*SCHAUM'S OUTLINE OF*

# THEORY AND PROBLEMS

*of Differential and Integral*

# CALCULUS

SECOND EDITION

•

BY

## FRANK AYRES, JR., Ph.D.

*Formerly Professor and Head,*
*Department of Mathematics*
*Dickinson College*

# SCHAUM'S OUTLINE SERIES

McGRAW-HILL BOOK COMPANY

*New York, St. Louis, San Francisco, Toronto, Sydney*

ISBN 07-002653-X

44 45 SH SH 8 7 6

# Preface

The purpose of this book, as in the case of the first edition, is to provide the beginning student of elementary calculus with a collection of carefully solved representative problems. The book will also be found helpful to students of science and engineering who feel the need for a review of fundamental theory and problem work in the subject. Moreover, since this edition includes proofs of theorems, derivations of differentiation and integration formulas, and an ample supply of supplementary problems, it may be used as a text for a formal course.

The plan of the book is essentially that of the previous edition. Each chapter begins with statements of pertinent definitions, principles, and theorems. The illustrative material and solved problems which follow have been selected not only to amplify the theory but also to provide practice in the formulation and solution of problems, to furnish the necessary repetition of basic principles for effective learning, to anticipate difficulties which normally beset the beginner, and to illustrate a wide variety of applications of the calculus. Numerous proofs of theorems and derivations of basic results are included among the solved problems. An effective use of the book, either as a supplement to a text or as the text itself, requires something more than a casual study of the solved problems. There is something to be learned from each, and this can best be accomplished by a step-by-step reproduction of them. When this has been done, no great difficulty should be encountered in solving most of the supplementary problems.

The increase by approximately fifty percent in the size of this edition is due only in part to the additions noted above. Of other changes and additions, attention is called to the fuller treatment given the limit concept, continuity of functions, and infinite series, and to a rather extensive introduction to both plane and space vectors.

In order that the more elementary work in integration, areas, volumes, etc., may be introduced out of the order in which they appear here, these chapters have been arranged so that ample portions of each may be considered at any point after Chapter 6. Likewise, those who use this book as a supplement will find little difficulty on this score in fitting it to their needs.

The author wishes to avail himself of the opportunity to express his gratitude to the Schaum Publishing Company for their unfailing cooperation in this endeavor.

FRANK AYRES, JR.

Carlisle, Penna.
March, 1964

# CONTENTS

# CONTENTS

# Chapter 1

## Variables and Functions

**THE SET OF REAL NUMBERS** consists of the rational numbers (the positive and negative integers, zero, and the common fractions $a/b$, where $a$ and $b$ are integers) and the irrational numbers (endless decimals as $\sqrt{2} = 1.4142\ldots$ and $\pi = 3.14159\ldots$ which are not ratios of integers).

The imaginary numbers of algebra will play no role here and the term will be used only in pointing out their exclusion. Since no confusion can result, the term number will be used hereinafter to mean a *real* number.

**THE ABSOLUTE OR NUMERICAL VALUE** ($|N|$) of a (real) number $N$ is defined as:

$$|N| = N \text{ if } N \text{ is zero or a positive number,}$$
$$|N| = -N \text{ if } N \text{ is a negative number.}$$

For example,

$$|3| = |-3| = 3, \quad |3-5| = |5-3| = 2,$$
$$|x-a| = x-a \text{ if } x \geq a \quad \text{and} \quad |x-a| = a-x \text{ if } x < a.$$

In general, if $a$ and $b$ are any two numbers,

$$-|a| \leq a \leq |a|$$

$$|a \pm b| = |b \pm a|; \quad |ab| = |a| \cdot |b|; \quad \left|\frac{a}{b}\right| = \frac{|a|}{|b|}, \quad b \neq 0;$$

$$|a+b| \geq |a| - |b|; \quad |a-b| \leq |a| + |b|;$$
$$|a+b| \leq |a| + |b|; \quad |a-b| \geq |a| - |b|.$$

**A NUMBER SCALE** is a graphical representation of the real numbers by the points of a straight line. To each number corresponds one and only one point, and conversely. As a consequence, the terms number and point (on a number scale) will be used interchangeably.

To set up a number scale on a given line: (i) select any point of the line as *origin* (corresponding to 0), (ii) choose a sense of positive direction (indicated by an arrow tip), and (iii) with any convenient unit of measure, locate the point $+1$ a unit's distance from 0. The numbers (points) $N$ and $-N$ are then on opposite sides of 0 and $|N|$ units from it.

1

If $a$ and $b$ are two distinct numbers, then $a < b$ means that $a$ is to the left of $b$ on the scale while $a > b$ means that $a$ is to the right of $b$.

The directed distance from $a$ to $b$ is given by $b - a$, being negative if $a > b$ and positive if $a < b$. In either case, $b$ is at the undirected distance $|b - a| = |a - b|$ from $a$.

**FINITE INTERVALS.** Let $a$ and $b$ be two numbers such that $a < b$. The set of all numbers $x$ between $a$ and $b$ is called the *open interval* from $a$ to $b$ and is written $a < x < b$. The points $a$ and $b$ are called the *endpoints* of the interval. An open interval does not contain its endpoints.

The open interval $a < x < b$ together with its endpoints $a$ and $b$ is called the *closed interval* from $a$ to $b$ and is written $a \leqq x \leqq b$.

open interval: $a < x < b$             closed interval: $a \leqq x \leqq b$

**INFINITE INTERVALS.** Let $a$ be any number. The set of all numbers $x$ satisfying $x < a$ is called an *infinite interval*. Other infinite intervals are defined by $x \leqq a$, $x > a$, and $x \geqq a$.

See Problems 1-2.

**CONSTANT AND VARIABLE.** In the definition of the interval $a < x < b$:

(i)    each of the symbols $a$ and $b$ represents a single number and is called a *constant*.

(ii)    the symbol $x$ represents any one of a set (collection) of numbers and is called a *variable*.

The *range* of a variable is another name for the set of numbers which it represents. For example:

(1)    $x$ is a volume of a ten-volume set of books; the range of $x$ is the set of integers $1, 2, 3, \ldots, 10$.

(2)    $x$ is a day in July; the range of $x$ is the set $1, 2, 3, \ldots, 31$.

(3)    $x$ is the quantity (in gallons) of water which can be removed from a filled ten gallon tank; the range of $x$ is the interval $0 \leqq x \leqq 10$.

**INEQUALITIES** such as $2x - 3 > 0$ and $x^2 - 5x - 24 \leqq 0$ define intervals on the number scale.

**Example 1:** Solve the inequality   (a) $2x - 3 > 0$,   (b) $x^2 - 5x - 24 \leqq 0$.

(a)   Set $2x - 3 = 0$, obtain $x = 3/2$, and consider the intervals $x < 3/2$ and $x > 3/2$. For any value of $x$ on the interval $x < 3/2$, say $x = 0$, $2x - 3 < 0$; for any value of $x$ on the interval $x > 3/2$, say $x = 3$, $2x - 3 > 0$. Thus, $2x - 3 > 0$ for all $x$ on the interval $x > 3/2$.

(b)   Set $x^2 - 5x - 24 = (x + 3)(x - 8) = 0$, obtain $x = -3$ and $x = 8$, and consider the intervals $x < -3$, $-3 < x < 8$, $x > 8$. Now $x^2 - 5x - 24 > 0$ on the intervals $x < -3$ and $x > 8$, while $x^2 - 5x - 24 < 0$ on the interval $-3 < x < 8$. Thus, $x^2 - 5x - 24 \leqq 0$ on the interval $-3 \leqq x \leqq 8$.

See Problem 3.

## FUNCTION OF A VARIABLE.

A variable $y$ is said to be a *function* of another variable $x$ if there is given a rule or device which associates with each value of $x$ in its range *one* value of $y$. The variable $y$, whose value depends upon the chosen value of $x$, is called the *dependent variable* while $x$ is called the *independent variable*. The rule or device may be a table of corresponding values (a table of logarithms), a graph, or an equation.

**Example 2:**

The equation $x^2 - y = 10$, with $x$ the independent variable, associates one value of $y$ with each value of $x$. The function defined is $y = x^2 - 10$. The same equation, with $y$ taken as independent variable, generally associates two values of $x$ with each value of $y$. Thus, two functions of $y$ are defined: $x = \sqrt{10 + y}$ and $x = -\sqrt{10 + y}$.

Some authors define $y$ to be a function of $x$ when, to each value of $x$ in its range, one or more values of $y$ are associated. Then, in Example 2, $y$ is called a *single-valued* function of $x$ while $x$ is called a *multi-valued* (more precisely, a *double-valued*) function of $y$. However, in the Calculus, one must think of a multi-valued function as consisting of several single-valued functions. Our definition of the term function is fashioned then to imply this property of single-valuedness.

The symbol $f(x)$, read "the $f$-function of $x$" or "$f$ of $x$" but never "$f$ times $x$", is used to denote a given function of $x$. If in the same problem another function of $x$ occurs, another letter is used to denote it: $g(x), h(x), F(x), \theta(x), \ldots$.

In the study of a function $y = f(x)$ it is necessary always to know the range of the independent variable, also called *the domain of definition* of the function.

**Example 3:**

(a) The function $f(x) = 18x - 3x^2$ is defined for every number $x$; that is, without exception $18x - 3x^2$ is a real number whenever $x$ is a real number. Thus, the range of $x$ or the domain of definition of the function is the set of all real numbers.

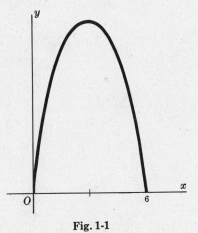

**Fig. 1-1**

(b) The area ($y$) of a certain rectangle, one of whose sides is $x$, is given by $y = 18x - 3x^2$. Here, both $x$ and $18x - 3x^2$ must be positive. From the adjoining figure or from Problem 3($a$) it is clear that the domain of definition is the interval $0 < x < 6$.

(c) The domain of definition of the function $y = x^2 - 10$ of Example 2 is the set of all real numbers. For the functions $x = \sqrt{10 + y}$ and $x = -\sqrt{10 + y}$, it is necessary that $10 + y \geqq 0$; hence, the domain of definition of each is $y \geqq -10$.

A function $f(x)$ will be said to be defined *on* an interval if it is defined for *every* point of the interval.

If $f(x)$ is a given function of $x$ and if $a$ is in its domain of definition, then by $f(a)$ is meant the number obtained by replacing $x$ by $a$ in $f(x)$ or the value assumed by $f(x)$ when $x = a$.

**Example 4:** If $f(x) = x^3 - 4x + 2$, then

$$f(1) = (1)^3 - 4(1) + 2 = 1 - 4 + 2 = -1,$$
$$f(-2) = (-2)^3 - 4(-2) + 2 = -8 + 8 + 2 = 2,$$
$$f(a) = a^3 - 4a + 2, \text{ etc.}$$

See Problems 4-13.

**AN INFINITE SEQUENCE** is a function of a variable (usually denoted by $n$) whose range is restricted to the set of positive integers. For example, when $n$ is given in turn the values $1, 2, 3, 4, \ldots$, the function $\dfrac{1}{n+1}$ yields the succession or sequence of terms $\frac{1}{2}, \frac{1}{3}, \frac{1}{4}, \frac{1}{5}, \ldots$. The sequence is called an *infinite sequence* to indicate that there is no last term.

The function, as $\dfrac{1}{n+1}$ in the paragraph above, is called the *general* or $n$th *term* of the sequence. An infinite sequence is denoted by enclosing the general term in braces, as $\left\{\dfrac{1}{n+1}\right\}$, or by displaying several terms of the sequence, as $\frac{1}{2}, \frac{1}{3}, \frac{1}{4}, \frac{1}{5}, \ldots$, $\dfrac{1}{n+1}, \ldots$.

See Problems 14-15.

# Solved Problems

1.   Describe and diagram the intervals: (a) $-3 < x < 5$, (b) $2 \leqq x \leqq 6$, (c) $-4 < x \leqq 0$, (d) $x > 5$, (e) $x \leqq 2$.

   (a)  All numbers greater than $-3$ and less than 5.

   (b)  All numbers equal to or greater than 2 and less than or equal to 6.

   (c)  All numbers greater than $-4$ and less than or equal to 0.

   　　This finite interval contains one but not both of its endpoints. It is called a *half-open* interval.

   (d)  All numbers greater than 5.

   (e)  All numbers less than or equal to 2.

2.   Describe and diagram the intervals:

   (a) $|x| < 2$;  (b) $|x| > 3$;  (c) $|x - 3| < 1$;  (d) $|x - 2| < \delta, \, \delta > 0$;  (e) $0 < |x + 3| < \delta, \, \delta > 0$.

   (a)  This is the open interval $-2 < x < 2$.

   (b)  Two infinite intervals are defined: $x < -3$ and $x > 3$.

   (c)  This is an open interval about the point 3. To find the endpoints, set $x - 3 = 1$ to obtain $x = 4$ and set $3 - x = 1$ to obtain $x = 2$. (Recall that $|x - 3| = x - 3$ or $3 - x$ depending upon the value of $x$.) The endpoints are 2 and 4; the interval is $2 < x < 4$. Note that the interval consists of all points whose distance from 3 is less than 1.

   (d)  Think of $\delta$ as a given positive number. The interval $2 - \delta < x < 2 + \delta$ consists of all points whose distance from 2 is less than $\delta$. It is called the $\delta$-*neighborhood* of 2.

   (e)  The inequality $|x + 3| < \delta$ defines the interval $-3 - \delta < x < -3 + \delta$ which includes the point $-3$. The additional condition $0 < |x + 3|$ requires $x \neq -3$. Thus the range of $x$ is the two open intervals $-3 - \delta < x < -3$ and $-3 < x < -3 + \delta$. The two intervals constitute the *deleted* $\delta$-*neighborhood* of $-3$.

3.  Solve the inequalities:   (a) $18x - 3x^2 > 0$,   (b) $(x+3)(x-2)(x-4) < 0$,   (c) $(x+1)^2(x-3) > 0$.

(a)  Set $18x - 3x^2 = 3x(6-x) = 0$, obtain $x = 0$ and $x = 6$, and determine the sign of $18x - 3x^2$ on each of the intervals $x < 0$, $0 < x < 6$, and $x > 6$. The inequality is satisfied by all $x$ on the interval $0 < x < 6$.

(b)  After determining the sign of $(x+3)(x-2)(x-4)$ on each of the intervals $x < -3$, $-3 < x < 2$, $2 < x < 4$, and $x > 4$, we conclude that the inequality is satisfied by all $x$ on the intervals $x < -3$ and $2 < x < 4$.

(c)  The intervals to be examined are $x < -1$, $-1 < x < 3$, and $x > 3$. The inequality is satisfied when $x > 3$. Note that since $(x+1)^2 > 0$ for all $x$, the factor may be ignored. Could a factor $(x+1)^3$ be ignored?

4.  Given   $f(x) = \dfrac{x-1}{x^2+2}$,   find $f(0)$, $f(-1)$, $f(2a)$, $f(1/x)$, $f(x+h)$.

$$f(0) = \frac{0-1}{0+2} = -\frac{1}{2}, \qquad f(-1) = \frac{-1-1}{1+2} = -\frac{2}{3}, \qquad f(2a) = \frac{2a-1}{4a^2+2},$$

$$f(1/x) = \frac{1/x - 1}{1/x^2 + 2} = \frac{x - x^2}{1 + 2x^2}, \qquad f(x+h) = \frac{x+h-1}{(x+h)^2 + 2} = \frac{x+h-1}{x^2 + 2hx + h^2 + 2}$$

5.  If $f(x) = 2^x$, show that   (a) $f(x+3) - f(x-1) = \dfrac{15}{2} f(x)$   and   (b) $\dfrac{f(x+3)}{f(x-1)} = f(4)$.

(a)  $f(x+3) - f(x-1) = 2^{x+3} - 2^{x-1} = 2^x(2^3 - \tfrac{1}{2}) = \dfrac{15}{2} f(x)$        (b)  $\dfrac{f(x+3)}{f(x-1)} = \dfrac{2^{x+3}}{2^{x-1}} = 2^4 = f(4)$

6.  If $f(x) = \log_a 1/x$, show that   (a) $f(a^3) = -3$   and   (b) $f(a^{-1/z}) = 1/z$.

(a)  $f(a^3) = \log_a 1/a^3 = \log_a a^{-3} = -3$        (b)  $f(a^{-1/z}) = \log_a 1/a^{-1/z} = \log_a a^{1/z} = 1/z$

7.  If $f(x) = \log_a x$ and $F(z) = a^z$, show that $F(f(x)) = f(F(x))$.

$F(f(x)) = F(\log_a x) = a^{\log_a x} = x = \log_a a^x = f(a^x) = f(F(x))$.

8.  Determine the range of the independent variable $x$, given:

(a)  $y = \sqrt{4 - x^2}$,   (b) $y = \sqrt{x^2 - 16}$,   (c) $y = \dfrac{1}{x-2}$,   (d) $y = \dfrac{1}{x^2 - 9}$,   (e) $y = \dfrac{x}{x^2 + 4}$.

(a)  Since $y$ must be real, $4 - x^2 \geq 0$ or $x^2 \leq 4$; the range of $x$ is the interval $-2 \leq x \leq 2$ or $|x| \leq 2$. In other words, $f(x) = \sqrt{4 - x^2}$ is *defined* on the interval $-2 \leq x \leq 2$ and only on this interval.

(b)  Here $x^2 - 16 \geq 0$ or $x^2 \geq 16$; the range of $x$ consists of the intervals $x \leq -4$ and $x \geq 4$, or $|x| \geq 4$.

(c)  The function is defined for every value of $x$ except $x = 2$. The range of $x$ may be given as $x < 2$, $x > 2$ or as $x \neq 2$.

(d)  The function is defined for $x \neq \pm 3$.

(e)  Since $x^2 + 4 \neq 0$ for all $x$, the range of $x$ is the set of real numbers.

9.  Sketch the graph of the function defined as follows:

$$f(x) = 5 \quad \text{when } 0 < x \leq 1 \qquad f(x) = 10 \quad \text{when } 1 < x \leq 2$$
$$f(x) = 15 \quad \text{when } 2 < x \leq 3 \qquad f(x) = 20 \quad \text{when } 3 < x \leq 4 \qquad \text{etc.}$$

Determine the range of $x$ and of $y = f(x)$.

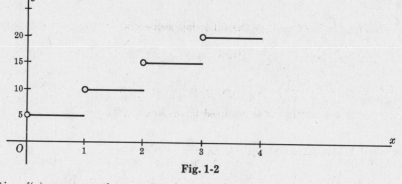

Fig. 1-2

The function $f(x)$ expresses the cost (in cents) of postage for first class domestic mail weighing $x$ oz. The range of $x$ is the interval $x > 0$ and the range of $y = f(x)$ is the set of integers $5, 10, 15, 20, \ldots$.

10.  A rectangular plot requires 2000 ft of fencing to enclose it. If one of the dimensions is $x$ ft, express the area $y$ (sq. ft.) as a function of $x$. Determine the range of $x$.

Since one dimension is $x$, the other is $\frac{1}{2}(2000 - 2x) = 1000 - x$.

The area is $y = x(1000 - x)$ and the range of $x$ is $0 < x < 1000$.

11.  Express the length $l$ of a chord of a circle of radius 8 in. as a function of its distance $x$ in. from the center of the circle. Determine the range of $x$.

From Fig. 1-3, it is seen that $\frac{1}{2}l = \sqrt{64 - x^2}$ and $l = 2\sqrt{64 - x^2}$.

The range of $x$ is the interval $0 \leqq x < 8$.

Fig. 1-3

12.  From each corner of a square of tin, 12 in. on a side, small squares of side $x$ in. are removed and the edges turned up to form an open box. Express the volume $V$ (cu. in.) as a function of $x$ and examine for the range of each variable.

The box has a square base of side $(12 - 2x)$ in. and a height of $x$ in. The volume of the box is then $V = x(12 - 2x)^2 = 4x(6 - x)^2$. The range of $x$ is the interval $0 < x < 6$.

As $x$ increases over its range, $V$ increases for a time and then decreases thereafter. Thus, among the boxes which may be constructed there is one of greatest volume, say $M$. To determine $M$, it is necessary to locate the precise point (value of $x$) at which $V$ ceases to increase. This problem will be studied in a later chapter.

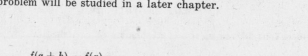

Fig. 1-4

13.  If $f(x) = x^2 + 2x$, find $\dfrac{f(a + h) - f(a)}{h}$ and interpret the result.

$$\frac{f(a + h) - f(a)}{h} = \frac{[(a + h)^2 + 2(a + h)] - (a^2 + 2a)}{h}$$

$$= 2a + 2 + h$$

On the graph of the function (Fig. 1-5) locate points $P$ and $Q$ whose respective abscissas are $a$ and $(a + h)$. The ordinate of $P$ is $f(a)$ and that of $Q$ is $f(a + h)$. Then

$$\frac{f(a + h) - f(a)}{h} = \frac{\text{difference of ordinates}}{\text{difference of abscissas}}$$

$$= \text{slope of } PQ$$

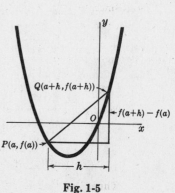

Fig. 1-5

14. Write the first five terms of each of the following sequences.

(a) $\left\{1 - \dfrac{1}{2n}\right\}$.   Set  $s_n = 1 - \dfrac{1}{2n}$;  then  $s_1 = 1 - \dfrac{1}{2 \cdot 1} = \dfrac{1}{2}$,

$$s_2 = 1 - \dfrac{1}{2 \cdot 2} = \dfrac{3}{4}, \quad s_3 = 1 - \dfrac{1}{2 \cdot 3} = \dfrac{5}{6}, \quad s_4 = 1 - \dfrac{1}{2 \cdot 4} = \dfrac{7}{8},$$

and $s_5 = 9/10$.  The required terms are $1/2, 3/4, 5/6, 7/8, 9/10$.

(b) $\left\{(-1)^{n+1} \dfrac{1}{3n - 1}\right\}$.   Here  $s_1 = (-1)^2 \dfrac{1}{3 \cdot 1 - 1} = 1/2$,

$$s_2 = (-1)^3 \dfrac{1}{3 \cdot 2 - 1} = -1/5, \quad s_3 = (-1)^4 \dfrac{1}{3 \cdot 3 - 1} = 1/8,$$

$s_4 = -1/11$,  $s_5 = 1/14$.  The required terms are $1/2, -1/5, 1/8, -1/11, 1/14$.

(c) $\left\{\dfrac{2n}{1 + n^2}\right\}$.   The terms are $1, 4/5, 3/5, 8/17, 5/13$.

(d) $\left\{(-1)^{n+1} \dfrac{n}{(n+1)(n+2)}\right\}$.   The terms are $\dfrac{1}{2 \cdot 3}, \dfrac{-2}{3 \cdot 4}, \dfrac{3}{4 \cdot 5}, \dfrac{-4}{5 \cdot 6}, \dfrac{5}{6 \cdot 7}$.

(e) $\{\tfrac{1}{2}[(-1)^n + 1]\}$.   The terms are $0, 1, 0, 1, 0$.

15. Write the general term of each of the following sequences.

(a) $1, 1/3, 1/5, 1/7, 1/9, \dots$.
   The terms are the reciprocals of the odd positive integers.  The general term is $\dfrac{1}{2n - 1}$.

(b) $1, -1/2, 1/3, -1/4, 1/5, \dots$.
   Apart from sign these are the reciprocals of the positive integers.  The general term is $(-1)^{n+1} \dfrac{1}{n}$ or $(-1)^{n-1} \dfrac{1}{n}$.

(c) $1, 1/4, 1/9, 1/16, 1/25, \dots$.
   The terms are the reciprocals of the squares of the positive integers.  The general term is $1/n^2$.

(d) $\dfrac{1}{2}, \dfrac{1 \cdot 3}{2 \cdot 4}, \dfrac{1 \cdot 3 \cdot 5}{2 \cdot 4 \cdot 6}, \dfrac{1 \cdot 3 \cdot 5 \cdot 7}{2 \cdot 4 \cdot 6 \cdot 8}, \dots$   The general term is $\dfrac{1 \cdot 3 \cdot 5 \dots (2n - 1)}{2 \cdot 4 \cdot 6 \dots (2n)}$.

(e) $1/2, -4/9, 9/28, -16/65, \dots$.
   Apart from sign the numerators are the squares of positive integers and the denominators are the cubes of these integers increased by 1.  The general term is $(-1)^{n+1} \dfrac{n^2}{n^3 + 1}$.

16. Prove:  If $a$ and $b$ are any two numbers,  $|a + b| \leqq |a| + |b|$.
   Consider the following cases: (a) $a$ and $b$ both non-negative, (b) $a$ and $b$ both negative, (c) one of $a, b$ positive and the other negative.

(a) Since $|a| = a$, $|b| = b$, and $a + b$ is zero or a positive number, then

$$|a + b| = a + b = |a| + |b|$$

(b) Since $|a| = -a$, $|b| = -b$, and $a + b$ is negative, then

$$|a + b| = -(a + b) = -a + (-b) = |a| + |b|$$

(c) Take $a > 0$ and $b < 0$; then $|a| = a$ and $|b| = -b$.

If $|a| > |b|$, then $|a + b| = a + b < a - b = |a| + |b|$.
If $|a| < |b|$, then $|a + b| = -a - b < a - b = |a| + |b|$.
If $|a| = |b|$, then $|a + b| = 0 < |a| + |b|$.

Thus, if $a > 0$ and $b < 0$ or if $a < 0$ and $b > 0$, then $|a + b| < |a| + |b|$.

# Supplementary Problems

**17.** Diagram each of the following intervals.

(a) $-5 < x < 0$    (c) $-2 \leqq x < 3$    (e) $|x| < 3$    (g) $|x - 2| < \frac{1}{2}$    (i) $0 < |x - 2| < 1$    (k) $|x - 2| \geqq 1$

(b) $x \leqq 0$    (d) $x \geqq 1$    (f) $|x| \geqq 5$    (h) $|x + 3| > 1$    (j) $0 < |x + 3| < \frac{1}{4}$

**18.** If $f(x) = x^2 - 4x + 6$, find (a) $f(0)$, (b) $f(3)$, (c) $f(-2)$.      *Ans.* (a) 6, (b) 3, (c) 18

Show that $f(\frac{1}{2}) = f(7/2)$ and $f(2 - h) = f(2 + h)$.

**19.** If $f(x) = \dfrac{x - 1}{x + 1}$, find (a) $f(0)$, (b) $f(1)$, (c) $f(-2)$.      *Ans.* (a) $-1$, (b) 0, (c) 3

Show that $f(1/x) = -f(x)$ and $f(-1/x) = -1/f(x)$.

**20.** If $f(x) = x^2 - x$, show that $f(x + 1) = f(-x)$.

**21.** If $f(x) = 1/x$, show that $f(a) - f(b) = f\left(\dfrac{ab}{b - a}\right)$.

**22.** If $y = f(x) = (5x + 3)/(4x - 5)$, show that $x = f(y)$.

**23.** Determine the domain of definition of each of the following functions.

(a) $y = x^2 + 4$      (c) $y = \sqrt{x^2 - 4}$      (e) $y = \dfrac{2x}{(x - 2)(x + 1)}$      (g) $y = \dfrac{x^2 - 1}{x^2 + 1}$

(b) $y = \sqrt{x^2 + 4}$      (d) $y = \dfrac{x}{x + 3}$      (f) $y = \dfrac{1}{\sqrt{9 - x^2}}$      (h) $y = \sqrt{\dfrac{x}{2 - x}}$.

*Ans.* (a), (b), (g) all values of $x$; (c) $|x| \geqq 2$; (d) $x \neq -3$; (e) $x \neq -1, 2$; (f) $-3 < x < 3$; (h) $0 \leqq x < 2$

**24.** Compute $\dfrac{f(a + h) - f(a)}{h}$, given: (a) $f(x) = \dfrac{1}{x - 2}$ when $a \neq 2$, $a + h \neq 2$; (b) $f(x) = \sqrt{x - 4}$ when

$a \geqq 4$, $a + h \geqq 4$; (c) $f(x) = \dfrac{x}{x + 1}$ when $a \neq -1$, $a + h \neq -1$.

*Ans.* (a) $\dfrac{-1}{(a - 2)(a + h - 2)}$, (b) $\dfrac{1}{\sqrt{a + h - 4} + \sqrt{a - 4}}$, (c) $\dfrac{1}{(a + 1)(a + h + 1)}$

**25.** Write the first five terms of each sequence.

(a) $\left\{1 + \dfrac{1}{n}\right\}$      (c) $\{a + (n - 1)d\}$      (e) $\left\{\dfrac{n}{\sqrt{1 + n^2}}\right\}$      (g) $\left\{(-1)^{n+1}\dfrac{n!}{n^n}\right\}$

(b) $\left\{\dfrac{1}{n(n + 1)}\right\}$      (d) $\{(-1)^{n+1} ar^{n-1}\}$      (f) $\left\{\dfrac{\sqrt{n + 1}}{n}\right\}$      (h) $\left\{\dfrac{(2n)!}{3^n 5^{n-1}}\right\}$

*Ans.* (a) 2, 3/2, 4/3, 5/4, 6/5          (e) $1/\sqrt{2}$, $2/\sqrt{5}$, $3/\sqrt{10}$, $4/\sqrt{17}$, $5/\sqrt{26}$

       (b) 1/2, 1/6, 1/12, 1/20, 1/30        (f) $\sqrt{2}$, $\frac{1}{2}\sqrt{3}$, 2/3, $\frac{1}{4}\sqrt{5}$, $\sqrt{6}/5$

       (c) $a$, $a + d$, $a + 2d$, $a + 3d$, $a + 4d$      (g) 1, $-1/2$, 2/9, $-3/32$, 24/625

       (d) $a$, $-ar$, $ar^2$, $-ar^3$, $ar^4$        (h) $\dfrac{2}{3}$, $\dfrac{2^3}{3 \cdot 5}$, $\dfrac{2^4}{3 \cdot 5}$, $\dfrac{7 \cdot 2^7}{3^2 \cdot 5^2}$, $\dfrac{7 \cdot 2^8}{3 \cdot 5^2}$

**26.** Determine the general term of each sequence.

(a) 1/2, 2/3, 3/4, 4/5, 5/6, ...          (d) $1/5^3$, $3/5^5$, $5/5^7$, $7/5^9$, $9/5^{11}$, ...

(b) 1/2, $-1/6$, 1/12, $-1/20$, 1/30, ...      (e) 1/2!, $-1/4!$, 1/6!, $-1/8!$, 1/10!, ...

(c) 1/2, 1/12, 1/30, 1/56, 1/90, ...

*Ans.* (a) $\dfrac{n}{n + 1}$, (b) $(-1)^{n-1}\dfrac{1}{n^2 + n}$, (c) $\dfrac{1}{(2n - 1)2n}$, (d) $\dfrac{2n - 1}{5^{2n+1}}$, (e) $(-1)^{n-1}\dfrac{1}{(2n)!}$

**27.** "Whenever $|x - 4| < 1$ then $|f(x)| > 1$" means "whenever $x$ is between 3 and 5 then $f(x)$ is either less than $-1$ or greater than $+1$". Interpret each of the following:

(a) Whenever $|x - 1| < 2$ then $|f(x)| < 10$.        (c) Whenever $0 < |x - 6| < 1$ then $|f(x)| > 0$.

(b) Whenever $|x - 5| < 2$ then $|f(x)| > 0$.        (d) Whenever $|x - 3| < 2$ then $|f(x) - 9| < 4$.

**28.** Take $y = f(x) = 6x - x^2$, construct the graph, and determine which of the statements $(a) - (d)$ of Problem 27 are true and which are false.      *Ans.* (b) is false.

**29.** Prove that when $a$ and $b$ are any two numbers: $|a \pm b| = |b \pm a|$; $|ab| = |a| \cdot |b|$; $|a/b| = |a| / |b|$, $b \neq 0$; $|a + b| \geqq |a| - |b|$; $|a - b| \leqq |a| + |b|$; $|a - b| \geqq |a| - |b|$.

# Chapter 2

## Limits

**LIMIT OF A SEQUENCE.** As consecutive points, given by the terms of the sequence

$$1, \ 3/2, \ 5/3, \ 7/4, \ 9/5, \ \ldots, \ 2 - 1/n, \ \ldots \tag{1}$$

are located on a number scale, it is noted that they cluster about the point 2 in such a way that there are points of the sequence whose distance from 2 is less than any

preassigned positive number, however small. For example, the point 2001/1001 and all subsequent points are at a distance $< 1/1000$ from 2, the point 20000001/10000001 and all subsequent points are at a distance $< 1/10000000$ from 2, and so on. This state of affairs is indicated by saying that *the limit of the sequence is* 2.

If $x$ is a variable whose range is the sequence (1), we say that $x$ *approaches* 2 *as limit* or $x$ *tends to* 2 *as limit* and write $x \to 2$.

The sequence (1) does not contain its limit 2 as a term. On the other hand, the sequence $1, 1/2, 1, 3/4, 1, 5/6, 1, \ldots$ has 1 as limit and every odd numbered term is 1. Thus, a sequence may or may not reach its limit. Hereinafter, the statement $x \to a$ will be understood to imply $x \neq a$, that is, *it is to be understood that any given arbitrary sequence does not contain its limit as a term.*

**LIMIT OF A FUNCTION.** Let $x \to 2$ over the sequence (1); then $f(x) = x^2 \to 4$ over the sequence $1, 9/4, 25/9, 49/16, \ldots, (2 - 1/n)^2, \ldots$. Now let $x \to 2$ over the sequence

$$2.1, \ 2.01, \ 2.001, \ 2.0001, \ \ldots, \ 2 + 1/10^n, \ \ldots \tag{2}$$

then $x^2 \to 4$ over the sequence $4.41, 4.0401, 4.004001, \ldots, (2 + 1/10^n)^2, \ldots$. It would seem reasonable to expect that $x^2$ would approach 4 as limit however $x$ may approach 2 as limit. Under this assumption, we say "the limit, as $x$ approaches 2, of $x^2$ is 4" and write $\lim_{x \to 2} x^2 = 4$.

See Problems 1-2.

**RIGHT AND LEFT LIMITS.** As $x \to 2$ over the sequence (1), its value is always less than 2. We say that $x$ *approaches* 2 *from the left* and write $x \to 2^-$. Similarly, as $x \to 2$ over the sequence (2), its value is always greater than 2. We say that $x$ *approaches* 2 *from the right* and write $x \to 2^+$. Clearly, the statement $\lim_{x \to a} f(x)$ exists implies that both the *left limit* $\lim_{x \to a^-} f(x)$ and the *right limit* $\lim_{x \to a^+} f(x)$ exist and are equal. However, the existence of a right (left) limit does not imply the existence of the left (right) limit.

9

**Example 1:**

The function $f(x) = \sqrt{9 - x^2}$ has the interval $-3 \le x \le 3$ as domain of definition. If $a$ is any number on the open interval $-3 < x < 3$, then $\lim_{x \to a} \sqrt{9 - x^2}$ exists and is equal to $\sqrt{9 - a^2}$. Now consider $a = 3$. First, let $x$ approach 3 from the left; then $\lim_{x \to 3^-} \sqrt{9 - x^2} = 0$. Next, let $x$ approach 3 from the right; then $\lim_{x \to 3^+} \sqrt{9 - x^2}$ does not exist since for $x > 3$, $\sqrt{9 - x^2}$ is imaginary. Thus, $\lim_{x \to 3} \sqrt{9 - x^2}$ does not exist.

Similarly, $\lim_{x \to -3^+} \sqrt{9 - x^2}$ exists and is equal to 0 but $\lim_{x \to -3^-} \sqrt{9 - x^2}$ and thus $\lim_{x \to -3} \sqrt{9 - x^2}$ do not exist.

**THEOREMS ON LIMITS.** The following theorems on limits are listed for future reference.

**I.**  If $f(x) = c$, a constant, then $\lim_{x \to a} f(x) = c$.

If $\lim_{x \to a} f(x) = A$ and $\lim_{x \to a} g(x) = B$, then:

**II.**  $\lim_{x \to a} k \cdot f(x) = kA$, $k$ being any constant.

**III.**  $\lim_{x \to a} [f(x) \pm g(x)] = \lim_{x \to a} f(x) \pm \lim_{x \to a} g(x) = A \pm B$.

**IV.**  $\lim_{x \to a} [f(x) \cdot g(x)] = \lim_{x \to a} f(x) \cdot \lim_{x \to a} g(x) = A \cdot B$.

**V.**  $\lim_{x \to a} \dfrac{f(x)}{g(x)} = \dfrac{\lim_{x \to a} f(x)}{\lim_{x \to a} g(x)} = \dfrac{A}{B}$, provided $B \ne 0$.

**VI.**  $\lim_{x \to a} \sqrt[n]{f(x)} = \sqrt[n]{\lim_{x \to a} f(x)} = \sqrt[n]{A}$, provided $\sqrt[n]{A}$ is a real number.

**INFINITY.**  Let the range of the variable $x$ be the sequence $s_1, s_2, s_3, s_4, \ldots, s_n, \ldots$; then

**(i)**  $x$ is said *to become positively infinite* $[x \to +\infty]$ if it eventually becomes and thereafter remains greater than any preassigned positive number, however large. For example, $x \to +\infty$ over the sequence $1, 2, 3, 4, \ldots$.

**(ii)**  $x$ is said *to become negatively infinite* $[x \to -\infty]$ if it eventually becomes and thereafter remains less than any preassigned negative number, however small. For example, $x \to -\infty$ over the sequence $-2, -4, -6, -8, \ldots$.

**(iii)**  $x$ is said *to become infinite* $[x \to \infty]$ if $|x| \to +\infty$, that is, if $x \to +\infty$ or $x \to -\infty$.

A function $f(x)$ is said *to become positively infinite* as $x \to a$, $\left[ \lim_{x \to a} f(x) = +\infty \right]$, if, as $x$ approaches its limit $a$ (without assuming the value $a$), $f(x)$ eventually becomes and thereafter remains greater than any preassigned positive number, however large.

A function $f(x)$ is said *to become negatively infinite* as $x \to a$, $\left[ \lim_{x \to a} f(x) = -\infty \right]$, if, as $x$ approaches its limit $a$ (without assuming the value $a$), $f(x)$ eventually becomes and thereafter remains less than any preassigned negative number, however small.

A function $f(x)$ is said *to become infinite* as $x \to a$, $\left[ \lim_{x \to a} f(x) = \infty \right]$, if $\lim_{x \to a} |f(x)| = +\infty$.

**Example 2:**

(a) As $x \to 2$ over the sequence (1), $f(x) = \dfrac{1}{2-x} \to +\infty$ over the sequence $1, 2, 3, 4, \ldots$ . In general, if $x \to 2^-$ then $\dfrac{1}{2-x} \to +\infty$ and we write $\lim\limits_{x \to 2^-} \dfrac{1}{2-x} = +\infty$.

(b) As $x \to 2$ over the sequence (2), $f(x) = \dfrac{1}{2-x} \to -\infty$ over the sequence $-10, -100, -1000, -10000,$ $\ldots$ . In general, if $x \to 2^+$ then $\dfrac{1}{2-x} \to -\infty$ and we write $\lim\limits_{x \to 2^+} \dfrac{1}{2-x} = -\infty$.

(c) As $x \to 2$ over (1) and (2), $|f(x)| = \left| \dfrac{1}{2-x} \right| \to +\infty$ and we write $\lim\limits_{x \to 2} \dfrac{1}{2-x} = \infty$.

*Note.* The symbols $+\infty, -\infty, \infty$ are *not* new numbers to be adjoined to the set of real numbers. These symbols are introduced to indicate a certain type of behavior of a variable or a function. When a variable or a function constantly increases in value but never exceeds a certain number $M$, the variable or function approaches $M$ or some smaller number as limit. When no such number $M$ exists, the variable or function is said to become infinite. In this latter case, *no limit exists*; the limit notation is used merely because of its convenience.

See Problems 3-12.

**THE STATEMENT** $\lim\limits_{x \to a} f(x) = A$ has been established by checking the behavior of $f(x)$ as $x \to a$ over a number of sequences. Having found that $f(x) \to A$ in each case, it was then concluded that the same result would be obtained for all other (unchecked) sequences having $a$ as limit. Now as $x \to a$ over each of the several sequences, $x$ must eventually come close to $a$. The essential notion of the limit concept is that whenever $x$ comes close to but remains different from $a$ then $f(x)$ comes close to $A$. This may be stated in precise terms as follows:

**A.** $\lim\limits_{x \to a} f(x) = A$ if for any chosen positive number $\epsilon$, however small, there exists a positive number $\delta$ such that whenever $0 < |x - a| < \delta$ then $|f(x) - A| < \epsilon$.

The two inequalities establish intervals:

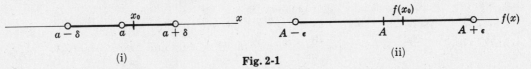

(i)                         Fig. 2-1                         (ii)

The gist of the definition is that after $\epsilon$ has been chosen (interval (ii) has been set up), $\delta$ can be found (interval (i) can be determined) so that whenever $x \neq a$ is on interval (i), say at $x_0$, then $f(x)$ is on the interval (ii).

**Example 3:**

Use the precise definition to show that $\lim\limits_{x \to 2} (x^2 + 3x) = 10$.

Let $\epsilon$ be chosen. We must produce a $\delta > 0$ such that whenever $0 < |x - 2| < \delta$ then $|(x^2 + 3x) - 10| < \epsilon$. We first note that whenever $0 < |x - 2| < \lambda < 1$, then $|x - 2|^n < \lambda$ for $n$ any positive integer. Then

$$|(x^2 + 3x) - 10| = |(x - 2)^2 + 7(x - 2)| \leq |x - 2|^2 + 7|x - 2| < \lambda + 7\lambda = 8\lambda$$

Now $8\lambda < \epsilon$ requires $\lambda < \epsilon/8$. Hence, any positive number smaller than both 1 and $\epsilon/8$ is an effective $\delta$ and the limit is established.

See Problems 13-14.

**OTHER TYPES OF LIMITS.** We define

**B.** $\lim\limits_{x \to a} f(x) = \infty$ if for any positive number $M$, *however large*, there exists a positive number $\delta$ such that whenever $0 < |x - a| < \delta$ then $|f(x)| > M$.

When $f(x) > M$, $\lim\limits_{x \to a} f(x) = +\infty$; when $f(x) < -M$, $\lim\limits_{x \to a} f(x) = -\infty$.

**C.** $\lim\limits_{x \to \infty} f(x) = A$   if for any positive number $\epsilon$, *however small*, there exists a positive number $M$ such that whenever $|x| > M$ then $|f(x) - A| < \epsilon$.

**D.** $\lim\limits_{x \to \infty} f(x) = \infty$   if for any positive number $M$, *however large*, there exists a positive number $P$ such that whenever $|x| > P$ then $|f(x)| > M$.

See Problem 15.

When $\lim\limits_{x \to \infty} f(x)$ and $\lim\limits_{x \to \infty} g(x)$ exist, the theorems on limits of this chapter remain valid. They must not be used, however, when $\lim\limits_{x \to a} f(x) = \infty$ and $\lim\limits_{x \to a} g(x) = \infty$ or when $\lim\limits_{x \to \infty} f(x) = \infty$ and $\lim\limits_{x \to \infty} g(x) = \infty$. For example, $\lim\limits_{x \to 1} \dfrac{x}{1-x} = \infty$ and $\lim\limits_{x \to 1} \dfrac{1}{1-x^2} = \infty$ but $\lim\limits_{x \to 1} \left( \dfrac{x}{1-x} \Big/ \dfrac{1}{1-x^2} \right) = \lim\limits_{x \to 1} x(1+x) = 2$. Also, $\lim\limits_{x \to +\infty} (x^2 + 5) = +\infty$ and $\lim\limits_{x \to +\infty} (2 - x^2) = -\infty$ but $\lim\limits_{x \to +\infty} \{ (x^2 + 5) + (2 - x^2) \} = \lim\limits_{x \to +\infty} 7 = 7$.

# Solved Problems

**1.** Determine the limit of each of the following sequences:

    (a)   1, 1/2, 1/3, 1/4, 1/5, ...      (c)   2, 5/2, 8/3, 11/4, 14/5, ...      (e)   1/2, 1/4, 1/8, 1/16, 1/32, ...

    (b)   1, 1/4, 1/9, 1/16, 1/25, ...      (d)   5, 4, 11/3, 7/2, 17/5, ...      (f)   .9, .99, .999, .9999, .99999, ...

    (a)   The general term is $1/n$. As $n$ takes on the values $1, 2, 3, 4, \ldots$ in turn, $1/n$ decreases but remains positive. The limit is 0.

    (b)   The general term is $(1/n)^2$; the limit is 0.

    (c)   The general term is $3 - 1/n$; the limit is 3.

    (d)   The general term is $3 + 2/n$; the limit is 3.

    (e)   The general term is $1/2^n$; as in (a) the limit is 0.

    (f)   The general term is $1 - 1/10^n$; the limit is 1.

**2.** Describe the behavior of $y = x + 2$ as $x$ ranges over the values of each of the sequences of Prob. 1.

    (a)   $y \to 2$ over the sequence   $3, 5/2, 7/3, 9/4, 11/5, \ldots, 2 + 1/n, \ldots$

    (b)   $y \to 2$ over the sequence   $3, 9/4, 19/9, 33/16, 51/25, \ldots, 2 + 1/n^2, \ldots$

    (c)   $y \to 5$ over the sequence   $4, 9/2, 14/3, 19/4, 24/5, \ldots, 5 - 1/n, \ldots$

    (d)   $y \to 5$ over the sequence   $7, 6, 17/3, 11/2, 27/5, \ldots, 5 + 2/n, \ldots$

    (e)   $y \to 2$ over the sequence   $5/2, 9/4, 17/8, 33/16, 65/32, \ldots, 2 + 1/2^n, \ldots$

    (f)   $y \to 3$ over the sequence   $2.9, 2.99, 2.999, 2.9999, \ldots, 3 - \dfrac{1}{10^n}, \ldots$

**3.** Evaluate:

    (a)   $\lim\limits_{x \to 2} 5x = 5 \lim\limits_{x \to 2} x = 5 \cdot 2 = 10$

    (b)   $\lim\limits_{x \to 2} (2x + 3) = 2 \lim\limits_{x \to 2} x + \lim\limits_{x \to 2} 3$
                        $= 2 \cdot 2 + 3 = 7$

    (c)   $\lim\limits_{x \to 2} (x^2 - 4x + 1) = 4 - 8 + 1 = -3$

    (d)   $\lim\limits_{x \to 3} \dfrac{x-2}{x+2} = \dfrac{\lim\limits_{x \to 3} (x-2)}{\lim\limits_{x \to 3} (x+2)} = \dfrac{1}{5}$

    (e)   $\lim\limits_{x \to -2} \dfrac{x^2 - 4}{x^2 + 4} = \dfrac{4-4}{4+4} = 0$

    (f)   $\lim\limits_{x \to 4} \sqrt{25 - x^2} = \sqrt{\lim\limits_{x \to 4} (25 - x^2)} = \sqrt{9} = 3$

*Note.* Do not assume from these problems that $\lim\limits_{x \to a} f(x)$ is invariably $f(a)$. By $f(a)$ is meant the value of $f(x)$ *when* $x = a$; by the convention of Page 9, $x$ is *never* equal to $a$ as $x \to a$.

4.   Examine the behavior of $f(x) = (-1)^x$ as $x$ ranges over the sequences

$\qquad$ (a)  1/3, 1/5, 1/7, 1/9, ...      and      (b)  2/3, 2/5, 2/7, 2/9, ...

What can be said concerning $\lim\limits_{x \to 0} (-1)^x$ and $f(0)$?

$\qquad$ (a)       $(-1)^x \to -1$  over the sequence  $-1, -1, -1, -1, \ldots$

$\qquad$ (b)       $(-1)^x \to +1$  over the sequence  $+1, +1, +1, +1, \ldots$

Since $(-1)^x$ approaches different limits over the two sequences, $\lim\limits_{x \to 0} (-1)^x$ does not exist; $f(0) = (-1)^0 = +1$.

5.   Evaluate:

(a)  $\lim\limits_{x \to 4} \dfrac{x-4}{x^2-x-12} = \lim\limits_{x \to 4} \dfrac{x-4}{(x+3)(x-4)} = \lim\limits_{x \to 4} \dfrac{1}{x+3} = \dfrac{1}{7}$

$\qquad$ The division by $(x - 4)$ before passing to the limit is valid since, by the convention on Page 9, $x \neq 4$ as $x \to 4$; hence, $x - 4$ is never zero.

(b)  $\lim\limits_{x \to 3} \dfrac{x^3-27}{x^2-9} = \lim\limits_{x \to 3} \dfrac{(x-3)(x^2+3x+9)}{(x-3)(x+3)} = \lim\limits_{x \to 3} \dfrac{x^2+3x+9}{x+3} = \dfrac{9}{2}$

(c)  $\lim\limits_{h \to 0} \dfrac{(x+h)^2-x^2}{h} = \lim\limits_{h \to 0} \dfrac{x^2+2hx+h^2-x^2}{h} = \lim\limits_{h \to 0} \dfrac{2hx+h^2}{h} = \lim\limits_{h \to 0} (2x+h) = 2x$

$\qquad$ Here, and again in Problems 7 and 8, $h$ is a variable so that it could be argued that we are in reality dealing with functions of two variables. However, the fact that $x$ is a variable plays no role in these problems; we may then for the moment consider $x$ to be a constant, that is, some one of the values of its range. The gist of this problem, as we shall see in Chapter 4, is that if $x$ is any value, say $x = x_0$, in the domain of $y = x^2$ then

$$\lim\limits_{h \to 0} \dfrac{(x+h)^2 - x^2}{h} \quad \text{is always twice the selected value of } x.$$

(d)  $\lim\limits_{x \to 2} \dfrac{4-x^2}{3-\sqrt{x^2+5}} = \lim\limits_{x \to 2} \dfrac{(4-x^2)(3+\sqrt{x^2+5})}{(3-\sqrt{x^2+5})(3+\sqrt{x^2+5})} = \lim\limits_{x \to 2} \dfrac{(4-x^2)(3+\sqrt{x^2+5})}{4-x^2}$

$\qquad\qquad\qquad$ $= \lim\limits_{x \to 2} (3+\sqrt{x^2+5}) = 6$

(e)  $\lim\limits_{x \to 1} \dfrac{x^2+x-2}{(x-1)^2} = \lim\limits_{x \to 1} \dfrac{(x-1)(x+2)}{(x-1)^2} = \lim\limits_{x \to 1} \dfrac{x+2}{x-1} = \infty$ ; no limit exists.

6.   Evaluate the following, first dividing numerator and denominator by the highest power of $x$ present and then using $\lim\limits_{x \to \infty} 1/x = 0$.

(a)  $\lim\limits_{x \to \infty} \dfrac{3x-2}{9x+7} = \lim\limits_{x \to \infty} \dfrac{3-2/x}{9+7/x} = \dfrac{3-0}{9+0} = \dfrac{1}{3}$

(b)  $\lim\limits_{x \to \infty} \dfrac{6x^2+2x+1}{6x^2-3x+4} = \lim\limits_{x \to \infty} \dfrac{6+2/x+1/x^2}{6-3/x+4/x^2} = \dfrac{6+0+0}{6-0+0} = 1$

(c)  $\lim\limits_{x \to \infty} \dfrac{x^2+x-2}{4x^3-1} = \lim\limits_{x \to \infty} \dfrac{1/x+1/x^2-2/x^3}{4-1/x^3} = \dfrac{0}{4} = 0$

(d)  $\lim\limits_{x \to \infty} \dfrac{2x^3}{x^2+1} = \lim\limits_{x \to \infty} \dfrac{2}{1/x+1/x^3} = \infty$ ; no limit exists.

7.   Given $f(x) = x^2 - 3x$, find $\lim\limits_{h \to 0} \dfrac{f(x+h) - f(x)}{h}$.

$\qquad$ Since $f(x) = x^2 - 3x$, $f(x+h) = (x+h)^2 - 3(x+h)$ and

$\lim\limits_{h \to 0} \dfrac{f(x+h)-f(x)}{h} = \lim\limits_{h \to 0} \dfrac{(x^2+2hx+h^2-3x-3h)-(x^2-3x)}{h} = \lim\limits_{h \to 0} \dfrac{2hx+h^2-3h}{h}$

$\qquad\qquad\qquad\qquad = \lim\limits_{h \to 0} (2x+h-3) = 2x - 3$

8.  Given $f(x) = \sqrt{5x+1}$, find $\lim\limits_{h \to 0} \dfrac{f(x+h)-f(x)}{h}$ when $x > -1/5$.

$$\lim_{h \to 0} \frac{f(x+h)-f(x)}{h} = \lim_{h \to 0} \frac{\sqrt{5x+5h+1} - \sqrt{5x+1}}{h}$$

$$= \lim_{h \to 0} \frac{\sqrt{5x+5h+1} - \sqrt{5x+1}}{h} \cdot \frac{\sqrt{5x+5h+1} + \sqrt{5x+1}}{\sqrt{5x+5h+1} + \sqrt{5x+1}}$$

$$= \lim_{h \to 0} \frac{(5x+5h+1) - (5x+1)}{h(\sqrt{5x+5h+1} + \sqrt{5x+1})}$$

$$= \lim_{h \to 0} \frac{5}{\sqrt{5x+5h+1} + \sqrt{5x+1}} = \frac{5}{2\sqrt{5x+1}}$$

9.  Determine the points $x = a$ for which each denominator is zero. Then examine $y$ as $x \to a^-$ and $x \to a^+$.

(a) $y = f(x) = 2/x$. The denominator is zero when $x = 0$. As $x \to 0^-$, $y \to -\infty$; as $x \to 0^+$, $y \to +\infty$.

(b) $y = f(x) = \dfrac{x-1}{(x+3)(x-2)}$. The denominator is zero for $x = -3$ and $x = 2$. As $x \to -3^-$, $y \to -\infty$; as $x \to -3^+$, $y \to +\infty$. As $x \to 2^-$, $y \to -\infty$; as $x \to 2^+$, $y \to +\infty$.

(c) $y = f(x) = \dfrac{x-3}{(x+2)(x-1)}$. The denominator is zero for $x = -2$ and $x = 1$. As $x \to -2^-$, $y \to -\infty$; as $x \to -2^+$, $y \to +\infty$. As $x \to 1^-$, $y \to +\infty$; as $x \to 1^+$, $y \to -\infty$.

(d) $y = f(x) = \dfrac{(x+2)(x-1)}{(x-3)^2}$. The denominator is zero for $x = 3$. As $x \to 3^-$, $y \to +\infty$; as $x \to 3^+$, $y \to +\infty$.

(e) $y = f(x) = \dfrac{(x+2)(1-x)}{x-3}$. The denominator is zero for $x = 3$. As $x \to 3^-$, $y \to +\infty$; as $x \to 3^+$, $y \to -\infty$.

10. Examine (a) $\lim\limits_{x \to 0} \dfrac{1}{3 + 2^{1/x}}$, (b) $\lim\limits_{x \to 0} \dfrac{1 + 2^{1/x}}{3 + 2^{1/x}}$.

(a) Let $x \to 0^-$; then $1/x \to -\infty$, $2^{1/x} \to 0$, and $\lim\limits_{x \to 0^-} \dfrac{1}{3 + 2^{1/x}} = 1/3$.

Let $x \to 0^+$; then $1/x \to +\infty$, $2^{1/x} \to +\infty$, and $\lim\limits_{x \to 0^+} \dfrac{1}{3 + 2^{1/x}} = 0$.

Thus $\lim\limits_{x \to 0} \dfrac{1}{3 + 2^{1/x}}$ does not exist.

(b) Let $x \to 0^-$; then $2^{1/x} \to 0$ and $\lim\limits_{x \to 0^-} \dfrac{1 + 2^{1/x}}{3 + 2^{1/x}} = \dfrac{1}{3}$.

Let $x \to 0^+$. For $x \neq 0$, $\dfrac{1 + 2^{1/x}}{3 + 2^{1/x}} = \dfrac{2^{-1/x} + 1}{3 \cdot 2^{-1/x} + 1}$ and since $\lim\limits_{x \to 0^+} 2^{-1/x} = 0$, $\lim\limits_{x \to 0^+} \dfrac{2^{-1/x} + 1}{3 \cdot 2^{-1/x} + 1} = 1$.
Thus, $\lim\limits_{x \to 0} \dfrac{1 + 2^{1/x}}{3 + 2^{1/x}}$ does not exist.

11. For each of the functions of Problem 9, examine $y$ as $x \to -\infty$ and as $x \to +\infty$.

(a) When $|x|$ is large, $|y|$ is small.

    For $x = -1000$, $y < 0$; as $x \to -\infty$, $y \to 0^-$. For $x = +1000$, $y > 0$; as $x \to +\infty$, $y \to 0^+$.

(b), (c)  Same as (a).

(d) When $|x|$ is large, $|y|$ is approximately 1.

    For $x = -1000$, $y < 1$; as $x \to -\infty$, $y \to 1^-$. For $x = +1000$, $y > 1$; as $x \to +\infty$, $y \to 1^+$.

(e) When $|x|$ is large, $|y|$ is large.

    For $x = -1000$, $y > 0$; as $x \to -\infty$, $y \to +\infty$. For $x = +1000$, $y < 0$; as $x \to +\infty$, $y \to -\infty$.

12. Examine the function of Prob. 9, Chap. 1, as $x \to a^-$ and as $x \to a^+$ when $a$ is any positive integer.

Consider $a = 2$. As $x \to 2^-$ over the sequence $(1)$, $f(x) \to 10$ over the sequence $5, 10, 10, 10, \ldots$; as $x \to 2^+$ over the sequence $(2)$, $f(x) \to 15$. Thus, $\lim\limits_{x \to 2} f(x)$ and, hence, $\lim\limits_{x \to a} f(x)$ does not exist.

13. Use the precise definition to show:

(a) $\lim\limits_{x \to 1} (4x^3 + 3x^2 - 24x + 22) = 5$,   (b) $\lim\limits_{x \to -1} (-2x^3 + 9x + 4) = -3$

(a) Let $\epsilon$ be chosen.  For $0 < |x - 1| < \lambda < 1$,

$$\begin{aligned}
|(4x^3 + 3x^2 - 24x + 22) - 5| &= |4(x-1)^3 + 15x^2 - 36x + 21| = |4(x-1)^3 + 15(x-1)^2 - 6(x-1)| \\
&\le 4|x-1|^3 + 15|x-1|^2 + 6|x-1| \\
&< 4\lambda + 15\lambda + 6\lambda = 25\lambda
\end{aligned}$$

Now $|(4x^3 + 3x^2 - 24x + 22) - 5| < \epsilon$ for $\lambda < \epsilon/25$; hence, any positive number smaller than both 1 and $\epsilon/25$ is an effective $\delta$ and the limit is established.

(b) Let $\epsilon$ be chosen.  For $0 < |x + 1| < \lambda < 1$,

$$\begin{aligned}
|(-2x^3 + 9x + 4) + 3| &= |-2(x+1)^3 + 6(x+1)^2 + 3(x+1)| \\
&\le 2|x+1|^3 + 6|x+1|^2 + 3|x+1| < 11\lambda
\end{aligned}$$

Any positive number smaller than both 1 and $\epsilon/11$ is an effective $\delta$ and the limit is established.

14. Given $\lim\limits_{x \to a} f(x) = A$ and $\lim\limits_{x \to a} g(x) = B$, prove

(a) $\lim\limits_{x \to a} \{f(x) + g(x)\} = A + B$,   (b) $\lim\limits_{x \to a} \{f(x) \cdot g(x)\} = AB$,   (c) $\lim\limits_{x \to a} \dfrac{f(x)}{g(x)} = \dfrac{A}{B}$, $B \ne 0$

Since $\lim\limits_{x \to a} f(x) = A$ and $\lim\limits_{x \to a} g(x) = B$, it follows by the precise definition that for numbers $\epsilon_1 > 0$ and $\epsilon_2 > 0$, however small, there exist numbers $\delta_1 > 0$ and $\delta_2 > 0$ such that:

(i)   whenever $0 < |x - a| < \delta_1$ then $|f(x) - A| < \epsilon_1$, and

(ii)  whenever $0 < |x - a| < \delta_2$ then $|g(x) - B| < \epsilon_2$.

Let $\lambda$ denote the smaller of $\delta_1$ and $\delta_2$; now

(iii) whenever $0 < |x - a| < \lambda$ then $|f(x) - A| < \epsilon_1$ and $|g(x) - B| < \epsilon_2$.

(a) Let $\epsilon$ be chosen.  We are required to produce a $\delta > 0$ such that

whenever   $0 < |x - a| < \delta$   then   $|\{f(x) + g(x)\} - \{A + B\}| < \epsilon$

Now $|\{f(x) + g(x)\} - \{A + B\}| = |\{f(x) - A\} + \{g(x) - B\}| \le |f(x) - A| + |g(x) - B|$.  By (iii), $|f(x) - A| < \epsilon_1$ whenever $0 < |x - a| < \lambda$ and $|g(x) - A| < \epsilon_2$ whenever $0 < |x - a| < \lambda$ where $\lambda$ is the smaller of $\delta_1$ and $\delta_2$.  Thus,

$$|\{f(x) + g(x)\} - \{A + B\}| < \epsilon_1 + \epsilon_2 \quad \text{whenever} \quad 0 < |x - a| < \lambda$$

Take $\epsilon_1 = \epsilon_2 = \frac{1}{2}\epsilon$ and $\delta = \lambda$ for this choice of $\epsilon_1$ and $\epsilon_2$; then, as required,

$$|\{f(x) + g(x)\} - \{A + B\}| < \tfrac{1}{2}\epsilon + \tfrac{1}{2}\epsilon = \epsilon \quad \text{whenever} \quad 0 < |x - a| < \delta$$

(b) Let $\epsilon$ be chosen.  We are required to produce a $\delta > 0$ such that

whenever   $0 < |x - a| < \delta$   then   $|f(x) \cdot g(x) - AB| < \epsilon$

Now
$$\begin{aligned}
|f(x) \cdot g(x) - AB| &= |\{f(x) - A\} \cdot \{g(x) - B\} + B\{f(x) - A\} + A\{g(x) - B\}| \\
&\le |f(x) - A| \cdot |g(x) - B| + |B| \cdot |f(x) - A| + |A| \cdot |g(x) - B|
\end{aligned}$$

so that, by (iii),    $|f(x) \cdot g(x) - AB| < \epsilon_1 \epsilon_2 + |B|\epsilon_1 + |A|\epsilon_2$ whenever $0 < |x - a| < \lambda$.

Take $\epsilon_1$ and $\epsilon_2$ such that $\epsilon_1 \epsilon_2 < \frac{1}{3}\epsilon$, $\epsilon_1 < \dfrac{1}{3}\dfrac{\epsilon}{|B|}$, and $\epsilon_2 < \dfrac{1}{3}\dfrac{\epsilon}{|A|}$ are simultaneously satisfied and $\delta = \lambda$ for this choice of $\epsilon_1$ and $\epsilon_2$.  Then, as required,

$$|f(x) \cdot g(x) - AB| < \epsilon/3 + \epsilon/3 + \epsilon/3 = \epsilon \quad \text{whenever} \quad 0 < |x - a| < \delta$$

*(c)* Since $\dfrac{f(x)}{g(x)} = f(x) \cdot \dfrac{1}{g(x)}$, the theorem follows from *(b)* provided we can show that $\lim\limits_{x \to a} \dfrac{1}{g(x)} = \dfrac{1}{B}$, $B \neq 0$.

Let $\epsilon$ be chosen. We are required to produce a $\delta > 0$ such that

$$\text{whenever } \quad 0 < |x - a| < \delta \quad \text{ then } \quad \left| \frac{1}{g(x)} - \frac{1}{B} \right| < \epsilon.$$

Now $\left| \dfrac{1}{g(x)} - \dfrac{1}{B} \right| = \left| \dfrac{B - g(x)}{B \cdot g(x)} \right| = \dfrac{|g(x) - B|}{|B| \cdot |g(x)|} = \dfrac{|g(x) - B|}{|B|} \cdot \dfrac{1}{|g(x)|}$. By (ii),

$$|g(x) - B| < \epsilon_2 \quad \text{whenever} \quad 0 < |x - a| < \delta_2$$

However, we are also dealing with $\dfrac{1}{g(x)}$ so we must be sure $\delta_2$ is sufficiently small that the interval $a - \delta_2 < x < a + \delta_2$ does not contain a root of $g(x) = 0$. Let $\delta_3 \leq \delta_2$ meet this requirement so that $|g(x) - B| < \epsilon_2$ and $|g(x)| > 0$ whenever $0 < |x - a| < \delta_3$. Now $|g(x)| > 0$ on the interval implies $|g(x)| > b > 0$ and $\dfrac{1}{|g(x)|} < \dfrac{1}{b}$ on the interval. Thus, we have

$$\left| \frac{1}{g(x)} - \frac{1}{B} \right| < \frac{\epsilon_2}{|B|} \cdot \frac{1}{b} \quad \text{whenever} \quad 0 < |x - a| < \delta_3$$

Take $\epsilon_2 < \epsilon b |B|$ so that $\dfrac{\epsilon_2}{|B| \cdot b} < \epsilon$ and $\delta = \delta_3$ for this choice of $\epsilon_2$. Then, as required,

$$\left| \frac{1}{g(x)} - \frac{1}{B} \right| < \epsilon \quad \text{whenever} \quad 0 < |x - a| < \delta$$

15. Prove: *(a)* $\lim\limits_{x \to 2} \dfrac{1}{(x-2)^3} = \infty$, *(b)* $\lim\limits_{x \to \infty} \dfrac{x}{x+1} = 1$, *(c)* $\lim\limits_{x \to \infty} \dfrac{x^2}{x-1} = \infty$.

*(a)* Let $M$ be chosen. For all $x$ such that $0 < |x - 2| < \delta$,

$$\left| \frac{1}{(x-2)^3} \right| > \frac{1}{\delta^3}. \quad \text{Then} \quad \left| \frac{1}{(x-2)^3} \right| > M \quad \text{when} \quad \frac{1}{\delta^3} > M \quad \text{or} \quad \delta < \frac{1}{\sqrt[3]{M}}.$$

*(b)* Let $\epsilon$ be chosen. For all $x$ such that $|x| > M$, $\left| \dfrac{x}{x+1} - 1 \right| = \dfrac{1}{|x+1|} \leq \dfrac{1}{|x|-1} < \dfrac{1}{M-1}$.

Then $\left| \dfrac{x}{x+1} - 1 \right| < \epsilon$ when $\dfrac{1}{M-1} < \epsilon$ or $M > 1 + \dfrac{1}{\epsilon}$.

*(c)* Let $M$, sufficiently large, be chosen. For all $x$ such that $|x| > P > 1$,

$$\left| \frac{x^2}{x-1} \right| \geq \frac{x^2}{|x|+1} > \frac{x^2}{2|x|} = \frac{1}{2}|x| > \frac{1}{2}P. \quad \text{Then} \quad \left| \frac{x^2}{x-1} \right| > M \quad \text{when} \quad P > 2M.$$

## Supplementary Problems

16. Describe the behavior of $y = 2x + 1$ as $x$ ranges over each sequence of Problem 1.
    *Ans.* *(a)* $y \to 1$, *(b)* $y \to 1$, *(c)* $y \to 7$, *(d)* $y \to 7$, *(e)* $y \to 1$, *(f)* $y \to 3$

17. Evaluate:

*(a)* $\lim\limits_{x \to 2} (x^2 - 4x)$   *(e)* $\lim\limits_{x \to 2} \dfrac{x-1}{x^2-1}$   *(i)* $\lim\limits_{x \to 2} \dfrac{x-2}{\sqrt{x^2-4}}$

*(b)* $\lim\limits_{x \to -1} (x^3 + 2x^2 - 3x - 4)$   *(f)* $\lim\limits_{x \to 2} \dfrac{x^2-4}{x^2-5x+6}$   *(j)* $\lim\limits_{x \to 2} \dfrac{\sqrt{x-2}}{x^2-4}$

*(c)* $\lim\limits_{x \to 1} \dfrac{(3x-1)^2}{(x+1)^3}$   *(g)* $\lim\limits_{x \to -1} \dfrac{x^2+3x+2}{x^2+4x+3}$   *(k)* $\lim\limits_{h \to 0} \dfrac{(x+h)^3 - x^3}{h}$

*(d)* $\lim\limits_{x \to 0} \dfrac{3^x - 3^{-x}}{3^x + 3^{-x}}$   *(h)* $\lim\limits_{x \to 2} \dfrac{x-2}{x^2-4}$   *(l)* $\lim\limits_{x \to 1} \dfrac{x-1}{\sqrt{x^2+3}-2}$

*Ans.* *(a)* $-4$; *(b)* $0$; *(c)* $\frac{1}{2}$; *(d)* $0$; *(e)* $\frac{1}{3}$; *(f)* $-4$; *(g)* $\frac{1}{2}$; *(h)* $\frac{1}{4}$; *(i)* $0$; *(j)* $\infty$, no limit; *(k)* $3x^2$; *(l)* $2$

18. Evaluate:

   (a) $\lim\limits_{x \to \infty} \dfrac{2x+3}{4x-5}$     (c) $\lim\limits_{x \to \infty} \dfrac{x}{x^2+5}$     (e) $\lim\limits_{x \to \infty} \dfrac{x+3}{x^2+5x+6}$     (g) $\lim\limits_{x \to -\infty} \dfrac{3^x - 3^{-x}}{3^x + 3^{-x}}$

   (b) $\lim\limits_{x \to \infty} \dfrac{2x^2+1}{6+x-3x^2}$     (d) $\lim\limits_{x \to \infty} \dfrac{x^2+5x+6}{x+1}$     (f) $\lim\limits_{x \to +\infty} \dfrac{3^x - 3^{-x}}{3^x + 3^{-x}}$

   *Ans.* (a) $\frac{1}{2}$; (b) $-2/3$; (c) 0; (d) $\infty$, no limit; (e) 0; (f) 1; (g) $-1$

19. Find $\lim\limits_{h \to 0} \dfrac{f(a+h) - f(a)}{h}$ for each function of Problem 24, Chapter 1.

   *Ans.* (a) $\dfrac{-1}{(a-2)^2}$, (b) $\dfrac{1}{2\sqrt{a-4}}$, (c) $\dfrac{1}{(a+1)^2}$

20. What can be said about $\lim\limits_{x \to \infty} \dfrac{a_0 x^m + a_1 x^{m-1} + \cdots + a_m}{b_0 x^n + b_1 x^{n-1} + \cdots + b_n}$ , where $a_0 b_0 \neq 0$ and $m, n$ are positive integers,

   when (a) $m > n$, (b) $m = n$, (c) $m < n$?     *Ans.* (a) no limit; (b) $a_0/b_0$; (c) 0

21. Investigate the behavior of $f(x) = |x|$ as $x \to 0$. Draw a graph.

   Hint: Examine $\lim\limits_{x \to 0^-} f(x)$ and $\lim\limits_{x \to 0^+} f(x)$.     *Ans.* $\lim\limits_{x \to 0} |x| = 0$

22. Investigate the behavior of $\begin{cases} f(x) = x, & x > 0 \\ f(x) = x+1, & x \leq 0 \end{cases}$ as $x \to 0$. Draw a graph.

   *Ans.* $\lim\limits_{x \to 0} f(x)$ does not exist.

23. (a) Use Theorem IV and mathematical induction to prove

$$\lim\limits_{x \to a} x^n = a^n, \text{ for } n \text{ a positive integer.}$$

   (b) Use Theorem III and mathematical induction to prove

$$\lim\limits_{x \to a} \{f_1(x) + f_2(x) + \cdots + f_n(x)\} = \lim\limits_{x \to a} f_1(x) + \lim\limits_{x \to a} f_2(x) + \cdots + \lim\limits_{x \to a} f_n(x)$$

24. Use Theorem II and the results of Problem 23 to prove

$$\lim\limits_{x \to a} P(x) = P(a), \text{ where } P(x) \text{ is any polynomial in } x.$$

25. For $f(x) = 5x - 6$, find a $\delta > 0$ such that whenever $0 < |x-4| < \delta$ then $|f(x) - 14| < \epsilon$, when
   (a) $\epsilon = \frac{1}{2}$, (b) $\epsilon = 0.001$.     *Ans.* (a) $1/10$, (b) $0.0002$

26. Use the precise definition to prove:
   (a) $\lim\limits_{x \to 3} 5x = 15$,   (b) $\lim\limits_{x \to 2} x^2 = 4$,   (c) $\lim\limits_{x \to 2} (x^2 - 3x + 5) = 3$.

27. Use the precise definition to prove:
   (a) $\lim\limits_{x \to 0} \dfrac{1}{x} = \infty$,   (b) $\lim\limits_{x \to 1} \dfrac{x}{x-1} = \infty$,   (c) $\lim\limits_{x \to \infty} \dfrac{x}{x-1} = 1$,   (d) $\lim\limits_{x \to \infty} \dfrac{x^2}{x+1} = \infty$.

   *Ans.* (a) $\delta < 1/M$,   (b) $\delta < \dfrac{1}{M+1}$,   (c) $M > 1 + \dfrac{1}{\epsilon}$,   (d) $P > 2M$

28. Prove: If $f(x)$ is defined for all $x$ near $x = a$ and has a limit as $x \to a$, that limit is unique.

   Hint: Assume $\lim\limits_{x \to a} f(x) = A$, $\lim\limits_{x \to a} f(x) = B$, and $B \neq A$. Choose $\epsilon_1, \epsilon_2 < \frac{1}{2}|A - B|$. Determine $\delta_1$ and $\delta_2$ for the two limits and take $\delta$ the smaller of $\delta_1$ and $\delta_2$. Show that then $|A - B| = |\{A - f(x)\} + \{f(x) - B\}| < |A - B|$, a contradiction.

29. Let $f(x), g(x), h(x)$ be such that (i) $f(x) \leq g(x) \leq h(x)$ for all values of $x$ near $x = a$ and (ii) $\lim\limits_{x \to a} f(x) = \lim\limits_{x \to a} h(x) = A$. Show that $\lim\limits_{x \to a} g(x) = A$.

   Hint: For a given $\epsilon > 0$, however small, there exists a $\delta > 0$ such that whenever $0 < |x - a| < \delta$ then $|f(x) - A| < \epsilon$ and $|h(x) - A| < \epsilon$ or $A - \epsilon < f(x) \leq g(x) \leq h(x) < A + \epsilon$.

30. Prove: If $f(x) \leq M$ for all $x$ and if $\lim\limits_{x \to a} f(x) = A$, then $A \leq M$.

   Hint: Suppose $A > M$. Choose $\epsilon = \frac{1}{2}(A - M)$ and obtain a contradiction.

# Chapter 3

## Continuity

**A FUNCTION** $f(x)$ is said to be *continuous* at $x = x_0$ if

(i) $f(x_0)$ is defined,    (ii) $\lim\limits_{x \to x_0} f(x)$ exists,    (iii) $\lim\limits_{x \to x_0} f(x) = f(x_0)$

For example, $f(x) = x^2 + 1$ is continuous at $x = 2$ since $\lim\limits_{x \to 2} f(x) = 5 = f(2)$. The condition (i) implies that a function can be continuous only at points on its domain of definition. Thus, $f(x) = \sqrt{4 - x^2}$ is not continuous at $x = 3$ since $f(3)$ is imaginary, i.e., is not defined.

A function which is continuous at every point of an interval (open or closed) is said to be continuous on that interval. A function $f(x)$ is called *continuous* if it is continuous at every point on its domain of definition. Thus, $f(x) = x^2 + 1$ and all other polynomials in $x$ are continuous functions; other examples are $e^x$, $\sin x$, $\cos x$.

If the domain of definition of a function is a closed interval $a \leqq x \leqq b$, condition (ii) fails at the endpoints $a$ and $b$. We shall call such a function continuous if it is continuous on the open interval $a < x < b$, if $\lim\limits_{x \to a^+} f(x) = f(a)$, and if $\lim\limits_{x \to b^-} f(x) = f(b)$. Thus, $f(x) = \sqrt{9 - x^2}$ will be called a continuous function (see Example 1, Chapter 2). The functions of elementary calculus are continuous on their domains of definition with the possible exception of a number of isolated points.

**A FUNCTION** $f(x)$ is said to be *discontinuous* at $x = x_0$ if one or more of the conditions for continuity fail there. The several types of discontinuity will be illustrated by examples:

(a)   $f(x) = \dfrac{1}{x - 2}$ is discontinuous at $x = 2$ since

    (i)   $f(2)$ is not defined (has zero as denominator)

    (ii)   $\lim\limits_{x \to 2} f(x)$ does not exist (equals $\infty$).

The function is continuous everywhere except at $x = 2$ where it is said to have an *infinite discontinuity*. See Fig. 3-1.

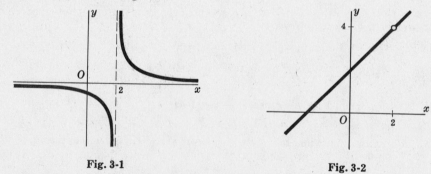

**Fig. 3-1**             **Fig. 3-2**

(b)   $f(x) = \dfrac{x^2 - 4}{x - 2}$ is discontinuous at $x = 2$ since

    (i)   $f(2)$ is not defined (both numerator and denominator are zero).

    (ii)   $\lim\limits_{x \to 2} f(x) = 4$.

The discontinuity here is called *removable* since it may be removed by redefining the function as $f(x) = \dfrac{x^2 - 4}{x - 2}$, $x \neq 2$; $f(2) = 4$. (Note that the discontinuity in ($a$) cannot be so removed since the limit also does not exist.) The graphs of $f(x) = \dfrac{x^2 - 4}{x - 2}$ and $g(x) = x + 2$ are identical except at $x = 2$ where the former has a 'hole'. Removing the discontinuity consists simply of properly filling the 'hole'.

($c$) $f(x) = \dfrac{x^3 - 27}{x - 3}$, $x \neq 3$; $f(3) = 9$ is discontinuous at $x = 3$ since

$$\textbf{(i)} \ \ f(3) = 9, \quad \textbf{(ii)} \ \ \lim_{x \to 3} f(x) = 27, \quad \textbf{(iii)} \ \ \lim_{x \to 3} f(x) \neq f(3)$$

The discontinuity may be removed by redefining the function as $f(x) = \dfrac{x^3 - 27}{x - 3}$, $x \neq 3$; $f(3) = 27$.

($d$) The function of Problem 9, Chapter 1, is defined for all $x > 0$ but has discontinuities at $x = 1, 2, 3, \ldots$ (see Problem 12, Chapter 2) arising from the fact that

$$\lim_{x \to s^-} f(x) \neq \lim_{x \to s^+} f(x) \qquad (s \text{ any positive integer})$$

These are called *jump discontinuities*.                        See Problems 1-2.

**PROPERTIES OF CONTINUOUS FUNCTIONS.** The theorems on limits of Chapter 2 lead readily to theorems on continuous functions. In particular, if $f(x)$ and $g(x)$ are continuous at $x = a$, so also are $f(x) \pm g(x)$, $f(x) \cdot g(x)$, and $f(x)/g(x)$, provided in the latter that $g(a) \neq 0$. Hence, polynomials in $x$ are everywhere continuous while rational functions of $x$ are everywhere continuous except at values of $x$ for which the denominator is zero.

The reader has used certain properties of continuous functions in the study of algebra:

($a$) In sketching the graph of a polynomial $y = f(x)$, any two points $(a, f(a))$ and $(b, f(b))$ were joined by an unbroken arc.

($b$) If $f(a)$ and $f(b)$ have opposite signs, the graph of $y = f(x)$ crosses the $x$-axis at least once and the equation $f(x) = 0$ has at least one root between $x = a$ and $x = b$.

The property of continuous functions used here is

**I.** If $f(x)$ is continuous on the interval $a \leqq x \leqq b$ and if $f(a) \neq f(b)$, then for any number $c$ between $f(a)$ and $f(b)$ there is at least one value of $x$, say $x = x_0$, for which $f(x_0) = c$.

Figures 3-3$a$ and 3-3$b$ illustrate the two applications of this property while Figures 3-4$a$ and 3-4$b$ show that continuity throughout the interval is essential.

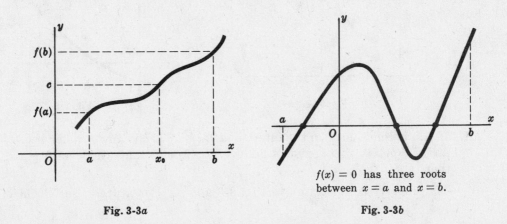

$f(x) = 0$ has three roots between $x = a$ and $x = b$.

Fig. 3-3$a$                                    Fig. 3-3$b$

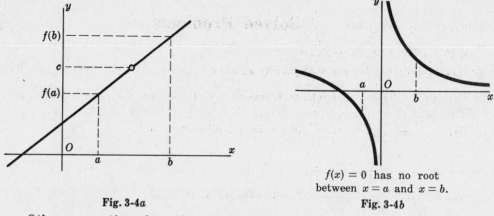

Fig. 3-4a

$f(x) = 0$ has no root
between $x = a$ and $x = b$.

Fig. 3-4b

Other properties of continuous functions are:

**II.** If $f(x)$ is continuous on the interval $a \leqq x \leqq b$, then $f(x)$ takes on a least value $m$ and a greatest value $M$ on the interval.

Although a proof of Property II is beyond the scope of this book, the property will be used freely in later chapters. The figures below merely establish the property intuitively. In Fig. 3-5a the function is continuous on $a \leqq x \leqq b$; the least value $m$ and the greatest value $M$ occur at $x = c$ and $x = d$ respectively, both points being within the interval. In Fig. 3-5b the function is continuous on $a \leqq x \leqq b$; the least value occurs at the endpoint $x = a$ while the greatest value occurs at $x = c$ within the interval. In Fig. 3-5c there is a discontinuity at $x = c$, where $a < c < b$; the function has a least value at $x = a$ but no greatest value.

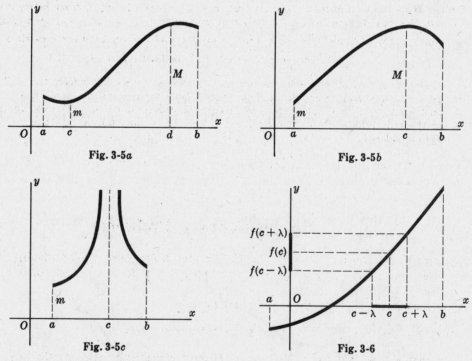

Fig. 3-5a

Fig. 3-5b

Fig. 3-5c

Fig. 3-6

**III.** If $f(x)$ is continuous on the interval $a \leqq x \leqq b$, if $c$ is any number between $a$ and $b$, and if $f(c) > 0$, there exists a number $\lambda > 0$ such that whenever $c - \lambda < x < c + \lambda$, then $f(x) > 0$.

This property is illustrated in Fig. 3-6 above. For a proof, see Problem 4.

# Solved Problems

1.  From Problem 9, Chapter 2, follow:

    (a)  $f(x) = 2/x$  has an infinite discontinuity at  $x = 0$.

    (b)  $f(x) = \dfrac{x-1}{(x+3)(x-2)}$  has infinite discontinuities at  $x = -3$  and  $x = 2$.

    (c)  $f(x) = \dfrac{(x+2)(x-1)}{(x-3)^2}$  has an infinite discontinuity at  $x = 3$.

2.  From Problem 5, Chapter 2, follow:

    (a)  $f(x) = \dfrac{x^3 - 27}{x^2 - 9}$  has a removable discontinuity at  $x = 3$.

    There is also an infinite discontinuity at  $x = -3$.

    (b)  $f(x) = \dfrac{4 - x^2}{3 - \sqrt{x^2 + 5}}$  has a removable discontinuity at  $x = 2$.

    There is also a removable discontinuity at  $x = -2$.

    (c)  $f(x) = \dfrac{x^2 + x - 2}{(x-1)^2}$  has an infinite discontinuity at  $x = 1$.

3.  Show that the existence of  $\lim\limits_{h \to 0} \dfrac{f(a+h) - f(a)}{h}$  implies  $f(x)$  is continuous at  $x = a$.

    The existence of the limit implies that  $f(a+h) - f(a) \to 0$  as  $h \to 0$.  Thus,  $\lim\limits_{h \to 0} f(a+h) = f(a)$  and  $f(x)$  is continuous at  $x = a$.

4.  Prove: If  $f(x)$  is continuous on the interval  $a \le x \le b$,  if  $c$  is any number between  $a$  and  $b$, and if  $f(c) > 0$  there exists a number  $\lambda > 0$  such that whenever  $c - \lambda < x < c + \lambda$,  then  $f(x) > 0$.

    Since  $f(x)$  is continuous at  $x = c$,  $\lim\limits_{x \to c} f(x) = f(c)$  and for any  $\epsilon > 0$  there exists a  $\delta > 0$  such that

    (i)   whenever   $0 < |x - c| < \delta$   then   $|f(x) - f(c)| < \epsilon$.

    Now  $f(x) > 0$  at all points on the interval  $c - \delta < x < c + \delta$  for which  $f(x) \ge f(c)$.  At all other points of the interval  $f(x) < f(c)$  so that  $|f(x) - f(c)| = f(c) - f(x) < \epsilon$  and  $f(x) > f(c) - \epsilon$.  Thus, at these points,  $f(x) > 0$  unless  $\epsilon \ge f(c)$.  Hence, to determine an interval meeting the requirements of the theorem, select  $\epsilon < f(c)$,  determine  $\delta$  satisfying (i), and take  $\lambda < \delta$.  See Problem 10 for the companion theorem.

# Supplementary Problems

5.  Examine the functions of Problem 17(a)-(h), Chapter 2, for points of discontinuity.
    Ans.  (a), (b), (d) none;  (c)  $x = -1$;  (e)  $x = \pm 1$;  (f)  $x = 2, 3$;  (g)  $x = -1, -3$;  (h)  $x = \pm 2$

6.  Show that  $f(x) = |x|$  is everywhere continuous.

7.  Show that  $f(x) = \dfrac{1 - 2^{1/x}}{1 + 2^{1/x}}$  has a jump discontinuity at  $x = 0$.

8.  Show that at  $x = 0$   (a)  $f(x) = \dfrac{1}{3^{1/x} + 1}$  has a jump discontinuity and  (b)  $f(x) = \dfrac{x}{3^{1/x} + 1}$  has a removable discontinuity.

9.  In Fig. 3-4a, the graph of  $f(x) = \dfrac{x^2 - 4x - 21}{x - 7}$,  take  $a = 3$  and  $b = 11$.  Show that  $c = 10$.

10. Prove: If  $f(x)$  is continuous on the interval  $a \le x \le b$,  if  $c$  is any number between  $a$  and  $b$, and if  $f(c) < 0$,  there exists a number  $\lambda > 0$  such that  whenever  $c - \lambda < x < c + \lambda$  then  $f(x) < 0$.

# Chapter 4

# The Derivative

**INCREMENTS.** The *increment* $\Delta x$ of a variable $x$ is the change in $x$ as it increases or decreases from one value $x = x_0$ to another value $x = x_1$ in its range. Here, $\Delta x = x_1 - x_0$ and we may write $x_1 = x_0 + \Delta x$.

If the variable $x$ is given an increment $\Delta x$ from $x = x_0$ (that is, if $x$ changes from $x = x_0$ to $x = x_0 + \Delta x$) and a function $y = f(x)$ is thereby given an increment $\Delta y = f(x_0 + \Delta x) - f(x_0)$ from $y = f(x_0)$, the quotient

$$\frac{\Delta y}{\Delta x} = \frac{\text{change in } y}{\text{change in } x}$$

is called the *average rate of change* of the function on the interval between $x = x_0$ and $x = x_0 + \Delta x$.

**Example 1:**

When $x$ is given the increment $\Delta x = 0.5$ from $x_0 = 1$, the function $y = f(x) = x^2 + 2x$ is given the increment $\Delta y = f(1 + 0.5) - f(1) = 5.25 - 3 = 2.25$. Thus, the average rate of change of $y$ on the interval between $x = 1$ and $x = 1.5$ is $\dfrac{\Delta y}{\Delta x} = \dfrac{2.25}{0.5} = 4.5$.

See Problems 1-2.

**THE DERIVATIVE** of a function $y = f(x)$ with respect to $x$ at the point $x = x_0$ is defined as

$$\lim_{\Delta x \to 0} \frac{\Delta y}{\Delta x} = \lim_{\Delta x \to 0} \frac{f(x_0 + \Delta x) - f(x_0)}{\Delta x}$$

provided the limit exists. This limit is also called the *instantaneous rate of change* (or simply, the *rate of change*) of $y$ with respect to $x$ at $x = x_0$.

**Example 2:**

Find the derivative of $y = f(x) = x^2 + 3x$ with respect to $x$ at $x = x_0$. Use this to find the value of the derivative at (a) $x_0 = 2$ and (b) $x_0 = -4$.

$$y_0 = f(x_0) = x_0^2 + 3x_0$$
$$y_0 + \Delta y = f(x_0 + \Delta x) = (x_0 + \Delta x)^2 + 3(x_0 + \Delta x)$$
$$= x_0^2 + 2x_0\Delta x + (\Delta x)^2 + 3x_0 + 3\Delta x$$
$$\Delta y = f(x_0 + \Delta x) - f(x_0) = 2x_0\Delta x + 3\Delta x + (\Delta x)^2$$
$$\frac{\Delta y}{\Delta x} = \frac{f(x_0 + \Delta x) - f(x_0)}{\Delta x} = 2x_0 + 3 + \Delta x$$

The derivative at $x = x_0$ is

$$\lim_{\Delta x \to 0} \frac{\Delta y}{\Delta x} = \lim_{\Delta x \to 0} (2x_0 + 3 + \Delta x) = 2x_0 + 3$$

(a) At $x_0 = 2$, the value of the derivative is $2 \cdot 2 + 3 = 7$.

(b) At $x_0 = -4$, the value of the derivative is $2(-4) + 3 = -5$.

**IN FINDING DERIVATIVES** it is customary to drop the subscript 0 and obtain the derivative of $y = f(x)$ *with respect to x* as

$$\lim_{\Delta x \to 0} \frac{\Delta y}{\Delta x} = \lim_{\Delta x \to 0} \frac{f(x + \Delta x) - f(x)}{\Delta x}$$

See the note following Problem 5(c), Chapter 2.

The derivative of $y = f(x)$ with respect to $x$ may be indicated by any one of the symbols

$$\frac{d}{dx} y, \quad \frac{dy}{dx}, \quad D_x y, \quad y', \quad f'(x), \quad \text{or} \quad \frac{d}{dx} f(x)$$

See Problems 3-8.

# Solved Problems

1.  Given $y = f(x) = x^2 + 5x - 8$, find $\Delta y$ and $\Delta y/\Delta x$ as $x$ changes

    (a) from $x_0 = 1$ to $x_1 = x_0 + \Delta x = 1.2$   and   (b) from $x_0 = 1$ to $x_1 = 0.8$.

    (a) $\Delta x = x_1 - x_0 = 1.2 - 1 = 0.2$

    $\Delta y = f(x_0 + \Delta x) - f(x_0) = f(1.2) - f(1) = -0.56 - (-2) = 1.44$   and   $\dfrac{\Delta y}{\Delta x} = \dfrac{1.44}{0.2} = 7.2$

    (b) $\Delta x = 0.8 - 1 = -0.2$

    $\Delta y = f(0.8) - f(1) = -3.36 - (-2) = -1.36$   and   $\dfrac{\Delta y}{\Delta x} = \dfrac{-1.36}{-0.2} = 6.8$

    Geometrically, $\dfrac{\Delta y}{\Delta x}$ in (a) is the slope of the secant line joining the points $(1, -2)$ and $(1.2, -0.56)$ of the parabola $y = x^2 + 5x - 8$, and in (b) is the slope of the secant line joining the points $(0.8, -3.36)$ and $(1, -2)$ of the same parabola.


2.  When $s$(ft) is the approximate distance a body falls freely from rest in $t$(sec), $s = 16t^2$. Find $\Delta s/\Delta t$ as $t$ changes from $t_0$ to $t_0 + \Delta t$. Use this to find $\Delta s/\Delta t$ as $t$ changes
    (a) from 3 to 3.5,  (b) from 3 to 3.2, and  (c) from 3 to 3.1.

    $$\frac{\Delta s}{\Delta t} = \frac{16(t_0 + \Delta t)^2 - 16t_0^2}{\Delta t} = \frac{32t_0 \cdot \Delta t + 16(\Delta t)^2}{\Delta t} = 32t_0 + 16\,\Delta t$$

    (a) Here $t_0 = 3$, $\Delta t = 0.5$, and $\Delta s/\Delta t = 32(3) + 16(0.5) = 104$ ft/sec.

    (b) Here $t_0 = 3$, $\Delta t = 0.2$, and $\Delta s/\Delta t = 32(3) + 16(0.2) = 99.2$ ft/sec.

    (c) Here $t_0 = 3$, $\Delta t = 0.1$, and $\Delta s/\Delta t = 97.6$ ft/sec.

    Since $\Delta s$ is the displacement of the body from time $t = t_0$ to $t = t_0 + \Delta t$,

    $$\frac{\Delta s}{\Delta t} = \frac{\text{displacement}}{\text{time}} = \text{average velocity of the body over the time interval}$$

3. Find $dy/dx$, given $y = x^3 - x^2 - 4$. Find also the value of $dy/dx$ when  (a) $x = 4$,  (b) $x = 0$,  (c) $x = -1$.

   (1)    $y + \Delta y = (x + \Delta x)^3 - (x + \Delta x)^2 - 4$

                     $= x^3 + 3x^2(\Delta x) + 3x(\Delta x)^2 + (\Delta x)^3 - x^2 - 2x(\Delta x) - (\Delta x)^2 - 4$

   (2)        $\Delta y = (3x^2 - 2x) \cdot \Delta x + (3x - 1)(\Delta x)^2 + (\Delta x)^3$

   (3)        $\dfrac{\Delta y}{\Delta x} = 3x^2 - 2x + (3x - 1) \cdot \Delta x + (\Delta x)^2$

   (4)        $\dfrac{dy}{dx} = \lim\limits_{\Delta x \to 0} \{3x^2 - 2x + (3x - 1) \cdot \Delta x + (\Delta x)^2\} = 3x^2 - 2x$

(a) $\dfrac{dy}{dx}\Big|_{x=4} = 3(4)^2 - 2(4) = 40$,   (b) $\dfrac{dy}{dx}\Big|_{x=0} = 3(0)^2 - 2(0) = 0$,   (c) $\dfrac{dy}{dx}\Big|_{x=-1} = 3(-1)^2 - 2(-1) = 5$

4. Find the derivative of $y = x^2 + 3x + 5$.

   (1)    $y + \Delta y = (x + \Delta x)^2 + 3(x + \Delta x) + 5 = x^2 + 2x\,\Delta x + \Delta x^2 + 3x + 3\Delta x + 5$

   (2)        $\Delta y = (2x + 3)\Delta x + \Delta x^2$

   (3)        $\dfrac{\Delta y}{\Delta x} = \dfrac{(2x + 3)\Delta x + \Delta x^2}{\Delta x} = 2x + 3 + \Delta x$

   (4)        $\dfrac{dy}{dx} = \lim\limits_{\Delta x \to 0} (2x + 3 + \Delta x) = 2x + 3$

5. Find the derivative of $y = \dfrac{1}{x - 2}$ at $x = 1$ and $x = 3$. Show that the derivative does not exist at $x = 2$, where the function is discontinuous.

   (1)    $y + \Delta y = \dfrac{1}{x + \Delta x - 2}$

   (2)        $\Delta y = \dfrac{1}{x + \Delta x - 2} - \dfrac{1}{x - 2} = \dfrac{(x - 2) - (x + \Delta x - 2)}{(x - 2)(x + \Delta x - 2)} = \dfrac{-\Delta x}{(x - 2)(x + \Delta x - 2)}$

   (3)        $\dfrac{\Delta y}{\Delta x} = \dfrac{-1}{(x - 2)(x + \Delta x - 2)}$

   (4)        $\dfrac{dy}{dx} = \lim\limits_{\Delta x \to 0} \dfrac{-1}{(x - 2)(x + \Delta x - 2)} = \dfrac{-1}{(x - 2)^2}$

At $x = 1$, $\dfrac{dy}{dx} = \dfrac{-1}{(1 - 2)^2} = -1$, and at $x = 3$, $\dfrac{dy}{dx} = \dfrac{-1}{(3 - 2)^2} = -1$.

At $x = 2$, $\dfrac{dy}{dx}$ does not exist since the denominator is zero.

6. Find the derivative of $f(x) = \dfrac{2x - 3}{3x + 4}$. Examine the derivative at $x = -\frac{4}{3}$, where the function is discontinuous.

   (1)             $f(x + \Delta x) = \dfrac{2(x + \Delta x) - 3}{3(x + \Delta x) + 4}$

   (2)    $f(x + \Delta x) - f(x) = \dfrac{2x + 2\Delta x - 3}{3x + 3\Delta x + 4} - \dfrac{2x - 3}{3x + 4}$

                         $= \dfrac{(3x + 4)[(2x - 3) + 2\Delta x] - (2x - 3)[(3x + 4) + 3\Delta x]}{(3x + 4)(3x + 3\Delta x + 4)}$

                         $= \dfrac{(6x + 8 - 6x + 9)\Delta x}{(3x + 4)(3x + 3\Delta x + 4)} = \dfrac{17\Delta x}{(3x + 4)(3x + 3\Delta x + 4)}$

(3) $\qquad \dfrac{f(x + \Delta x) - f(x)}{\Delta x} \;=\; \dfrac{17}{(3x + 4)(3x + 3\Delta x + 4)}$

(4) $\qquad\qquad f'(x) \;=\; \lim\limits_{\Delta x \to 0} \dfrac{17}{(3x + 4)(3x + 3\Delta x + 4)} \;=\; \dfrac{17}{(3x + 4)^2}$

At $x = -4/3$, the derivative does not exist since the denominator is zero. In general, the *derivative of a function does not exist at a point of discontinuity of the function.*

7.  Find the derivative of $y = \sqrt{2x + 1}$.

(1) $\quad y + \Delta y \;=\; (2x + 2\Delta x + 1)^{1/2}$

(2) $\qquad \Delta y \;=\; (2x + 2\Delta x + 1)^{1/2} - (2x + 1)^{1/2}$

$\qquad\qquad =\; [(2x + 2\Delta x + 1)^{1/2} - (2x + 1)^{1/2}] \,\dfrac{(2x + 2\Delta x + 1)^{1/2} + (2x + 1)^{1/2}}{(2x + 2\Delta x + 1)^{1/2} + (2x + 1)^{1/2}}$

$\qquad\qquad =\; \dfrac{(2x + 2\Delta x + 1) - (2x + 1)}{(2x + 2\Delta x + 1)^{1/2} + (2x + 1)^{1/2}} \;=\; \dfrac{2\Delta x}{(2x + 2\Delta x + 1)^{1/2} + (2x + 1)^{1/2}}$

(3) $\qquad \dfrac{\Delta y}{\Delta x} \;=\; \dfrac{2}{(2x + 2\Delta x + 1)^{1/2} + (2x + 1)^{1/2}}$

(4) $\qquad \dfrac{dy}{dx} \;=\; \lim\limits_{\Delta x \to 0} \dfrac{2}{(2x + 2\Delta x + 1)^{1/2} + (2x + 1)^{1/2}} \;=\; \dfrac{1}{(2x + 1)^{1/2}}$

For the function $f(x) = \sqrt{2x + 1}$, $\lim\limits_{x \to -\frac{1}{2}^{+}} f(x) = 0 = f(-\frac{1}{2})$ while $\lim\limits_{x \to -\frac{1}{2}^{-}} f(x)$ does not exist; the function has *right hand continuity* at $x = -\frac{1}{2}$. At $x = -\frac{1}{2}$, the derivative is infinite.

8.  Find the derivative of $f(x) = x^{1/3}$. Examine $f'(0)$.

(1) $\qquad\quad f(x + \Delta x) \;=\; (x + \Delta x)^{1/3}$

(2) $\quad f(x + \Delta x) - f(x) \;=\; (x + \Delta x)^{1/3} - x^{1/3}$

$\qquad\qquad =\; \dfrac{[(x + \Delta x)^{1/3} - x^{1/3}][(x + \Delta x)^{2/3} + x^{1/3}(x + \Delta x)^{1/3} + x^{2/3}]}{(x + \Delta x)^{2/3} + x^{1/3}(x + \Delta x)^{1/3} + x^{2/3}}$

$\qquad\qquad =\; \dfrac{x + \Delta x - x}{(x + \Delta x)^{2/3} + x^{1/3}(x + \Delta x)^{1/3} + x^{2/3}}$

(3) $\quad \dfrac{f(x + \Delta x) - f(x)}{\Delta x} \;=\; \dfrac{1}{(x + \Delta x)^{2/3} + x^{1/3}(x + \Delta x)^{1/3} + x^{2/3}}$

(4) $\qquad\qquad f'(x) \;=\; \lim\limits_{\Delta x \to 0} \dfrac{1}{(x + \Delta x)^{2/3} + x^{1/3}(x + \Delta x)^{1/3} + x^{2/3}} \;=\; \dfrac{1}{3x^{2/3}}$

The derivative does not exist at $x = 0$ since the denominator is zero. Note that the function is continuous at $x = 0$. This together with the remark at the end of Problem 7 illustrates: *If the derivative of a function exists at $x = a$ then the function is continuous there, but not conversely.*

9.  Interpret $dy/dx$ geometrically.

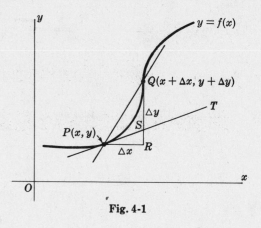

Fig. 4-1

From Fig. 4-1 it is seen that $\Delta y/\Delta x$ is the slope of the secant line joining an arbitrary but fixed point $P(x,y)$ and a nearby point $Q(x+\Delta x, y+\Delta y)$ of the curve. As $\Delta x \to 0$, $P$ remains fixed while $Q$ moves along the curve toward $P$, and the line $PQ$ revolves about $P$ toward its limiting position, the tangent line $PT$ to the curve at $P$. Thus, $dy/dx$ gives the slope of the tangent at $P$ to the curve $y = f(x)$.

For example, from Problem 3, the slope of the cubic $y = x^3 - x^2 - 4$ is $m = 40$ at the point $x = 4$, is $m = 0$ at the point $x = 0$, and is $m = 5$ at the point $x = -1$.

10. Find $ds/dt$ for the function of Problem 2 and interpret.

Here $\dfrac{\Delta s}{\Delta t} = 32t_0 + 16\,\Delta t$ and $\dfrac{ds}{dt} = \lim\limits_{\Delta t \to 0} (32t_0 + 16\,\Delta t) = 32t_0$.

As $\Delta t \to 0$, $\Delta s/\Delta t$ gives the average velocity of the body for shorter and shorter time intervals $\Delta t$. We define $ds/dt$ to be the instantaneous velocity $v$ of the body at time $t = t_0$. For example, at $t = 3$, $v = 32(3) = 96$ ft/sec.

11. Find $f'(x)$, given $f(x) = |x|$.

The function is continuous for all values of $x$. For $x < 0$, $f(x) = -x$ and $f'(x) = -1$; for $x > 0$, $f(x) = x$ and $f'(x) = 1$.

At $x = 0$, $f(x) = 0$ and $\lim\limits_{\Delta x \to 0} \dfrac{f(x+\Delta x) - f(x)}{\Delta x} = \lim\limits_{\Delta x \to 0} \dfrac{f(0+\Delta x) - f(0)}{\Delta x} = \lim\limits_{\Delta x \to 0} \dfrac{|\Delta x|}{\Delta x}$.

As $\Delta x \to 0^-$, $\dfrac{|\Delta x|}{\Delta x} \to -1$ while as $\Delta x \to 0^+$, $\dfrac{|\Delta x|}{\Delta x} \to 1$.

Hence, the derivative does not exist at $x = 0$.

12. Compute $\epsilon = \dfrac{\Delta y}{\Delta x} - \dfrac{dy}{dx}$ for the function of (a) Prob. 3 and (b) Prob. 5. Verify that $\epsilon \to 0$ as $\Delta x \to 0$.

(a)  $\epsilon = \{3x^2 - 2x + (3x-1)\Delta x + (\Delta x)^2\} - \{3x^2 - 2x\} = (3x - 1 + \Delta x)\Delta x$.

(b)  $\epsilon = \dfrac{-1}{(x-2)(x+\Delta x - 2)} - \dfrac{-1}{(x-2)^2} = \dfrac{-(x-2) + (x+\Delta x - 2)}{(x-2)^2(x+\Delta x - 2)} = \dfrac{1}{(x-2)^2(x+\Delta x - 2)}\,\Delta x$

13. Interpret $\Delta y = \dfrac{dy}{dx} \cdot \Delta x + \epsilon \cdot \Delta x$ of Problem 12 geometrically.

In the figure of Problem 9, $\Delta y = RQ$ and $\dfrac{dy}{dx} \cdot \Delta x = PR \cdot \tan \angle TPR = RS$; thus, $\epsilon \cdot \Delta x = SQ$. For a change $\Delta x$ in $x$ from $P(x, y)$, $\Delta y$ is the corresponding change in $y$ *along the curve* while $\dfrac{dy}{dx}\Delta x$ is the corresponding change in $y$ *along the tangent line $PT$*. Since their difference $\epsilon \cdot \Delta x = (\ldots)(\Delta x)^2 \to 0$ faster than does $\Delta x$, $\dfrac{dy}{dx} \cdot \Delta x$ will be used in Chapter 23 as an approximation of $\Delta y$ when $|\Delta x|$ is small.

# Supplementary Problems

**14.** Find $\Delta y$ and $\Delta y/\Delta x$, given:

    (a) $y = 2x - 3$ and $x$ changes from 3.3 to 3.5.

    (b) $y = x^2 + 4x$ and $x$ changes from 0.7 to 0.85.

    (c) $y = 2/x$ and $x$ changes from 0.75 to 0.5.

    *Ans.* (a) 0.4; 2,  (b) 0.8325; 5.55,  (c) 4/3; $-16/3$

**15.** Find $\Delta y$, given $y = x^2 - 3x + 5$, $x = 5$, and $\Delta x = -0.01$. What then is the value of $y$ when $x = 4.99$?
    *Ans.* $\Delta y = -0.0699$; $y = 14.9301$

**16.** Find the average velocity, given:

    (a) $s = (3t^2 + 5)$ ft and $t$ changes from 2 to 3 sec.

    (b) $s = (2t^2 + 5t - 3)$ ft and $t$ changes from 2 to 5 sec.

    *Ans.* (a) 15 ft/sec,  (b) 19 ft/sec.

**17.** Find the increase in the volume of a spherical balloon when its radius is increased  (a) from $r$ to $r + \Delta r$ in.,  (b) from 2 to 3 in.

    *Ans.* (a) $\dfrac{4\pi}{3}(3r^2 + 3r \cdot \Delta r + \Delta r^2) \cdot \Delta r$ cu in.,   (b) $\frac{76}{3}\pi$ cu. in.

**18.** Find the derivative of each of the following:

    (a) $y = 4x - 3$　　　　(d) $y = 1/x^2$　　　　(g) $y = \sqrt{x}$　　　　(i) $y = \sqrt{1 + 2x}$

    (b) $y = 4 - 3x$　　　　(e) $y = (2x - 1)/(2x + 1)$　　(h) $y = 1/\sqrt{x}$　　(j) $y = 1/\sqrt{2 + x}$

    (c) $y = x^2 + 2x - 3$　(f) $y = (1 + 2x)/(1 - 2x)$

    *Ans.* (a) 4

        (b) $-3$

        (c) $2(x + 1)$

        (d) $-2/x^3$

    (e) $\dfrac{4}{(2x + 1)^2}$

    (f) $\dfrac{4}{(1 - 2x)^2}$

    (g) $\dfrac{1}{2\sqrt{x}}$

    (h) $-\dfrac{1}{2x\sqrt{x}}$

    (i) $\dfrac{1}{\sqrt{1 + 2x}}$

    (j) $-\dfrac{1}{2(2 + x)^{3/2}}$

**19.** Find the slope of the following curves at the point $x = 1$:

    (a) $y = 8 - 5x^2$,  (b) $y = \dfrac{4}{x + 1}$,  (c) $y = \dfrac{2}{x + 3}$.

    *Ans.* (a) $-10$,  (b) $-1$,  (c) $-1/8$.

**20.** Find the coordinates of the vertex of the parabola $y = x^2 - 4x + 1$ by making use of the fact that at the vertex the slope of the tangent is zero.　　　*Ans.* $V(2, -3)$.

**21.** Find the slope of the tangents to the parabola $y = -x^2 + 5x - 6$ at its points of intersection with the $x$-axis.　　　*Ans.* At $x = 2$, $m = 1$; at $x = 3$, $m = -1$.

**22.** When $s$ is measured in feet and $t$ in seconds, find the velocity at time $t = 2$ of the following motions:
    (a) $s = t^2 + 3t$,  (b) $s = t^3 - 3t^2$,  (c) $s = \sqrt{t + 2}$.
    *Ans.* (a) 7 ft/sec,  (b) 0 ft/sec,  (c) $\frac{1}{4}$ ft/sec.

**23.** Show that the instantaneous rate of change of the volume of a cube with respect to its edge $x$(in.) is 12 in³/in when $x = 2$ in.

# Differentiation of Algebraic Functions

**A FUNCTION** is said to be *differentiable at* $x = x_0$ if it has a derivative there. A function is said to be *differentiable on an interval* if it is differentiable at every point of the interval.

The functions of elementary calculus are differentiable, except possibly at certain isolated points, on their intervals of definition.

**DIFFERENTIATION FORMULAS.** In these formulas $u$, $v$ and $w$ are differentiable functions of $x$.

1. $\dfrac{d}{dx}(c) = 0$, where $c$ is any constant

2. $\dfrac{d}{dx}(x) = 1$

3. $\dfrac{d}{dx}(u + v + \cdots) = \dfrac{d}{dx}(u) + \dfrac{d}{dx}(v) + \cdots$

4. $\dfrac{d}{dx}(cu) = c\dfrac{d}{dx}(u)$

5. $\dfrac{d}{dx}(uv) = u\dfrac{d}{dx}(v) + v\dfrac{d}{dx}(u)$

6. $\dfrac{d}{dx}(uvw) = uv\dfrac{d}{dx}(w) + uw\dfrac{d}{dx}(v) + vw\dfrac{d}{dx}(u)$

7. $\dfrac{d}{dx}\left(\dfrac{u}{c}\right) = \dfrac{1}{c} \cdot \dfrac{d}{dx}(u)$, $c \neq 0$

8. $\dfrac{d}{dx}\left(\dfrac{c}{u}\right) = c\dfrac{d}{dx}\left(\dfrac{1}{u}\right) = -\dfrac{c}{u^2} \cdot \dfrac{d}{dx}(u)$, $u \neq 0$

9. $\dfrac{d}{dx}\left(\dfrac{u}{v}\right) = \dfrac{v\dfrac{d}{dx}(u) - u\dfrac{d}{dx}(v)}{v^2}$, $v \neq 0$

10. $\dfrac{d}{dx}(x^m) = mx^{m-1}$

11. $\dfrac{d}{dx}(u^m) = mu^{m-1}\dfrac{d}{dx}(u)$

See Problems 1-13.

**INVERSE FUNCTIONS.** Let $y = f(x)$ be differentiable on the interval $a \leqq x \leqq b$ and suppose that $dy/dx$ does not change sign on the interval. Then from Fig. 5-1$a$ and 5-1$b$ the function assumes once and only once every value between $f(a) = c$ and $f(b) = d$. Thus, for each value of $y$ on the respective interval, there corresponds one and only one value of $x$ and $x$ is a function of $y$, say $x = g(y)$. The functions $y = f(x)$ and $x = g(y)$ are called *inverse functions*.

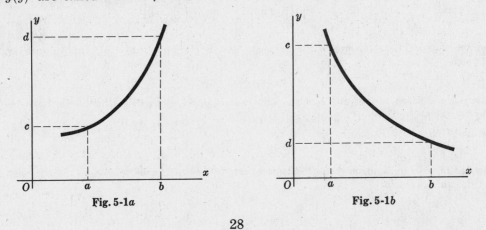

Fig. 5-1$a$          Fig. 5-1$b$

**Example 1:**

(a) $y = f(x) = 3x + 2$　and　$x = g(y) = \frac{1}{3}(y - 2)$　are inverse functions.

(b) When $x \leqq 2$ and $y \geqq -1$,　$y = x^2 - 4x + 3$　and　$x = 2 - \sqrt{y+1}$　are inverse functions.　When $x \geqq 2$ and $y \geqq -1$,　$y = x^2 - 4x + 3$　and　$x = 2 + \sqrt{y+1}$　are inverse functions.

　　　　To find $dy/dx$, given $x = g(y)$:

(a) Solve for $y$, when possible, and differentiate with respect to $x$; or

(b) Differentiate $x = g(y)$ with respect to $y$ and use

$$12. \qquad \frac{dy}{dx} = \frac{1}{\dfrac{dx}{dy}}$$

**Example 2:**

Find $dy/dx$, given $x = \sqrt{y} + 5$.

Using (a): $y = (x - 5)^2$ and $dy/dx = 2(x - 5)$.

Using (b): $\dfrac{dx}{dy} = \frac{1}{2}y^{-1/2} = \dfrac{1}{2\sqrt{y}}$;　then　$\dfrac{dy}{dx} = 2\sqrt{y} = 2(x - 5)$.

See Problems 14-15.

**DIFFERENTIATION OF A FUNCTION OF A FUNCTION.** If $y = f(u)$ and $u = g(x)$, then $y = f\{g(x)\}$ is a function of $x$. If $y$ is a differentiable function of $u$ and if $u$ is a differentiable function of $x$, then $y = f\{g(x)\}$ is a differentiable function of $x$ and the derivative $dy/dx$ may be obtained by one of the following procedures:

(a) Express $y$ explicitly in terms of $x$ and differentiate.

**Example 3:**

If $y = u^2 + 3$ and $u = 2x + 1$, then $y = (2x + 1)^2 + 3$ and $dy/dx = 8x + 4$.

(b) Differentiate each function with respect to the independent variable and use the formula (*the chain rule*)

$$13. \qquad \frac{dy}{dx} = \frac{dy}{du} \cdot \frac{du}{dx}$$

**Example 4:**

If $y = u^2 + 3$ and $u = 2x + 1$, then $\dfrac{dy}{du} = 2u$, $\dfrac{du}{dx} = 2$ and $\dfrac{dy}{dx} = \dfrac{dy}{du} \cdot \dfrac{du}{dx} = 4u = 8x + 4$.

See Problems 16-20.

**HIGHER DERIVATIVES.** Let $y = f(x)$ be a differentiable function of $x$ and let its derivative be called the *first derivative* of the function. If the first derivative is differentiable, its derivative is called the *second derivative* of the (original) function and is denoted by one of the symbols $\dfrac{d^2y}{dx^2}$, $y''$, or $f''(x)$. In turn, the derivative of the second derivative is called the *third derivative* of the function and is denoted by one of the symbols $\dfrac{d^3y}{dx^3}$, $y'''$, or $f'''(x)$; ....

*Note.* The derivative of a given order at a point can exist only when the function and all derivatives of lower order are differentiable at the point.

See Problems 21-23.

# Solved Problems

1. Prove: (a) $\dfrac{d}{dx}(c) = 0$, where $c$ is any constant;   (b) $\dfrac{d}{dx}(x) = 1$;   (c) $\dfrac{d}{dx}(cx) = c$, where $c$ is any constant;   (d) $\dfrac{d}{dx}(x^n) = nx^{n-1}$, when $n$ is a positive integer.

Since $\dfrac{d}{dx} f(x) = \lim\limits_{\Delta x \to 0} \dfrac{f(x + \Delta x) - f(x)}{\Delta x}$,

(a) $\dfrac{d}{dx}(c) = \lim\limits_{\Delta x \to 0} \dfrac{c - c}{\Delta x} = \lim\limits_{\Delta x \to 0} 0 = 0$

(b) $\dfrac{d}{dx}(x) = \lim\limits_{\Delta x \to 0} \dfrac{(x + \Delta x) - x}{\Delta x} = \lim\limits_{\Delta x \to 0} \dfrac{\Delta x}{\Delta x} = \lim\limits_{\Delta x \to 0} 1 = 1$

(c) $\dfrac{d}{dx}(cx) = \lim\limits_{\Delta x \to 0} \dfrac{c(x + \Delta x) - cx}{\Delta x} = \lim\limits_{\Delta x \to 0} c = c$

(d) $\dfrac{d}{dx}(x^n) = \lim\limits_{\Delta x \to 0} \dfrac{(x + \Delta x)^n - x^n}{\Delta x} = \lim\limits_{\Delta x \to 0} \dfrac{\left\{ x^n + nx^{n-1}\Delta x + \dfrac{n(n-1)}{1 \cdot 2} x^{n-2}(\Delta x)^2 + \cdots + (\Delta x)^n \right\} - x^n}{\Delta x}$

$\qquad = \lim\limits_{\Delta x \to 0} \left\{ nx^{n-1} + \dfrac{n(n-1)}{1 \cdot 2} x^{n-2}\Delta x + \cdots + (\Delta x)^{n-1} \right\} = nx^{n-1}$

2. Let $u$ and $v$ be differentiable functions of $x$. Prove: (a) $\dfrac{d}{dx}(u + v) = \dfrac{d}{dx}(u) + \dfrac{d}{dx}(v)$

(b) $\dfrac{d}{dx}(u \cdot v) = u \cdot \dfrac{d}{dx}(v) + v \cdot \dfrac{d}{dx}(u)$      (c) $\dfrac{d}{dx}\left(\dfrac{u}{v}\right) = \dfrac{v \cdot \dfrac{d}{dx}(u) - u \cdot \dfrac{d}{dx}(v)}{v^2}$,   $v \neq 0$

(a) Set $f(x) = u + v = u(x) + v(x)$; then

$\dfrac{f(x + \Delta x) - f(x)}{\Delta x} = \dfrac{u(x + \Delta x) + v(x + \Delta x) - u(x) - v(x)}{\Delta x} = \dfrac{u(x + \Delta x) - u(x)}{\Delta x} + \dfrac{v(x + \Delta x) - v(x)}{\Delta x}$

Taking the limit as $\Delta x \to 0$,   $\dfrac{d}{dx} f(x) = \dfrac{d}{dx}(u + v) = \dfrac{d}{dx} u(x) + \dfrac{d}{dx} v(x) = \dfrac{d}{dx}(u) + \dfrac{d}{dx}(v)$.

(b) Set $f(x) = u \cdot v = u(x) \cdot v(x)$; then

$\dfrac{f(x + \Delta x) - f(x)}{\Delta x} = \dfrac{u(x + \Delta x) \cdot v(x + \Delta x) - u(x) \cdot v(x)}{\Delta x}$

$\qquad = \dfrac{[u(x + \Delta x) \cdot v(x + \Delta x) - v(x) \cdot u(x + \Delta x)] + [v(x) \cdot u(x + \Delta x) - u(x) \cdot v(x)]}{\Delta x}$

$\qquad = u(x + \Delta x) \dfrac{v(x + \Delta x) - v(x)}{\Delta x} + v(x) \dfrac{u(x + \Delta x) - u(x)}{\Delta x}$

and   $\dfrac{d}{dx} f(x) = \dfrac{d}{dx}(u \cdot v) = u(x) \dfrac{d}{dx} v(x) + v(x) \dfrac{d}{dx} u(x) = u \dfrac{d}{dx}(v) + v \dfrac{d}{dx}(u)$.

(c) Set $f(x) = \dfrac{u}{v} = \dfrac{u(x)}{v(x)}$; then

$\dfrac{f(x + \Delta x) - f(x)}{\Delta x} = \dfrac{\dfrac{u(x + \Delta x)}{v(x + \Delta x)} - \dfrac{u(x)}{v(x)}}{\Delta x} = \dfrac{u(x + \Delta x) \cdot v(x) - u(x) \cdot v(x + \Delta x)}{\Delta x \{v(x) \cdot v(x + \Delta x)\}}$

$\qquad = \dfrac{[u(x + \Delta x) \cdot v(x) - u(x) \cdot v(x)] - [u(x) \cdot v(x + \Delta x) - u(x) \cdot v(x)]}{\Delta x \{v(x) \cdot v(x + \Delta x)\}}$

$\qquad = \dfrac{v(x) \dfrac{u(x + \Delta x) - u(x)}{\Delta x} - u(x) \dfrac{v(x + \Delta x) - v(x)}{\Delta x}}{v(x) \cdot v(x + \Delta x)}$

and   $\dfrac{d}{dx} f(x) = \dfrac{d}{dx}\left(\dfrac{u}{v}\right) = \dfrac{v(x) \dfrac{d}{dx} u(x) - u(x) \dfrac{d}{dx} v(x)}{\{v(x)\}^2} = \dfrac{v \dfrac{d}{dx}(u) - u \dfrac{d}{dx}(v)}{v^2}$.

Differentiate each of the following.

3.  $y = 4 + 2x - 3x^2 - 5x^3 - 8x^4 + 9x^5$

$$\frac{dy}{dx} = 0 + 2(1) - 3(2x) - 5(3x^2) - 8(4x^3) + 9(5x^4) = 2 - 6x - 15x^2 - 32x^3 + 45x^4$$

4.  $y = \dfrac{1}{x} + \dfrac{3}{x^2} + \dfrac{2}{x^3} = x^{-1} + 3x^{-2} + 2x^{-3}$

$$\frac{dy}{dx} = -x^{-2} + 3(-2x^{-3}) + 2(-3x^{-4}) = -x^{-2} - 6x^{-3} - 6x^{-4} = -\frac{1}{x^2} - \frac{6}{x^3} - \frac{6}{x^4}$$

5.  $y = 2x^{1/2} + 6x^{1/3} - 2x^{3/2}$

$$\frac{dy}{dx} = 2\left(\frac{1}{2}x^{-1/2}\right) + 6\left(\frac{1}{3}x^{-2/3}\right) - 2\left(\frac{3}{2}x^{1/2}\right) = x^{-1/2} + 2x^{-2/3} - 3x^{1/2} = \frac{1}{x^{1/2}} + \frac{2}{x^{2/3}} - 3x^{1/2}$$

6.  $y = \dfrac{2}{x^{1/2}} + \dfrac{6}{x^{1/3}} - \dfrac{2}{x^{3/2}} - \dfrac{4}{x^{3/4}} = 2x^{-1/2} + 6x^{-1/3} - 2x^{-3/2} - 4x^{-3/4}$

$$\frac{dy}{dx} = 2\left(-\frac{1}{2}x^{-3/2}\right) + 6\left(-\frac{1}{3}x^{-4/3}\right) - 2\left(-\frac{3}{2}x^{-5/2}\right) - 4\left(-\frac{3}{4}x^{-7/4}\right)$$

$$= -x^{-3/2} - 2x^{-4/3} + 3x^{-5/2} + 3x^{-7/4} = -\frac{1}{x^{3/2}} - \frac{2}{x^{4/3}} + \frac{3}{x^{5/2}} + \frac{3}{x^{7/4}}$$

7.  $y = \sqrt[3]{3x^2} - \dfrac{1}{\sqrt{5x}} = (3x^2)^{1/3} - (5x)^{-1/2}$

$$\frac{dy}{dx} = \frac{1}{3}(3x^2)^{-2/3}\cdot 6x - \left(-\frac{1}{2}\right)(5x)^{-3/2}\cdot 5 = \frac{2x}{(9x^4)^{1/3}} + \frac{5}{2(5x)(5x)^{1/2}} = \frac{2}{\sqrt[3]{9x}} + \frac{1}{2x\sqrt{5x}}$$

8.  $s = (t^2 - 3)^4$

$$\frac{ds}{dt} = 4(t^2 - 3)^3(2t) = 8t(t^2 - 3)^3$$

9.  $z = \dfrac{3}{(a^2 - y^2)^2} = 3(a^2 - y^2)^{-2}$

$$\frac{dz}{dy} = 3(-2)(a^2 - y^2)^{-3}\cdot\frac{d}{dy}(a^2 - y^2) = 3(-2)(a^2 - y^2)^{-3}(-2y) = \frac{12y}{(a^2 - y^2)^3}$$

10.  $f(x) = \sqrt{x^2 + 6x + 3} = (x^2 + 6x + 3)^{1/2}$

$$f'(x) = \frac{1}{2}(x^2 + 6x + 3)^{-1/2}\cdot\frac{d}{dx}(x^2 + 6x + 3) = \frac{1}{2}(x^2 + 6x + 3)^{-1/2}(2x + 6) = \frac{x + 3}{\sqrt{x^2 + 6x + 3}}$$

11.  $y = (x^2 + 4)^2(2x^3 - 1)^3$

$$y' = (x^2 + 4)^2\cdot\frac{d}{dx}(2x^3 - 1)^3 + (2x^3 - 1)^3\cdot\frac{d}{dx}(x^2 + 4)^2$$

$$= (x^2 + 4)^2\cdot 3(2x^3 - 1)^2\cdot\frac{d}{dx}(2x^3 - 1) + (2x^3 - 1)^3\cdot 2(x^2 + 4)\cdot\frac{d}{dx}(x^2 + 4)$$

$$= (x^2 + 4)^2\cdot 3(2x^3 - 1)^2\cdot 6x^2 + (2x^3 - 1)^3\cdot 2(x^2 + 4)\cdot 2x = 2x(x^2 + 4)(2x^3 - 1)^2(13x^3 + 36x - 2)$$

12.  $y = \dfrac{3 - 2x}{3 + 2x}$

$$y' = \frac{(3 + 2x)\cdot\frac{d}{dx}(3 - 2x) - (3 - 2x)\cdot\frac{d}{dx}(3 + 2x)}{(3 + 2x)^2} = \frac{(3 + 2x)(-2) - (3 - 2x)(2)}{(3 + 2x)^2} = \frac{-12}{(3 + 2x)^2}$$

**13.** $y = \dfrac{x^2}{\sqrt{4-x^2}} = \dfrac{x^2}{(4-x^2)^{1/2}}$

$$\frac{dy}{dx} = \frac{(4-x^2)^{1/2} \cdot \frac{d}{dx}(x^2) - x^2 \cdot \frac{d}{dx}(4-x^2)^{1/2}}{4-x^2} = \frac{(4-x^2)^{1/2}(2x) - x^2 \cdot \frac{1}{2}(4-x^2)^{-1/2}(-2x)}{4-x^2}$$

$$= \frac{(4-x^2)^{1/2}(2x) + x^3(4-x^2)^{-1/2}}{4-x^2} \cdot \frac{(4-x^2)^{1/2}}{(4-x^2)^{1/2}} = \frac{2x(4-x^2) + x^3}{(4-x^2)^{3/2}} = \frac{8x - x^3}{(4-x^2)^{3/2}}$$

**14.** Find $dy/dx$, given $x = y\sqrt{1-y^2}$.

$$\frac{dx}{dy} = (1-y^2)^{1/2} + \frac{1}{2}y(1-y^2)^{-1/2}(-2y) = \frac{1-2y^2}{\sqrt{1-y^2}} \quad \text{and} \quad \frac{dy}{dx} = \frac{1}{dx/dy} = \frac{\sqrt{1-y^2}}{1-2y^2}$$

**15.** Find the slope of the curve $x = y^2 - 4y$ at the points where it crosses the $y$-axis.

The points of crossing are $(0,0)$ and $(0,4)$.

$$\frac{dx}{dy} = 2y - 4, \quad \text{and} \quad \frac{dy}{dx} = \frac{1}{dx/dy} = \frac{1}{2y-4}. \quad \text{At } (0,0) \text{ the slope is } -\tfrac{1}{4}, \text{ and at } (0,4) \text{ the slope is } \tfrac{1}{4}.$$

## THE CHAIN RULE

**16.** Derive the chain rule $\dfrac{dy}{dx} = \dfrac{dy}{du} \cdot \dfrac{du}{dx}$.

Let $\Delta u$ and $\Delta y$ respectively be the increments given to $u$ and $y$ when $x$ is given an increment $\Delta x$. Now, provided $\Delta u \neq 0$,

$$\frac{\Delta y}{\Delta x} = \frac{\Delta y}{\Delta u} \cdot \frac{\Delta u}{\Delta x}$$

and, provided $\Delta u \neq 0$ as $\Delta x \to 0$, $\dfrac{dy}{dx} = \dfrac{dy}{du} \cdot \dfrac{du}{dx}$ as required.

The restriction on $\Delta u$ can usually be met by taking $|\Delta x|$ sufficiently small. When this is not possible, the chain rule may be established as follows:

Set $\Delta y = \dfrac{dy}{du} \cdot \Delta u + \epsilon \cdot \Delta u$ where $\epsilon \to 0$ as $\Delta x \to 0$. (See Problem 13, Chapter 4.) Then

$$\frac{\Delta y}{\Delta x} = \frac{dy}{du} \cdot \frac{\Delta u}{\Delta x} + \epsilon \frac{\Delta u}{\Delta x}$$

and, taking the limits as $\Delta x \to 0$, $\dfrac{dy}{dx} = \dfrac{dy}{du} \cdot \dfrac{du}{dx} + 0\dfrac{du}{dx} = \dfrac{dy}{du} \cdot \dfrac{du}{dx}$ as before.

**17.** Find $dy/dx$, given $y = \dfrac{u^2 - 1}{u^2 + 1}$ and $u = \sqrt[3]{x^2 + 2}$.

$$\frac{dy}{du} = \frac{4u}{(u^2+1)^2} \quad \text{and} \quad \frac{du}{dx} = \frac{2x}{3(x^2+2)^{2/3}} = \frac{2x}{3u^2}$$

Then $\qquad \dfrac{dy}{dx} = \dfrac{dy}{du} \cdot \dfrac{du}{dx} = \dfrac{4u}{(u^2+1)^2} \cdot \dfrac{2x}{3u^2} = \dfrac{8x}{3u(u^2+1)^2}$

**18.** A point moves along the curve $y = x^3 - 3x + 5$ so that $x = \frac{1}{2}\sqrt{t} + 3$, where $t$ is time. At what rate is $y$ changing when $t = 4$?

We are to find the value of $dy/dt$ when $t = 4$.

$$\frac{dy}{dx} = 3(x^2 - 1), \qquad \frac{dx}{dt} = \frac{1}{4\sqrt{t}}, \qquad \frac{dy}{dt} = \frac{dy}{dx} \cdot \frac{dx}{dt} = \frac{3(x^2-1)}{4\sqrt{t}}$$

When $t = 4$, $x = \frac{1}{2}\sqrt{4} + 3 = 4$, and $\dfrac{dy}{dt} = \dfrac{3(16-1)}{4 \cdot 2} = \dfrac{45}{8}$ units per unit of time.

**19.** A point moves in the plane according to the law $x = t^2 + 2t$, $y = 2t^3 - 6t$. Find $dy/dx$ when $t = 0, 2, 5$.

Since the first relation may be solved for $t$ and this result substituted for $t$ in the second relation, $y$ is clearly a function of $x$.

$$\frac{dy}{dt} = 6t^2 - 6, \quad \frac{dx}{dt} = 2t + 2, \quad \frac{dt}{dx} = \frac{1}{2t+2}, \quad \text{and} \quad \frac{dy}{dx} = \frac{dy}{dt} \cdot \frac{dt}{dx} = 6(t^2 - 1) \cdot \frac{1}{2(t+1)} = 3(t-1).$$

The required values of $dy/dx$ are $-3$ at $t = 0$, $3$ at $t = 2$, and $12$ at $t = 5$.

**20.** If $y = x^2 - 4x$ and $x = \sqrt{2t^2 + 1}$, find $dy/dt$ when $t = \sqrt{2}$.

$$\frac{dy}{dx} = 2(x-2), \qquad \frac{dx}{dt} = \frac{2t}{(2t^2+1)^{1/2}}, \qquad \frac{dy}{dt} = \frac{dy}{dx} \cdot \frac{dx}{dt} = \frac{4t(x-2)}{(2t^2+1)^{1/2}}$$

When $t = \sqrt{2}$, $x = \sqrt{5}$ and $\dfrac{dy}{dt} = \dfrac{4\sqrt{2}\,(\sqrt{5}-2)}{\sqrt{5}} = \dfrac{4\sqrt{2}}{5}(5 - 2\sqrt{5})$.

**21.** Show that the function $f(x) = x^3 + 3x^2 - 8x + 2$ has derivatives of all orders at $x = a$.

$$
\begin{array}{lll}
f'(x) = 3x^2 + 6x - 8 & \text{and} & f'(a) = 3a^2 + 6a - 8 \\
f''(x) = 6x + 6 & \text{and} & f''(a) = 6a + 6 \\
f'''(x) = 6 & \text{and} & f'''(a) = 6
\end{array}
$$

All derivatives of higher order are identically zero.

**22.** Investigate the successive derivatives of $f(x) = x^{4/3}$ at $x = 0$.

$$f'(x) = \frac{4}{3}x^{1/3} \qquad \text{and} \qquad f'(0) = 0$$

$$f''(x) = \frac{4}{9x^{2/3}} \qquad \text{and} \qquad f''(0) \text{ does not exist}$$

Thus the first derivative, but no derivative of higher order, exists at $x = 0$.

**23.** Given $f(x) = \dfrac{2}{1-x} = 2(1-x)^{-1}$, find $f^{(n)}(x)$.

We find

$$
\begin{array}{lll}
f'(x) &= 2(-1)(1-x)^{-2}(-1) = 2(1-x)^{-2} &= 2 \cdot 1! \,(1-x)^{-2} \\
f''(x) &= 2(1!)(-2)(1-x)^{-3}(-1) = 2 \cdot 2! \,(1-x)^{-3} \\
f'''(x) &= 2(2!)(-3)(1-x)^{-4}(-1) = 2 \cdot 3! \,(1-x)^{-4}
\end{array}
$$

which suggests $f^{(n)}(x) = 2 \cdot n! \,(1-x)^{-(n+1)}$.

This may be established by mathematical induction by showing that if $f^{(k)}(x) = 2 \cdot k! \,(1-x)^{-(k+1)}$, then

$$f^{(k+1)}(x) = -2 \cdot k! \,(k+1)(1-x)^{-(k+2)}(-1) = 2 \cdot (k+1)! \,(1-x)^{-(k+2)}$$

# Supplementary Problems

**24.** Establish formula 10 for $m = -1/n$, $n$ a positive integer, by using formula 9 to compute $\dfrac{d}{dx}\left(\dfrac{1}{x^n}\right)$. (For the case $m = p/q$, $p$ and $q$ integers, see Problem 4, Chapter 6.)

In Problems 25-43, find the derivative.

**25.** $y = x^5 + 5x^4 - 10x^2 + 6$  $\qquad$ *Ans.* $dy/dx = 5x(x^3 + 4x^2 - 4)$

**26.** $y = 3x^{1/2} - x^{3/2} + 2x^{-1/2}$  $\qquad$ *Ans.* $\dfrac{dy}{dx} = \dfrac{3}{2\sqrt{x}} - \dfrac{3}{2}\sqrt{x} - \dfrac{1}{x^{3/2}}$

**27.** $y = \dfrac{1}{2x^2} + \dfrac{4}{\sqrt{x}} = \dfrac{1}{2}x^{-2} + 4x^{-1/2}$  $\qquad$ *Ans.* $\dfrac{dy}{dx} = -\dfrac{1}{x^3} - \dfrac{2}{x^{3/2}}$

**28.** $y = \sqrt{2x} + 2\sqrt{x}$  $\qquad$ *Ans.* $y' = \dfrac{1 + \sqrt{2}}{\sqrt{2x}}$

**29.** $f(t) = \dfrac{2}{\sqrt{t}} + \dfrac{6}{\sqrt[3]{t}}$  $\qquad$ *Ans.* $f'(t) = -\dfrac{t^{1/2} + 2t^{2/3}}{t^2}$

**30.** $y = (1 - 5x)^6$  $\qquad$ *Ans.* $y' = -30(1 - 5x)^5$

**31.** $f(x) = (3x - x^3 + 1)^4$  $\qquad$ *Ans.* $f'(x) = 12(1 - x^2)(3x - x^3 + 1)^3$

**32.** $y = (3 + 4x - x^2)^{1/2}$  $\qquad$ *Ans.* $y' = \dfrac{2 - x}{y}$

33. $\theta = \dfrac{3r+2}{2r+3}$ $\qquad$ Ans. $\dfrac{d\theta}{dr} = \dfrac{5}{(2r+3)^2}$

34. $y = \left(\dfrac{x}{1+x}\right)^5$ $\qquad$ Ans. $y' = \dfrac{5x^4}{(1+x)^6}$

35. $y = 2x^2\sqrt{2-x}$ $\qquad$ Ans. $y' = \dfrac{x(8-5x)}{\sqrt{2-x}}$

36. $f(x) = x\sqrt{3-2x^2}$ $\qquad$ Ans. $f'(x) = \dfrac{3-4x^2}{\sqrt{3-2x^2}}$

37. $y = (x-1)\sqrt{x^2-2x+2}$ $\qquad$ Ans. $\dfrac{dy}{dx} = \dfrac{2x^2-4x+3}{\sqrt{x^2-2x+2}}$

38. $z = \dfrac{w}{\sqrt{1-4w^2}}$ $\qquad$ Ans. $\dfrac{dz}{dw} = \dfrac{1}{(1-4w^2)^{3/2}}$

39. $y = \sqrt{1+\sqrt{x}}$ $\qquad$ Ans. $y' = \dfrac{1}{4\sqrt{x+x\sqrt{x}}}$

40. $f(x) = \sqrt{\dfrac{x-1}{x+1}}$ $\qquad$ Ans. $f'(x) = \dfrac{1}{(x+1)\sqrt{x^2-1}}$

41. $y = (x^2+3)^4\,(2x^3-5)^3$ $\qquad$ Ans. $y' = 2x(x^2+3)^3\,(2x^3-5)^2\,(17x^3+27x-20)$

42. $s = \dfrac{t^2+2}{3-t^2}$ $\qquad$ Ans. $\dfrac{ds}{dt} = \dfrac{10t}{(3-t^2)^2}$

43. $y = \left(\dfrac{x^3-1}{2x^3+1}\right)^4$ $\qquad$ Ans. $y' = \dfrac{36x^2(x^3-1)^3}{(2x^3+1)^5}$

44. Compute $dy/dx$ by two different methods and check that the results are the same: (a) $x = (1+2y)^3$, (b) $x = 1/(2+y)$.

## Use the chain rule to find $dy/dx$ in Problems 45-48.

45. $y = \dfrac{u-1}{u+1}$, $u = \sqrt{x}$ $\qquad$ Ans. $\dfrac{dy}{dx} = \dfrac{1}{\sqrt{x}\,(1+\sqrt{x}\,)^2}$

46. $y = u^3+4$, $u = x^2+2x$ $\qquad$ Ans. $\dfrac{dy}{dx} = 6x^2(x+2)^2\,(x+1)$

47. $y = \sqrt{1+u}$, $u = \sqrt{x}$ $\qquad$ Ans. See Problem 39

48. $y = \sqrt{u}$, $u = v(3-2v)$, $v = x^2$ $\qquad$ Hint: $\dfrac{dy}{dx} = \dfrac{dy}{du}\cdot\dfrac{du}{dv}\cdot\dfrac{dv}{dx}$. $\qquad$ Ans. See Problem 36.

## In Problems 49-52, find the indicated derivative.

49. $y = 3x^4-2x^2+x-5$; $y'''$ $\qquad$ Ans. $y''' = 72x$

50. $y = 1/\sqrt{x}$; $y^{(iv)}$ $\qquad$ Ans. $y^{(iv)} = \dfrac{105}{16x^{9/2}}$

51. $f(x) = \sqrt{2-3x^2}$; $f''(x)$ $\qquad$ Ans. $f''(x) = \dfrac{-6}{(2-3x^2)^{3/2}}$

52. $y = x/\sqrt{x-1}$, $y''$ $\qquad$ Ans. $y'' = \dfrac{4-x}{4(x-1)^{5/2}}$

## In Problems 53-54, find the $n$th derivative.

53. $y = 1/x^2$ $\qquad$ Ans. $y^{(n)} = \dfrac{(-1)^n\,(n+1)!}{x^{n+2}}$

54. $f(x) = 1/(3x+2)$ $\qquad$ Ans. $f^n(x) = (-1)^n\,\dfrac{3^n\cdot n!}{(3x+2)^{n+1}}$

55. If $y = f(u)$ and $u = g(x)$, show:

(a) $\dfrac{d^2y}{dx^2} = \dfrac{dy}{du}\cdot\dfrac{d^2u}{dx^2} + \dfrac{d^2y}{du^2}\left(\dfrac{du}{dx}\right)^2$ $\quad$ (b) $\dfrac{d^3y}{dx^3} = \dfrac{dy}{du}\cdot\dfrac{d^3u}{dx^3} + 3\dfrac{d^2y}{du^2}\cdot\dfrac{d^2u}{dx^2}\cdot\dfrac{du}{dx} + \dfrac{d^3y}{du^3}\left(\dfrac{du}{dx}\right)^3$

56. From $\dfrac{dx}{dy} = \dfrac{1}{y'}$, derive $\dfrac{d^2x}{dy^2} = -\dfrac{y''}{(y')^3}$ and $\dfrac{d^3x}{dy^3} = \dfrac{3(y'')^2-y'y'''}{(y')^5}$.

# Chapter 6

## Implicit Differentiation

**IMPLICIT FUNCTIONS.** An equation $f(x, y) = 0$, on perhaps certain restricted ranges of the variables, is said to define $y$ *implicitly* as a function of $x$.

**Example 1:**

(a) The equation $xy + x - 2y - 1 = 0$, with $x \neq 2$, defines the function $y = \dfrac{1-x}{x-2}$.

(b) The equation $4x^2 + 9y^2 - 36 = 0$ defines the function $y = \frac{2}{3}\sqrt{9-x^2}$ when $|x| \leq 3$ and $y \geq 0$ and the function $y = -\frac{2}{3}\sqrt{9-x^2}$ when $|x| \leq 3$ and $y \leq 0$. Note that the ellipse is to be thought of as consisting of two arcs joined at the points $(-3, 0)$ and $(3, 0)$.

The derivative $y'$ may be obtained by one of the following procedures:

(a) Solve, when possible, for $y$ and differentiate with respect to $x$. Except for very simple equations, this procedure is to be avoided.

(b) Thinking of $y$ as a function of $x$, differentiate the given equation with respect to $x$ and solve the resulting relation for $y'$. This differentiation process is known as *implicit differentiation*.

**Example 2:**

(a) Find $y'$, given $xy + x - 2y - 1 = 0$.

We have $\quad x \cdot \dfrac{d}{dx}(y) + y \cdot \dfrac{d}{dx}(x) + \dfrac{d}{dx}(x) - 2 \cdot \dfrac{d}{dx}(y) - \dfrac{d}{dx}(1) = \dfrac{d}{dx}(0)$

or $\quad xy' + y + 1 - 2y' = 0$; then $\quad y' = \dfrac{1+y}{2-x}$.

(b) Find $y'$, when $x = \sqrt{5}$, given $4x^2 + 9y^2 - 36 = 0$.

We have $\quad 4 \cdot \dfrac{d}{dx}(x^2) + 9 \cdot \dfrac{d}{dx}(y^2) = 8x + 9 \cdot \dfrac{d}{dy}(y^2)\dfrac{dy}{dx} = 8x + 18yy' = 0$ and $y' = -\dfrac{4x}{9y}$.

When $x = \sqrt{5}$, $y = \pm 4/3$. At the point $(\sqrt{5}, 4/3)$ on the upper arc of the ellipse, $y' = -\sqrt{5}/3$ and at the point $(\sqrt{5}, -4/3)$ on the lower arc, $y' = \sqrt{5}/3$.

**DERIVATIVES OF HIGHER ORDERS** may be obtained by one of the procedures:

(a) Differentiate implicitly the derivative of one lower order and replace $y'$ by the relation previously found.

**Example 3:** From Example 2(a), $y' = \dfrac{1+y}{2-x}$. Then

$$\dfrac{d}{dx}(y') = y'' = \dfrac{d}{dx}\left(\dfrac{1+y}{2-x}\right) = \dfrac{(2-x)y' + 1 + y}{(2-x)^2} = \dfrac{(2-x)\left(\dfrac{1+y}{2-x}\right) + 1 + y}{(2-x)^2} = \dfrac{2+2y}{(2-x)^2}$$

(b) Differentiate implicitly the given equation as many times as necessary to produce the required derivative and eliminate all derivatives of lower order. This procedure is recommended only when a derivative of higher order at a given point is required.

**Example 4:** Find the value of $y''$ at the point $(-1, 1)$ of the curve $x^2y + 3y - 4 = 0$.

Differentiate implicitly with respect to $x$ twice:

$$x^2y' + 2xy + 3y' = 0 \quad \text{and} \quad x^2y'' + 2xy' + 2xy' + 2y + 3y'' = 0$$

Substitute $x = -1$, $y = 1$ in the first relation; then $y' = \frac{1}{2}$.

Substitute $x = -1$, $y = 1$, $y' = \frac{1}{2}$ in the second relation; then $y'' = 0$.

# Solved Problems

1. Find $y'$, given $x^2y - xy^2 + x^2 + y^2 = 0$.

$$\frac{d}{dx}(x^2y) - \frac{d}{dx}(xy^2) + \frac{d}{dx}(x^2) + \frac{d}{dx}(y^2) = 0$$

$$x^2\frac{d}{dx}(y) + y\frac{d}{dx}(x^2) - x\frac{d}{dx}(y^2) - y^2\frac{d}{dx}(x) + \frac{d}{dx}(x^2) + \frac{d}{dx}(y^2) = 0$$

$$x^2y' + 2xy - 2xyy' - y^2 + 2x + 2yy' = 0 \quad \text{and} \quad y' = \frac{y^2 - 2x - 2xy}{x^2 + 2y - 2xy}$$

2. Find $y'$ and $y''$, given $x^2 - xy + y^2 = 3$.

$$\frac{d}{dx}(x^2) - \frac{d}{dx}(xy) + \frac{d}{dx}(y^2) = 2x - xy' - y + 2yy' = 0 \quad \text{and} \quad y' = \frac{2x - y}{x - 2y}$$

$$y'' = \frac{(x-2y)\dfrac{d}{dx}(2x-y) - (2x-y)\dfrac{d}{dx}(x-2y)}{(x-2y)^2} = \frac{(x-2y)(2-y') - (2x-y)(1-2y')}{(x-2y)^2}$$

$$= \frac{3xy' - 3y}{(x-2y)^2} = \frac{3x\left(\dfrac{2x-y}{x-2y}\right) - 3y}{(x-2y)^2} = \frac{6(x^2-xy+y^2)}{(x-2y)^3} = \frac{18}{(x-2y)^3}$$

3. Find $y'$ and $y''$, given $x^3y + xy^3 = 2$ and $x = 1$.

$$x^3y' + 3x^2y + 3xy^2y' + y^3 = 0$$

and  $$x^3y'' + 3x^2y' + 3x^2y' + 6xy + 3xy^2y'' + 6xy(y')^2 + 3y^2y' + 3y^2y' = 0$$

When $x = 1$, $y = 1$; substituting in the first derived relation, $y' = -1$.

Substituting $x = 1$, $y = 1$, $y' = -1$ in the second relation, $y'' = 0$.

# Supplementary Problems

4. Establish Formula 10, Chapter 5, for $m = p/q$, $p$ and $q$ integers, by writing $y = x^{p/q}$ as $y^q = x^p$ and differentiating with respect to $x$.

5. Find $y''$, given: (a) $x + xy + y = 2$, (b) $x^3 - 3xy + y^3 = 1$.  $Ans.$ $y'' = \dfrac{2(1+y)}{(1+x)^2}$, (b) $y'' = -\dfrac{4xy}{(y^2-x)^3}$

6. Find $y'$, $y''$, and $y'''$ at (a) the point $(2,1)$ on $x^2 - y^2 - x = 1$, (b) the point $(1,1)$ on $x^3 + 3x^2y - 6xy^2 + 2y^3 = 0$.  $Ans.$ (a) $3/2$, $-5/4$, $45/8$; (b) $1, 0, 0$

7. Find the slope at the point $(x_0, y_0)$ of (a) $b^2x^2 + a^2y^2 = a^2b^2$, (b) $b^2x^2 - a^2y^2 = a^2b^2$, (c) $x^3 + y^3 - 6x^2y = 0$.

 $Ans.$ (a) $-\dfrac{b^2x_0}{a^2y_0}$, (b) $\dfrac{b^2x_0}{a^2y_0}$, (c) $\dfrac{4x_0y_0 - x_0^2}{y_0^2 - 2x_0^2}$

8. Prove that the curves $5y - 2x + y^3 - x^2y = 0$ and $2y + 5x + x^4 - x^3y^2 = 0$ intersect at right angles at the origin.

9. (a) The total surface area of a rectangular parallelepiped of square base $y$ on a side and height $x$ is given by $S = 2y^2 + 4xy$. If $S$ is constant, find $dy/dx$ without solving for $y$.

  (b) The total surface area of a right circular cylinder of radius $r$ and height $h$ is given by $S = 2\pi r^2 + 2\pi rh$. If $S$ is constant, find $dr/dh$.  $Ans.$ (a) $-\dfrac{y}{x+y}$; (b) $-\dfrac{r}{2r+h}$

10. For the circle $x^2 + y^2 = r^2$, show that $\left|\dfrac{y''}{\{1+(y')^2\}^{3/2}}\right| = \dfrac{1}{r}$.

11. Given $S = \pi x(x + 2y)$ and $V = \pi x^2y$, show that $dS/dx = 2\pi(x - y)$ when $V$ is a constant and $dV/dx = -\pi x(x - y)$ when $S$ is a constant.

# Chapter 7

## Tangents and Normals

**IF THE FUNCTION** $f(x)$ has a finite derivative $f'(x_0)$ at $x = x_0$, the curve $y = f(x)$ has a tangent at $P_0(x_0, y_0)$ whose slope is

$$m = \tan \theta = f'(x_0)$$

If $m = 0$, the curve has a horizontal tangent of equation $y = y_0$ at $P_0$, as at $A$, $C$, and $E$ of Fig. 7-1. Otherwise, the equation of the tangent is

$$y - y_0 = m(x - x_0)$$

If $f(x)$ is continuous at $x = x_0$ but $\lim_{x \to x_0} f'(x) = \infty$, the curve has a vertical tangent of equation $x = x_0$, as at $B$ and $D$ of Fig. 7-1.

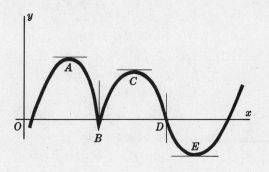

**Fig. 7-1**

The *normal* to a curve at one of its points is the line which passes through the point and is perpendicular to the tangent at the point. The equation of the normal at $P_0(x_0, y_0)$ is

$$x = x_0, \text{ if the tangent is horizontal}$$
$$y = y_0, \text{ if the tangent is vertical;}$$

otherwise,

$$y - y_0 = -\frac{1}{m}(x - x_0)$$

See Problems 1-9.

**THE ANGLE OF INTERSECTION** of two curves is defined as the angle between the tangents to the curve at their point of intersection.

To determine the angles of intersection of two curves:

(1) Solve the equations simultaneously for the points of intersection.

(2) Find the slopes $m_1$ and $m_2$ of the tangents to the two curves at each point of intersection.

(3) If $m_1 = m_2$, the angle of intersection is $\phi = 0°$,
if $m_1 = -1/m_2$, the angle of intersection is $\phi = 90°$;

otherwise, $$\tan \phi = \frac{m_1 - m_2}{1 + m_1 m_2}$$

$\phi$ is the *acute* angle of intersection when $\tan \phi > 0$, and $180° - \phi$ is the acute angle of intersection when $\tan \phi < 0$.

See Problems 10-12.

**LENGTH OF TANGENT, NORMAL, SUBTANGENT, AND SUBNORMAL.** The *length of the tangent* of a curve at one of its points is defined as the length of the portion of the tangent between its point of contact and the $x$-axis. The length of the projection of this segment on the $x$-axis is called the *length of the subtangent*.

The *length of the normal* is defined as the length of the portion of the normal between the point of contact of the tangent and the $x$-axis. The length of the projection of this segment on the $x$-axis is called the *length of the subnormal*.

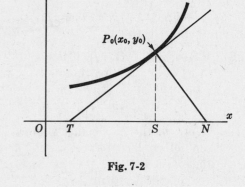

**Fig. 7-2**

Length of subtangent $= TS = y_0/m$

Length of subnormal $= SN = my_0$

Length of tangent

$$= TP_0 = \sqrt{(TS)^2 + (SP_0)^2}$$

Length of normal

$$= P_0N = \sqrt{(SN)^2 + (SP_0)^2}$$

*Note.* The lengths of the subtangent and subnormal are directed lengths. Certain authors prefer undirected lengths $|y_0/m|$ and $|my_0|$ respectively. In this event, the signs in the answers below are to be ignored.

See Problem 13.

# Solved Problems

1. Find the points of tangency of horizontal and vertical tangents to the curve $x^2 - xy + y^2 = 27$.

Differentiating, $y' = \dfrac{y - 2x}{2y - x}$.

For horizontal tangents: Set the numerator of $y'$ equal to zero and obtain $y = 2x$. The points of tangency are the points of intersection of the line $y = 2x$ and the given curve. Solve simultaneously the two equations for the points $(3, 6)$ and $(-3, -6)$.

For vertical tangents: Set the denominator of $y'$ equal to zero and obtain $x = 2y$. The points of tangency are the points of intersection of the line $x = 2y$ and the given curve. Solve simultaneously the two equations for the points $(6, 3)$ and $(-6, -3)$.

2. Find the equations of the tangent and normal to $y = x^3 - 2x^2 + 4$ at $(2, 4)$.

$f'(x) = 3x^2 - 4x$; the slope of the tangent at $(2, 4)$ is $m = f'(2) = 4$.

The equation of the tangent is $y - 4 = 4(x - 2)$ or $y = 4x - 4$.

The equation of the normal is $y - 4 = -\frac{1}{4}(x - 2)$ or $x + 4y = 18$.

3. Find the equations of the tangent and normal to $x^2 + 3xy + y^2 = 5$ at $(1, 1)$.

$\dfrac{dy}{dx} = -\dfrac{2x + 3y}{3x + 2y}$. The slope of the tangent at $(1, 1)$ is $m = -1$.

The equation of the tangent is $y - 1 = -1(x - 1)$ or $x + y = 2$.

The equation of the normal is $y - 1 = 1(x - 1)$ or $x - y = 0$.

4. Find the equations of the tangents with slope $m = -2/9$ to the ellipse $4x^2 + 9y^2 = 40$.

Let $P_0(x_0, y_0)$ be the point of tangency of a required tangent. Then

(a) $4x_0^2 + 9y_0^2 = 40$, since the point $P_0$ is on the ellipse.

(b) $\dfrac{dy}{dx} = -\dfrac{4x}{9y}$. At $(x_0, y_0)$, $m = -\dfrac{4x_0}{9y_0} = -\dfrac{2}{9}$ and $y_0 = 2x_0$.

(c) The points of tangency are the simultaneous solutions $(1, 2)$ and $(-1, -2)$ of the equations of $(a)$ and $(b)$.

The equation of the tangent at $(1, 2)$ is $y - 2 = -\frac{2}{9}(x - 1)$ or $2x + 9y = 20$.

The equation of the tangent at $(-1, -2)$ is $y + 2 = -\frac{2}{9}(x + 1)$ or $2x + 9y = -20$.

5.　Find the equation of the tangent, through the point $(2, -2)$, to the hyperbola $x^2 - y^2 = 16$.

Let $P_0(x_0, y_0)$ be the point of tangency of the required tangent. Then

(a) $x_0^2 - y_0^2 = 16$, since the point $P_0$ is on the hyperbola.

(b) $\dfrac{dy}{dx} = \dfrac{x}{y}$. At $(x_0, y_0)$, $m = \dfrac{x_0}{y_0} = \dfrac{y_0 + 2}{x_0 - 2} =$ slope of the line joining $P_0$ and $(2, -2)$; then

$$2x_0 + 2y_0 = x_0^2 - y_0^2 = 16 \quad \text{or} \quad x_0 + y_0 = 8$$

(c) The point of tangency is the simultaneous solution $(5, 3)$ of the equations of (a) and (b). Thus the equation of the tangent is $y - 3 = \frac{5}{3}(x - 5)$ or $5x - 3y = 16$.

6.　Find the equations of the vertical lines which meet the curves　(1) $y = x^3 + 2x^2 - 4x + 5$　and (2) $3y = 2x^3 + 9x^2 - 3x - 3$ in points at which the tangents to the respective curves are parallel.

Let $x = x_0$ be such a vertical line.

$$\text{For } (1): \quad y' = 3x^2 + 4x - 4; \text{ at } x = x_0, \ m = 3x_0^2 + 4x_0 - 4.$$
$$\text{For } (2): \quad 3y' = 6x^2 + 18x - 3; \text{ at } x = x_0, \ m = 2x_0^2 + 6x_0 - 1.$$

Since $3x_0^2 + 4x_0 - 4 = 2x_0^2 + 6x_0 - 1$, $x_0 = -1$ and 3. The lines are $x = -1$ and $x = 3$.

7.　(a) Show that the equation of the tangent of slope $m \neq 0$ to the parabola $y^2 = 4px$ is $y = mx + p/m$.

(b) Show that the equation of the tangent to the ellipse $b^2x^2 + a^2y^2 = a^2b^2$ at the point $P_0(x_0, y_0)$ on it is $b^2x_0x + a^2y_0y = a^2b^2$.

(a) $y' = 2p/y$. Let $P_0(x_0, y_0)$ be the point of tangency; then $y_0^2 = 4px_0$ and $m = 2p/y_0$. Hence, $y_0 = 2p/m$, $x_0 = \frac{1}{4}y_0^2/p = p/m^2$ and the equation of the tangent is $y - 2p/m = m(x - p/m^2)$ or $y = mx + p/m$.

(b) $y' = -\dfrac{b^2x}{a^2y}$. At $P_0$, $m = -\dfrac{b^2x_0}{a^2y_0}$ and the equation of the tangent is $y - y_0 = -\dfrac{b^2x_0}{a^2y_0}(x - x_0)$ or $b^2x_0x + a^2y_0y = b^2x_0^2 + a^2y_0^2 = a^2b^2$.

8.　Show that at a point $P_0(x_0, y_0)$ on the hyperbola $b^2x^2 - a^2y^2 = a^2b^2$, the tangent bisects the angle included by the focal radii of $P_0$.

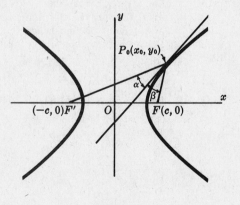

At $P_0$ the slope of the tangent to the hyperbola is $b^2x_0/a^2y_0$ and the slopes of the focal radii $P_0F'$ and $P_0F$ are $y_0/(x_0 + c)$ and $y_0/(x_0 - c)$ respectively. Now

$$\tan \alpha = \frac{\dfrac{b^2x_0}{a^2y_0} - \dfrac{y_0}{x_0 + c}}{1 + \dfrac{b^2x_0}{a^2y_0} \cdot \dfrac{y_0}{x_0 + c}}$$

$$= \frac{(b^2x_0^2 - a^2y_0^2) + b^2cx_0}{(a^2 + b^2)x_0y_0 + a^2cy_0}$$

$$= \frac{a^2b^2 + b^2cx_0}{c^2x_0y_0 + a^2cy_0} = \frac{b^2(a^2 + cx_0)}{cy_0(a^2 + cx_0)} = \frac{b^2}{cy_0}$$

**Fig. 7-3**

since $b^2x_0^2 - a^2y_0^2 = a^2b^2$ and $a^2 + b^2 = c^2$, and

$$\tan \beta = \frac{\dfrac{y_0}{x_0 - c} - \dfrac{b^2x_0}{a^2y_0}}{1 + \dfrac{b^2x_0}{a^2y_0} \cdot \dfrac{y_0}{x_0 - c}} = \frac{b^2cx_0 - (b^2x_0^2 - a^2y_0^2)}{(a^2 + b^2)x_0y_0 - a^2cy_0} = \frac{b^2cx_0 - a^2b^2}{c^2x_0y_0 - a^2cy_0} = \frac{b^2}{cy_0}$$

Hence, since $\tan \alpha = \tan \beta$, $\alpha = \beta$.

9.  **Prove:** The chord joining the points of contact of the tangents to the ellipse $b^2x^2 + a^2y^2 = a^2b^2$ through any point on a directrix passes through the associated focus.

Let $P_0(x_0, y_0)$ be any point through which two tangents can be drawn to the ellipse and let $P_1(x_1, y_1)$ and $P_2(x_2, y_2)$ be the corresponding points of contact. The equations of the tangents at $P_1$ and $P_2$ are $b^2x_1x + a^2y_1y = a^2b^2$ and $b^2x_2x + a^2y_2y = a^2b^2$. Since they pass through $P_0$, $b^2x_1x_0 + a^2y_1y_0 = a^2b^2$ and $b^2x_2x_0 + a^2y_2y_0 = a^2b^2$. The line $b^2x_0x + a^2y_0y = a^2b^2$ which passes through $P_1$ and $P_2$ is the chord of contact. Let $P(a^2/c, \bar{y})$ be a point on the right hand directrix. The chord of contact through $P$ has equation $(b^2a^2/c)x + a^2\bar{y}y = a^2b^2$ and, as can be readily checked, passes through the corresponding focus $F(c, 0)$.

10. Find the acute angles of intersection of the curves  $(1)$ $y^2 = 4x$  and  $(2)$ $2x^2 = 12 - 5y$.

(a) The points of intersection of the curves are $P_1(1, 2)$ and $P_2(4, -4)$.

(b) For $(1)$, $y' = 2/y$; for $(2)$, $y' = -4x/5$.

At $P_1(1, 2)$, $m_1 = 1$ and $m_2 = -4/5$; at $P_2(4, -4)$, $m_1 = -1/2$ and $m_2 = -16/5$.

(c) At $P_1$:  $\tan \phi = \dfrac{m_1 - m_2}{1 + m_1m_2} = \dfrac{1 + 4/5}{1 - 4/5} = 9$  and  $\phi = 83°40'$ is the acute angle of intersection.

At $P_2$:  $\tan \phi = \dfrac{-1/2 + 16/5}{1 + 8/5} = 1.0385$  and  $\phi = 46°5'$ is the acute angle of intersection.

11. Find the acute angles of intersection of the curves  $(1)$ $2x^2 + y^2 = 20$  and  $(2)$ $4y^2 - x^2 = 8$.

The points of intersection are $(\pm 2\sqrt{2}, 2)$ and $(\pm 2\sqrt{2}, -2)$.

For $(1)$, $y' = -2x/y$, and for $(2)$, $y' = x/4y$.

At the point $(2\sqrt{2}, 2)$, $m_1 = -2\sqrt{2}$ and $m_2 = \frac{1}{4}\sqrt{2}$. Since $m_1m_2 = -1$, the angle of intersection is $\phi = 90°$ (i.e., the curves are *orthogonal*). By symmetry, the curves are orthogonal at each of their points of intersection.

12. The cable of a certain suspension bridge is attached to supporting pillars 250 feet apart. If it hangs in the form of a parabola with the lowest point 50 feet below the point of suspension, find the angle between the cable and the pillar.

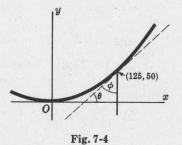

Take the origin at the vertex of the parabola as in Fig. 7-4. The equation of the parabola is $y = \dfrac{2}{625}x^2$, and $y' = \dfrac{4x}{625}$. At $(125, 50)$, $m = 4(125)/625 = .8000$ and $\theta = 38°40'$. The required angle is $\phi = 90° - \theta = 51°20'$.

**Fig. 7-4**

13. Find the length of the subtangent, subnormal, tangent, and normal of $xy + 2x - y = 5$ at the point $(2, 1)$.

$\dfrac{dy}{dx} = \dfrac{2 + y}{1 - x}$; at point $(2, 1)$, $m = -3$.

Length of subtangent $= y_0/m = -1/3$. Length of subnormal $= my_0 = -3$.

Length of tangent $= \sqrt{1/9 + 1} = \sqrt{10}/3$. Length of normal $= \sqrt{9 + 1} = \sqrt{10}$.

# Supplementary Problems

14. Examine $x^2 + 4xy + 16y^2 = 27$ for horizontal and vertical tangents.

    *Ans.* H.T. at $(3, -3/2)$ and $(-3, 3/2)$
    
    　　　 V.T. at $(6, -3/4)$ and $(-6, 3/4)$

15. Find the equations of the tangent and normal to $x^2 - y^2 = 7$ at the point $(4, -3)$.

    *Ans.* $4x + 3y = 7$;　$3x - 4y = 24$

16. At what point is the tangent to the curve $y = x^3 + 5$ (a) parallel to the line $12x - y = 17$, (b) perpendicular to the line $x + 3y = 2$?　　*Ans.* (a) $(2, 13)$, $(-2, -3)$;　(b) $(1, 6)$, $(-1, 4)$

17. Find the equations of the tangents to $9x^2 + 16y^2 = 52$ parallel to the line $9x - 8y = 1$.

    *Ans.* $9x - 8y = \pm 26$

18. Find the equations of the tangents to the hyperbola $xy = 1$ through the point $(-1, 1)$.

    *Ans.* $y = (2\sqrt{2} - 3)x + 2\sqrt{2} - 2$;　$y = -(2\sqrt{2} + 3)x - 2\sqrt{2} - 2$

19. For the parabola $y^2 = 4px$, show that the equation of the tangent at one of its points $P(x_0, y_0)$ is $yy_0 = 2p(x + x_0)$.

20. For the ellipse $b^2x^2 + a^2y^2 = a^2b^2$, show that the equations of the tangents of slope $m$ are $y = mx \pm \sqrt{a^2m^2 + b^2}$.

21. For the hyperbola $b^2x^2 - a^2y^2 = a^2b^2$, show (a) the equation of the tangent at one of its points $P(x_0, y_0)$ is $b^2x_0x - a^2y_0y = a^2b^2$ and (b) the equations of the tangents of slope $m$ are $y = mx \pm \sqrt{a^2m^2 - b^2}$.

22. Show that the normal to a parabola at any of its points $P_0$ bisects the angle included by the focal radius of $P_0$ and the line through $P_0$ parallel to the axis of the parabola.

23. Prove: Any tangent to a parabola, except at the vertex, intersects the directrix and the latus rectum (produced if necessary) in points equidistant from the focus.

24. Prove: The chord joining the points of contact of the tangents to a parabola through any point on its directrix passes through the focus.

25. Prove: The normal to an ellipse at any of its points $P_0$ bisects the angle included by the focal radii of $P_0$.

26. Prove: The chord joining the points of contact of the tangents to the hyperbola through any point on a directrix passes through the associated focus.

27. Prove: The point of contact of a tangent of a hyperbola is the midpoint of the segment of the tangent included between the asymptotes.

28. Prove: The slope of the tangent at any extremity of either latus rectum of a hyperbola or ellipse is numerically equal to its eccentricity.

29. Prove: (a) The sum of the intercepts on the coordinate axes of any tangent to $\sqrt{x} + \sqrt{y} = \sqrt{a}$ is a constant. (b) The sum of the squares of the intercepts on the coordinate axes of any tangent to $x^{2/3} + y^{2/3} = a^{2/3}$ is a constant.

30. Find the acute angles of intersection of the circles $x^2 - 4x + y^2 = 0$ and $x^2 + y^2 = 8$.　　*Ans.* $45°$

31. Show that the curves $y = x^3 + 2$ and $y = 2x^2 + 2$ have a common tangent at the point $(0, 2)$ and intersect at an angle $\phi = \text{Arc tan } 4/97$ at the point $(2, 10)$.

32. Show that the ellipse $4x^2 + 9y^2 = 45$ and the hyperbola $x^2 - 4y^2 = 5$ are orthogonal.

33. Find the equations of the tangent and normal, and the lengths of the subtangent, subnormal, tangent, and normal to the parabola $y = 4x^2$ at the point $(-1, 4)$.

    *Ans.* $y + 8x + 4 = 0$,　$8y - x - 33 = 0$;　$-\frac{1}{2}$, $-32$, $\frac{1}{2}\sqrt{65}$, $4\sqrt{65}$

34. Find the length of the subtangent, subnormal, tangent, and normal to the hyperbola $3x^2 - 2y^2 = 10$ at the point $(-2, 1)$.　　*Ans.* $-1/3$, $-3$, $\sqrt{10}/3$, $\sqrt{10}$

35. At what points on the curve $y = 2x^3 + 13x^2 + 5x + 9$ will the tangent pass through the origin?

    *Ans.* $x = -3$, $-1$, $3/4$.

# Chapter 8

## Maximum and Minimum Values

**INCREASING AND DECREASING FUNCTIONS.** A function $f(x)$ is said to be *increasing* at $x = x_0$ if for $h$, positive and sufficiently small, $f(x_0 - h) < f(x_0) < f(x_0 + h)$. A function $f(x)$ is said to be *decreasing* at $x = x_0$ if for $h$, positive and sufficiently small, $f(x_0 - h) > f(x_0) > f(x_0 + h)$.

If $f'(x_0) > 0$, then $f(x)$ is an increasing function at $x = x_0$; if $f'(x_0) < 0$, then $f(x)$ is a decreasing function at $x = x_0$. (For a proof, see Problem 17.) If $f'(x_0) = 0$, then $f(x)$ is said to be *stationary* at $x = x_0$.

A non-constant function is said to be an increasing (decreasing) function over an interval if it is increasing (decreasing) or stationary at every point of the interval.

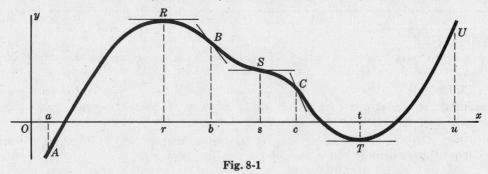

**Fig. 8-1**

In Fig. 8-1, the curve $y = f(x)$ is rising (the function is increasing) on the intervals $a < x < r$ and $t < x < u$; the curve is falling (the function is decreasing) on the interval $r < x < t$. The function is stationary at $x = r$, $x = s$, and $x = t$; the curve has a horizontal tangent at the points $R$, $S$, and $T$. The values of $x$, ($r$, $s$, and $t$), for which the function $f(x)$ is stationary ($f'(x) = 0$) are more frequently called *critical values* for the function and the corresponding points ($R$, $S$, and $T$) of the graph are called *critical points* of the curve.

**RELATIVE MAXIMUM AND MINIMUM VALUES OF A FUNCTION.** A function $y = f(x)$ is said to have a *relative maximum* (*relative minimum*) value at $x = x_0$ if $f(x_0)$ is greater (smaller) than immediately preceding *and* succeeding values of the function. See Problem 1.

In Fig. 8-1, $R(r, f(r))$ is a relative maximum point of the curve since $f(r) > f(x)$ on any sufficiently small neighborhood $0 < |x - r| < \delta$. We shall say that $y = f(x)$ has a *relative maximum value* ($= f(r)$) when $x = r$. In the same figure, $T(t, f(t))$ is a relative minimum point of the curve since $f(t) < f(x)$ on any sufficiently small neighborhood $0 < |x - t| < \delta$. We shall say that $y = f(x)$ has a *relative minimum value* ($= f(t)$) when $x = t$. Note that $R$ joins an arc $AR$ which is rising ($f'(x) > 0$) and an arc $RB$ which is falling ($f'(x) < 0$), while $T$ joins an arc $CT$ which is falling $[f'(x) < 0]$ and an arc $TU$ which is rising $[f'(x) > 0]$. At $S$ two arcs $BS$ and $SC$ both of which are falling are joined; $S$ is neither a relative maximum nor a relative minimum point of the curve.

42

If $y = f(x)$ is differentiable on $a \leqq x \leqq b$ and if $f(x)$ has a relative maximum (minimum) value at $x = x_0$, where $a < x_0 < b$, then $f'(x_0) = 0$. For a proof, see Prob. 18.

To find the relative maximum (minimum) values (hereinafter called maximum (minimum) values) of functions $f(x)$ which, together with their first derivatives, are continuous:

## FIRST DERIVATIVE TEST

1. Solve $f'(x) = 0$ for the critical values.
2. Locate the critical values on a number scale, thereby establishing a number of intervals.
3. Determine the sign of $f'(x)$ on each interval.
4. Let $x$ *increase* through each critical value $x = x_0$; then

$f(x)$ has a maximum value $(= f(x_0))$ if $f'(x)$ changes from $+$ to $-$,

$f(x)$ has a minimum value $(= f(x_0))$ if $f'(x)$ changes from $-$ to $+$,

$f(x)$ has neither a maximum nor a minimum value at $x = x_0$ if $f'(x)$ does not change sign.

See Problems 2-5.

**A FUNCTION** $y = f(x)$, necessarily less simple than those of Problems 2-5, may have a maximum or minimum value $(= f(x_0))$ although $f'(x_0)$ does not exist. The values $x = x_0$ for which $f(x)$ is defined but $f'(x)$ does not exist will also be called critical values for the function. They, together with the values for which $f'(x) = 0$ are to be used in determining the intervals of Step 2 above.

See Problems 6-8.

A final case in which $f(x_0)$ is a maximum (minimum) value although there is no interval $x_0 - \delta < x < x_0$ on which $f'(x)$ is positive (negative) and no interval $x_0 < x < x_0 + \delta$ on which $f'(x)$ is negative (positive) will not be treated here.

**DIRECTION OF BENDING.** An arc of a curve $y = f(x)$ is called *concave upward* if, at each of its points, the arc lies above the tangent at the point. As $x$ increases, $f'(x)$ either is of the same sign and increasing (as on the interval $b < x < s$ of Fig. 8-1) or changes sign from negative to positive (as on the interval $c < x < u$). In either case, the slope $f'(x)$ is increasing and $f''(x) > 0$.

An arc of a curve $y = f(x)$ is called *concave downward* if, at each of its points, the arc lies below the tangent at the point. As $x$ increases, $f'(x)$ either is of the same sign and decreasing (as on the interval $s < x < c$) or changes sign from positive to negative (as on the interval $a < x < b$). In either case the slope $f'(x)$ is decreasing and $f''(x) < 0$.

**A POINT OF INFLECTION** is a point at which a curve is changing from concave upward to concave downward, or vice versa. In Fig. 8-1, the points of inflection are $B$, $S$, and $C$.

A curve $y = f(x)$ has one of its points $x = x_0$ as an inflection point

if $f''(x_0) = 0$ or is not defined and

if $f''(x)$ changes sign as $x$ increases through $x = x_0$.

The latter condition may be replaced by $f'''(x_0) \neq 0$ when $f'''(x_0)$ exists.

See Problems 9-13.

## A SECOND TEST FOR MAXIMA AND MINIMA. SECOND DERIVATIVE TEST

1. Solve $f'(x) = 0$ for the critical values.

2. For a critical value $x = x_0$:

$f(x)$ has a maximum value $(= f(x_0))$ if $f''(x_0) < 0$,
$f(x)$ has a minimum value $(= f(x_0))$ if $f''(x_0) > 0$,
the test fails if $f''(x_0) = 0$ or becomes infinite.

In the latter case, the first derivative method must be used.

See Problems 14-16.

# Solved Problems

1.  (a)  $y = -x^2$ has a relative maximum value $(= 0)$ when $x = 0$, since $y = 0$ when $x = 0$ and $y < 0$ when $x \neq 0$.

   (b)  $y = (x - 3)^2$ has a relative minimum value $(= 0)$ when $x = 3$, since $y = 0$ when $x = 3$ and $y > 0$ when $x \neq 3$.

   (c)  $y = \sqrt{25 - 4x^2}$ has a relative maximum value $(= 5)$ when $x = 0$, since $y = 5$ when $x = 0$ and $y < 5$ when $-1 < x < 1$.

   (d)  $y = \sqrt{x - 4}$ has neither a relative maximum nor a relative minimum value. [Some authors define relative maximum (minimum) values so that this function has a relative minimum at $x = 4$. See Problem 30.]

2.  Given  $y = \frac{1}{3}x^3 + \frac{1}{2}x^2 - 6x + 8$,  find

   (a)  the critical points
   (b)  the intervals on which $y$ is increasing and decreasing, and
   (c)  the maximum and minimum values of $y$.

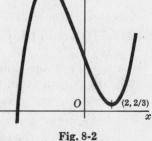
Fig. 8-2

   (a)  $y' = x^2 + x - 6 = (x + 3)(x - 2)$

   Setting $y' = 0$ gives the critical values $x = -3, 2$.

   The critical points are $(-3, 43/2)$, $(2, 2/3)$.

   (b)  When $y'$ is positive, $y$ increases; when $y'$ is negative, $y$ decreases.

   When $x < -3$, say $x = -4$,       $y' = (-)(-) = +$,  and $y$ is increasing.
   When $-3 < x < 2$, say $x = 0$,     $y' = (+)(-) = -$,  and $y$ is decreasing.
   When $x > 2$, say $x = 3$,          $y' = (+)(+) = +$,  and $y$ is increasing.

   This is illustrated by the following diagram.

|  | Max. |  | Min. |  |
| :---: | :---: | :---: | :---: | :---: |
| $x < -3$ | $x = -3$ | $-3 < x < 2$ | $x = 2$ | $x > 2$ |
| $y' = +$ |  | $y' = -$ |  | $y' = +$ |
| $y$ increases |  | $y$ decreases |  | $y$ increases |

   (c)  Test the critical values  $x = -3, 2$  for maxima and minima.

   As $x$ increases through $-3$, $y'$ changes sign from $+$ to $-$; hence at $x = -3$, $y$ has a maximum value 43/2.

   As $x$ increases through 2, $y'$ changes sign from $-$ to $+$; hence at $x = 2$, $y$ has a minimum value 2/3.

3. Given $y = x^4 + 2x^3 - 3x^2 - 4x + 4$, find

(a) the intervals on which $y$ is increasing and decreasing, and

(b) the maximum and minimum values of $y$.

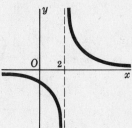

Fig. 8-3

$$y' = 4x^3 + 6x^2 - 6x - 4 = 2(x+2)(2x+1)(x-1)$$

Setting $y' = 0$ gives the critical values $x = -2, -\frac{1}{2}, 1$.

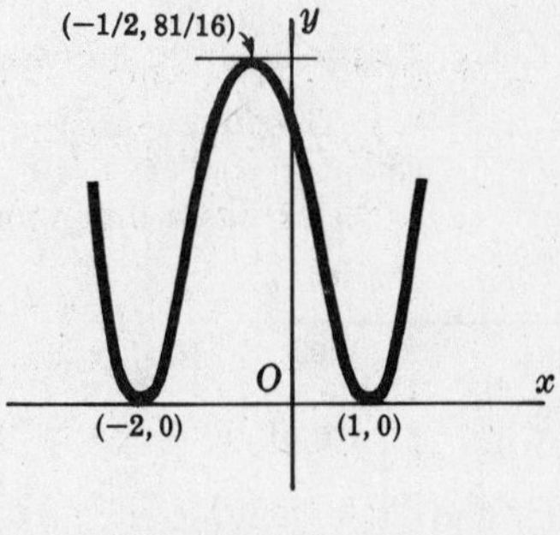

(a) When $x < -2$,　　　　$y' = 2(-)(-)(-) = -$, and $y$ is decreasing.

When $-2 < x < -\frac{1}{2}$,　$y' = 2(+)(-)(-) = +$, and $y$ is increasing.

When $-\frac{1}{2} < x < 1$,　$y' = 2(+)(+)(-) = -$, and $y$ is decreasing.

When $x > 1$,　　　　　$y' = 2(+)(+)(+) = +$, and $y$ is increasing.

This is illustrated by the following diagram.

| | Min. | | Max. | | Min. | |
|---|---|---|---|---|---|---|
| $x < -2$ | $x = -2$ | $-2 < x < -\frac{1}{2}$ | $x = -\frac{1}{2}$ | $-\frac{1}{2} < x < 1$ | $x = 1$ | $x > 1$ |
| $y' = -$ | | $y' = +$ | | $y' = -$ | | $y' = +$ |
| $y$ decreases | | $y$ increases | | $y$ decreases | | $y$ increases |

(b) Test the critical values $x = -2, -\frac{1}{2}, 1$ for maxima and minima.

As $x$ increases through $-2$, $y'$ changes from $-$ to $+$; hence at $x = -2$, $y$ has a minimum value 0.

As $x$ increases through $-\frac{1}{2}$, $y'$ changes from $+$ to $-$; hence at $x = -\frac{1}{2}$, $y$ has a maximum value 81/16.

As $x$ increases through 1, $y'$ changes from $-$ to $+$; hence at $x = 1$, $y$ has a minimum value 0.

4. Show that the curve $y = x^3 - 8$ has no maximum or minimum value.

$y' = 3x^2$. Setting $y' = 0$ gives the critical value $x = 0$.

When $x < 0$ and when $x > 0$, $y' > 0$. Then $y$ has no maximum or minimum value.

At $x = 0$ the curve has a point of inflection.

5. Examine $y = f(x) = \dfrac{1}{x-2}$ for maxima and minima, and locate the intervals on which the function is increasing and decreasing.

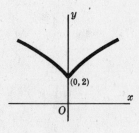

Fig. 8-4

$f'(x) = -\dfrac{1}{(x-2)^2}$. Since $f(2)$ is not defined (i.e. $f(x)$ becomes infinite as $x$ approaches 2), there is no critical value. However, $x = 2$ is employed to locate intervals on which $f(x)$ is increasing and decreasing.

$f'(x) < 0$ for all $x \ne 2$. Hence $f(x)$ is decreasing on the intervals $x < 2$ and $x > 2$.

6. Locate the maximum and minimum values of $f(x) = 2 + x^{2/3}$ and the intervals on which the function is increasing and decreasing.

$f'(x) = \dfrac{2}{3x^{1/3}}$. The critical value is $x = 0$, since $f'(x)$ becomes infinite as $x$ approaches 0.

When $x < 0$, $f'(x) = -$, and $f(x)$ is decreasing.

When $x > 0$, $f'(x) = +$, and $f(x)$ is increasing.

Hence at $x = 0$ the function has a minimum value 2.

Fig. 8-5

7. Examine $y = x^{4/3}(1-x)^{1/3}$ for maximum and minimum values.

Here $y' = \dfrac{x^{1/3}(4-5x)}{3(1-x)^{2/3}}$ and the critical values are $x = 0$, 4/5, and 1. When $x < 0$, $y' < 0$; when $0 < x < 4/5$, $y' > 0$; when $4/5 < x < 1$, $y' < 0$; when $x > 1$, $y' < 0$. The function has a minimum value $(= 0)$ when $x = 0$ and a maximum value $(= \frac{4}{25}\sqrt[3]{20})$ when $x = 4/5$.

8.    Examine $y = |x|$ for maximum and minimum values.

The function is everywhere defined and has a derivative for all $x$ except $x = 0$. (See Problem 11, Chapter 4.) Thus, $x = 0$ is a critical value. For $x < 0$, $f'(x) = -1$ while for $x > 0$, $f'(x) = +1$. The function has a minimum $(= 0)$ when $x = 0$. This result is immediate from a figure.

9.    Examine $y = 3x^4 - 10x^3 - 12x^2 + 12x - 7$ for directions of bending and points of inflection.

$$y' = 12x^3 - 30x^2 - 24x + 12$$
$$y'' = 36x^2 - 60x - 24 = 12(3x + 1)(x - 2)$$

Set $y'' = 0$ and solve to obtain the possible points of inflection $x = -1/3, 2$.

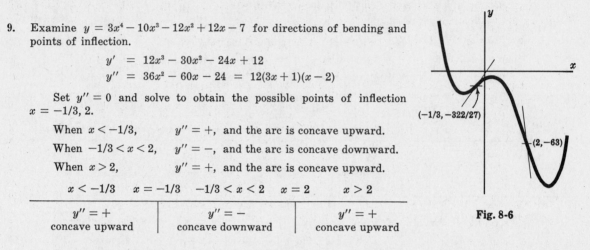

When $x < -1/3$,         $y'' = +$, and the arc is concave upward.

When $-1/3 < x < 2$,    $y'' = -$, and the arc is concave downward.

When $x > 2$,             $y'' = +$, and the arc is concave upward.

| $x < -1/3$ | $x = -1/3$ | $-1/3 < x < 2$ | $x = 2$ | $x > 2$ |
|---|---|---|---|---|
| $y'' = +$ concave upward | | $y'' = -$ concave downward | | $y'' = +$ concave upward |

Fig. 8-6

The points of inflection are $(-1/3, -322/27)$ and $(2, -63)$, since $y''$ changes sign at $x = -1/3$ and $x = 2$.

10.    Examine $y = x^4 - 6x + 2$ for directions of bending and points of inflection. See Fig. 8-7.

$y'' = 12x^2$. The possible point of inflection is at $x = 0$.

On the intervals $x < 0$ and $x > 0$, $y'' = +$, and the arcs are concave upward. Point $(0, 2)$ is not a point of inflection.

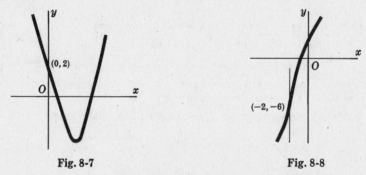

Fig. 8-7                              Fig. 8-8

11.    Examine $y = 3x + (x + 2)^{3/5}$ for directions of bending and points of inflection. See Fig. 8-8.

$$y' = 3 + \frac{3}{5(x + 2)^{2/5}} \qquad y'' = \frac{-6}{25(x + 2)^{7/5}}$$

The possible point of inflection is at $x = -2$.

When $x > -2$, $y'' = -$, and the arc is concave downward.

When $x < -2$, $y'' = +$, and the arc is concave upward.

Point $(-2, -6)$ is a point of inflection.

12.    Find the equations of the tangents at the points of inflection of

$$y = f(x) = x^4 - 6x^3 + 12x^2 - 8x$$

A point of inflection exists at $x = x_0$ when $f''(x_0) = 0$ and $f'''(x_0) \neq 0$.

$$\begin{aligned}
f'(x) &= 4x^3 - 18x^2 + 24x - 8 \\
f''(x) &= 12x^2 - 36x + 24 = 12(x-1)(x-2) \\
f'''(x) &= 24x - 36 = 12(2x-3)
\end{aligned}$$

The possible points of inflection are at $x = 1, 2$. Since $f'''(1) \neq 0$ and $f'''(2) \neq 0$, points $(1, -1)$ and $(2, 0)$ are points of inflection.

At $(1, -1)$, slope $m = f'(1) = 2$, and the equation of the tangent is

$$y - y_1 = m(x - x_1) \quad \text{or} \quad y + 1 = 2(x-1) \quad \text{or} \quad y = 2x - 3$$

At $(2, 0)$, slope $= f'(2) = 0$, and the equation of the tangent is $y = 0$.

13. Show that the points of inflection of $y = \dfrac{a-x}{x^2 + a^2}$ lie on a straight line and find its equation.

$$y' = \frac{x^2 - 2ax - a^2}{(x^2 + a^2)^2} \quad \text{and} \quad y'' = -2\frac{x^3 - 3ax^2 - 3a^2 x + a^3}{(x^2 + a^2)^3}$$

Now $x^3 - 3ax^2 - 3a^2 x + a^3 = 0$ when $x = -a, a(2 \pm \sqrt{3})$; and the points of inflection are $(-a, 1/a)$, $(a(2 + \sqrt{3}), (1 - \sqrt{3})/4a)$, $(a(2 - \sqrt{3}), (1 + \sqrt{3})/4a)$. The slope of the line joining any two of the points is $-1/4a^2$ and the equation of the line of inflection points is $x + 4a^2 y = 3a$.

14. Examine $f(x) = x(12 - 2x)^2$ for maxima and minima using the second derivative method.

(a) $f'(x) = 12(x^2 - 8x + 12) = 12(x-2)(x-6)$. The critical values are $x = 2, 6$.

(b) $f''(x) = 12(2x - 8) = 24(x - 4)$.

(c) $f''(2) < 0$. Hence $f(x)$ has a maximum value 128 at $x = 2$.
   $f''(6) > 0$. Hence $f(x)$ has a minimum value 0 at $x = 6$.

15. Examine $y = x^2 + \dfrac{250}{x}$ for maxima and minima using the second derivative method.

(a) $y' = 2x - \dfrac{250}{x^2} = \dfrac{2(x^3 - 125)}{x^2}$. The critical value is $x = 5$.

(b) $y'' = 2 + \dfrac{500}{x^3}$

(c) $y'' > 0$ at $x = 5$. Hence $y$ has a minimum value 75 at $x = 5$.

16. Examine $y = (x - 2)^{2/3}$ for maximum and minimum values.

(a) $y' = \dfrac{2}{3}(x-2)^{-1/3} = \dfrac{2}{3(x-2)^{1/3}}$. The critical value is $x = 2$.

(b) $y'' = -\dfrac{2}{9}(x-2)^{-4/3} = -\dfrac{2}{9(x-2)^{4/3}}$

(c) $y''$ becomes infinite as $x$ approaches 2. Hence the test fails.

Employ the first derivative method. When $x < 2$, $y' = -$; when $x > 2$, $y' = +$. Hence $y$ has a relative minimum value 0 at $x = 2$.

17. A function $f(x)$ is said to be increasing at $x = x_0$ if for $h > 0$ and sufficiently small, $f(x_0 - h) < f(x_0) < f(x_0 + h)$. Prove: If $f'(x_0) > 0$, then $f(x)$ is increasing at $x = x_0$.

Since $\lim\limits_{\Delta x \to 0} \dfrac{f(x_0 + \Delta x) - f(x_0)}{\Delta x} = f'(x_0) > 0$, we have $\dfrac{f(x_0 + \Delta x) - f(x_0)}{\Delta x} > 0$ for sufficiently small $|\Delta x|$ by Problem 4, Chapter 3.

If $\Delta x < 0$, then $f(x_0 + \Delta x) - f(x_0) < 0$ and, setting $\Delta x = -h$, $f(x_0 - h) < f(x_0)$. If $\Delta x > 0$, say $\Delta x = h$, then $f(x_0 + h) > f(x_0)$. Hence, $f(x_0 - h) < f(x_0) < f(x_0 + h)$ as required in the definition. See Problem 33 for a companion theorem.

18. Prove: If $y = f(x)$ is differentiable on $a \leq x \leq b$ and if $f(x)$ has a relative maximum at $x = x_0$, where $a < x_0 < b$, then $f'(x_0) = 0$.

   Since $f(x)$ has a relative maximum at $x = x_0$ then for every $\Delta x$ with $|\Delta x|$ sufficiently small,

   $$f(x_0 + \Delta x) < f(x_0) \quad \text{and} \quad f(x_0 + \Delta x) - f(x_0) < 0$$

   Now when $\Delta x < 0$, $\dfrac{f(x_0 + \Delta x) - f(x_0)}{\Delta x} > 0$ and $f'(x_0) = \lim\limits_{\Delta x \to 0^-} \dfrac{f(x_0 + \Delta x) - f(x_0)}{\Delta x} \geq 0$.

   Also when $\Delta x > 0$, $\dfrac{f(x_0 + \Delta x) - f(x_0)}{\Delta x} < 0$ and $f'(x_0) = \lim\limits_{\Delta x \to 0^+} \dfrac{f(x_0 + \Delta x) - f(x_0)}{\Delta x} \leq 0$.

   Thus $0 \leq f'(x_0) \leq 0$ and $f'(x_0) = 0$, as was to be proved. See Problem 34 for a companion theorem.

19. Prove the second derivative test for maximum and minimum: If $f(x)$ and $f'(x)$ are differentiable on $a \leq x \leq b$, if $x = x_0$ where $a < x_0 < b$ is a critical value for $f(x)$, and if $f''(x_0) > 0$, then $f(x)$ has a relative minimum value at $x = x_0$.

   Since $f''(x_0) > 0$, $f'(x)$ is increasing at $x = x_0$ and there exists an $h > 0$ such that $f'(x_0 - h) < f'(x_0) < f'(x_0 + h)$. Thus, when $x$ is near to but $< x_0$, $f'(x) < f'(x_0)$; when $x$ is near to but $> x_0$, $f'(x) > f'(x_0)$. Now since $f'(x_0) = 0$, $f'(x) < 0$ when $x < x_0$ and $f'(x) > 0$ when $x > x_0$. These are the conditions (see Problem 18) which assure that $f(x)$ has a relative minimum at $x = x_0$. It is left for the reader to consider the companion theorem for relative maximum.

20. Consider the problem of locating the point $(X, Y)$ on the hyperbola $x^2 - y^2 = 1$ nearest the given point $P(a, 0)$ where $a > 0$. We have $D^2 = (X - a)^2 + Y^2$ for the square of the distance between the two points and $X^2 - Y^2 = 1$ since $(X, Y)$ is on the hyperbola.

   Expressing $D^2$ as a function of $X$ alone, we obtain

   $$f(X) = (X - a)^2 + X^2 - 1 = 2X^2 - 2aX + a^2 - 1$$

   with critical value $X = \frac{1}{2}a$.

   Take $a = \frac{1}{2}$. No point is found since $Y$ is imaginary for the critical value $X = \frac{1}{4}$. From a figure, however, it is clear that the point on the hyperbola nearest $P(\frac{1}{4}, 0)$ is $V(1, 0)$. The trouble here is we have overlooked the fact that $f(X) = (X - \frac{1}{2})^2 + X^2 - 1$ is to be minimized subject to the restriction $X \geq 1$. (Note that this restriction does not arise from $f(X)$ itself. The function $f(X)$, with $X$ unrestricted has indeed a relative minimum at $X = \frac{1}{4}$.) On the interval $X \geq 1$, $f(X)$ has an absolute minimum at the endpoint $X = 1$ but no relative minimum. It is left as an exercise to examine the cases (i) $a = \sqrt{2}$ and (ii) $a = 3$.

## Supplementary Problems

21. Examine each function of Prob. 1 and determine the intervals on which it is increasing and decreasing. *Ans.* (a) Inc. $x < 0$; Dec. $x > 0$. (b) Inc. $x > 3$; Dec. $x < 3$. (c) Inc. $-5/2 < x < 0$; Dec. $0 < x < 5/2$. (d) Inc. $x > 4$

22. (a) Show that $y = x^5 + 20x - 6$ is an increasing function for all values of $x$.
    (b) Show that $y = 1 - x^3 - x^7$ is a decreasing function for all values of $x$.

23. Examine each of the following for relative maximum and minimum values, using the first derivative method:

   (a) $f(x) = x^2 + 2x - 3$         *Ans.* $x = -1$ yields rel. min. $= -4$

   (b) $f(x) = 3 + 2x - x^2$         *Ans.* $x = 1$ yields rel. max. $= 4$

   (c) $f(x) = x^3 + 2x^2 - 4x - 8$      *Ans.* $x = \frac{2}{3}$ yields rel. min. $= -256/27$
                                            $x = -2$ yields rel. max. $= 0$

   (d) $f(x) = x^3 - 6x^2 + 9x - 8$      *Ans.* $x = 1$ yields rel. max. $= -4$
                                              $x = 3$ yields rel. min. $= -8$

   (e) $f(x) = (2 - x)^3$             *Ans.* Neither rel. max. nor min.

   (f) $f(x) = (x^2 - 4)^2$           *Ans.* $x = 0$ yields rel. max. $= 16$
                                              $x = \pm 2$ yields rel. min. $= 0$

(g)  $f(x) = (x-4)^4 (x+3)^3$      Ans. $x = 0$ yields rel. max. $= 6912$
$$x = 4 \text{ yields rel. min.} = 0$$
$$x = -3 \text{ yields neither}$$

(h)  $f(x) = x^3 + 48/x$      Ans. $x = -2$ yields rel. max. $= -32$
$$x = 2 \text{ yields rel. min.} = 32$$

(i)  $f(x) = (x-1)^{1/3}(x+2)^{2/3}$      Ans. $x = -2$ yields rel. max. $= 0$
$$x = 0 \text{ yields rel. min.} = -\sqrt[3]{4}$$
$$x = 1 \text{ yields neither}$$

24. Examine the functions of Problems 23(a)-(f) for relative maximum and minimum values using the second derivative method. Also determine the points of inflection and the intervals on which the curve is concave upward and concave downward.

Ans. (a) No P.I.; concave upward everywhere.

    (b) No P.I.; concave downward everywhere.

    (c) P.I., $x = -2/3$; conc. up, $x > -2/3$; conc. down, $x < -2/3$

    (d) P.I., $x = 2$; conc. up, $x > 2$; conc. down, $x < 2$

    (e) P.I., $x = 2$; conc. down, $x > 2$; conc. up, $x < 2$

    (f) P.I., $x = \pm 2\sqrt{3}/3$;   conc. up, $x > 2\sqrt{3}/3$ and $x < -2\sqrt{3}/3$
$$\text{conc. down, } -2\sqrt{3}/3 < x < 2\sqrt{3}/3$$

25. Show that $y = \dfrac{ax+b}{cx+d}$ has neither a relative maximum nor minimum.

26. Examine $y = x^3 - 3px + q$ for relative maximum and minimum values.
Ans. Min. $= q - 2p^{3/2}$, Max. $= q + 2p^{3/2}$ if $p > 0$; otherwise, neither.

27. Show that $y = (a_1 - x)^2 + (a_2 - x)^2 + \cdots + (a_n - x)^2$ has a relative minimum when $x = (a_1 + a_2 + \cdots + a_n)/n$.

28. Prove: If $f''(x_0) = 0$ and $f'''(x_0) \neq 0$, then there is a point of inflection at $x = x_0$.

29. Prove: If $y = ax^3 + bx^2 + cx + d$ has 2 critical points, they are bisected by the point of inflection. If the curve has just one critical point, it is the point of inflection.

30. A function $f(x)$ is said to have an absolute maximum (minimum) value at $x = x_0$ provided $f(x_0)$ is greater (less) than or equal to every other value of the function on its domain of definition. Use graphs to verify: (a) $y = -x^2$ has an absolute maximum at $x = 0$; (b) $y = (x-3)^2$ has an absolute minimum $(= 0)$ at $x = 3$; (c) $y = \sqrt{25 - 4x^2}$ has an absolute maximum $(= 5)$ at $x = 0$ and an absolute minimum $(= 0)$ at $x = \pm 5/2$; (d) $y = \sqrt{x-4}$ has an absolute minimum $(= 0)$ at $x = 4$.

31. Examine for absolute maximum and minimum values on the given interval only:

(a)  $y = -x^2$ on $-2 < x < 2$      Ans. Max. $(= 0)$ at $x = 0$

(b)  $y = (x-3)^2$ on $0 \leq x \leq 4$      Ans. Max. $(= 9)$ at $x = 0$
$$\text{Min. } (= 0) \text{ at } x = 3$$

(c)  $y = \sqrt{25 - 4x^2}$ on $-2 \leq x \leq 2$      Ans. Max. $(= 5)$ at $x = 0$
$$\text{Min. } (= 3) \text{ at } x = \pm 2$$

(d)  $y = \sqrt{x-4}$ on $4 \leq x \leq 29$      Ans. Max. $(= 5)$ at $x = 29$
$$\text{Min. } (= 0) \text{ at } x = 4$$

    Note. These are the greatest and least values of Property II, Chapter 3, of continuous functions.

32. Verify: A function $f(x)$ is increasing (decreasing) at $x = x_0$ if the angle of inclination of the tangent at $x = x_0$ to the curve $y = f(x)$ is acute (obtuse).

33. State and prove the companion of Problem 17 for a decreasing function.

34. State and prove the companion of Problem 18 for a relative minimum.

35. Examine $2x^2 - 4xy + 3y^2 - 8x + 8y - 1 = 0$ for maximum and minimum points.
Ans. Max. at $(5, 3)$; min. at $(-1, -3)$

36. An electric current, when flowing in a circular coil of radius $r$, exerts a force $F = \dfrac{kx}{(x^2 + r^2)^{5/2}}$ on a small magnet located a distance $x$ above the center of the coil. Show that $F$ is maximum when $x = \frac{1}{2}r$.

37. The work done by a voltaic cell of constant electromotive force $E$ and constant internal resistance $r$ in passing a steady current through an external resistance $R$ is proportional to $E^2 R/(r+R)^2$. Show that the work done is maximum when $R = r$.

# Chapter 9

## Applied Problems in Maxima and Minima

**PROBLEMS IN MAXIMA AND MINIMA.** In the simpler applications it is rarely necessary to prove a relative maximum or minimum. The proper choice of the critical value can be made from a study of the problem.

A relative maximum or minimum may at times be an *absolute* maximum or minimum (i.e., the greatest or smallest value) of a function. In such cases, the terms *nearest, greatest,* etc., are justified in the statement of the problem.

## Solved Problems

1. Divide the number 120 into two parts such that the product $P$ of one part and the square of the other is a maximum.

   Let $x =$ one part; $120 - x =$ the other part. Then $P = (120 - x)x^2$.

   $dP/dx = 3x(80 - x)$. The critical values are $x = 0$ and $x = 80$.

   The critical value $x = 0$ is readily excluded. The required parts are $x = 80$ and $120 - x = 40$.

2. A sheet of paper for a poster contains 18 square feet. The margins at the top and bottom are 9 in. and at the sides 6 in. What are the dimensions if the printed area is a maximum?

   Let $x =$ length of poster and $18/x =$ width of poster, in feet. See Figure 9-1 below.

   The printed area in square feet is $A = (x - 1)\left(\dfrac{18}{x} - \dfrac{3}{2}\right)$.

   $\dfrac{dA}{dx} = \dfrac{18}{x^2} - \dfrac{3}{2}$. Solving $\dfrac{dA}{dx} = 0$, the critical value is $x = 2\sqrt{3}$.

   The overall dimensions of the poster are $x = 2\sqrt{3}$ ft and $18/x = 3\sqrt{3}$ ft.

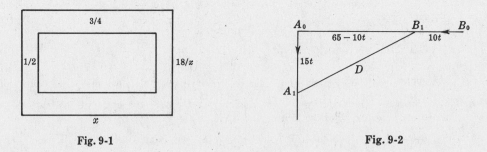

   Fig. 9-1                    Fig. 9-2

3. At 9 A.M. ship $B$ was 65 miles due east of another ship $A$. Ship $B$ was then sailing due west at 10 mi/hr, and $A$ was sailing due south at 15 mi/hr. If they continue their respective courses, when will they be nearest one another and how near? See Fig. 9-2 above.

   Let $A_0$ and $B_0$ be the positions of the ships at 9 A.M., and $A_1$ and $B_1$ be their positions $t$ hours later. Distance covered in $t$ hours by $A = 15t$ miles; by $B = 10t$ miles.

   The distance $D$ between the ships is given by $D^2 = (15t)^2 + (65 - 10t)^2$.

$\dfrac{dD}{dt} = \dfrac{325t - 650}{D}$. Solving $\dfrac{dD}{dt} = 0$ gives the critical value $t = 2$ for which $D$ is a minimum.

Putting $t = 2$ in $D^2 = (15t)^2 + (65 - 10t)^2$ gives $D = 15\sqrt{13}$ miles.

The ships are nearest at 11 A.M., at which time they are $15\sqrt{13}$ miles apart.

4.  A cylindrical container with circular base is to hold 64 cubic inches.  Find the dimensions so that the amount (surface area) of metal required is a minimum when (*a*) the container is an open cup and (*b*) a closed can.

    Let $r$ and $h$ respectively be the radius of the base and height in inches, $A$ the amount of metal, and $V$ the volume of the container.

(*a*)  $V = \pi r^2 h = 64$, and $A = 2\pi rh + \pi r^2$.

    To express $A$ as a function of one variable, solve for $h$ in the first relation (since it is easier) and substitute in the second to obtain $A = 2\pi r(64/\pi r^2) + \pi r^2 = 128/r + \pi r^2$.

$$\frac{dA}{dr} = -\frac{128}{r^2} + 2\pi r = \frac{2(\pi r^3 - 64)}{r^2}, \quad \text{and the critical value is } r = \frac{4}{\sqrt[3]{\pi}}.$$

    Then $h = 64/\pi r^2 = 4/\sqrt[3]{\pi}$.  Thus, $r = h = 4/\sqrt[3]{\pi}$ in.

(*b*)  $V = \pi r^2 h = 64$, and $A = 2\pi rh + 2\pi r^2 = 2\pi r(64/\pi r^2) + 2\pi r^2 = 128/r + 2\pi r^2$.

$$\frac{dA}{dr} = -\frac{128}{r^2} + 4\pi r = \frac{4(\pi r^3 - 32)}{r^2}, \quad \text{and the critical value is } r = 2\sqrt[3]{4/\pi}$$

    Then $h = 64/\pi r^2 = 4\sqrt[3]{4/\pi}$.  Thus, $h = 2r = 4\sqrt[3]{4/\pi}$ in.

5.  The total cost of producing $x$ radio sets per day is $\$(\tfrac14 x^2 + 35x + 25)$ and the price per set at which they may be sold is $\$(50 - \tfrac12 x)$.

(*a*)  What should be the daily output to obtain a maximum total profit?

(*b*)  Show that the cost of producing a set is a relative minimum.

(*a*)  Profit on sale of $x$ sets per day is $P = x(50 - \tfrac12 x) - (\tfrac14 x^2 + 35x + 25)$.

$$\frac{dP}{dx} = 15 - \frac{3x}{2}. \quad \text{Solving } dP/dx = 0 \text{ gives the critical value } x = 10.$$

    Thus the production to give maximum profit is 10 sets/day.

(*b*)  Cost of producing a set is $C = \dfrac{\$(\tfrac14 x^2 + 35x + 25)}{x} = \$\left(\dfrac14 x + 35 + \dfrac{25}{x}\right)$.

$$\frac{dC}{dx} = \frac14 - \frac{25}{x^2}. \quad \text{Solving } dC/dx = 0 \text{ gives } x = 10, \text{ a minimum.}$$

6.  The cost of fuel in running a locomotive is proportional to the square of the speed and is \$25 per hour for a speed of 25 mi/hr.  Other costs amount to \$100 per hour, regardless of the speed.  Find the speed which will make the cost per mile a minimum.

    Let $v =$ required speed, and $C =$ total cost per mile.

    Fuel cost per hour $= kv^2$, where constant $k$ is to be determined.  When $v = 25$ mi/hr, $kv^2 = 625k = 25$; hence $k = 1/25$.

$$C \text{ (in \$/mile)} = \frac{\text{cost in \$/hr}}{\text{speed in mi/hr}} = \frac{v^2/25 + 100}{v} = \frac{v}{25} + \frac{100}{v}$$

$$\frac{dC}{dv} = \frac{1}{25} - \frac{100}{v^2} = \frac{(v - 50)(v + 50)}{25v^2}. \quad \text{Since } v > 0, \text{ the only relevant critical value is } v = 50. \quad \text{Thus,}$$

the most economical speed is 50 mi/hr.

7.  A man in a rowboat at $P$, 5 miles from the nearest point $A$ on a straight shore, wishes to reach a point $B$, 6 miles from $A$ along the shore, in the shortest time.  Where should he land if he can row 2 mi/hr and walk 4 mi/hr?

    Let $C$ be the point between $A$ and $B$ at which the man lands, and let $AC = x$.

    The distance rowed is $PC = \sqrt{25 + x^2}$ and the time required is

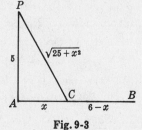

**Fig. 9-3**

$t_1 = \dfrac{\text{distance}}{\text{speed}} = \dfrac{\sqrt{25 + x^2}}{2}$. The distance walked is $CB = 6 - x$ and the time required is $t_2 = (6 - x)/4$.

The total time required is $t = t_1 + t_2 = \frac{1}{2}\sqrt{25 + x^2} + \frac{1}{4}(6 - x)$.

$\dfrac{dt}{dx} = \dfrac{x}{2\sqrt{25 + x^2}} - \dfrac{1}{4} = \dfrac{2x - \sqrt{25 + x^2}}{4\sqrt{25 + x^2}}$ and the critical value, from $2x - \sqrt{25 + x^2} = 0$, is

$x = \frac{5}{3}\sqrt{3} = 2.89$. Thus, he should land at a point 2.89 miles from $A$ toward $B$.

8. A rectangular field to contain a given area is to be fenced off along a straight river. If no fencing is needed along the river, show that the least amount of fencing will be required when the length of the field is twice its width.

Let $x = $ length and $y = $ width of field. Area of field, $A = xy$. Fencing required, $F = x + 2y$.

$dF/dx = 1 + 2\,dy/dx$. When $dF/dx = 0$, $dy/dx = -\frac{1}{2}$.

$dA/dx = 0 = y + x\,dy/dx$. Then $y - \frac{1}{2}x = 0$, and $x = 2y$ as required.

9. Find the dimensions of the right circular cone of minimum volume which can be circumscribed about a sphere of radius 8 inches.

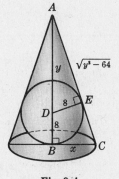

Let $x = $ radius of base of cone, and $y + 8 = $ altitude of cone. From similar right triangles $ABC$ and $AED$, we have

$\dfrac{x}{8} = \dfrac{y + 8}{\sqrt{y^2 - 64}}$. Then $x^2 = \dfrac{64(y + 8)^2}{y^2 - 64} = \dfrac{64(y + 8)}{y - 8}$.

Volume of cone, $V = \dfrac{(\pi x^2)(y + 8)}{3} = \dfrac{64\pi(y + 8)^2}{3(y - 8)}$.

$\dfrac{dV}{dy} = \dfrac{64\pi(y + 8)(y - 24)}{3(y - 8)^2}$. The pertinent critical value is $y = 24$.

Altitude of cone $= y + 8 = 32$ in.;   radius of base $= x = 8\sqrt{2}$ in.

**Fig. 9-4**

10. Find the dimensions of the rectangle of maximum area that can be inscribed in the portion of the parabola $y^2 = 4px$ intercepted by the line $x = a$.

Let $PBB'P'$ be the rectangle, and $(x, y)$ the coordinates of $P$. See Fig. 9-5 below.

Area of rectangle, $A = 2y(a - x) = 2y(a - y^2/4p) = 2ay - y^3/2p$.

$dA/dy = 2a - 3y^2/2p$. Solving $dA/dy = 0$, the critical value is $y = \sqrt{4ap/3}$.

Dimensions of rectangle are $2y = \frac{4}{3}\sqrt{3ap}$ and $a - x = a - y^2/4p = 2a/3$.

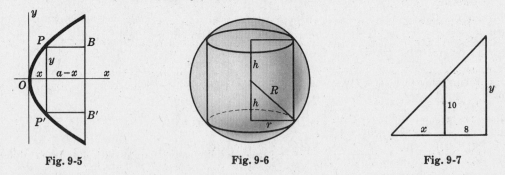

**Fig. 9-5**          **Fig. 9-6**          **Fig. 9-7**

11. Find the height of the right circular cylinder of maximum volume $V$ which can be inscribed in a sphere of radius $R$. See Fig. 9-6 above.

Let $r$ be the radius of the base and $2h$ the height of the cylinder.

$V = 2\pi r^2 h$ and $r^2 + h^2 = R^2$. Then $dV/dr = 2\pi(r^2\,dh/dr + 2rh)$ and $2r + 2h\,dh/dr = 0$.

From the last relation above, $dh/dr = -r/h$. Then $dV/dr = 2\pi(-r^3/h + 2rh)$.

When $V$ is a maximum, $dV/dr = 2\pi(-r^3/h + 2rh) = 0$ and $r^2 = 2h^2$.

Since $r^2 + h^2 = R^2$, $2h^2 + h^2 = R^2$ and $h = R/\sqrt{3}$. Height of cylinder $= 2h = 2R/\sqrt{3}$.

12. The side wall of a building is to be braced by a beam which must pass over a parallel wall 10 feet high and 8 feet from the building.  Find the length $L$ of the shortest beam that can be used.

   Let $x$ be the distance from the foot of the beam to the foot of the parallel wall and $y$ be the distance from the ground to the top of the beam, in feet.  See Fig. 9-7 above.

   $L = \sqrt{(x+8)^2 + y^2}$.  From similar triangles,  $\dfrac{y}{10} = \dfrac{x+8}{x}$  and  $y = \dfrac{10(x+8)}{x}$.

   Then  $L = \sqrt{(x+8)^2 + \dfrac{100(x+8)^2}{x^2}} = \dfrac{x+8}{x}\sqrt{x^2 + 100}$  and

$$\frac{dL}{dx} = \frac{x[(x^2+100)^{1/2} + x(x+8)(x^2+100)^{-1/2}] - (x+8)(x^2+100)^{1/2}}{x^2} = \frac{x^3 - 800}{x^2\sqrt{x^2+100}}$$

   The relevant critical value is $x = 2\sqrt[3]{100}$.  The length of the shortest beam is

$$\frac{2\sqrt[3]{100}+8}{2\sqrt[3]{100}}\sqrt{4\sqrt[3]{10{,}000}+100} = (\sqrt[3]{100}+4)^{3/2} \text{ ft}$$

# Supplementary Problems

13. The sum of two positive numbers is 20.  Find the numbers (a) if their product is a maximum, (b) if the sum of their squares is a minimum, (c) if the product of the square of one and the cube of the other is a maximum.     Ans. (a) 10, 10;  (b) 10, 10;  (c) 8, 12

14. The product of two positive numbers is 16.  Find the numbers (a) if their sum is least, (b) if the sum of one and the square of the other is least.     Ans. (a) 4, 4;  (b) 8, 2

15. An open rectangular box with square ends to hold 6400 cu. ft. is to be built at a cost of 75¢ per sq. ft. for the base and 25¢ per sq. ft. for the sides.  Find the most economical dimensions.
   Ans.  $20 \times 20 \times 16$ ft.

16. A wall 8 ft. high is $3\frac{3}{8}$ ft. from a house.  Find the shortest ladder which will reach from the ground to the house when leaning over the wall.     Ans. $15\frac{5}{8}$ ft.

17. A company offers the following schedule of charges: \$30 per thousand for orders of 50,000 or less, with the charge per thousand decreased by $37\frac{1}{2}$¢ for each thousand above 50,000.  Find the order which will make the company's receipts a maximum.     Ans. 65,000

18. Find the equation of the line through the point $(3, 4)$ which cuts from the first quadrant a triangle of minimum area.     Ans. $4x + 3y - 24 = 0$

19. At what first quadrant point on the parabola $y = 4 - x^2$ does the tangent together with the coordinate axes determine a triangle of minimum area.     Ans. $(2\sqrt{3}/3, 8/3)$

20. Find the minimum distance from the point $(4, 2)$ to the parabola $y^2 = 8x$.     Ans. $2\sqrt{2}$ units

21. A tangent is drawn to the ellipse $x^2/25 + y^2/16 = 1$ so that the part intercepted by the coordinate axes is a minimum.  Show that the length is 9 units.

22. A rectangle is inscribed in the ellipse $x^2/400 + y^2/225 = 1$ with its sides parallel to the axes of the ellipse.  Find the dimensions of the rectangle of (a) maximum area and (b) maximum perimeter which can be so inscribed.     Ans. (a) $20\sqrt{2} \times 15\sqrt{2}$,  (b) $32 \times 18$

23. Find the radius $R$ of the right circular cone of maximum volume which can be inscribed in a sphere of radius $r$.     Ans. $R = \frac{2}{3}r\sqrt{2}$

24. A right circular cylinder is inscribed in a right circular cone of radius $r$.  Find the radius $R$ of the cylinder (a) if its volume is a maximum, (b) if its lateral area is a maximum.
   Ans. (a) $R = \frac{2}{3}r$,  (b) $R = \frac{1}{2}r$

25. Show that a conical tent of given capacity will require the least amount of material when its height = $\sqrt{2}$(the radius of the base).

26. Show that the equilateral triangle of altitude $3r$ is the isosceles triangle of least area circumscribing a circle of radius $r$.

27. Determine the dimensions of the right circular cylinder of maximum lateral surface which can be inscribed in a sphere of radius 8 in.     Ans. $h = 2r = 8\sqrt{2}$ in.

28. Investigate the possibility of inscribing a right circular cylinder of maximum total area in a right circular cone of radius $r$ and height $h$.     Ans. If $h > 2r$, radius of cylinder = $\frac{1}{2}hr/(h-r)$.

# Chapter 10

## Rectilinear and Circular Motion

### RECTILINEAR MOTION

*The Motion of a Particle P* along a straight line is completely described by the equation $s = f(t)$, where $t \geq 0$ is time and $s$ is the distance of $P$ from a fixed point $O$ in its path.

*The Velocity* of $P$ at time $t$ is $v = \dfrac{ds}{dt}$.

If $v > 0$, $P$ is moving in the direction of increasing $s$.

If $v < 0$, $P$ is moving in the direction of decreasing $s$.

If $v = 0$, $P$ is instantaneously at rest.

*The Acceleration* of $P$ at time $t$ is $a = \dfrac{dv}{dt} = \dfrac{d^2s}{dt^2}$.

If $a > 0$, $v$ is increasing; if $a < 0$, $v$ is decreasing.

If $v$ and $a$ have the same sign, the speed of $P$ is increasing.

If $v$ and $a$ have opposite signs, the speed of $P$ is decreasing.

See Problems 1-5.

### CIRCULAR MOTION

*The Motion of a Particle P* along a circle is completely described by the equation $\theta = f(t)$, where $\theta$ is the central angle (in radians) swept over in time $t$ by a line joining $P$ to the center of the circle.

*The Angular Velocity* of $P$ at time $t$ is $\omega = \dfrac{d\theta}{dt}$.

*The Angular Acceleration* of $P$ at time $t$ is $\alpha = \dfrac{d\omega}{dt} = \dfrac{d^2\theta}{dt^2}$.

If $\alpha = $ constant for all $t$, $P$ moves with constant angular acceleration.

If $\alpha = 0$ for all $t$, $P$ moves with constant angular velocity.

See Problem 6.

## Solved Problems

In the following problems on straight line motion, distance $s$ is in feet and time $t$ in seconds.

1.  A body moves along a straight line according to the law $s = \frac{1}{2}t^3 - 2t$. Determine its velocity and acceleration at the end of 2 seconds.

$$v = \frac{ds}{dt} = \frac{3}{2}t^2 - 2 \qquad \text{When } t = 2, \quad v = \frac{3}{2}(2)^2 - 2 = 4 \text{ ft/sec.}$$

$$a = \frac{dv}{dt} = 3t \qquad \text{When } t = 2, \quad a = 3(2) = 6 \text{ ft/sec}^2.$$

2.  The path of a particle moving in a straight line is given by $s = t^3 - 6t^2 + 9t + 4$.

    (a)  Find $s$ and $a$ when $v = 0$.    (d)  When is $v$ increasing?

    (b)  Find $s$ and $v$ when $a = 0$.    (e)  When does the direction of motion change?

    (c)  When is $s$ increasing?

$$v = ds/dt = 3t^2 - 12t + 9 = 3(t-1)(t-3) \qquad a = dv/dt = 6(t-2)$$

(a)  When  $v=0$,  $t=1$  and  3.  When  $t=1$,  $s=8$  and  $a=-6$.  When  $t=3$,  $s=4$  and  $a=6$.

(b)  When  $a=0$,  $t=2$.  At  $t=2$,  $s=6$  and  $v=-3$.

(c)  $s$  is increasing when  $v>0$,  i.e.  when  $t<1$  and  $t>3$.

(d)  $v$  is increasing when  $a>0$,  i.e.  when  $t>2$.

(e)  The direction of motion changes when  $v=0$  and  $a \ne 0$.  From  (a)  the direction changes when  $t=1$  and  $t=3$.

3.  A body moves along a horizontal line according to the law  $s = f(t) = t^3 - 9t^2 + 24t$.

(a)  When is  $s$  increasing and when decreasing?

(b)  When is  $v$  increasing and when decreasing?

(c)  When is the speed of the body increasing and when decreasing?

(d)  Find the total distance traveled in the first 5 seconds of motion.

$$v = ds/dt = 3t^2 - 18t + 24 = 3(t-2)(t-4) \qquad a = dv/dt = 6(t-3)$$

(a)  $s$  is increasing when  $v>0$,  that is, when  $t<2$  and  $t>4$.
$s$  is decreasing when  $v<0$,  that is, when  $2<t<4$.

(b)  $v$  is increasing when  $a>0$,  that is, when  $t>3$.
$v$  is decreasing when  $a<0$,  that is, when  $t<3$.

(c)  The speed is increasing when  $v$  and  $a$  have the same sign and is decreasing when  $v$  and  $a$  have opposite signs.  Since  $v$  may change sign when  $t=2$  and  $t=4$  while  $a$  may change sign at  $t=3$,  their signs are to be compared on the intervals  $t<2$,  $2<t<3$,  $3<t<4$,  and  $t>4$.

On the interval  $t<2$,  $v>0$  and  $a<0$;  the speed is decreasing.

On the interval  $2<t<3$,  $v<0$  and  $a<0$;  the speed is increasing.

On the interval  $3<t<4$,  $v<0$  and  $a>0$;  the speed is decreasing.

On the interval  $t>4$,  $v>0$  and  $a>0$;  the speed is increasing.

(d)  When  $t=0$,  $s=0$  and the body is at  $O$.  The initial motion is to the right  $(v>0)$  for the first two seconds, at which time the body is  $s=f(2)=20$ ft  from  $O$.

During the next two seconds it moves to the left, at the end of which time the body is  $s=f(4)=16$ ft  from  $O$.

It then moves to the right and after 5 sec of motion, in all, is  $s=f(5)=20$ ft  from  $O$.

The total distance traveled is  $20+4+4 = 28$ ft.

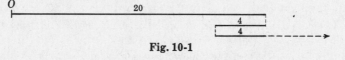

**Fig. 10-1**

4.  A particle moves in a horizontal line according to the law  $s = f(t) = t^4 - 6t^3 + 12t^2 - 10t + 3$.

(a)  When is the speed increasing and when decreasing?

(b)  When does the direction of motion change?

(c)  Find the total distance traveled in the first 3 seconds of motion.

$$v = ds/dt = 4t^3 - 18t^2 + 24t - 10 = 2(t-1)^2(2t-5) \qquad a = dv/dt = 12(t-1)(t-2)$$

(a)  $v$  may change sign when  $t=1$  and  $t=2.5$;  $a$  may change sign when  $t=1$  and  $t=2$.

On the interval  $t<1$,  $v<0$  and  $a>0$;  the speed is decreasing.

On the interval  $1<t<2$,  $v<0$  and  $a<0$;  the speed is increasing.

On the interval  $2<t<2.5$,  $v<0$  and  $a>0$;  the speed is decreasing.

On the interval  $t>2.5$,  $v>0$  and  $a>0$;  the speed is increasing.

(b)  The direction of motion changes at  $t=2.5$  since  $v=0$,  $a \ne 0$;  but it does not change at  $t=1$  since  $v$  does not change sign as  $t$  increases through  $t=1$.  Note that when  $t=1$,  $v=0$  and  $a=0$  so that no information is given.

(c)  When $t = 0$, $s = 3$ and the particle is 3 ft to the right of $O$.

The motion is to the left for the first 2.5 sec, at which time the particle is 27/16 ft to the left of $O$. When $t = 3$, $s = 0$; the particle has moved 27/16 ft to the right.

The total distance traveled is  $3 + 27/16 + 27/16 = 51/8$ ft.

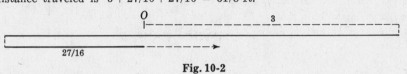

**Fig. 10-2**

5.  A stone, projected vertically upward with initial velocity 112 ft/sec, moves according to the law $s = 112t - 16t^2$, where $s$ is the distance from the starting point. Compute (a) the velocity and acceleration when $t = 3$ and when $t = 4$, and (b) the greatest height reached. (c) When will its height be 96 ft?

$$v = ds/dt = 112 - 32t \qquad a = dv/dt = -32$$

(a)  At $t = 3$, $v = 16$ and $a = -32$. The stone is rising at 16 ft/sec.
    At $t = 4$, $v = -16$ and $a = -32$. The stone is falling at 16 ft/sec.

(b)  At the highest point of the motion, $v = 0$.
    Solving $v = 0 = 112 - 32t$, $t = 3.5$. At this time, $s = 196$ ft.

(c)  $96 = 112t - 16t^2$, $t^2 - 7t + 6 = 0$, $(t-1)(t-6) = 0$, $t = 1, 6$.

At the end of 1 sec of motion the stone is at a height of 96 ft and is rising since $v > 0$. At the end of 6 sec it is at the same height but is falling since $v < 0$.

6.  A particle rotates counterclockwise from rest according to the law $\theta = t^3/50 - t$, where $\theta$ is in radians and $t$ in seconds. Calculate the angular displacement $\theta$, the angular velocity $\omega$, and the angular acceleration $\alpha$ at the end of 10 seconds.

$$\theta = t^3/50 - t = 10 \text{ rad}, \quad \omega = d\theta/dt = 3t^2/50 - 1 = 5 \text{ rad/sec}, \alpha = d\omega/dt = 6t/50 = 6/5 \text{ rad/sec}^2$$

# Supplementary Problems

7.  A particle moves in a straight line according to the law $s = t^3 - 6t^2 + 9t$, the units being feet and seconds. Locate the particle with respect to its initial position $(t = 0)$ at $O$; find its direction and velocity, and determine whether its speed is increasing or decreasing when (a) $t = 1/2$, (b) $t = 3/2$, (c) $t = 5/2$, (d) $t = 4$.
    Ans. (a)  25/8 ft to the right of $O$; moving to the right with $v = 15/4$ ft/sec; decreasing.
        (b)  27/8 ft to the right of $O$; moving to the left with $v = -9/4$ ft/sec; increasing.
        (c)  5/8 ft to the right of $O$; moving to the left with $v = -9/4$ ft/sec; decreasing.
        (d)  4 ft to the right of $O$; moving to the right with $v = 9$ ft/sec; increasing.

8.  The distance a locomotive is from a fixed point on a straight track at time $t$ is given by $s = 3t^4 - 44t^3 + 144t^2$. When was it in reverse?     Ans. $3 < t < 8$

9.  Examine, as in Problem 2, each of the following straight line motions:
    (a) $s = t^3 - 9t^2 + 24t$, (b) $s = t^3 - 3t^2 + 3t + 3$, (c) $s = 2t^3 - 12t^2 + 18t - 5$, (d) $s = 3t^4 - 28t^3 + 90t^2 - 108t$.
    Ans. (a)  Stops at $t = 2$ and $t = 4$ with change of direction.
        (b)  Stops at $t = 1$ without change of direction.
        (c)  Stops at $t = 1$ and $t = 3$ with change of direction.
        (d)  Stops at $t = 1$ with change and $t = 3$ without change of direction.

10.  A body rises vertically from the earth according to the law $s = 64t - 16t^2$. Show that it has lost one-half its velocity in its first 48 ft of rise.

11.  A ball is thrown vertically upward from the edge of a roof in such a manner that it eventually falls in the street 112 ft below. If it moves so that its distance $s$ ft from the roof at time $t$ sec is given by $s = 96t - 16t^2$, find (a) the position of the ball, its velocity, and the direction of motion when $t = 2$, and (b) its velocity when it strikes the street.
    Ans. (a)  240 ft above the street, 32 ft/sec, upward. (b) −128 ft/sec.

12.  A wheel turns through an angle $\theta$ radians in time $t$ sec so that $\theta = 128t - 12t^2$. Find the angular velocity and acceleration at the end of 3 sec.     Ans. $\omega = 56$ rad/sec, $\alpha = -24$ rad/sec$^2$

13.  Examine Problems 2 and 9 to conclude that stops with reversal of direction occur at values of $t$ for which $s = f(t)$ has a maximum or minimum value while stops without reversal of direction occur at inflection points.

# Chapter 11

## Related Rates

**RELATED RATES.** If a variable $x$ is a function of time $t$, the *time rate of change* of $x$ is given by $dx/dt$.

When two or more variables, all functions of $t$, are related by an equation, the relation between their rates of change may be obtained by differentiating the equation with respect to $t$.

## Solved Problems

1. Gas is escaping from a spherical balloon at the rate of 2 cubic feet per minute (2 ft³/min). How fast is the surface area shrinking when the radius is 12 ft?

   At time $t$ the sphere has radius $r$, volume $V = \frac{4}{3}\pi r^3$, and surface $S = 4\pi r^2$.

   Then $\dfrac{dV}{dt} = 4\pi r^2 \dfrac{dr}{dt}$, $\dfrac{dS}{dt} = 8\pi r \dfrac{dr}{dt}$, $\dfrac{dS/dt}{dV/dt} = \dfrac{2}{r}$, and $\dfrac{dS}{dt} = \dfrac{2}{r}\left(\dfrac{dV}{dt}\right) = \dfrac{2}{12}(-2) = -\dfrac{1}{3}$ ft²/min.

2. Water is running out of a conical funnel at the rate of 1 cubic inch per sec (1 in³/sec). If the radius of the base of the funnel is 4 in. and the altitude is 8 in., find the rate at which the water level is dropping when it is 2 in. from the top.

   Let $r$ be the radius and $h$ the height of the surface of the water at time $t$, and $V$ the volume of water in the cone.

   By similar triangles, $r/4 = h/8$ or $r = \frac{1}{2}h$.

   $V = \frac{1}{3}\pi r^2 h = \frac{1}{12}\pi h^3$ and $dV/dt = \frac{1}{4}\pi h^2\, dh/dt$.

   When $dV/dt = -1$ and $h = 8 - 2 = 6$, then $dh/dt = -1/9\pi$ in/sec.

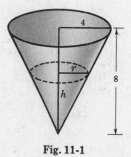

**Fig. 11-1**

3. Sand falling from a chute forms a conical pile whose altitude is always equal to 4/3 the radius of the base. (*a*) How fast is the volume increasing when the radius of the base is 3 feet and is increasing at the rate of 3 in/min? (*b*) How fast is the radius increasing when it is 6 ft and the volume is increasing at the rate of 24 ft³/min?

   Let $r$ = radius of base and $h$ = height of pile at time $t$.

   Since $h = \dfrac{4}{3}r$, $V = \dfrac{1}{3}\pi r^2 h = \dfrac{4}{9}\pi r^3$, and $\dfrac{dV}{dt} = \dfrac{4}{3}\pi r^2 \dfrac{dr}{dt}$.

   (*a*) When $r = 3$ and $\dfrac{dr}{dt} = \dfrac{1}{4}$, $\dfrac{dV}{dt} = 3\pi$ ft³/min.　　(*b*) When $r = 6$ and $\dfrac{dV}{dt} = 24$, $\dfrac{dr}{dt} = \dfrac{1}{2\pi}$ ft/min.

4. One ship $A$ is sailing due south at 16 mi/hr and a second ship $B$, 32 miles south of $A$, is sailing due east at 12 mi/hr. (*a*) At what rate are they approaching or separating at the end of 1 hr? (*b*) at the end of 2 hr? (*c*) When do they cease to approach each other and how far apart are they at that time?

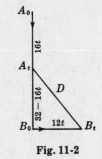

Let $A_o$ and $B_o$ be the initial positions of the ships, and $A_t$ and $B_t$ their positions $t$ hours later. Let $D = $ distance between them $t$ hours later.

$$D^2 = (32 - 16t)^2 + (12t)^2 \quad \text{and} \quad \frac{dD}{dt} = \frac{400t - 512}{D}$$

(a) When $t = 1$, $D = 20$ and $dD/dt = -5.6$. They are approaching at 5.6 mi/hr.

(b) When $t = 2$, $D = 24$ and $dD/dt = 12$. They are separating at 12 mi/hr.

(c) They will cease to approach each other when $dD/dt = 0$, i.e. when $t = 512/400 = 1.28$ hr, at which time they are $D = 19.2$ miles apart.

**Fig. 11-2**

5. Two parallel sides of a rectangle are being lengthened at the rate of 2 in/sec, while the other two sides are shortened in such a way that the figure remains a rectangle with constant area $A$ of 50 square inches. What is the rate of change of the perimeter $P$ when the length of an increasing side is (a) 5 in.? (b) 10 in.? (c) What are the dimensions when the perimeter ceases to decrease?

Let $x = $ length of sides which are being lengthened   and   $y = $ length of other sides at time $t$.

$$P = 2(x + y) \quad \text{and} \quad \frac{dP}{dt} = 2\left(\frac{dx}{dt} + \frac{dy}{dt}\right). \quad A = xy = 50 \quad \text{and} \quad x\frac{dy}{dt} + y\frac{dx}{dt} = 0.$$

(a) When $x = 5$, $y = 10$ and $dx/dt = 2$.

Then $5\dfrac{dy}{dt} + 10(2) = 0$ or $\dfrac{dy}{dt} = -4$, and $\dfrac{dP}{dt} = 2(2 - 4) = -4$ in/sec (decreasing).

(b) When $x = 10$, $y = 5$ and $dx/dt = 2$.

Then $10\dfrac{dy}{dt} + 5(2) = 0$ or $\dfrac{dy}{dt} = -1$, and $\dfrac{dP}{dt} = 2(2 - 1) = 2$ in/sec (increasing).

(c) The perimeter will cease to decrease when $dP/dt = 0$, i.e. when $dy/dt = -dx/dt = -2$.

Then $x(-2) + y(2) = 0$, and the rectangle is a square of side $x = y = 5\sqrt{2}$ in.

6. The radius of a sphere is $r$ in. at time $t$ sec. Find the radius when the rates of increase of the surface area and the radius are numerically equal.

Surface area of sphere, $S = 4\pi r^2$.   $\dfrac{dS}{dt} = 8\pi r\dfrac{dr}{dt}$.

When $\dfrac{dS}{dt} = \dfrac{dr}{dt}$, $\dfrac{dr}{dt} = 8\pi r\dfrac{dr}{dt}$ and the radius is $r = \dfrac{1}{8\pi}$ in.

7. A weight $W$ is attached to a rope 50 ft long which passes over a pulley at $P$, 20 ft above the ground. The other end of the rope is attached to a truck at a point $A$, 2 ft above the ground as shown in Fig. 11-3. If the truck moves off at the rate of 9 ft/sec, how fast is the weight rising when it is 6 ft above the ground?

Let $x$ denote the distance the weight has been raised and $y$ the horizontal distance from the point $A$ where the rope is attached to the truck to the vertical line passing through the pulley at time $t$.

We must find $\dfrac{dx}{dt}$ when $\dfrac{dy}{dt} = 9$ and $x = 6$.

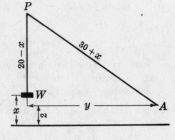

**Fig. 11-3**

Now $y^2 = (30 + x)^2 - (18)^2$ and $\dfrac{dy}{dt} = \dfrac{30 + x}{y} \cdot \dfrac{dx}{dt}$.

When $x = 6$, $y = 18\sqrt{3}$ and $\dfrac{dy}{dt} = 9$. Then $9 = \dfrac{30 + 6}{18\sqrt{3}} \cdot \dfrac{dx}{dt}$ from which $\dfrac{dx}{dt} = \dfrac{9}{2}\sqrt{3}$ ft/sec.

8. A light $L$ hangs $H$ ft above a street. An object $h$ ft tall at $O$, directly under the light, is moved in a straight line along the street at $v$ ft/sec. Investigate the velocity $V$ of the tip of the shadow on the street after $t$ sec. See Fig. 11-4 below.

After $t$ sec the object has been moved a distance $vt$. Let $y =$ distance of tip of shadow from $O$.

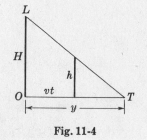

Fig. 11-4

$$\frac{y - vt}{y} = \frac{h}{H} \quad \text{or} \quad y = \frac{Hvt}{H - h} \quad \text{and} \quad V = \frac{dy}{dt} = \frac{Hv}{H - h} = \frac{1}{1 - h/H}\, v$$

Thus the velocity of the tip of the shadow is proportional to the velocity of the object, the factor of proportionality depending upon the ratio $h/H$. As $h \to 0$, $V \to v$, while as $h \to H$, $V$ increases ever more rapidly.

# Supplementary Problems

9. A rectangular trough is 8 ft long, 2 ft across the top, and 4 ft deep. If water flows in at the rate 2 ft³/min, how fast is the surface rising when the water is 1 ft deep?    *Ans.* 1/8 ft/min

10. A liquid is flowing into a vertical cylindrical tank of radius 6 ft at the rate 8 ft³/min. How fast is the surface rising?    *Ans.* $2/9\pi$ ft/min

11. A man 5 ft tall walks at the rate 4 ft/sec directly away from a street light which is 20 ft above the street. (*a*) At what rate is the tip of his shadow changing? (*b*) At what rate is the length of his shadow changing?    *Ans.* (*a*) 16/3 ft/sec, (*b*) 4/3 ft/sec

12. A balloon is rising vertically over a point $A$ on the ground at the rate 15 ft/sec. A point $B$ on the ground is level with and 30 ft from $A$. When the balloon is 40 ft from $A$, at what rate is its distance from $B$ changing?    *Ans.* 12 ft/sec

13. A ladder 20 ft long leans against a house. Find the rate at which (*a*) the top of the ladder is moving downward if its foot is 12 ft from the house and moving away at the rate 2 ft/sec, (*b*) the slope of the ladder decreases.    *Ans.* (*a*) 3/2 ft/sec, (*b*) 25/72 per sec

14. Water is being withdrawn from a conical reservoir 3 ft in radius and 10 ft deep at 4 ft³/min. How fast is the surface falling when the depth of the water is 6 ft? How fast is the radius of this surface diminishing?    *Ans.* $100/81\pi$ ft/min, $10/27\pi$ ft/min

15. A barge whose deck is 10 ft below the level of a dock, is being drawn in by means of a cable attached to the deck and passing through a ring on the dock. When the barge is 24 ft away and approaching the dock at 3/4 ft/sec, how fast is the cable being pulled in? (Neglect any sag in the cable.)    *Ans.* 9/13 ft/sec

16. A boy is flying a kite at a height 150 ft. If the kite moves horizontally away from the boy at the rate 20 ft/sec, how fast is the string being paid out when the kite is 250 ft from him?    *Ans.* 16 ft/sec

17. A train, starting at 11 A.M., travels east at 45 mi/hr while another, starting at noon from the same point, travels south at 60 mi/hr. How fast are they separating at 3 P.M.?    *Ans.* $105\sqrt{2}/2$ mi/hr

18. A light is at the top of a pole 80 ft high. A ball is dropped at the same height from a point 20 ft from the light. Assuming that the ball falls according to the law $s = 16t^2$, how fast is the shadow of the ball moving along the ground one second later?    *Ans.* 200 ft/sec

19. Ship $A$ is 15 miles east of $O$ and moving west at 20 mi/hr; ship $B$ is 60 miles south of $O$ and moving north at 15 mi/hr. (*a*) Are they approaching or separating after 1 hr and at what rate? (*b*) Same, after 3 hrs? (*c*) When are they nearest one another?
*Ans.* (*a*) App., $115/\sqrt{82}$ mi/hr; (*b*) Sep., $9\sqrt{10}/2$ mi/hr; (*c*) 1 hr 55 min

20. Water, at the rate 10 ft³/min, is pouring into a leaky cistern whose shape is a cone 16′ deep and 8′ in diameter at the top. At the time the water is 12′ deep, the water level is observed to be rising 4″/min. How fast is the water leaking away?    *Ans.* $(10 - 3\pi)$ ft³/min

21. A solution is passing through a conical filter 24″ deep and 16″ across the top into a cylindrical vessel of diameter 12″. At what rate is the level of the solution in the cylinder rising if when the depth of the solution in the filter is 12″ its level is falling at the rate 1″/min?    *Ans.* $\frac{4}{9}$ in/min

# Chapter 12

# Differentiation of Trigonometric Functions

**RADIAN MEASURE.** Let $s$ denote the length of arc $AB$ intercepted by the central angle $AOB$ on a circle of radius $r$ and let $S$ denote the area of the sector $AOB$. (If $s$ is 1/360 of the circumference, $\angle AOB = 1°$; if $s = r$, $\angle AOB = 1$ radian.) Suppose $\angle AOB$ is measured as $\alpha$ degrees; then

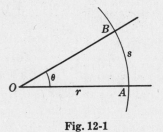

**Fig. 12-1**

(i) $\qquad s = \dfrac{\pi}{180}\alpha r \qquad$ and $\qquad S = \dfrac{\pi}{360}\alpha r^2$

Suppose next that $\angle AOB$ is measured as $\theta$ radians; then

(ii) $\qquad s = \theta r \qquad$ and $\qquad S = \tfrac{1}{2}\theta r^2$

A comparison of (i) and (ii) will make clear one of the advantages of radian measure.

**TRIGONOMETRIC FUNCTIONS.** Let $\theta$ be any real number. Construct the angle whose measure is $\theta$ radians with vertex at the origin of a rectangular coordinate system and initial side along the positive $x$-axis. Take $P(x,y)$ on the terminal side of the angle a unit distance from $O$; then $\sin\theta = y$ and $\cos\theta = x$. The domain of definition of both $\sin\theta$ and $\cos\theta$ is the set of real numbers; the range of $\sin\theta$ is $-1 \le y \le 1$ and the range of $\cos\theta$ is $-1 \le x \le 1$. From

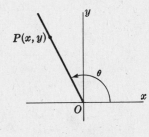

**Fig. 12-2**

$$\tan\theta = \frac{\sin\theta}{\cos\theta} \qquad \text{and} \qquad \sec\theta = \frac{1}{\cos\theta}$$

it follows that the range of both $\tan\theta$ and $\sec\theta$ is the set of real numbers while the domain of definition $(\cos\theta \ne 0)$ is $\theta \ne \pm\dfrac{2n-1}{2}\pi$, $(n = 1,2,3,\dots)$. It is left as an exercise for the reader to consider the functions $\cot\theta$ and $\csc\theta$.

In Problem 1, we prove

$$\lim_{\theta \to 0}\frac{\sin\theta}{\theta} = 1$$

(Had the angle been measured in degrees, the limit would have been $\pi/180$. For this reason, radian measure is always used in the calculus.)

**RULES OF DIFFERENTIATION.** Let $u$ be a differentiable function of $x$; then

14. $\dfrac{d}{dx}(\sin u) = \cos u\dfrac{du}{dx}$

15. $\dfrac{d}{dx}(\cos u) = -\sin u\dfrac{du}{dx}$

16. $\dfrac{d}{dx}(\tan u) = \sec^2 u\dfrac{du}{dx}$

17. $\dfrac{d}{dx}(\cot u) = -\csc^2 u\dfrac{du}{dx}$

18. $\dfrac{d}{dx}(\sec u) = \sec u\tan u\dfrac{du}{dx}$

19. $\dfrac{d}{dx}(\csc u) = -\csc u\cot u\dfrac{du}{dx}$

See Problems 2-23.

# Solved Problems

**1.** Prove: $\displaystyle\lim_{\theta\to 0}\frac{\sin\theta}{\theta}=1$.

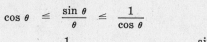

$OC=\cos\theta,\ \ CB=\sin\theta$

**Fig. 12-3**

Since $\dfrac{\sin(-\theta)}{-\theta}=\dfrac{\sin\theta}{\theta}$, we need consider only $\displaystyle\lim_{\theta\to 0^+}\frac{\sin\theta}{\theta}$.

In Fig. 12-3, let $\theta=\angle AOB$ be a small positive central angle of a circle of radius $OA=1$. Denote by $C$ the foot of the perpendicular dropped from $B$ on $OA$ and by $D$ the intersection of $OB$ and an arc of radius $OC$. Now

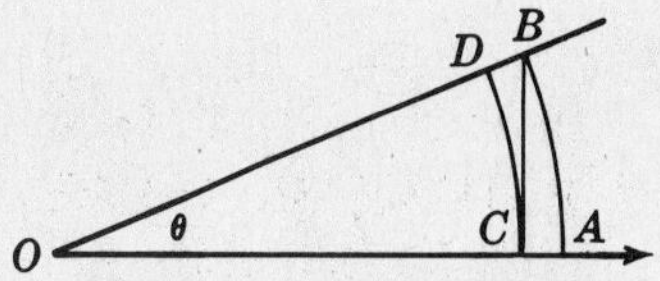

$$\text{sector } COD \ \le\ \triangle COB \ \le\ \text{sector } AOB$$

so that
$$\tfrac{1}{2}\theta\cos^2\theta \ \le\ \tfrac{1}{2}\sin\theta\cos\theta \ \le\ \tfrac{1}{2}\theta$$

Dividing by $\tfrac{1}{2}\theta\cos\theta>0$, we have

$$\cos\theta \ \le\ \frac{\sin\theta}{\theta} \ \le\ \frac{1}{\cos\theta}$$

Let $\theta\to 0^+$; then $\cos\theta\to 1$, $\dfrac{1}{\cos\theta}\to 1$, and $1\le\displaystyle\lim_{\theta\to 0^+}\frac{\sin\theta}{\theta}\le 1$; hence, $\displaystyle\lim_{\theta\to 0^+}\frac{\sin\theta}{\theta}=1$.

**2.** Derive $\dfrac{d}{dx}(\sin u)=\cos u\dfrac{du}{dx}$, $u$ being a differentiable function of $x$.

Let
$$y=\sin u$$
then
$$y+\Delta y=\sin(u+\Delta u)$$
$$\Delta y=\sin(u+\Delta u)-\sin u=2\cos(u+\tfrac{1}{2}\Delta u)\sin\tfrac{1}{2}\Delta u$$

$$\frac{\Delta y}{\Delta u}=\cos(u+\tfrac{1}{2}\Delta u)\frac{\sin\tfrac{1}{2}\Delta u}{\tfrac{1}{2}\Delta u}$$

$$\frac{dy}{du}=\lim_{\Delta u\to 0}\frac{\Delta y}{\Delta u}=\lim_{\Delta u\to 0}\cos(u+\tfrac{1}{2}\Delta u)\cdot\lim_{\Delta u\to 0}\frac{\sin\tfrac{1}{2}\Delta u}{\tfrac{1}{2}\Delta u}=\cos u$$

and, using the chain rule,

$$\frac{d}{dx}(\sin u)=\frac{d}{du}(\sin u)\cdot\frac{du}{dx}=\cos u\frac{du}{dx}$$

**3.** $\dfrac{d}{dx}(\cos u)=\dfrac{d}{dx}[\sin(\tfrac{1}{2}\pi-u)]=\dfrac{d}{du}[\sin(\tfrac{1}{2}\pi-u)]\dfrac{du}{dx}=-\cos(\tfrac{1}{2}\pi-u)\dfrac{du}{dx}=-\sin u\dfrac{du}{dx}.$

**4.** $\dfrac{d}{dx}(\tan u)=\dfrac{d}{dx}\left(\dfrac{\sin u}{\cos u}\right)=\dfrac{\cos u\cdot\cos u\dfrac{du}{dx}-\sin u\left(-\sin u\dfrac{du}{dx}\right)}{\cos^2 u}=\dfrac{1}{\cos^2 u}\dfrac{du}{dx}=\sec^2 u\dfrac{du}{dx}.$

In Problems 5-12, find the first derivative.

**5.** $y=\sin 3x+\cos 2x.$     $y'=\cos 3x\dfrac{d}{dx}(3x)-\sin 2x\dfrac{d}{dx}(2x)=3\cos 3x-2\sin 2x$

**6.** $y=\tan x^2.$     $y'=\sec^2 x^2\dfrac{d}{dx}(x^2)=2x\sec^2 x^2$

**7.** $y=\tan^2 x=(\tan x)^2.$     $y'=2\tan x\dfrac{d}{dx}(\tan x)=2\tan x\sec^2 x$

**8.** $y=\cot(1-2x^2).$     $y'=-\csc^2(1-2x^2)\dfrac{d}{dx}(1-2x^2)=4x\csc^2(1-2x^2)$

**9.** $y=\sec^3\sqrt{x}=\sec^3 x^{1/2}.$

$$y'=3\sec^2 x^{1/2}\frac{d}{dx}(\sec x^{1/2})=3\sec^2 x^{1/2}\cdot\sec x^{1/2}\tan x^{1/2}\cdot\frac{d}{dx}(x^{1/2})=\frac{3}{2\sqrt{x}}\sec^3\sqrt{x}\tan\sqrt{x}$$

**10.** $\rho=\sqrt{\csc 2\theta}=(\csc 2\theta)^{1/2}.$

$$\rho'=\tfrac{1}{2}(\csc 2\theta)^{-1/2}\frac{d}{dx}(\csc 2\theta)=-\tfrac{1}{2}(\csc 2\theta)^{-1/2}\cdot\csc 2\theta\cot 2\theta\cdot 2=-\sqrt{\csc 2\theta}\cdot\cot 2\theta$$

**11.** $f(x) = x^2 \sin x.$     $f'(x) = x^2 \dfrac{d}{dx}(\sin x) + \sin x \dfrac{d}{dx}(x^2) = x^2 \cos x + 2x \sin x$

**12.** $f(x) = \dfrac{\cos x}{x}.$     $f'(x) = \dfrac{x \dfrac{d}{dx}(\cos x) - \cos x \dfrac{d}{dx}(x)}{x^2} = \dfrac{-x \sin x - \cos x}{x^2}$

In Problems 13-16, find the indicated derivative.

**13.** $y = x \sin x;\ y'''.$

$$y' = x \cos x + \sin x$$
$$y'' = x(-\sin x) + \cos x + \cos x = -x \sin x + 2 \cos x$$
$$y''' = -x \cos x - \sin x - 2 \sin x = -x \cos x - 3 \sin x$$

**14.** $y = \tan^2 (3x - 2);\ y''.$

$$y' = 2 \tan (3x - 2) \sec^2 (3x - 2) \cdot 3 = 6 \tan (3x - 2) \sec^2 (3x - 2)$$
$$y'' = 6[\tan (3x - 2) \cdot 2 \sec (3x - 2) \cdot \sec (3x - 2) \tan (3x - 2) \cdot 3 + \sec^2 (3x - 2) \sec^2 (3x - 2) \cdot 3]$$
$$= 36 \tan^2 (3x - 2) \sec^2 (3x - 2) + 18 \sec^4 (3x - 2)$$

**15.** $y = \sin (x + y);\ y'.$     $y' = \cos (x + y) \cdot (1 + y')$   and   $y' = \dfrac{\cos (x + y)}{1 - \cos (x + y)}$

**16.** $\sin y + \cos x = 1;\ y''.$

$$\cos y \cdot y' - \sin x = 0 \quad \text{and} \quad y' = (\sin x)/(\cos y)$$

$$y'' = \frac{\cos y \cos x - \sin x (-\sin y) \cdot y'}{\cos^2 y} = \frac{\cos x \cos y + \sin x \sin y \cdot y'}{\cos^2 y}$$

$$= \frac{\cos x \cos y + \sin x \sin y (\sin x)/(\cos y)}{\cos^2 y} = \frac{\cos x \cos^2 y + \sin^2 x \sin y}{\cos^3 y}$$

**17.** Find $f'(\pi/3),\ f''(\pi/3),\ f'''(\pi/3)$, given $f(x) = \sin x \cos 3x$.

$$f'(x) = -3 \sin x \sin 3x + \cos 3x \cos x$$
$$= (\cos 3x \cos x - \sin 3x \sin x) - 2 \sin x \sin 3x$$
$$= \cos 4x - 2 \sin x \sin 3x. \qquad f'(\pi/3) = -\tfrac{1}{2} - 2(\sqrt{3}/2)(0) = -\tfrac{1}{2}$$

$$f''(x) = -4 \sin 4x - 2(3 \sin x \cos 3x + \sin 3x \cos x)$$
$$= -4 \sin 4x - 2(\sin x \cos 3x + \sin 3x \cos x) - 4 \sin x \cos 3x$$
$$= -6 \sin 4x - 4f(x). \qquad f''(\pi/3) = -6(-\sqrt{3}/2) - 4(\sqrt{3}/2)(-1) = 5\sqrt{3}$$

$$f'''(x) = -24 \cos 4x - 4f'(x). \qquad f'''(\pi/3) = -24(-\tfrac{1}{2}) - 4(-\tfrac{1}{2}) = 14$$

**18.** Find the acute angles of intersection of the curves
(1) $y = 2 \sin^2 x$ and (2) $y = \cos 2x$ on the interval $0 < x < 2\pi$.

(a) Solve $2 \sin^2 x = \cos 2x = 1 - 2 \sin^2 x$ to obtain $\pi/6,\ 5\pi/6,\ 7\pi/6$, and $11\pi/6$ as the abscissas of the points of intersection.

(b) $y' = 4 \sin x \cos x$ for (1), and
$y' = -2 \sin 2x$ for (2).

At the point $\pi/6$, $m_1 = \sqrt{3}$ and $m_2 = -\sqrt{3}$.

(c) $\tan \phi = \dfrac{\sqrt{3} + \sqrt{3}}{1 - 3} = -\sqrt{3}$; the acute angle of intersection is $60°$. At each of the remaining points the acute angle of intersection is $60°$.

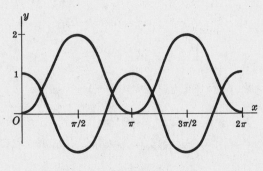

Fig. 12-4

19. A rectangular plot of ground has two adjacent sides along Highways 20 and 32. In the plot is a small lake, one end of which is 256 ft from Highway 20 and 108 ft from Highway 32. Find the length of the shortest straight path which cuts across the plot from one highway to the other and passes by the end of the lake.

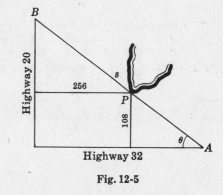

Let $s$ be the length of the path and $\theta$ be the angle which it makes with Highway 32.

$$s = AP + PB = 108 \csc \theta + 256 \sec \theta$$
$$ds/d\theta = -108 \csc \theta \cot \theta + 256 \sec \theta \tan \theta$$
$$= \frac{-108 \cos^3 \theta + 256 \sin^3 \theta}{\sin^2 \theta \cos^2 \theta}$$

Fig. 12-5

From $-108 \cos^3 \theta + 256 \sin^3 \theta = 0$, $\tan^3 \theta = 27/64$ and the critical value is $\theta = \arctan 3/4$. Then $s = 108 \csc \theta + 256 \sec \theta = 108(5/3) + 256(5/4) = 500$ ft.

20. Discuss the curve $y = f(x) = 4 \sin x - 3 \cos x$ on the interval $[0, 2\pi]$.

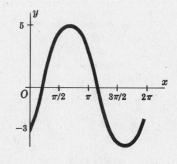

When $x = 0$, $y = f(0) = 4(0) - 3(1) = -3$.

$f(x) = 4 \sin x - 3 \cos x$. Setting $f(x) = 0$ gives $\tan x = 3/4$, and the $x$-intercepts are $x = .64$ (radian) and $x = \pi + .64 = 3.78$ (radians).

$f'(x) = 4 \cos x + 3 \sin x$. Setting $f'(x) = 0$ gives $\tan x = -4/3$, and the critical values are $x = \pi - .93 = 2.21$ and $x = 2\pi - .93 = 5.35$.

$f''(x) = -4 \sin x + 3 \cos x$. Setting $f''(x) = 0$ gives $\tan x = 3/4$, and the possible points of inflection are $x = .64$ and $x = \pi + .64 = 3.78$.

$f'''(x) = -4 \cos x - 3 \sin x$.

Fig. 12-6

(a) When $x = 2.21$, $\sin x = 4/5$ and $\cos x = -3/5$; $f''(x) < 0$ and $x = 2.21$ yields a relative maximum $= 5$. $x = 5.35$ yields a relative minimum $= -5$.

(b) $f'''(.64) \neq 0$ and $f'''(3.78) \neq 0$. The points of inflection are $(.64, 0)$ and $(3.78, 0)$.

(c) The curve is concave upward from $x = 0$ to $x = .64$; concave downward from $x = .64$ to $3.78$; and concave upward from $x = 3.78$ to $2\pi$.

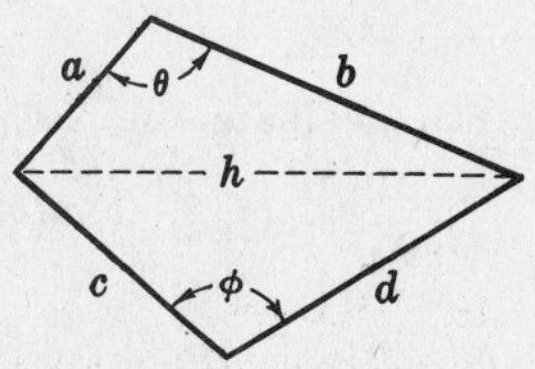

21. Four bars of lengths $a, b, c, d$ are hinged together to form a quadrilateral. Show that the area $A$ is greatest when the opposite angles are supplementary.

Denote by $\theta$ the angle included by the bars of lengths $a$ and $b$, by $\phi$ the opposite angle, and by $h$ the length of the diagonal opposite these angles.

It is required to maximize

Fig. 12-7

(1)  $A = \frac{1}{2}ab \sin \theta + \frac{1}{2}cd \sin \phi$  subject to the condition

(2)  $h^2 = a^2 + b^2 - 2ab \cos \theta = c^2 + d^2 - 2cd \cos \phi$.  Differentiate with respect to $\theta$:

$$(1')\ \frac{dA}{d\theta} = \frac{1}{2}ab \cos \theta + \frac{1}{2}cd \cos \phi \frac{d\phi}{d\theta} = 0 \quad \text{and} \quad (2')\ ab \sin \theta = cd \sin \phi \frac{d\phi}{d\theta}$$

Solve for $d\phi/d\theta$ in (2') and substitute in (1') to obtain

$$ab \cos \theta + cd \cos \phi \frac{ab \sin \theta}{cd \sin \phi} = 0 \quad \text{or} \quad \sin \phi \cos \theta + \cos \phi \sin \theta = \sin(\phi + \theta) = 0$$

Then $\phi + \theta = 0$ or $\pi$, the first of which is easily rejected.

22. A bombardier is sighting on a target on the ground directly ahead.  If the
bomber is flying 2 miles above the ground at 240 mi/hr, how fast must the
sighting instrument be turning when the angle between the path of the
bomber and the line of sight is 30°?

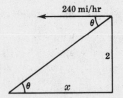

$$\frac{dx}{dt} = -240 \text{ mi/hr}, \quad \theta = 30°, \quad \text{and} \quad x = 2 \cot \theta.$$

$$\frac{dx}{dt} = -2 \csc^2 \theta \, \frac{d\theta}{dt} \quad \text{or} \quad -240 = -2(4)\frac{d\theta}{dt}, \quad \text{and} \quad \frac{d\theta}{dt} = 30 \text{ rad/hr} = \frac{3}{2\pi} \text{ deg/sec}.$$

**Fig. 12-8**

23. A ray of light passes through the air with velocity $v_1$ from a
point $P$, $a$ units above the surface of a body of water, to some
point $O$ on the surface and then with velocity $v_2$ to a point $Q$,
$b$ units below the surface.  If $OP$ and $OQ$ make angles $\theta_1$ and $\theta_2$
with a perpendicular to the surface, show that the passage from

$P$ to $Q$ is most rapid when $\dfrac{\sin \theta_1}{\sin \theta_2} = \dfrac{v_1}{v_2}$

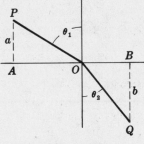

Let $t$ denote the length of time for the passage from $P$ to $Q$,
and $c$ denote the distance from $A$ to $B$; then

$$t = \frac{a \sec \theta_1}{v_1} + \frac{b \sec \theta_2}{v_2} \quad \text{and} \quad c = a \tan \theta_1 + b \tan \theta_2$$

**Fig. 12-9**

Differentiating with respect to $\theta_1$,

$$\frac{dt}{d\theta_1} = \frac{a \sec \theta_1 \tan \theta_1}{v_1} + \frac{b \tan \theta_2 \sec \theta_2}{v_2} \cdot \frac{d\theta_2}{d\theta_1} \quad \text{and} \quad 0 = a \sec^2 \theta_1 + b \sec^2 \theta_2 \cdot \frac{d\theta_2}{d\theta_1}$$

From the last equation, $\dfrac{d\theta_2}{d\theta_1} = -\dfrac{a \sec^2 \theta_1}{b \sec^2 \theta_2}$.  For $t$ to be a minimum, it is necessary that

$$\frac{dt}{d\theta_1} = \frac{a \sec \theta_1 \tan \theta_1}{v_1} + \frac{b \sec \theta_2 \tan \theta_2}{v_2}\left(-\frac{a \sec^2 \theta_1}{b \sec^2 \theta_2}\right) = 0$$

from which the required relation follows.

# Supplementary Problems

24. Evaluate:  (a) $\displaystyle\lim_{x \to 0} \frac{\sin 2x}{x} = 2 \lim_{x \to 0} \frac{\sin 2x}{2x}$,  (b) $\displaystyle\lim_{x \to 0} \frac{\sin ax}{\sin bx}$,  (c) $\displaystyle\lim_{x \to 0} \frac{\sin^3 2x}{x \sin^2 3x}$  (d) $\displaystyle\lim_{x \to 0} \frac{1 - \cos x}{x}$.

Ans.  (a) 2,  (b) $a/b$,  (c) 8/9,  (d) 0

25. Derive the differentiation formula 17, using  (a) $\cot u = \dfrac{\cos u}{\sin u}$  and  (b) $\cot u = \dfrac{1}{\tan u}$.  Derive
also the differentiation formulas 18 and 19.

In Problems 26–45, find the derivative $dy/dx$ or $d\rho/d\theta$.

26. $y = 3 \sin 2x$          Ans.  $6 \cos 2x$

27. $y = 4 \cos \frac{1}{2}x$          Ans.  $-2 \sin \frac{1}{2}x$

28. $y = 4 \tan 5x$          Ans.  $20 \sec^2 5x$

29. $y = \frac{1}{4} \cot 8x$          Ans.  $-2 \csc^2 8x$

30. $y = 9 \sec \frac{1}{3}x$          Ans.  $3 \sec \frac{1}{3}x \tan \frac{1}{3}x$

31. $y = \frac{1}{4} \csc 4x$          Ans.  $y = -\csc 4x \cot 4x$

**32.** $y = \sin x - x \cos x + x^2 + 4x + 3$ $\qquad$ *Ans.* $x \sin x + 2x + 4$

**33.** $\rho = \sqrt{\sin \theta}$ $\qquad$ *Ans.* $(\cos \theta)/(2\sqrt{\sin \theta})$

**34.** $y = \sin 2/x$ $\qquad$ *Ans.* $(-2 \cos 2/x)/x^2$

**35.** $y = \cos (1 - x^2)$ $\qquad$ *Ans.* $2x \sin (1 - x^2)$

**36.** $y = \cos (1 - x)^2$ $\qquad$ *Ans.* $y = 2(1 - x) \sin (1 - x)^2$

**37.** $y = \sin^2 (3x - 2)$ $\qquad$ *Ans.* $3 \sin (6x - 4)$

**38.** $y = \sin^3 (2x - 3)$ $\qquad$ *Ans.* $-\frac{3}{2} \{\cos (6x - 9) - \cos (2x - 3)\}$

**39.** $y = \frac{1}{2} \tan x \sin 2x$ $\qquad$ *Ans.* $\sin 2x$

**40.** $\rho = \dfrac{1}{(\sec 2\theta - 1)^{3/2}}$ $\qquad$ *Ans.* $\dfrac{-3 \sec 2\theta \tan 2\theta}{(\sec 2\theta - 1)^{5/2}}$

**41.** $\rho = \dfrac{\tan 2\theta}{1 - \cot 2\theta}$ $\qquad$ *Ans.* $2 \dfrac{\sec^2 2\theta - 4 \csc 4\theta}{(1 - \cot 2\theta)^2}$

**42.** $y = x^2 \sin x + 2x \cos x - 2 \sin x$ $\qquad$ *Ans.* $x^2 \cos x$

**43.** $\sin y = \cos 2x$ $\qquad$ *Ans.* $-\dfrac{2 \sin 2x}{\cos y}$

**44.** $\cos 3y = \tan 2x$ $\qquad$ *Ans.* $-\dfrac{2 \sec^2 2x}{3 \sin 3y}$

**45.** $x \cos y = \sin (x + y)$ $\qquad$ *Ans.* $\dfrac{\cos y - \cos (x + y)}{x \sin y + \cos (x + y)}$

**46.** If $x = A \sin kt + B \cos kt$, where $A, B, k$ are constants, show that $\dfrac{d^2x}{dt^2} = -k^2x$ and $\dfrac{d^{2n}x}{dt^{2n}} = (-1)^n k^{2n} x$.

**47.** Show: (a) $y'' + 4y = 0$ when $y = 3 \sin (2x + 3)$, (b) $y''' + y'' + y' + y = 0$ when $y = \sin x + 2 \cos x$.

**48.** Discuss and sketch on the interval $0 \leqq x < 2\pi$:

$\quad$ (a) $y = \frac{1}{2} \sin 2x$ $\qquad$ (c) $y = x - 2 \sin x$ $\qquad$ (e) $y = 4 \cos^3 x - 3 \cos x$

$\quad$ (b) $y = \cos^2 x - \cos x$ $\qquad$ (d) $y = \sin x (1 + \cos x)$

$\quad$ *Ans.* (a) Max. at $x = \pi/4, 5\pi/4$; min. at $x = 3\pi/4, 7\pi/4$; P.I. at $x = 0, \pi/2, \pi, 3\pi/2$

$\qquad$ (b) Max. at $x = 0, \pi$; min. at $x = \pi/3, 5\pi/3$; P.I. at $x = 32°32', 126°23', 233°37', 327°28'$

$\qquad$ (c) Max. at $x = 5\pi/3$; min. at $x = \pi/3$; P.I. at $x = 0, \pi$

$\qquad$ (d) Max. at $x = \pi/3$; min. at $x = 5\pi/3$; P.I. at $x = 0, \pi, 104°29', 255°31'$

$\qquad$ (e) Max. at $x = 0, 2\pi/3, 4\pi/3$; min. at $x = \pi/3, \pi, 5\pi/3$; P.I. at $x = \pi/2, 3\pi/2, \pi/6, 5\pi/6, 7\pi/6, 11\pi/6$

**49.** If the angle of elevation of the sun is 45° and is decreasing at $\frac{1}{4}$ rad/hr, how fast is the shadow cast on level ground by a pole 50 ft tall lengthening? $\qquad$ *Ans.* 25 ft/hr

**50.** A kite, 120 ft above the ground, is moving horizontally at the rate 10 ft/sec. At what rate is the inclination of the string to the horizontal diminishing when 240 ft of string are out? *Ans.* 1/48 rad/sec

**51.** A revolving beacon is situated 3600 ft off a straight shore. If the beacon turns at $4\pi$ rad/min, how fast does the beam sweep along the shore at (a) its nearest point, (b) at a point 4800 ft from the nearest point? $\qquad$ *Ans.* (a) $240\pi$ ft/sec, (b) $2000\pi/3$ ft/sec

**52.** Two sides of a triangle are 15 and 20 ft long, respectively. (a) How fast is the third side increasing if the angle between the given sides is 60° and is increasing at the rate 2° per sec? (b) How fast is the area increasing? $\qquad$ *Ans.* (a) $\pi/\sqrt{39}$ ft/sec, (b) $\frac{5}{6}\pi$ ft²/sec

# Chapter 13

## Differentiation of Inverse Trigonometric Functions

**THE INVERSE TRIGONOMETRIC FUNCTIONS.** If $x = \sin y$, the inverse function is written $y = \arcsin x$. The domain of definition of arc sin $x$ is $-1 \leq x \leq 1$, the range of sin $y$; the range of arc sin $x$ is the set of real numbers, the domain of definition of sin $y$. The domain of definition and the range of the remaining inverse trigonometric functions may be established in a similar manner.

The inverse trigonometric functions are multi-valued. In order that there be agreement on separating the graph into single-valued arcs, we define below one such arc (called the *principal branch*) for each function. In the accompanying graphs, the principal branch is indicated by a thickening of the line.

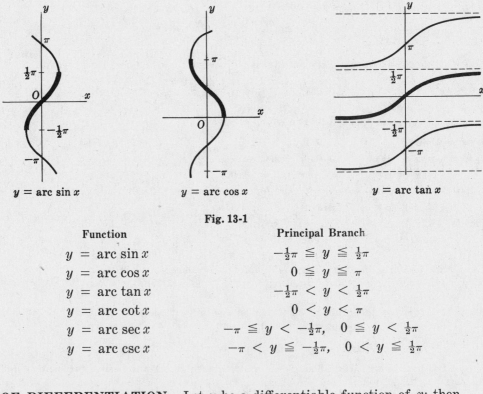

$y = \arcsin x$     $y = \arccos x$     $y = \arctan x$

**Fig. 13-1**

| Function | Principal Branch |
|----------|------------------|
| $y = \arcsin x$ | $-\tfrac{1}{2}\pi \leq y \leq \tfrac{1}{2}\pi$ |
| $y = \arccos x$ | $0 \leq y \leq \pi$ |
| $y = \arctan x$ | $-\tfrac{1}{2}\pi < y < \tfrac{1}{2}\pi$ |
| $y = \operatorname{arc\ cot} x$ | $0 < y < \pi$ |
| $y = \operatorname{arc\ sec} x$ | $-\pi \leq y < -\tfrac{1}{2}\pi, \quad 0 \leq y < \tfrac{1}{2}\pi$ |
| $y = \operatorname{arc\ csc} x$ | $-\pi < y \leq -\tfrac{1}{2}\pi, \quad 0 < y \leq \tfrac{1}{2}\pi$ |

**RULES OF DIFFERENTIATION.** Let $u$ be a differentiable function of $x$; then

**20.** $\dfrac{d}{dx}(\arcsin u) = \dfrac{1}{\sqrt{1-u^2}}\dfrac{du}{dx}$

**23.** $\dfrac{d}{dx}(\operatorname{arc\ cot} u) = -\dfrac{1}{1+u^2}\dfrac{du}{dx}$

**21.** $\dfrac{d}{dx}(\arccos u) = -\dfrac{1}{\sqrt{1-u^2}}\dfrac{du}{dx}$

**24.** $\dfrac{d}{dx}(\operatorname{arc\ sec} u) = \dfrac{1}{u\sqrt{u^2-1}}\dfrac{du}{dx}$

**22.** $\dfrac{d}{dx}(\arctan u) = \dfrac{1}{1+u^2}\dfrac{du}{dx}$

**25.** $\dfrac{d}{dx}(\operatorname{arc\ csc} u) = -\dfrac{1}{u\sqrt{u^2-1}}\dfrac{du}{dx}$

# Solved Problems

1. Derive:   (a) $\dfrac{d}{dx}(\text{arc sin } u) = \dfrac{1}{\sqrt{1-u^2}}\dfrac{du}{dx}$,   (b) $\dfrac{d}{dx}(\text{arc sec } u) = \dfrac{1}{u\sqrt{u^2-1}}\dfrac{du}{dx}$.

(a) Let $y = \text{arc sin } u$, where $u$ is a differentiable function of $x$. Then $u = \sin y$ and

$$\frac{du}{dx} = \frac{d}{dx}(\sin y) = \frac{d}{dy}(\sin y)\frac{dy}{dx} = \cos y \frac{dy}{dx} = \sqrt{1-u^2}\frac{dy}{dx}$$

the sign being positive since $\cos y \geq 0$ on the interval $-\frac{1}{2}\pi \leq y \leq \frac{1}{2}\pi$. Thus, $\dfrac{dy}{dx} = \dfrac{1}{\sqrt{1-u^2}}\dfrac{du}{dx}$.

(b) Let $y = \text{arc sec } u$, where $u$ is a differentiable function of $x$. Then $u = \sec y$ and

$$\frac{du}{dx} = \frac{d}{dy}(\sec y)\frac{dy}{dx} = \sec y \tan y \frac{dy}{dx} = u\sqrt{u^2-1}\frac{dy}{dx}$$

the sign being positive since $\tan y \geq 0$ on the intervals $0 \leq y < \frac{1}{2}\pi$ and $-\pi \leq y < -\frac{1}{2}\pi$. Thus,

$$\frac{d}{dx}(\text{arc sec } u) = \frac{1}{u\sqrt{u^2-1}}\frac{du}{dx}.$$

In Problems 2-9, find the first derivative.

2. $y = \text{arc sin }(2x-3)$

$$\frac{dy}{dx} = \frac{1}{\sqrt{1-(2x-3)^2}}\frac{d}{dx}(2x-3) = \frac{1}{\sqrt{3x-x^2-2}}$$

3. $y = \text{arc cos } x^2$

$$\frac{dy}{dx} = -\frac{1}{\sqrt{1-x^4}}\frac{d}{dx}(x^2) = -\frac{2x}{\sqrt{1-x^4}}$$

4. $y = \text{arc tan } 3x^2$

$$\frac{dy}{dx} = \frac{1}{1+(3x^2)^2}\frac{d}{dx}(3x^2) = \frac{6x}{1+9x^4}$$

5. $f(x) = \text{arc cot }\dfrac{1+x}{1-x}$

$$f'(x) = -\frac{1}{1+\left(\dfrac{1+x}{1-x}\right)^2}\frac{d}{dx}\left(\frac{1+x}{1-x}\right) = -\frac{1}{1+\left(\dfrac{1+x}{1-x}\right)^2}\cdot\frac{(1-x)-(1+x)(-1)}{(1-x)^2} = -\frac{1}{1+x^2}$$

6. $f(x) = x\sqrt{a^2-x^2} + a^2\text{ arc sin }\dfrac{x}{a}$

$$f'(x) = x\cdot\tfrac{1}{2}(a^2-x^2)^{-1/2}(-2x) + (a^2-x^2)^{1/2} + a^2\frac{1}{\sqrt{1-(x/a)^2}}\cdot\frac{1}{a} = 2\sqrt{a^2-x^2}$$

7. $y = x\text{ arc csc}\dfrac{1}{x} + \sqrt{1-x^2}$

$$y' = x\left[\frac{-1}{\dfrac{1}{x}\sqrt{\dfrac{1}{x^2}-1}}\cdot\frac{d}{dx}\left(\frac{1}{x}\right)\right] + \text{arc csc}\frac{1}{x}\cdot\frac{d}{dx}(x) + \tfrac{1}{2}(1-x^2)^{-1/2}(-2x) = \text{arc csc}\frac{1}{x}$$

8. $y = \dfrac{1}{ab}\text{ arc tan}\left(\dfrac{b}{a}\tan x\right)$

$$y' = \frac{1}{ab}\frac{1}{1+\left(\dfrac{b}{a}\tan x\right)^2}\cdot\frac{d}{dx}\left(\frac{b}{a}\tan x\right)\bigg] = \frac{1}{ab}\cdot\frac{a^2}{a^2+b^2\tan^2 x}\cdot\frac{b}{a}\sec^2 x$$

$$= \frac{\sec^2 x}{a^2+b^2\tan^2 x} = \frac{1}{a^2\cos^2 x + b^2\sin^2 x}$$

9. $y^2\sin x + y = \text{arc tan } x$     $2yy'\sin x + y^2\cos x + y' = \dfrac{1}{1+x^2}$

$$y'(2y\sin x + 1) = \frac{1}{1+x^2} - y^2\cos x \quad \text{and} \quad y' = \frac{1-(1+x^2)y^2\cos x}{(1+x^2)(2y\sin x + 1)}$$

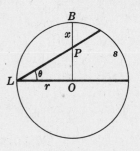

10. In a circular arena there is a light at $L$. A boy starting from $B$ runs at the rate of 10 ft/sec toward the center $O$. At what rate will his shadow be moving along the side when he is half-way from $B$ to $O$?

Let $P$, $x$ feet from $B$, be the position of the boy at time $t$; denote by $r$ the radius of the arena, by $\theta$ the angle $OLP$, and by $s$ the arc intercepted by $\theta$.

$$s = r(2\theta), \quad \text{and} \quad \theta = \text{arc tan } OP/LO = \text{arc tan } (r-x)/r.$$

$$\frac{ds}{dt} = 2r\frac{d\theta}{dt} = 2r \cdot \frac{1}{1+[(r-x)/r]^2} \cdot \left(-\frac{1}{r}\right) \cdot \frac{dx}{dt} = \frac{-2r^2}{x^2 - 2rx + 2r^2} \cdot \frac{dx}{dt}.$$

**Fig. 13-2**

When $x = \frac{1}{2}r$ and $dx/dt = 10$, then $ds/dt = -16$ ft/sec.

The shadow is moving along the wall at 16 ft/sec.

11. The lower edge of a mural, 12 ft high, is 6 ft above an observer's eyes. Under the assumption that the most favorable view is obtained when the angle subtended by the mural at the eye is a maximum, at what distance from the wall should the observer stand?

Let $\theta$ denote the subtended angle and $x$ the distance from the wall. From Fig. 13-3, $\tan(\theta+\phi) = 18/x$, $\tan\phi = 6/x$, and

$$\tan\theta = \tan\{(\theta+\phi)-\phi\} = \frac{\tan(\theta+\phi)-\tan\phi}{1+\tan(\theta+\phi)\tan\phi} = \frac{18/x-6/x}{1+(18/x)(6/x)} = \frac{12x}{x^2+108}$$

$$\theta = \text{arc tan}\frac{12x}{x^2+108} \quad \text{and} \quad \frac{d\theta}{dx} = \frac{12(-x^2+108)}{x^4+360x^2+11,664}. \quad \text{The critical value is}$$

**Fig. 13-3**

$x = 6\sqrt{3} = 10.4$. The observer should stand 10.4 feet in front of the wall.

# Supplementary Problems

12. Derive the differentiation formulas 21, 22, 23, and 25.

In Problems 13-20, find $dy/dx$.

13. $y = \text{arc sin } 3x$      *Ans.* $\dfrac{3}{\sqrt{1-9x^2}}$      17. $y = x^2 \text{ arc cos } 2/x$     *Ans.* $2x\left(\text{arc cos }\dfrac{2}{x} + \dfrac{1}{\sqrt{x^2-4}}\right)$

14. $y = \text{arc cos }\frac{1}{2}x$      *Ans.* $-\dfrac{1}{\sqrt{4-x^2}}$    18. $y = \dfrac{x}{\sqrt{a^2-x^2}} - \text{arc sin}\dfrac{x}{a}$    *Ans.* $\dfrac{x^2}{(a^2-x^2)^{3/2}}$

15. $y = \text{arc tan } 3/x$      *Ans.* $-\dfrac{3}{x^2+9}$      19. $y = (x-a)\sqrt{2ax-x^2} + a^2 \text{ arc sin}\dfrac{x-a}{a}$    *Ans.* $2\sqrt{2ax-x^2}$

16. $y = \text{arc sin }(x-1)$    *Ans.* $\dfrac{1}{\sqrt{2x-x^2}}$    20. $y = \dfrac{\sqrt{x^2-4}}{x^2} + \dfrac{1}{2}\text{ arc sec}\dfrac{x}{2}$    *Ans.* $\dfrac{8}{x^3\sqrt{x^2-4}}$

21. A light is to be placed directly above the center of a circular plot, of radius 30 ft, at such a height that the edge of the plot will get maximum illumination. Find the height if the intensity at any point on the edge is directly proportional to the cosine of the angle of incidence (angle between the ray of light and the vertical) and inversely proportional to the square of the distance from the source.

Hint: Let $x$ be the required height, $y$ the distance from the light to a point on the edge, and $\theta$ the angle of incidence. Then $I = k\dfrac{\cos\theta}{y^2} = \dfrac{kx}{(x^2+900)^{3/2}}.$    *Ans.* $15\sqrt{2}$ ft.

22. Two ships sail from $A$ at the same time. One sails south at 15 mi/hr; the other sails east at 25 mi/hr for 1 hour and then turns north. Find the rate of rotation of the line joining them after 3 hr.
*Ans.* 20/193 rad/hr

# Chapter 14

# Differentiation of
# Exponential and Logarithmic Functions

**THE NUMBER** $e = \lim\limits_{h \to +\infty} \left(1 + \dfrac{1}{h}\right)^{h} = \lim\limits_{k \to 0} (1+k)^{1/k}$

$$= 1 + 1 + \frac{1}{2!} + \frac{1}{3!} + \cdots + \frac{1}{n!} + \cdots = 2.71828\ldots$$

See Problem 1.

**NOTATION.** If $a > 0$ and $a \neq 1$, and if $a^{y} = x$, then $y = \log_a x$.

$$y = \log_e x = \ln x \qquad y = \log_{10} x = \log x$$

The domain of definition is $x > 0$; the range is the set of real numbers.

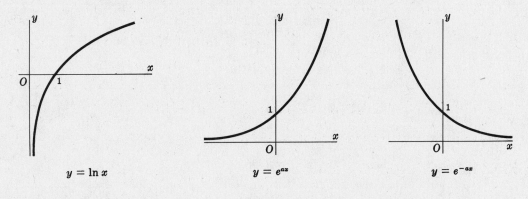

$$y = \ln x \qquad\qquad y = e^{ax} \qquad\qquad y = e^{-ax}$$

**Fig. 14-1**

**RULES OF DIFFERENTIATION.** If $u$ is a differentiable function of $x$,

26. $\dfrac{d}{dx}(\log_a u) = \dfrac{1}{u} \log_a e \dfrac{du}{dx}, \quad (a > 0, a \neq 1)$  28. $\dfrac{d}{dx}(a^u) = a^u \ln a \dfrac{du}{dx}, \quad (a > 0)$

27. $\dfrac{d}{dx}(\ln u) = \dfrac{1}{u} \dfrac{du}{dx}$  29. $\dfrac{d}{dx}(e^u) = e^u \dfrac{du}{dx}$

See Problems 2-17.

**LOGARITHMIC DIFFERENTIATION.** If a differentiable function $y = f(x)$ is the product of several factors, the process of differentiation may be simplified by taking the natural logarithm of the function before differentiating or, what is the same thing, by using the formula

30.
$$\frac{d}{dx}(y) = y \frac{d}{dx}(\ln y)$$

See Problems 18-19.

69

# Solved Problems

1. Verify: $2 < \lim\limits_{n \to +\infty} \left(1 + \dfrac{1}{n}\right)^n < 3$.

   By the binomial theorem, for $n$ a positive integer,

   (i) $\left(1 + \dfrac{1}{n}\right)^n = 1 + n\left(\dfrac{1}{n}\right) + \dfrac{n(n-1)}{1 \cdot 2}\left(\dfrac{1}{n}\right)^2 + \dfrac{n(n-1)(n-2)}{1 \cdot 2 \cdot 3}\left(\dfrac{1}{n}\right)^3$

   $\qquad\qquad\qquad + \cdots + \dfrac{n(n-1)(n-2)\ldots 1}{1 \cdot 2 \cdot 3 \ldots n}\left(\dfrac{1}{n}\right)^n$

   $\qquad\qquad = 1 + 1 + \left(1 - \dfrac{1}{n}\right) \cdot \dfrac{1}{2!} + \left(1 - \dfrac{1}{n}\right)\left(1 - \dfrac{2}{n}\right) \cdot \dfrac{1}{3!}$

   $\qquad\qquad\qquad + \cdots + \left(1 - \dfrac{1}{n}\right)\left(1 - \dfrac{2}{n}\right)\ldots\left(1 - \dfrac{n-1}{n}\right) \cdot \dfrac{1}{n!}$

   Clearly, for every value of $n \neq 1, \left(1 + \dfrac{1}{n}\right)^n > 2$. Also, if in (i) each difference $\left(1 - \dfrac{1}{n}\right), \left(1 - \dfrac{2}{n}\right),$
   ... is replaced by the larger number 1, we have

   $$\left(1 + \dfrac{1}{n}\right)^n < 2 + \dfrac{1}{2!} + \dfrac{1}{3!} + \cdots + \dfrac{1}{n!}$$

   $$< 2 + \dfrac{1}{2} + \dfrac{1}{2^2} + \dfrac{1}{2^3} + \cdots + \dfrac{1}{2^{n-1}} \quad \left(\text{since } \dfrac{1}{n!} < \dfrac{1}{2^{n-1}}\right)$$

   $$< 3 \quad \left(\text{since } \dfrac{1}{2} + \dfrac{1}{2^2} + \dfrac{1}{2^3} + \cdots + \dfrac{1}{2^{n-1}} < 1\right)$$

   Hence, $2 < \left(1 + \dfrac{1}{n}\right)^n < 3$.

   Let $n \to \infty$ through positive integral values; then

   $$1 - \dfrac{1}{n} \to 1, \; 1 - \dfrac{2}{n} \to 1, \; \ldots, \quad \text{and} \quad \left(1 - \dfrac{1}{n}\right)\left(1 - \dfrac{2}{n}\right)\ldots\left(1 - \dfrac{k}{n}\right) \cdot \dfrac{1}{k!} \to \dfrac{1}{k!}$$

   This suggests $\lim\limits_{n \to +\infty} \left(1 + \dfrac{1}{n}\right)^n = 1 + 1 + \dfrac{1}{2!} + \dfrac{1}{3!} + \cdots + \dfrac{1}{k!} + \cdots = 2.71828\ldots$

2. Derive $\dfrac{d}{dx}(\log_a u) = \dfrac{1}{u}\log_a e \dfrac{du}{dx}$ and $\dfrac{d}{dx}\ln u = \dfrac{1}{u}\dfrac{du}{dx}$.

   Let $y = \log_a u$, where $u$ is a differentiable function of $x$. Then

   $$y + \Delta y = \log_a(u + \Delta u)$$
   $$\Delta y = \log_a(u + \Delta u) - \log_a u = \log_a \dfrac{u + \Delta u}{u} = \log_a\left(1 + \dfrac{\Delta u}{u}\right)$$

   $$\dfrac{\Delta y}{\Delta u} = \dfrac{1}{\Delta u}\log_a\left(1 + \dfrac{\Delta u}{u}\right) = \dfrac{1}{u} \cdot \dfrac{u}{\Delta u}\log_a\left(1 + \dfrac{\Delta u}{u}\right) = \dfrac{1}{u}\log_a\left(1 + \dfrac{\Delta u}{u}\right)^{u/\Delta u}$$

   and

   $$\dfrac{dy}{du} = \dfrac{1}{u}\lim_{\Delta u \to 0}\log_a\left(1 + \dfrac{\Delta u}{u}\right)^{u/\Delta u} = \dfrac{1}{u}\log_a\left\{\lim_{\Delta u \to 0}\left(1 + \dfrac{\Delta u}{u}\right)^{u/\Delta u}\right\} = \dfrac{1}{u}\log_a e$$

   Thus, using the chain rule, we have $\dfrac{d}{dx}(\log_a u) = \dfrac{d}{du}(\log_a u)\dfrac{du}{dx} = \dfrac{1}{u}\log_a e \dfrac{du}{dx}$.

   When $a = e$, $\log_a e = \log_e e = 1$ and $\dfrac{d}{dx}(\ln u) = \dfrac{1}{u}\dfrac{du}{dx}$.

3. $y = \log_a(3x^2 - 5)$ $\qquad\qquad \dfrac{dy}{dx} = \dfrac{1}{3x^2 - 5}\log_a e \cdot \dfrac{d}{dx}(3x^2 - 5) = \dfrac{6x}{3x^2 - 5}\log_a e$

4. $y = \ln(x + 3)^2 = 2\ln(x + 3)$ $\qquad \dfrac{dy}{dx} = 2\dfrac{1}{x + 3} \cdot \dfrac{d}{dx}(x + 3) = \dfrac{2}{x + 3}$

5. $y = \ln^2(x + 3)$

   $$y' = 2\ln(x + 3) \cdot \dfrac{d}{dx}[\ln(x + 3)] = 2\ln(x + 3) \cdot \dfrac{1}{x + 3} \cdot \dfrac{d}{dx}(x + 3) = \dfrac{2\ln(x + 3)}{x + 3}$$

**6.** $y = \ln(x^3 + 2)(x^2 + 3) = \ln(x^3 + 2) + \ln(x^2 + 3)$

$$y' = \frac{1}{x^3 + 2} \cdot \frac{d}{dx}(x^3 + 2) + \frac{1}{x^2 + 3} \cdot \frac{d}{dx}(x^2 + 3) = \frac{3x^2}{x^3 + 2} + \frac{2x}{x^2 + 3}$$

**7.** $f(x) = \ln\frac{x^4}{(3x - 4)^2} = \ln x^4 - \ln(3x - 4)^2 = 4\ln x - 2\ln(3x - 4)$

$$f'(x) = 4\frac{1}{x}\frac{d}{dx}(x) - 2\frac{1}{3x - 4}\frac{d}{dx}(3x - 4) = \frac{4}{x} - \frac{6}{3x - 4}$$

**8.** $y = \ln\sin 3x \qquad y' = \frac{1}{\sin 3x} \cdot \frac{d}{dx}(\sin 3x) = 3\frac{\cos 3x}{\sin 3x} = 3\cot 3x$

**9.** $y = \ln(x + \sqrt{1 + x^2})$

$$y' = \frac{1 + \frac{1}{2}(1 + x^2)^{-1/2}(2x)}{x + (1 + x^2)^{1/2}} = \frac{1 + x(1 + x^2)^{-1/2}}{x + (1 + x^2)^{1/2}} \cdot \frac{(1 + x^2)^{1/2}}{(1 + x^2)^{1/2}} = \frac{1}{\sqrt{1 + x^2}}$$

**10.** Derive $\dfrac{d}{dx}(a^u) = a^u \ln a \dfrac{du}{dx}$ and $\dfrac{d}{dx}(e^u) = e^u \dfrac{du}{dx}$.

Let $y = a^u$, where $u$ is a differentiable function of $x$. Then $\ln y = u \ln a$,

$$\frac{d}{dx}(\ln y) = \frac{1}{y}\frac{dy}{dx} = \ln a\frac{du}{dx}, \qquad \frac{dy}{dx} = y\ln a\frac{du}{dx} \qquad \text{or} \qquad \frac{d}{dx}(a^u) = a^u\ln a\frac{du}{dx}$$

When $a = e$, $\ln a = \ln(e^u) = 1$, and we have $\dfrac{d}{dx}(e^u) = e^u\dfrac{du}{dx}$.

**11.** $y = e^{-\frac{1}{2}x} \qquad y' = e^{-\frac{1}{2}x}\dfrac{d}{dx}(-\tfrac{1}{2}x) = -\tfrac{1}{2}e^{-\frac{1}{2}x}$

**12.** $y = e^{x^2} \qquad y' = e^{x^2} \cdot \dfrac{d}{dx}(x^2) = 2xe^{x^2}$

**13.** $y = a^{3x^2} \qquad y' = a^{3x^2}\ln a \cdot \dfrac{d}{dx}(3x^2) = 6xa^{3x^2}\ln a$

**14.** $y = x^2 3^x \qquad y' = x^2 \cdot \dfrac{d}{dx}(3^x) + 3^x \cdot \dfrac{d}{dx}(x^2) = x^2 3^x \ln 3 + 3^x 2x = x3^x(x\ln 3 + 2)$

**15.** $y = \dfrac{e^{ax} - e^{-ax}}{e^{ax} + e^{-ax}}$

$$y' = \frac{(e^{ax} + e^{-ax})\dfrac{d}{dx}(e^{ax} - e^{-ax}) - (e^{ax} - e^{-ax})\dfrac{d}{dx}(e^{ax} + e^{-ax})}{(e^{ax} + e^{-ax})^2}$$

$$= \frac{(e^{ax} + e^{-ax})[a(e^{ax} + e^{-ax})] - (e^{ax} - e^{-ax})[a(e^{ax} - e^{-ax})]}{(e^{ax} + e^{-ax})^2}$$

$$= a\frac{(e^{2ax} + 2 + e^{-2ax}) - (e^{2ax} - 2 + e^{-2ax})}{(e^{ax} + e^{-ax})^2} = \frac{4a}{(e^{ax} + e^{-ax})^2}$$

**16.** Find $y''$, given $y = e^{-x}\ln x$.

$$y' = e^{-x}\frac{d}{dx}(\ln x) + \ln x\frac{d}{dx}(e^{-x}) = \frac{e^{-x}}{x} - e^{-x}\ln x = \frac{e^{-x}}{x} - y$$

$$y'' = \frac{x\dfrac{d}{dx}(e^{-x}) - e^{-x}\dfrac{d}{dx}(x)}{x^2} - y' = \frac{-xe^{-x} - e^{-x}}{x^2} - \frac{e^{-x}}{x} + e^{-x}\ln x = -e^{-x}\left(\frac{2}{x} + \frac{1}{x^2} - \ln x\right)$$

**17.** Find $y''$, given $y = e^{-2x}\sin 3x$.

$$y' = e^{-2x}\frac{d}{dx}(\sin 3x) + \sin 3x\frac{d}{dx}(e^{-2x}) = 3e^{-2x}\cos 3x - 2e^{-2x}\sin 3x = 3e^{-2x}\cos 3x - 2y$$

$$y'' = 3e^{-2x}\frac{d}{dx}(\cos 3x) + 3\cos 3x\frac{d}{dx}(e^{-2x}) - 2y'$$

$$= -9e^{-2x}\sin 3x - 6e^{-2x}\cos 3x - 2(3e^{-2x}\cos 3x - 2e^{-2x}\sin 3x)$$

$$= -e^{-2x}(12\cos 3x + 5\sin 3x)$$

Use logarithmic differentiation to find the first derivative.

**18.** $y = (x^2 + 2)^3 (1 - x^3)^4$      $\ln y = \ln (x^2 + 2)^3 (1 - x^3)^4 = 3 \ln (x^2 + 2) + 4 \ln (1 - x^3)$

$$y' = y \frac{d}{dx} [3 \ln (x^2 + 2) + 4 \ln (1 - x^3)] = (x^2 + 2)^3 (1 - x^3)^4 \left[ \frac{6x}{x^2 + 2} - \frac{12x^2}{1 - x^3} \right]$$

$$= 6x(x^2 + 2)^2 (1 - x^3)^3 (1 - 4x - 3x^3)$$

**19.** $y = \dfrac{x(1 - x^2)^2}{(1 + x^2)^{1/2}}$      $\ln y = \ln x + 2 \ln (1 - x^2) - \frac{1}{2} \ln (1 + x^2)$

$$y' = \frac{x(1 - x^2)^2}{(1 + x^2)^{1/2}} \left[ \frac{1}{x} - \frac{4x}{1 - x^2} - \frac{x}{1 + x^2} \right] = \frac{(1 - x^2)^2}{(1 + x^2)^{1/2}} - \frac{4x^2(1 - x^2)}{(1 + x^2)^{1/2}} - \frac{x^2(1 - x^2)^2}{(1 + x^2)^{3/2}}$$

$$= \frac{(1 - 5x^2 - 4x^4)(1 - x^2)}{(1 + x^2)^{3/2}}$$

**20.** Locate (a) the relative maximum and minimum points and (b) the points of inflection of the curve $y = f(x) = x^2 e^x$.

$f'(x) = 2xe^x + x^2 e^x = xe^x(2 + x)$

$f''(x) = 2e^x + 4xe^x + x^2 e^x = e^x(2 + 4x + x^2)$

$f'''(x) = 6e^x + 6xe^x + x^2 e^x = e^x(6 + 6x + x^2)$

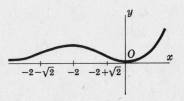

**Fig. 14-2**

(a) Solving $f'(x) = 0$ gives the critical values $x = 0$ and $x = -2$.

$f''(0) > 0$ and $(0, 0)$ is a relative minimum point.

$f''(-2) < 0$ and $(-2, 4/e^2)$ is a relative maximum point.

(b) Solving $f''(x) = 0$ gives the possible points of inflection at $x = -2 \pm \sqrt{2}$.

$f'''(-2 - \sqrt{2}) \neq 0$ and $f'''(-2 + \sqrt{2}) \neq 0$; the points $x = -2 \pm \sqrt{2}$ are points of inflection.

**21.** Discuss the probability curve $y = ae^{-b^2 x^2}$, $a > 0$.

(a) The curve lies entirely above the $x$-axis, since $e^{-b^2 x^2} > 0$ for all $x$. As $x \to \pm\infty$, $y \to 0$ and the $x$-axis is a horizontal asymptote.

(b) $y' = -2ab^2 xe^{-b^2 x^2}$  and  $y'' = 2ab^2(2b^2 x^2 - 1)e^{-b^2 x^2}$.

When $y' = 0$, $x = 0$; and when $x = 0$, $y'' < 0$. The point $(0, a)$ is a maximum point of the curve.

**Fig. 14-3**

(c) When $y'' = 0$, $2b^2 x^2 - 1 = 0$, and $x = \pm\sqrt{2}/2b$ are possible points of inflection.

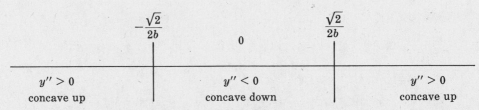

| $-\dfrac{\sqrt{2}}{2b}$ | 0 | $\dfrac{\sqrt{2}}{2b}$ |
|---|---|---|
| $y'' > 0$ | $y'' < 0$ | $y'' > 0$ |
| concave up | concave down | concave up |

The points $(\pm\sqrt{2}/2b, ae^{-1/2})$ are points of inflection.

**22.** The equilibrium constant $K$ of a balanced chemical reaction changes with the absolute temperature $T$ according to the law $K = K_0\, e^{-\frac{1}{2}q(T-T_0)/T_0 T}$, where $K_0$, $q$, and $T_0$ are constants. Find the percentage rate of change of $K$ per degree of change of $T$.

The percentage rate of change of $K$ per degree of change of $T$ is given by $\dfrac{1}{K}\dfrac{dK}{dT} = \dfrac{d(\ln K)}{dT}$.

Then $\ln K = \ln K_0 - \frac{1}{2}q\dfrac{T-T_0}{T_0 T}$ and $\dfrac{d(\ln K)}{dT} = -\dfrac{q}{2T^2} = -\dfrac{50q}{T^2}\,\%$.

**23.** Discuss the damped vibration curve $y = f(t) = e^{-\frac{1}{2}t}\sin 2\pi t$.

(a) When $t = 0$, $y = 0$. The $y$-intercept is 0.

When $y = 0$, $\sin 2\pi t = 0$ and $t = \ldots, -\frac{3}{2}, -1, -\frac{1}{2}, 0, \frac{1}{2},$ $1, \frac{3}{2}, \ldots$. These are the $t$-intercepts.

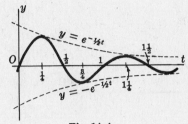

(b) When $t = \ldots, -\frac{7}{4}, -\frac{3}{4}, \frac{1}{4}, \frac{5}{4}, \ldots,$ $\sin 2\pi t = 1$ and $y = e^{-\frac{1}{2}t}$.

When $t = \ldots, -\frac{5}{4}, -\frac{1}{4}, \frac{3}{4}, \frac{7}{4}, \ldots,$ $\sin 2\pi t = -1$ and $y = -e^{\frac{1}{2}t}$.

The given curve oscillates between the two curves $y = e^{-\frac{1}{2}t}$ and $y = -e^{-\frac{1}{2}t}$, touching them at the above mentioned points.

**Fig. 14-4**

(c)
$$y' = f'(t) = e^{-\frac{1}{2}t}(2\pi\cos 2\pi t - \tfrac{1}{2}\sin 2\pi t)$$
$$y'' = f''(t) = e^{-\frac{1}{2}t}\{(\tfrac{1}{4} - 4\pi^2)\sin 2\pi t - 2\pi\cos 2\pi t\}$$

When $y' = 0$, $2\pi\cos 2\pi t - \frac{1}{2}\sin 2\pi t = 0$, that is, $\tan 2\pi t = 4\pi$.

If $t = \xi = .237$ is the smallest positive angle satisfying this relation, then $t = \cdots, \xi - \frac{3}{2}, \xi - 1,$ $\xi - \frac{1}{2}, \xi, \xi + \frac{1}{2}, \xi + 1, \ldots$ are the critical values.

Since, for $n = 0, 1, 2, \ldots$, $f''(\xi \pm \frac{1}{2}n)$ and $f''\left(\xi \pm \dfrac{n+1}{2}\right)$ have opposite signs and $f''(\xi \pm \frac{1}{2}n)$ and $f''\left(\xi \pm \dfrac{n+2}{2}\right)$ have the same sign, the critical values yield alternate maximum and minimum points of the curve. These points are slightly to the left of the points of contact with the curves $y = e^{-\frac{1}{2}t}$ and $y = -e^{-\frac{1}{2}t}$.

(d) When $y'' = 0$, $\tan 2\pi t = \dfrac{2\pi}{\frac{1}{4} - 4\pi^2} = \dfrac{8\pi}{1 - 16\pi^2}$.

If $t = \eta = .475$ is the smallest positive angle satisfying this relation, then $t = \cdots, \eta - 1,$ $\eta - \frac{1}{2}, \eta, \eta + \frac{1}{2}, \eta + 1, \ldots$ are the possible points of inflection. These points, located slightly to the left of the points of intersection of the curve and the $t$-axis, are points of inflection.

**24.** The equation $s = ce^{-bt}\sin(kt + \theta)$, where $c, b, k,$ and $\theta$ are constants, represents damped (slowed down) vibratory motion. Show that $a = -2bv - (k^2 + b^2)s$.

$$v = ds/dt = ce^{-bt}[-b\sin(kt+\theta) + k\cos(kt+\theta)]$$

$$\begin{aligned}
a = dv/dt &= ce^{-bt}[(b^2 - k^2)\sin(kt+\theta) - 2bk\cos(kt+\theta)]\\
&= ce^{-bt}[-2b\{-b\sin(kt+\theta) + k\cos(kt+\theta)\} - (k^2+b^2)\sin(kt+\theta)]\\
&= -2bv - (k^2 + b^2)s
\end{aligned}$$

# Supplementary Problems

In Problems 25-35, find $dy/dx$.

**25.** $y = \ln(4x - 5)$    Ans. $4/(4x - 5)$

**26.** $y = \ln\sqrt{3 - x^2}$    Ans. $x/(x^2 - 3)$

**27.** $y = \ln 3x^5$    Ans. $5/x$

**28.** $y = \ln(x^2 + x - 1)^3$    Ans. $(6x + 3)/(x^2 + x - 1)$

**29.** $y = x \cdot \ln x - x$    Ans. $\ln x$

**30.** $y = \ln(\sec x + \tan x)$    Ans. $\sec x$

**31.** $y = \ln(\ln\tan x)$    Ans. $2/(\sin 2x \ln\tan x)$

**32.** $y = (\ln x^2)/x^2$    Ans. $(2 - 4\ln x)/x^3$

**33.** $y = \frac{1}{5}x^5(\ln x - \frac{1}{5})$    Ans. $x^4\ln x$

**34.** $y = x(\sin\ln x - \cos\ln x)$    Ans. $2\sin\ln x$

**35.** $y = x\ln(4 + x^2) + 4\arctan\frac{1}{2}x - 2x$
    Ans. $\ln(4 + x^2)$

**36.** Find the equation of the tangent to $y = \ln x$ at any one of its points $(x_0, y_0)$. Use the $y$-intercept of the tangent to obtain a simple construction for the tangent line.

**37.** Discuss and sketch: $y = x^2\ln x$.    Ans. Min. at $x = 1/\sqrt{e}$, P.I. at $x = 1/e^{3/2}$

**38.** Show that the angle of intersection of the curves $y = \ln(x - 2)$ and $y = x^2 - 4x + 3$ at the point $(3, 0)$ is $\phi = \arctan 1/3$.

In Problems 39-46, find $dy/dx$.

**39.** $y = e^{5x}$    Ans. $5e^{5x}$

**40.** $y = e^{x^3}$    Ans. $3x^2e^{x^3}$

**41.** $y = e^{\sin 3x}$    Ans. $3e^{\sin 3x}\cos 3x$

**42.** $y = 3^{-x^2}$    Ans. $-2x \cdot 3^{-x^2}\ln 3$

**43.** $y = e^{-x}\cos x$    Ans. $-e^{-x}(\cos x + \sin x)$

**44.** $y = \arcsin e^x$    Ans. $e^x/\sqrt{1 - e^{2x}}$

**45.** $y = \tan^2 e^{3x}$    Ans. $6e^{3x}\tan e^{3x}\sec^2 e^{3x}$

**46.** $y = e^{e^x}$    Ans. $e^{(x + e^x)}$

**47.** If $y = x^2e^x$, show that $y''' = (x^2 + 6x + 6)e^x$.

**48.** If $y = e^{-2x}(\sin 2x + \cos 2x)$, show that $y'' + 4y' + 8y = 0$.

**49.** Discuss and sketch:  (a) $y = x^2e^{-x}$,  (b) $y = x^2e^{-x^2}$.
    Ans. (a) Max. at $x = 2$; min. at $x = 0$;  P.I. at $x = 2 \pm \sqrt{2}$
       (b) Max. at $x = \pm 1$; min. at $x = 0$;  P.I. at $x = \pm 1.51$, $x = \pm 0.47$

**50.** Find the rectangle of maximum area, having one edge along the $x$-axis, under the curve $y = e^{-x^2}$. Hint: $A = 2xy = 2xe^{-x^2}$, where $P(x, y)$ is a vertex of the rectangle on the curve.    Ans. $A = \sqrt{2/e}$

**51.** Show that the curves $y = e^{ax}$ and $y = e^{ax}\cos ax$ are tangent at the points for which $x = 2n\pi/a$, $(n = 1,2,3,...)$ and that the curves $y = e^{-ax}/a^2$ and $y = e^{ax}\cos ax$ are mutually perpendicular at the same points.

**52.** For the curve $y = xe^x$, show  (a) $(-1, -1/e)$ is a relative minimum point,  (b) $(-2, -2/e^2)$ is a point of inflection, and  (c) the curve is concave downward to the left and concave upward to the right of the point of inflection.

In Problems 53-56, use logarithmic differentiation to find $dy/dx$.

**53.** $y = x^x$    Ans. $x^x(1 + \ln x)$

**55.** $y = x^2 e^{2x}\cos 3x$    Ans. $x^2 e^{2x}\cos 3x \{2/x + 2 - 3\tan 3x\}$

**54.** $y = x^{\ln x}$    Ans. $2x^{(\ln x - 1)}\ln x$

**56.** $y = x^{e^{-x^2}}$    Ans. $e^{-x^2}x^{e^{-x^2}}(1/x - 2x\ln x)$

**57.** Show  (a) $\dfrac{d^n}{dx^n}(xe^x) = (x + n)e^x$,  (b) $\dfrac{d^n}{dx^n}(x^{n-1}\ln x) = \dfrac{(n-1)!}{x}$.

# Chapter 15

## Differentiation of Hyperbolic Functions

**DEFINITIONS OF HYPERBOLIC FUNCTIONS.** For $u$ any real number, except where noted:

$$\sinh u = \frac{e^u - e^{-u}}{2}$$

$$\coth u = \frac{1}{\tanh u} = \frac{e^u + e^{-u}}{e^u - e^{-u}}, \quad (u \neq 0)$$

$$\cosh u = \frac{e^u + e^{-u}}{2}$$

$$\operatorname{sech} u = \frac{1}{\cosh u} = \frac{2}{e^u + e^{-u}}$$

$$\tanh u = \frac{\sinh u}{\cosh u} = \frac{e^u - e^{-u}}{e^u + e^{-u}}$$

$$\operatorname{csch} u = \frac{1}{\sinh u} = \frac{2}{e^u - e^{-u}}, \quad (u \neq 0)$$

**DIFFERENTIATION FORMULAS.** If $u$ is a differentiable function of $x$,

31. $\dfrac{d}{dx}(\sinh u) = \cosh u \dfrac{du}{dx}$

34. $\dfrac{d}{dx}(\coth u) = -\operatorname{csch}^2 u \dfrac{du}{dx}$

32. $\dfrac{d}{dx}(\cosh u) = \sinh u \dfrac{du}{dx}$

35. $\dfrac{d}{dx}(\operatorname{sech} u) = -\operatorname{sech} u \tanh u \dfrac{du}{dx}$

33. $\dfrac{d}{dx}(\tanh u) = \operatorname{sech}^2 u \dfrac{du}{dx}$

36. $\dfrac{d}{dx}(\operatorname{csch} u) = -\operatorname{csch} u \coth u \dfrac{du}{dx}$

See Problems 1-12.

**DEFINITIONS OF INVERSE HYPERBOLIC FUNCTIONS.**

$$\sinh^{-1} u = \ln(u + \sqrt{1 + u^2}), \quad \text{all } u \qquad \coth^{-1} u = \tfrac{1}{2} \ln \frac{u+1}{u-1}, \qquad (u^2 > 1)$$

$$\cosh^{-1} u = \ln(u + \sqrt{u^2 - 1}), \quad (u \geq 1) \qquad \operatorname{sech}^{-1} u = \ln \frac{1 + \sqrt{1 - u^2}}{u}, \quad (0 < u \leq 1)$$

$$\tanh^{-1} u = \tfrac{1}{2} \ln \frac{1+u}{1-u}, \qquad (u^2 < 1) \qquad \operatorname{csch}^{-1} u = \ln \left( \frac{1}{u} + \frac{\sqrt{1 + u^2}}{|u|} \right), \quad (u \neq 0)$$

(Only principal values of $\cosh^{-1} x$ and $\operatorname{sech}^{-1} x$ are included here.)

**DIFFERENTIATION FORMULAS.** If $u$ is a differentiable function of $x$,

37. $\dfrac{d}{dx}(\sinh^{-1} u) = \dfrac{1}{\sqrt{1 + u^2}} \dfrac{du}{dx}$

40. $\dfrac{d}{dx}(\coth^{-1} u) = \dfrac{1}{1 - u^2} \dfrac{du}{dx}, \qquad (u^2 > 1)$

38. $\dfrac{d}{dx}(\cosh^{-1} u) = \dfrac{1}{\sqrt{u^2 - 1}} \dfrac{du}{dx}, \quad (u > 1)$

41. $\dfrac{d}{dx}(\operatorname{sech}^{-1} u) = \dfrac{-1}{u\sqrt{1 - u^2}} \dfrac{du}{dx}, \quad (0 < u < 1)$

39. $\dfrac{d}{dx}(\tanh^{-1} u) = \dfrac{1}{1 - u^2} \dfrac{du}{dx}, \quad (u^2 < 1)$

42. $\dfrac{d}{dx}(\operatorname{csch}^{-1} u) = \dfrac{-1}{|u|\sqrt{1 + u^2}} \dfrac{du}{dx}, \qquad (u \neq 0)$

See Problems 13-19.

# Solved Problems

**1.** Prove: $\cosh^2 u - \sinh^2 u = 1$.

$$\cosh^2 u - \sinh^2 u = \left(\frac{e^u + e^{-u}}{2}\right)^2 - \left(\frac{e^u - e^{-u}}{2}\right)^2 = \tfrac{1}{4}(e^{2u} + 2 + e^{-2u}) - \tfrac{1}{4}(e^{2u} - 2 + e^{-2u}) = 1$$

**2.** Derive: $\dfrac{d}{dx}(\sinh u) = \cosh u \dfrac{du}{dx}$, when $u$ is a differentiable function of $x$.

$$\frac{d}{dx}(\sinh u) = \frac{d}{dx}\left(\frac{e^u - e^{-u}}{2}\right) = \frac{e^u + e^{-u}}{2}\frac{du}{dx} = \cosh u \frac{du}{dx}$$

In Problems 3-12, find $\dfrac{dy}{dx}$.

**3.** $y = \sinh 3x$

$$\frac{dy}{dx} = \cosh 3x \cdot \frac{d}{dx}(3x) = 3\cosh 3x$$

**4.** $y = \cosh \tfrac{1}{2}x$

$$\frac{dy}{dx} = \sinh \tfrac{1}{2}x \cdot \frac{d}{dx}(\tfrac{1}{2}x) = \tfrac{1}{2}\sinh \tfrac{1}{2}x$$

**5.** $y = \tanh(1 + x^2)$

$$\frac{dy}{dx} = \operatorname{sech}^2(1 + x^2) \cdot \frac{d}{dx}(1 + x^2) = 2x \operatorname{sech}^2(1 + x^2)$$

**6.** $y = \coth\dfrac{1}{x}$

$$\frac{dy}{dx} = -\operatorname{csch}^2\frac{1}{x} \cdot \frac{d}{dx}\left(\frac{1}{x}\right) = \frac{1}{x^2}\operatorname{csch}^2\frac{1}{x}$$

**7.** $y = x \operatorname{sech} x^2$

$$\frac{dy}{dx} = x \cdot \frac{d}{dx}(\operatorname{sech} x^2) + \operatorname{sech} x^2 \cdot \frac{d}{dx}(x)$$
$$= x(-\operatorname{sech} x^2 \tanh x^2)2x + \operatorname{sech} x^2$$
$$= -2x^2 \operatorname{sech} x^2 \tanh x^2 + \operatorname{sech} x^2$$

**8.** $y = \operatorname{csch}^2(x^2 + 1)$

$$\frac{dy}{dx} = 2\operatorname{csch}(x^2 + 1) \cdot \frac{d}{dx}[\operatorname{csch}(x^2 + 1)]$$
$$= 2\operatorname{csch}(x^2 + 1)[-\operatorname{csch}(x^2 + 1)\coth(x^2 + 1) \cdot 2x]$$
$$= -4x \operatorname{csch}^2(x^2 + 1)\coth(x^2 + 1)$$

**9.** $y = \tfrac{1}{4}\sinh 2x - \tfrac{1}{2}x$

$$\frac{dy}{dx} = \tfrac{1}{4}(\cosh 2x)2 - \tfrac{1}{2} = \tfrac{1}{2}(\cosh 2x - 1) = \sinh^2 x$$

**10.** $y = \ln \tanh 2x$

$$\frac{dy}{dx} = \frac{1}{\tanh 2x}(2\operatorname{sech}^2 2x) = \frac{2}{\sinh 2x \cosh 2x} = 4\operatorname{csch} 4x$$

**11.** Find the coordinates of the minimum point of the catenary $y = a\cosh\dfrac{x}{a}$.

$$f'(x) = \frac{1}{a}\left(a\sinh\frac{x}{a}\right) = \sinh\frac{x}{a}, \qquad f''(x) = \frac{1}{a}\cosh\frac{x}{a} = \frac{1}{a}\left(\frac{e^{x/a} + e^{-x/a}}{2}\right)$$

When $f'(x) = \dfrac{e^{x/a} - e^{-x/a}}{2} = 0$, $x = 0$; $f''(0) > 0$. The point $(0, a)$ is the minimum point.

**12.** Examine (*a*) $y = \sinh x$, (*b*) $y = \cosh x$, (*c*) $y = \tanh x$ for points of inflection.

    (*a*) $f'(x) = \cosh x$, $f''(x) = \sinh x$, and $f'''(x) = \cosh x$.

        $f''(x) = \sinh x = 0$, when $x = 0$; $f'''(0) \neq 0$. The point $(0,0)$ is a point of inflection.

    (*b*) $f'(x) = \sinh x$, $f''(x) = \cosh x \neq 0$ for all values of $x$. There is no point of inflection.

    (*c*) $f'(x) = \operatorname{sech}^2 x$, $f''(x) = -2\operatorname{sech}^2 x \tanh x = -2\dfrac{\sinh x}{\cosh^3 x}$, and $f'''(x) = \dfrac{4\sinh^2 x - 2}{\cosh^4 x}$.

        $f''(x) = 0$ when $x = 0$; $f'''(0) \neq 0$. The point $(0,0)$ is a point of inflection.

**13.** Derive:  (*a*)  $\sinh^{-1} x = \ln(x + \sqrt{x^2 + 1})$, all $x$

        (*b*)  $\operatorname{sech}^{-1} x = \cosh^{-1}\dfrac{1}{x} = \ln\dfrac{1 + \sqrt{1 - x^2}}{x}$, $\quad 0 < x \leq 1$.

    (*a*) Let $\sinh^{-1} x = y$; then $x = \sinh y = \frac{1}{2}(e^y - e^{-y})$ or $e^{2y} - 2xe^y - 1 = 0$.

        Solving for $e^y$: $e^y = x + \sqrt{x^2 + 1}$, since $e^y > 0$. Thus, $y = \ln(x + \sqrt{x^2 + 1})$.

    (*b*) Let $\operatorname{sech}^{-1} x = y$; then $x = \operatorname{sech} y = \dfrac{1}{\cosh y}$, $\cosh y = \dfrac{1}{x}$, and $y = \cosh^{-1}\left(\dfrac{1}{x}\right) = \operatorname{sech}^{-1} x$.

        Also, $x = \operatorname{sech} y = \dfrac{2}{e^y + e^{-y}}$ or $e^{2y} x - 2e^y + x = 0$.

        Solving for $e^y$: $e^y = \dfrac{1 + \sqrt{1 - x^2}}{x}$, when $y \geq 0$. Thus, $y = \ln\dfrac{1 + \sqrt{1 - x^2}}{x}$, $\quad 0 < x \leq 1$.

**14.** Derive: $\dfrac{d}{dx}(\sinh^{-1} u) = \dfrac{1}{\sqrt{1 + u^2}}\dfrac{du}{dx}$.

    Let $y = \sinh^{-1} u$, where $u$ is a differentiable function of $x$. Then $\sinh y = u$,

$$\cosh y\,\frac{dy}{dx} = \frac{du}{dx} \quad\text{and}\quad \frac{dy}{dx} = \frac{1}{\cosh y}\frac{du}{dx} = \frac{1}{\sqrt{1 + \sinh^2 y}}\frac{du}{dx} = \frac{1}{\sqrt{1 + u^2}}\frac{du}{dx}$$

In Problems 15–19, find $dy/dx$.

**15.** $y = \sinh^{-1} 3x$

    $\dfrac{dy}{dx} = \dfrac{1}{\sqrt{(3x)^2 + 1}} \cdot \dfrac{d}{dx}(3x) = \dfrac{3}{\sqrt{9x^2 + 1}}$

**16.** $y = \cosh^{-1} e^x$

    $\dfrac{dy}{dx} = \dfrac{1}{\sqrt{e^{2x} - 1}} \cdot \dfrac{d}{dx}(e^x) = \dfrac{e^x}{\sqrt{e^{2x} - 1}}$

**17.** $y = 2\tanh^{-1}(\tan \frac{1}{2}x)$

    $\dfrac{dy}{dx} = 2\dfrac{1}{1 - \tan^2 \frac{1}{2}x} \cdot \dfrac{d}{dx}(\tan \frac{1}{2}x)$

    $= 2\dfrac{1}{1 - \tan^2 \frac{1}{2}x}\sec^2 \frac{1}{2}x \cdot \frac{1}{2} = \dfrac{\sec^2 \frac{1}{2}x}{1 - \tan^2 \frac{1}{2}x} = \sec x$

**18.** $y = \coth^{-1}\dfrac{1}{x}$

    $\dfrac{dy}{dx} = \dfrac{1}{1 - \left(\dfrac{1}{x}\right)^2} \cdot \dfrac{d}{dx}\left(\dfrac{1}{x}\right) = \dfrac{-\dfrac{1}{x^2}}{1 - \dfrac{1}{x^2}} = \dfrac{-1}{x^2 - 1}$

**19.** $y = \operatorname{sech}^{-1}(\cos x)$

    $\dfrac{dy}{dx} = \dfrac{-1}{\cos x\sqrt{1 - \cos^2 x}} \cdot \dfrac{d}{dx}(\cos x) = \dfrac{\sin x}{\cos x\sqrt{1 - \cos^2 x}} = \sec x$

# Supplementary Problems

20. (a) Sketch $y = e^x$ and $y = -e^{-x}$ and average the ordinates of the two curves for various values of $x$ to obtain points on $y = \sinh x$. Complete the curve.

     (b) Proceed as in (a), using $y = e^x$ and $y = e^{-x}$ to obtain the graph of $y = \cosh x$.

21. For the hyperbola $x^2 - y^2 = 1$, show: (a) $P(\cosh u, \sinh u)$ is a point on the hyperbola, (b) the tangent line at $A$ intersects the line $OP$ in $T(1, \tanh u)$. Refer to Fig. 15-1.

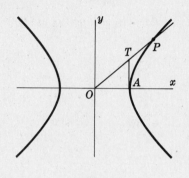

22. Show: (a) $\sinh (x + y) = \sinh x \cosh y + \cosh x \sinh y$

     (b) $\cosh (x + y) = \cosh x \cosh y + \sinh x \sinh y$

     (c) $\sinh 2x = 2 \sinh x \cosh x$

     (d) $\cosh 2x = \cosh^2 x + \sinh^2 x = 2 \cosh^2 x - 1$
                   $= 2 \sinh^2 x + 1$

**Fig. 15-1**

     (e) $\tanh 2x = \dfrac{2 \tanh x}{1 + \tanh^2 x}$

In Problems 23-28, find $dy/dx$.

23. $y = \sinh \frac{1}{4}x$          *Ans.* $\frac{1}{4} \cosh \frac{1}{4}x$

24. $y = \cosh^2 3x$          *Ans.* $3 \sinh 6x$

25. $y = \tanh 2x$          *Ans.* $2 \operatorname{sech}^2 2x$

26. $y = \ln \cosh x$          *Ans.* $\tanh x$

27. $y = \arctan \sinh x$          *Ans.* $\operatorname{sech} x$

28. $y = \ln \sqrt{\tanh 2x}$          *Ans.* $2 \operatorname{csch} 4x$

29. Show: (a) If $y = a \cosh \dfrac{x}{a}$, then $y'' = \dfrac{1}{a} \sqrt{1 + (y')^2}$.

     (b) If $y = A \cosh bx + B \sinh bx$, where $b, A, B$ are constants, then $y'' = b^2 y$.

30. Show: (a) $\cosh^{-1} u = \ln (u + \sqrt{u^2 - 1}), \ u \geqq 1$

     (b) $\tanh^{-1} u = \frac{1}{2} \ln \dfrac{1 + u}{1 - u}, \ u^2 < 1$.

31. (a) Trace the curve $y = \sinh^{-1} x$ by reflecting in the 45° line the curve $y = \sinh x$.

     (b) Trace the principal branch of $y = \cosh^{-1} x$ by reflecting in the 45° line the right half of $y = \cosh x$.

32. Derive the differentiation formulas 32-36 and 38-40, 42.

In Problems 33-36, find $dy/dx$.

33. $y = \sinh^{-1} \frac{1}{2}x$          *Ans.* $1/\sqrt{x^2 + 4}$

34. $y = \cosh^{-1} (1/x)$          *Ans.* $-1/x\sqrt{1 - x^2}$

35. $y = \tanh^{-1} (\sin x)$          *Ans.* $\sec x$

36. $x = a \operatorname{sech}^{-1} (y/a) - \sqrt{a^2 - y^2}$      *Ans.* $-y/\sqrt{a^2 - y^2}$

# Chapter 16

## Parametric Representation of Curves

**PARAMETRIC EQUATIONS.** If the coordinates $(x, y)$ of a point $P$ on a curve are given as functions $x = f(u)$, $y = g(u)$ of a third variable or *parameter u*, the equations $x = f(u)$, $y = g(u)$ are called *parametric equations* of the curve.

**Example:**

(a) $x = \cos\theta$, $y = 4\sin^2\theta$ are parametric equations, with parameter $\theta$, of the parabola $4x^2 + y = 4$, since $4x^2 + y = 4\cos^2\theta + 4\sin^2\theta = 4$.

(b) $x = \frac{1}{2}t$, $y = 4 - t^2$ is another parametric representation, with parameter $t$, of the same curve.

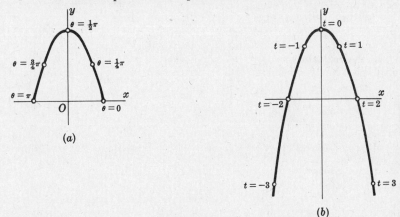

(a)

(b)

Fig. 16-1

It should be noted that the first set of parametric equations represents only a portion of the parabola, whereas the second represents the entire curve.

**THE FIRST DERIVATIVE** $\dfrac{dy}{dx}$ is given by $\quad \dfrac{dy}{dx} = \dfrac{dy/du}{dx/du}.$

**THE SECOND DERIVATIVE** $\dfrac{d^2y}{dx^2}$ is given by $\quad \dfrac{d^2y}{dx^2} = \dfrac{d}{du}\left(\dfrac{dy}{dx}\right) \cdot \dfrac{du}{dx}.$

## Solved Problems

1. Find $\dfrac{dy}{dx}$ and $\dfrac{d^2y}{dx^2}$, given $x = \theta - \sin\theta$, $y = 1 - \cos\theta$.

$$\frac{dx}{d\theta} = 1 - \cos\theta, \quad \frac{dy}{d\theta} = \sin\theta, \quad \text{and} \quad \frac{dy}{dx} = \frac{dy/d\theta}{dx/d\theta} = \frac{\sin\theta}{1 - \cos\theta}$$

$$\frac{d^2y}{dx^2} = \frac{d}{d\theta}\left(\frac{\sin\theta}{1 - \cos\theta}\right) \cdot \frac{d\theta}{dx} = \frac{\cos\theta - 1}{(1 - \cos\theta)^2} \cdot \frac{1}{1 - \cos\theta} = -\frac{1}{(1 - \cos\theta)^2}$$

2.  Find $\dfrac{dy}{dx}$ and $\dfrac{d^2y}{dx^2}$, given $x = e^t \cos t$, $y = e^t \sin t$.

$$\frac{dx}{dt} = e^t(\cos t - \sin t), \quad \frac{dy}{dt} = e^t(\sin t + \cos t), \quad \frac{dy}{dx} = \frac{dy/dt}{dx/dt} = \frac{\sin t + \cos t}{\cos t - \sin t}$$

$$\frac{d^2y}{dx^2} = \frac{d}{dt}\left(\frac{\sin t + \cos t}{\cos t - \sin t}\right) \cdot \frac{dx}{dt} = \frac{2}{(\cos t - \sin t)^2} \cdot \frac{1}{e^t(\cos t - \sin t)} = \frac{2}{e^t(\cos t - \sin t)^3}$$

3.  Find the equation of the tangent to $x = \sqrt{t}$, $y = t - 1/\sqrt{t}$ at the point where $t = 4$.

$$\frac{dx}{dt} = \frac{1}{2\sqrt{t}}, \quad \frac{dy}{dt} = 1 + \frac{1}{2t\sqrt{t}}, \quad \text{and} \quad \frac{dy}{dx} = \frac{dy/dt}{dx/dt} = 2\sqrt{t} + \frac{1}{t}$$

At $t = 4$:  $x = 2$, $y = 7/2$, and $m = dy/dx = 17/4$.

The equation of the tangent is $(y - 7/2) = (17/4)(x - 2)$ or $17x - 4y = 20$.

4.  The position of a particle, moving along a curve, is given at time $t$ by the parametric equations $x = 2 - 3 \cos t$, $y = 3 + 2 \sin t$, where $x$ and $y$ are measured in feet and $t$ in seconds. Find the time rate and direction of change  (a) of the abscissa when $t = \pi/3$, (b) of the ordinate when $t = 5\pi/3$, (c) of $\theta$, the angle of inclination of the tangent, when $t = 2\pi/3$.

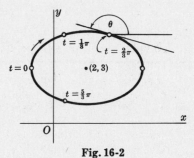

**Fig. 16-2**

$dx/dt = 3 \sin t$,  $dy/dt = 2 \cos t$,  $\tan \theta = dy/dx = \frac{2}{3} \cot t$

(a)  When $t = \pi/3$, $dx/dt = 3\sqrt{3}/2$. The abscissa is increasing at $3\sqrt{3}/2$ ft/sec.

(b)  When $t = 5\pi/3$, $dy/dt = 2(\frac{1}{2}) = 1$. The ordinate is increasing at the rate 1 ft/sec.

(c)  $\theta = \arctan (\frac{2}{3} \cot t)$, and $\dfrac{d\theta}{dt} = \dfrac{-6 \csc^2 t}{9 + 4 \cot^2 t}$. When $t = 2\pi/3$, $\dfrac{d\theta}{dt} = \dfrac{-6(2/\sqrt{3})^2}{9 + 4(-1/\sqrt{3})^2} = -\dfrac{24}{31}$. The angle of inclination of the tangent is decreasing at the rate 24/31 radians/second.

## Supplementary Problems

In Problems 5-9, find $dy/dx$ and $d^2y/dx^2$.

5.  $x = 2 + t$, $y = 1 + t^2$      Ans.  $dy/dx = 2t$, $d^2y/dx^2 = 2$

6.  $x = t + 1/t$, $y = t + 1$      Ans.  $dy/dx = t^2/(t^2 - 1)$, $d^2y/dx^2 = -2t^3/(t^2 - 1)^3$

7.  $x = 2 \sin t$, $y = \cos 2t$      Ans.  $dy/dx = -2 \sin t$, $d^2y/dx^2 = -1$

8.  $x = \cos^3 \theta$, $y = \sin^3 \theta$      Ans.  $dy/dx = -\tan \theta$, $d^2y/dx^2 = 1/(3 \cos^4 \theta \sin \theta)$

9.  $x = a(\cos \phi + \phi \sin \phi)$, $y = a(\sin \phi - \phi \cos \phi)$      Ans.  $dy/dx = \tan \phi$, $d^2y/dx^2 = 1/(a\phi \cos^3 \phi)$

10.  Find the slope of the curve $x = e^{-t} \cos 2t$, $y = e^{-2t} \sin 2t$ at the point $t = 0$.      Ans.  $-2$

11.  Find the rectangular coordinates of the highest point of the curve $x = 96t$, $y = 96t - 16t^2$.  Hint: Find $t$ for maximum $y$.      Ans.  $(288, 144)$

12.  Find the equation of the tangent and the normal to the curve  (a) $x = 3e^t$, $y = 5e^{-t}$ at $t = 0$, (b) $x = a \cos^4 \theta$, $y = a \sin^4 \theta$ at $\theta = \frac{1}{4}\pi$.
Ans.  (a) $5x + 3y - 30 = 0$, $3x - 5y + 16 = 0$;  (b) $2x + 2y - a = 0$, $x - y = 0$

13.  Find the equation of the tangent at any point $P(x, y)$ of the curve $x = a \cos^3 t$, $y = a \sin^3 t$. Show that the length of the segment of the tangent intercepted by the coordinate axes is $a$.
Ans.  $x \sin t + y \cos t = \frac{1}{2}a \sin 2t$

14.  For the curve $x = t^2 - 1$, $y = t^3 - t$, locate the points where the tangent line is  (a) horizontal and (b) vertical.  Show that at the point where the curve crosses itself, the two tangents are mutually perpendicular.      Ans.  (a) $t = \pm\sqrt{3}/3$, (b) $t = 0$

# Chapter 17

## Curvature

**DERIVATIVE OF ARC LENGTH.** Let $y = f(x)$ be a function having a continuous first derivative. Let $A$ (see Fig. 17-1) be a fixed point on the graph and denote by $s$ the arc length measured from $A$ to any other point on the curve. Let $P(x, y)$ be an arbitrary point and $Q(x + \Delta x, y + \Delta y)$ be a neighboring point on the curve. Denote by $\Delta s$ the arc length from $P$ to $Q$. The rate of change of the arc $s$ $(= AP)$ per unit change in $x$ and the rate of change per unit change in $y$ are given respectively by

$$\frac{ds}{dx} = \lim_{\Delta x \to 0} \frac{\Delta s}{\Delta x} = \pm \sqrt{1 + \left(\frac{dy}{dx}\right)^2}, \qquad \frac{ds}{dy} = \lim_{\Delta y \to 0} \frac{\Delta s}{\Delta y} = \pm \sqrt{1 + \left(\frac{dx}{dy}\right)^2}$$

The plus or minus sign is to be taken in the first formula according as $s$ increases or decreases as $x$ increases, and in the second formula according as $s$ increases or decreases as $y$ increases.

When a curve is given by the parametric equations $x = f(u)$, $y = g(u)$, the rate of change of $s$ with respect to $u$ is given by $\dfrac{ds}{du} = \pm \sqrt{\left(\dfrac{dx}{du}\right)^2 + \left(\dfrac{dy}{du}\right)^2}$. Here the plus or minus sign is to be taken according as $s$ increases or decreases as $u$ increases.

To avoid the repetition of ambiguous signs, we shall assume hereafter that direction on each arc considered has been established so that the derivative of arc length will be positive.

See Problems 1-5.

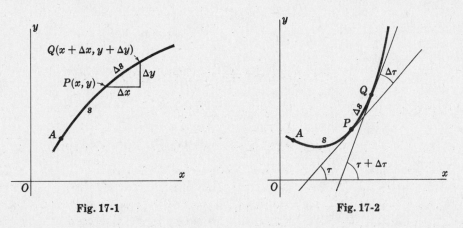

**Fig. 17-1**          **Fig. 17-2**

**CURVATURE.** The curvature $K$ of a curve $y = f(x)$, at any point $P$ on it, is the rate of change in direction (i.e., the angle of inclination $\tau$ of the tangent line at $P$) per unit of arc length $s$. (See Fig. 17-2.) Thus,

$$K = \frac{d\tau}{ds} = \lim_{\Delta s \to 0} \frac{\Delta \tau}{\Delta s} = \frac{\dfrac{d^2 y}{dx^2}}{\left\{ 1 + \left(\dfrac{dy}{dx}\right)^2 \right\}^{3/2}}; \qquad K = \frac{-\dfrac{d^2 x}{dy^2}}{\left\{ 1 + \left(\dfrac{dx}{dy}\right)^2 \right\}^{3/2}}$$

81

From the first of these formulas, it is clear that $K$ is positive when $P$ is on an arc which is concave upward and is negative when $P$ is on an arc which is concave downward.

The reader will find that $K$ is sometimes defined so as to be positive, that is, as the numerical value of that given by the above formulas. With this latter definition, the sign of $K$ in the answers below should be ignored.

**THE RADIUS OF CURVATURE** $R$ for a point $P$ on a curve is given by $R = |1/K|$, provided $K \neq 0$.

**THE CIRCLE OF CURVATURE** or Osculating Circle of a curve at a point $P$ on it is the circle of radius $R$ lying on the concave side of the curve and tangent to it at $P$.

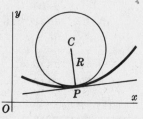

To construct the circle of curvature: On the concave side of the curve construct the normal at $P$ and on it lay off $PC = R$. The point $C$ is the center of the required circle.

**Fig. 17-3**

**THE CENTER OF CURVATURE** for a point $P(x, y)$ of a curve is the center $C$ of the circle of curvature at $P$. The coordinates $(\alpha, \beta)$ of the center of curvature are given by

$$\alpha = x - \frac{\dfrac{dy}{dx}\left[1 + \left(\dfrac{dy}{dx}\right)^2\right]}{\dfrac{d^2y}{dx^2}}, \qquad \beta = y + \frac{1 + \left(\dfrac{dy}{dx}\right)^2}{\dfrac{d^2y}{dx^2}}$$

or

$$\alpha = x + \frac{1 + \left(\dfrac{dx}{dy}\right)^2}{\dfrac{d^2x}{dy^2}}, \qquad \beta = y - \frac{\dfrac{dx}{dy}\left[1 + \left(\dfrac{dx}{dy}\right)^2\right]}{\dfrac{d^2x}{dy^2}}$$

**THE EVOLUTE** of a curve is the locus of the centers of curvature of the given curve. See Problems 6-13.

# Solved Problems

1. Derive: $\left(\dfrac{ds}{dx}\right)^2 = 1 + \left(\dfrac{dy}{dx}\right)^2$.

Refer to Fig. 17-1. On the curve $y = f(x)$, where $f(x)$ has a continuous derivative, let $s$ denote the arc length from a fixed point $A$ to a variable point $P(x, y)$. Denote by $\Delta s$ the arc length from $P$ to a neighboring point $Q(x + \Delta x, y + \Delta y)$ of the curve and by $PQ$ the length of the chord joining $P$ and $Q$.

Now $\dfrac{\Delta s}{\Delta x} = \dfrac{\Delta s}{PQ} \cdot \dfrac{PQ}{\Delta x}$ and, since $(PQ)^2 = (\Delta x)^2 + (\Delta y)^2$,

$$\left(\frac{\Delta s}{\Delta x}\right)^2 = \left(\frac{\Delta s}{PQ}\right)^2\left(\frac{PQ}{\Delta x}\right)^2 = \left(\frac{\Delta s}{PQ}\right)^2 \frac{(\Delta x)^2 + (\Delta y)^2}{(\Delta x)^2} = \left(\frac{\Delta s}{PQ}\right)^2\left\{1 + \left(\frac{\Delta y}{\Delta x}\right)^2\right\}$$

As $Q$ approaches $P$ along the curve, $\Delta x \to 0$, $\Delta y \to 0$, and $\dfrac{\Delta s}{PQ} = \dfrac{\text{arc } PQ}{\text{chord } PQ} \to 1$. (For a proof of the latter, see Chapter 41, Problem 22.) Then

$$\left(\frac{ds}{dx}\right)^2 = \lim_{\Delta x \to 0}\left(\frac{\Delta s}{\Delta x}\right)^2 = \lim_{\Delta x \to 0}\left\{1 + \left(\frac{\Delta y}{\Delta x}\right)^2\right\} = 1 + \left(\frac{dy}{dx}\right)^2$$

2. Find $\dfrac{ds}{dx}$ at $P(x, y)$ on the parabola $y = 3x^2$. $\dfrac{ds}{dx} = \sqrt{1 + \left(\dfrac{dy}{dx}\right)^2} = \sqrt{1 + (6x)^2} = \sqrt{1 + 36x^2}$

3.  Find $\dfrac{ds}{dx}$ and $\dfrac{ds}{dy}$ at $P(x, y)$ on the ellipse $x^2 + 4y^2 = 8$.

(a) $\dfrac{dy}{dx} = -\dfrac{x}{4y}; \ \ 1 + \left(\dfrac{dy}{dx}\right)^2 = 1 + \dfrac{x^2}{16y^2} = \dfrac{x^2 + 16y^2}{16y^2} = \dfrac{32 - 3x^2}{32 - 4x^2}$ and $\ \dfrac{ds}{dx} = \sqrt{\dfrac{32 - 3x^2}{32 - 4x^2}}$

(b) $\dfrac{dx}{dy} = -\dfrac{4y}{x}; \ \ 1 + \left(\dfrac{dx}{dy}\right)^2 = 1 + \dfrac{16y^2}{x^2} = \dfrac{x^2 + 16y^2}{x^2} = \dfrac{2 + 3y^2}{2 - y^2}$ and $\ \dfrac{ds}{dy} = \sqrt{\dfrac{2 + 3y^2}{2 - y^2}}$

4.  Find $\dfrac{ds}{d\theta}$ at $P(\theta)$ on the curve $x = \sec \theta,\ y = \tan \theta$.

$$\frac{ds}{d\theta} = \sqrt{\left(\frac{dx}{d\theta}\right)^2 + \left(\frac{dy}{d\theta}\right)^2} = \sqrt{\sec^2\theta \tan^2\theta + \sec^4\theta} = |\sec\theta|\sqrt{\tan^2\theta + \sec^2\theta}$$

5.  The coordinates $(x, y)$ in feet of a moving particle $P$ are given by $x = \cos t - 1$, $y = 2 \sin t + 1$, where $t$ is the time in seconds. At what rate is $P$ moving along the curve when (a) $t = 5\pi/6$, (b) $t = 5\pi/3$, and (c) $P$ is moving at extreme speeds?

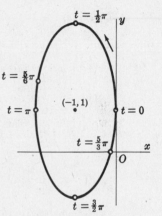

$$\frac{ds}{dt} = \sqrt{\left(\frac{dx}{dt}\right)^2 + \left(\frac{dy}{dt}\right)^2} = \sqrt{\sin^2 t + 4\cos^2 t} = \sqrt{1 + 3\cos^2 t}$$

(a)  When $t = 5\pi/6$, $ds/dt = \sqrt{1 + 3(\frac{3}{4})} = \sqrt{13}/2$ ft/sec.

(b)  When $t = 5\pi/3$, $ds/dt = \sqrt{1 + 3(\frac{1}{4})} = \sqrt{7}/2$ ft/sec.

(c)  Let $S = \dfrac{ds}{dt} = \sqrt{1 + 3\cos^2 t}$. Then $\dfrac{dS}{dt} = \dfrac{-3\cos t \sin t}{S}$.

Solving $dS/dt = 0$ gives the critical values $t = 0,\ \pi/2,\ \pi,\ 3\pi/2$.

When $t = 0$ and $\pi$, the rate $ds/dt = \sqrt{1 + 3(1)} = 2$ ft/sec is fastest.

When $t = \pi/2$ and $3\pi/2$, the rate $ds/dt = \sqrt{1 + 3(0)} = 1$ ft/sec is slowest.

Fig. 17-4

6.  Find the curvature of the parabola $y^2 = 12x$ at the points (a) $(3, 6)$, (b) $(\frac{3}{4}, -3)$, (c) $(0, 0)$.

$$\frac{dy}{dx} = \frac{6}{y}, \qquad 1 + \left(\frac{dy}{dx}\right)^2 = 1 + \frac{36}{y^2}, \qquad \text{and} \qquad \frac{d^2y}{dx^2} = -\frac{6}{y^2} \cdot \frac{dy}{dx} = -\frac{36}{y^3}$$

(a)  At $(3, 6)$: $\dfrac{dy}{dx} = 1$, $1 + \left(\dfrac{dy}{dx}\right)^2 = 2$, $\dfrac{d^2y}{dx^2} = -\dfrac{1}{6}$, and $K = \dfrac{-1/6}{2^{3/2}} = -\dfrac{\sqrt{2}}{24}$.

(b)  At $(\frac{3}{4}, -3)$: $\dfrac{dy}{dx} = -2$, $1 + \left(\dfrac{dy}{dx}\right)^2 = 5$, $\dfrac{d^2y}{dx^2} = \dfrac{4}{3}$, and $K = \dfrac{4/3}{5^{3/2}} = \dfrac{4\sqrt{5}}{75}$.

(c)  At $(0, 0)$, $\dfrac{dy}{dx}$ is undefined. But $\dfrac{dx}{dy} = \dfrac{y}{6} = 0$, $1 + \left(\dfrac{dx}{dy}\right)^2 = 1$, $\dfrac{d^2x}{dy^2} = \dfrac{1}{6}$, and $K = -\dfrac{1}{6}$.

7.  Find the curvature of the cycloid $x = \theta - \sin \theta$, $y = 1 - \cos \theta$, at the highest point of an arch.

To find the highest point on the interval $0 < x < 2\pi$: $dy/d\theta = \sin \theta$ and the critical value on the interval is $x = \pi$. Since $d^2y/d\theta^2 = \cos \theta < 0$ when $\theta = \pi$, the point $\theta = \pi$ is a relative maximum point and the highest point of the curve on the interval.

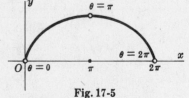

Fig. 17-5

To find the curvature:

$$\frac{dx}{d\theta} = 1 - \cos \theta, \qquad \frac{dy}{d\theta} = \sin \theta, \qquad \frac{dy}{dx} = \frac{\sin \theta}{1 - \cos \theta}, \qquad \frac{d^2y}{dx^2} = \frac{d}{d\theta}\left(\frac{\sin \theta}{1 - \cos \theta}\right) \cdot \frac{d\theta}{dx} = \frac{-1}{(1 - \cos \theta)^2}$$

At $\theta = \pi$, $dy/dx = 0$, $d^2y/dx^2 = -1/4$, and $K = -1/4$.

8.   Find the curvature of the cissoid $y^2(2-x) = x^3$ at the point $(1,1)$.

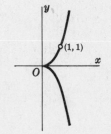

Fig. 17-6

Differentiating the given equation implicitly with respect to $x$, we have

$$(a) \quad -y^2 + (2-x)2yy' = 3x^2 \quad \text{and}$$

$$(b) \quad -2yy' + (2-x)2yy'' + (2-x)2(y')^2 - 2yy' = 6x$$

From $(a)$, for $x = y = 1$, $-1 + 2y' = 3$ and $y' = 2$.

From $(b)$, for $x = y = 1$ and $y' = 2$, $-4 + 2y'' + 8 - 4 = 6$ and $y'' = 3$.

Then $K = 3/(1+4)^{3/2} = 3\sqrt{5}/25$.

9.   Find the point of greatest curvature on the curve $y = \ln x$.

$$\frac{dy}{dx} = \frac{1}{x}, \qquad \frac{d^2y}{dx^2} = -\frac{1}{x^2}, \qquad \text{and} \qquad K = \frac{-x}{(1+x^2)^{3/2}}$$

$\dfrac{dK}{dx} = \dfrac{2x^2-1}{(1+x^2)^{5/2}}$ and the critical value is $x = 1/\sqrt{2}$. The required point is $(1/\sqrt{2}, -\tfrac{1}{2}\ln 2)$.

10.  Locate the center of curvature $C$ for the curve $y = f(x)$ at one of its points $P(x, y)$ at which $y' \neq 0$.
     (See Fig. 17-3.)

The center of curvature $C(\alpha, \beta)$ lies (1) on the normal line at $P$ and (2) at a distance $R$ from $P$ measured toward the concave side of the curve. Thus,

$$(1) \quad \beta - y = -\frac{1}{y'}(\alpha - x) \qquad \text{and} \qquad (2) \quad (\alpha - x)^2 + (\beta - y)^2 = R^2 = \frac{[1+(y')^2]^3}{(y'')^2}$$

From $(1)$, $\alpha - x = -y'(\beta - y)$; substituting in $(2)$,

$$(\beta - y)^2 [1 + (y')^2] = \frac{[1+(y')^2]^3}{(y'')^2} \qquad \text{and} \qquad \beta - y = \pm \frac{1 + (y')^2}{y''}$$

To determine the correct sign, note that when the curve is concave upward $y'' > 0$ and, since $C$ then lies above $P$, $\beta - y > 0$. Thus, the proper sign in this case is $+$. (The reader will show the sign is $+$ when $y'' < 0$.) Thus,

$$\beta = y + \frac{1 + (y')^2}{y''} \quad \text{and from } (1), \quad \alpha = x - \frac{y'[1+(y')^2]}{y''}$$

11.  Find the equation of the circle of curvature of $2xy + x + y = 4$ at the point $(1,1)$.

$2y + 2xy' + 1 + y' = 0$; at $(1,1)$, $y' = -1$. Then $1 + (y')^2 = 2$.

$4y' + 2xy'' + y'' = 0$; at $(1,1)$, $y'' = 4/3$.

$K = \dfrac{4/3}{2\sqrt{2}}$ and $R = \dfrac{3\sqrt{2}}{2}$.   $\alpha = 1 - \dfrac{-1(2)}{4/3} = \dfrac{5}{2}$, $\beta = 1 + \dfrac{2}{4/3} = \dfrac{5}{2}$.

The required equation is $(x-\alpha)^2 + (y-\beta)^2 = R^2$ or $(x - 5/2)^2 + (y - 5/2)^2 = 9/2$.

12.  Find the equation of the evolute of the parabola $y^2 = 12x$.

At $P(x, y)$: $\dfrac{dy}{dx} = \dfrac{6}{y} = \dfrac{\sqrt{3}}{\sqrt{x}}$, $1 + \left(\dfrac{dy}{dx}\right)^2 = 1 + \dfrac{36}{y^2} = 1 + \dfrac{3}{x}$, and $\dfrac{d^2y}{dx^2} = -\dfrac{36}{y^3} = -\dfrac{\sqrt{3}}{2x^{3/2}}$.

$$\alpha = x - \frac{\sqrt{3/x}\,(1 + 3/x)}{-\sqrt{3}/2x^{3/2}} = x + \frac{2\sqrt{3}\,(x+3)}{\sqrt{3}} = 3x + 6$$

$$\beta = y + \frac{1 + 36/y^2}{-36/y^3} = y - \frac{y^3 + 36y}{36} = -\frac{y^3}{36}$$

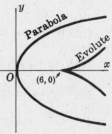

Fig. 17-7

The equations $\alpha = 3x + 6$, $\beta = -y^3/36$ may be regarded as parametric equations of the evolute with $x$ and $y$, connected by the equation of the parabola, as parameters. However, it is relatively simple in this problem to eliminate the parameters. Thus: $x = (\alpha - 6)/3$, $y = -\sqrt[3]{36\beta}$, and, substituting in the equation of the parabola, we have

$$(36\beta)^{2/3} = 4(\alpha - 6) \qquad \text{or} \qquad 81\beta^2 = 4(\alpha - 6)^3.$$

13. Find the equation of the evolute of the curve $x = \cos\theta + \theta\sin\theta$, $y = \sin\theta - \theta\cos\theta$.

At $P(x,y)$: $\dfrac{dx}{d\theta} = \theta\cos\theta$, $\dfrac{dy}{d\theta} = \theta\sin\theta$, $\dfrac{dy}{dx} = \tan\theta$,

and $\dfrac{d^2y}{dx^2} = \dfrac{\sec^2\theta}{\theta\cos\theta} = \dfrac{\sec^3\theta}{\theta}$

$$\alpha = x - \frac{\tan\theta\sec^2\theta}{(\sec^3\theta)/\theta} = x - \theta\sin\theta = \cos\theta$$

$$\beta = y + \frac{\sec^2\theta}{(\sec^3\theta)/\theta} = y + \theta\cos\theta = \sin\theta$$

$\alpha = \cos\theta$, $\beta = \sin\theta$ are parametric equations of the evolute.

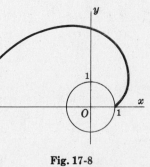

**Fig. 17-8**

# Supplementary Problems

In Problems 14-19, find the indicated derivative of arc length.

14. $x^2 + y^2 = 25$      *Ans.* $ds/dx = 5/\sqrt{25 - x^2}$, $ds/dy = 5/\sqrt{25 - y^2}$

15. $y^2 = x^3$      *Ans.* $ds/dx = \frac{1}{2}\sqrt{4 + 9x}$, $ds/dy = \sqrt{4 + 9y^{2/3}}/3y^{1/3}$

16. $x^{2/3} + y^{2/3} = a^{2/3}$      *Ans.* $ds/dx = (a/x)^{1/3}$, $ds/dy = (a/y)^{1/3}$

17. $6xy = x^4 + 3$      *Ans.* $ds/dx = (x^4 + 1)/2x^2$

18. $27ay^2 = 4(x - a)^3$      *Ans.* $ds/dx = \sqrt{(x + 2a)/3a}$

19. $y = a\cosh x/a$      *Ans.* $ds/dx = \cosh x/a$

20. For the curve $x = f(u)$, $y = g(u)$, derive $(ds/du)^2 = (dx/du)^2 + (dy/du)^2$.

In Problems 21-24 find $ds/dt$.

21. $x = t^2$, $y = t^3$      *Ans.* $t\sqrt{4 + 9t^2}$      23. $x = 2\cos t$, $y = 3\sin t$      *Ans.* $\sqrt{4 + 5\cos^2 t}$

22. $x = \cos t$, $y = \sin t$      *Ans.* 1      24. $x = \cos^3 t$, $y = \sin^3 t$      *Ans.* $\frac{3}{2}\sin 2t$

25. Use $dy/dx = \tan\tau$, to obtain $dx/ds = \cos\tau$, $dy/ds = \sin\tau$.

26. Use $\tau = \arctan\left(\dfrac{dy}{dx}\right)$ to obtain $K = \dfrac{d\tau}{ds} = \dfrac{d\tau}{dx}\cdot\dfrac{dx}{ds} = \dfrac{y''}{\{1 + (y')^2\}^{3/2}}$.

27. Find the curvature of each curve at the given point.

    (a) $y = x^3/3$ at $x = 0$, $x = 1$, $x = -2$      *Ans.* $0, \sqrt{2}/2, -4\sqrt{17}/289$

    (b) $x^2 = 4ay$ at $x = 0$, $x = 2a$      *Ans.* $\dfrac{1}{2a}, \dfrac{\sqrt{2}}{8a}$

    (c) $y = \sin x$ at $x = 0$, $x = \frac{1}{2}\pi$      *Ans.* $0, -1$

    (d) $y = e^{-x^2}$ at $x = 0$      *Ans.* $-2$

28. Show (a) the curvature of a straight line is 0 and (b) the curvature of a circle is numerically the reciprocal of its radius.

29. Find the points of maximum curvature: (a) $y = e^x$, (b) $y = x^3/3$.      *Ans.* (a) $x = \frac{1}{2}\ln\frac{1}{2}$, (b) $x = 1/\sqrt[4]{5}$

30. In a certain coordinate system a railroad track follows a portion of the negative $x$-axis to the origin $O$, then passes along a transition curve $y = \frac{1}{4}x^4$ to $A(1, \frac{1}{4})$, and then along an arc of the circle $144x^2 + 144y^2 - 96x - 264y + 9 = 0$. Verify (a) that the transition curve is tangent to the straight and circular track at the joins and (b) that the curvature of the transition curve is zero at $O$ and the reciprocal of the radius of the circular section at $A$.

31. Find the radius of curvature of (a) $x^3 + xy^2 - 6y^2 = 0$ at $(3,3)$, (b) $x = a\operatorname{sech}^{-1}y/a - \sqrt{a^2 - y^2}$ at $(x,y)$, (c) $x = 2a\tan\theta$, $y = a\tan^2\theta$, (d) $x = a\cos^4\theta$, $y = a\sin^4\theta$.      *Ans.* (a) $5\sqrt{5}$, (b) $a\sqrt{a^2 - y^2}/|y|$, (c) $2a|\sec^3\theta|$, (d) $2a(\sin^4\theta + \cos^4\theta)^{3/2}$

32. Find the center of curvature of (a) Problem 31(a), (b) $y = \sin x$ at a maximum point.      *Ans.* (a) $C(-7, 8)$, (b) $C(\frac{1}{2}\pi, 0)$

33. Find the equation of the circle of curvature of the parabola $y^2 = 12x$ at the points $(0,0)$ and $(3,6)$.      *Ans.* $(x - 6)^2 + y^2 = 36$, $(x - 15)^2 + (y + 6)^2 = 288$

34. Find the equation of the evolute of

    (a) $b^2x^2 + a^2y^2 = a^2b^2$, (b) $x^{2/3} + y^{2/3} = a^{2/3}$, (c) $x = 2\cos t + \cos 2t$, $y = 2\sin t + \sin 2t$.

    *Ans.* (a) $(a\alpha)^{2/3} + (b\beta)^{2/3} = (a^2 - b^2)^{2/3}$      (b) $(\alpha + \beta)^{2/3} + (\alpha - \beta)^{2/3} = 2a^{2/3}$

        (c) $\alpha = \frac{1}{3}(2\cos t - \cos 2t)$, $\beta = \frac{1}{3}(2\sin t - \sin 2t)$

# Chapter 18

# Plane Vectors

**SCALARS AND VECTORS.** Quantities such as time, temperature, and speed, which have magnitude only, are called scalar quantities or *scalars*. Scalars, being merely numbers, obey all the laws of ordinary algebra, e.g. 5 sec + 3 sec = 8 sec.

Quantities such as force, velocity, acceleration, and momentum, which have both magnitude and direction, are called vector quantities or *vectors*. Vectors are represented geometrically by directed line segments (arrows). The direction of the arrow (the angle which it makes with some fixed line of the plane) is the direction of the vector, and the length of the arrow (in terms of a chosen unit of measure) represents the magnitude of the vector. Scalars will be denoted here by letters $a, b, c, \ldots$ in ordinary type; vectors will be denoted by letters $\mathbf{a}, \mathbf{b}, \mathbf{c}, \ldots$ in bold faced type or as **OP** [see Fig. 18-1(a)]. The magnitude of a vector **a** or **OP** will be denoted by $|\mathbf{a}|$ or $|\mathbf{OP}|$.

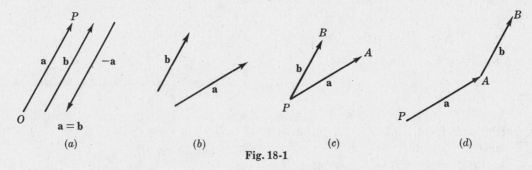

$$(a) \qquad\qquad (b) \qquad\qquad (c) \qquad\qquad (d)$$

**Fig. 18-1**

Two vectors **a** and **b** are called *equal*, $\mathbf{a} = \mathbf{b}$, if they have the same magnitude and the same direction. A vector whose magnitude is that of **a** but whose direction is opposite that of **a**, is defined as the negative of **a** and is denoted by $-\mathbf{a}$. More generally, if **a** is a vector and $k$ is a scalar, then $k\mathbf{a}$ is a vector whose direction is that of **a** and whose magnitude is $k$ times that of **a** if $k$ is positive but whose direction is opposite that of **a** and whose magnitude is $|k|$ times that of **a** if $k$ is negative.

Unless indicated otherwise, a given vector has no fixed position in the plane and so may be moved under parallel displacement at will. In particular, if **a** and **b** are two vectors (Fig. 18-1(b)) they may be placed so as to have a common initial or beginning point $P$ (see Fig. 18-1(c)) or so that the initial point of **b** coincides with the terminal or end point of **a** (see Fig. 18-1(d)).

**SUM AND DIFFERENCE OF TWO VECTORS.** If **a** and **b** are the vectors of Fig. 18-1(b), their *sum* or *resultant* $\mathbf{a} + \mathbf{b}$ is found

(i)   by taking the vectors as in Fig. 18-1(c) and completing the parallelogram $PAQB$ of Fig. 18-2(e). The vector **PQ** is the required sum.

(ii)  by taking the vectors as in Fig. 18-1(d) and completing the triangle $PAB$ of Fig. 18-2(f). Here, the vector **PB** is the required sum.

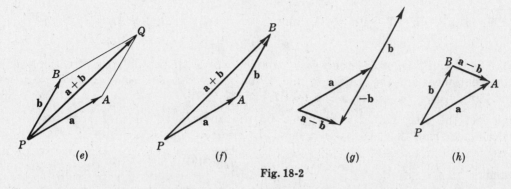

**Fig. 18-2**

It follows from Fig. 18-2(*f*) that three vectors may be displaced to form a triangle provided one of them is either the sum or the negative of the sum of the other two.

If **a** and **b** are the vectors of Fig. 18-1(*b*), their difference **a** − **b** is found

(iii)   from the relation   **a** − **b** = **a** + (−**b**)   as in Fig. 18-2(*g*).

(iv)   by taking the vectors as in Fig. 18-1(*c*) and completing the triangle.   In Fig. 18-2(*h*), the vector   **BA** = **a** − **b**.

If **a**, **b**, **c** are vectors and *k* is a scalar, then

1.   **a** + **b** = **b** + **a**              (Commutative law)
2.   **a** + (**b** + **c**) = (**a** + **b**) + **c**       (Associative law)
3.   *k*(**a** + **b**) = *k***a** + *k***b**           (Distributive law)

See Problems 1-4.

**COMPONENTS OF A VECTOR.**   In Fig. 18-3(*i*), let **a** = **PQ** be a given vector, let *PM* and *PN* be any two other lines (directions) through *P*, and construct the parallelogram *PAQB*.   Now

$$\mathbf{a} = \mathbf{PA} + \mathbf{PB}$$

and **a** is said to be *resolved* in the directions *PM* and *PN*.   We shall call **PA** and **PB** the *vector components of* **a** *in the pair of directions PM and PN*.

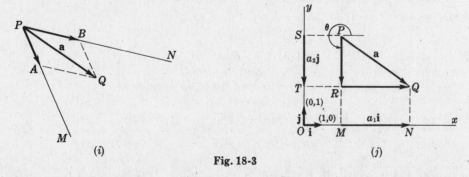

(*i*)

**Fig. 18-3**

(*j*)

Consider next the vector **a** in a rectangular coordinate system [Fig. 18-3(*j*)] having equal units of measure on the two axes.   Denote by **i** the vector from $(0,0)$ to $(1,0)$ and by **j** the vector from $(0,0)$ to $(0,1)$.   The direction of **i** is that of the positive *x*-axis, the direction of **j** is that of the positive *y*-axis, and both are *unit vectors*, that is, vectors of magnitude 1.

From the initial point *P* and the terminal point *Q* of **a** drop perpendiculars to the *x*-axis meeting it in *M* and *N* respectively and to the *y*-axis meeting it in *S* and *T* respectively.   Now **MN** = $a_1$**i**, with $a_1$ positive, and **ST** = $a_2$**j**, with $a_2$ negative.   Then

$$\mathbf{MN} = \mathbf{RQ} = a_1\mathbf{i}, \quad \mathbf{ST} = \mathbf{PR} = a_2\mathbf{j}, \quad \text{and}$$

$$\mathbf{a} = a_1\mathbf{i} + a_2\mathbf{j} \tag{1}$$

We shall call $a_1\mathbf{i}$ and $a_2\mathbf{j}$ the *vector components* of $\mathbf{a}$ (the pair of directions need not be mentioned) and the scalars $a_1$ and $a_2$ the *scalar components* or *x- and y-components* or simply *components* of $\mathbf{a}$.

Let the direction of $\mathbf{a}$ be given by the angle $\theta$, $0 \leqq \theta < 2\pi$, measured counterclockwise from the positive $x$-axis to the vector. Then

$$|\mathbf{a}| = \sqrt{a_1^2 + a_2^2} \tag{2}$$

and

$$\tan \theta = a_2/a_1 \tag{3}$$

with the quadrant of $\theta$ being determined by

$$a_1 = |\mathbf{a}| \cos \theta, \quad a_2 = |\mathbf{a}| \sin \theta$$

If $\mathbf{a} = a_1\mathbf{i} + a_2\mathbf{j}$ and $\mathbf{b} = b_1\mathbf{i} + b_2\mathbf{j}$, then

4.  $\mathbf{a} = \mathbf{b}$ if and only if $a_1 = b_1$ and $a_2 = b_2$     6.  $\mathbf{a} + \mathbf{b} = (a_1 + b_1)\mathbf{i} + (a_2 + b_2)\mathbf{j}$

5.  $k\mathbf{a} = ka_1\mathbf{i} + ka_2\mathbf{j}$     7.  $\mathbf{a} - \mathbf{b} = (a_1 - b_1)\mathbf{i} + (a_2 - b_2)\mathbf{j}$

See Problem 5.

**SCALAR OR DOT PRODUCT.** The scalar or dot product of two vectors $\mathbf{a}$ and $\mathbf{b}$ is defined by

$$\mathbf{a} \cdot \mathbf{b} = |\mathbf{a}| |\mathbf{b}| \cos \theta \tag{4}$$

where $\theta$ is the smaller angle between the two vectors when drawn with a common initial point (see Fig. 18-4).

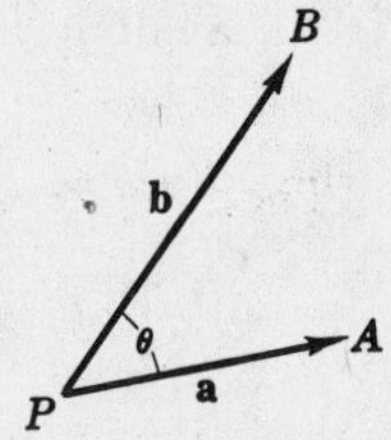

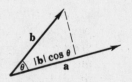

Fig. 18-4

From (4) follow

8.  $\mathbf{a} \cdot \mathbf{b} = \mathbf{b} \cdot \mathbf{a}$      (Commutative law)

9.  $\mathbf{a} \cdot \mathbf{a} = |\mathbf{a}| |\mathbf{a}| = |\mathbf{a}|^2$ or $|\mathbf{a}| = \sqrt{\mathbf{a} \cdot \mathbf{a}}$

10.  $\mathbf{a} \cdot \mathbf{b} = 0$ if (i) $\mathbf{a} = 0$, or (ii) $\mathbf{b} = 0$, or (iii) $\mathbf{a}$ is perpendicular to $\mathbf{b}$

11.  $\mathbf{i} \cdot \mathbf{i} = \mathbf{j} \cdot \mathbf{j} = 1$; $\mathbf{i} \cdot \mathbf{j} = 0$

12.  $\mathbf{a} \cdot \mathbf{b} = (a_1\mathbf{i} + a_2\mathbf{j}) \cdot (b_1\mathbf{i} + b_2\mathbf{j}) = a_1 b_1 + a_2 b_2$

13.  $\mathbf{a} \cdot (\mathbf{b} + \mathbf{c}) = \mathbf{a} \cdot \mathbf{b} + \mathbf{a} \cdot \mathbf{c}$      (Distributive law)

14.  $(\mathbf{a} + \mathbf{b}) \cdot (\mathbf{c} + \mathbf{d}) = \mathbf{a} \cdot \mathbf{c} + \mathbf{a} \cdot \mathbf{d} + \mathbf{b} \cdot \mathbf{c} + \mathbf{b} \cdot \mathbf{d}$

**SCALAR AND VECTOR PROJECTIONS.** In equation (1), the scalar $a_1$ may be called the *scalar projection* of $\mathbf{a}$ on any vector whose direction is that of the positive $x$-axis while the vector $a_1\mathbf{i}$ may be called the *vector projection* of $\mathbf{a}$ on any vector whose direction is that of the positive $x$-axis. In Problem 7, the scalar projection $\mathbf{a} \cdot \dfrac{\mathbf{b}}{|\mathbf{b}|}$ and the vector projection $\left(\mathbf{a} \cdot \dfrac{\mathbf{b}}{|\mathbf{b}|}\right)\dfrac{\mathbf{b}}{|\mathbf{b}|}$ of a vector $\mathbf{a}$ on another vector $\mathbf{b}$ are found. (Note that when $\mathbf{b}$ has the direction of the positive $x$-axis, then $\dfrac{\mathbf{b}}{|\mathbf{b}|} = \mathbf{i}$.)

Fig. 18-5

There follows

15.  $\mathbf{a} \cdot \mathbf{b} =$ product of length of $\mathbf{a}$ and the scalar projection of $\mathbf{b}$ on $\mathbf{a}$

        $=$ product of length of $\mathbf{b}$ and the scalar projection of $\mathbf{a}$ on $\mathbf{b}$. (See Fig. 18-5)

See Problems 8-9.

**DIFFERENTIATION OF VECTORS.** Let the curve of Fig. 18-6 be given by the parametric equations

$$x = f(u), \quad y = g(u)$$

The vector

$$\mathbf{r} = x\mathbf{i} + y\mathbf{j} = \mathbf{i}f(u) + \mathbf{j}g(u)$$

joining the origin to the point $P(x, y)$ of the curve is called the *position vector* or *radius vector* of $P$. (Hereinafter, the letter $\mathbf{r}$ will be used exclusively to denote position vectors; thus, $\mathbf{a} = 3\mathbf{i} + 4\mathbf{j}$ is a 'free' vector while $\mathbf{r} = 3\mathbf{i} + 4\mathbf{j}$ is *the vector* joining the origin to $P(3, 4)$.)

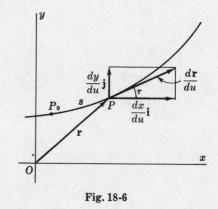

Fig. 18-6

The derivative of $\mathbf{r}$ with respect to $u$ is given by

$$\frac{d\mathbf{r}}{du} = \frac{dx}{du}\mathbf{i} + \frac{dy}{du}\mathbf{j} \tag{5}$$

Let $s$ denote the arc length measured from a fixed point $P_0$ of the curve so that $s$ increases with $u$. If $\tau$ is the angle which $d\mathbf{r}/du$ makes with the positive $x$-axis,

$$\tan \tau = \frac{dy/du}{dx/du} = \frac{dy}{dx} = \text{slope of curve at } P$$

Then $d\mathbf{r}/du$ is a vector of magnitude

$$\left| \frac{d\mathbf{r}}{du} \right| = \sqrt{\left(\frac{dx}{du}\right)^2 + \left(\frac{dy}{du}\right)^2} = \frac{ds}{du}$$

and direction that of the tangent to the curve at $P$. It is customary to show this vector with $P$ as initial point.

If now the scalar variable $u$ is the length of arc $s$, (5) becomes

$$\mathbf{t} = \frac{d\mathbf{r}}{ds} = \frac{dx}{ds}\mathbf{i} + \frac{dy}{ds}\mathbf{j} \tag{6}$$

Here, the direction of $\mathbf{t}$ is $\tau$ as before, while the magnitude is $\sqrt{(dx/ds)^2 + (dy/ds)^2} = 1$. Thus, $\mathbf{t} = d\mathbf{r}/ds$ is the *unit tangent* to the curve at $P$.

Since $\mathbf{t}$ is a unit vector, $\mathbf{t}$ and $d\mathbf{t}/ds$ are perpendicular (see Problem 11). Denote by $\mathbf{n}$ a unit vector at $P$ having the direction of $d\mathbf{t}/ds$. As $P$ moves along the curve shown in Fig. 18-7, the magnitude of $\mathbf{t}$ remains constant; hence, $d\mathbf{t}/ds$ measures the rate of change of the direction of $\mathbf{t}$.

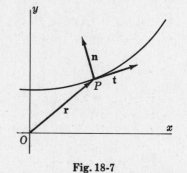

Fig. 18-7

is the numerical value of the curvature at $P$, e.g. $|dt/ds| = |K|$. Thus, the magnitude of $d\mathbf{t}/ds$ at $P$ and

$$\frac{d\mathbf{t}}{ds} = |K|\,\mathbf{n} \tag{7}$$

See Problems 10-13.

# Solved Problems

1.  Prove:  $\mathbf{a} + \mathbf{b} = \mathbf{b} + \mathbf{a}$

    From Fig. 18-8,  $\mathbf{a} + \mathbf{b} = \mathbf{PQ} = \mathbf{b} + \mathbf{a}$

2.  Prove:  $(\mathbf{a} + \mathbf{b}) + \mathbf{c} = \mathbf{a} + (\mathbf{b} + \mathbf{c})$

    From Fig. 18-9,  $\mathbf{PC} = \mathbf{PB} + \mathbf{BC} = (\mathbf{a} + \mathbf{b}) + \mathbf{c}$.   Also,  $\mathbf{PC} = \mathbf{PA} + \mathbf{AC} = \mathbf{a} + (\mathbf{b} + \mathbf{c})$.

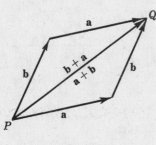

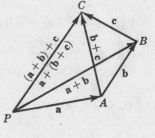

Fig. 18-8                                                         Fig. 18-9

3.  Let  $\mathbf{a}, \mathbf{b}, \mathbf{c}$  be three vectors issuing from  $P$  such that their end points  $A, B, C$  lie on a line as shown in Fig. 18-10. If  $C$  divides  $BA$  in the ratio  $x : y$  where  $x + y = 1$,  then

    $$\mathbf{c} = \mathbf{PB} + \mathbf{BC} = \mathbf{b} + x(\mathbf{a} - \mathbf{b}) = x\mathbf{a} + (1 - x)\mathbf{b} = x\mathbf{a} + y\mathbf{b}$$

    For example, if  $C$  bisects  $BA$,  then  $\mathbf{c} = \tfrac{1}{2}(\mathbf{a} + \mathbf{b})$  and  $\mathbf{BC} = \tfrac{1}{2}(\mathbf{a} - \mathbf{b})$.

4.  Prove:   The diagonals of a parallelogram bisect each other.

    Let the diagonals intersect at  $Q$,  as in Fig. 18-11.   Since

    $$\mathbf{PB} = \mathbf{PQ} + \mathbf{QB} = \mathbf{PQ} - \mathbf{BQ} \quad \text{or} \quad \mathbf{b} = x(\mathbf{a} + \mathbf{b}) - y(\mathbf{a} - \mathbf{b}) = (x - y)\mathbf{a} + (x + y)\mathbf{b}$$

    $x + y = 1$  and  $x - y = 0$.   Then  $x = y = \tfrac{1}{2}$  and  $Q$  is the midpoint of each diagonal.

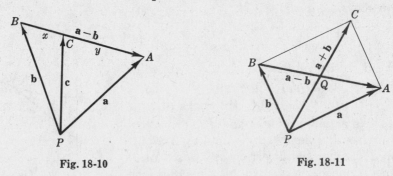

Fig. 18-10                                                         Fig. 18-11

5.  For the vectors  $\mathbf{a} = 3\mathbf{i} + 4\mathbf{j}$  and  $\mathbf{b} = 2\mathbf{i} - \mathbf{j}$,  find the magnitude and direction of  (a) $\mathbf{a}$ and $\mathbf{b}$, (b) $\mathbf{a} + \mathbf{b}$,  (c) $\mathbf{b} - \mathbf{a}$.

    (a)  For  $\mathbf{a} = 3\mathbf{i} + 4\mathbf{j}$:  $|\mathbf{a}| = \sqrt{a_1^2 + a_2^2} = \sqrt{3^2 + 4^2} = 5$;  $\tan\theta = a_2/a_1 = 4/3$,  $a_1 = |\mathbf{a}|\cos\theta$,  and  $\cos\theta = 3/5$.   Then  $\theta$,  a first quadrant angle, is  $53°8'$.

    For  $\mathbf{b} = 2\mathbf{i} - \mathbf{j}$:  $|\mathbf{b}| = \sqrt{4 + 1} = \sqrt{5}$;  $\tan\theta = -1/2$,  $\cos\theta = 2/\sqrt{5}$;  $\theta = 360° - 26°34' = 333°26'$.

    (b)  $\mathbf{a} + \mathbf{b} = (3\mathbf{i} + 4\mathbf{j}) + (2\mathbf{i} - \mathbf{j}) = 5\mathbf{i} + 3\mathbf{j}$.

    $|\mathbf{a} + \mathbf{b}| = \sqrt{5^2 + 3^2} = \sqrt{34}$;  $\tan\theta = 3/5$,  $\cos\theta = 5/\sqrt{34}$;  $\theta = 30°58'$.

    (c)  $\mathbf{b} - \mathbf{a} = (2\mathbf{i} - \mathbf{j}) - (3\mathbf{i} + 4\mathbf{j}) = -\mathbf{i} - 5\mathbf{j}$.

    $|\mathbf{b} - \mathbf{a}| = \sqrt{26}$;  $\tan\theta = 5$,  $\cos\theta = -1/\sqrt{26}$;  $\theta = 258°41'$.

6. Prove: The median to the base of an isosceles triangle is perpendicular to the base. (In Fig. 18-12, $|\mathbf{a}| = |\mathbf{b}|$.)

From Problem 3, since $\mathbf{m}$ bisects the base,

$$\mathbf{m} = \tfrac{1}{2}(\mathbf{a} + \mathbf{b})$$

Then
$$\mathbf{m} \cdot (\mathbf{b} - \mathbf{a}) = \tfrac{1}{2}(\mathbf{a} + \mathbf{b}) \cdot (\mathbf{b} - \mathbf{a})$$
$$= \tfrac{1}{2}(\mathbf{a} \cdot \mathbf{b} - \mathbf{a} \cdot \mathbf{a} + \mathbf{b} \cdot \mathbf{b} - \mathbf{b} \cdot \mathbf{a}) = \tfrac{1}{2}(\mathbf{b} \cdot \mathbf{b} - \mathbf{a} \cdot \mathbf{a}) = 0$$

as was to be proved.

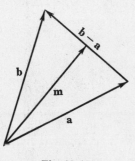

Fig. 18-12

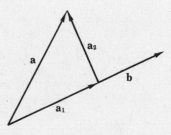

Fig. 18-13

7. Resolve $\mathbf{a}$ into components $\mathbf{a}_1$ and $\mathbf{a}_2$ respectively parallel and perpendicular to $\mathbf{b}$.

In Fig. 18-13 above: $\mathbf{a} = \mathbf{a}_1 + \mathbf{a}_2$, $\mathbf{a}_1 = c\mathbf{b}$, and $\mathbf{a}_2 \cdot \mathbf{b} = 0$. Now

$$\mathbf{a}_2 = \mathbf{a} - \mathbf{a}_1 = \mathbf{a} - c\mathbf{b}, \qquad \mathbf{a}_2 \cdot \mathbf{b} = (\mathbf{a} - c\mathbf{b}) \cdot \mathbf{b} = \mathbf{a} \cdot \mathbf{b} - c|\mathbf{b}|^2 = 0$$

and $c = \dfrac{\mathbf{a} \cdot \mathbf{b}}{|\mathbf{b}|^2}$. Thus, $\mathbf{a}_1 = \dfrac{\mathbf{a} \cdot \mathbf{b}}{|\mathbf{b}|^2}\mathbf{b}$ and $\mathbf{a}_2 = \mathbf{a} - c\mathbf{b} = \mathbf{a} - \dfrac{\mathbf{a} \cdot \mathbf{b}}{|\mathbf{b}|^2}\mathbf{b}$.

The scalar $\mathbf{a} \cdot \dfrac{\mathbf{b}}{|\mathbf{b}|}$ is the scalar projection of $\mathbf{a}$ on $\mathbf{b}$; the vector $\left(\mathbf{a} \cdot \dfrac{\mathbf{b}}{|\mathbf{b}|}\right)\dfrac{\mathbf{b}}{|\mathbf{b}|}$ is the vector projection of $\mathbf{a}$ on $\mathbf{b}$.

8. Resolve $\mathbf{a} = 4\mathbf{i} + 3\mathbf{j}$ into components $\mathbf{a}_1$ and $\mathbf{a}_2$ parallel and perpendicular to $\mathbf{b} = 3\mathbf{i} + \mathbf{j}$.

From Prob. 7, $c = \dfrac{\mathbf{a} \cdot \mathbf{b}}{|\mathbf{b}|^2} = \dfrac{12 + 3}{10} = \dfrac{3}{2}$. Then $\mathbf{a}_1 = c\mathbf{b} = \tfrac{9}{2}\mathbf{i} + \tfrac{3}{2}\mathbf{j}$ and $\mathbf{a}_2 = \mathbf{a} - \mathbf{a}_1 = -\tfrac{1}{2}\mathbf{i} + \tfrac{3}{2}\mathbf{j}$.

9. Find the work done in moving an object along a vector $\mathbf{a} = 3\mathbf{i} + 4\mathbf{j}$ if the force applied is $\mathbf{b} = 2\mathbf{i} + \mathbf{j}$.

Work done $=$ (magnitude of $\mathbf{b}$ in the direction of $\mathbf{a}$)(distance moved)
$$= (|\mathbf{b}| \cos \theta)\,|\mathbf{a}| = \mathbf{b} \cdot \mathbf{a} = (2\mathbf{i} + \mathbf{j}) \cdot (3\mathbf{i} + 4\mathbf{j}) = 10$$

10. If $\mathbf{a} = \mathbf{i}\,f_1(u) + \mathbf{j}\,f_2(u)$ and $\mathbf{b} = \mathbf{i}\,g_1(u) + \mathbf{j}\,g_2(u)$, show that $\dfrac{d}{du}(\mathbf{a} \cdot \mathbf{b}) = \dfrac{d\mathbf{a}}{du} \cdot \mathbf{b} + \mathbf{a} \cdot \dfrac{d\mathbf{b}}{du}$.

$$\mathbf{a} \cdot \mathbf{b} = (\mathbf{i}f_1 + \mathbf{j}f_2) \cdot (\mathbf{i}g_1 + \mathbf{j}g_2) = f_1 g_1 + f_2 g_2,$$

$$\frac{d}{du}(\mathbf{a} \cdot \mathbf{b}) = f_1' g_1 + f_1 g_1' + f_2' g_2 + f_2 g_2' \qquad \left(f_1' = \frac{df_1(u)}{du}\right)$$

$$= (f_1' g_1 + f_2' g_2) + (f_1 g_1' + f_2 g_2')$$

$$= (\mathbf{i}f_1' + \mathbf{j}f_2') \cdot (\mathbf{i}g_1 + \mathbf{j}g_2) + (\mathbf{i}f_1 + \mathbf{j}f_2) \cdot (\mathbf{i}g_1' + \mathbf{j}g_2') = \frac{d\mathbf{a}}{du} \cdot \mathbf{b} + \mathbf{a} \cdot \frac{d\mathbf{b}}{du}$$

11. If $\mathbf{a} = \mathbf{i}\,f_1(u) + \mathbf{j}\,f_2(u)$ is of constant magnitude, show that $\mathbf{a}$ and $d\mathbf{a}/du$ are perpendicular.

From $\mathbf{a} \cdot \mathbf{a} = $ constant $\neq 0$, we obtain by Problem 10, $\dfrac{d}{du}(\mathbf{a} \cdot \mathbf{a}) = \dfrac{d\mathbf{a}}{du} \cdot \mathbf{a} + \mathbf{a} \cdot \dfrac{d\mathbf{a}}{du} = 2\mathbf{a} \cdot \dfrac{d\mathbf{a}}{du} = 0$.

Then $\mathbf{a} \cdot \dfrac{d\mathbf{a}}{du} = 0$ so that $\mathbf{a}$ and $\dfrac{d\mathbf{a}}{du}$ are perpendicular.

Thus, the tangent to a circle at one of its points $P$ is perpendicular to the radius drawn to $P$.

12. Given $\mathbf{r} = \mathbf{i} \cos^2 \theta + \mathbf{j} \sin^2 \theta$, find $\mathbf{t}$.

$$d\mathbf{r}/d\theta = -\mathbf{i} \sin 2\theta + \mathbf{j} \sin 2\theta$$

$$\frac{ds}{d\theta} = \left| \frac{d\mathbf{r}}{d\theta} \right| = \sqrt{\frac{d\mathbf{r}}{d\theta} \cdot \frac{d\mathbf{r}}{d\theta}} = \sqrt{2} \sin 2\theta \quad \text{and} \quad \mathbf{t} = \frac{d\mathbf{r}}{ds} = \frac{d\mathbf{r}}{d\theta} \cdot \frac{d\theta}{ds} = -\frac{1}{\sqrt{2}}\mathbf{i} + \frac{1}{\sqrt{2}}\mathbf{j}$$

13. Given $x = a \cos^3 \theta$, $y = a \sin^3 \theta$, find $\mathbf{t}$ and $\mathbf{n}$ when $\theta = \frac{1}{4}\pi$.

$$\mathbf{r} = a\mathbf{i} \cos^3 \theta + a\mathbf{j} \sin^3 \theta$$

$$d\mathbf{r}/d\theta = -3a\mathbf{i} \cos^2 \theta \sin \theta + 3a\mathbf{j} \sin^2 \theta \cos \theta$$

$$ds/d\theta = |d\mathbf{r}/d\theta| = 3a \sin \theta \cos \theta$$

$$\mathbf{t} = \frac{d\mathbf{r}}{ds} = \frac{d\mathbf{r}}{d\theta}\frac{d\theta}{ds} = -\mathbf{i} \cos \theta + \mathbf{j} \sin \theta$$

$$\frac{d\mathbf{t}}{ds} = (\mathbf{i} \sin \theta + \mathbf{j} \cos \theta)\frac{d\theta}{ds} = \frac{1}{3a \cos \theta} \mathbf{i} + \frac{1}{3a \sin \theta} \mathbf{j}$$

At $\theta = \frac{1}{4}\pi$:   $\mathbf{t} = -\frac{1}{\sqrt{2}}\mathbf{i} + \frac{1}{\sqrt{2}}\mathbf{j}$,   $\frac{d\mathbf{t}}{ds} = \frac{\sqrt{2}}{3a}\mathbf{i} + \frac{\sqrt{2}}{3a}\mathbf{j}$,   $|K| = \left|\frac{d\mathbf{t}}{ds}\right| = \frac{2}{3a}$,   $\mathbf{n} = \frac{1}{|K|}\frac{d\mathbf{t}}{ds} = \frac{1}{\sqrt{2}}\mathbf{i} + \frac{1}{\sqrt{2}}\mathbf{j}$.

14. Show that the vector $\mathbf{a} = a\mathbf{i} + b\mathbf{j}$ is perpendicular to the line $ax + by + c = 0$.

Let $P_1(x_1, y_1)$ and $P_2(x_2, y_2)$ be two distinct points on the line. Then $ax_1 + by_1 + c = 0$, $ax_2 + by_2 + c = 0$, and subtracting the first from the second,

$$a(x_2 - x_1) + b(y_2 - y_1) = 0$$

Now

$$a(x_2 - x_1) + b(y_2 - y_1) = (a\mathbf{i} + b\mathbf{j}) \cdot [(x_2 - x_1)\mathbf{i} + (y_2 - y_1)\mathbf{j}]$$
$$= \mathbf{a} \cdot \mathbf{P_1P_2} = 0$$

Thus, $\mathbf{a}$ is perpendicular (normal) to the line.

15. Use vector methods to find

    (a) the equation of the line through $P_1(2, 3)$ and perpendicular to the line $x + 2y + 5 = 0$;

    (b) the equation of the line through $P_1(2, 3)$ and $P_2(5, -1)$.

        Take $P(x, y)$ any other point on the required line.

    (a) By Problem 14, the vector $\mathbf{a} = \mathbf{i} + 2\mathbf{j}$ is normal to $x + 2y + 5 = 0$. Then $\mathbf{P_1P} = (x - 2)\mathbf{i} + (y - 3)\mathbf{j}$ is parallel to $\mathbf{a}$, that is,

$$(x - 2)\mathbf{i} + (y - 3)\mathbf{j} = k(\mathbf{i} + 2\mathbf{j}) \quad (k, \text{ a scalar variable})$$

Equating components, we have $x - 2 = k$, $y - 3 = 2k$. Eliminating $k$, the required equation is $y - 3 = 2(x - 2)$ or $2x - y - 1 = 0$.

    (b) We have          $\mathbf{P_1P} = (x - 2)\mathbf{i} + (y - 3)\mathbf{j}$   and   $\mathbf{P_1P_2} = 3\mathbf{i} - 4\mathbf{j}$

Now $\mathbf{a} = 4\mathbf{i} + 3\mathbf{j}$ is perpendicular to $\mathbf{P_1P_2}$ and, hence, to $\mathbf{P_1P}$. The required equation is

$$\mathbf{a} \cdot \mathbf{P_1P} = (4\mathbf{i} + 3\mathbf{j}) \cdot [(x - 2)\mathbf{i} + (y - 3)\mathbf{j}] = 0 \quad \text{or} \quad 4x + 3y - 17 = 0$$

16. Use vector methods to find the distance of the point $P_1(2, 3)$ from the line $3x + 4y - 12 = 0$.

At any convenient point, say $A(4, 0)$, on the line construct the vector $\mathbf{a} = 3\mathbf{i} + 4\mathbf{j}$ perpendicular to the line. The required distance is

$$d = |\mathbf{AP_1}| \cos \theta$$

Now $\mathbf{a} \cdot \mathbf{AP_1} = |\mathbf{a}| |\mathbf{AP_1}| \cos \theta = |\mathbf{a}| d$; hence

$$d = \frac{\mathbf{a} \cdot \mathbf{AP_1}}{|\mathbf{a}|} = \frac{(3\mathbf{i} + 4\mathbf{j}) \cdot (-2\mathbf{i} + 3\mathbf{j})}{5} = \frac{-6 + 12}{5} = \frac{6}{5}$$

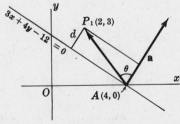

Fig. 18-14

# Supplementary Problems

**17.** Given the vectors $\mathbf{a}, \mathbf{b}, \mathbf{c}$ (see Fig. 18-15), construct
   (*a*) $2\mathbf{a}$           (*d*) $\mathbf{a} + \mathbf{b} - \mathbf{c}$
   (*b*) $-3\mathbf{b}$         (*e*) $\mathbf{a} - 2\mathbf{b} + 3\mathbf{c}$
   (*c*) $\mathbf{a} + 2\mathbf{b}$

                                                                **Fig. 18-15**

**18.** Prove: The line joining the midpoints of two sides of a triangle is parallel to and one-half the length of the third side. See Fig. 18-16.

**19.** If $\mathbf{a}, \mathbf{b}, \mathbf{c}, \mathbf{d}$ are consecutive sides of a quadrilateral, show that $\mathbf{a} + \mathbf{b} + \mathbf{c} + \mathbf{d} = 0$. See Fig. 18-17.
*Hint:* Let $P$ and $Q$ be two non-consecutive vertices. Express $\mathbf{PQ}$ in two ways.

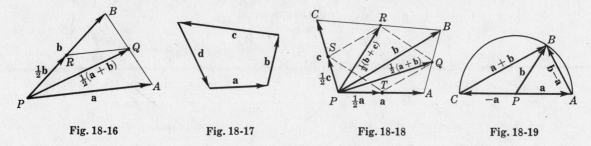

     **Fig. 18-16**            **Fig. 18-17**              **Fig. 18-18**            **Fig. 18-19**

**20.** Prove: If the midpoints of the consecutive sides of any quadrilateral are joined, the resulting quadrilateral is a parallelogram. See Fig. 18-18.

**21.** Using the figure in which $|\mathbf{a}| = |\mathbf{b}|$ is the radius of a circle, prove that the angle inscribed in a semicircle is a right angle. See Fig. 18-19.

**22.** Find the length of each vector and the angle which it makes with the positive $x$-axis:
   (*a*) $\mathbf{i} + \mathbf{j}$,  (*b*) $-\mathbf{i} + \mathbf{j}$,  (*c*) $\mathbf{i} + \sqrt{3}\,\mathbf{j}$,  (*d*) $\mathbf{i} - \sqrt{3}\,\mathbf{j}$.
   *Ans.* (*a*) $\sqrt{2}$; $\theta = \frac{1}{4}\pi$,  (*b*) $\sqrt{2}$; $\theta = 3\pi/4$,  (*c*) $2$; $\theta = \pi/3$,  (*d*) $2$; $\theta = 5\pi/3$

**23.** Prove: If $\mathbf{u}$ is obtained by rotating the unit vector $\mathbf{i}$ counterclockwise about the origin through the angle $\theta$, then $\mathbf{u} = \mathbf{i}\cos\theta + \mathbf{j}\sin\theta$.

**24.** Use the law of cosines for triangles to obtain $\mathbf{a} \cdot \mathbf{b} = |\mathbf{a}|\,|\mathbf{b}|\cos\theta = \frac{1}{2}\{|\mathbf{a}|^2 + |\mathbf{b}|^2 - |\mathbf{c}|^2\}$.

**25.** Write each of the following vectors in the form $a\mathbf{i} + b\mathbf{j}$:
   (*a*) joining the origin to $P(2, -3)$;  (*b*) joining $P_1(2, 3)$ to $P_2(4, 2)$;  (*c*) joining $P_2(4, 2)$ to $P_1(2, 3)$;  (*d*) unit vector in the direction of $3\mathbf{i} + 4\mathbf{j}$;  (*e*) having magnitude 6 and direction $120°$.
   *Ans.* (*a*) $2\mathbf{i} - 3\mathbf{j}$,  (*b*) $2\mathbf{i} - \mathbf{j}$,  (*c*) $-2\mathbf{i} + \mathbf{j}$,  (*d*) $\frac{3}{5}\mathbf{i} + \frac{4}{5}\mathbf{j}$,  (*e*) $-3\mathbf{i} + 3\sqrt{3}\,\mathbf{j}$

**26.** Using vector methods, derive the formula for the distance between $P_1(x_1, y_1)$ and $P_2(x_2, y_2)$.

**27.** Given $O(0, 0)$, $A(3, 1)$ and $B(1, 5)$ as vertices of the parallelogram $OAPB$, find the coordinates of $P$.

**28.** (*a*) Find $k$ so that $\mathbf{a} = 3\mathbf{i} - 2\mathbf{j}$ and $\mathbf{b} = \mathbf{i} + k\mathbf{j}$ are perpendicular.
   (*b*) Write a vector perpendicular to $\mathbf{a} = 2\mathbf{i} + 5\mathbf{j}$.

**29.** Prove the properties 8-15 of the dot product.

**30.** Find the vector projection and the scalar projection of $\mathbf{b}$ on $\mathbf{a}$, given:  (*a*) $\mathbf{a} = \mathbf{i} - 2\mathbf{j}$,  $\mathbf{b} = -3\mathbf{i} + \mathbf{j}$;
   (*b*) $\mathbf{a} = 2\mathbf{i} + 3\mathbf{j}$, $\mathbf{b} = 10\mathbf{i} + 2\mathbf{j}$.    *Ans.* (*a*) $-\mathbf{i} + 2\mathbf{j}$, $-\sqrt{5}$   (*b*) $4\mathbf{i} + 6\mathbf{j}$, $2\sqrt{13}$

**31.** Prove: Three vectors $\mathbf{a}, \mathbf{b}, \mathbf{c}$ will after parallel displacement form a triangle provided (*a*) some one is the sum of the other two or (*b*) $\mathbf{a} + \mathbf{b} + \mathbf{c} = 0$.

**32.** Show that $\mathbf{a} = 3\mathbf{i} - 6\mathbf{j}$, $\mathbf{b} = 4\mathbf{i} + 2\mathbf{j}$, $\mathbf{c} = -7\mathbf{i} + 4\mathbf{j}$ are the sides of a right triangle. Verify that the midpoint of the hypotenuse is equidistant from the vertices.

**33.** Find the unit tangent vector $\mathbf{t} = d\mathbf{r}/ds$, given:
   (*a*) $\mathbf{r} = 4\mathbf{i}\cos\theta + 4\mathbf{j}\sin\theta$;  (*b*) $\mathbf{r} = e^{\theta}\mathbf{i} + e^{-\theta}\mathbf{j}$;  (*c*) $\mathbf{r} = \theta\mathbf{i} + \theta^2\mathbf{j}$.

   *Ans.* (*a*) $-\mathbf{i}\sin\theta + \mathbf{j}\cos\theta$,  (*b*) $\dfrac{e^{\theta}\mathbf{i} - e^{-\theta}\mathbf{j}}{\sqrt{e^{2\theta} + e^{-2\theta}}}$,  (*c*) $\dfrac{\mathbf{i} + 2\theta\mathbf{j}}{\sqrt{1 + 4\theta^2}}$

**34.** (*a*) Find $\mathbf{n}$ for the curve of Problem 33(*a*).   (*b*) Find $\mathbf{n}$ for the curve of Problem 33(*c*).
   (*c*) Find $\mathbf{t}$ and $\mathbf{n}$ given $x = \cos\theta + \theta\sin\theta$, $y = \sin\theta - \theta\cos\theta$.

   *Ans.* (*a*) $-\mathbf{i}\cos\theta - \mathbf{j}\sin\theta$,  (*b*) $\dfrac{-2\theta}{\sqrt{1 + 4\theta^2}}\mathbf{i} + \dfrac{1}{\sqrt{1 + 4\theta^2}}\mathbf{j}$,  (*c*) $\mathbf{t} = \mathbf{i}\cos\theta + \mathbf{j}\sin\theta$, $\mathbf{n} = -\mathbf{i}\sin\theta + \mathbf{j}\cos\theta$

# Chapter 19

## Curvilinear Motion

**VELOCITY IN CURVILINEAR MOTION.**  Consider a point $P(x, y)$ moving along a curve
of equation
$$x = f(t), \ y = g(t)$$

where $t$ is time.  By differentiating the position
vector
$$\mathbf{r} = \mathbf{i}x + \mathbf{j}y \qquad (1)$$

with respect to $t$, we obtain the *velocity vector*

$$\mathbf{v} = \frac{d\mathbf{r}}{dt} = \mathbf{i}\frac{dx}{dt} + \mathbf{j}\frac{dy}{dt} = \mathbf{i}v_x + \mathbf{j}v_y \qquad (2)$$

where $v_x = dx/dt$ and $v_y = dy/dt$.

The magnitude of $\mathbf{v}$ is given by

$$|\mathbf{v}| = \sqrt{\mathbf{v} \cdot \mathbf{v}} = \sqrt{v_x^2 + v_y^2} = \frac{ds}{dt}$$

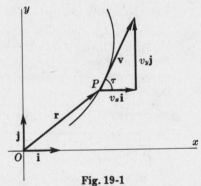

**Fig. 19-1**

The direction of $\mathbf{v}$ at $P$ is along the tangent to the path at $P$ as shown in Fig. 19-1.  If
$\tau$ denotes the direction of $\mathbf{v}$ (angle between $\mathbf{v}$ and the positive $x$-axis) then $\tan \tau = v_y/v_x$,
with the quadrant being determined by $v_x = |\mathbf{v}| \cos \tau$ and $v_y = |\mathbf{v}| \sin \tau$.

**ACCELERATION IN CURVILINEAR MOTION.**  Differentiating $(2)$ with respect to $t$,
we obtain the *acceleration vector*

$$\mathbf{a} = \frac{d\mathbf{v}}{dt} = \frac{d^2\mathbf{r}}{dt^2} = \mathbf{i}\frac{d^2x}{dt^2} + \mathbf{j}\frac{d^2y}{dt^2} = \mathbf{i}a_x + \mathbf{j}a_y \qquad (3)$$

where $a_x = d^2x/dt^2$ and $a_y = d^2y/dt^2$.

The magnitude of $\mathbf{a}$ is given by

$$|\mathbf{a}| = \sqrt{\mathbf{a} \cdot \mathbf{a}} = \sqrt{a_x^2 + a_y^2}$$

The direction $\phi$ of $\mathbf{a}$ is given by $\tan \phi = a_y/a_x$ with
the quadrant being determined by $a_x = |\mathbf{a}| \cos \phi$
and $a_y = |\mathbf{a}| \sin \phi$.  See Fig. 19-2.

In Problems 1-3, two solutions are offered.
One uses the position vector $(1)$, the velocity vector $(2)$, and the acceleration vector $(3)$.  This
solution requires a parametric representation of
the path.  The other and more popular solution
makes use only of the $x$ and $y$ components of these
vectors.  A parametric representation of the path
is not necessary. The two solutions are, of course,
basically the same.

See Problems 1-3.

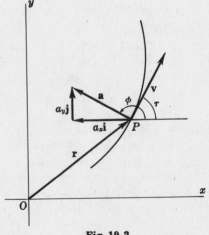

**Fig. 19-2**

94

**TANGENTIAL AND NORMAL COMPONENTS OF ACCELERATION.** By (6), Chapter 18, we have

$$\mathbf{v} = \frac{d\mathbf{r}}{dt} = \frac{d\mathbf{r}}{ds}\frac{ds}{dt} = \mathbf{t}\frac{ds}{dt} \tag{4}$$

Then

$$\mathbf{a} = \frac{d\mathbf{v}}{dt} = \mathbf{t}\frac{d^2s}{dt^2} + \frac{d\mathbf{t}}{dt}\frac{ds}{dt} \tag{5}$$

$$= \mathbf{t}\frac{d^2s}{dt^2} + \frac{d\mathbf{t}}{ds}\left(\frac{ds}{dt}\right)^2 = \mathbf{t}\frac{d^2s}{dt^2} + |K|\,\mathbf{n}\left(\frac{ds}{dt}\right)^2$$

by (7), Chapter 18.

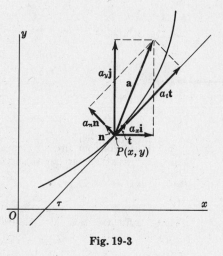

Now (5) gives the resolution of the acceleration vector at $P$ along the tangent and normal there. Denoting the components by $a_t$ and $a_n$ respectively, we have for their magnitudes

$$|a_t| = \left|\frac{d^2s}{dt^2}\right| \quad \text{and} \quad |a_n| = \frac{(ds/dt)^2}{R} = \frac{|\mathbf{v}|^2}{R}$$

where $R$ is the radius of curvature of the path at $P$. See Fig. 19-3.

Since $|\mathbf{a}|^2 = a_x^2 + a_y^2 = a_t^2 + a_n^2$, we have

$$a_n^2 = |\mathbf{a}|^2 - a_t^2$$

as a second procedure for determining $|a_n|$.

See Problems 4-8.

**Fig. 19-3**

# Solved Problems

1. Discuss the motion given by the equations $x = \cos 2\pi t$, $y = 3\sin 2\pi t$. Find the magnitude and direction of the velocity and acceleration vectors when (a) $t = 1/6$ and (b) $t = 2/3$.

    The motion is along the ellipse $9x^2 + y^2 = 9$. Beginning $(t = 0)$ at $(1, 0)$, the moving point traverses the curve counterclockwise.

    *First Solution.*

$$\mathbf{r} = \mathbf{i}x + \mathbf{j}y = \mathbf{i}\cos 2\pi t + 3\mathbf{j}\sin 2\pi t$$
$$\mathbf{v} = d\mathbf{r}/dt = \mathbf{i}v_x + \mathbf{j}v_y = -2\pi\mathbf{i}\sin 2\pi t + 6\pi\mathbf{j}\cos 2\pi t$$
$$\mathbf{a} = d\mathbf{v}/dt = \mathbf{i}a_x + \mathbf{j}a_y = -4\pi^2\mathbf{i}\cos 2\pi t - 12\pi^2\mathbf{j}\sin 2\pi t$$

    (a) At $t = 1/6$:

$$\mathbf{v} = -\sqrt{3}\pi\mathbf{i} + 3\pi\mathbf{j} \quad \text{and} \quad \mathbf{a} = -2\pi^2\mathbf{i} - 6\sqrt{3}\,\pi^2\mathbf{j}$$

$$|\mathbf{v}| = \sqrt{\mathbf{v}\cdot\mathbf{v}} = \sqrt{(-\sqrt{3}\pi)^2 + (3\pi)^2} = 2\sqrt{3}\pi$$

$$\tan\tau = v_y/v_x = -\sqrt{3}, \quad \cos\tau = v_x/|\mathbf{v}| = -1/2, \quad \text{and} \quad \tau = 120°$$

$$|\mathbf{a}| = \sqrt{\mathbf{a}\cdot\mathbf{a}} = \sqrt{(-2\pi^2)^2 + (-6\sqrt{3}\,\pi^2)^2} = 4\sqrt{7}\,\pi^2$$

$$\tan\phi = a_y/a_x = 3\sqrt{3}, \quad \cos\phi = a_x/|\mathbf{a}| = -1/2\sqrt{7}, \quad \text{and} \quad \phi = 259°6'$$

    (b) At $t = 2/3$:

$$\mathbf{v} = \sqrt{3}\,\pi\mathbf{i} - 3\pi\mathbf{j} \quad \text{and} \quad \mathbf{a} = 2\pi^2\mathbf{i} + 6\sqrt{3}\,\pi^2\mathbf{j}$$

$$|\mathbf{v}| = 2\sqrt{3}\,\pi; \quad \tan\tau = -\sqrt{3}, \quad \cos\tau = 1/2, \quad \text{and} \quad \tau = 5\pi/3$$

$$|\mathbf{a}| = 4\sqrt{7}\,\pi^2; \quad \tan\phi = 3\sqrt{3}, \quad \cos\phi = 1/2\sqrt{7}, \quad \text{and} \quad \phi = 79°6'$$

*Second Solution.*

$$x = \cos 2\pi t, \quad v_x = dx/dt = -2\pi \sin 2\pi t, \quad a_x = d^2x/dt^2 = -4\pi^2 \cos 2\pi t$$
$$y = 3 \sin 2\pi t, \quad v_y = dy/dt = 6\pi \cos 2\pi t, \quad a_y = d^2y/dt^2 = -12\pi^2 \sin 2\pi t$$

(a)  At  $t = 1/6$:

$$v_x = -\sqrt{3}\,\pi, \quad v_y = 3\pi, \quad |\mathbf{v}| = \sqrt{v_x^2 + v_y^2} = 2\sqrt{3}\,\pi$$
$$\tan \tau = v_y/v_x = -\sqrt{3}, \quad \cos \tau = v_x/|\mathbf{v}| = -1/2, \quad \text{and} \quad \tau = 120°$$
$$a_x = -2\pi^2, \quad a_y = -6\sqrt{3}\,\pi^2, \quad |\mathbf{a}| = \sqrt{a_x^2 + a_y^2} = 4\sqrt{7}\,\pi^2$$
$$\tan \phi = a_y/a_x = 3\sqrt{3}, \quad \cos \phi = a_x/|\mathbf{a}| = -1/2\sqrt{7}, \quad \text{and} \quad \phi = 259°6'$$

(b)  At  $t = 2/3$:

$$v_x = \sqrt{3}\,\pi, \quad v_y = -3\pi, \quad |\mathbf{v}| = 2\sqrt{3}\,\pi$$
$$\tan \tau = -\sqrt{3}, \quad \cos \tau = \tfrac{1}{2}, \quad \text{and} \quad \tau = 5\pi/3$$
$$a_x = 2\pi^2, \quad a_y = 6\sqrt{3}\,\pi^2, \quad |\mathbf{a}| = 4\sqrt{7}\,\pi^2$$
$$\tan \phi = 3\sqrt{3}, \quad \cos \phi = 1/2\sqrt{7}, \quad \text{and} \quad \phi = 79°6'$$

2.  A point travels counterclockwise about the circle  $x^2 + y^2 = 625$  at the rate  $|\mathbf{v}| = 15$.  Find  $\tau$,  $|\mathbf{a}|$, and  $\phi$  at  (a) the point  $(20, 15)$  and  (b) the point  $(5, -10\sqrt{6})$.   Refer to Fig. 19-4.

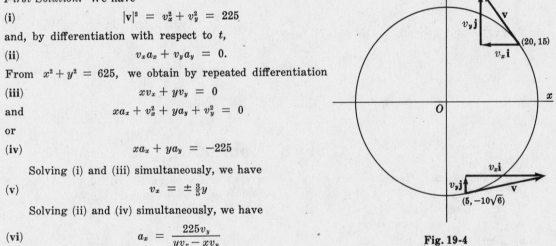

*First Solution.*  We have

(i)                          $|\mathbf{v}|^2 = v_x^2 + v_y^2 = 225$

and, by differentiation with respect to $t$,

(ii)                         $v_x a_x + v_y a_y = 0$.

From  $x^2 + y^2 = 625$,  we obtain by repeated differentiation

(iii)                        $x v_x + y v_y = 0$

and                          $x a_x + v_x^2 + y a_y + v_y^2 = 0$

or

(iv)                         $x a_x + y a_y = -225$

Solving (i) and (iii) simultaneously, we have

(v)                          $v_x = \pm \tfrac{3}{5} y$

Solving (ii) and (iv) simultaneously, we have

(vi)                         $a_x = \dfrac{225 v_y}{y v_x - x v_y}$

**Fig. 19-4**

(a)  From Fig. 19-4, $v_x < 0$ at $(20, 15)$. From (v), $v_x = -9$; from (iii), $v_y = 12$. Then  $\tan \tau = -4/3$, $\cos \tau = -3/5$, and $\tau = 126°52'$. From (vi), $a_x = -36/5$; from (iv), $a_y = -27/5$; and $|\mathbf{a}| = 9$. Then $\tan \phi = 3/4$, $\cos \phi = -4/5$, and $\phi = 216°52'$.

(b)  From the figure, $v_x > 0$ at $(5, -10\sqrt{6})$. From (v), $v_x = 6\sqrt{6}$; from (iii), $v_y = 3$. Then $\tan \tau = \sqrt{6}/12$, $\sin \tau = 1/5$, and $\tau = 11°32'$. From (vi), $a_x = -9/5$; from (iv), $a_y = 18\sqrt{6}/5$; and $|\mathbf{a}| = 9$. Then $\tan \phi = -2\sqrt{6}$, $\cos \phi = -1/5$, and $\phi = 101°32'$.

*Second Solution.*

Using the parametric equations  $x = 25 \cos \theta, \; y = 25 \sin \theta$,  we have at $P(x, y)$

$$\mathbf{r} = 25\mathbf{i} \cos \theta + 25\mathbf{j} \sin \phi$$

$$\mathbf{v} = \frac{d\mathbf{r}}{dt} = (-25\mathbf{i} \sin \theta + 25\mathbf{j} \cos \theta)\frac{d\theta}{dt} = -15\mathbf{i} \sin \theta + 15\mathbf{j} \cos \theta$$

$$\mathbf{a} = \frac{d\mathbf{v}}{dt} = (-15\mathbf{i} \cos \theta - 15\mathbf{j} \sin \theta)\frac{d\theta}{dt} = -9\mathbf{i} \cos \theta - 9\mathbf{j} \sin \theta$$

since  $|\mathbf{v}| = 15$  is equivalent to a constant angular speed of  $d\theta/dt = 3/5$.

(a) At the point $(20, 15)$, $\sin \theta = 3/5$ and $\cos \theta = 4/5$.

$$\mathbf{v} = -9\mathbf{i} + 12\mathbf{j}; \quad \tan \tau = -4/3, \quad \cos \tau = -3/5, \quad \text{and} \quad \tau = 126°52'$$
$$\mathbf{a} = -\tfrac{36}{5}\mathbf{i} - \tfrac{27}{5}\mathbf{j}; \quad |\mathbf{a}| = 9; \quad \tan \phi = \tfrac{3}{4}, \quad \cos \phi = -\tfrac{4}{5}, \quad \text{and} \quad \phi = 216°52'.$$

(b) At the point $(5, -10\sqrt{6})$, $\sin \theta = -\tfrac{2}{5}\sqrt{6}$ and $\cos \theta = 1/5$.

$$\mathbf{v} = 6\sqrt{6}\mathbf{i} + 3\mathbf{j}; \quad \tan \tau = \sqrt{6}/12, \quad \cos \tau = \tfrac{2}{5}\sqrt{6}, \quad \text{and} \quad \tau = 11°32'$$
$$\mathbf{a} = -\tfrac{9}{5}\mathbf{i} + \tfrac{18}{5}\sqrt{6}\mathbf{j}; \quad |\mathbf{a}| = 9; \quad \tan \phi = -2\sqrt{6}, \quad \cos \phi = -1/5, \quad \text{and} \quad \phi = 101°32'$$

3. A particle moves on the first quadrant arc of $x^2 = 8y$ so that $v_y = 2$. Find $|\mathbf{v}|$, $\tau$, $|\mathbf{a}|$, and $\phi$ at the point $(4, 2)$.

*First Solution.*

Differentiating $x^2 = 8y$ twice with respect to $t$ and using $v_y = 2$, we have

$$2xv_x = 8v_y = 16 \quad \text{or} \quad xv_x = 8, \quad \text{and} \quad xa_x + v_x^2 = 0$$

At $(4, 2)$: $v_x = 8/x = 2$; $|\mathbf{v}| = 2\sqrt{2}$; $\tan \tau = 1$, $\cos \tau = \tfrac{1}{2}\sqrt{2}$, and $\tau = \tfrac{1}{4}\pi$.

$a_x = -1$; $a_y = 0$; $|\mathbf{a}| = 1$; $\tan \phi = 0$, $\cos \phi = -1$, and $\phi = \pi$.

*Second Solution.*

Using the parametric equations $x = 4\theta$, $y = 2\theta^2$, we have

$$\mathbf{r} = 4\mathbf{i}\theta + 2\mathbf{j}\theta^2$$

$$\mathbf{v} = 4\mathbf{i}\frac{d\theta}{dt} + 4\mathbf{j}\theta\frac{d\theta}{dt} = \frac{2}{\theta}\mathbf{i} + 2\mathbf{j}, \quad \text{since} \quad v_y = 4\theta\frac{d\theta}{dt} = 2 \quad \text{and} \quad \frac{d\theta}{dt} = \frac{1}{2\theta}. \quad \mathbf{a} = -\frac{1}{\theta^3}\mathbf{i}.$$

At the point $(4, 2)$, $\theta = 1$. Then

$$\mathbf{v} = 2\mathbf{i} + 2\mathbf{j}; \quad |\mathbf{v}| = 2\sqrt{2}; \quad \tan \tau = 1, \quad \cos \tau = \tfrac{1}{2}\sqrt{2}, \quad \text{and} \quad \tau = \tfrac{1}{4}\pi.$$
$$\mathbf{a} = -\mathbf{i}; \quad |\mathbf{a}| = 1; \quad \tan \phi = 0, \quad \cos \phi = -1, \quad \text{and} \quad \phi = \pi.$$

4. Find the magnitudes of the tangential and normal components of acceleration for the motion $x = e^t \cos t$, $y = e^t \sin t$ at any time $t$.

$$\mathbf{r} = \mathbf{i}x + \mathbf{j}y = \mathbf{i}e^t \cos t + \mathbf{j}e^t \sin t$$
$$\mathbf{v} = \mathbf{i}e^t (\cos t - \sin t) + \mathbf{j}e^t (\sin t + \cos t)$$
$$\mathbf{a} = -2\mathbf{i}e^t \sin t + 2\mathbf{j}e^t \cos t$$

Then $|\mathbf{a}| = 2e^t$; $ds/dt = |\mathbf{v}| = \sqrt{2}\,e^t$ and $|a_t| = |d^2s/dt^2| = \sqrt{2}\,e^t$; $|a_n| = \sqrt{|\mathbf{a}|^2 - a_t^2} = \sqrt{2}\,e^t$.

5. A particle moves from left to right along the parabola $y = x^2$ with constant speed 5. Find the magnitude of the tangential and normal components of acceleration at the point $(1, 1)$.

Since the speed is constant, $|a_t| = |d^2s/dt^2| = 0$. At $(1, 1)$, $y' = 2x = 2$ and $y'' = 2$.

The radius of curvature at $(1, 1)$ is $R = \dfrac{[1 + (y')^2]^{3/2}}{|y''|} = \dfrac{5\sqrt{5}}{2}$ and $|a_n| = \dfrac{|\mathbf{v}|^2}{R} = 2\sqrt{5}$.

6. The centrifugal force $F$ lb exerted by a particle of weight $W$ lb at a point in its path is $F = \dfrac{W}{g}|a_n|$.

Find the centrifugal force exerted by a particle, weighing 5 lb, at the ends of the major and minor axes as it traverses the elliptical path $x = 20 \cos t$, $y = 15 \sin t$, the measurements being in feet and seconds. Use $g = 32$ ft/sec².

$$\mathbf{r} = 20\mathbf{i} \cos t + 15\mathbf{j} \sin t$$
$$\mathbf{v} = -20\mathbf{i} \sin t + 15\mathbf{j} \cos t$$
$$\mathbf{a} = -20\mathbf{i} \cos t - 15\mathbf{j} \sin t$$

$$\frac{ds}{dt} = |\mathbf{v}| = \sqrt{400 \sin^2 t + 225 \cos^2 t}, \qquad \frac{d^2s}{dt^2} = \frac{175 \sin t \cos t}{\sqrt{400 \sin^2 t + 225 \cos^2 t}}$$

At the ends of the major axis ($t = 0$ or $t = \pi$):

$$|\mathbf{a}| = 20, \quad |a_t| = |d^2s/dt^2| = 0, \quad |a_n| = 20, \quad \text{and} \quad F = (5/32)(20) = 25/8 \text{ lb}$$

At the ends of the minor axis ($t = \pi/2$ or $t = 3\pi/2$):

$$|\mathbf{a}| = 15, \quad |a_t| = 0, \quad |a_n| = 15, \quad \text{and} \quad F = (5/32)(15) = 75/32 \text{ lb}$$

7.  Assuming the equations of motion of a projectile to be $x = v_0 t \cos\psi$, $y = v_0 t \sin\psi - \frac{1}{2}gt^2$, where $v_0$ is the initial velocity, $\psi$ is the angle of projection, $g = 32$ ft/sec$^2$, and $x$ and $y$ are measured in feet and $t$ in seconds, find:

(a) the equation of motion in rectangular coordinates;

(b) the range;   (c) the angle of projection for maximum range;

(d) the speed and direction of the projectile after 5 sec of flight if $v_0 = 500$ ft/sec and $\psi = 45°$.

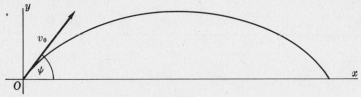

Fig. 19-5

(a) Solve the first of the equations for $t = \dfrac{x}{v_0 \cos\psi}$ and substitute in the second:

$$y = v_0\left(\frac{x}{v_0 \cos\psi}\right)\sin\psi - \frac{1}{2}g\left(\frac{x}{v_0 \cos\psi}\right)^2 = x\tan\psi - \frac{gx^2}{2v_0^2 \cos^2\psi}$$

(b) When $y = v_0 t \sin\psi - \frac{1}{2}gt^2 = 0$, $t = 0$ and $t = (2v_0 \sin\psi)/g$.

For $t = \dfrac{2v_0 \sin\psi}{g}$, the range $= x = v_0 \cos\psi\left(\dfrac{2v_0 \sin\psi}{g}\right) = \dfrac{v_0^2 \sin 2\psi}{g}$.

(c) For $x$, a maximum: $\dfrac{dx}{d\psi} = \dfrac{2v_0^2 \cos 2\psi}{g} = 0$, $\cos 2\psi = 0$, and $\psi = \frac{1}{4}\pi$.

(d) When $v_0 = 500$ and $\psi = \frac{1}{4}\pi$, $x = 250\sqrt{2}\,t$ and $y = 250\sqrt{2}\,t - 16t^2$. Then

$$v_x = 250\sqrt{2} \quad \text{and} \quad v_y = 250\sqrt{2} - 32t$$

When $t = 5$: $v_x = 250\sqrt{2}$ and $v_y = 250\sqrt{2} - 160$. Then

$$\tan\tau = v_y/v_x = 0.5475, \quad \tau = 28°42' \quad \text{and} \quad |\mathbf{v}| = \sqrt{v_x^2 + v_y^2} = 403 \text{ ft/sec}.$$

8.  A point $P$ moves on a circle $x = r\cos\beta$, $y = r\sin\beta$ with constant speed $v$. Show that, if the radius vector to $P$ moves with angular velocity $\omega$ and angular acceleration $\alpha$, (a) $v = r\omega$ and (b) $a = r\sqrt{\omega^4 + \alpha^2}$.

(a) $v_x = -r\sin\beta \cdot \dfrac{d\beta}{dt} = -r\sin\beta \cdot \omega$ and $v_y = r\cos\beta \cdot \dfrac{d\beta}{dt} = r\cos\beta \cdot \omega$.

$$v = \sqrt{v_x^2 + v_y^2} = \sqrt{(r^2 \sin^2\beta + r^2 \cos^2\beta)\omega^2} = r\omega$$

(b) $a_x = \dfrac{dv_x}{dt} = -r\omega\cos\beta \cdot \dfrac{d\beta}{dt} - r\sin\beta \cdot \dfrac{d\omega}{dt} = -r\omega^2 \cos\beta - r\alpha\sin\beta$.

$a_y = \dfrac{dv_y}{dt} = -r\omega\sin\beta \cdot \dfrac{d\beta}{dt} + r\cos\beta \cdot \dfrac{d\omega}{dt} = -r\omega^2 \sin\beta + r\alpha\cos\beta$.

$$a^2 = a_x^2 + a_y^2 = r^2(\omega^4 + \alpha^2) \quad \text{and} \quad a = r\sqrt{\omega^4 + \alpha^2}$$

# Supplementary Problems

9.  Find the magnitude and direction of velocity and acceleration, given

    (a)  $x = e^t$,  $y = e^{2t} - 4e^t + 3$  at $t = 0$

    (b)  $x = 2 - t$,  $y = 2t^3 - t$  at $t = 1$

    (c)  $x = \cos 3t$,  $y = \sin t$  at $t = \frac{1}{4}\pi$

    (d)  $x = e^t \cos t$,  $y = e^t \sin t$  at $t = 0$

    Ans.  (a)  $|\mathbf{v}| = \sqrt{5}$, $\tau = 296°34'$; $|\mathbf{a}| = 1$, $\phi = 0$

        (b)  $|\mathbf{v}| = \sqrt{26}$, $\tau = 101°19'$; $|\mathbf{a}| = 12$, $\phi = \frac{1}{2}\pi$

        (c)  $|\mathbf{v}| = \sqrt{5}$, $\tau = 161°34'$; $|\mathbf{a}| = \sqrt{41}$, $\phi = 353°40'$

        (d)  $|\mathbf{v}| = \sqrt{2}$, $\tau = \frac{1}{4}\pi$; $|\mathbf{a}| = 2$, $\phi = \frac{1}{2}\pi$

10. A particle moves on the first quadrant arc of the parabola $y^2 = 12x$ with $v_x = 15$. Find $v_y$, $|\mathbf{v}|$, $\tau$; $a_x$, $a_y$, $|\mathbf{a}|$, and $\phi$ at $(3, 6)$.
    Ans.  $v_y = 15$, $|\mathbf{v}| = 15\sqrt{2}$, $\tau = \frac{1}{4}\pi$; $a_x = 0$, $a_y = -75/2$, $|\mathbf{a}| = 75/2$, $\phi = 3\pi/2$.

11. A particle moves along the curve $y = x^3/3$ with $v_x = 2$ at all times. Find the magnitude and direction of the velocity and acceleration when $x = 3$.    Ans.  $|\mathbf{v}| = 2\sqrt{82}$, $\tau = 83°40'$; $|\mathbf{a}| = 24$, $\phi = \frac{1}{2}\pi$

12. A particle moves around a circle of radius 6 ft at the constant speed of 4 ft/sec. Determine the magnitude of its acceleration at any position.    Ans.  $|a_t| = 0$, $|\mathbf{a}| = |a_n| = 8/3$ ft/sec²

13. Find the magnitude and direction of the velocity and acceleration, and the magnitudes of the tangential and normal components of acceleration for the motion

    (a)  $x = 3t$,  $y = 9t - 3t^2$,  when $t = 2$

    (b)  $x = \cos t + t \sin t$,  $y = \sin t - t \cos t$  when $t = 1$.

    Ans.  (a)  $|\mathbf{v}| = 3\sqrt{2}$, $\tau = 7\pi/4$; $|\mathbf{a}| = 6$, $\phi = 3\pi/2$; $|a_t| = |a_n| = 3\sqrt{2}$

        (b)  $|\mathbf{v}| = 1$, $\tau = 1$; $|\mathbf{a}| = \sqrt{2}$, $\phi = 102°18'$; $|a_t| = |a_n| = 1$

14. A particle moves along the curve $y = \frac{1}{2}x^2 - \frac{1}{4}\ln x$ so that $x = \frac{1}{2}t^2$, $t > 0$. Find $v_x$, $v_y$, $|\mathbf{v}|$, $\tau$; $a_x$, $a_y$, $|\mathbf{a}|$, $\phi$; $|a_t|$ and $|a_n|$ when $t = 1$.
    Ans.  $v_x = 1$, $v_y = 0$, $|\mathbf{v}| = 1$, $\tau = 0$; $a_x = 1$, $a_y = 2$, $|\mathbf{a}| = \sqrt{5}$, $\phi = 63°26'$; $|a_t| = 1$, $|a_n| = 2$.

15. A particle moves along the path $y = 2x - x^2$ with $v_x = 4$ at all times. Find the magnitudes of the tangential and normal components of acceleration at the position  (a) $(1, 1)$ and  (b) $(2, 0)$.
    Ans.  (a)  $|a_t| = 0$, $|a_n| = 32$;  (b)  $|a_t| = 64/\sqrt{5}$, $|a_n| = 32/\sqrt{5}$

# Chapter 20

## Polar Coordinates

**THE POSITION OF A POINT** $P$ in a given plane, relative to a fixed point $O$ of the plane, may be described by giving the projections of the vector **OP** on two mutually perpendicular lines of the plane through $O$. This, in essence, is the rectangular coordinate system. Another description may be made by giving the directed distance $\rho = OP$ and the angle $\theta$ which $OP$ makes with a fixed half-line $OX$ through $O$. This is the *polar coordinate system*.

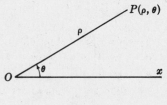

**Fig. 20-1**

To each number pair $(\rho, \theta)$ there corresponds one and only one point. The converse is not true; for example, the point $P$ in the figure may be described as $(\rho, \theta \pm 2n\pi)$ and $[-\rho, \theta \pm (2n+1)\pi]$, where $n$ is any positive integer including 0. In particular, the polar coordinates of the pole may be given as $(0, \theta)$ with $\theta$ perfectly arbitrary.

The curve, whose equation in polar coordinates is $\rho = f(\theta)$ or $F(\rho, \theta) = 0$, consists of the totality of distinct points $(\rho, \theta)$ which satisfy the equation.

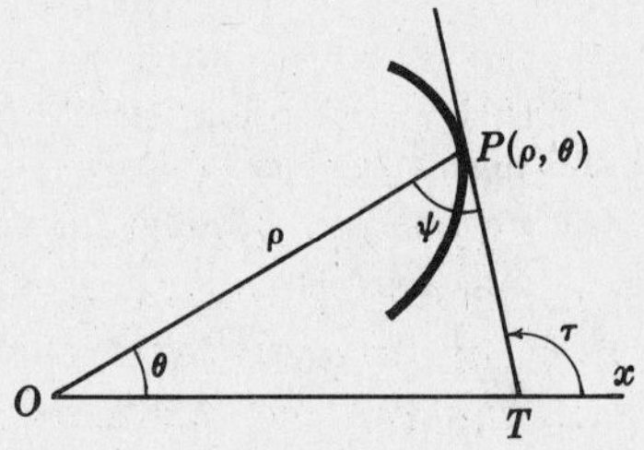

**THE ANGLE** $\psi$ from the radius vector $OP$ to the tangent $PT$ to a curve, at a point $P(\rho, \theta)$ on it, is given by

$$\tan \psi = \rho \frac{d\theta}{d\rho} = \frac{\rho}{\rho'}, \quad \text{where} \quad \rho' = \frac{d\rho}{d\theta}$$

Tan $\psi$ plays a role in polar coordinates somewhat similar to that of the slope of the tangent in rectangular coordinates.

See Problems 1-3.

**Fig. 20-2**

**THE ANGLE OF INCLINATION** $\tau$ of the tangent to a curve at a point $P(\rho, \theta)$ on it is given by

$$\tan \tau = \frac{\rho \cos \theta + \rho' \sin \theta}{-\rho \sin \theta + \rho' \cos \theta}$$

See Problems 4-10.

**THE POINTS OF INTERSECTION** of two curves whose equations are

$$\rho = f_1(\theta) \quad \text{and} \quad \rho = f_2(\theta)$$

may frequently be found by solving

$$f_1(\theta) = f_2(\theta) \tag{1}$$

**Example 1:**

Find the points of intersection of $\rho = 1 + \sin \theta$ and $\rho = 5 - 3 \sin \theta$.

Setting $1 + \sin \theta = 5 - 3 \sin \theta$, we have $\sin \theta = 1$.

Then $\theta = \frac{1}{2}\pi$ and $(2, \frac{1}{2}\pi)$ is the only point of intersection.

When the pole is a point of intersection, it may not appear among the solutions of (1). The pole is a point of intersection provided there are values of $\theta$, say $\theta_1$ and $\theta_2$, such that $f_1(\theta_1) = 0$ and $f_2(\theta_2) = 0$.

**Example 2:**

Find the points of intersection of $\rho = \sin\theta$ and $\rho = \cos\theta$.

From $$\sin\theta = \cos\theta \qquad (1)$$

we obtain the point of intersection $(\frac{1}{2}\sqrt{2}, \frac{1}{4}\pi)$. Now the curves are circles passing through the pole. The pole is not obtained as a point of intersection from (1), since on $\rho = \sin\theta$ it has coordinate $(0, 0)$ while on $\rho = \cos\theta$ it has coordinate $(0, \frac{1}{2}\pi)$.

**Example 3:**

Find the points of intersection of $\rho = \cos 2\theta$ and $\rho = \cos\theta$.

Setting $\cos 2\theta = 2\cos^2\theta - 1 = \cos\theta$, we find $(\cos\theta - 1)(2\cos\theta + 1) = 0$.

Then $\theta = 0, 2\pi/3, 4\pi/3$; and we have as points of intersection $(1, 0)$, $(-\frac{1}{2}, 2\pi/3)$, $(-\frac{1}{2}, 4\pi/3)$. The pole is also a point of intersection.

**THE ANGLE OF INTERSECTION** $\phi$ of two curves at a common point $P(\rho, \theta)$, not the pole, is given by

$$\tan\phi = \frac{\tan\psi_1 - \tan\psi_2}{1 + \tan\psi_1 \tan\psi_2}$$

where $\psi_1$ and $\psi_2$ are the angles from the radius vector $OP$ to the respective tangents to the curves at $P$.

The procedure here is similar to that in the case of curves given in rectangular coordinates, the use of the tangents of the angles from the radius vector to the tangent instead of the slopes of the tangents being a matter of convenience in computing.

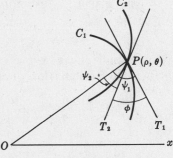

**Fig. 20-3**

**Example 4:**

Find the (acute) angles of intersection of $\rho = \cos\theta$ and $\rho = \cos 2\theta$.

The points of intersection were found in Example 3.

At the pole: On $\rho = \cos\theta$, the pole is given by $\theta = \frac{1}{2}\pi$; on $\rho = \cos 2\theta$, the pole is given by $\theta = \pi/4$ and $3\pi/4$. Thus, at the pole there are two intersections, the acute angle being $\frac{1}{4}\pi$ for each.

For $\rho = \cos\theta$      For $\rho = \cos 2\theta$
$\tan\psi_1 = -\cot\theta$      $\tan\psi_2 = -\frac{1}{2}\cot 2\theta$

At the point $(1, 0)$: $\tan\psi_1 = -\cot 0 = \infty$ and $\tan\psi_2 = \infty$. Then $\psi_1 = \psi_2 = \frac{1}{2}\pi$ and $\phi = 0$.

At the point $(-\frac{1}{2}, 2\pi/3)$: $\tan\psi_1 = \sqrt{3}/3$ and $\tan\psi_2 = -\sqrt{3}/6$. $\tan\phi = \dfrac{\sqrt{3}/3 + \sqrt{3}/6}{1 - 1/6} = 3\sqrt{3}/5$ and the acute angle of intersection is $\phi = 46°6'$.

By symmetry, this is also the acute angle of intersection at the point $(-\frac{1}{2}, 4\pi/3)$.

See Problems 11-13.

**THE DERIVATIVE OF ARC LENGTH** is given by $\dfrac{ds}{d\theta} = \sqrt{\rho^2 + (\rho')^2}$, where $\rho' = \dfrac{d\rho}{d\theta}$

and with the understanding that $s$ increases as $\theta$ increases.

See Problems 14-16.

**THE CURVATURE** of a curve is given by $K = \dfrac{\rho^2 + 2(\rho')^2 - \rho\rho''}{\{\rho^2 + (\rho')^2\}^{3/2}}$.

See Problems 17-19.

**CURVILINEAR MOTION.** Suppose a particle $P$ moves along a curve whose equation is given in polar coordinates, $\rho = f(\theta)$. By means of a parametric representation

$$x = \rho \cos \theta = g(\theta), \qquad y = \rho \sin \theta = h(\theta)$$

of the curve, the position vector of $P$ is

$$\mathbf{r} = \mathbf{OP} = x\mathbf{i} + y\mathbf{j} = \rho\mathbf{i}\cos\theta + \rho\mathbf{j}\sin\theta = \rho(\mathbf{i}\cos\theta + \mathbf{j}\sin\theta)$$

and the motion may be studied as in Chapter 19.

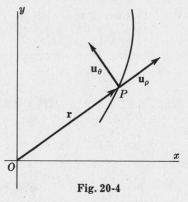

Fig. 20-4

An alternate procedure is to express $\mathbf{r}$ and, thus, $\mathbf{v}$ and $\mathbf{a}$ in terms of unit vectors along and perpendicular to the radius vector of $P$. For this purpose, we define the unit vector

$$\mathbf{u}_\rho = \mathbf{i}\cos\theta + \mathbf{j}\sin\theta$$

along $\mathbf{r}$ in the direction of increasing $\rho$ and the unit vector

$$\mathbf{u}_\theta = -\mathbf{i}\sin\theta + \mathbf{j}\cos\theta$$

perpendicular to $\mathbf{r}$ and in the direction of increasing $\theta$. An easy calculation yields

$$\frac{d\mathbf{u}_\rho}{dt} = \frac{d\mathbf{u}_\rho}{d\theta}\frac{d\theta}{dt} = \mathbf{u}_\theta\frac{d\theta}{dt} \qquad \text{and} \qquad \frac{d\mathbf{u}_\theta}{dt} = -\mathbf{u}_\rho\frac{d\theta}{dt}$$

From

$$\mathbf{r} = \rho\,\mathbf{u}_\rho$$

we obtain in Problem 20,

$$\mathbf{v} = \frac{d\mathbf{r}}{dt} = \mathbf{u}_\rho\frac{d\rho}{dt} + \rho\mathbf{u}_\theta\frac{d\theta}{dt} = v_\rho\mathbf{u}_\rho + v_\theta\mathbf{u}_\theta$$

and

$$\mathbf{a} = \frac{d\mathbf{v}}{dt} = \mathbf{u}_\rho\left[\frac{d^2\rho}{dt^2} - \rho\left(\frac{d\theta}{dt}\right)^2\right] + \mathbf{u}_\theta\left[\rho\frac{d^2\theta}{dt^2} + 2\frac{d\rho}{dt}\frac{d\theta}{dt}\right]$$

$$= a_\rho\mathbf{u}_\rho + a_\theta\mathbf{u}_\theta$$

Here $v_\rho = \dfrac{d\rho}{dt}$ and $v_\theta = \rho\dfrac{d\theta}{dt}$ are respectively the components of $\mathbf{v}$ along and perpendicular to the radius vector and $a_\rho = \dfrac{d^2\rho}{dt^2} - \rho\left(\dfrac{d\theta}{dt}\right)^2$ and $a_\theta = \rho\dfrac{d^2\theta}{dt^2} + 2\dfrac{d\rho}{dt}\dfrac{d\theta}{dt}$ are the corresponding components of $\mathbf{a}$.

See Problem 21.

# Solved Problems

1. Derive $\tan\psi = \rho\dfrac{d\theta}{d\rho}$, where $\psi$ is the angle measured from the radius vector $OP$ of a point $P(\rho, \theta)$ on the curve of equation $\rho = f(\theta)$ to the tangent $PT$.

In Fig. 20-5 below, $Q(\rho + \Delta\rho, \theta + \Delta\theta)$ is a point on the curve near $P$. From the right triangle $PSQ$,

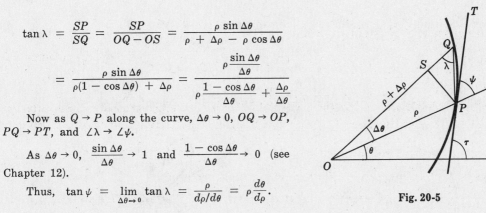

$$\tan \lambda = \frac{SP}{SQ} = \frac{SP}{OQ - OS} = \frac{\rho \sin \Delta\theta}{\rho + \Delta\rho - \rho \cos \Delta\theta}$$

$$= \frac{\rho \sin \Delta\theta}{\rho(1 - \cos \Delta\theta) + \Delta\rho} = \frac{\rho \dfrac{\sin \Delta\theta}{\Delta\theta}}{\rho \dfrac{1 - \cos \Delta\theta}{\Delta\theta} + \dfrac{\Delta\rho}{\Delta\theta}}$$

Now as $Q \to P$ along the curve, $\Delta\theta \to 0$, $OQ \to OP$, $PQ \to PT$, and $\angle\lambda \to \angle\psi$.

As $\Delta\theta \to 0$, $\dfrac{\sin \Delta\theta}{\Delta\theta} \to 1$ and $\dfrac{1 - \cos \Delta\theta}{\Delta\theta} \to 0$ (see Chapter 12).

Thus, $\tan \psi = \lim\limits_{\Delta\theta \to 0} \tan \lambda = \dfrac{\rho}{d\rho/d\theta} = \rho \dfrac{d\theta}{d\rho}$.

**Fig. 20-5**

In Problems 2-3, find $\tan \psi$ for the given curve at the given point.

2.   $\rho = 2 + \cos \theta$;  $\theta = \pi/3$.  See Fig. 20-6 below.

   At  $\theta = \pi/3$:  $\rho = 2 + \frac{1}{2} = 5/2$,  $\rho' = -\sin \theta = -\sqrt{3}/2$, and $\tan \psi = \rho/\rho' = -5/\sqrt{3}$.

| | |
|---|---|
| **Fig. 20-6** | **Fig. 20-7** |

3.   $\rho = 2 \sin 3\theta$;  $\theta = \pi/4$.  See Fig. 20-7 above.

   At  $\theta = \pi/4$:  $\rho = 2(1/\sqrt{2}) = \sqrt{2}$,   $\rho' = 6 \cos 3\theta = 6(-1/\sqrt{2}) = -3\sqrt{2}$, and  $\tan \psi = \rho/\rho' = -1/3$.

4.   Derive   $\tan \tau = \dfrac{\rho \cos \theta + \rho' \sin \theta}{-\rho \sin \theta + \rho' \cos \theta}$.

   From the figure of Problem 1,  $\tau = \psi + \theta$  and

$$\tan \tau = \tan(\psi + \theta) = \frac{\tan \psi + \tan \theta}{1 - \tan \psi \tan \theta} = \frac{\rho \dfrac{d\theta}{d\rho} + \dfrac{\sin \theta}{\cos \theta}}{1 - \rho \dfrac{d\theta}{d\rho} \dfrac{\sin \theta}{\cos \theta}}$$

$$= \frac{\rho \cos \theta + \dfrac{d\rho}{d\theta} \sin \theta}{\dfrac{d\rho}{d\theta} \cos \theta - \rho \sin \theta} = \frac{\rho \cos \theta + \rho' \sin \theta}{-\rho \sin \theta + \rho' \cos \theta}$$

5.   Show that, if $\rho = f(\theta)$ passes through the pole and $\theta_1$ is such that $f(\theta_1) = 0$, the direction of the tangent to the curve at the pole $(0, \theta_1)$ is $\theta_1$.

   At $(0, \theta_1)$:  $\rho = 0$ and $\rho' = f'(\theta_1)$.

   If $\rho' \neq 0$:  $\tan \tau = \dfrac{\rho \cos \theta + \rho' \sin \theta}{-\rho \sin \theta + \rho' \cos \theta}$

$$= \frac{0 + \sin \theta_1 \cdot f'(\theta_1)}{0 + \cos \theta_1 \cdot f'(\theta_1)} = \tan \theta_1$$

   If $\rho' = 0$:  $\tan \tau = \lim\limits_{\theta \to \theta_1} \dfrac{\sin \theta \cdot f'(\theta)}{\cos \theta \cdot f'(\theta)} = \tan \theta_1$

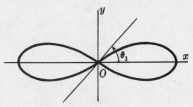

**Fig. 20-8**

In Problems 6-8, find the slope of the given curve at the given point.

**6.** $\rho = 1 - \cos\theta;\ \theta = \frac{1}{2}\pi.$  See Fig. 20-9 below.

At $\theta = \frac{1}{2}\pi$:  $\sin\theta = 1,\ \cos\theta = 0,\ \rho = 1,\ \rho' = \sin\theta = 1,$  and

$$\tan\tau = \frac{\rho\cos\theta + \rho'\sin\theta}{-\rho\sin\theta + \rho'\cos\theta} = \frac{1\cdot 0 + 1\cdot 1}{-1\cdot 1 + 1\cdot 0} = -1$$

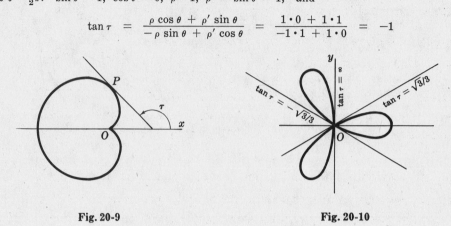

Fig. 20-9                      Fig. 20-10

**7.** $\rho = \cos 3\theta;$ pole.  See Fig. 20-10 above.

When $\rho = 0,\ \cos 3\theta = 0.$  Then $3\theta = \pi/2,\ 3\pi/2,\ 5\pi/2,$  and $\theta = \pi/6,\ \pi/2,\ 5\pi/6.$

By Problem 5, $\tan\tau = 1/\sqrt3,\ \infty,$  and $-1/\sqrt3.$

**8.** $\rho\theta = a;\ \theta = \pi/3.$

At $\theta = \pi/3$:  $\sin\theta = \sqrt3/2,\ \cos\theta = \frac{1}{2},\ \rho = 3a/\pi,$  and $\rho' = -a/\theta^2 = -9a/\pi^2.$

$$\tan\tau = \frac{\rho\cos\theta + \rho'\sin\theta}{-\rho\sin\theta + \rho'\cos\theta} = -\frac{\pi - 3\sqrt3}{\sqrt3\,\pi + 3}$$

**9.** Investigate $\rho = 1 + \sin\theta$ for horizontal and vertical tangents.

At $P(\rho,\theta)$:  $\tan\tau = \dfrac{(1 + \sin\theta)\cos\theta + \cos\theta\sin\theta}{-(1 + \sin\theta)\sin\theta + \cos^2\theta}$

$$= -\frac{\cos\theta\,(1 + 2\sin\theta)}{(\sin\theta + 1)(2\sin\theta - 1)}$$

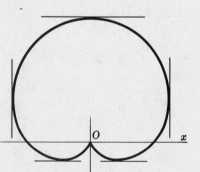

Fig. 20-11

(a)  Set $\cos\theta\,(1 + 2\sin\theta) = 0$ and solve:
$\theta = \pi/2,\ 3\pi/2,\ 7\pi/6,$ and $11\pi/6.$
Set $(\sin\theta + 1)(2\sin\theta - 1) = 0$ and solve:
$\theta = 3\pi/2,\ \pi/6,$ and $5\pi/6.$

(b)  For $\theta = \pi/2$:  There is a horizontal tangent at $(2, \pi/2).$
For $\theta = 7\pi/6$ and $11\pi/6$:  There are horizontal tangents at $(\frac{1}{2}, 7\pi/6)$ and $(\frac{1}{2}, 11\pi/6).$
For $\theta = \pi/6$ and $5\pi/6$:  There are vertical tangents at $(3/2, \pi/6)$ and $(3/2, 5\pi/6).$
For $\theta = 3\pi/2$:  By Problem 5, there is a vertical tangent at the pole.

**10.** Show that the angle which the radius vector to any point of the cardioid $\rho = a(1 - \cos\theta)$ makes with the curve is one-half that which the radius vector makes with the polar axis.

At any point $P(\rho,\theta)$ on the cardioid:

$$\rho' = a\sin\theta,\ \ \text{and}\ \ \tan\psi = \rho/\rho' = (1 - \cos\theta)/\sin\theta = \tan\tfrac{1}{2}\theta\ \ \text{or}\ \ \psi = \tfrac{1}{2}\theta$$

In Problems 11-13, find the angles of intersection of the pairs of curves.

11.  $\rho = 3 \cos \theta$, $\rho = 1 + \cos \theta$.

(a) Solve $\rho = 3 \cos \theta = 1 + \cos \theta$ for the points of intersection $(3/2, \pi/3)$ and $(3/2, 5\pi/3)$. The curves intersect also at the pole.

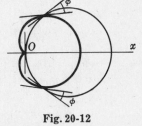

(b) For $\rho = 3 \cos \theta$:    $\rho' = -3 \sin \theta$ and $\tan \psi_1 = -\cot \theta$.

For $\rho = 1 + \cos \theta$: $\rho' = -\sin \theta$ and $\tan \psi_2 = -\dfrac{1 + \cos \theta}{\sin \theta}$.

**Fig. 20-12**

(c) At $\theta = \pi/3$: $\tan \psi_1 = -1/\sqrt{3}$, $\tan \psi_2 = -\sqrt{3}$, and $\tan \phi = 1/\sqrt{3}$.

The acute angle of intersection at $(3/2, \pi/3)$ and, by symmetry, at $(3/2, 5\pi/3)$ is $\pi/6$.

At the pole:  From a diagram or by using the result of Problem 5, the curves are orthogonal.

12.  $\rho = \sec^2 \frac{1}{2}\theta$, $\rho = 3 \csc^2 \frac{1}{2}\theta$.

(a) Solve $\rho = \sec^2 \frac{1}{2}\theta = 3 \csc^2 \frac{1}{2}\theta$ for the points of intersection $(4, 2\pi/3)$ and $(4, 4\pi/3)$.

(b) For $\rho = \sec^2 \frac{1}{2}\theta$:    $\rho' = \sec^2 \frac{1}{2}\theta \tan \frac{1}{2}\theta$ and $\tan \psi_1 = \cot \frac{1}{2}\theta$.
For $\rho = 3 \csc^2 \frac{1}{2}\theta$: $\rho' = -3 \csc^2 \frac{1}{2}\theta \cot \frac{1}{2}\theta$ and $\tan \psi_2 = -\tan \frac{1}{2}\theta$.

(c) At $\theta = 2\pi/3$: $\tan \psi_1 = 1/\sqrt{3}$, $\tan \psi_2 = -\sqrt{3}$, and $\phi = \frac{1}{2}\pi$. The curves are orthogonal.

13.  $\rho = \sin 2\theta$, $\rho = \cos \theta$.

(a) The curves intersect at the points $(\sqrt{3}/2, \pi/6)$, $(-\sqrt{3}/2, 5\pi/6)$, and the pole.

(b) For $\rho = \sin 2\theta$:    $\rho' = 2 \cos 2\theta$ and $\tan \psi_1 = \frac{1}{2} \tan 2\theta$.
For $\rho = \cos \theta$:    $\rho' = -\sin \theta$ and $\tan \psi_2 = -\cot \theta$.

(c) At $\theta = \pi/6$: $\tan \psi_1 = \sqrt{3}/2$, $\tan \psi_2 = -\sqrt{3}$, and $\tan \phi = -3\sqrt{3}$. The acute angle of intersection at the points $(\sqrt{3}/2, \pi/6)$ and $(-\sqrt{3}/2, 5\pi/6)$ is  $\phi = \arctan 3\sqrt{3} = 79°6'$.

At the pole the angles of intersection are $0°$ and $\frac{1}{2}\pi$.

In Problems 14-16, find $ds/d\theta$ at the point $P(\rho, \theta)$.

14.  $\rho = \cos 2\theta$.

$\rho' = -2 \sin 2\theta$ and $ds/d\theta = \sqrt{\rho^2 + (\rho')^2} = \sqrt{\cos^2 2\theta + 4 \sin^2 2\theta} = \sqrt{1 + 3 \sin^2 2\theta}$.

15.  $\rho(1 + \cos \theta) = 4$.

$-\rho \sin \theta + \rho'(1 + \cos \theta) = 0$. Then  $\rho' = \dfrac{\rho \sin \theta}{1 + \cos \theta} = \dfrac{4 \sin \theta}{(1 + \cos \theta)^2}$  and

$$\frac{ds}{d\theta} = \sqrt{\rho^2 + (\rho')^2} = \frac{4\sqrt{2}}{(1 + \cos \theta)^{3/2}}$$

16.  $\rho = \sin^3 \frac{1}{3}\theta$. Evaluate $ds/d\theta$ at $\theta = \frac{1}{2}\pi$.

$\rho' = \sin^2 \frac{1}{3}\theta \cos \frac{1}{3}\theta$ and $ds/d\theta = \sqrt{\sin^6 \frac{1}{3}\theta + \sin^4 \frac{1}{3}\theta \cos^2 \frac{1}{3}\theta} = \sin^2 \frac{1}{3}\theta$.

At $\theta = \frac{1}{2}\pi$, $ds/d\theta = \sin^2 \frac{1}{6}\pi = 1/4$.

17.  Derive  $K = \dfrac{\rho^2 + 2(\rho')^2 - \rho\rho''}{\{\rho^2 + (\rho')^2\}^{3/2}}$ .

By definition, $K = d\tau/ds$. Now $\tau = \theta + \psi$  and

$$\frac{d\tau}{ds} = \frac{d\theta}{ds} + \frac{d\psi}{ds} = \frac{d\theta}{ds} + \frac{d\psi}{d\theta}\frac{d\theta}{ds} = \frac{d\theta}{ds}\left(1 + \frac{d\psi}{d\theta}\right), \quad \text{where } \psi = \arctan \frac{\rho}{\rho'}$$

$$\frac{d\psi}{d\theta} = \frac{[(\rho')^2 - \rho\rho'']/(\rho')^2}{1 + (\rho/\rho')^2} = \frac{(\rho')^2 - \rho\rho''}{\rho^2 + (\rho')^2} \quad \text{and} \quad 1 + \frac{d\psi}{d\theta} = 1 + \frac{(\rho')^2 - \rho\rho''}{\rho^2 + (\rho')^2} = \frac{\rho^2 + 2(\rho')^2 - \rho\rho''}{\rho^2 + (\rho')^2}$$

Thus, $\quad K = \dfrac{d\theta}{ds}\left(1 + \dfrac{d\psi}{d\theta}\right) = \dfrac{1 + d\psi/d\theta}{ds/d\theta} = \dfrac{1 + d\psi/d\theta}{\sqrt{\rho^2 + (\rho')^2}} = \dfrac{\rho^2 + 2(\rho')^2 - \rho\rho''}{\{\rho^2 + (\rho')^2\}^{3/2}}.$

18.  $\rho = 2 + \sin\theta$.  Find the curvature at the point $P(\rho, \theta)$.

$$K = \frac{\rho^2 + 2(\rho')^2 - \rho\rho''}{\{\rho^2 + (\rho')^2\}^{3/2}} = \frac{(2 + \sin\theta)^2 + 2\cos^2\theta + (\sin\theta)(2 + \sin\theta)}{\{(2 + \sin\theta)^2 + \cos^2\theta\}^{3/2}} = \frac{6(1 + \sin\theta)}{(5 + 4\sin\theta)^{3/2}}.$$

19.  $\rho(1 - \cos\theta) = 1$.  Find the curvature at $\theta = \pi/2$ and at $\theta = 4\pi/3$.

$$\rho' = \frac{-\sin\theta}{(1 - \cos\theta)^2}, \quad \rho'' = \frac{-\cos\theta}{(1 - \cos\theta)^2} + \frac{2\sin^2\theta}{(1 - \cos\theta)^3}, \quad \text{and} \quad K = \sin^3 \tfrac{1}{2}\theta.$$

At $\theta = \pi/2$, $K = (1/\sqrt{2})^3 = \sqrt{2}/4$; at $\theta = 4\pi/3$, $K = (\sqrt{3}/2)^3 = 3\sqrt{3}/8$.

20.  From $\mathbf{r} = \rho\mathbf{u}_\rho$, derive formulas for $\mathbf{v}$ and $\mathbf{a}$ in terms of $\mathbf{u}_\rho$ and $\mathbf{u}_\theta$.

$$\mathbf{r} = \rho\mathbf{u}_\rho$$

$$\mathbf{v} = \frac{d\mathbf{r}}{dt} = \mathbf{u}_\rho\frac{d\rho}{dt} + \rho\frac{d\mathbf{u}_\rho}{dt} = \mathbf{u}_\rho\frac{d\rho}{dt} + \rho\mathbf{u}_\theta\frac{d\theta}{dt}$$

$$\mathbf{a} = \frac{d\mathbf{v}}{dt} = \mathbf{u}_\rho\frac{d^2\rho}{dt^2} + \mathbf{u}_\theta\frac{d\rho}{dt}\frac{d\theta}{dt} + \rho\mathbf{u}_\theta\frac{d^2\theta}{dt^2} + \mathbf{u}_\theta\frac{d\rho}{dt}\frac{d\theta}{dt} - \rho\mathbf{u}_\rho\left(\frac{d\theta}{dt}\right)^2$$

$$= \mathbf{u}_\rho\left[\frac{d^2\rho}{dt^2} - \rho\left(\frac{d\theta}{dt}\right)^2\right] + \mathbf{u}_\theta\left[\rho\frac{d^2\theta}{dt^2} + 2\frac{d\rho}{dt}\frac{d\theta}{dt}\right]$$

21.  A particle moves counterclockwise along $\rho = 4\sin 2\theta$ with $d\theta/dt = \frac{1}{2}$ rad/sec.  (a) Express $\mathbf{v}$ and $\mathbf{a}$ in terms of $\mathbf{u}_\rho$ and $\mathbf{u}_\theta$.  (b) Find $|\mathbf{v}|$ and $|\mathbf{a}|$ when $\theta = \pi/6$.

$$\mathbf{r} = 4\sin 2\theta\,\mathbf{u}_\rho, \quad d\rho/dt = 8\cos 2\theta\,d\theta/dt = 4\cos 2\theta, \quad d^2\rho/dt^2 = -4\sin 2\theta$$

(a) $\quad \mathbf{v} = \mathbf{u}_\rho\dfrac{d\rho}{dt} + \rho\mathbf{u}_\theta\dfrac{d\theta}{dt} = 4\mathbf{u}_\rho\cos 2\theta + 2\mathbf{u}_\theta\sin 2\theta$

$\quad \mathbf{a} = \mathbf{u}_\rho\left[\dfrac{d^2\rho}{dt^2} - \rho\left(\dfrac{d\theta}{dt}\right)^2\right] + \mathbf{u}_\theta\left[\rho\dfrac{d^2\theta}{dt^2} + 2\dfrac{d\rho}{dt}\dfrac{d\theta}{dt}\right]$

$\qquad = -5\mathbf{u}_\rho\sin 2\theta + 4\mathbf{u}_\theta\cos 2\theta$

(b) At $\theta = \pi/6$:  $\mathbf{u}_\rho = \dfrac{\sqrt{3}}{2}\mathbf{i} + \dfrac{1}{2}\mathbf{j}$,  $\mathbf{u}_\theta = -\dfrac{1}{2}\mathbf{i} + \dfrac{\sqrt{3}}{2}\mathbf{j}$.  $\mathbf{v} = \dfrac{\sqrt{3}}{2}\mathbf{i} + \dfrac{5}{2}\mathbf{j}$;  $\mathbf{a} = -\dfrac{19}{4}\mathbf{i} - \dfrac{\sqrt{3}}{4}\mathbf{j}$.
$|\mathbf{v}| = \sqrt{7}$,  $|\mathbf{a}| = \frac{1}{2}\sqrt{91}$.

# Supplementary Problems

In Problems 22-25, find $\tan\psi$ at the given point.

22.  $\rho = 3 - \sin\theta$ at $\theta = 0$, $\theta = 3\pi/4$ $\qquad$ *Ans.* $-3$, $3\sqrt{2} - 1$

23.  $\rho = a(1 - \cos\theta)$ at $\theta = \pi/4$, $\theta = 3\pi/2$ $\qquad$ *Ans.* $\sqrt{2} - 1$, $-1$

24.  $\rho(1 - \cos\theta) = a$ at $\theta = \pi/3$, $\theta = 5\pi/4$ $\qquad$ *Ans.* $-\sqrt{3}/3$, $1 + \sqrt{2}$

25.  $\rho^2 = 4\sin 2\theta$ at $\theta = 5\pi/12$, $\theta = 2\pi/3$ $\qquad$ *Ans.* $-1/\sqrt{3}$, $\sqrt{3}$

In Problems 26-29, find $\tan \tau$.

**26.** $\rho = 2 + \sin \theta$ at $\theta = \pi/6$    *Ans.* $-3\sqrt{3}$        **28.** $\rho = \sin^3 \theta/3$ at $\theta = \pi/2$    *Ans.* $-\sqrt{3}$

**27.** $\rho^2 = 9 \cos 2\theta$ at $\theta = \pi/6$    *Ans.* 0        **29.** $2\rho(1 - \sin \theta) = 3$ at $\theta = \pi/4$    *Ans.* $1 + \sqrt{2}$

**30.** Investigate $\rho = \sin 2\theta$ for horizontal and vertical tangents.
   *Ans.* H.T. at $\theta = 0$, $\pi$, $54°44'$, $125°16'$, $234°44'$, $305°16'$
   V.T. at $\theta = \pi/2$, $3\pi/2$, $35°16'$, $144°44'$, $215°16'$, $324°44'$

In Problems 31-33, find the acute angles of intersection of each pair of curves.

**31.** $\rho = \sin \theta$, $\rho = \sin 2\theta$        *Ans.* $\phi = 79°6'$ at $\theta = \pi/3$ and $5\pi/3$; $\phi = 0$ at the pole.

**32.** $\rho = \sqrt{2} \sin \theta$, $\rho^2 = \cos 2\theta$        *Ans.* $\phi = \pi/3$ at $\theta = \pi/6$, $5\pi/6$; $\phi = \pi/4$ at the pole.

**33.** $\rho^2 = 16 \sin 2\theta$, $\rho^2 = 4 \csc 2\theta$    *Ans.* $\phi = \pi/3$ at each intersection.

**34.** Show that each of the following pairs of curves intersect at right angles at all points of intersection.
   (a) $\rho = 4 \cos \theta$, $\rho = 4 \sin \theta$        (c) $\rho^2 \cos 2\theta = 4$, $\rho^2 \sin 2\theta = 9$
   (b) $\rho = e^\theta$, $\rho = e^{-\theta}$        (d) $\rho = 1 + \cos \theta$, $\rho = 1 - \cos \theta$

**35.** Find the angle of intersection of the tangents to $\rho = 2 - 4 \sin \theta$ at the pole.    *Ans.* $2\pi/3$

**36.** Find the curvature of each curve at $P(\rho, \theta)$:
   (a) $\rho = e^\theta$,  (b) $\rho = \sin \theta$,  (c) $\rho^2 = 4 \cos 2\theta$,  (d) $\rho = 3 \sin \theta + 4 \cos \theta$.
   *Ans.* (a) $1/(\sqrt{2}\, e^\theta)$,  (b) 2,  (c) $\frac{3}{2}\sqrt{\cos 2\theta}$,  (d) 2/5

**37.** Let $\rho = f(\theta)$ be the polar equation of a curve and let $s$ be the arc length along the curve. Using
$x = \rho \cos \theta$, $y = \rho \sin \theta$ and recalling $\left(\dfrac{ds}{d\theta}\right)^2 = \left(\dfrac{dx}{d\theta}\right)^2 + \left(\dfrac{dy}{d\theta}\right)^2$, derive $\left(\dfrac{ds}{d\theta}\right)^2 = \rho^2 + (\rho')^2$.

**38.** Find $ds/d\theta$ for each of the following, assuming $s$ increases in the direction of increasing $\theta$:
   (a) $\rho = a \cos \theta$,  (b) $\rho = a(1 + \cos \theta)$,  (c) $\rho = \cos 2\theta$.
   *Ans.* (a) $a$,  (b) $a\sqrt{2 + 2 \cos \theta}$,  (c) $\sqrt{1 + 3 \sin^2 2\theta}$

**39.** Suppose a particle moves along a curve $\rho = f(\theta)$ with its position at any time $t$ given by $\rho = g(t)$, $\theta = h(t)$.
   (a) Multiply the relation obtained in Prob. 37 by $\left(\dfrac{d\theta}{dt}\right)^2$ to obtain $v^2 = \left(\dfrac{ds}{dt}\right)^2 = \rho^2 \left(\dfrac{d\theta}{dt}\right)^2 + \left(\dfrac{d\rho}{dt}\right)^2$.

   (b) From $\tan \psi = \rho \dfrac{d\theta}{d\rho} = \rho \dfrac{d\theta/dt}{d\rho/dt}$, obtain $\sin \psi = \dfrac{\rho}{v} \dfrac{d\theta}{dt}$ and $\cos \psi = \dfrac{1}{v} \dfrac{d\rho}{dt}$.

**40.** Show that $\dfrac{d\mathbf{u}_\rho}{dt} = \mathbf{u}_\theta \dfrac{d\theta}{dt}$ and $\dfrac{d\mathbf{u}_\theta}{dt} = -\mathbf{u}_\rho \dfrac{d\theta}{dt}$.

**41.** A particle moves counterclockwise about the cardioid $\rho = 4(1 + \cos \theta)$ with $d\theta/dt = \pi/6$ rad/sec. Express $\mathbf{v}$ and $\mathbf{a}$ in terms of $\mathbf{u}_\rho$ and $\mathbf{u}_\theta$.
   *Ans.* $\mathbf{v} = -\dfrac{2\pi}{3} \mathbf{u}_\rho \sin \theta + \dfrac{2\pi}{3} \mathbf{u}_\theta (1 + \cos \theta)$, $\mathbf{a} = -\dfrac{\pi^2}{9} \mathbf{u}_\rho (1 + 2 \cos \theta) - \dfrac{2\pi^2}{9} \mathbf{u}_\theta \sin \theta$

**42.** A particle moves counterclockwise on $\rho = 8 \cos \theta$ with constant speed 4 units/sec. Express $\mathbf{v}$ and $\mathbf{a}$ in terms of $\mathbf{u}_\rho$ and $\mathbf{u}_\theta$.
   *Ans.* $\mathbf{v} = -4\mathbf{u}_\rho \sin \theta + 4\mathbf{u}_\theta \cos \theta$, $\mathbf{a} = -4\mathbf{u}_\rho \cos \theta - 4\mathbf{u}_\theta \sin \theta$

**43.** If a particle of mass $m$ moves along a path under a force $\mathbf{F}$ which is always directed toward the origin, we have $\mathbf{F} = m\mathbf{a}$ or $\mathbf{a} = \dfrac{1}{m} \mathbf{F}$ so that $a_\theta = 0$. Show that when $a_\theta = 0$, then $\rho^2 \dfrac{d\theta}{dt} = k$, a constant, and the radius vector sweeps over area at a constant rate.

**44.** A particle moves along $\rho = \dfrac{2}{1 - \cos \theta}$ with $a_\theta = 0$. Show that $a_\rho = -\dfrac{k^2}{2} \dfrac{1}{\rho^2}$, where $k$ is defined in Problem 43.

# Chapter 21

# The Law of the Mean

**ROLLE'S THEOREM.** If $f(x)$ is continuous on the interval $a \leqq x \leqq b$, if $f(a) = f(b) = 0$, and if $f'(x)$ exists everywhere on the interval except possibly at the endpoints; then $f'(x) = 0$ for at least one value of $x$, say $x = x_0$, between $a$ and $b$.

Geometrically, this means that if a continuous curve intersects the $x$-axis at $x = a$ and $x = b$, and has a tangent at every point between $a$ and $b$, then there is at least one point $x = x_0$ between $a$ and $b$ where the tangent is parallel to the $x$-axis. See Fig. 21-1 below.

For a proof, see Problem 11.

*Corollary.* If $f(x)$ satisfies the conditions of Rolle's Theorem, except that $f(a) = f(b) \neq 0$, then $f'(x) = 0$ for at least one value of $x$, say $x = x_0$, between $a$ and $b$. See Fig. 21-2 below.

See Problems 1-2.

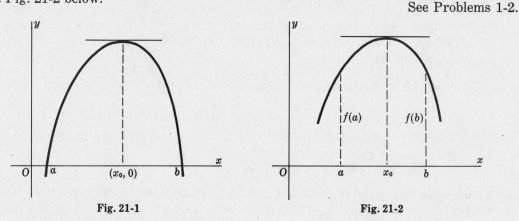

Fig. 21-1                    Fig. 21-2

**THE LAW OF THE MEAN.** If $f(x)$ is continuous on the interval $a \leqq x \leqq b$ and if $f'(x)$ exists everywhere on the interval except possibly at the endpoints, then there is at least one value of $x$, say $x = x_0$, between $a$ and $b$ such that

$$\frac{f(b) - f(a)}{b - a} = f'(x_0)$$

Geometrically, this means that if $P_1$ and $P_2$ are two points of a continuous curve, having a tangent at each intervening point, then there exists at least one point of the curve between $P_1$ and $P_2$ at which the slope of the curve is equal to the slope of $P_1P_2$. See Fig. 21-3.

For a proof see Problem 12.

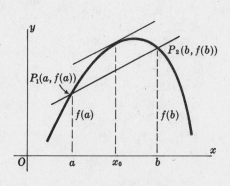

Fig. 21-3

The Law of the Mean may be put in several useful forms:

(I)                  $f(b) = f(a) + (b - a) \cdot f'(x_0)$,    $x_0$ between $a$ and $b$.

By a simple change of letter, this becomes

(II)                  $f(x) = f(a) + (x - a) \cdot f'(x_0)$,    $x_0$ between $a$ and $x$.

It is clear from Fig. 21-4 that $x_0 = a + \theta(b - a)$, where $0 < \theta < 1$. With this replacement, (I) takes the form

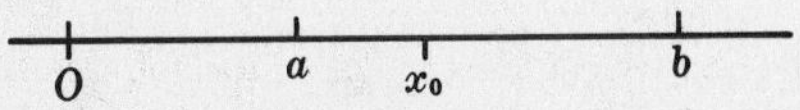

Fig. 21-4

(III)                  $f(b) = f(a) + (b - a) \cdot f'[a + \theta(b - a)]$,    $0 < \theta < 1$

Writing $(b - a) = h$, (III) becomes

(IV)                  $f(a + h) = f(a) + h \cdot f'(a + \theta h)$,    $0 < \theta < 1$

Finally, writing $a = x$ and $h = \Delta x$, (IV) becomes

(V)                  $f(x + \Delta x) = f(x) + \Delta x \cdot f'(x + \theta \cdot \Delta x)$,    $0 < \theta < 1$

See Problems 3-9.

**GENERALIZED LAW OF THE MEAN.** If $f(x)$ and $g(x)$ are continuous on the interval $a \leqq x \leqq b$, if $f'(x)$ and $g'(x)$ exist and $g'(x) \neq 0$ everywhere on the interval, except possibly at the endpoints, then there exists at least one value of $x$, say $x = x_0$, between $a$ and $b$ such that

$$\frac{f(b) - f(a)}{g(b) - g(a)} = \frac{f'(x_0)}{g'(x_0)}$$

For the case $g(x) = x$, this becomes the Law of the Mean.

For a proof, see Problem 13.

**EXTENDED LAW OF THE MEAN.** If $f(x)$, together with its first $(n - 1)$ derivatives are continuous on the interval $a \leqq x \leqq b$ and if $f^{(n)}(x)$ exists everywhere on the interval except possibly at the endpoints, then there is at least one value of $x$, say $x = x_0$, between $a$ and $b$ such that

(VI)     $f(b) = f(a) + \dfrac{f'(a)}{1!}(b - a) + \dfrac{f''(a)}{2!}(b - a)^2 + \cdots$

$$+ \frac{f^{(n-1)}(a)}{(n-1)!}(b - a)^{n-1} + \frac{f^{(n)}(x_0)}{n!}(b - a)^n$$

For a proof, see Problem 15.

When $b$ is replaced by the variable $x$, (VI) becomes

(VII)     $f(x) = f(a) + \dfrac{f'(a)}{1!}(x - a) + \dfrac{f''(a)}{2!}(x - a)^2 + \cdots$

$$+ \frac{f^{(n-1)}(a)}{(n-1)!}(x - a)^{n-1} + \frac{f^{(n)}(x_0)}{n!}(x - a)^n, \quad x_0 \text{ between } a \text{ and } x.$$

When $a$ is replaced by 0, (VII) becomes

(VIII)     $f(x) = f(0) + \dfrac{f'(0)}{1!}x + \dfrac{f''(0)}{2!}x^2 + \cdots + \dfrac{f^{(n-1)}(0)}{(n-1)!}x^{n-1} + \dfrac{f^{(n)}(x_0)}{n!}x^n$

$x_0$ between 0 and $x$.

# Solved Problems

1.   Find the value of $x_0$ prescribed in Rolle's Theorem for $f(x) = x^3 - 12x$ on the interval $0 \leqq x \leqq 2\sqrt{3}$.
     $f'(x) = 3x^2 - 12 = 0$ when $x = \pm 2$; then $x_0 = 2$ is the prescribed value.

2.   Does Rolle's Theorem apply to the functions $(a)$ $f(x) = \dfrac{x^2 - 4x}{x - 2}$ and $(b)$ $f(x) = \dfrac{x^2 - 4x}{x + 2}$ ?

     $(a)$ $f(x) = 0$ when $x = 0, 4$. Since $f(x)$ is discontinuous at $x = 2$, a point on the interval $0 \leqq x \leqq 4$, the theorem does not apply.

     $(b)$ $f(x) = 0$ when $x = 0, 4$. Here $f(x)$ is discontinuous at $x = -2$, a point not on the interval $0 \leqq x \leqq 4$.

     $f'(x) = (x^2 + 4x - 8)/(x + 2)^2$ exists everywhere except at $x = -2$. Hence, the theorem applies and $x_0 = 2(\sqrt{3} - 1)$, the positive root of $x^2 + 4x - 8 = 0$.

3.   Find the value of $x_0$ prescribed by the Law of the Mean, given $f(x) = 3x^2 + 4x - 3$, $a = 1$, $b = 3$.

     Using (I) with $f(a) = f(1) = 4$, $f(b) = f(3) = 36$, $f'(x_0) = 6x_0 + 4$, and $b - a = 2$, we have $36 = 4 + 2(6x_0 + 4) = 12x_0 + 12$ and $x_0 = 2$.

4.   Use the Law of the Mean to approximate $\sqrt[6]{65}$.

     Let $f(x) = \sqrt[6]{x}$, $a = 64$, $b = 65$, and use (I).

     Then $f(65) = f(64) + (65 - 64)/6x_0^{5/6}$, $64 < x_0 < 65$. Since $x_0$ is not known, take $x_0 = 64$; then approximately, $\sqrt[6]{65} = \sqrt[6]{64} + 1/(6\sqrt[6]{64^5}) = 2 + 1/192 = 2.00521$.

5.   A circular hole 4 in. in diameter and 1 ft deep in a metal block is rebored to increase the diameter to 4.12 in. Estimate the amount of metal removed.

     The volume (in.³) of a circular hole of radius $x$ in. and depth 12 in. is given by $V = f(x) = 12\pi x^2$. We are to estimate $f(2.06) - f(2)$. By the Law of the Mean,

     $$f(2.06) - f(2) = .06\, f'(x_0) = .06(24\pi x_0), \qquad 2 < x_0 < 2.06$$

     Take $x_0 = 2$; then, approximately, $f(2.06) - f(2) = .06(24\pi \cdot 2) = 2.88\pi$ in.³

6.   Apply the Law of the Mean to $y = f(x)$, $a = x$, $b = x + \Delta x$ with all conditions satisfied to show that $\Delta y = f'(x) \cdot \Delta x$ approximately.

     We have $\Delta y = f(x + \Delta x) - f(x) = [x + \Delta x - x] \cdot f'(x_0)$, $x < x_0 < x + \Delta x$.

     Take $x_0 = x$; then approximately $\Delta y = f'(x) \cdot \Delta x$.

7.   Use the Law of the Mean to show $\sin x < x$ for $x > 0$.

     Since $\sin x \leqq 1$, $\sin x < x$ when $x > 1$. Take $f(x) = \sin x$ with $0 \leqq x \leqq 1$ and use (II); then
     $$\sin x = \sin 0 + x \cos x_0 = x \cos x_0, \qquad 0 < x_0 < x$$
     Now on this interval $\cos x_0 < 1$ and $x \cos x_0 < x$; hence, $\sin x < x$.

8.   Use the Law of the Mean to show $\dfrac{x}{1 + x} < \ln(1 + x) < x$ for $-1 < x < 0$ and for $x > 0$.

     Use (IV) with $f(x) = \ln x$, $a = 1$ and $h = x$; then
     $$\ln(1 + x) = \ln 1 + x\frac{1}{1 + \theta x} = \frac{x}{1 + \theta x}, \qquad 0 < \theta < 1$$

     When $x > 0$, $1 < 1 + \theta x < 1 + x$; hence, $1 > \dfrac{1}{1 + \theta x} > \dfrac{1}{1 + x}$ and $x > \dfrac{x}{1 + \theta x} > \dfrac{x}{1 + x}$.

     When $-1 < x < 0$, $1 > 1 + \theta x > 1 + x$; hence, $1 < \dfrac{1}{1 + \theta x} < \dfrac{1}{1 + x}$ and $x > \dfrac{x}{1 + \theta x} > \dfrac{x}{1 + x}$.

     In each case, $\dfrac{x}{1 + \theta x} < x$ and $\ln(1 + x) = \dfrac{x}{1 + \theta x} < x$; also, $\dfrac{x}{1 + \theta x} > \dfrac{x}{1 + x}$ and $\ln(1 + x) = \dfrac{x}{1 + \theta x} > \dfrac{x}{1 + x}$. Hence, $\dfrac{x}{1 + x} < \ln(1 + x) < x$ when $-1 < x < 0$ and when $x > 0$.

9.  Use the Law of the Mean to show $\sqrt{1+x} < 1 + \frac{1}{2}x$ for $-1 < x < 0$ and for $x > 0$.

    Take $f(x) = \sqrt{x}$ and use (IV) with $a = 1$, $h = x$; then

    $$\sqrt{1+x} = 1 + \frac{x}{2\sqrt{1+\theta x}}, \quad 0 < \theta < 1$$

    When $x > 0$, $\sqrt{1+\theta x} < \sqrt{1+x}$ and $\dfrac{x}{2\sqrt{1+\theta x}} > \dfrac{x}{2\sqrt{1+x}}$; when $-1 < x < 0$, $\sqrt{1+\theta x} > \sqrt{1+x}$ and $\dfrac{x}{2\sqrt{1+\theta x}} > \dfrac{x}{2\sqrt{1+x}}$.

    In each case, $\sqrt{1+x} = 1 + \dfrac{x}{2\sqrt{1+\theta x}} > 1 + \dfrac{x}{2\sqrt{1+x}}$. Multiplying the inequality by $\sqrt{1+x} > 0$, we have $1 + x > \sqrt{1+x} + \frac{1}{2}x$ and $\sqrt{1+x} < 1 + \frac{1}{2}x$.

10. Find a value $x_0$ as prescribed by the Generalized Law of the Mean, given $f(x) = 3x + 2$, $g(x) = x^2 + 1$, $1 \le x \le 4$.

    We are to find $x_0$ so that

    $$\frac{f(b) - f(a)}{g(b) - g(a)} = \frac{f(4) - f(1)}{g(4) - g(1)} = \frac{14 - 5}{17 - 2} = \frac{3}{5} = \frac{f'(x_0)}{g'(x_0)} = \frac{3}{2x_0}$$

    Then $2x_0 = 5$ and $x_0 = 5/2$.

11. Prove Rolle's Theorem: If $f(x)$ is continuous on the interval $a \le x \le b$, if $f(a) = f(b) = 0$, and if $f'(x)$ exists everywhere on the interval except possibly at the endpoints, then $f'(x) = 0$ for at least one value of $x$, say $x = x_0$, between $a$ and $b$.

    If $f(x) = 0$ throughout the interval so also is $f'(x) = 0$ and the theorem is proved. Otherwise, if $f(x)$ is positive (negative) somewhere on the interval, it has a relative maximum (minimum) at some $x = x_0$, $a < x_0 < b$, (see Property II, Chapter 3) and $f'(x_0) = 0$.

12. Prove the Law of the Mean: If $f(x)$ is continuous on the interval $a \le x \le b$ and if $f'(x)$ exists everywhere on the interval except possibly at the endpoints, then there is a value of $x$, say $x = x_0$, between $a$ and $b$ such that

    $$\frac{f(b) - f(a)}{b - a} = f'(x_0)$$

    Refer to Fig. 21-3. The equation of the secant line $P_1 P_2$ is $y = f(b) + K(x - b)$ where $K = \dfrac{f(b) - f(a)}{b - a}$. At any point $x$ on the interval $a < x < b$, the vertical distance from the secant line to the curve is $F(x) = f(x) - f(b) - K(x - b)$. Now $F(x)$ satisfies the conditions of Rolle's Theorem (check this); hence, $F'(x) = f'(x) - K = 0$ for some $x = x_0$ between $a$ and $b$. Thus,

    $$K = f'(x_0) = \frac{f(b) - f(a)}{b - a}$$

    as was to be proved.

13. Prove the Generalized Law of the Mean: If $f(x)$ and $g(x)$ are continuous on the interval $a \le x \le b$, if $f'(x)$ and $g'(x)$ exist and $g'(x) \ne 0$ everywhere on the interval, except possibly at the endpoints, then there exists at least one value of $x$, say $x = x_0$, between $a$ and $b$ such that

    $$\frac{f(b) - f(a)}{g(b) - g(a)} = \frac{f'(x_0)}{g'(x_0)}$$

    Suppose $g(b) = g(a)$; then by the Corollary to Rolle's Theorem, $g'(x) = 0$ for some $x$ between $a$ and $b$. But this is contrary to the hypothesis; thus, $g(b) \ne g(a)$. Set $\dfrac{f(b) - f(a)}{g(b) - g(a)} = K$, a constant, and form the function

    $$F(x) = f(x) - f(b) - K[g(x) - g(b)]$$

    Now this function satisfies the conditions of Rolle's Theorem (check this) so that $F'(x) = f'(x) - K g'(x) = 0$ for at least one value of $x$, say $x = x_0$, between $a$ and $b$. Thus,

    $$K = \frac{f'(x_0)}{g'(x_0)} = \frac{f(b) - f(a)}{g(b) - g(a)} \qquad \text{as was to be proved.}$$

**14.** An arc $PQ$ of a curve $y = f(x)$ is concave upward when it lies below the chord $PQ$ and is concave downward when it lies above the chord. Prove: If $f(x)$ and $f'(x)$ are continuous on $a \leqq x \leqq b$, and if $f'(x)$ has the same sign on $a < x < b$, then

(**i**) $f(x)$ is concave upward on $a < x < b$ when $f''(x) > 0$,

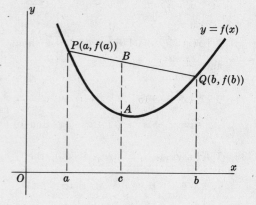

(**ii**) $f(x)$ is concave downward on $a < x < b$ when $f''(x) < 0$.

The equation of the chord $PQ$ joining $P[a, f(a)]$ and $Q[b, f(b)]$ is $y = f(a) + (x - a)\dfrac{f(b) - f(a)}{b - a}$. Let $A$ and $B$ be points on the arc and chord respectively having abscissa $x = c$, $a < c < b$. The corresponding ordinates are $f(c)$ and

$$f(a) + (c - a)\frac{f(b) - f(a)}{b - a} = \frac{(b - c) \cdot f(a) + (c - a) \cdot f(b)}{b - a}$$

(i) We are to prove

$$f(c) < \frac{(b - c) \cdot f(a) + (c - a) \cdot f(b)}{b - a}$$

**Fig. 21-5**

when $f''(x) > 0$. By the Law of the Mean, $\dfrac{f(c) - f(a)}{c - a} = f'(\xi)$, where $\xi$ is between $a$ and $c$, and $\dfrac{f(b) - f(c)}{b - c} = f'(\eta)$, where $\eta$ is between $c$ and $b$. Since $f''(x) > 0$ on $a < x < b$, $f'(x)$ is an increasing function on the interval and $f'(\xi) < f'(\eta)$. Thus $\dfrac{f(c) - f(a)}{c - a} < \dfrac{f(b) - f(c)}{b - c}$, from which it follows that

$$f(c) < \frac{(b - c) \cdot f(a) + (c - a) \cdot f(b)}{b - a}$$

as required.

Part (ii) is left as an exercise for the reader.

**15.** Prove: If $f(x)$, together with its first $(n - 1)$ derivatives, are continuous on the interval $a \leqq x \leqq b$ and if $f^{(n)}(x)$ exists everywhere on the interval except possibly at the endpoints, then there is a value of $x$, say $x = x_0$, between $a$ and $b$ such that

$$f(b) = f(a) + \frac{f'(a)}{1!}(b - a) + \frac{f''(a)}{2!}(b - a)^2 + \cdots + \frac{f^{(n-1)}(a)}{(n-1)!}(b - a)^{n-1} + \frac{f^{(n)}(x_0)}{n!}(b - a)^n$$

For the case $n = 1$, this becomes the Law of the Mean. The proof below parallels that of Prob. 12. Let $K$ be defined by

(**i**)      $$f(b) = f(a) + \frac{f'(a)}{1!}(b - a) + \frac{f''(a)}{2!}(b - a)^2 + \cdots + \frac{f^{(n-1)}(a)}{(n-1)!}(b - a)^{n-1} + K(b - a)^n$$

and consider

$$F(x) = f(x) - f(b) + \frac{f'(x)}{1!}(b - x) + \frac{f''(x)}{2!}(b - x)^2 + \cdots + \frac{f^{(n-1)}(x)}{(n-1)!}(b - x)^{n-1} + K(b - x)^n$$

Now $F(a) = 0$ by (i) and $F(b) = 0$. By Rolle's Theorem, there exists an $x = x_0$, where $a < x_0 < b$, such that

$$F'(x_0) = f'(x_0) + \{f''(x_0) \cdot (b - x_0) - f'(x_0)\} + \left\{\frac{f'''(x_0)}{2!}(b - x_0)^2 - f''(x_0) \cdot (b - x_0)\right\}$$

$$+ \cdots + \left\{\frac{f^{(n)}(x_0)}{(n-1)!}(b - x_0)^{n-1} - \frac{f^{(n-1)}(x_0)}{(n-2)!}(b - x_0)^{n-2}\right\} - Kn(b - x_0)^{n-1}$$

$$= \frac{f^{(n)}(x_0)}{(n-1)!}(b - x_0)^{n-1} - Kn(b - x_0)^{n-1} = 0$$

Then $K = \dfrac{f^{(n)}(x_0)}{n!}$ and (i) becomes

$$f(b) = f(a) + \frac{f'(a)}{1!}(b - a) + \frac{f''(a)}{2!}(b - a)^2 + \cdots + \frac{f^{(n-1)}(a)}{(n-1)!}(b - a)^{n-1} + \frac{f^{(n)}(x_0)}{n!}(b - a)^n$$

# Supplementary Problems

16. Find a value for $x_0$ as prescribed by Rolle's Theorem, given:

    (a) $f(x) = x^2 - 4x + 3$, $1 \leq x \leq 3$      *Ans.* $x_0 = 2$

    (b) $f(x) = \sin x$, $0 \leq x \leq \pi$      *Ans.* $x_0 = \frac{1}{2}\pi$

    (c) $f(x) = \cos x$, $\pi/2 < x < 3\pi/2$      *Ans.* $x_0 = \pi$

17. Find a value for $x_0$ as prescribed by the Law of the Mean, given:

    (a) $y = x^3$, $0 \leq x \leq 6$      *Ans.* $x_0 = 2\sqrt{3}$

    (b) $y = ax^2 + bx + c$, $x_1 \leq x \leq x_2$      *Ans.* $x_0 = \frac{1}{2}(x_1 + x_2)$

    (c) $y = \ln x$, $1 \leq x \leq 2e$      *Ans.* $x_0 = \dfrac{2e - 1}{1 + \ln 2}$

18. Use the Law of the Mean to approximate   (a) $\sqrt{15}$,   (b) $(3.001)^3$,   (c) $1/999$.
    *Ans.* (a) 3.875, (b) 27.027, (c) 0.001001

19. Use the Law of the Mean to prove:

    (a) $\tan x > x$, $0 < x < \frac{1}{2}\pi$;   (b) $\dfrac{x}{1 + x^2} < \text{Arc} \tan x < x$, $x > 0$;   (c) $x < \text{Arc} \sin x < \dfrac{x}{\sqrt{1 - x^2}}$, $0 < x < 1$.

20. Show that   $|f(x) - f(x_1)| \leq |x - x_1|$,   $x_1$ being any number, when   (a) $f(x) = \sin x$,   (b) $f(x) = \cos x$.

21. Use the Law of the Mean to prove:

    (a) If $f'(x) = 0$ everywhere on the interval $a \leq x \leq b$, then $f(x) = f(a) = c$, a constant, everywhere on the interval.

    (b) On a given interval $a \leq x \leq b$, $f(x)$ increases as $x$ increases if $f'(x) > 0$ throughout the interval.
    *Hint*: Let $x_1 < x_2$ be two points on the interval; then $f(x_2) = f(x_1) + (x_2 - x_1)f'(x_0)$, $x_1 < x_0 < x_2$.

22. Use the theorem of Problem 21(a) to prove: If $f(x)$ and $g(x)$ are different but $f'(x) = g'(x)$ throughout an interval, then $f(x) - g(x) = c \neq 0$, a constant, on the interval.

23. Define a *bend point* of $f(x)$ to be a critical point $x = x_0$ for which $f'(x)$ changes sign as $x$ increases through $x = x_0$. Let $x_1 < x_2 < \cdots < x_{m-1} < x_m$ be the distinct bend points of $f(x)$. Show that $f(x) = 0$ has at most one real root on each of the intervals $x < x_1$, $x_1 < x < x_2$, ..., $x_{m-1} < x < x_m$, $x > x_m$.

24. Prove: If $f(x) = 0$ of degree $n$ has $n$ simple real roots, then $f'(x) = 0$ has exactly $n - 1$ simple real roots.

25. Show that $x^3 + px + q = 0$ has   (a) just one real root if $p > 0$,   (b) three real roots if $4p^3 + 27q^2 < 0$.

26. Find a value $x_0$ as prescribed by the Generalized Law of the Mean, given:

    (a) $f(x) = x^2 + 2x - 3$, $g(x) = x^2 - 4x + 6$; $a = 0$, $b = 1$.      *Ans.* $\frac{1}{2}$

    (b) $f(x) = \sin x$, $g(x) = \cos x$; $a = \pi/6$, $b = \pi/3$.      *Ans.* $\frac{1}{4}\pi$

27. Use the Extended Law of the Mean (VIII) to show:

    (a) $\sin x$ can be approximated by $x$ with allowable error 0.005 for $x < 0.31$.
    *Hint*: For $n = 3$, $\sin x = x - \frac{1}{6}x^3 \cos x_0$. Set $\frac{1}{6}|x^3 \cos x_0| \leq \frac{1}{6}|x^3| < 0.005$.

    (b) $\sin x$ can be approximated by $x - x^3/6$ with allowable error 0.00005 for $x < 0.359$.

# Chapter 22

## Indeterminate Forms

**IN FINDING THE DERIVATIVE** of a differentiable function $f(x)$ by the step rule of Chapter 4, we were led to consider

$$(a) \qquad \lim_{\Delta x \to 0} \frac{f(x + \Delta x) - f(x)}{(x + \Delta x) - x} \;=\; \lim_{\Delta x \to 0} \frac{F(\Delta x)}{G(\Delta x)}$$

Since the limit of both the numerator and the denominator of the fraction is zero, it is customary to call (a) *indeterminate* of the type 0/0. Other examples are found in Problem 5, Chapter 2.

Similarly, it is customary to call $\lim\limits_{x \to \infty} \dfrac{3x - 2}{9x + 7}$ (see Problem 6, Chapter 2) indeterminate of the type $\infty/\infty$. These symbols 0/0, $\infty/\infty$, and others $(0 \cdot \infty, \; \infty - \infty, \; 0^0, \; \infty^0$ and $1^\infty)$ to be introduced later must not be taken literally; they are merely convenient labels for distinguishing types.

### TYPE 0/0.

*L'Hospital's Rule.* If $a$ is a number, if $f(x)$ and $g(x)$ are differentiable and $g(x) \neq 0$ for all $x$ on some interval $0 < |x - a| < \delta$, if $\lim\limits_{x \to a} f(x) = 0$ and $\lim\limits_{x \to a} g(x) = 0$; then, when $\lim\limits_{x \to a} \dfrac{f'(x)}{g'(x)}$ exists or is infinite,

$$\lim_{x \to a} \frac{f(x)}{g(x)} \;=\; \lim_{x \to a} \frac{f'(x)}{g'(x)}$$

**Example 1:** $\lim\limits_{x \to 3} \dfrac{x^4 - 81}{x - 3}$ is indeterminate of type 0/0. Since

$$\lim_{x \to 3} \frac{\dfrac{d}{dx}(x^4 - 81)}{\dfrac{d}{dx}(x - 3)} \;=\; \lim_{x \to 3} 4x^3 \;=\; 108, \quad \text{then} \quad \lim_{x \to 3} \frac{x^4 - 81}{x - 3} \;=\; 108$$

See Problems 1-7.

*Note.* L'Hospital's rule implies $\lim\limits_{x \to a^+} f(x)/g(x) = \lim\limits_{x \to a^-} f(x)/g(x)$. In certain of the problems (Problem 8, for example), the existence of one but not both of these limits is indicated.

### TYPE $\infty/\infty$.

The conclusion of l'Hospital's rule is unchanged if one or both of the following changes are made in the hypotheses:

(i)   "$\lim\limits_{x \to a} f(x) = 0$ and $\lim\limits_{x \to a} g(x) = 0$" is replaced by "$\lim\limits_{x \to a} f(x) = \infty$ and $\lim\limits_{x \to a} g(x) = \infty$",

(ii)  "$a$ is a number" is replaced by "$a = +\infty, \; -\infty,$ or $\infty$" and "$0 < |x - a| < \delta$" is replaced by "$|x| > M$."

**Example 2:** $\lim\limits_{x \to +\infty} \dfrac{x^2}{e^x}$ is indeterminate of type $\infty/\infty$. Then

$$\lim_{x \to +\infty} \frac{x^2}{e^x} \;=\; \lim_{x \to +\infty} \frac{2x}{e^x} \;=\; \lim_{x \to +\infty} \frac{2}{e^x} \;=\; 0$$

See Problems 9-11.

**TYPES** $0 \cdot \infty$ and $\infty - \infty$.

These may be handled by first transforming to one of the types $0/0$ or $\infty/\infty$. For example:

$$\lim_{x \to +\infty} x^2 e^{-x} \text{ is of type } 0 \cdot \infty \text{ and } \lim_{x \to +\infty} \frac{x^2}{e^x} \text{ is of type } \infty/\infty;$$

$$\lim_{x \to 0} \left( \csc x - \frac{1}{x} \right) \text{ is of type } \infty - \infty \text{ and } \lim_{x \to 0} \left( \frac{x - \sin x}{x \sin x} \right) \text{ is of type } 0/0.$$

See Problems 13-16.

**TYPES** $0^0$, $\infty^0$, and $1^\infty$.

If $\lim y$ is one of these types, then $\lim (\ln y)$ is of type $0 \cdot \infty$.

**Example 3:**   Evaluate $\lim_{x \to 0} (\sec^3 2x)^{\cot^2 3x}$.

This is of the type $1^\infty$. Let $y = (\sec^3 2x)^{\cot^2 3x}$; then $\ln y = \cot^2 3x \ln \sec^3 2x = \dfrac{3 \ln \sec 2x}{\tan^2 3x}$ and $\lim_{x \to 0} \ln y$ is of the type $0/0$.

Now $\lim_{x \to 0} \dfrac{3 \ln \sec 2x}{\tan^2 3x} = \lim_{x \to 0} \dfrac{6 \tan 2x}{6 \tan 3x \sec^2 3x} = \lim_{x \to 0} \dfrac{\tan 2x}{\tan 3x}$, since $\lim_{x \to 0} \sec^2 3x = 1$, is of type $0/0$.

Then $\lim_{x \to 0} \dfrac{\tan 2x}{\tan 3x} = \lim_{x \to 0} \dfrac{2 \sec^2 2x}{3 \sec^2 3x} = \dfrac{2}{3}$.

Since $\lim_{x \to 0} \ln y = 2/3$,  $\lim_{x \to 0} y = \lim_{x \to 0} (\sec^3 2x)^{\cot^2 3x} = e^{2/3}$.

See Problems 17-19.

# Solved Problems

1.   Prove l'Hospital's Rule: If $a$ is a number, if $f(x)$ and $g(x)$ are differentiable and $g(x) \neq 0$ for all $x$ on some interval $0 < |x - a| < \delta$, if $\lim_{x \to a} f(x) = 0$ and $\lim_{x \to a} g(x) = 0$; then

$$\text{if } \lim_{x \to a} \frac{f'(x)}{g'(x)} \text{ exists, } \quad \lim_{x \to a} \frac{f(x)}{g(x)} = \lim_{x \to a} \frac{f'(x)}{g'(x)}$$

When $b$ is replaced by $x$ in the Generalized Law of the Mean (Chapter 21) we have, since $f(a) = g(a) = 0$,

$$\frac{f(x) - f(a)}{g(x) - g(a)} = \frac{f(x)}{g(x)} = \frac{f'(x_0)}{g'(x_0)}$$

where $x_0$ is between $a$ and $x$.   Now $x_0 \to a$ as $x \to a$; hence,

$$\lim_{x \to a} \frac{f(x)}{g(x)} = \lim_{x_0 \to a} \frac{f'(x_0)}{g'(x_0)} = \lim_{x \to a} \frac{f'(x)}{g'(x)}$$

2.   Evaluate $\lim_{x \to 2} \dfrac{x^2 + x - 6}{x^2 - 4}$.      When $x \to 2$, both numerator and denominator approach 0.

Hence the rule applies and  $\lim_{x \to 2} \dfrac{x^2 + x - 6}{x^2 - 4} = \lim_{x \to 2} \dfrac{2x + 1}{2x} = \dfrac{5}{4}$.

3.   Evaluate $\lim_{x \to 0} \dfrac{x + \sin 2x}{x - \sin 2x}$.      When $x \to 0$, both numerator and denominator approach 0.

Hence the rule applies and  $\lim_{x \to 0} \dfrac{x + \sin 2x}{x - \sin 2x} = \lim_{x \to 0} \dfrac{1 + 2 \cos 2x}{1 - 2 \cos 2x} = \dfrac{1 + 2}{1 - 2} = -3$.

4.   Evaluate $\lim_{x \to 0} \dfrac{e^x - 1}{x^2}$.      $\lim_{x \to 0} \dfrac{e^x - 1}{x^2} = \lim_{x \to 0} \dfrac{e^x}{2x} = \infty$.

**5.** Evaluate $\lim\limits_{x\to 0}\dfrac{e^x + e^{-x} - x^2 - 2}{\sin^2 x - x^2}$.

When $x \to 0$, both numerator and denominator approach 0. Hence the rule applies and

$$\lim_{x\to 0}\frac{e^x + e^{-x} - x^2 - 2}{\sin^2 x - x^2} = \lim_{x\to 0}\frac{e^x - e^{-x} - 2x}{\sin 2x - 2x}$$

Since the resulting function is indeterminate of type 0/0, we apply the rule to it:

$$\lim_{x\to 0}\frac{e^x + e^{-x} - x^2 - 2}{\sin^2 x - x^2} = \lim_{x\to 0}\frac{e^x - e^{-x} - 2x}{\sin 2x - 2x} = \lim_{x\to 0}\frac{e^x + e^{-x} - 2}{2\cos 2x - 2}$$

Again, the resulting function is indeterminate of type 0/0. With the understanding that each equality is justified, we obtain:

$$\lim_{x\to 0}\frac{e^x + e^{-x} - x^2 - 2}{\sin^2 x - x^2} = \lim_{x\to 0}\frac{e^x - e^{-x} - 2x}{\sin 2x - 2x} = \lim_{x\to 0}\frac{e^x + e^{-x} - 2}{2\cos 2x - 2}$$

$$= \lim_{x\to 0}\frac{e^x - e^{-x}}{-4\sin 2x} = \lim_{x\to 0}\frac{e^x + e^{-x}}{-8\cos 2x} = -\frac{1}{4}$$

**6.** Criticize: $\lim\limits_{x\to 2}\dfrac{x^3 - x^2 - x - 2}{x^3 - 3x^2 + 3x - 2} = \lim\limits_{x\to 2}\dfrac{3x^2 - 2x - 1}{3x^2 - 6x + 3} = \lim\limits_{x\to 2}\dfrac{6x - 2}{6x - 6} = \lim\limits_{x\to 2}\dfrac{6}{6} = 1.$

The given function is indeterminate of the form 0/0 and the rule applies. But the resulting function is not indeterminate (the limit is 7/3), and the succeeding applications of the rule are not justified. This is a fairly common error.

**7.** Criticize: $\lim\limits_{x\to 1}\dfrac{x^3 - x^2 - x + 1}{x^3 - 2x^2 + x} = \dfrac{3x^2 - 2x - 1}{3x^2 - 4x + 1} = \dfrac{6x - 2}{6x - 4} = 2.$

The correct statement is $\lim\limits_{x\to 1}\dfrac{x^3 - x^2 - x + 1}{x^3 - 2x^2 + x} = \lim\limits_{x\to 1}\dfrac{3x^2 - 2x - 1}{3x^2 - 4x + 1} = \lim\limits_{x\to 1}\dfrac{6x - 2}{6x - 4} = 2.$

The fact that the limit is correct does not justify the series of incorrect statements in obtaining it.

**8.** Evaluate $\lim\limits_{x\to \pi^+}\dfrac{\sin x}{\sqrt{x - \pi}}$.

$$\lim_{x\to \pi^+}\frac{\sin x}{\sqrt{x - \pi}} = \lim_{x\to \pi^+}\frac{\cos x}{\frac{1}{2}(x - \pi)^{-1/2}} = \lim_{x\to \pi^+} 2(x - \pi)^{1/2}\cos x = 0$$

Here the approach must be from the right, since otherwise $(x - \pi)^{1/2}$ is imaginary.

**9.** Evaluate $\lim\limits_{x\to +\infty}\dfrac{\ln x}{x}$.

When $x \to +\infty$, both numerator and denominator approach $+\infty$. Then $\lim\limits_{x\to +\infty}\dfrac{\ln x}{x} = \lim\limits_{x\to +\infty}\dfrac{1/x}{1} = 0.$

**10.** Evaluate $\lim\limits_{x\to 0^+}\dfrac{\ln \sin x}{\ln \tan x}$.　　$\lim\limits_{x\to 0^+}\dfrac{\ln \sin x}{\ln \tan x} = \lim\limits_{x\to 0^+}\dfrac{\cos x / \sin x}{\sec^2 x / \tan x} = \lim\limits_{x\to 0^+}\cos^2 x = 1.$

**11.** Evaluate $\lim\limits_{x\to 0}\dfrac{\cot x}{\cot 2x}$.　　$\lim\limits_{x\to 0}\dfrac{\cot x}{\cot 2x} = \lim\limits_{x\to 0}\dfrac{\csc^2 x}{2\csc^2 2x} = \lim\limits_{x\to 0}\dfrac{\csc^2 x \cot x}{4\csc^2 2x \cot 2x}$

Here each application of the rule results in an indeterminate form of type $\infty/\infty$. We try a trigonometric substitution:

$$\lim_{x\to 0}\frac{\cot x}{\cot 2x} = \lim_{x\to 0}\frac{\tan 2x}{\tan x} = \lim_{x\to 0}\frac{2\sec^2 2x}{\sec^2 x} = 2$$

**12.** Let $\lim\limits_{x \to +\infty} f(x) = 0$ and $\lim\limits_{x \to +\infty} g(x) = 0$. Prove: If $\lim\limits_{x \to +\infty} \dfrac{f'(x)}{g'(x)} = L$, then $\lim\limits_{x \to +\infty} \dfrac{f(x)}{g(x)} = L$.

Let $x = 1/y$. As $x \to +\infty$, $y \to 0^+$ and $\lim\limits_{x \to +\infty} \dfrac{f(x)}{g(x)} = \lim\limits_{y \to 0^+} \dfrac{f(1/y)}{g(1/y)}$. Then

$$L = \lim_{x \to +\infty} \frac{f'(x)}{g'(x)} = \lim_{x \to 0^+} \frac{f'(1/y)}{g'(1/y)} = \lim_{y \to 0^+} \frac{-f'(1/y) \cdot y^{-2}}{-g'(1/y) \cdot y^{-2}}$$

$$= \lim_{y \to 0^+} \frac{\dfrac{d}{dy} f(1/y)}{\dfrac{d}{dy} g(1/y)} = \lim_{y \to 0^+} \frac{f(1/y)}{g(1/y)} = \lim_{x \to +\infty} \frac{f(x)}{g(x)}$$

**13.** Evaluate $\lim\limits_{x \to 0^+} (x^2 \ln x)$.

As $x \to 0^+$, $x^2 \to 0$ and $\ln x \to -\infty$. Then $\dfrac{\ln x}{1/x^2}$ is indeterminate of type $\infty/\infty$.

$$\lim_{x \to 0^+} (x^2 \ln x) = \lim_{x \to 0^+} \frac{\ln x}{1/x^2} = \lim_{x \to 0^+} \frac{1/x}{-2/x^3} = \lim_{x \to 0^+} (-\tfrac{1}{2}x^2) = 0$$

**14.** $\lim\limits_{x \to \pi/4} (1 - \tan x) \sec 2x = \lim\limits_{x \to \pi/4} \dfrac{1 - \tan x}{\cos 2x} = \lim\limits_{x \to \pi/4} \dfrac{-\sec^2 x}{-2 \sin 2x} = 1.$

**15.** $\lim\limits_{x \to 0} \left( \dfrac{1}{x} - \dfrac{1}{e^x - 1} \right) = \lim\limits_{x \to 0} \dfrac{e^x - 1 - x}{x(e^x - 1)} = \lim\limits_{x \to 0} \dfrac{e^x - 1}{xe^x + e^x - 1} = \lim\limits_{x \to 0} \dfrac{e^x}{xe^x + 2e^x} = \dfrac{1}{2}.$

**16.** $\lim\limits_{x \to 0} (\csc x - \cot x) = \lim\limits_{x \to 0} \dfrac{1 - \cos x}{\sin x} = \lim\limits_{x \to 0} \dfrac{\sin x}{\cos x} = 0.$

**17.** Evaluate $\lim\limits_{x \to 1} x^{1/(x-1)}$. (This is type $1^\infty$.)

Let $y = x^{1/(x-1)}$. Then $\ln y = \dfrac{\ln x}{x - 1}$ is indeterminate of type $\dfrac{0}{0}$.

$$\lim_{x \to 1} \ln y = \lim_{x \to 1} \frac{\ln x}{x - 1} = \lim_{x \to 1} \frac{1/x}{1} = 1$$

Since $\ln y \to 1$ as $x \to 1$, $y \to e$. Thus the required limit is $e$.

**18.** Evaluate $\lim\limits_{x \to \frac{1}{2}\pi^-} (\tan x)^{\cos x}$. (This is type $\infty^0$.)

Let $y = (\tan x)^{\cos x}$. Then $\ln y = \cos x \ln \tan x = \dfrac{\ln \tan x}{\sec x}$ is of type $\dfrac{\infty}{\infty}$.

$$\lim_{x \to \frac{1}{2}\pi^-} \ln y = \lim_{x \to \frac{1}{2}\pi^-} \frac{\ln \tan x}{\sec x} = \lim_{x \to \frac{1}{2}\pi^-} \frac{\sec^2 x / \tan x}{\sec x \tan x} = \lim_{x \to \frac{1}{2}\pi^-} \frac{\cos x}{\sin^2 x} = 0$$

Since $\ln y \to 0$ as $x \to \frac{1}{2}\pi^-$, $y \to 1$. Thus, the required limit is $1$.

**19.** Evaluate $\lim\limits_{x \to 0^+} x^{\sin x}$. (This is type $0^0$.)

Let $y = x^{\sin x}$. Then $\ln y = \sin x \ln x = \dfrac{\ln x}{\csc x}$ is indeterminate of type $\dfrac{\infty}{\infty}$.

$$\lim_{x \to 0^+} \ln y = \lim_{x \to 0^+} \frac{\ln x}{\csc x} = \lim_{x \to 0^+} \frac{1/x}{-\csc x \cot x} = \lim_{x \to 0^+} \frac{\sin^2 x}{-x \cos x} = \lim_{x \to 0^+} \frac{2 \sin x \cos x}{x \sin x - \cos x} = 0$$

Since $\ln y \to 0$ as $x \to 0^+$, $y \to 1$. Thus, the required limit is $1$.

**20.** Evaluate $\lim\limits_{x \to +\infty} \dfrac{\sqrt{2 + x^2}}{x}$.  $\quad \lim\limits_{x \to +\infty} \dfrac{\sqrt{2 + x^2}}{x} = \lim\limits_{x \to +\infty} \dfrac{x}{\sqrt{2 + x^2}} = \lim\limits_{x \to +\infty} \dfrac{\sqrt{2 + x^2}}{x}$, etc.

However, $\lim\limits_{x \to +\infty} \dfrac{\sqrt{2 + x^2}}{x} = \lim\limits_{x \to +\infty} \sqrt{\dfrac{2 + x^2}{x^2}} = \lim\limits_{x \to +\infty} \sqrt{\dfrac{2}{x^2} + 1} = 1.$

21. The current in a coil containing a resistance $R$, an inductance $L$, and a constant electromotive force $E$ at time $t$ is given by $i = \dfrac{E}{R}(1 - e^{-Rt/L})$. Obtain a suitable formula to be used when $R$ is very small.

$$\lim_{R \to 0} i = \lim_{R \to 0} \frac{E(1 - e^{-Rt/L})}{R} = \lim_{R \to 0} E \frac{t}{L} e^{-Rt/L} = \frac{Et}{L}$$

# Supplementary Problems

Evaluate

22. $\displaystyle\lim_{x \to 4} \frac{x^4 - 256}{x - 4} = 256$

23. $\displaystyle\lim_{x \to 4} \frac{x^4 - 256}{x^2 - 16} = 32$

24. $\displaystyle\lim_{x \to 3} \frac{x^2 - 3x}{x^2 - 9} = 1/2$

25. $\displaystyle\lim_{x \to 2} \frac{e^x - e^2}{x - 2} = e^2$

26. $\displaystyle\lim_{x \to 0} \frac{xe^x}{1 - e^x} = -1$

27. $\displaystyle\lim_{x \to 0} \frac{e^x - 1}{\tan 2x} = 1/2$

28. $\displaystyle\lim_{x \to -1} \frac{\ln (2 + x)}{x + 1} = 1$

29. $\displaystyle\lim_{x \to 0} \frac{\cos x - 1}{\cos 2x - 1} = 1/4$

30. $\displaystyle\lim_{x \to 0} \frac{e^{2x} - e^{-2x}}{\sin x} = 4$

31. $\displaystyle\lim_{x \to 0} \frac{8^x - 2^x}{4x} = \frac{1}{2}\ln 2$

32. $\displaystyle\lim_{x \to 0} \frac{2 \arctan x - x}{2x - \arcsin x} = 1$

33. $\displaystyle\lim_{x \to 0} \frac{\ln \sec 2x}{\ln \sec x} = 4$

34. $\displaystyle\lim_{x \to 0} \frac{\ln \cos x}{x^2} = -1/2$

35. $\displaystyle\lim_{x \to 0} \frac{\cos 2x - \cos x}{\sin^2 x} = -3/2$

36. $\displaystyle\lim_{x \to +\infty} \frac{\ln x}{\sqrt{x}} = 0$

37. $\displaystyle\lim_{x \to \frac{1}{2}\pi} \frac{\csc 6x}{\csc 2x} = 1/3$

38. $\displaystyle\lim_{x \to +\infty} \frac{5x + 2\ln x}{x + 3\ln x} = 5$

39. $\displaystyle\lim_{x \to +\infty} \frac{x^4 + x^2}{e^x + 1} = 0$

40. $\displaystyle\lim_{x \to 0^+} \frac{\ln \cot x}{e^{\csc^2 x}} = 0$

41. $\displaystyle\lim_{x \to +\infty} \frac{e^x + 3x^3}{4e^x + 2x^2} = 1/4$

42. $\displaystyle\lim_{x \to 0} (e^x - 1) \cot x = 1$

43. $\displaystyle\lim_{x \to -\infty} x^2 e^x = 0$

44. $\displaystyle\lim_{x \to 0} x \csc x = 1$

45. $\displaystyle\lim_{x \to 1} \csc \pi x \ln x = -1/\pi$

46. $\displaystyle\lim_{x \to \frac{1}{2}\pi^-} e^{-\tan x} \sec^2 x = 0$

47. $\displaystyle\lim_{x \to 0} (x - \arcsin x) \csc^3 x = -1/6$

48. $\displaystyle\lim_{x \to 2} \left( \frac{4}{x^2 - 4} - \frac{1}{x - 2} \right) = -1/4$

49. $\displaystyle\lim_{x \to 0} \left( \frac{1}{x} - \frac{1}{\sin x} \right) = 0$

50. $\displaystyle\lim_{x \to \frac{1}{2}\pi} (\sec^3 x - \tan^3 x) = \infty$

51. $\displaystyle\lim_{x \to 1} \left( \frac{1}{\ln x} - \frac{x}{x - 1} \right) = -1/2$

52. $\displaystyle\lim_{x \to 0} \left( \frac{4}{x^2} - \frac{2}{1 - \cos x} \right) = -1/3$

53. $\displaystyle\lim_{x \to +\infty} \left( \frac{\ln x}{x} - \frac{1}{\sqrt{x}} \right) = 0$

54. $\displaystyle\lim_{x \to 0^+} x^x = 1$

55. $\displaystyle\lim_{x \to 0} (\cos x)^{1/x} = 1$

56. $\displaystyle\lim_{x \to 0} (e^x + 3x)^{1/x} = e^4$

57. $\displaystyle\lim_{x \to +\infty} (1 - e^{-x})^{e^x} = 1/e$

58. $\displaystyle\lim_{x \to \frac{1}{2}\pi} (\sin x - \cos x)^{\tan x} = 1/e$

59. $\displaystyle\lim_{x \to \frac{1}{2}\pi^-} (\tan x)^{\cos x} = 1$

60. $\displaystyle\lim_{x \to 1} x^{\tan \frac{1}{2}\pi x} = e^{-2/\pi}$

61. $\displaystyle\lim_{x \to +\infty} (1 + 1/x)^x = e$

62. Evaluate:  (a) $\displaystyle\lim_{x \to 0} \frac{e^x(1 - e^x)}{(1 + x)\ln(1 - x)} = \lim_{x \to 0} \frac{e^x}{1 + x} \cdot \lim_{x \to 0} \frac{1 - e^x}{\ln(1 - x)} = 1,$

(b) $\displaystyle\lim_{x \to +\infty} \frac{2^x}{3^{x^2}} = 0$,   (c) $\displaystyle\lim_{x \to 0^+} \frac{e^{-3/x}}{x^2} = 0$

63. Evaluate $\displaystyle\lim_{x \to +\infty} \frac{\ln^5 x}{x^2} = 0$. Also, $\displaystyle\lim_{x \to +\infty} \frac{\ln^{1000} x}{x^5}$.

# Chapter 23

## Differentials

**DIFFERENTIALS.** For the function $y = f(x)$, we define:

(a) $dx$, called *differential of* $x$, by the relation $dx = \Delta x$.

(b) $dy$, called *differential of* $y$, by the relation $dy = f'(x)\,dx$.

The differential of the independent variable is by definition equal to the increment of the variable, but the differential of the dependent variable is *not* equal to the increment of that variable. See Fig. 23-1 below.

**Example 1:**

When $y = x^2$, $dy = 2x \cdot dx$ while $\Delta y = (x + \Delta x)^2 - x^2 = 2x \cdot \Delta x + (\Delta x)^2 = 2x\,dx + (dx)^2$. A geometric interpretation is given in Fig. 23-2. It will be seen that $\Delta y$ and $dy$ differ by the small square of area $(dx)^2$.

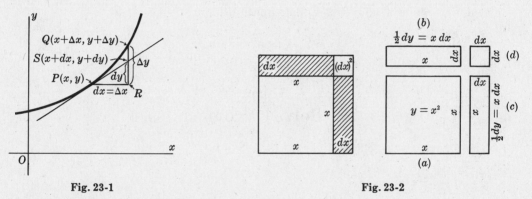

Fig. 23-1          Fig. 23-2

**THE DIFFERENTIAL** $dy$ may be found by using the definition $dy = f'(x)\,dx$ or by means of rules obtained readily from the rules for finding derivatives. Some of these are:

$$d(c) = 0, \quad d(cu) = c\,du, \quad d(uv) = u\,dv + v\,du,$$

$$d\left(\frac{u}{v}\right) = \frac{v\,du - u\,dv}{v^2}, \quad d(\sin u) = \cos u\,du, \quad d(\ln u) = \frac{du}{u}, \quad \text{etc.}$$

**Example 2:** Find $dy$ for each of the following:

(a) $y = x^3 + 4x^2 - 5x + 6$

$\quad dy = d(x^3) + d(4x^2) - d(5x) + d(6) = (3x^2 + 8x - 5)\,dx$

(b) $y = (2x^3 + 5)^{3/2}$

$\quad dy = \frac{3}{2}(2x^3 + 5)^{1/2}\,d(2x^3 + 5) = \frac{3}{2}(2x^3 + 5)^{1/2} \cdot 6x^2\,dx = 9x^2(2x^3 + 5)^{1/2}\,dx$

See Problems 1-5.

**APPROXIMATIONS BY DIFFERENTIALS.** If $dx = \Delta x$ is relatively small when compared with $x$, $dy$ is a fairly good approximation of $\Delta y$.

**Example 3:**

Take $y = x^2 + x + 1$ and let $x$ change from $x = 2$ to $x = 2.01$. The actual change in $y$ is $\Delta y = \{(2.01)^2 + 2.01 + 1\} - \{2^2 + 2 + 1\} = .0501$. The approximate change in $y$, obtained by taking $x = 2$ and $dx = .01$, is $dy = f'(x)\,dx = (2x + 1)\,dx = \{2(2) + 1\}\,.01 = .05$.

See Problems 6-10.

**APPROXIMATIONS OF ROOTS OF EQUATIONS.** Let $x = x_1$ be a fairly close approximation of a root $r$ of the equation $y = f(x) = 0$ and let $f(x_1) = y_1 \neq 0$. Then $y_1$ differs from 0 by a small amount. Now if $x_1$ were changed to $r$, the corresponding change in $f(x_1)$ would be $\Delta y_1 = -y_1$. An approximation of this change in $x_1$ is given by $f'(x_1)\,dx_1 = -y_1$ or $dx_1 = -\dfrac{y_1}{f'(x_1)}$.

Thus, a second and better approximation of the root $r$ is

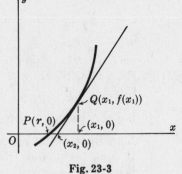

**Fig. 23-3**

$$x_2 = x_1 + dx_1 = x_1 - \frac{y_1}{f'(x_1)} = x_1 - \frac{f(x_1)}{f'(x_1)}$$

A third approximation is $x_3 = x_2 + dx_2 = x_2 - \dfrac{f(x_2)}{f'(x_2)}$, and so on.

When $x_1$ is not a sufficiently close approximation of a root, it will be found that $x_2$ differs materially from $x_1$. While at times the process is self-correcting, it will be simpler to make a new first approximation.

See Problems 11-12.

# Solved Problems

1. Find $dy$ for each of the following:

(a) $y = \dfrac{x^3 + 2x + 1}{x^2 + 3}$.

$$dy = \frac{(x^2 + 3)\cdot d(x^3 + 2x + 1) - (x^3 + 2x + 1)\cdot d(x^2 + 3)}{(x^2 + 3)^2}$$

$$= \frac{(x^2 + 3)(3x^2 + 2)\,dx - (x^3 + 2x + 1)(2x)\,dx}{(x^2 + 3)^2} = \frac{x^4 + 7x^2 - 2x + 6}{(x^2 + 3)^2}\,dx$$

(b) $y = \cos^2 2x + \sin 3x$.

$$dy = 2\cos 2x\, d(\cos 2x) + d(\sin 3x) = 2\cos 2x(-2\sin 2x\, dx) + 3\cos 3x\, dx$$

$$= -4\sin 2x \cos 2x\, dx + 3\cos 3x\, dx = (-2\sin 4x + 3\cos 3x)\,dx$$

(c) $y = e^{3x} + \arcsin 2x$.       $dy = (3e^{3x} + 2/\sqrt{1 - 4x^2})\,dx$

Differentiate each of Problems 2-5, using differentials, and obtain $dy/dx$.

2. $xy + x - 2y = 5$.

$$d(xy) + d(x) - d(2y) = d(5).$$

$$x\,dy + y\,dx + dx - 2\,dy = 0 \quad \text{or} \quad (x - 2)\,dy + (y + 1)\,dx = 0. \quad \text{Then } \frac{dy}{dx} = -\frac{y + 1}{x - 2}.$$

3. $x^3y^2 - 2x^2y + 3xy^2 - 8xy = 6$.

$$2x^3y\,dy + 3x^2y^2\,dx - 2x^2\,dy - 4xy\,dx + 6xy\,dy + 3y^2\,dx - 8x\,dy - 8y\,dx = 0$$

$$\frac{dy}{dx} = \frac{8y - 3y^2 + 4xy - 3x^2y^2}{2x^3y - 2x^2 + 6xy - 8x}$$

4. $\dfrac{2x}{y} - \dfrac{3y}{x} = 8$.       $2\left(\dfrac{y\,dx - x\,dy}{y^2}\right) - 3\left(\dfrac{x\,dy - y\,dx}{x^2}\right) = 0$    and    $\dfrac{dy}{dx} = \dfrac{2x^2y + 3y^3}{3xy^2 + 2x^3}$

**5.**  $x = 3\cos\theta - \cos 3\theta$,  $y = 3\sin\theta - \sin 3\theta$.

$dx = (-3\sin\theta + 3\sin 3\theta)\,d\theta$,  $dy = (3\cos\theta - 3\cos 3\theta)\,d\theta$, and  $\dfrac{dy}{dx} = \dfrac{\cos\theta - \cos 3\theta}{-\sin\theta + \sin 3\theta}$

**6.**  Use differentials to approximate: (a) $\sqrt[3]{124}$, (b) $\sin 60°1'$.

(a) For $y = x^{1/3}$, $dy = \dfrac{1}{3x^{2/3}}\,dx$. Take $x = 125 = 5^3$ and $dx = -1$. Then $dy = \dfrac{1}{3(125)^{2/3}}(-1) = \dfrac{-1}{75} = -0.0133$ and, approximately, $\sqrt[3]{124} = y + dy = 5 - 0.0133 = 4.9867$.

(b) For $x = 60°$ and $dx = 1' = 0.0003$ rad, $y = \sin x = \sqrt{3}/2 = 0.86603$ and $dy = \cos x\,dx = \frac{1}{2}(0.0003) = 0.00015$. Then, approximately, $\sin 60°1' = y + dy = 0.86603 + 0.00015 = 0.86618$.

**7.**  Compute $\Delta y$, $dy$, and $\Delta y - dy$, given $y = \frac{1}{2}x^2 + 3x$, $x = 2$, and $dx = 0.5$.

$\Delta y = \{\frac{1}{2}(2.5)^2 + 3(2.5)\} - \{\frac{1}{2}(2)^2 + 3(2)\} = 2.625$.

$dy = (x+3)dx = (2+3)(0.5) = 2.5$   $\Delta y - dy = 2.625 - 2.5 = 0.125$.

**8.**  Find the approximate change in the volume of a cube of side $x$ in. caused by increasing the sides by 1%.

$V = x^3$ and $dV = 3x^2\,dx$. When $dx = 0.01x$, $dV = 3x^2(0.01x) = 0.03x^3$ in.$^3$

**9.**  Find the approximate weight of an 8 ft length of copper tubing if the inside diameter is 1 in. and the thickness is 1/8 in. The specific weight of copper is 550 lb/ft.$^3$

First find the change in volume when the radius $r = 1/24$ ft is changed by $dr = 1/96$ ft.

$$V = 8\pi r^2 \quad \text{and} \quad dV = 16\pi r\,dr = 16\pi(1/24)(1/96) = \pi/144 \text{ ft}^3$$

The required weight is $550(\pi/144) = 12$ lb.

**10.**  For what values of $x$ may $\sqrt[5]{x}$ be used in place of $\sqrt[5]{x+1}$, if an allowable error must be less than 0.001?

When $y = x^{1/5}$ and $dx = 1$, $dy = \frac{1}{5}x^{-4/5}\,dx = \frac{1}{5}x^{-4/5}$.

If $\frac{1}{5}x^{-4/5} < 10^{-3}$, then $x^{-4/5} < 5 \cdot 10^{-3}$ and $x^{-4} < 5^5 \cdot 10^{-15}$.

If $x^{-4} < 10 \cdot 5^5 \cdot 10^{-16}$, then $x^4 > \dfrac{10^{16}}{31250}$ and $x > \dfrac{10^4}{\sqrt[4]{31250}} = 752.1$.

**11.**  Approximate the (real) roots of $x^3 + 2x - 5 = 0$ or $x^3 = 5 - 2x$.

(a) On the same axes, construct the graphs of $y = x^3$ and $y = 5 - 2x$.

The abscissas of the points of intersection of the curves are roots of the given equation.

From the graph, it is seen that there is one root whose approximate value is $x_1 = 1.3$.

(b) A second approximation of this root is

$$x_2 = x_1 - \frac{f(x_1)}{f'(x_1)} = 1.3 - \frac{(1.3)^3 + 2(1.3) - 5}{3(1.3)^2 + 2} = 1.3 - \frac{-.203}{7.07} = 1.3 + .03 = 1.33$$

The division is carried out to yield two decimal places, since there is but one zero immediately following the decimal point. This is in accord with the theorem: If in a division, $k$ zeros immediately follow the decimal point in the quotient, the division can be carried out to yield $2k$ decimal places.

(c) A third and fourth approximation are:

$$x_3 = x_2 - \frac{f(x_2)}{f'(x_2)} = 1.33 - \frac{(1.33)^3 + 2(1.33) - 5}{3(1.33)^2 + 2} = 1.33 - .0017 = 1.3283$$

$$x_4 = x_3 - \frac{f(x_3)}{f'(x_3)} = 1.3283 - .00003114 = 1.32826886$$

12. Approximate the roots of $2 \cos x - x^2 = 0$.

   (a) The curves $y = 2 \cos x$ and $y = x^2$ intersect in two points whose abscissas are approximately 1 and $-1$. Note that if $r$ is one root, $-r$ is the other.

   (b) Using $x_1 = 1$:    $x_2 = 1 - \dfrac{2 \cos 1 - 1}{-2 \sin 1 - 2} = 1 + \dfrac{2(.5403) - 1}{2(.8415) + 2} = 1 + .02 = 1.02$.

   (c)   $x^3 = 1.02 - \dfrac{2 \cos (1.02) - (1.02)^2}{-2 \sin (1.02) - 2(1.02)} = 1.02 + \dfrac{.0064}{3.7442} = 1.02 + .0017 = 1.0217$.

   Thus, to four decimal places, the roots are 1.0217 and $-1.0217$.

# Supplementary Problems

13. Find $dy$ for each of the following.

   (a) $y = (5 - x)^3$     *Ans.* $-3(5 - x)^2 \, dx$        (d) $y = \cos bx^2$     *Ans.* $-2bx \sin bx^2 \, dx$

   (b) $y = e^{4x^2}$       *Ans.* $8xe^{4x^2} dx$            (e) $y = \arccos 2x$     *Ans.* $\dfrac{-2}{\sqrt{1 - 4x^2}} dx$

   (c) $y = (\sin x)/x$    *Ans.* $\dfrac{x \cos x - \sin x}{x^2} \, dx$      (f) $y = \ln \tan x$     *Ans.* $\dfrac{2 \, dx}{\sin 2x}$

14. Find $dy/dx$ as in Problems 2-5.

   (a) $2xy^3 + 3x^2 y = 1$    *Ans.* $-\dfrac{2y(y^2 + 3x)}{3x(2y^2 + x)}$    (c) $\arctan \dfrac{y}{x} = \ln (x^2 + y^2)$    *Ans.* $\dfrac{2x + y}{x - 2y}$

   (b) $xy = \sin (x - y)$    *Ans.* $\dfrac{\cos (x - y) - y}{\cos (x - y) + x}$    (d) $x^2 \ln y + y^2 \ln x = 2$    *Ans.* $-\dfrac{(2x^2 \ln y + y^2)y}{(2y^2 \ln x + x^2)x}$

15. Use differentials to approximate:   (a) $\sqrt[4]{17}$,   (b) $\sqrt[5]{1020}$,   (c) $\cos 59°$,   (d) $\tan 44°$.
   *Ans.* (a) 2.03125,   (b) 3.99688,   (c) 0.5151,   (d) 0.9651

16. Use differentials to approximate the change in:   (a) $x^3$ as $x$ changes from 5 to 5.01;   (b) $1/x$ as $x$ changes from 1 to 0.98.     *Ans.* (a) 0.75,   (b) 0.02

17. A circular plate expands under the influence of heat so that its radius increases from 5 in. to 5.06 in. Find the approximate increase in area.     *Ans.* $0.6\pi = 1.88 \text{ in}^2$

18. A sphere of ice of radius 10 in. shrinks to radius 9.8 in. Approximate the decrease in (a) volume and (b) surface area.     *Ans.* (a) $80\pi \text{ in}^3$, (b) $16\pi \text{ in}^2$

19. The velocity ($v$ ft/sec) attained by a body falling freely from rest a distance $h$ ft is given by $v = \sqrt{64.4h}$. Find the error in $v$ due to an error of 0.5 ft when $h$ is measured as 100 ft.     *Ans.* 0.2 ft/sec

20. If an aviator flies around the world at a distance 2 miles above the equator, how many more miles will he travel than a person who travels along the equator?     *Ans.* 12.6 mi

21. The radius of a circle is to be measured and its area computed. If the radius can be measured to 0.001 in. and the area must be accurate to 0.1 in², find the maximum radius for which this process can be used.     *Ans.* Approx. 16 in.

22. If $pV = 20$ and $p$ is measured as $5 \pm 0.02$, find $V$.     *Ans.* $V = 4 \mp 0.016$

23. If $F = 1/r^2$ and $F$ is measured as $4 \pm 0.05$, find $r$.     *Ans.* $0.5 \mp 0.003$

24. Find the change in the total surface of a right circular cone when (a) the radius remains constant while the altitude changes by a small amount, (b) the altitude remains constant while the radius changes by a small amount.

   *Ans.* (a) $\dfrac{\pi rh \, dh}{\sqrt{r^2 + h^2}}$,   (b) $\pi \left\{ \dfrac{h^2 + 2r^2}{\sqrt{r^2 + h^2}} + 2r \right\} dr$

25. Find to 4 decimal places (a) the real root of $x^3 + 3x + 1 = 0$, (b) the smallest root of $e^{-x} = \sin x$, (c) the root of $x^2 + \ln x = 2$, (d) the root of $x - \cos x = 0$.
   *Ans.* (a) $-0.3222$, (b) 0.5885, (c) 1.3141, (d) 0.7391

# Chapter 24

## Curve Tracing

**A PLANE ALGEBRAIC CURVE** is one whose equation may be written in the form

$$ay^n + (bx + c)y^{n-1} + (dx^2 + ex + f)y^{n-2} + \cdots u_n(x) = 0$$

where $u_n(x)$ is a polynomial in $x$ of degree $n$. The properties of an algebraic curve are discussed below.

**SYMMETRY.** A curve is symmetric with respect to
(1)  the $x$-axis, if its equation is unchanged when $y$ is replaced by $-y$.
(2)  the $y$-axis, if its equation is unchanged when $x$ is replaced by $-x$.
(3)  the origin, if its equation is unchanged when $x$ is replaced by $-x$ and $y$ by $-y$ simultaneously.
(4)  the line $y = x$, if its equation is unchanged when $x$ and $y$ are interchanged.

**INTERCEPTS.** The $x$-intercepts are obtained by setting $y = 0$ in the equation and solving for $x$. The $y$-intercepts are obtained by setting $x = 0$ and solving for $y$.

**EXTENT.** The *horizontal extent* is given by the range of $x$, i.e. the intervals of $x$ for which the curve exists.

The *vertical extent* of a curve is given by the range of $y$.

A point $(x_0, y_0)$ is called an *isolated point* of the curve if its coordinates satisfy the equation of the curve while those of no other nearby point do.

**MAXIMUM AND MINIMUM POINTS,** Points of Inflection, and Concavity. These are discussed in Chapter 8.

**ASYMPTOTES.** An *asymptote* of a curve of infinite extent is a line whose position is approached as a limit by a secant of the curve as two of its points of intersection with the curve recede indefinitely along the curve.

A curve will have *vertical asymptotes* if, when its equation is written in the form above, the coefficient of the highest power of $y$ is a non-constant function of $x$ having one or more (real) linear factors. To each such factor, there corresponds a vertical asymptote.

A curve will have *horizontal asymptotes* if, when its equation is written in the form $ax^n + (by + c)x^{n-1} + (dy^2 + ey + f)x^{n-2} + \cdots = 0$, the coefficient of the highest power of $x$ is a non-constant function of $y$ having one or more (real) linear factors. To each such factor, there corresponds a horizontal asymptote.

To obtain the equations of the *oblique asymptotes*:
(1)  Replace $y$ by $mx + b$ in the equation of the curve and arrange the result in the form

$$a_0 x^n + a_1 x^{n-1} + a_2 x^{n-2} + \cdots + a_{n-1} x + a_n = 0$$

(2)  Solve simultaneously the equations $a_0 = 0$ and $a_1 = 0$ for $m$ and $b$.
(3)  For each pair of solutions $m$ and $b$, write the equation of an asymptote $y = mx + b$.

If $a_1 = 0$, irrespective of the value of $b$, the equations $a_0 = 0$ and $a_2 = 0$ are to be used in (3).

**SINGULAR POINTS.** A *singular point* of an algebraic curve is a point for which $dy/dx$ has the indeterminate form $0/0$.

To locate the singular points of a curve, obtain $\dfrac{dy}{dx} = \dfrac{g(x)}{h(x)}$, without simplifying by the cancellation of common factors, and find the common roots of $g(x) = 0$ and $h(x) = 0$.

If $(x_0, y_0)$ is a singular point of a curve, further study is simplified by the substitution of $x = x' + x_0$, $y = y' + y_0$. The singular point is now the point $(0, 0)$ in the new coordinate system.

**SINGULAR POINT AT THE ORIGIN.** When the origin is a point on a curve, its equation may be put in the form

$$(a_1 x + b_1 y) + (a_2 x^2 + b_2 xy + c_2 y^2) + (a_3 x^3 + b_3 x^2 y + c_3 xy^2 + d_3 y^3) + \cdots = 0$$

If $a_1 = b_1 = 0$, the origin is a singular point of the curve.

If $a_1 = b_1 = 0$, but not all of $a_2, b_2, c_2$ are zero, the singular point is called a *double point*.

If $a_1 = b_1 = a_2 = b_2 = c_2 = 0$, but not all of $a_3, b_3, c_3, d_3$ are zero, the singular point is called a *triple point*, and so on.

### CLASSIFICATION OF A DOUBLE POINT AT THE ORIGIN

**A.** Case: $c_2 \neq 0$.

(1) Replace $y$ by $mx$ in the terms $a_2 x^2 + b_2 xy + c_2 y^2$ to obtain $(c_2 m^2 + b_2 m + a_2)x^2$.

(2) Solve $c_2 m^2 + b_2 m + a_2 = 0$ for $m$.

If the roots $m_1, m_2$ are real and distinct, the curve has two distinct tangents $y = m_1 x$ and $y = m_2 x$ at the origin and the double point is a *node*.

If the roots are real and equal, the curve has generally a single tangent at the origin and the double point is called

(a) a *cusp*, provided the curve does not continue through the origin.

(b) a *tacnode*, provided the curve continues through the origin.

In exceptional cases, the origin may be an isolated point.

If the roots are imaginary, the origin is an *isolated double point*.

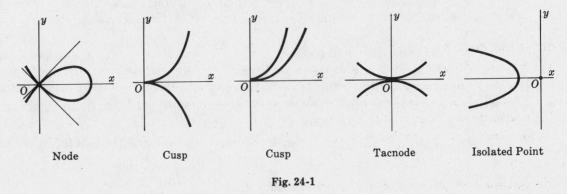

| Node | Cusp | Cusp | Tacnode | Isolated Point |

**Fig. 24-1**

**B.** Case: $c_2 = 0$, $a_2 \neq 0$.

Replace $x$ by $ny$ in the terms $a_2 x^2 + b_2 xy$ and proceed as in *A*.

**C.** Case: $a_2 = c_2 = 0$, $b_2 \neq 0$.

The origin is a node, the two tangents there being the coordinate axes.

# Solved Problems

**ASYMPTOTES**

1.  Find the equations of the asymptotes of $y^2(1+x) = x^2(1-x)$.

The coefficient of the highest power of $y$ is $(1+x)$; the line $x+1=0$ is a vertical asymptote. There is no horizontal asymptote since the coefficient of the highest power of $x$ is a constant.

For the oblique asymptotes, replace $y$ by $mx+b$ to obtain

$$(m^2+1)x^3 + (m^2+2mb-1)x^2 + b(b+2m)x + b^2 = 0 \tag{1}$$

The simultaneous solutions of the coefficients of the two highest powers of $x$ equated to zero,

$$m^2+1=0 \quad \text{and} \quad m^2+2mb-1=0$$

are imaginary. There are no oblique asymptotes. (See Fig. 24-2 below.)

2.  Find the equations of the asymptotes of $x^3+y^3-6x^2=0$.

There are neither horizontal nor vertical asymptotes since the coefficients of the highest powers of $x$ and $y$ are constants.

For the oblique asymptotes, replace $y$ by $mx+b$ to obtain

$$(m^3+1)x^3 + 3(m^2b-2)x^2 + 3mb^2x + b^3 = 0 \tag{1}$$

Solve simultaneously $m^3+1=0$ and $m^2b-2=0$: $m=-1$, $b=2$. The equation of the asymptote is $y=-x+2$.

When $m=-1$ and $b=2$ are substituted in (1), the equation becomes $-12x+8=0$. Then $x=2/3$ is the abscissa of the finite point of intersection of the curve and its asymptote. (See Fig. 24-3 below.)

3.  Find the equations of the asymptotes of $y^2(x-1)-x^3=0$.

The coefficient of the highest power of $y$ is $(x-1)$; the line $x-1=0$ is a vertical asymptote. There are no horizontal asymptotes.

For the oblique asymptotes, replace $y$ by $mx+b$ to obtain

$$(m^2-1)x^3 + m(2b-m)x^2 + b(b-2m)x - b^2 = 0 \tag{1}$$

Solve simultaneously $m^2-1=0$ and $m(2b-m)=0$: $m=1$, $b=\frac{1}{2}$ and $m=-1$, $b=-\frac{1}{2}$. The equations of the asymptotes are $y=x+\frac{1}{2}$ and $y=-x-\frac{1}{2}$.

The asymptote $y=x+\frac{1}{2}$ intersects the curve in a finite point whose abscissa is given by $\frac{1}{2}(\frac{1}{2}-2)x-\frac{1}{4}=0$, i.e. $x=-\frac{1}{3}$. The abscissa of the finite point of intersection of the curve and the asymptote $y=-x-\frac{1}{2}$ is also $-\frac{1}{3}$. (See Fig. 24-4 below.)

**SINGULAR POINTS**

4.  Examine $y^2(1+x) = x^2(1-x)$ for singular points.

The terms of lowest degree are of the second degree; the origin is a double point.

Since $c_2 \neq 0$, that is, the term in $y^2$ is present, replace $y$ by $mx$ in the terms $y^2-x^2$ and equate to zero the coefficient of $x^2$ to obtain $m^2-1=0$.

Then $m=\pm1$ and the lines $y=x$ and $y=-x$ are tangent to the curve at the origin. The origin is a node. (See Fig. 24-2 below.)

5.  Examine $x^3+y^3-6x^2=0$ for singular points.

The terms of lowest degree are of the second degree; the origin is a double point.

Since $c_2=0$, replace $x$ by $ny$ in the terms of lowest degree and equate to zero the coefficient of $y^2$ to obtain $n^2=0$. There is a single tangent, $x=0$, to the curve at the origin.

The double point is a cusp since, when $y=-\xi$, the equation $x^3-6x^2-\xi^3=0$ has, by Descartes rule of signs, one positive and two imaginary roots and the curve does not continue through the origin. (See Fig. 24-3 below.)

**6.** Examine $y^2(x-1) - x^3 = 0$ for singular points.

The terms of lowest degree are of the second degree; the origin is a double point.

Since $c_2 \neq 0$, replace $y$ by $mx$ in the terms of lowest degree and equate to zero the coefficient of $x^2$ to obtain $m^2 = 0$. The origin is a cusp since, for $x < 0$, $y$ is defined, but for $0 < x < 1$, $y$ is imaginary. (See Fig. 24-4 below.)

**7.** Examine $y^2(x^2 - 4) = x^4$ for (a) singular points and (b) asymptotes.

(a) The origin is a double point. Since $a_2 = b_2 = 0$ and $c_2 \neq 0$, the result of substituting $y = mx$ and equating to zero is $m^2 = 0$. The origin is an isolated double point since for $x$ near 0, $y$ is imaginary.

(b) The lines $x = 2$ and $x = -2$ are vertical asymptotes.

For the oblique asymptotes, replace $y$ by $mx + b$ to obtain

$$(m^2 - 1)x^4 + 2mbx^3 + (b^2 - 4m^2)x^2 - 8mbx - 4b^2 = 0$$

Solve simultaneously $m^2 - 1 = 0$ and $mb = 0$: $m = 1$, $b = 0$ and $m = -1$, $b = 0$. The equations of the asymptotes are $y = x$ and $y = -x$.

The oblique asymptotes intersect the curve at the origin. (See Fig. 24-5 below.)

## CURVE TRACING

**8.** Discuss and sketch the curve $y^2(1+x) = x^2(1-x)$.

*Symmetry.* The curve is symmetric with respect to the $x$-axis.

*Intercepts.* The $x$-intercepts are $x = 0$ and $x = 1$. The $y$-intercept is $y = 0$.

*Extent.* The curve exists on the interval $-1 < x \leqq 1$ and for all values of $y$.

*Maximum and Minimum Points, etc.* The curve consists of two branches $y = \dfrac{x\sqrt{1-x}}{\sqrt{1+x}}$ and $y = -\dfrac{x\sqrt{1-x}}{\sqrt{1+x}}$. For the first of these

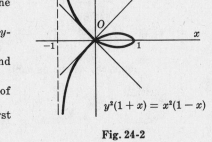

$$y^2(1+x) = x^2(1-x)$$

**Fig. 24-2**

$$\frac{dy}{dx} = \frac{1 - x - x^2}{(1+x)^{3/2}(1-x)^{1/2}} \quad \text{and} \quad \frac{d^2y}{dx^2} = \frac{x - 2}{(1+x)^{5/2}(1-x)^{3/2}}$$

The critical values are $x = 1$ and $(-1 + \sqrt{5})/2$. The point $\left(\dfrac{-1+\sqrt{5}}{2}, \dfrac{(-1+\sqrt{5})\sqrt{\sqrt{5}-2}}{2}\right)$ is a maximum point. There is no point of inflection. The branch is concave downward.

By symmetry, there is a minimum point at $\left(\dfrac{-1+\sqrt{5}}{2}, -\dfrac{(-1+\sqrt{5})\sqrt{\sqrt{5}-2}}{2}\right)$ and the second branch is concave upward.

*Asymptotes.* From Problem 1, the line $x = -1$ is a vertical asymptote.

*Singular Points.* From Problem 4, the origin is a node, the (double point or nodal) tangents being the lines $y = x$ and $y = -x$.

**9.** Discuss and sketch the curve $y^3 - x^2(6-x) = 0$. See Fig. 24-3 below.

*Symmetry.* There is no symmetry.

*Intercepts.* The intercepts are $x = 0$, $x = 6$, and $y = 0$.

*Extent.* The curve exists for all values of $x$ and $y$.

*Maximum and Minimum Points, etc.* $\dfrac{dy}{dx} = \dfrac{4-x}{x^{1/3}(6-x)^{2/3}}$ and $\dfrac{d^2y}{dx^2} = \dfrac{-8}{x^{4/3}(6-x)^{5/3}}$.

The critical values are $x = 0, 4, 6$; $(0, 0)$ is a minimum point and $(4, 2\sqrt[3]{4})$ is a maximum point. The point $(6, 0)$ is a point of inflection, the curve being concave downward to the left and concave upward to the right.

*Asymptotes.* From Problem 2, the line $y = -x + 2$ is an asymptote.

*Singular Points.* From Problem 5, the origin is a cusp, the (cuspidal) tangent being the $y$-axis.

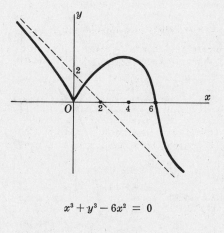

$$x^3 + y^3 - 6x^2 = 0$$

Fig. 24-3

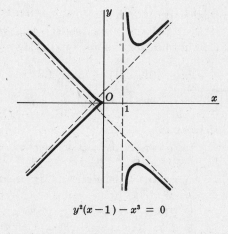

$$y^2(x-1) - x^3 = 0$$

Fig. 24-4

10. Discuss and sketch the curve $y^2(x-1) - x^3 = 0$.  See Fig. 24-4 above.

*Symmetry.* The curve is symmetric with respect to the $x$-axis.

*Intercepts.* The intercepts are $x = 0$ and $y = 0$.

*Extent.* The curve exists on the intervals $-\infty < x \le 0$ and $x > 1$, and for all values of $y$.

*Maximum and Minimum Points, etc.* For the branch $y = x\sqrt{\dfrac{x}{x-1}}$,

$$\frac{dy}{dx} = \frac{(2x-3)x^{1/2}}{2(x-1)^{3/2}} \quad \text{and} \quad \frac{d^2y}{dx^2} = \frac{3}{4x^{1/2}(x-1)^{5/2}}$$

The critical values are $x = 0$ and $3/2$.  The point $(3/2, 3\sqrt{3}/2)$ is a minimum point.  There is no point of inflection.  The branch is concave upward.  By symmetry, there is a maximum point $(3/2, -3\sqrt{3}/2)$ on the branch $y = -x\sqrt{\dfrac{x}{x-1}}$ and the branch is concave downward.

*Asymptotes.* From Problem 3, the lines $x = 1$, $y = x + \frac{1}{2}$, and $y = -x - \frac{1}{2}$ are asymptotes.
*Singular Points.* From Prob. 6, the origin is a cusp, the line $y = 0$ being the (cuspidal) tangent.

11. Discuss and sketch the curve $y^2(x^2 - 4) = x^4$.

*Symmetry.* The curve is symmetric with respect to the coordinate axes and the origin.

*Intercepts.* The intercepts are $x = 0$ and $y = 0$.

*Extent.* The curve exists on the interval $-\infty < x < -2$ and $2 < x < +\infty$ and on the intervals $-\infty < y \le -4$ and $4 \le y < +\infty$.  The point $(0, 0)$ is an isolated point.

*Maximum and Minimum Points, etc.*

For the portion $y = \dfrac{x^2}{\sqrt{x^2 - 4}}$, $x > 2$,

$$\frac{dy}{dx} = \frac{x^3 - 8x}{(x^2 - 4)^{3/2}} \quad \text{and} \quad \frac{d^2y}{dx^2} = \frac{4x^2 + 32}{(x^2 - 4)^{5/2}}$$

The critical value is $x = 2\sqrt{2}$.  The portion is concave upward and $(2\sqrt{2}, 4)$ is a minimum point.

By symmetry, there is a minimum point at $(-2\sqrt{2}, 4)$ and maximum points at $(2\sqrt{2}, -4)$ and $(-2\sqrt{2}, -4)$.

*Asymptotes, Singular Points.* See Problem 7.

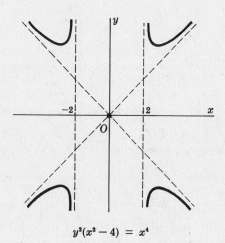

$$y^2(x^2 - 4) = x^4$$

Fig. 24-5

12. Discuss and sketch the curve $(x+3)(x^2+y^2) = 4$.

We first locate the singular point, if any, and translate to the singular point as new origin before making the analysis.

$$\frac{dy}{dx} = -\frac{(x+2)(x+2+\sqrt{3})(x+2-\sqrt{3})}{(x+3)^2 y} \ . \text{ When } x = -2, y = 0$$

and $\frac{dy}{dx}$ has the indeterminate form $\frac{0}{0}$. The point $(-2, 0)$ is a singular point.

Under the transformation $x = x'-2$, $y = y'$, the equation becomes $y'^2(x'+1) + x'^3 - 3x'^2 = 0$.

*Symmetry.* The curve is symmetric with respect to the $x'$-axis.

*Intercepts.* The intercepts are $x' = 0$, $x' = 3$, and $y' = 0$.

*Extent.* The curve is defined on the interval $-1 < x' \leqq 3$ and for all values of $y'$.

*Maximum and Minimum Points, etc.*

For the branch $y' = \dfrac{x'\sqrt{3-x'}}{\sqrt{x'+1}}$

$$\frac{dy'}{dx'} = \frac{3-x'^2}{(3-x')^{1/2}(x'+1)^{3/2}} \quad \text{and} \quad \frac{d^2y'}{dx'^2} = \frac{-12}{(3-x')^{3/2}(x'+1)^{5/2}} .$$

The critical values are $x' = \sqrt{3}$ and $3$. The point $\left(\sqrt{3}, \sqrt{6\sqrt{3}-9}\right)$ is a maximum point. The branch is concave downward.

By symmetry, $\left(\sqrt{3}, \sqrt{6\sqrt{3}-9}\right)$ is a minimum point on the other branch which is concave upward.

*Asymptotes.* The line $x' = -1$ is a vertical asymptote. For the oblique asymptotes, replace $y'$ by $mx'+b$ to obtain $(m^2+1)x'^3 + \cdots = 0$. There are no oblique asymptotes. Why?

*Singular Points.* The origin is a double point. When $y'$ is replaced by $mx'$ in the terms of lowest degree $y'^2 - 3x'^2$, the result is $(m^2-3)x'^2$. From $m^2-3=0$, $m = \pm\sqrt{3}$ and the (nodal) tangents are $y' = \pm\sqrt{3}\,x'$.

In the original coordinates, $\left(\sqrt{3}-2, \sqrt{6\sqrt{3}-9}\right)$ is a maximum point and $\left(\sqrt{3}-2, -\sqrt{6\sqrt{3}-9}\right)$ is a minimum point. The line $x = -3$ is a vertical asymptote. The point $(-2, 0)$ is a node, the equations of the (nodal) tangents being $y = \pm\sqrt{3}\,(x+2)$.

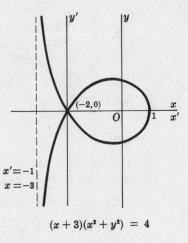

$$(x+3)(x^2+y^2) = 4$$

**Fig. 24-6**

# Supplementary Problems

Discuss and sketch each of the following.

13. $(x-2)(x-6)y = 2x^2$
14. $x(3-x^2)y = 1$
15. $(1-x^2)y = x^4$
16. $xy = (x^2-9)^2$
17. $2xy = (x^2-1)^3$
18. $x(x^2-4)y = x^2-6$
19. $y^2 = x(x^2-4)$
20. $y^2 = (x^2-1)(x^2-4)$
21. $xy^2 = x^2+3x+2$
22. $(x^2-2x-3)y^2 = 2x+3$

23. $x(x-1)y = x^2-4$
24. $(x+1)(x+4)^2y^2 = x(x^2-4)$
25. $y^2 = 4x^2(4-x^2)$
26. $y^2 = 5x^4+4x^5$
27. $y^3 = x^2(8-x^2)$
28. $y^3 = x^2(3-x)$
29. $(x^2-1)y^3 = x^2$
30. $(x-3)y^3 = x^4$
31. $(x-6)y^2 = x^2(x-4)$
32. $(x^2-16)y^2 = x^3(x-2)$

33. $(x^2+y^2)^2 = 8xy$
34. $(x^2+y^2)^3 = 4x^2y^2$
35. $y^4 - 4xy^2 = x^4$
36. $(x^2+y^2)^3 = 4xy(x^2-y^2)$
37. $y^2 = x(x-3)^2$
38. $y^2 = x(x-2)^3$
39. $3y^4 = x(x^2-9)^3$
40. $x^3y^3 = (x-3)^2$

# Chapter 25

## Fundamental Integration Formulas

IF $F(x)$ IS A FUNCTION whose derivative $F'(x) = f(x)$ on a certain interval of the $x$-axis, then $F(x)$ is called an *anti-derivative* or *indefinite integral* of $f(x)$. The indefinite integral of a given function is not unique; for example, $x^2$, $x^2 + 5$, $x^2 - 4$ are indefinite integrals of $f(x) = 2x$ since $\frac{d}{dx}(x^2) = \frac{d}{dx}(x^2 + 5) = \frac{d}{dx}(x^2 - 4) = 2x$. All indefinite integrals of $f(x) = 2x$ are then included in $x^2 + C$ where $C$, called the *constant of integration*, is an arbitrary constant.

The symbol $\int f(x)\,dx$ is used to indicate that the indefinite integral of $f(x)$ is to be found. Thus we write $\int 2x\,dx = x^2 + C$.

FUNDAMENTAL INTEGRATION FORMULAS. A number of the formulas below follow immediately from the standard differentiation formulas of earlier chapters while Formula 25, for example, may be checked by showing that

$$\frac{d}{du}\left\{ \tfrac{1}{2}u\sqrt{a^2 - u^2} + \tfrac{1}{2}a^2 \arcsin \frac{u}{a} + C \right\} = \sqrt{a^2 - u^2}$$

Absolute value signs appear in certain of the formulas. For example, we write

$$5. \qquad \int \frac{du}{u} = \ln|u| + C$$

instead of

$$5(a). \quad \int \frac{du}{u} = \ln u + C, \quad u > 0 \qquad 5(b). \quad \int \frac{du}{u} = \ln(-u) + C, \quad u < 0$$

and

$$10. \qquad \int \tan u\,du = \ln|\sec u| + C$$

instead of

$$10(a). \qquad \int \tan u\,du = \ln \sec u + C, \quad \text{all } u \text{ such that } \sec u \geqq 1$$

$$10(b). \qquad \int \tan u\,du = \ln(-\sec u) + C, \quad \text{all } u \text{ such that } \sec u \leqq -1$$

1. $\int \frac{d}{dx}[f(x)]\,dx = f(x) + C$

2. $\int (u + v)\,dx = \int u\,dx + \int v\,dx$

3. $\int au\,dx = a\int u\,dx, \quad a \text{ any constant}$

4. $\int u^m\,du = \frac{u^{m+1}}{m+1} + C, \quad m \neq -1$

5. $\int \frac{du}{u} = \ln|u| + C$

6. $\int a^u\,du = \frac{a^u}{\ln a} + C, \quad a > 0, a \neq 1$

7. $\int e^u\,du = e^u + C$

8. $\int \sin u\,du = -\cos u + C$

9. $\int \cos u\,du = \sin u + C$

10. $\int \tan u\,du = \ln|\sec u| + C$

11. $\int \cot u\,du = \ln|\sin u| + C$

12. $\int \sec u\,du = \ln|\sec u + \tan u| + C$

13. $\int \csc u\,du = \ln|\csc u - \cot u| + C$

14. $\int \sec^2 u\,du = \tan u + C$

15. $\int \csc^2 u\,du = -\cot u + C$

16. $\int \sec u \tan u\,du = \sec u + C$

17. $\displaystyle\int \csc u \cot u \, du = -\csc u + C$　　　23. $\displaystyle\int \frac{du}{\sqrt{u^2 + a^2}} = \ln\left(u + \sqrt{u^2 + a^2}\right) + C$

18. $\displaystyle\int \frac{du}{\sqrt{a^2 - u^2}} = \arcsin \frac{u}{a} + C$　　　24. $\displaystyle\int \frac{du}{\sqrt{u^2 - a^2}} = \ln\left|u + \sqrt{u^2 - a^2}\right| + C$

19. $\displaystyle\int \frac{du}{a^2 + u^2} = \frac{1}{a}\arctan \frac{u}{a} + C$　　　25. $\displaystyle\int \sqrt{a^2 - u^2}\, du = \tfrac{1}{2}u\sqrt{a^2 - u^2} + \tfrac{1}{2}a^2 \arcsin \frac{u}{a} + C$

20. $\displaystyle\int \frac{du}{u\sqrt{u^2 - a^2}} = \frac{1}{a}\operatorname{arc\,sec} \frac{u}{a} + C$　　　26. $\displaystyle\int \sqrt{u^2 + a^2}\, du = \tfrac{1}{2}u\sqrt{u^2 + a^2}$
$$+ \tfrac{1}{2}a^2 \ln\left(u + \sqrt{u^2 + a^2}\right) + C$$

21. $\displaystyle\int \frac{du}{u^2 - a^2} = \frac{1}{2a}\ln\left|\frac{u - a}{u + a}\right| + C$

22. $\displaystyle\int \frac{du}{a^2 - u^2} = \frac{1}{2a}\ln\left|\frac{a + u}{a - u}\right| + C$　　　27. $\displaystyle\int \sqrt{u^2 - a^2}\, du = \tfrac{1}{2}u\sqrt{u^2 - a^2}$
$$- \tfrac{1}{2}a^2 \ln\left|u + \sqrt{u^2 - a^2}\right| + C$$

# Solved Problems

1. $\displaystyle\int x^5 \, dx = \frac{x^6}{6} + C$　　　3. $\displaystyle\int \sqrt[3]{z} \, dz = \int z^{1/3} \, dz = \frac{z^{4/3}}{4/3} + C = \frac{3}{4}z^{4/3} + C$

2. $\displaystyle\int \frac{dx}{x^2} = \int x^{-2} \, dx = \frac{x^{-1}}{-1} + C = -\frac{1}{x} + C$　　　4. $\displaystyle\int \frac{dx}{\sqrt[3]{x^2}} = \int x^{-2/3} \, dx = \frac{x^{1/3}}{1/3} + C = 3x^{1/3} + C$

5. $\displaystyle\int (2x^2 - 5x + 3)\, dx = 2\int x^2 \, dx - 5\int x \, dx + 3\int dx = \frac{2x^3}{3} - \frac{5x^2}{2} + 3x + C$

6. $\displaystyle\int (1 - x)\sqrt{x} \, dx = \int (x^{1/2} - x^{3/2})\, dx = \int x^{1/2} \, dx - \int x^{3/2} \, dx = \tfrac{2}{3}x^{3/2} - \tfrac{2}{5}x^{5/2} + C$

7. $\displaystyle\int (3s + 4)^2 \, ds = \int (9s^2 + 24s + 16)\, ds = 9(\tfrac{1}{3}s^3) + 24(\tfrac{1}{2}s^2) + 16s + C = 3s^3 + 12s^2 + 16s + C$

8. $\displaystyle\int \frac{x^3 + 5x^2 - 4}{x^2}\, dx = \int (x + 5 - 4x^{-2})\, dx = \frac{1}{2}x^2 + 5x - \frac{4x^{-1}}{-1} + C = \frac{1}{2}x^2 + 5x + \frac{4}{x} + C$

9. Evaluate:　(a) $\displaystyle\int (x^3 + 2)^2 \cdot 3x^2 \, dx$,　(b) $\displaystyle\int (x^3 + 2)^{1/2}x^2 \, dx$,　(c) $\displaystyle\int \frac{8x^2 \, dx}{(x^3 + 2)^3}$,　(d) $\displaystyle\int \frac{x^2 \, dx}{\sqrt[4]{(x^3 + 2)}}$.

　　Let $x^3 + 2 = u$; then $du = 3x^2 \, dx$.

(a) $\displaystyle\int (x^3 + 2)^2 \cdot 3x^2 \, dx = \int u^2 \, du = \tfrac{1}{3}u^3 + C = \tfrac{1}{3}(x^3 + 2)^3 + C$

(b) $\displaystyle\int (x^3 + 2)^{1/2}x^2 \, dx = \frac{1}{3}\int (x^3 + 2)^{1/2} \cdot 3x^2 \, dx = \frac{1}{3}\int u^{1/2} \, du = \frac{1}{3}\cdot\frac{u^{3/2}}{3/2} + C = \frac{2}{9}(x^3 + 2)^{3/2} + C$

(c) $\displaystyle\int \frac{8x^2 \, dx}{(x^3 + 2)^3} = 8\cdot\frac{1}{3}\int (x^3 + 2)^{-3}\, 3x^2 \, dx = \frac{8}{3}\int u^{-3} \, du = -\frac{8}{3}\left(\frac{1}{2}u^{-2}\right) + C = -\frac{4}{3(x^3 + 2)^2} + C$

(d) $\displaystyle\int \frac{x^2}{\sqrt[4]{x^3 + 2}}\, dx = \frac{1}{3}\int (x^3 + 2)^{-1/4}\, 3x^2 \, dx = \frac{1}{3}\int u^{-1/4} \, du = \frac{1}{3}\cdot\frac{4}{3}u^{3/4} + C = \frac{4}{9}(x^3 + 2)^{3/4} + C$

10. Evaluate $\displaystyle\int 3x\sqrt{1 - 2x^2} \, dx$.　　Let $1 - 2x^2 = u$; then $du = -4x \, dx$.

$$\int 3x\sqrt{1 - 2x^2} \, dx = 3\left(-\frac{1}{4}\right)\int (1 - 2x^2)^{1/2}(-4x \, dx) = -\frac{3}{4}\int u^{1/2} \, du$$
$$= -\frac{3}{4}\cdot\frac{2}{3}u^{3/2} + C = -\frac{1}{2}(1 - 2x^2)^{3/2} + C$$

11. Evaluate $\displaystyle\int \frac{(x + 3)\, dx}{(x^2 + 6x)^{1/3}}$.　　Let $x^2 + 6x = u$; then $du = (2x + 6)\, dx$.

$$\int \frac{(x + 3)\, dx}{(x^2 + 6x)^{1/3}} = \frac{1}{2}\int (x^2 + 6x)^{-1/3}(2x + 6)\, dx = \frac{1}{2}\int u^{-1/3} \, du$$
$$= \frac{1}{2}\cdot\frac{3}{2}u^{2/3} + C = \frac{3}{4}(x^2 + 6x)^{2/3} + C$$

**12.** $\int \sqrt[3]{1-x^2}\, x\, dx \;=\; -\frac{1}{2}\int (1-x^2)^{1/3}\,(-2x\,dx) \;=\; -\frac{1}{2}\cdot\frac{3}{4}(1-x^2)^{4/3} + C \;=\; -\frac{3}{8}(1-x^2)^{4/3} + C$

**13.** $\int \sqrt{x^2-2x^4}\, dx \;=\; \int (1-2x^2)^{1/2}x\, dx \;=\; -\frac{1}{4}\int (1-2x^2)^{1/2}\,(-4x\,dx)$

$\qquad\qquad =\; -\frac{1}{4}\cdot\frac{2}{3}(1-2x^2)^{3/2} + C \;=\; -\frac{1}{6}(1-2x^2)^{3/2} + C$

**14.** $\int \frac{(1+x)^2}{\sqrt{x}}\, dx \;=\; \int \frac{1+2x+x^2}{x^{1/2}}\, dx \;=\; \int (x^{-1/2}+2x^{1/2}+x^{3/2})\, dx \;=\; 2x^{1/2}+\frac{4}{3}x^{3/2}+\frac{2}{5}x^{5/2}+C$

**15.** $\int \frac{x^2+2x}{(x+1)^2}\, dx \;=\; \int \left\{1-\frac{1}{(x+1)^2}\right\}dx \;=\; x+\frac{1}{x+1}+C' \;=\; \frac{x^2}{x+1}+1+C' \;=\; \frac{x^2}{x+1}+C$

## FORMULAS 5-7

**16.** $\int \frac{dx}{x} \;=\; \ln|x| + C$        **17.** $\int \frac{dx}{x+2} \;=\; \int \frac{d(x+2)}{x+2} \;=\; \ln|x+2| + C$

**18.** $\int \frac{dx}{2x-3} \;=\; \frac{1}{2}\int \frac{du}{u} \;=\; \frac{1}{2}\ln|u| + C \;=\; \frac{1}{2}\ln|2x-3| + C,$   where   $u=2x-3$   and   $du=2\,dx,$    or

$\qquad\qquad\qquad \int \frac{dx}{2x-3} \;=\; \frac{1}{2}\int \frac{d(2x-3)}{2x-3} \;=\; \frac{1}{2}\ln|2x-3| + C$

**19.** $\int \frac{x\,dx}{x^2-1} \;=\; \frac{1}{2}\int \frac{2x\,dx}{x^2-1} \;=\; \frac{1}{2}\ln|x^2-1| + C \;=\; \frac{1}{2}\ln|x^2-1| + \ln c \;=\; \ln c\sqrt{|x^2-1|}$

**20.** $\int \frac{x^2\,dx}{1-2x^3} \;=\; -\frac{1}{6}\int \frac{-6x^2\,dx}{1-2x^3} \;=\; -\frac{1}{6}\ln|1-2x^3| + C \;=\; \ln\frac{c}{\sqrt[6]{|1-2x^3|}}$

**21.** $\int \frac{x+2}{x+1}\, dx \;=\; \int \left(1+\frac{1}{x+1}\right)dx \;=\; x+\ln|x+1| + C$

**22.** $\int e^{-x}\, dx \;=\; -\int e^{-x}\,(-dx) \;=\; -e^{-x} + C$      **24.** $\int e^{3x}\, dx \;=\; \frac{1}{3}\int e^{3x}\,(3\,dx) \;=\; \frac{e^{3x}}{3} + C$

**23.** $\int a^{2x}\, dx \;=\; \frac{1}{2}\int a^{2x}\,(2\,dx) \;=\; \frac{1}{2}\left(\frac{a^{2x}}{\ln a}\right) + C$    **25.** $\int \frac{e^{1/x}\,dx}{x^2} \;=\; -\int e^{1/x}\left(-\frac{dx}{x^2}\right) \;=\; -e^{1/x} + C$

**26.** $\int (e^x+1)^3\, e^x\, dx \;=\; \int u^3\, du \;=\; \frac{u^4}{4} + C \;=\; \frac{(e^x+1)^4}{4} + C$   where $u=e^x+1$ and $du=e^x\,dx,$    or

$\qquad\qquad\qquad \int (e^x+1)^3\, e^x\, dx \;=\; \int (e^x+1)^3\, d(e^x+1) \;=\; \frac{(e^x+1)^4}{4} + C$

**27.** $\int \frac{dx}{e^x+1} \;=\; \int \frac{e^{-x}\,dx}{1+e^{-x}} \;=\; -\int \frac{-e^{-x}\,dx}{1+e^{-x}} \;=\; -\ln(1+e^{-x}) + C \;=\; \ln\frac{e^x}{1+e^x} + C$

$\qquad\qquad\qquad\qquad\qquad =\; x-\ln(1+e^x) + C$

The absolute value sign is not needed here since $1+e^{-x} > 0$ for all values of $x$.

## FORMULAS 8-17

**28.** $\int \sin\tfrac{1}{2}x\, dx \;=\; 2\int \sin\tfrac{1}{2}x\cdot\tfrac{1}{2}dx \;=\; -2\cos\tfrac{1}{2}x + C$

**29.** $\int \cos 3x\, dx \;=\; \frac{1}{3}\int \cos 3x\cdot 3\,dx \;=\; \frac{1}{3}\sin 3x + C$

**30.** $\int \sin^2 x\cos x\, dx \;=\; \int \sin^2 x\,(\cos x\,dx) \;=\; \int \sin^2 x\, d(\sin x) \;=\; \frac{\sin^3 x}{3} + C$

**31.** $\int \tan x\, dx \;=\; \int \frac{\sin x}{\cos x}\, dx \;=\; -\int \frac{-\sin x\,dx}{\cos x} \;=\; -\ln|\cos x| + C \;=\; \ln|\sec x| + C$

**32.** $\int \tan 2x\, dx \;=\; \frac{1}{2}\int \tan 2x\cdot 2\,dx \;=\; \frac{1}{2}\ln|\sec 2x| + C$

**33.** $\int x\cot x^2\, dx \;=\; \frac{1}{2}\int \cot x^2\cdot 2x\,dx \;=\; \frac{1}{2}\ln|\sin x^2| + C$

**34.** $\displaystyle\int \sec x \, dx \;=\; \int \frac{\sec x \, (\sec x + \tan x)}{\sec x + \tan x} \, dx \;=\; \frac{\sec x \tan x + \sec^2 x}{\sec x + \tan x} \, dx \;=\; \ln |\sec x + \tan x | + C$

**35.** $\displaystyle\int \sec \sqrt{x} \, \frac{dx}{\sqrt{x}} \;=\; 2 \int \sec x^{1/2} \cdot \frac{1}{2} x^{-1/2} \, dx \;=\; 2 \ln |\sec \sqrt{x} + \tan \sqrt{x}| + C$

**36.** $\displaystyle\int \sec^2 2ax \, dx \;=\; \frac{1}{2a} \int \sec^2 2ax \cdot 2a \, dx \;=\; \frac{\tan 2ax}{2a} + C$

**37.** $\displaystyle\int \frac{\sin x + \cos x}{\cos x} \, dx \;=\; \int (\tan x + 1) \, dx \;=\; \ln |\sec x| + x + C$

**38.** $\displaystyle\int \frac{\sin y \, dy}{\cos^2 y} \;=\; \int \tan y \sec y \, dy \;=\; \sec y + C$

**39.** $\displaystyle\int (1 + \tan x)^2 \, dx \;=\; \int (1 + 2 \tan x + \tan^2 x) \, dx \;=\; \int (\sec^2 x + 2 \tan x) \, dx$
$$=\; \tan x + 2 \ln |\sec x| + C$$

**40.** $\displaystyle\int e^x \cos e^x \, dx \;=\; \int \cos e^x \cdot e^x \, dx \;=\; \sin e^x + C$

**41.** $\displaystyle\int e^{3 \cos 2x} \sin 2x \, dx \;=\; -\frac{1}{6} \int e^{3 \cos 2x} (-6 \sin 2x \, dx) \;=\; -\frac{e^{3 \cos 2x}}{6} + C$

**42.** $\displaystyle\int \frac{dx}{1 + \cos x} \;=\; \int \frac{1 - \cos x}{1 - \cos^2 x} \, dx \;=\; \int \frac{1 - \cos x}{\sin^2 x} \, dx \;=\; \int (\csc^2 x - \cot x \csc x) \, dx$
$$=\; -\cot x + \csc x + C$$

**43.** $\displaystyle\int (\tan 2x + \sec 2x)^2 \, dx \;=\; \int (\tan^2 2x + 2 \tan 2x \sec 2x + \sec^2 2x) \, dx$
$$=\; \int (2 \sec^2 2x + 2 \tan 2x \sec 2x - 1) \, dx \;=\; \tan 2x + \sec 2x - x + C$$

**44.** $\displaystyle\int \csc u \, du \;=\; \int \frac{du}{\sin u} \;=\; \int \frac{du}{2 \sin \frac{1}{2}u \cos \frac{1}{2}u} \;=\; \int \frac{\sec^2 \frac{1}{2}u \cdot \frac{1}{2} du}{\tan \frac{1}{2}u} \;=\; \ln |\tan \frac{1}{2}u| + C$

**45.** $\displaystyle\int (\sec 4x - 1)^2 \, dx \;=\; \int (\sec^2 4x - 2 \sec 4x + 1) \, dx \;=\; \frac{1}{4} \tan 4x - \frac{1}{2} \ln |\sec 4x + \tan 4x| + x + C$

**46.** $\displaystyle\int \frac{\sec x \tan x \, dx}{a + b \sec x} \;=\; \frac{1}{b} \int \frac{\sec x \tan x \cdot b \, dx}{a + b \sec x} \;=\; \frac{1}{b} \ln |a + b \sec x| + C$

**47.** $\displaystyle\int \frac{dx}{\csc 2x - \cot 2x} \;=\; \int \frac{\sin 2x \, dx}{1 - \cos 2x} \;=\; \frac{1}{2} \int \frac{\sin 2x \cdot 2 \, dx}{1 - \cos 2x} \;=\; \frac{1}{2} \ln (1 - \cos 2x) + C'$
$$=\; \tfrac{1}{2} \ln (2 \sin^2 x) + C' \;=\; \tfrac{1}{2}(\ln 2 + 2 \ln |\sin x|) + C' \;=\; \ln |\sin x| + C$$

## FORMULAS 18-20

**48.** $\displaystyle\int \frac{dx}{\sqrt{1 - x^2}} \;=\; \arcsin x + C$        **51.** $\displaystyle\int \frac{dx}{\sqrt{4 - x^2}} \;=\; \arcsin \frac{x}{2} + C$

**49.** $\displaystyle\int \frac{dx}{1 + x^2} \;=\; \arctan x + C$        **52.** $\displaystyle\int \frac{dx}{9 + x^2} \;=\; \frac{1}{3} \arctan \frac{x}{3} + C$

**50.** $\displaystyle\int \frac{dx}{x\sqrt{x^2 - 1}} \;=\; \operatorname{arc sec} x + C$        **53.** $\displaystyle\int \frac{dx}{\sqrt{25 - 16x^2}} \;=\; \frac{1}{4} \int \frac{4 \, dx}{\sqrt{5^2 - (4x)^2}} \;=\; \frac{1}{4} \arcsin \frac{4x}{5} + C$

**54.** $\displaystyle\int \frac{dx}{4x^2 + 9} \;=\; \frac{1}{2} \int \frac{2 \, dx}{(2x)^2 + 3^2} \;=\; \frac{1}{6} \arctan \frac{2x}{3} + C$

**55.** $\displaystyle\int \frac{dx}{x\sqrt{4x^2 - 9}} \;=\; \int \frac{2 \, dx}{2x\sqrt{(2x)^2 - 3^2}} \;=\; \frac{1}{3} \operatorname{arc sec} \frac{2x}{3} + C$

**56.** $\displaystyle\int \frac{x^2 \, dx}{\sqrt{1 - x^6}} \;=\; \frac{1}{3} \int \frac{3x^2 \, dx}{\sqrt{1 - (x^3)^2}} \;=\; \frac{1}{3} \arcsin x^3 + C$

**57.** $\displaystyle\int \frac{x \, dx}{x^4 + 3} \;=\; \frac{1}{2} \int \frac{2x \, dx}{(x^2)^2 + 3} \;=\; \frac{1}{2} \cdot \frac{1}{\sqrt{3}} \arctan \frac{x^2}{\sqrt{3}} + C \;=\; \frac{\sqrt{3}}{6} \arctan \frac{x^2 \sqrt{3}}{3} + C$

**58.** $\displaystyle\int \frac{dx}{x\sqrt{x^4-1}} \;=\; \frac{1}{2}\int \frac{2x\,dx}{x^2\sqrt{(x^2)^2-1}} \;=\; \frac{1}{2}\arccsc x^2 + C \;=\; \frac{1}{2}\arccos\frac{1}{x^2} + C$

**59.** $\displaystyle\int \frac{dx}{\sqrt{4-(x+2)^2}} \;=\; \arcsin\frac{x+2}{2} + C$　　**60.** $\displaystyle\int \frac{dx}{e^x+e^{-x}} \;=\; \int \frac{e^x\,dx}{e^{2x}+1} \;=\; \arctan e^x + C$

**61.** $\displaystyle\int \frac{3x^3-4x^2+3x}{x^2+1}\,dx \;=\; \int \left(3x-4+\frac{4}{x^2+1}\right)dx \;=\; \frac{3x^2}{2} - 4x + 4\arctan x + C$

**62.** $\displaystyle\int \frac{\sec x\tan x\,dx}{9+4\sec^2 x} \;=\; \frac{1}{2}\int \frac{2\sec x\tan x\,dx}{3^2+(2\sec x)^2} \;=\; \frac{1}{6}\arctan\frac{2\sec x}{3} + C$

**63.** $\displaystyle\int \frac{(x+3)\,dx}{\sqrt{1-x^2}} \;=\; \int \frac{x\,dx}{\sqrt{1-x^2}} + 3\int \frac{dx}{\sqrt{1-x^2}} \;=\; -\sqrt{1-x^2} + 3\arcsin x + C$

**64.** $\displaystyle\int \frac{(2x-7)\,dx}{x^2+9} \;=\; \int \frac{2x\,dx}{x^2+9} - 7\int \frac{dx}{x^2+9} \;=\; \ln(x^2+9) - \frac{7}{3}\arctan\frac{x}{3} + C$

**65.** $\displaystyle\int \frac{dy}{y^2+10y+30} \;=\; \int \frac{dy}{(y^2+10y+25)+5} \;=\; \int \frac{dy}{(y+5)^2+5} \;=\; \frac{\sqrt{5}}{5}\arctan\frac{(y+5)\sqrt{5}}{5} + C$

**66.** $\displaystyle\int \frac{dx}{\sqrt{20+8x-x^2}} \;=\; \int \frac{dx}{\sqrt{36-(x^2-8x+16)}} \;=\; \int \frac{dx}{\sqrt{36-(x-4)^2}} \;=\; \arcsin\frac{x-4}{6} + C$

**67.** $\displaystyle\int \frac{dx}{2x^2+2x+5} \;=\; \int \frac{2\,dx}{4x^2+4x+10} \;=\; \int \frac{2\,dx}{(2x+1)^2+9} \;=\; \frac{1}{3}\arctan\frac{2x+1}{3} + C$

**68.** $\displaystyle\int \frac{x+1}{x^2-4x+8}\,dx \;=\; \frac{1}{2}\int \frac{2x+2}{x^2-4x+8}\,dx \;=\; \frac{1}{2}\int \frac{(2x-4)+6}{x^2-4x+8}\,dx \;=\; \frac{1}{2}\int \frac{(2x-4)\,dx}{x^2-4x+8} + 3\int \frac{dx}{x^2-4x+8}$

$$= \;\frac{1}{2}\int \frac{(2x-4)\,dx}{x^2-4x+8} + 3\int \frac{dx}{(x-2)^2+4} \;=\; \frac{1}{2}\ln(x^2-4x+8) + \frac{3}{2}\arctan\frac{x-2}{2} + C$$

The absolute value sign is not needed here since $x^2-4x+8 > 0$ for all values of $x$.

**69.** $\displaystyle\int \frac{dx}{\sqrt{28-12x-x^2}} \;=\; \int \frac{dx}{\sqrt{64-(x^2+12x+36)}} \;=\; \int \frac{dx}{\sqrt{64-(x+6)^2}} \;=\; \arcsin\frac{x+6}{8} + C$

**70.** $\displaystyle\int \frac{x+3}{\sqrt{5-4x-x^2}}\,dx \;=\; -\frac{1}{2}\int \frac{-2x-6}{\sqrt{5-4x-x^2}}\,dx \;=\; -\frac{1}{2}\int \frac{(-2x-4)-2}{\sqrt{5-4x-x^2}}\,dx$

$$= \;-\frac{1}{2}\int \frac{-2x-4}{\sqrt{5-4x-x^2}}\,dx + \int \frac{dx}{\sqrt{5-4x-x^2}}$$

$$= \;-\frac{1}{2}\int \frac{-2x-4}{\sqrt{5-4x-x^2}}\,dx + \int \frac{dx}{\sqrt{9-(x+2)^2}}$$

$$= \;-\sqrt{5-4x-x^2} + \arcsin\frac{x+2}{3} + C$$

**71.** $\displaystyle\int \frac{2x+3}{9x^2-12x+8}\,dx \;=\; \frac{1}{9}\int \frac{18x+27}{9x^2-12x+8}\,dx \;=\; \frac{1}{9}\int \frac{(18x-12)+39}{9x^2-12x+8}\,dx$

$$= \;\frac{1}{9}\int \frac{18x-12}{9x^2-12x+8}\,dx + \frac{13}{3}\int \frac{dx}{(3x-2)^2+4}$$

$$= \;\frac{1}{9}\ln(9x^2-12x+8) + \frac{13}{18}\arctan\frac{3x-2}{2} + C$$

**72.** $\displaystyle\int \frac{x+2}{\sqrt{4x-x^2}}\,dx \;=\; -\frac{1}{2}\int \frac{-2x-4}{\sqrt{4x-x^2}}\,dx \;=\; -\frac{1}{2}\int \frac{(-2x+4)-8}{\sqrt{4x-x^2}}\,dx$

$$= \;-\frac{1}{2}\int \frac{4-2x}{\sqrt{4x-x^2}}\,dx + 4\int \frac{dx}{\sqrt{4-(x-2)^2}} \;=\; -\sqrt{4x-x^2} + 4\arcsin\frac{x-2}{2} + C$$

## FORMULAS 21-24

**73.** $\int \dfrac{dx}{x^2 - 1} = \dfrac{1}{2} \ln \left| \dfrac{x-1}{x+1} \right| + C$      **76.** $\int \dfrac{dx}{9 - x^2} = \dfrac{1}{6} \ln \left| \dfrac{3+x}{3-x} \right| + C$

**74.** $\int \dfrac{dx}{1 - x^2} = \dfrac{1}{2} \ln \left| \dfrac{1+x}{1-x} \right| + C$      **77.** $\int \dfrac{dx}{\sqrt{x^2 + 1}} = \ln (x + \sqrt{x^2 + 1}) + C$

**75.** $\int \dfrac{dx}{x^2 - 4} = \dfrac{1}{4} \ln \left| \dfrac{x-2}{x+2} \right| + C$      **78.** $\int \dfrac{dx}{\sqrt{x^2 - 1}} = \ln | x + \sqrt{x^2 - 1} | + C$

**79.** $\int \dfrac{dx}{\sqrt{4x^2 + 9}} = \dfrac{1}{2} \int \dfrac{2\,dx}{\sqrt{(2x)^2 + 3^2}} = \dfrac{1}{2} \ln (2x + \sqrt{4x^2 + 9}) + C$

**80.** $\int \dfrac{dz}{\sqrt{9z^2 - 25}} = \dfrac{1}{3} \int \dfrac{3\,dz}{\sqrt{9z^2 - 25}} = \dfrac{1}{3} \ln | 3z + \sqrt{9z^2 - 25} | + C$

**81.** $\int \dfrac{dx}{9x^2 - 16} = \dfrac{1}{3} \int \dfrac{3\,dx}{(3x)^2 - 16} = \dfrac{1}{24} \ln \left| \dfrac{3x-4}{3x+4} \right| + C$

**82.** $\int \dfrac{dy}{25 - 16y^2} = \dfrac{1}{4} \int \dfrac{4\,dy}{25 - (4y)^2} = \dfrac{1}{40} \ln \left| \dfrac{5+4y}{5-4y} \right| + C$

**83.** $\int \dfrac{dx}{x^2 + 6x + 8} = \int \dfrac{dx}{(x+3)^2 - 1} = \dfrac{1}{2} \ln \left| \dfrac{(x+3)-1}{(x+3)+1} \right| + C = \dfrac{1}{2} \ln \left| \dfrac{x+2}{x+4} \right| + C$

**84.** $\int \dfrac{dx}{4x - x^2} = \int \dfrac{dx}{4 - (x-2)^2} = \dfrac{1}{4} \ln \left| \dfrac{2+(x-2)}{2-(x-2)} \right| + C = \dfrac{1}{4} \ln \left| \dfrac{x}{4-x} \right| + C$

**85.** $\int \dfrac{ds}{\sqrt{4s + s^2}} = \int \dfrac{ds}{\sqrt{(s+2)^2 - 4}} = \ln | s + 2 + \sqrt{4s + s^2} | + C$

**86.** $\int \dfrac{x+2}{\sqrt{x^2 + 9}}\,dx = \dfrac{1}{2} \int \dfrac{2x+4}{\sqrt{x^2 + 9}}\,dx = \dfrac{1}{2} \int \dfrac{2x\,dx}{\sqrt{x^2 + 9}} + 2 \int \dfrac{dx}{\sqrt{x^2 + 9}}$

$\qquad\qquad = \sqrt{x^2 + 9} + 2 \ln (x + \sqrt{x^2 + 9}) + C$

**87.** $\int \dfrac{2x-3}{4x^2 - 11}\,dx = \dfrac{1}{4} \int \dfrac{8x-12}{4x^2 - 11}\,dx = \dfrac{1}{4} \int \dfrac{8x\,dx}{4x^2 - 11} - \dfrac{3}{2} \int \dfrac{2\,dx}{4x^2 - 11}$

$\qquad\qquad = \dfrac{1}{4} \ln |4x^2 - 11| - \dfrac{3\sqrt{11}}{44} \ln \left| \dfrac{2x - \sqrt{11}}{2x + \sqrt{11}} \right| + C$

**88.** $\int \dfrac{x+2}{\sqrt{x^2 + 2x - 3}}\,dx = \dfrac{1}{2} \int \dfrac{2x+4}{\sqrt{x^2 + 2x - 3}}\,dx = \dfrac{1}{2} \int \dfrac{2x+2}{\sqrt{x^2 + 2x - 3}}\,dx + \int \dfrac{dx}{\sqrt{(x+1)^2 - 4}}$

$\qquad\qquad = \sqrt{x^2 + 2x - 3} + \ln | x + 1 + \sqrt{x^2 + 2x - 3} | + C$

**89.** $\int \dfrac{2-x}{4x^2 + 4x - 3}\,dx = -\dfrac{1}{8} \int \dfrac{8x-16}{4x^2 + 4x - 3}\,dx = -\dfrac{1}{8} \int \dfrac{8x+4}{4x^2 + 4x - 3}\,dx + \dfrac{5}{2} \int \dfrac{dx}{(2x+1)^2 - 4}$

$\qquad\qquad = -\dfrac{1}{8} \ln | 4x^2 + 4x - 3 | + \dfrac{5}{16} \ln \left| \dfrac{2x-1}{2x+3} \right| + C$

## FORMULAS 25-27

**90.** $\int \sqrt{25 - x^2}\,dx = \dfrac{1}{2} x \sqrt{25 - x^2} + \dfrac{25}{2} \arcsin \dfrac{x}{5} + C$

**91.** $\int \sqrt{3 - 4x^2}\,dx = \dfrac{1}{2} \int \sqrt{3 - 4x^2} \cdot 2\,dx = \dfrac{1}{2} \left( \dfrac{2x}{2} \sqrt{3 - 4x^2} + \dfrac{3}{2} \arcsin \dfrac{2x}{\sqrt{3}} \right) + C$

$\qquad\qquad = \dfrac{1}{2} x \sqrt{3 - 4x^2} + \dfrac{3}{4} \arcsin \dfrac{2x\sqrt{3}}{3} + C$

**92.** $\int \sqrt{x^2 - 36}\,dx = \dfrac{1}{2} x \sqrt{x^2 - 36} - 18 \ln | x + \sqrt{x^2 - 36} | + C$

93. $\displaystyle\int \sqrt{3x^2+5}\,dx = \frac{1}{\sqrt{3}}\int \sqrt{3x^2+5}\cdot\sqrt{3}\,dx = \frac{1}{\sqrt{3}}\left[\frac{\sqrt{3}}{2}x\sqrt{3x^2+5} + \frac{5}{2}\ln(\sqrt{3}\,x + \sqrt{3x^2+5}\,)\right] + C$

$\displaystyle\qquad\qquad = \frac{1}{2}x\sqrt{3x^2+5} + \frac{5\sqrt{3}}{6}\ln(\sqrt{3}\,x + \sqrt{3x^2+5}\,) + C$

94. $\displaystyle\int \sqrt{3-2x-x^2}\,dx = \int \sqrt{4-(x+1)^2}\,dx = \frac{x+1}{2}\sqrt{3-2x-x^2} + 2\arcsin\frac{x+1}{2} + C$

95. $\displaystyle\int \sqrt{4x^2-4x+5}\,dx = \frac{1}{2}\int \sqrt{(2x-1)^2+4}\cdot 2\,dx$

$\displaystyle\qquad\qquad = \frac{1}{2}\left[\frac{2x-1}{2}\sqrt{4x^2-4x+5} + 2\ln(2x-1+\sqrt{4x^2-4x+5}\,)\right] + C$

$\displaystyle\qquad\qquad = \frac{2x-1}{4}\sqrt{4x^2-4x+5} + \ln(2x-1+\sqrt{4x^2-4x+5}\,) + C$

# Supplementary Problems

Perform the following integrations.

96. $\displaystyle\int (4x^3+3x^2+2x+5)\,dx = x^4+x^3+x^2+5x+C$

97. $\displaystyle\int (3-2x-x^4)\,dx = 3x-x^2-x^5/5+C$

98. $\displaystyle\int (2-3x+x^3)\,dx = 2x-3x^2/2+x^4/4+C$ 　　99. $\displaystyle\int (x^2-1)^2\,dx = x^5/5-2x^3/3+x+C$

100. $\displaystyle\int (\sqrt{x}-\tfrac{1}{2}x+2/\sqrt{x})\,dx = \tfrac{2}{3}x^{3/2}-\tfrac{1}{4}x^2+4x^{1/2}+C$

101. $\displaystyle\int (a+x)^3\,dx = \tfrac{1}{4}(a+x)^4+C$ 　　112. $\displaystyle\int (x^3+3)x^2\,dx = \tfrac{1}{6}(x^3+3)^2+C$

102. $\displaystyle\int (x-2)^{3/2}\,dx = \tfrac{2}{5}(x-2)^{5/2}+C$ 　　113. $\displaystyle\int (4-x^2)^2x^2\,dx = \tfrac{16}{3}x^3-\tfrac{8}{5}x^5+\tfrac{1}{7}x^7+C$

103. $\displaystyle\int \frac{dx}{x^3} = -\frac{1}{2x^2}+C$ 　　114. $\displaystyle\int \frac{dy}{(2-y)^3} = \frac{1}{2(2-y)^2}+C$

104. $\displaystyle\int \frac{dx}{(x-1)^3} = -\frac{1}{2(x-1)^2}+C$ 　　115. $\displaystyle\int \frac{x\,dx}{(x^2+4)^3} = -\frac{1}{4(x^2+4)^2}+C$

105. $\displaystyle\int \frac{dx}{\sqrt{x+3}} = 2\sqrt{x+3}+C$ 　　116. $\displaystyle\int (1-x^3)^2\,dx = x-\tfrac{1}{2}x^4+\tfrac{1}{7}x^7+C$

106. $\displaystyle\int \sqrt{3x-1}\,dx = \tfrac{2}{9}(3x-1)^{3/2}+C$ 　　117. $\displaystyle\int (1-x^3)^2x\,dx = \tfrac{1}{2}x^2-\tfrac{2}{5}x^5+\tfrac{1}{8}x^8+C$

107. $\displaystyle\int \sqrt{2-3x}\,dx = -\tfrac{2}{9}(2-3x)^{3/2}+C$ 　　118. $\displaystyle\int (1-x^3)^2x^2\,dx = -\tfrac{1}{9}(1-x^3)^3+C$

108. $\displaystyle\int (2x^2+3)^{1/3}x\,dx = \tfrac{3}{16}(2x^2+3)^{4/3}+C$ 　　119. $\displaystyle\int (x^2-x)^4(2x-1)\,dx = \tfrac{1}{5}(x^2-x)^5+C$

109. $\displaystyle\int (x-1)^2x\,dx = \tfrac{1}{4}x^4-\tfrac{2}{3}x^3+\tfrac{1}{2}x^2+C$ 　　120. $\displaystyle\int \frac{3t\,dt}{\sqrt[3]{t^2+3}} = \frac{9}{4}(t^2+3)^{2/3}+C$

110. $\displaystyle\int (x^2-1)x\,dx = \tfrac{1}{4}(x^2-1)^2+C$ 　　121. $\displaystyle\int \frac{(x+1)\,dx}{\sqrt{x^2+2x-4}} = \sqrt{x^2+2x-4}+C$

111. $\displaystyle\int \sqrt{1+y^4}\,y^3\,dy = \tfrac{1}{6}(1+y^4)^{3/2}+C$ 　　122. $\displaystyle\int \frac{dx}{(a+bx)^{1/3}} = \frac{3}{2b}(a+bx)^{2/3}+C$

123. $\int \frac{(1+\sqrt{x}\,)^2}{\sqrt{x}}\,dx \;=\; \frac{2}{3}(1+\sqrt{x}\,)^3 + C$

147. $\int \frac{dx}{x+x^{1/3}} \;=\; \frac{3}{2}\ln C(x^{2/3}+1), \;\; C>0$

124. $\int \sqrt{x}\,(3-5x)\,dx \;=\; 2x^{3/2}(1-x) + C$

148. $\int \sin 2x\,dx \;=\; -\frac{1}{2}\cos 2x + C$

125. $\int \frac{(x+1)(x-2)}{\sqrt{x}}\,dx \;=\; \frac{2}{5}x^{5/2} - \frac{2}{3}x^{3/2} - 4x^{1/2} + C$

149. $\int \cos \frac{1}{2}x\,dx \;=\; 2\sin \frac{1}{2}x + C$

126. $\int \frac{dx}{x-1} \;=\; \ln|x-1| + C$

150. $\int \sec 3x \tan 3x\,dx \;=\; \frac{1}{3}\sec 3x + C$

127. $\int \frac{dx}{3x+1} \;=\; \frac{1}{3}\ln|3x+1| + C$

151. $\int \csc^2 2x\,dx \;=\; -\frac{1}{2}\cot 2x + C$

128. $\int \frac{3x\,dx}{x^2+2} \;=\; \frac{3}{2}\ln(x^2+2) + C$

152. $\int x \sec^2 x^2\,dx \;=\; \frac{1}{2}\tan x^2 + C$

129. $\int \frac{x^2\,dx}{1-x^3} \;=\; -\frac{1}{3}\ln|1-x^3| + C$

153. $\int \tan^2 x\,dx \;=\; \tan x - x + C$

130. $\int \frac{x-1}{x+1}\,dx \;=\; x - 2\ln|x+1| + C$

154. $\int \tan \frac{1}{2}x\,dx \;=\; 2\ln|\sec \frac{1}{2}x| + C$

131. $\int \frac{x^2+2x+2}{x+2}\,dx \;=\; \frac{1}{2}x^2 + 2\ln|x+2| + C$

155. $\int \csc 3x\,dx \;=\; \frac{1}{3}\ln|\csc 3x - \cot 3x| + C$

132. $\int \frac{x+1}{x^2+2x+2}\,dx \;=\; \frac{1}{2}\ln(x^2+2x+2) + C$

156. $\int b \sec ax \tan ax\,dx \;=\; \frac{b}{a}\sec ax + C$

133. $\int \left(\frac{dx}{2x-1} - \frac{dx}{2x+1}\right) \;=\; \ln \sqrt{\left|\frac{2x-1}{2x+1}\right|} + C$

157. $\int (\cos x - \sin x)^2\,dx \;=\; x + \frac{1}{2}\cos 2x + C$

134. $\int a^{4x}\,dx \;=\; \frac{1}{4}\frac{a^{4x}}{\ln a} + C$

158. $\int \sin ax \cos ax\,dx \;=\; \frac{1}{2a}\sin^2 ax + C$

$$= -\frac{1}{2a}\cos^2 ax + C' \;=\; -\frac{1}{4a}\cos 2ax + K$$

135. $\int e^{4x}\,dx \;=\; \frac{1}{4}e^{4x} + C$

136. $\int \frac{e^{1/x^2}}{x^3}\,dx \;=\; -\frac{1}{2}e^{1/x^2} + C$

159. $\int \sin^3 x \cos x\,dx \;=\; \frac{1}{4}\sin^4 x + C$

137. $\int e^{-x^2+2}x\,dx \;=\; -\frac{1}{2}e^{-x^2+2} + C$

160. $\int \cos^4 x \sin x\,dx \;=\; -\frac{1}{5}\cos^5 x + C$

138. $\int x^2 e^{x^3}\,dx \;=\; \frac{1}{3}e^{x^3} + C$

161. $\int \tan^5 x \sec^2 x\,dx \;=\; \frac{1}{6}\tan^6 x + C$

139. $\int (e^x+1)^2\,dx \;=\; \frac{1}{2}e^{2x} + 2e^x + x + C$

162. $\int \cot^4 3x \csc^2 3x\,dx \;=\; -\frac{1}{15}\cot^5 3x + C$

140. $\int (e^x - x^e)\,dx \;=\; e^x - \frac{x^{e+1}}{e+1} + C$

163. $\int \frac{dx}{1-\sin \frac{1}{2}x} \;=\; 2(\tan \frac{1}{2}x + \sec \frac{1}{2}x) + C$

141. $\int (e^x+1)^2 e^x\,dx \;=\; \frac{1}{3}(e^x+1)^3 + C$

142. $\int \frac{e^{2x}}{e^{2x}+3}\,dx \;=\; \frac{1}{2}\ln(e^{2x}+3) + C$

164. $\int \frac{dx}{1+\cos 3x} \;=\; \frac{1-\cos 3x}{3\sin 3x} + C$

143. $\int \left(e^x + \frac{1}{e^x}\right)^2\,dx \;=\; \frac{1}{2}e^{2x} + 2x - \frac{1}{2e^{2x}} + C$

165. $\int \frac{dx}{1+\sec ax} \;=\; x + \frac{1}{a}(\cot ax - \csc ax) + C$

144. $\int \frac{e^x-1}{e^x+1}\,dx \;=\; \ln(e^x+1)^2 - x + C$

166. $\int \sec^2 \frac{x}{a} \tan \frac{x}{a}\,dx \;=\; \frac{1}{2}a \tan^2 \frac{x}{a} + C$

145. $\int \frac{e^{2x}-1}{e^{2x}+3}\,dx \;=\; \ln(e^{2x}+3)^{2/3} - \frac{1}{3}x + C$

167. $\int \frac{\sec^2 3x}{\tan 3x}\,dx \;=\; \frac{1}{3}\ln|\tan 3x| + C$

146. $\int \frac{dx}{\sqrt{x}(1-\sqrt{x}\,)} \;=\; \ln \frac{C}{(1-\sqrt{x}\,)^2}, \;\; C>0$

168. $\int \frac{\sec^5 x}{\csc x}\,dx \;=\; \frac{1}{4}\sec^4 x + C$

169. $\int e^{\tan 2x} \sec^2 2x \, dx = \frac{1}{2} e^{\tan 2x} + C$

176. $\int \frac{dx}{\sqrt{4 - 9x^2}} = \frac{1}{3} \arcsin \frac{3x}{2} + C$

170. $\int e^{2 \sin 3x} \cos 3x \, dx = \frac{1}{6} e^{2 \sin 3x} + C$

177. $\int \frac{dx}{9x^2 + 4} = \frac{1}{6} \arctan \frac{3x}{2} + C$

171. $\int \frac{dx}{\sqrt{5 - x^2}} = \arcsin \frac{x\sqrt{5}}{5} + C$

178. $\int \frac{\sin 8x}{9 + \sin^4 4x} \, dx = \frac{1}{12} \arctan \frac{\sin^2 4x}{3} + C$

172. $\int \frac{dx}{5 + x^2} = \frac{\sqrt{5}}{5} \arctan \frac{x\sqrt{5}}{5} + C$

179. $\int \frac{\sec^2 x \, dx}{\sqrt{1 - 4 \tan^2 x}} = \frac{1}{2} \arcsin (2 \tan x) + C$

173. $\int \frac{dx}{x\sqrt{x^2 - 5}} = \frac{\sqrt{5}}{5} \operatorname{arc\,sec} \frac{x\sqrt{5}}{5} + C$

180. $\int \frac{dx}{x\sqrt{4 - 9 \ln^2 x}} = \frac{1}{3} \arcsin \ln x^{3/2} + C$

174. $\int \frac{e^x \, dx}{\sqrt{1 - e^{2x}}} = \arcsin e^x + C$

181. $\int \frac{2x^4 - x^2}{2x^2 + 1} \, dx = \frac{1}{3} x^3 - x + \frac{\sqrt{2}}{2} \arctan x\sqrt{2} + C$

175. $\int \frac{e^{2x} \, dx}{1 + e^{4x}} = \frac{1}{2} \arctan e^{2x} + C$

182. $\int \frac{\cos 2x \, dx}{\sin^2 2x + 8} = \frac{\sqrt{2}}{8} \arctan \frac{\sin 2x}{2\sqrt{2}} + C$

183. $\int \frac{(2x - 3) \, dx}{x^2 + 6x + 13} = \int \frac{(2x + 6) \, dx}{x^2 + 6x + 13} - 9 \int \frac{dx}{x^2 + 6x + 13} = \ln (x^2 + 6x + 13) - \frac{9}{2} \arctan \frac{x + 3}{2} + C$

184. $\int \frac{(x - 1) \, dx}{3x^2 - 4x + 3} = \frac{1}{6} \int \frac{(6x - 4) \, dx}{3x^2 - 4x + 3} - \int \frac{dx}{9x^2 - 12x + 9} = \frac{1}{6} \ln (3x^2 - 4x + 3) - \frac{\sqrt{5}}{15} \arctan \frac{3x - 2}{\sqrt{5}} + C$

185. $\int \frac{x \, dx}{\sqrt{27 + 6x - x^2}} = -\sqrt{27 + 6x - x^2} + 3 \arcsin \frac{x - 3}{6} + C$

186. $\int \frac{(5 - 4x) \, dx}{\sqrt{12x - 4x^2 - 8}} = \sqrt{12x - 4x^2 - 8} - \frac{1}{2} \arcsin (2x - 3) + C$

187. $\int \frac{dx}{x^2 - 4} = \frac{1}{4} \ln \left| \frac{x - 2}{x + 2} \right| + C$

190. $\int \frac{dx}{25 - 9x^2} = \frac{1}{30} \ln \left| \frac{3x + 5}{3x - 5} \right| + C$

188. $\int \frac{dx}{4x^2 - 9} = \frac{1}{12} \ln \left| \frac{2x - 3}{2x + 3} \right| + C$

191. $\int \frac{dx}{\sqrt{x^2 + 4}} = \ln (x + \sqrt{x^2 + 4}) + C$

189. $\int \frac{dx}{9 - x^2} = \frac{1}{6} \ln \left| \frac{x + 3}{x - 3} \right| + C$

192. $\int \frac{dx}{\sqrt{4x^2 - 25}} = \frac{1}{2} \ln | 2x + \sqrt{4x^2 - 25} | + C$

193. $\int \sqrt{16 - 9x^2} \, dx = \frac{1}{2} x\sqrt{16 - 9x^2} + \frac{8}{3} \arcsin \frac{3x}{4} + C$

194. $\int \sqrt{x^2 - 16} \, dx = \frac{1}{2} x\sqrt{x^2 - 16} - 8 \ln | x + \sqrt{x^2 - 16} | + C$

195. $\int \sqrt{4x^2 + 9} \, dx = \frac{1}{2} x\sqrt{4x^2 + 9} + \frac{9}{4} \ln (2x + \sqrt{4x^2 + 9}) + C$

196. $\int \sqrt{x^2 - 2x - 3} \, dx = \frac{1}{2} (x - 1) \sqrt{x^2 - 2x - 3} - 2 \ln | x - 1 + \sqrt{x^2 - 2x - 3} | + C$

197. $\int \sqrt{12 + 4x - x^2} \, dx = \frac{1}{2} (x - 2) \sqrt{12 + 4x - x^2} + 8 \arcsin \frac{1}{4} (x - 2) + C$

198. $\int \sqrt{x^2 + 4x} \, dx = \frac{1}{2} (x + 2) \sqrt{x^2 + 4x} - 2 \ln | x + 2 + \sqrt{x^2 + 4x} | + C$

199. $\int \sqrt{x^2 - 8x} \, dx = \frac{1}{2} (x - 4) \sqrt{x^2 - 8x} - 8 \ln | x - 4 + \sqrt{x^2 - 8x} | + C$

200. $\int \sqrt{6x - x^2} \, dx = \frac{1}{2} (x - 3) \sqrt{6x - x^2} + \frac{9}{2} \arcsin \frac{x - 3}{3} + C$

# Chapter 26

## Integration by Parts

**INTEGRATION BY PARTS.** When $u$ and $v$ are differentiable functions of $x$,

$$d(uv) = u\,dv + v\,du$$
$$u\,dv = d(uv) - v\,du$$

(**i**)
$$\int u\,dv = uv - \int v\,du$$

To use (i) in effecting a required integration, the given integral must be separated into two parts, one part being $u$ and the other part, together with $dx$, being $dv$. (For this reason, integration by the use of (i) is called *integration by parts*.) Two general rules can be stated:

(*a*)  the part selected as $dv$ must be readily integrable.

(*b*)  $\displaystyle\int v\,du$  must not be more complex than  $\displaystyle\int u\,dv$.

**Example 1:**  Find $\displaystyle\int x^3 e^{x^2}\,dx$.

Take $u = x^2$ and $dv = e^{x^2} x\,dx$; then $du = 2x\,dx$ and $v = \frac{1}{2}e^{x^2}$. Now by the rule,

$$\int x^3 e^{x^2}\,dx = \tfrac{1}{2}x^2 e^{x^2} - \int x e^{x^2}\,dx = \tfrac{1}{2}x^2 e^{x^2} - \tfrac{1}{2}e^{x^2} + C$$

**Example 2:**  Find $\displaystyle\int \ln(x^2 + 2)\,dx$.

Take $u = \ln(x^2 + 2)$ and $dv = dx$; then $du = \dfrac{2x\,dx}{x^2 + 2}$ and $v = x$. By the rule,

$$\int \ln(x^2 + 2)\,dx = x\ln(x^2 + 2) - \int \frac{2x^2\,dx}{x^2 + 2} = x\ln(x^2 + 2) - \int\left(2 - \frac{4}{x^2 + 2}\right)dx$$
$$= x\ln(x^2 + 2) - 2x + 2\sqrt{2}\,\arctan x/\sqrt{2} + C$$

See Problems 1-10.

**REDUCTION FORMULAS.** The labor involved in successive applications (see Problem 9) of integration by parts to evaluate an integral may be materially reduced by the use of *reduction formulas*. In general, a reduction formula yields a new integral of the same form as the original but with an exponent increased or reduced. A reduction formula succeeds if ultimately it produces an integral which can be evaluated. Among the reduction formulas are:

(A)  $\displaystyle \int \frac{du}{(a^2 \pm u^2)^m} = \frac{1}{a^2}\left\{\frac{u}{(2m-2)(a^2 \pm u^2)^{m-1}} + \frac{2m-3}{2m-2}\int \frac{du}{(a^2 \pm u^2)^{m-1}}\right\}, \quad m \neq 1$

(B)  $\displaystyle \int (a^2 \pm u^2)^m\,du = \frac{u(a^2 \pm u^2)^m}{2m+1} + \frac{2ma^2}{2m+1}\int (a^2 \pm u^2)^{m-1}\,du, \quad m \neq -1/2$

(C)  $\displaystyle \int \frac{du}{(u^2 - a^2)^m} = -\frac{1}{a^2}\left\{\frac{u}{(2m-2)(u^2 - a^2)^{m-1}} + \frac{2m-3}{2m-2}\int \frac{du}{(u^2 - a^2)^{m-1}}\right\}, \quad m \neq 1$

(D)  $\displaystyle \int (u^2 - a^2)^m\,du = \frac{u(u^2 - a^2)^m}{2m+1} - \frac{2ma^2}{2m+1}\int (u^2 - a^2)^{m-1}\,du, \quad m \neq -1/2$

(E)  $\displaystyle \int u^m e^{au}\,du = \frac{1}{a}u^m e^{au} - \frac{m}{a}\int u^{m-1} e^{au}\,du$

(F) $\quad \displaystyle\int \sin^m u \, du \;=\; -\frac{\sin^{m-1} u \cos u}{m} + \frac{m-1}{m} \int \sin^{m-2} u \, du$

(G) $\quad \displaystyle\int \cos^m u \, du \;=\; \frac{\cos^{m-1} u \sin u}{m} + \frac{m-1}{m} \int \cos^{m-2} u \, du$

(H) $\quad \displaystyle\int \sin^m u \cos^n u \, du \;=\; \frac{\sin^{m+1} u \cos^{n-1} u}{m+n} + \frac{n-1}{m+n} \int \sin^m u \cos^{n-2} u \, du$

$$\;=\; -\frac{\sin^{m-1} u \cos^{n+1} u}{m+n} + \frac{m-1}{m+n} \int \sin^{m-2} u \cos^n u \, du \,, \quad m \neq -n$$

(I) $\quad \displaystyle\int u^m \sin bu \, du \;=\; -\frac{u^m}{b} \cos bu + \frac{m}{b} \int u^{m-1} \cos bu \, du$

(J) $\quad \displaystyle\int u^m \cos bu \, du \;=\; \frac{u^m}{b} \sin bu - \frac{m}{b} \int u^{m-1} \sin bu \, du$

See Problem 11.

# Solved Problems

1. Find $\displaystyle\int x \sin x \, dx$.

   We have the following choices:

   $\quad$ (a) $u = x \sin x, \; dv = dx$; $\quad$ (b) $u = \sin x, \; dv = x \, dx$; $\quad$ (c) $u = x, \; dv = \sin x \, dx$.

   (a) $u = x \sin x, \; dv = dx$. Then $du = (\sin x + x \cos x) \, dx, \; v = x,$ and

   $$\int x \sin x \, dx \;=\; x \cdot x \sin x - \int x(\sin x + x \cos x) \, dx$$

   The resulting integral is not as simple as the original and this choice is discarded.

   (b) $u = \sin x, \; dv = x \, dx$. Then $du = \cos x \, dx, \; v = \tfrac{1}{2}x^2,$ and

   $$\int x \sin x \, dx \;=\; \tfrac{1}{2}x^2 \sin x - \int \tfrac{1}{2}x^2 \cos x \, dx$$

   The resulting integral is not as simple as the original and this choice is discarded.

   (c) $u = x, \; dv = \sin x \, dx$. Then $du = dx, \; v = -\cos x,$ and

   $$\int x \sin x \, dx \;=\; -x \cos x - \int -\cos x \, dx \;=\; -x \cos x + \sin x + C$$

2. Find $\displaystyle\int xe^x \, dx$.

   Let $u = x, \; dv = e^x \, dx$. Then $du = dx, \; v = e^x,$ and

   $$\int xe^x \, dx \;=\; xe^x - \int e^x \, dx \;=\; xe^x - e^x + C$$

3. Find $\displaystyle\int x^2 \ln x \, dx$.

   Let $u = \ln x, \; dv = x^2 \, dx$. Then $du = \dfrac{dx}{x}, \; v = \dfrac{x^3}{3},$ and

   $$\int x^2 \ln x \, dx \;=\; \frac{x^3}{3} \ln x - \int \frac{x^3}{3} \cdot \frac{dx}{x} \;=\; \frac{x^3}{3} \ln x - \frac{1}{3} \int x^2 \, dx \;=\; \frac{x^3}{3} \ln x - \frac{1}{9} x^3 + C$$

4. Find $\displaystyle\int x\sqrt{1+x} \, dx$.

   Let $u = x, \; dv = \sqrt{1+x} \, dx$. Then $du = dx, \; v = \tfrac{2}{3}(1+x)^{3/2},$ and

   $$\int x\sqrt{1+x} \, dx \;=\; \tfrac{2}{3}x(1+x)^{3/2} - \tfrac{2}{3} \int (1+x)^{3/2} \, dx \;=\; \tfrac{2}{3}x(1+x)^{3/2} - \frac{4}{15}(1+x)^{5/2} + C$$

**5.**   Find $\int \arcsin x \, dx$.

Let $u = \arcsin x$, $dv = dx$. Then $du = dx/\sqrt{1-x^2}$, $v = x$, and

$$\int \arcsin x \, dx \;=\; x \arcsin x - \int \frac{x \, dx}{\sqrt{1-x^2}} \;=\; x \arcsin x + \sqrt{1-x^2} + C$$

**6.**   Find $\int \sin^2 x \, dx$.

Let $u = \sin x$, $dv = \sin x \, dx$. Then $du = \cos x \, dx$, $v = -\cos x$, and

$$\int \sin^2 x \, dx \;=\; -\sin x \cos x + \int \cos^2 x \, dx$$

$$=\; -\sin x \cos x + \int (1 - \sin^2 x) \, dx \;=\; -\tfrac{1}{2} \sin 2x + \int dx - \int \sin^2 x \, dx$$

Transposing the integral on the right,

$$2 \int \sin^2 x \, dx \;=\; -\tfrac{1}{2} \sin 2x + x + C' \quad \text{and} \quad \int \sin^2 x \, dx \;=\; \tfrac{1}{2}x - \tfrac{1}{4} \sin 2x + C$$

**7.**   Find $\int \sec^3 x \, dx$.

Let $u = \sec x$, $dv = \sec^2 x \, dx$. Then $du = \sec x \tan x \, dx$, $v = \tan x$, and

$$\int \sec^3 x \, dx \;=\; \sec x \tan x - \int \sec x \tan^2 x \, dx \;=\; \sec x \tan x - \int \sec x \, (\sec^2 x - 1) \, dx$$

$$=\; \sec x \tan x - \int \sec^3 x \, dx + \int \sec x \, dx$$

Then   $$2 \int \sec^3 x \, dx \;=\; \sec x \tan x + \int \sec x \, dx \;=\; \sec x \tan x + \ln|\sec x + \tan x| + C'$$

and   $$\int \sec^3 x \, dx \;=\; \tfrac{1}{2}\{\sec x \tan x + \ln|\sec x + \tan x|\} + C$$

**8.**   Find $\int x^2 \sin x \, dx$.

Let $u = x^2$, $dv = \sin x \, dx$. Then $du = 2x \, dx$, $v = -\cos x$, and

$$\int x^2 \sin x \, dx \;=\; -x^2 \cos x + 2 \int x \cos x \, dx$$

For the resulting integral, let $u = x$ and $dv = \cos x \, dx$. Then $du = dx$, $v = \sin x$, and

$$\int x^2 \sin x \, dx \;=\; -x^2 \cos x + 2 \left\{ x \sin x - \int \sin x \, dx \right\}$$

$$=\; -x^2 \cos x + 2x \sin x + 2 \cos x + C$$

**9.**   Find $\int x^3 e^{2x} \, dx$.

Let $u = x^3$, $dv = e^{2x} \, dx$. Then $du = 3x^2 \, dx$, $v = \tfrac{1}{2} e^{2x}$, and

$$\int x^3 e^{2x} \, dx \;=\; \frac{1}{2} x^3 e^{2x} - \frac{3}{2} \int x^2 e^{2x} \, dx$$

For the resulting integral, let $u = x^2$ and $dv = e^{2x} \, dx$. Then $du = 2x \, dx$, $v = \tfrac{1}{2} e^{2x}$, and

$$\int x^3 e^{2x} \, dx \;=\; \frac{1}{2} x^3 e^{2x} - \frac{3}{2} \left\{ \frac{1}{2} x^2 e^{2x} - \int x e^{2x} \, dx \right\} \;=\; \frac{1}{2} x^3 e^{2x} - \frac{3}{4} x^2 e^{2x} + \frac{3}{2} \int x e^{2x} \, dx$$

For the resulting integral, let $u = x$ and $dv = e^{2x} \, dx$. Then $du = dx$, $v = \tfrac{1}{2} e^{2x}$, and

$$\int x^3 e^{2x} \, dx \;=\; \frac{1}{2} x^3 e^{2x} - \frac{3}{4} x^2 e^{2x} + \frac{3}{2} \left\{ \frac{1}{2} x e^{2x} - \frac{1}{2} \int e^{2x} \, dx \right\}$$

$$=\; \frac{1}{2} x^3 e^{2x} - \frac{3}{4} x^2 e^{2x} + \frac{3}{4} x e^{2x} - \frac{3}{8} e^{2x} + C$$

**10.**   (a) Take $u = x$, $dv = \dfrac{x \, dx}{(a^2 \pm x^2)^m}$; then $du = dx$, $v = \dfrac{\mp 1}{(2m-2)(a^2 \pm x^2)^{m-1}}$, and

$$\int \frac{x^2 \, dx}{(a^2 \pm x^2)^m} \;=\; \frac{\mp x}{(2m-2)(a^2 \pm x^2)^{m-1}} \pm \frac{1}{2m-2} \int \frac{dx}{(a^2 \pm x^2)^{m-1}}$$

(b) Take $u = x$, $dv = x(a^2 \pm x^2)^{m-1} \, dx$; then $du = dx$, $v = \dfrac{\pm 1}{2m}(a^2 \pm x^2)^m$, and

$$\int x^2 (a^2 \pm x^2)^{m-1} \, dx \;=\; \frac{\pm x}{2m}(a^2 \pm x^2)^m \mp \frac{1}{2m} \int (a^2 \pm x^2)^m \, dx$$

11. Find:  (a) $\int \dfrac{dx}{(1+x^2)^{5/2}}$,  (b) $\int (9+x^2)^{3/2}\, dx$.

(a). Since Reduction Formula (A) reduces the exponent in the denominator by 1, we use this formula twice to obtain

$$\int \frac{dx}{(1+x^2)^{5/2}} \;=\; \frac{x}{3(1+x^2)^{3/2}} + \frac{2}{3}\int \frac{dx}{(1+x^2)^{3/2}} \;=\; \frac{x}{3(1+x^2)^{3/2}} + \frac{2}{3}\frac{x}{(1+x^2)^{1/2}} + C$$

(b) Using Reduction Formula (B),

$$\int (9+x^2)^{3/2}\, dx \;=\; \frac{1}{4}x(9+x^2)^{3/2} + \frac{27}{4}\int (9+x^2)^{1/2}\, dx$$

$$\;=\; \frac{1}{4}x(9+x^2)^{3/2} + \frac{27}{8}\{x(9+x^2)^{1/2} + 9\ln(x+\sqrt{9+x^2})\} + C$$

# Supplementary Problems

12. $\int x\cos x\, dx \;=\; x\sin x + \cos x + C$   13. $\int x\sec^2 3x\, dx \;=\; \dfrac{x}{3}\tan 3x - \dfrac{1}{9}\ln|\sec 3x| + C$

14. $\int \arccos 2x\, dx \;=\; x\arccos 2x - \dfrac{1}{2}\sqrt{1-4x^2} + C$

15. $\int \arctan x\, dx \;=\; x\arctan x - \ln\sqrt{1+x^2} + C$

16. $\int x^2\sqrt{1-x}\, dx \;=\; -\dfrac{2}{105}(1-x)^{3/2}(15x^2+12x+8) + C$   17. $\int \dfrac{xe^x\, dx}{(1+x)^2} \;=\; \dfrac{e^x}{1+x} + C$

18. $\int x\arctan x\, dx \;=\; \dfrac{1}{2}(x^2+1)\arctan x - \dfrac{1}{2}x + C$

19. $\int x^2 e^{-3x}\, dx \;=\; -\dfrac{1}{3}e^{-3x}(x^2+\tfrac{2}{3}x+\tfrac{2}{9}) + C$

20. $\int \sin^3 x\, dx \;=\; -\dfrac{2}{3}\cos^3 x - \sin^2 x\cos x + C$

21. $\int x^3\sin x\, dx \;=\; -x^3\cos x + 3x^2\sin x + 6x\cos x - 6\sin x + C$

22. $\int \dfrac{x\, dx}{\sqrt{a+bx}} \;=\; \dfrac{2(bx-2a)\sqrt{a+bx}}{3b^2} + C$   23. $\int \dfrac{x^2\, dx}{\sqrt{1+x}} \;=\; \dfrac{2}{15}(3x^2-4x+8)\sqrt{1+x} + C$

24. $\int x\arcsin x^2\, dx \;=\; \dfrac{1}{2}x^2\arcsin x^2 + \dfrac{1}{2}\sqrt{1-x^4} + C$

25. $\int \sin x\sin 3x\, dx \;=\; \dfrac{1}{8}\sin 3x\cos x - \dfrac{3}{8}\sin x\cos 3x + C$

26. $\int \sin(\ln x)\, dx \;=\; \dfrac{1}{2}x(\sin\ln x - \cos\ln x) + C$

27. $\int e^{ax}\cos bx\, dx \;=\; \dfrac{e^{ax}(b\sin bx + a\cos bx)}{a^2+b^2} + C$

28. $\int e^{ax}\sin bx\, dx \;=\; \dfrac{e^{ax}(a\sin bx - b\cos bx)}{a^2+b^2} + C$

29. (a) Write $\int \dfrac{a^2\, dx}{(a^2\pm x^2)^m} \;=\; \int \dfrac{(a^2\pm x^2)\mp x^2}{(a^2\pm x^2)^m}\, dx \;=\; \int \dfrac{dx}{(a^2\pm x^2)^{m-1}} \mp \int \dfrac{x^2\, dx}{(a^2\pm x^2)^m}$  and use the result of Problem 10(a) to obtain the reduction formula (A).

(b) Write $\int (a^2\pm x^2)^m\, dx \;=\; a^2\int (a^2\pm x^2)^{m-1}\, dx \pm \int x^2(a^2\pm x^2)^{m-1}\, dx$  and use the result of Problem 10(b) to obtain the reduction formula (B).

30. Derive the reduction formulas (C)–(J).

**31.** $\displaystyle \int \frac{dx}{(1-x^2)^3} = \frac{x(5-3x^2)}{8(1-x^2)^2} + \frac{3}{16}\ln\left|\frac{1+x}{1-x}\right| + C$     **32.** $\displaystyle \int \frac{dx}{(4+x^2)^{3/2}} = \frac{x}{4(4+x^2)^{1/2}} + C$

**33.** $\displaystyle \int (4-x^2)^{3/2}\,dx = \frac{1}{4}x(10-x^2)\sqrt{4-x^2} + 6\arcsin\frac{1}{2}x + C$

**34.** $\displaystyle \int \frac{dx}{(x^2-16)^3} = \frac{1}{2048}\left\{\frac{x(3x^2-80)}{(x^2-16)^2} + \frac{3}{8}\ln\left|\frac{x-4}{x+4}\right|\right\} + C$

**35.** $\displaystyle \int (x^2-1)^{5/2}\,dx = \frac{1}{48}x(8x^4-26x^2+33)\sqrt{x^2-1} - \frac{5}{16}\ln|x+\sqrt{x^2-1}| + C$

**36.** $\displaystyle \int \sin^4 x\,dx = \frac{3}{8}x - \frac{3}{8}\sin x\cos x - \frac{1}{4}\sin^3 x\cos x + C$

**37.** $\displaystyle \int \cos^5 x\,dx = \frac{1}{15}(3\cos^4 x + 4\cos^2 x + 8)\sin x + C$

**38.** $\displaystyle \int \sin^3 x\cos^2 x\,dx = -\frac{1}{5}\cos^3 x\,(\sin^2 x + \frac{2}{3}) + C$

**39.** $\displaystyle \int \sin^4 x\cos^5 x\,dx = \frac{1}{9}\sin^5 x\,(\cos^4 x + \frac{4}{7}\cos^2 x + \frac{8}{35}) + C$

*An alternate procedure* for some of the more tedious problems of this section can be found by noting that in (see Problem 9)

**(i)** $$\int x^3 e^{2x}\,dx = \frac{1}{2}x^3 e^{2x} - \frac{3}{4}x^2 e^{2x} + \frac{3}{4}xe^{2x} - \frac{3}{8}e^{2x} + C$$

the terms on the right, apart from the coefficients, are the different terms obtained by repeated differentiations of the integrand $x^3 e^{2x}$. Thus, we may write at once

**(ii)** $$\int x^3 e^{2x}\,dx = Ax^3 e^{2x} + Bx^2 e^{2x} + Dxe^{2x} + Ee^{2x} + C$$

and from it obtain by differentiation

$$x^3 e^{2x} = 2Ax^3 e^{2x} + (3A+2B)x^2 e^{2x} + (2B+2D)xe^{2x} + (D+2E)e^{2x}$$

Equating coefficients, we have

$$2A = 1, \quad 3A+2B = 0, \quad 2B+2D = 0, \quad D+2E = 0$$

so that $A = \frac{1}{2}$, $B = -\frac{3}{2}A = -\frac{3}{4}$, $D = -B = \frac{3}{4}$, $E = -\frac{1}{2}D = -\frac{3}{8}$. Substituting for $A, B, D, E$ in (ii), we obtain (i).

This procedure may be used for finding $\displaystyle \int f(x)\,dx$ whenever repeated differentiation of $f(x)$ yields only a finite number of different terms.

**40.** Find $\displaystyle \int e^{2x}\cos 3x\,dx = \frac{1}{13}e^{2x}(3\sin 3x + 2\cos 3x) + C$   using

$$\int e^{2x}\cos 3x\,dx = Ae^{2x}\sin 3x + Be^{2x}\cos 3x + C$$

**41.** Find $\displaystyle \int e^{3x}(2\sin 4x - 5\cos 4x)\,dx = \frac{1}{25}e^{3x}(-14\sin 4x - 23\cos 4x) + C$   using

$$\int e^{3x}(2\sin 4x - 5\cos 4x)\,dx = Ae^{3x}\sin 4x + Be^{3x}\cos 4x + C$$

**42.** Find $\displaystyle \int \sin 3x\cos 2x\,dx = -\frac{1}{5}(2\sin 3x\sin 2x + 3\cos 3x\cos 2x) + C$   using

$$\int \sin 3x\cos 2x\,dx = A\sin 3x\sin 2x + B\cos 3x\cos 2x$$
$$+ D\cos 3x\sin 2x + E\sin 3x\cos 2x + C$$

**43.** Find $\displaystyle \int e^{3x}x^2\sin x\,dx = \frac{e^{3x}}{250}[25x^2(3\sin x - \cos x) - 10x(4\sin x - 3\cos x) + 9\sin x - 13\cos x] + C$

# Chapter 27

# Trigonometric Integrals

**THE FOLLOWING IDENTITIES** are employed to find the trigonometric integrals of this chapter.

1. $\sin^2 x + \cos^2 x = 1$
2. $1 + \tan^2 x = \sec^2 x$
3. $1 + \cot^2 x = \csc^2 x$
4. $\sin^2 x = \frac{1}{2}(1 - \cos 2x)$
5. $\cos^2 x = \frac{1}{2}(1 + \cos 2x)$
6. $\sin x \cos x = \frac{1}{2} \sin 2x$

7. $\sin x \cos y = \frac{1}{2}[\sin(x-y) + \sin(x+y)]$
8. $\sin x \sin y = \frac{1}{2}[\cos(x-y) - \cos(x+y)]$
9. $\cos x \cos y = \frac{1}{2}[\cos(x-y) + \cos(x+y)]$
10. $1 - \cos x = 2 \sin^2 \frac{1}{2}x$
11. $1 + \cos x = 2 \cos^2 \frac{1}{2}x$
12. $1 \pm \sin x = 1 \pm \cos(\frac{1}{2}\pi - x)$

## Solved Problems

### SINES AND COSINES

1. $\displaystyle\int \sin^2 x \, dx = \int \frac{1}{2}(1 - \cos 2x) \, dx = \frac{1}{2}x - \frac{1}{4}\sin 2x + C$

2. $\displaystyle\int \cos^2 3x \, dx = \int \frac{1}{2}(1 + \cos 6x) \, dx = \frac{1}{2}x + \frac{1}{12}\sin 6x + C$

3. $\displaystyle\int \sin^3 x \, dx = \int \sin^2 x \sin x \, dx = \int (1 - \cos^2 x) \sin x \, dx = -\cos x + \frac{1}{3}\cos^3 x + C$

4. $\displaystyle\int \cos^5 x \, dx = \int \cos^4 x \cos x \, dx = \int (1 - \sin^2 x)^2 \cos x \, dx$

   $\displaystyle\qquad = \int \cos x \, dx - 2\int \sin^2 x \cos x \, dx + \int \sin^4 x \cos x \, dx$

   $\displaystyle\qquad = \sin x - \frac{2}{3}\sin^3 x + \frac{1}{5}\sin^5 x + C$

5. $\displaystyle\int \sin^2 x \cos^3 x \, dx = \int \sin^2 x \cos^2 x \cos x \, dx = \int \sin^2 x\,(1 - \sin^2 x) \cos x \, dx$

   $\displaystyle\qquad = \int \sin^2 x \cos x \, dx - \int \sin^4 x \cos x \, dx = \frac{1}{3}\sin^3 x - \frac{1}{5}\sin^5 x + C$

6. $\displaystyle\int \cos^4 2x \sin^3 2x \, dx = \int \cos^4 2x \sin^2 2x \sin 2x \, dx = \int \cos^4 2x\,(1 - \cos^2 2x) \sin 2x \, dx$

   $\displaystyle\qquad = \int \cos^4 2x \sin 2x \, dx - \int \cos^6 2x \sin 2x \, dx = -\frac{1}{10}\cos^5 2x + \frac{1}{14}\cos^7 2x + C$

7. $\displaystyle\int \sin^3 3x \cos^5 3x \, dx = \int (1 - \cos^2 3x) \cos^5 3x \sin 3x \, dx$

   $\displaystyle\qquad = \int \cos^5 3x \sin 3x \, dx - \int \cos^7 3x \sin 3x \, dx = -\frac{1}{18}\cos^6 3x + \frac{1}{24}\cos^8 3x + C$

   or

   $\displaystyle\int \sin^3 3x \cos^5 3x \, dx = \int \sin^3 3x\,(1 - \sin^2 3x)^2 \cos 3x \, dx$

   $\displaystyle\qquad = \int \sin^3 3x \cos 3x \, dx - 2\int \sin^5 3x \cos 3x \, dx + \int \sin^7 3x \cos 3x \, dx$

   $\displaystyle\qquad = \frac{1}{12}\sin^4 3x - \frac{1}{9}\sin^6 3x + \frac{1}{24}\sin^8 3x + C$

**8.** $\int \cos^3 \frac{x}{3}\, dx \;=\; \int \left(1 - \sin^2 \frac{x}{3}\right) \cos \frac{x}{3}\, dx \;=\; 3 \sin \frac{x}{3} - \sin^3 \frac{x}{3} + C$

**9.** $\int \sin^4 x\, dx \;=\; \int (\sin^2 x)^2\, dx \;=\; \frac{1}{4} \int (1 - \cos 2x)^2\, dx$

$\qquad =\; \frac{1}{4} \int dx - \frac{1}{2} \int \cos 2x\, dx + \frac{1}{4} \int \cos^2 2x\, dx$

$\qquad =\; \frac{1}{4} \int dx - \frac{1}{2} \int \cos 2x\, dx + \frac{1}{8} \int (1 + \cos 4x)\, dx$

$\qquad =\; \frac{1}{4} x - \frac{1}{4} \sin 2x + \frac{1}{8} x + \frac{1}{32} \sin 4x + C \;=\; \frac{3}{8} x - \frac{1}{4} \sin 2x + \frac{1}{32} \sin 4x + C$

**10.** $\int \sin^2 x \cos^2 x\, dx \;=\; \frac{1}{4} \int \sin^2 2x\, dx \;=\; \frac{1}{8} \int (1 - \cos 4x)\, dx \;=\; \frac{1}{8} x - \frac{1}{32} \sin 4x + C$

**11.** $\int \sin^4 3x \cos^2 3x\, dx \;=\; \int (\sin^2 3x \cos^2 3x) \sin^2 3x\, dx \;=\; \frac{1}{8} \int \sin^2 6x\, (1 - \cos 6x)\, dx$

$\qquad =\; \frac{1}{8} \int \sin^2 6x\, dx - \frac{1}{8} \int \sin^2 6x \cos 6x\, dx$

$\qquad =\; \frac{1}{16} \int (1 - \cos 12x)\, dx - \frac{1}{8} \int \sin^2 6x \cos 6x\, dx$

$\qquad =\; \frac{1}{16} x - \frac{1}{192} \sin 12x - \frac{1}{144} \sin^3 6x + C$

**12.** $\int \sin 3x \sin 2x\, dx \;=\; \int \frac{1}{2}\{\cos(3x - 2x) - \cos(3x + 2x)\}\, dx \;=\; \frac{1}{2} \int (\cos x - \cos 5x)\, dx$

$\qquad =\; \frac{1}{2} \sin x - \frac{1}{10} \sin 5x + C$

**13.** $\int \sin 3x \cos 5x\, dx \;=\; \int \frac{1}{2}\{\sin(3x - 5x) + \sin(3x + 5x)\}\, dx \;=\; \frac{1}{4} \cos 2x - \frac{1}{16} \cos 8x + C$

**14.** $\int \cos 4x \cos 2x\, dx \;=\; \frac{1}{2} \int (\cos 2x + \cos 6x)\, dx \;=\; \frac{1}{4} \sin 2x + \frac{1}{12} \sin 6x + C$

**15.** $\int \sqrt{1 - \cos x}\, dx \;=\; \sqrt{2} \int \sin \frac{1}{2} x\, dx \;=\; -2\sqrt{2} \cos \frac{1}{2} x + C$

**16.** $\int (1 + \cos 3x)^{3/2}\, dx \;=\; 2\sqrt{2} \int \cos^3 \frac{3}{2} x\, dx \;=\; 2\sqrt{2} \int (1 - \sin^2 \frac{3}{2} x) \cos \frac{3}{2} x\, dx$

$\qquad =\; 2\sqrt{2}\, (\frac{2}{3} \sin \frac{3}{2} x - \frac{2}{9} \sin^3 \frac{3}{2} x) + C$

**17.** $\int \frac{dx}{\sqrt{1 - \sin 2x}} \;=\; \int \frac{dx}{\sqrt{1 - \cos(\frac{1}{2}\pi - 2x)}} \;=\; \frac{\sqrt{2}}{2} \int \frac{dx}{\sin(\frac{1}{4}\pi - x)} \;=\; \frac{\sqrt{2}}{2} \int \csc(\frac{1}{4}\pi - x)\, dx$

$\qquad =\; -\frac{\sqrt{2}}{2} \ln |\csc(\frac{1}{4}\pi - x) - \cot(\frac{1}{4}\pi - x)| + C$

## TANGENTS, SECANTS, COTANGENTS, COSECANTS

**18.** $\int \tan^4 x\, dx \;=\; \int \tan^2 x \tan^2 x\, dx \;=\; \int \tan^2 x\, (\sec^2 x - 1)\, dx \;=\; \int \tan^2 x \sec^2 x\, dx - \int \tan^2 x\, dx$

$\qquad =\; \int \tan^2 x \sec^2 x\, dx - \int (\sec^2 x - 1)\, dx \;=\; \frac{1}{3} \tan^3 x - \tan x + x + C$

**19.** $\int \tan^5 x\, dx \;=\; \int \tan^3 x \tan^2 x\, dx \;=\; \int \tan^3 x\, (\sec^2 x - 1)\, dx$

$\qquad =\; \int \tan^3 x \sec^2 x\, dx - \int \tan^3 x\, dx \;=\; \int \tan^3 x \sec^2 x\, dx - \int \tan x\, (\sec^2 x - 1)\, dx$

$\qquad =\; \frac{1}{4} \tan^4 x - \frac{1}{2} \tan^2 x + \ln |\sec x| + C$

**20.** $\int \sec^4 2x\, dx \;=\; \int \sec^2 2x \sec^2 2x\, dx \;=\; \int \sec^2 2x\, (1 + \tan^2 2x)\, dx$

$\qquad =\; \int \sec^2 2x\, dx + \int \tan^2 2x \sec^2 2x\, dx \;=\; \frac{1}{2} \tan 2x + \frac{1}{6} \tan^3 2x + C$

**21.** $\int \tan^3 3x \sec^4 3x \, dx = \int \tan^3 3x \, (1 + \tan^2 3x) \sec^2 3x \, dx$

$\qquad = \int \tan^3 3x \sec^2 3x \, dx + \int \tan^5 3x \sec^2 3x \, dx = \frac{1}{12} \tan^4 3x + \frac{1}{18} \tan^6 3x + C$

**22.** $\int \tan^2 x \sec^3 x \, dx = \int (\sec^2 x - 1) \sec^3 x \, dx = \int \sec^5 x \, dx - \int \sec^3 x \, dx$

$\qquad = \frac{1}{4} \sec^3 x \tan x - \frac{1}{8} \sec x \tan x - \frac{1}{8} \ln |\sec x + \tan x| + C$, integrating by parts.

**23.** $\int \tan^3 2x \sec^3 2x \, dx = \int \tan^2 2x \sec^2 2x \cdot \sec 2x \tan 2x \, dx$

$\qquad = \int (\sec^2 2x - 1) \sec^2 2x \cdot \sec 2x \tan 2x \, dx$

$\qquad = \int \sec^4 2x \cdot \sec 2x \tan 2x \, dx - \int \sec^2 2x \cdot \sec 2x \tan 2x \, dx$

$\qquad = \frac{1}{10} \sec^5 2x - \frac{1}{6} \sec^3 2x + C$

**24.** $\int \cot^3 2x \, dx = \int \cot 2x \, (\csc^2 2x - 1) \, dx = -\frac{1}{4} \cot^2 2x + \frac{1}{2} \ln |\csc 2x| + C$

**25.** $\int \cot^4 3x \, dx = \int \cot^2 3x \, (\csc^2 3x - 1) \, dx = \int \cot^2 3x \csc^2 3x \, dx - \int \cot^2 3x \, dx$

$\qquad = \int \cot^2 3x \csc^2 3x \, dx - \int (\csc^2 3x - 1) \, dx = -\frac{1}{9} \cot^3 3x + \frac{1}{3} \cot 3x + x + C$

**26.** $\int \csc^6 x \, dx = \int \csc^2 x \, (1 + \cot^2 x)^2 \, dx$

$\qquad = \int \csc^2 x \, dx + 2 \int \cot^2 x \csc^2 x \, dx + \int \cot^4 x \csc^2 x \, dx$

$\qquad = -\cot x - \frac{2}{3} \cot^3 x - \frac{1}{5} \cot^5 x + C$

**27.** $\int \cot 3x \csc^4 3x \, dx = \int \cot 3x \, (1 + \cot^2 3x) \csc^2 3x \, dx$

$\qquad = \int \cot 3x \csc^2 3x \, dx + \int \cot^3 3x \csc^2 3x \, dx = -\frac{1}{6} \cot^2 3x - \frac{1}{12} \cot^4 3x + C$

**28.** $\int \cot^3 x \csc^5 x \, dx = \int \cot^2 x \csc^4 x \cdot \csc x \cot x \, dx = \int (\csc^2 x - 1) \csc^4 x \cdot \csc x \cot x \, dx$

$\qquad = \int \csc^6 x \cdot \csc x \cot x \, dx - \int \csc^4 x \cdot \csc x \cot x \, dx$

$\qquad = -\frac{1}{7} \csc^7 x + \frac{1}{5} \csc^5 x + C$

# Supplementary Problems

**29.** $\int \cos^2 x \, dx = \frac{1}{2}x + \frac{1}{4} \sin 2x + C$        **30.** $\int \sin^3 2x \, dx = \frac{1}{6} \cos^3 2x - \frac{1}{2} \cos 2x + C$

**31.** $\int \sin^4 2x \, dx = \frac{3}{8}x - \frac{1}{8} \sin 4x + \frac{1}{64} \sin 8x + C$

**32.** $\int \cos^4 \frac{1}{2}x \, dx = \frac{3}{8}x + \frac{1}{2} \sin x + \frac{1}{16} \sin 2x + C$

**33.** $\int \sin^7 x \, dx = \frac{1}{7} \cos^7 x - \frac{3}{5} \cos^5 x + \cos^3 x - \cos x + C$

**34.** $\int \cos^6 \frac{1}{2}x \, dx \;=\; \frac{5}{16}x + \frac{1}{2}\sin x + \frac{3}{32}\sin 2x - \frac{1}{24}\sin^3 x + C$

**35.** $\int \sin^2 x \cos^5 x \, dx \;=\; \frac{1}{3}\sin^3 x - \frac{2}{5}\sin^5 x + \frac{1}{7}\sin^7 x + C$

**36.** $\int \sin^3 x \cos^2 x \, dx \;=\; \frac{1}{5}\cos^5 x - \frac{1}{3}\cos^3 x + C$

**37.** $\int \sin^3 x \cos^3 x \, dx \;=\; \frac{1}{48}\cos^3 2x - \frac{1}{16}\cos 2x + C$

**38.** $\int \sin^4 x \cos^4 x \, dx \;=\; \frac{1}{128}\left(3x - \sin 4x + \frac{1}{8}\sin 8x\right) + C$

**39.** $\int \sin 2x \cos 4x \, dx \;=\; \frac{1}{4}\cos 2x - \frac{1}{12}\cos 6x + C$

**40.** $\int \cos 3x \cos 2x \, dx \;=\; \frac{1}{2}\sin x + \frac{1}{10}\sin 5x + C$

**41.** $\int \sin 5x \sin x \, dx \;=\; \frac{1}{8}\sin 4x - \frac{1}{12}\sin 6x + C$

**42.** $\int \dfrac{\cos^3 x \, dx}{1 - \sin x} \;=\; \sin x + \dfrac{1}{2}\sin^2 x + C$     **43.** $\int \dfrac{\cos^{2/3} x}{\sin^{8/3} x}\, dx \;=\; -\dfrac{3}{5}\cot^{5/3} x + C$

**44.** $\int \dfrac{\cos^3 x}{\sin^4 x}\, dx \;=\; \csc x - \dfrac{1}{3}\csc^3 x + C$

**45.** $\int x \left(\cos^3 x^2 - \sin^3 x^2\right) dx \;=\; \frac{1}{12}\left(\sin x^2 + \cos x^2\right)\left(4 + \sin 2x^2\right) + C$

**46.** $\int \tan^3 x \, dx \;=\; \frac{1}{2}\tan^2 x + \ln|\cos x| + C$

**47.** $\int \tan^3 3x \sec 3x \, dx \;=\; \frac{1}{9}\sec^3 3x - \frac{1}{3}\sec 3x + C$

**48.** $\int \tan^{3/2} x \sec^4 x \, dx \;=\; \frac{2}{5}\tan^{5/2} x + \frac{2}{9}\tan^{9/2} x + C$

**49.** $\int \tan^4 x \sec^4 x \, dx \;=\; \frac{1}{7}\tan^7 x + \frac{1}{5}\tan^5 x + C$     **53.** $\int \csc^4 2x \, dx \;=\; -\frac{1}{2}\cot 2x - \frac{1}{6}\cot^3 2x + C$

**50.** $\int \cot^3 x \, dx \;=\; -\frac{1}{2}\cot^2 x - \ln|\sin x| + C$     **54.** $\int \left(\dfrac{\sec x}{\tan x}\right)^4 dx \;=\; -\dfrac{1}{3\tan^3 x} - \dfrac{1}{\tan x} + C$

**51.** $\int \cot^3 x \csc^4 x \, dx \;=\; -\frac{1}{4}\cot^4 x - \frac{1}{6}\cot^6 x + C$     **55.** $\int \dfrac{\cot^3 x}{\csc x}\, dx \;=\; -\sin x - \csc x + C$

**52.** $\int \cot^3 x \csc^3 x \, dx \;=\; -\frac{1}{5}\csc^5 x + \frac{1}{3}\csc^3 x + C$     **56.** $\int \tan x \sqrt{\sec x} \, dx \;=\; 2\sqrt{\sec x} + C$

**57.** Use integration by parts to derive the reduction formulas

    (a) $\int \sec^m u \, du \;=\; \dfrac{1}{m-1}\sec^{m-2} u \tan u + \dfrac{m-2}{m-1}\int \sec^{m-2} u \, du$

    (b) $\int \csc^m u \, du \;=\; -\dfrac{1}{m-1}\csc^{m-2} u \cot u + \dfrac{m-2}{m-1}\int \csc^{m-2} u \, du$

## Use the reduction formulas of Problem 57 to evaluate Problems 58-60.

**58.** $\int \sec^3 x \, dx \;=\; \frac{1}{2}\sec x \tan x + \frac{1}{2}\ln|\sec x + \tan x| + C$

**59.** $\int \csc^5 x \, dx \;=\; -\frac{1}{4}\csc^3 x \cot x - \frac{3}{8}\csc x \cot x + \frac{3}{8}\ln|\csc x - \cot x| + C$

**60.** $\int \sec^6 x \, dx \;=\; \frac{1}{5}\sec^4 x \tan x + \frac{4}{15}\sec^2 x \tan x + \frac{8}{15}\tan x + C$

         $=\; \frac{1}{5}\tan^5 x + \frac{2}{3}\tan^3 x + \tan x + C$

# Chapter 28

## Trigonometric Substitutions

**AN INTEGRAND,** which contains one of the forms $\sqrt{a^2 - b^2 u^2}$, $\sqrt{a^2 + b^2 u^2}$, or $\sqrt{b^2 u^2 - a^2}$ but no other irrational factor, may be transformed into another involving trigonometric functions of a new variable as follows:

| For | use | to obtain |
|-----|-----|-----------|
| $\sqrt{a^2 - b^2 u^2}$ | $u = \dfrac{a}{b}\sin z$ | $a\sqrt{1 - \sin^2 z} = a\cos z$ |
| $\sqrt{a^2 + b^2 u^2}$ | $u = \dfrac{a}{b}\tan z$ | $a\sqrt{1 + \tan^2 z} = a\sec z$ |
| $\sqrt{b^2 u^2 - a^2}$ | $u = \dfrac{a}{b}\sec z$ | $a\sqrt{\sec^2 z - 1} = a\tan z$ |

In each case, integration yields an expression in the variable $z$. The corresponding expression in the original variable may be obtained by the use of a right triangle as shown in the solved problems below.

## Solved Problems

1. Find $\displaystyle\int \frac{dx}{x^2\sqrt{4 + x^2}}$.

   Let $x = 2\tan z$; then $dx = 2\sec^2 z\, dz$ and $\sqrt{4 + x^2} = 2\sec z$.

   $$\int \frac{dx}{x^2\sqrt{4 + x^2}} = \int \frac{2\sec^2 z\, dz}{(4\tan^2 z)(2\sec z)} = \frac{1}{4}\int \frac{\sec z}{\tan^2 z}\, dz$$
   $$= \frac{1}{4}\int \sin^{-2} z\cos z\, dz = -\frac{1}{4\sin z} + C = -\frac{\sqrt{4 + x^2}}{4x} + C$$

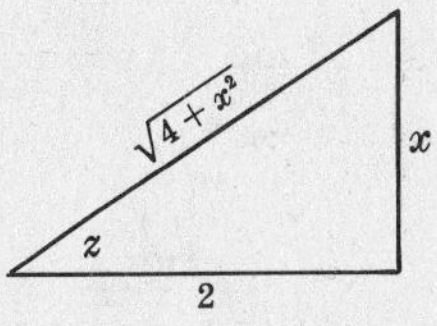

   **Fig. 28-1**

2. Find $\displaystyle\int \frac{x^2}{\sqrt{x^2 - 4}}\, dx$.

   Let $x = 2\sec z$; then $dx = 2\sec z\tan z\, dz$ and $\sqrt{x^2 - 4} = 2\tan z$.

   $$\int \frac{x^2}{\sqrt{x^2 - 4}}\, dx = \int \frac{4\sec^2 z}{2\tan z}(2\sec z\tan z\, dz) = 4\int \sec^3 z\, dz$$
   $$= 2\sec z\tan z + 2\ln|\sec z + \tan z| + C'$$
   $$= \tfrac{1}{2}x\sqrt{x^2 - 4} + 2\ln|x + \sqrt{x^2 - 4}| + C$$

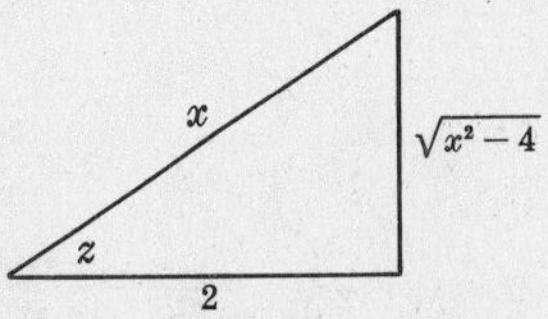

   **Fig. 28-2**

3. Find $\displaystyle\int \frac{\sqrt{9 - 4x^2}}{x}\, dx$.

   Let $x = \tfrac{3}{2}\sin z$; then $dx = \tfrac{3}{2}\cos z\, dz$ and $\sqrt{9 - 4x^2} = 3\cos z$.

147

$$\int \frac{\sqrt{9-4x^2}}{x}dx \;=\; \int \frac{3\cos z}{\frac{3}{2}\sin z}\left(\tfrac{3}{2}\cos z\,dz\right) \;=\; 3\int \frac{\cos^2 z}{\sin z}dz$$

$$=\; 3\int \frac{1-\sin^2 z}{\sin z}dz \;=\; 3\int \csc z\,dz \;-\; 3\int \sin z\,dz$$

$$=\; 3\ln\left|\csc z - \cot z\right| \;+\; 3\cos z \;+\; C'$$

$$=\; 3\ln\left|\frac{3-\sqrt{9-4x^2}}{x}\right| \;+\; \sqrt{9-4x^2} \;+\; C$$

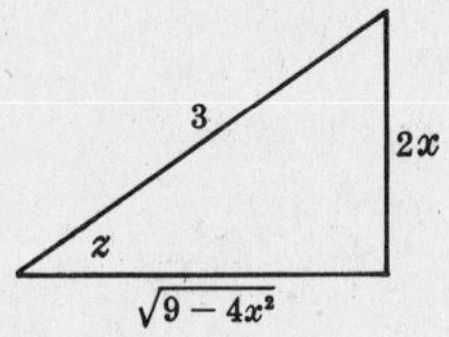

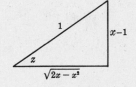

**Fig. 28-3**

4.   Find $\displaystyle\int \frac{dx}{x\sqrt{9+4x^2}}$.

Let $x = \tfrac{3}{2}\tan z$; then $dx = \tfrac{3}{2}\sec^2 z\,dz$ and $\sqrt{9+4x^2} = 3\sec z$.

$$\int \frac{dx}{x\sqrt{9+4x^2}} \;=\; \int \frac{\frac{3}{2}\sec^2 z\,dz}{\frac{3}{2}\tan z \cdot 3\sec z} \;=\; \frac{1}{3}\int \csc z\,dz$$

$$=\; \frac{1}{3}\ln\left|\csc z - \cot z\right| \;+\; C' \;=\; \frac{1}{3}\ln\left|\frac{\sqrt{9+4x^2}-3}{x}\right| \;+\; C$$

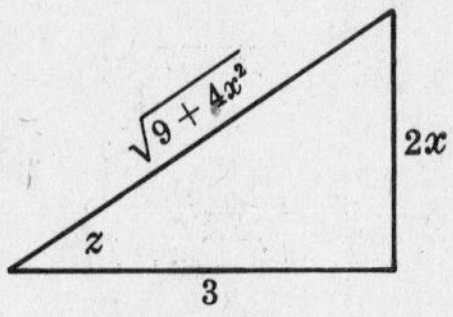

**Fig. 28-4**

5.   Find $\displaystyle\int \frac{(16-9x^2)^{3/2}}{x^6}\,dx$.

Let $x = \tfrac{4}{3}\sin z$; then $dx = \tfrac{4}{3}\cos z\,dz$ and $\sqrt{16-9x^2} = 4\cos z$.

$$\int \frac{(16-9x^2)^{3/2}}{x^6}\,dx \;=\; \int \frac{64\cos^3 z \cdot \frac{4}{3}\cos z\,dz}{\frac{4096}{729}\sin^6 z} \;=\; \frac{243}{16}\int \frac{\cos^4 z}{\sin^6 z}dz$$

$$=\; \frac{243}{16}\int \cot^4 z\,\csc^2 z\,dz \;=\; -\frac{243}{80}\cot^5 z \;+\; C$$

$$=\; -\frac{243}{80}\cdot\frac{(16-9x^2)^{5/2}}{243x^5} \;+\; C \;=\; -\frac{1}{80}\cdot\frac{(16-9x^2)^{5/2}}{x^5} \;+\; C$$

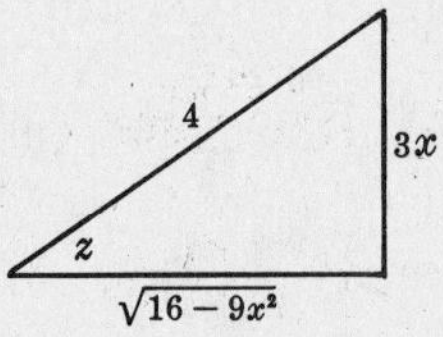

**Fig. 28-5**

6.   Find $\displaystyle\int \frac{x^2\,dx}{\sqrt{2x-x^2}} \;=\; \int \frac{x^2\,dx}{\sqrt{1-(x-1)^2}}$.

Let $x-1 = \sin z$; then $dx = \cos z\,dz$ and $\sqrt{2x-x^2} = \cos z$.

$$\int \frac{x^2\,dx}{\sqrt{2x-x^2}} \;=\; \int \frac{(1+\sin z)^2}{\cos z}\cos z\,dz \;=\; \int (1+\sin z)^2\,dz$$

$$=\; \int \left(\tfrac{3}{2}+2\sin z - \tfrac{1}{2}\cos 2z\right)dz \;=\; \tfrac{3}{2}z - 2\cos z - \tfrac{1}{4}\sin 2z \;+\; C$$

**Fig. 28-6**

$$=\; \tfrac{3}{2}\arcsin(x-1) - 2\sqrt{2x-x^2} - \tfrac{1}{2}(x-1)\sqrt{2x-x^2} \;+\; C$$

$$=\; \tfrac{3}{2}\arcsin(x-1) - \tfrac{1}{2}(x+3)\sqrt{2x-x^2} \;+\; C$$

7.   Find $\displaystyle\int \frac{dx}{(4x^2-24x+27)^{3/2}} \;=\; \int \frac{dx}{\{4(x-3)^2-9\}^{3/2}}$.

Let $x-3 = \tfrac{3}{2}\sec z$; then $dx = \tfrac{3}{2}\sec z\tan z\,dz$ and $\sqrt{4x^2-24x+27} = 3\tan z$.

$$\int \frac{dx}{(4x^2-24x+27)^{3/2}} \;=\; \int \frac{\frac{3}{2}\sec z\tan z\,dz}{27\tan^3 z}$$

$$=\; \frac{1}{18}\int \sin^{-2} z\,\cos z\,dz$$

$$=\; -\frac{1}{18}\csc z \;+\; C$$

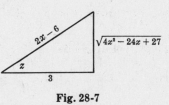

**Fig. 28-7**

$$=\; -\frac{1}{9}\frac{x-3}{\sqrt{4x^2-24x+27}} \;+\; C$$

# Supplementary Problems

8. $\int \dfrac{dx}{(4-x^2)^{3/2}} = \dfrac{x}{4\sqrt{4-x^2}} + C$

9. $\int \dfrac{\sqrt{25-x^2}}{x}\,dx = 5\ln\left|\dfrac{5-\sqrt{25-x^2}}{x}\right| + \sqrt{25-x^2} + C$

10. $\int \dfrac{dx}{x^2\sqrt{a^2-x^2}} = -\dfrac{\sqrt{a^2-x^2}}{a^2x} + C$

11. $\int \sqrt{x^2+4}\,dx = \tfrac{1}{2}x\sqrt{x^2+4} + 2\ln(x+\sqrt{x^2+4}) + C$

12. $\int \dfrac{x^2\,dx}{(a^2-x^2)^{3/2}} = \dfrac{x}{\sqrt{a^2-x^2}} - \arcsin\dfrac{x}{a} + C$

13. $\int \sqrt{x^2-4}\,dx = \tfrac{1}{2}x\sqrt{x^2-4} - 2\ln|x+\sqrt{x^2-4}| + C$

14. $\int \dfrac{\sqrt{x^2+a^2}}{x}\,dx = \sqrt{x^2+a^2} + \dfrac{a}{2}\ln\dfrac{\sqrt{a^2+x^2}-a}{\sqrt{a^2+x^2}+a} + C$

15. $\int \dfrac{x^2\,dx}{(4-x^2)^{5/2}} = \dfrac{x^3}{12(4-x^2)^{3/2}} + C$

16. $\int \dfrac{dx}{(a^2+x^2)^{3/2}} = \dfrac{x}{a^2\sqrt{a^2+x^2}} + C$

17. $\int \dfrac{dx}{x^2\sqrt{9-x^2}} = -\dfrac{\sqrt{9-x^2}}{9x} + C$

18. $\int \dfrac{x^2\,dx}{\sqrt{x^2-16}} = \tfrac{1}{2}x\sqrt{x^2-16} + 8\ln|x+\sqrt{x^2-16}| + C$

19. $\int x^3\sqrt{a^2-x^2}\,dx = \dfrac{1}{5}(a^2-x^2)^{5/2} - \dfrac{a^2}{3}(a^2-x^2)^{3/2} + C$

20. $\int \dfrac{dx}{\sqrt{x^2-4x+13}} = \ln(x-2+\sqrt{x^2-4x+13}) + C$

21. $\int \dfrac{dx}{(4x-x^2)^{3/2}} = \dfrac{x-2}{4\sqrt{4x-x^2}} + C$

22. $\int \dfrac{dx}{(9+x^2)^2} = \dfrac{1}{54}\arctan\dfrac{x}{3} + \dfrac{x}{18(9+x^2)} + C$

In Problems 23-24, integrate by parts and apply the method of this chapter.

23. $\int x\arcsin x\,dx = \tfrac{1}{4}(2x^2-1)\arcsin x + \tfrac{1}{4}x\sqrt{1-x^2} + C$

24. $\int x\arccos x\,dx = \tfrac{1}{4}(2x^2-1)\arccos x - \tfrac{1}{4}x\sqrt{1-x^2} + C$

# Chapter 29

# Integration by Partial Fractions

**A POLYNOMIAL IN** $x$ is a function of the form $a_0x^n + a_1x^{n-1} + \cdots + a_{n-1}x + a_n$, where the $a$'s are constants, $a_0 \neq 0$, and $n$ is a positive integer including zero.

If two polynomials of the same degree are equal for all values of the variable, the coefficients of the like powers of the variable in the two polynomials are equal.

Every polynomial with real coefficients can be expressed (at least, theoretically) as a product of real linear factors of the form $ax + b$ and real irreducible quadratic factors of the form $ax^2 + bx + c$.

**A FUNCTION** $F(x) = \dfrac{f(x)}{g(x)}$, where $f(x)$ and $g(x)$ are polynomials, is called a *rational fraction*.

If the degree of $f(x)$ is less than the degree of $g(x)$, $F(x)$ is called *proper*; otherwise, $F(x)$ is called *improper*.

An improper rational fraction can be expressed as the sum of a polynomial and a proper rational fraction. Thus, $\dfrac{x^3}{x^2+1} = x - \dfrac{x}{x^2+1}$.

Every proper rational fraction can be expressed (at least, theoretically) as a sum of simpler fractions (*partial fractions*) whose denominators are of the form $(ax + b)^n$ and $(ax^2 + bx + c)^n$, $n$ being a positive integer. Four cases, depending upon the nature of the factors of the denominator, arise.

## CASE I. DISTINCT LINEAR FACTORS

To each linear factor $ax + b$ occurring once in the denominator of a proper rational fraction, there corresponds a single partial fraction of the form $\dfrac{A}{ax+b}$, where $A$ is a constant to be determined.

See Problems 1-2.

## CASE II. REPEATED LINEAR FACTORS

To each linear factor $ax + b$ occurring $n$ times in the denominator of a proper rational fraction, there corresponds a sum of $n$ partial fractions of the form

$$\frac{A_1}{ax+b} + \frac{A_2}{(ax+b)^2} + \cdots + \frac{A_n}{(ax+b)^n}$$

where the $A$'s are constants to be determined.

See Problems 3-4.

## CASE III. DISTINCT QUADRATIC FACTORS

To each irreducible quadratic factor $ax^2 + bx + c$ occurring once in the denominator of a proper rational fraction, there corresponds a single partial fraction of the form $\dfrac{Ax + B}{ax^2 + bx + c}$, where $A$ and $B$ are constants to be determined.

See Problems 5-6.

## CASE IV.  REPEATED QUADRATIC FACTORS

To each irreducible quadratic factor $ax^2 + bx + c$ occurring $n$ times in the denominator of a proper rational fraction, there corresponds a sum of $n$ partial fractions of the form

$$\frac{A_1 x + B_1}{ax^2 + bx + c} + \frac{A_2 x + B_2}{(ax^2 + bx + c)^2} + \cdots + \frac{A_n x + B_n}{(ax^2 + bx + c)^n}$$

where the $A$'s and $B$'s are constants to be determined.

See Problems 7-8.

## Solved Problems

1.  Find $\int \dfrac{dx}{x^2 - 4}$.

(a)  Factor the denominator:  $x^2 - 4 = (x-2)(x+2)$.

Write  $\dfrac{1}{x^2 - 4} = \dfrac{A}{x-2} + \dfrac{B}{x+2}$  and clear of fractions to obtain

$$(1)\quad 1 = A(x+2) + B(x-2) \quad\text{or}\quad (2)\quad 1 = (A+B)x + (2A - 2B)$$

(b)  Determine the constants.

*General method.* Equate coefficients of like powers of $x$ in (2) and solve simultaneously for the constants. Thus, $A + B = 0$ and $2A - 2B = 1$; $A = \frac{1}{4}$ and $B = -\frac{1}{4}$.

*Short method.* Substitute in (1) the values $x = 2$ and $x = -2$ to obtain $1 = 4A$ and $1 = -4B$; then $A = \frac{1}{4}$ and $B = -\frac{1}{4}$, as before. (Note that the values of $x$ used are those for which the denominators of the partial fractions become 0.)

(c)  By either method:  $\dfrac{1}{x^2 - 4} = \dfrac{\frac{1}{4}}{x-2} - \dfrac{\frac{1}{4}}{x+2}$  and

$$\int \frac{dx}{x^2 - 4} = \frac{1}{4} \int \frac{dx}{x-2} - \frac{1}{4} \int \frac{dx}{x+2} = \frac{1}{4} \ln|x-2| - \frac{1}{4} \ln|x+2| + C = \frac{1}{4} \ln\left|\frac{x-2}{x+2}\right| + C$$

2.  Find $\int \dfrac{(x+1)\,dx}{x^3 + x^2 - 6x}$.

(a)  $x^3 + x^2 - 6x = x(x-2)(x+3)$.  Then  $\dfrac{x+1}{x^3 + x^2 - 6x} = \dfrac{A}{x} + \dfrac{B}{x-2} + \dfrac{C}{x+3}$  and

$$(1)\quad x + 1 = A(x-2)(x+3) + Bx(x+3) + Cx(x-2) \quad\text{or}$$

$$(2)\quad x + 1 = (A+B+C)x^2 + (A+3B-2C)x - 6A$$

(b)  *General method.*  Solve simultaneously the system of equations

$$A+B+C = 0, \quad A+3B-2C = 1, \quad\text{and}\quad -6A = 1$$

to obtain $A = -1/6$, $B = 3/10$, and $C = -2/15$.

*Short method.* Substitute in (1) the values $x = 0$, $x = 2$, and $x = -3$ to obtain $1 = -6A$ or $A = -1/6$, $3 = 10B$ or $B = 3/10$, and $-2 = 15C$ or $C = -2/15$.

(c)  $\int \dfrac{(x+1)\,dx}{x^3 + x^2 - 6x} = -\dfrac{1}{6} \int \dfrac{dx}{x} + \dfrac{3}{10} \int \dfrac{dx}{x-2} - \dfrac{2}{15} \int \dfrac{dx}{x+3}$

$$= -\frac{1}{6} \ln|x| + \frac{3}{10} \ln|x-2| - \frac{2}{15} \ln|x+3| + C = \ln \frac{|x-2|^{3/10}}{|x|^{1/6}\,|x+3|^{2/15}} + C$$

3.  Find $\int \dfrac{(3x+5)\,dx}{x^3 - x^2 - x + 1}$.

$x^3 - x^2 - x + 1 = (x+1)(x-1)^2$.  Then  $\dfrac{3x+5}{x^3 - x^2 - x + 1} = \dfrac{A}{x+1} + \dfrac{B}{x-1} + \dfrac{C}{(x-1)^2}$  and

$$3x + 5 = A(x-1)^2 + B(x+1)(x-1) + C(x+1)$$

For $x = -1$, $2 = 4A$ and $A = \frac{1}{2}$. For $x = 1$, $8 = 2C$ and $C = 4$. To determine the remaining constant, use any other value of $x$, say $x = 0$; for $x = 0$, $5 = A - B + C$ and $B = -\frac{1}{2}$. Thus

$$\int \frac{3x + 5}{x^3 - x^2 - x + 1}\, dx = \frac{1}{2} \int \frac{dx}{x + 1} - \frac{1}{2} \int \frac{dx}{x - 1} + 4 \int \frac{dx}{(x - 1)^2}$$

$$= \frac{1}{2} \ln|x + 1| - \frac{1}{2} \ln|x - 1| - \frac{4}{x - 1} + C = -\frac{4}{x - 1} + \frac{1}{2} \ln\left|\frac{x + 1}{x - 1}\right| + C$$

4.  Find $\int \dfrac{x^4 - x^3 - x - 1}{x^3 - x^2}\, dx$.

The integrand is an improper fraction. By division,

$$\frac{x^4 - x^3 - x - 1}{x^3 - x^2} = x - \frac{x + 1}{x^3 - x^2} = x - \frac{x + 1}{x^2(x - 1)}$$

We write $\dfrac{x + 1}{x^2(x - 1)} = \dfrac{A}{x} + \dfrac{B}{x^2} + \dfrac{C}{x - 1}$.  Then,

$$x + 1 = Ax(x - 1) + B(x - 1) + Cx^2$$

For $x = 0$, $1 = -B$ and $B = -1$. For $x = 1$, $2 = C$. For $x = 2$, $3 = 2A + B + 4C$ and $A = -2$. Thus

$$\int \frac{x^4 - x^3 - x - 1}{x^3 - x^2}\, dx = \int x\, dx + 2 \int \frac{dx}{x} + \int \frac{dx}{x^2} - 2 \int \frac{dx}{x - 1}$$

$$= \frac{1}{2}x^2 + 2 \ln|x| - \frac{1}{x} - 2 \ln|x - 1| + C = \frac{1}{2}x^2 - \frac{1}{x} + 2 \ln\left|\frac{x}{x - 1}\right| + C$$

5.  Find $\int \dfrac{x^3 + x^2 + x + 2}{x^4 + 3x^2 + 2}\, dx$.

$x^4 + 3x^2 + 2 = (x^2 + 1)(x^2 + 2)$.  We write $\dfrac{x^3 + x^2 + x + 2}{x^4 + 3x^2 + 2} = \dfrac{Ax + B}{x^2 + 1} + \dfrac{Cx + D}{x^2 + 2}$.  Then

$$x^3 + x^2 + x + 2 = (Ax + B)(x^2 + 2) + (Cx + D)(x^2 + 1)$$
$$= (A + C)x^3 + (B + D)x^2 + (2A + C)x + (2B + D)$$

Hence $A + C = 1$, $B + D = 1$, $2A + C = 1$, and $2B + D = 2$. Solving simultaneously, $A = 0$, $B = 1$, $C = 1$, $D = 0$. Thus

$$\int \frac{x^3 + x^2 + x + 2}{x^4 + 3x^2 + 2}\, dx = \int \frac{dx}{x^2 + 1} + \int \frac{x\, dx}{x^2 + 2} = \arctan x + \frac{1}{2} \ln(x^2 + 2) + C$$

6.  Solve the equation $\int \dfrac{x^2\, dx}{a^4 - x^4} = \int k\, dt$  which occurs in physical chemistry.

We write $\dfrac{x^2}{a^4 - x^4} = \dfrac{A}{a - x} + \dfrac{B}{a + x} + \dfrac{Cx + D}{a^2 + x^2}$.  Then

$$x^2 = A(a + x)(a^2 + x^2) + B(a - x)(a^2 + x^2) + (Cx + D)(a - x)(a + x)$$

For $x = a$, $a^2 = 4Aa^3$ and $A = 1/4a$. For $x = -a$, $a^2 = 4Ba^3$ and $B = 1/4a$. For $x = 0$, $0 = Aa^3 + Ba^3 + Da^2 = a^2/2 + Da^2$ and $D = -1/2$. For $x = 2a$, $4a^2 = 15Aa^3 - 5Ba^3 - 6Ca^3 - 3Da^2$ and $C = 0$. Thus

$$\int \frac{x^2\, dx}{a^4 - x^4} = \frac{1}{4a} \int \frac{dx}{a - x} + \frac{1}{4a} \int \frac{dx}{a + x} - \frac{1}{2} \int \frac{dx}{a^2 + x^2}$$

$$= -\frac{1}{4a} \ln|a - x| + \frac{1}{4a} \ln|a + x| - \frac{1}{2a} \arctan\frac{x}{a} + C$$

and

$$\int k\, dt = kt = \frac{1}{4a} \ln\left|\frac{a + x}{a - x}\right| - \frac{1}{2a} \arctan\frac{x}{a} + C$$

7.  Find $\int \dfrac{x^5 - x^4 + 4x^3 - 4x^2 + 8x - 4}{(x^2 + 2)^3}\, dx$.

We write $\dfrac{x^5 - x^4 + 4x^3 - 4x^2 + 8x - 4}{(x^2 + 2)^3} = \dfrac{Ax + B}{x^2 + 2} + \dfrac{Cx + D}{(x^2 + 2)^2} + \dfrac{Ex + F}{(x^2 + 2)^3}$.  Then

$$x^5 - x^4 + 4x^3 - 4x^2 + 8x - 4 = (Ax + B)(x^2 + 2)^2 + (Cx + D)(x^2 + 2) + Ex + F$$
$$= Ax^5 + Bx^4 + (4A + C)x^3 + (4B + D)x^2$$
$$+ (4A + 2C + E)x + (4B + 2D + F)$$

from which $A = 1$, $B = -1$, $C = 0$, $D = 0$, $E = 4$, $F = 0$. Thus the given integral is equal to

$$\int \frac{x-1}{x^2+2}\,dx \;+\; 4\int \frac{x\,dx}{(x^2+2)^3} \;=\; \tfrac{1}{2}\ln(x^2+2) \;-\; \frac{\sqrt{2}}{2}\arctan\frac{x}{\sqrt{2}} \;-\; \frac{1}{(x^2+2)^2} \;+\; C$$

**8.** Find $\displaystyle\int \frac{2x^2+3}{(x^2+1)^2}\,dx$.

We write $\displaystyle\frac{2x^2+3}{(x^2+1)^2} \;=\; \frac{Ax+B}{x^2+1} + \frac{Cx+D}{(x^2+1)^2}$. Then

$$2x^2+3 \;=\; (Ax+B)(x^2+1) + Cx + D \;=\; Ax^3 + Bx^2 + (A+C)x + (B+D)$$

from which $A = 0$, $B = 2$, $A + C = 0$, $B + D = 3$. Thus $A = 0$, $B = 2$, $C = 0$, $D = 1$ and

$$\int \frac{2x^2+3}{(x^2+1)^2}\,dx \;=\; \int \frac{2\,dx}{x^2+1} \;+\; \int \frac{dx}{(x^2+1)^2}$$

For the second integral on the right, let $x = \tan z$. Then

$$\int \frac{dx}{(x^2+1)^2} \;=\; \int \frac{\sec^2 z\,dz}{\sec^4 z} \;=\; \int \cos^2 z\,dz \;=\; \tfrac{1}{2}z + \tfrac{1}{4}\sin 2z + C$$

and $\displaystyle\int \frac{2x^2+3}{(x^2+1)^2}\,dx \;=\; 2\arctan x + \tfrac{1}{2}\arctan x + \frac{\tfrac{1}{2}x}{x^2+1} + C \;=\; \tfrac{5}{2}\arctan x + \frac{\tfrac{1}{2}x}{x^2+1} + C$

# Supplementary Problems

**9.** $\displaystyle\int \frac{dx}{x^2-9} \;=\; \tfrac{1}{6}\ln\left|\frac{x-3}{x+3}\right| + C$

**12.** $\displaystyle\int \frac{x^2+3x-4}{x^2-2x-8}\,dx \;=\; x + \ln|(x+2)(x-4)^4| + C$

**10.** $\displaystyle\int \frac{dx}{x^2+7x+6} \;=\; \tfrac{1}{5}\ln\left|\frac{x+1}{x+6}\right| + C$

**13.** $\displaystyle\int \frac{x^2-3x-1}{x^3+x^2-2x}\,dx \;=\; \ln\left|\frac{x^{1/2}(x+2)^{3/2}}{x-1}\right| + C$

**11.** $\displaystyle\int \frac{x\,dx}{x^2-3x-4} \;=\; \tfrac{1}{5}\ln|(x+1)(x-4)^4| + C$

**14.** $\displaystyle\int \frac{x\,dx}{(x-2)^2} \;=\; \ln|x-2| - \frac{2}{x-2} + C$

**15.** $\displaystyle\int \frac{x^4}{(1-x)^3}\,dx \;=\; -\tfrac{1}{2}x^2 - 3x - \ln(1-x)^6 - \frac{4}{1-x} + \frac{1}{2(1-x)^2} + C$

**16.** $\displaystyle\int \frac{dx}{x^3+x} \;=\; \ln\left|\frac{x}{\sqrt{x^2+1}}\right| + C$

**17.** $\displaystyle\int \frac{x^3+x^2+x+3}{(x^2+1)(x^2+3)}\,dx \;=\; \ln\sqrt{x^2+3} + \arctan x + C$

**18.** $\displaystyle\int \frac{x^4-2x^3+3x^2-x+3}{x^3-2x^2+3x}\,dx \;=\; \tfrac{1}{2}x^2 + \ln\left|\frac{x}{\sqrt{x^2-2x+3}}\right| + C$

**19.** $\displaystyle\int \frac{2x^3\,dx}{(x^2+1)^2} \;=\; \ln(x^2+1) + \frac{1}{x^2+1} + C$

**20.** $\displaystyle\int \frac{2x^3+x^2+4}{(x^2+4)^2}\,dx \;=\; \ln(x^2+4) + \tfrac{1}{2}\arctan\tfrac{1}{2}x + \frac{4}{x^2+4} + C$

**21.** $\displaystyle\int \frac{x^3+x-1}{(x^2+1)^2}\,dx \;=\; \ln\sqrt{x^2+1} - \tfrac{1}{2}\arctan x - \frac{1}{2}\left(\frac{x}{x^2+1}\right) + C$

**22.** $\displaystyle\int \frac{x^4+8x^3-x^2+2x+1}{(x^2+x)(x^3+1)}\,dx \;=\; \ln\left|\frac{x^3-x^2+x}{(x+1)^2}\right| - \frac{3}{x+1} + \frac{2}{\sqrt{3}}\arctan\frac{2x-1}{\sqrt{3}} + C$

**23.** $\displaystyle\int \frac{x^3+x^2-5x+15}{(x^2+5)(x^2+2x+3)}\,dx \;=\; \ln\sqrt{x^2+2x+3} + \frac{5}{\sqrt{2}}\arctan\frac{x+1}{\sqrt{2}} - \sqrt{5}\arctan\frac{x}{\sqrt{5}} + C$

**24.** $\displaystyle\int \frac{x^6+7x^5+15x^4+32x^3+23x^2+25x-3}{(x^2+x+2)^2(x^2+1)^2}\,dx \;=\; \frac{1}{x^2+x+2} - \frac{3}{x^2+1} + \ln\frac{x^2+1}{x^2+x+2} + C$

**25.** $\displaystyle\int \frac{dx}{e^{2x}-3e^x} \;=\; \frac{1}{3e^x} + \tfrac{1}{9}\ln\left|\frac{e^x-3}{e^x}\right| + C$    (Let $e^x = u$.)

**26.** $\displaystyle\int \frac{\sin x\,dx}{\cos x\,(1+\cos^2 x)} \;=\; \ln\left|\frac{\sqrt{1+\cos^2 x}}{\cos x}\right| + C$    (Let $\cos x = u$.)

**27.** $\displaystyle\int \frac{(2+\tan^2\theta)\sec^2\theta\,d\theta}{1+\tan^3\theta} \;=\; \ln|1+\tan\theta| + \frac{2}{\sqrt{3}}\arctan\frac{2\tan\theta-1}{\sqrt{3}} + C$

# Chapter 30

## Miscellaneous Substitutions

**IF THE INTEGRAND IS RATIONAL** except for a radical of the form:

1. $\sqrt[n]{au+b}$, the substitution $au+b = z^n$ will replace it by a rational integrand.

2. $\sqrt{q+pu+u^2}$, the substitution $q+pu+u^2 = (z-u)^2$ will replace it by a rational integrand.

3. $\sqrt{q+pu-u^2} = \sqrt{(\alpha+u)(\beta-u)}$, the substitution $q+pu-u^2 = (\alpha+u)^2 z^2$ or $q+pu-u^2 = (\beta-u)^2 z^2$ will replace it by a rational integrand.

See Problems 1-5.

**THE SUBSTITUTION** $u = 2 \arctan z$ will replace any rational function of $\sin u$ and $\cos u$ by a rational function of $z$, since

$$\sin u = \frac{2z}{1+z^2}, \qquad \cos u = \frac{1-z^2}{1+z^2}, \qquad \text{and} \qquad du = \frac{2\,dz}{1+z^2}$$

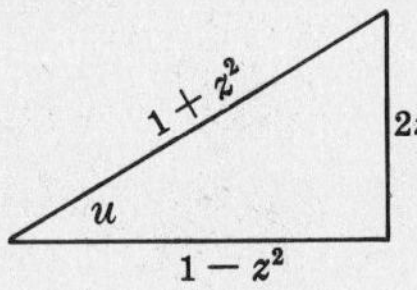

The first and second of these relations are obtained from the adjoining Fig. 30-1, and the third by differentiating

$$u = 2 \arctan z$$

**Fig. 30-1**

After integrating, use $z = \tan \tfrac{1}{2}u$ to return to the original variable.

See Problems 6-10.

**EFFECTIVE SUBSTITUTIONS** are often suggested by the form of the integrand function.

See Problems 11-12.

## Solved Problems

1. Find $\displaystyle\int \frac{dx}{x\sqrt{1-x}}$.   Let $1-x = z^2$. Then $x = 1-z^2$, $dx = -2z\,dz$, and

$$\int \frac{dx}{x\sqrt{1-x}} = \int \frac{-2z\,dz}{(1-z^2)z} = -2\int \frac{dz}{1-z^2} = -\ln\left|\frac{1+z}{1-z}\right| + C = \ln\left|\frac{1-\sqrt{1-x}}{1+\sqrt{1-x}}\right| + C$$

2. Find $\displaystyle\int \frac{dx}{(x-2)\sqrt{x+2}}$.   Let $x+2 = z^2$. Then $x = z^2-2$, $dx = 2z\,dz$, and

$$\int \frac{dx}{(x-2)\sqrt{x+2}} = \int \frac{2z\,dz}{z(z^2-4)} = 2\int \frac{dz}{z^2-4} = \frac{1}{2}\ln\left|\frac{z-2}{z+2}\right| + C = \frac{1}{2}\ln\left|\frac{\sqrt{x+2}-2}{\sqrt{x+2}+2}\right| + C$$

3. Find $\displaystyle\int \frac{dx}{x^{1/2}-x^{1/4}}$.   Let $x = z^4$. Then $dx = 4z^3\,dz$ and

$$\int \frac{dx}{x^{1/2}-x^{1/4}} = \int \frac{4z^3\,dz}{z^2-z} = 4\int \frac{z^2}{z-1}\,dz = 4\int \left(z+1+\frac{1}{z-1}\right) dz$$

$$= 4(\tfrac{1}{2}z^2 + z + \ln|z-1|) + C = 2\sqrt{x} + 4\sqrt[4]{x} + \ln(\sqrt[4]{x}-1)^4 + C$$

**4.** Find $\displaystyle\int \frac{dx}{x\sqrt{x^2+x+2}}$. 　Let　$x^2+x+2=(z-x)^2$. 　Then

$$x = \frac{z^2-2}{1+2z}, \quad dx = \frac{2(z^2+z+2)\,dz}{(1+2z)^2}, \quad \sqrt{x^2+x+2} = \frac{z^2+z+2}{1+2z}, \quad \text{and}$$

$$\int \frac{dx}{x\sqrt{x^2+x+2}} = \int \frac{\dfrac{2(z^2+z+2)}{(1+2z)^2}}{\dfrac{z^2-2}{1+2z}\cdot\dfrac{z^2+z+2}{1+2z}}\,dz = 2\int \frac{dz}{z^2-2} = \frac{1}{\sqrt{2}}\ln\left|\frac{z-\sqrt{2}}{z+\sqrt{2}}\right| + C$$

$$= \frac{1}{\sqrt{2}}\ln\left|\frac{\sqrt{x^2+x+2}+x-\sqrt{2}}{\sqrt{x^2+x+2}+x+\sqrt{2}}\right| + C$$

**5.** Find $\displaystyle\int \frac{x\,dx}{(5-4x-x^2)^{3/2}}$. 　Let　$5-4x-x^2=(5+x)(1-x)=(1-x)^2z^2$. 　Then

$$x = \frac{z^2-5}{1+z^2}, \quad dx = \frac{12z\,dz}{(1+z^2)^2}, \quad \sqrt{5-4x-x^2} = (1-x)z = \frac{6z}{1+z^2}, \quad \text{and}$$

$$\int \frac{x\,dx}{(5-4x-x^2)^{3/2}} = \int \frac{\dfrac{z^2-5}{1+z^2}\cdot\dfrac{12z}{(1+z^2)^2}}{\dfrac{216z^3}{(1+z^2)^3}}\,dz = \frac{1}{18}\int\left(1-\frac{5}{z^2}\right)dz$$

$$= \frac{1}{18}\left(z+\frac{5}{z}\right) + C = \frac{5-2x}{9\sqrt{5-4x-x^2}} + C$$

**6.** $\displaystyle\int \frac{dx}{1+\sin x-\cos x} = \int \frac{\dfrac{2\,dz}{1+z^2}}{1+\dfrac{2z}{1+z^2}-\dfrac{1-z^2}{1+z^2}} = \int \frac{dz}{z(1+z)} = \ln|z| - \ln|1+z| + C$

$$= \ln\left|\frac{z}{1+z}\right| + C = \ln\left|\frac{\tan\frac{1}{2}x}{1+\tan\frac{1}{2}x}\right| + C$$

**7.** $\displaystyle\int \frac{dx}{3-2\cos x} = \int \frac{\dfrac{2\,dz}{1+z^2}}{3-2\dfrac{1-z^2}{1+z^2}} = \int \frac{2\,dz}{1+5z^2} = \frac{2\sqrt{5}}{5}\arctan z\sqrt{5} + C$

$$= \frac{2\sqrt{5}}{5}\arctan(\sqrt{5}\tan\tfrac{1}{2}x) + C$$

**8.** $\displaystyle\int \sec x\,dx = \int \frac{1+z^2}{1-z^2}\cdot\frac{2\,dz}{1+z^2} = 2\int \frac{dz}{1-z^2} = \ln\left|\frac{1+z}{1-z}\right| + C = \ln\left|\frac{1+\tan\frac{1}{2}x}{1-\tan\frac{1}{2}x}\right| + C$

$$= \ln\left|\tan\left(\tfrac{1}{2}x+\tfrac{1}{4}\pi\right)\right| + C$$

**9.** $\displaystyle\int \frac{dx}{2+\cos x} = \int \frac{\dfrac{2\,dz}{1+z^2}}{2+\dfrac{1-z^2}{1+z^2}} = 2\int \frac{dz}{3+z^2} = \frac{2}{\sqrt{3}}\arctan\frac{z}{\sqrt{3}} + C$

$$= \frac{2\sqrt{3}}{3}\arctan\left(\frac{\sqrt{3}}{3}\tan\tfrac{1}{2}x\right) + C$$

**10.** $\displaystyle\int \frac{dx}{5+4\sin x} = \int \frac{\dfrac{2\,dz}{1+z^2}}{5+4\dfrac{2z}{1+z^2}} = \int \frac{2\,dz}{5+8z+5z^2} = \frac{2}{5}\int \frac{dz}{(z+\frac{4}{5})^2+\frac{9}{25}}$

$$= \frac{2}{3}\arctan\frac{z+4/5}{3/5} + C = \frac{2}{3}\arctan\frac{5\tan\frac{1}{2}x+4}{3} + C$$

**11.** Use the substitution $1-x^3=z^2$ to find $\displaystyle\int x^5\sqrt{1-x^3}\,dx$. 　　$x^3=1-z^2$, $3x^2\,dx=-2z\,dz$, 　and

$$\int x^5\sqrt{1-x^3}\,dx = \int x^3\sqrt{1-x^3}\cdot x^2\,dx = \int (1-z^2)z(-\tfrac{2}{3}z\,dz) = -\frac{2}{3}\int (1-z^2)z^2\,dz$$

$$= -\frac{2}{3}\left(\frac{z^3}{3}-\frac{z^5}{5}\right) + C = -\frac{2}{45}(1-x^3)^{3/2}(2+3x^3) + C$$

12. Use $x = \dfrac{1}{z}$ to find $\displaystyle\int \dfrac{\sqrt{x-x^2}}{x^4}\,dx$.    Then $dx = -\dfrac{dz}{z^2}$, $\sqrt{x-x^2} = \dfrac{1}{z}\sqrt{z-1}$, and

$$\int \frac{\sqrt{x-x^2}}{x^4}\,dx \;=\; \int \frac{\dfrac{1}{z}\sqrt{z-1}\left(-\dfrac{dz}{z^2}\right)}{1/z^4} \;=\; -\int z\sqrt{z-1}\,dz$$

Let $z - 1 = s^2$. Then

$$-\int z\sqrt{z-1}\,dz \;=\; -\int (s^2+1)s\cdot 2s\,ds \;=\; -2\left(\frac{s^5}{5} + \frac{s^3}{3}\right) + C$$

$$=\; -2\left(\frac{(z-1)^{5/2}}{5} + \frac{(z-1)^{3/2}}{3}\right) + C \;=\; -2\left(\frac{(1-x)^{5/2}}{5x^{5/2}} + \frac{(1-x)^{3/2}}{3x^{3/2}}\right) + C$$

# Supplementary Problems

13. $\displaystyle\int \frac{\sqrt{x}}{1+x}\,dx \;=\; 2\sqrt{x} - 2\arctan\sqrt{x} + C$      14. $\displaystyle\int \frac{dx}{\sqrt{x}(1+\sqrt{x})} \;=\; 2\ln(1+\sqrt{x}) + C$

15. $\displaystyle\int \frac{dx}{3+\sqrt{x+2}} \;=\; 2\sqrt{x+2} - 6\ln(3+\sqrt{x+2}) + C$

16. $\displaystyle\int \frac{1-\sqrt{3x+2}}{1+\sqrt{3x+2}}\,dx \;=\; -x + \frac{4}{3}\left\{\sqrt{3x+2} - \ln(1+\sqrt{3x+2})\right\} + C$

17. $\displaystyle\int \frac{dx}{\sqrt{x^2-x+1}} \;=\; \ln\left|2\sqrt{x^2-x+1} + 2x - 1\right| + C$

18. $\displaystyle\int \frac{dx}{x\sqrt{x^2+x-1}} \;=\; 2\arctan(\sqrt{x^2+x-1} + x) + C$

19. $\displaystyle\int \frac{dx}{\sqrt{6+x-x^2}} \;=\; \arcsin\frac{2x-1}{5} + C$     20. $\displaystyle\int \frac{\sqrt{4x-x^2}}{x^3}\,dx \;=\; -\frac{(4x-x^2)^{3/2}}{6x^3} + C$

21. $\displaystyle\int \frac{dx}{(x+1)^{1/2} + (x+1)^{1/4}} \;=\; 2(x+1)^{1/2} - 4(x+1)^{1/4} + 4\ln(1+(x+1)^{1/4}) + C$

22. $\displaystyle\int \frac{dx}{2+\sin x} \;=\; \frac{2}{\sqrt{3}}\arctan\frac{2\tan\frac{1}{2}x + 1}{\sqrt{3}} + C$

23. $\displaystyle\int \frac{dx}{1-2\sin x} \;=\; \frac{\sqrt{3}}{3}\ln\left|\frac{\tan\frac{1}{2}x - 2 - \sqrt{3}}{\tan\frac{1}{2}x - 2 + \sqrt{3}}\right| + C$

24. $\displaystyle\int \frac{dx}{3+5\sin x} \;=\; \frac{1}{4}\ln\left|\frac{3\tan\frac{1}{2}x + 1}{\tan\frac{1}{2}x + 3}\right| + C$     25. $\displaystyle\int \frac{dx}{\sin x - \cos x - 1} \;=\; \ln\left|\tan\frac{1}{2}x - 1\right| + C$

26. $\displaystyle\int \frac{dx}{5+3\sin x} \;=\; \frac{1}{2}\arctan\frac{5\tan\frac{1}{2}x + 3}{4} + C$     27. $\displaystyle\int \frac{\sin x\,dx}{1+\sin^2 x} \;=\; \frac{\sqrt{2}}{4}\ln\left|\frac{\tan^2\frac{1}{2}x + 3 - 2\sqrt{2}}{\tan^2\frac{1}{2}x + 3 + 2\sqrt{2}}\right| + C$

28. $\displaystyle\int \frac{dx}{1+\sin x + \cos x} \;=\; \ln\left|1 + \tan\frac{1}{2}x\right| + C$     29. $\displaystyle\int \frac{dx}{2-\cos x} \;=\; \frac{2}{\sqrt{3}}\arctan(\sqrt{3}\tan\frac{1}{2}x) + C$

30. $\displaystyle\int \sin\sqrt{x}\,dx \;=\; -2\sqrt{x}\cos\sqrt{x} + 2\sin\sqrt{x} + C$

31. $\displaystyle\int \frac{dx}{x\sqrt{3x^2+2x-1}} \;=\; -\arcsin\frac{1-x}{2x} + C$     Let $x = 1/z$.

32. $\displaystyle\int \frac{(e^x-2)e^x}{e^x+1}\,dx \;=\; e^x - 3\ln(e^x+1) + C$     Let $e^x + 1 = z$.

33. $\displaystyle\int \frac{\sin x \cos x}{1-\cos x}\,dx \;=\; \cos x + \ln(1-\cos x) + C$     Let $\cos x = z$.

34. $\displaystyle\int \frac{dx}{x^2\sqrt{4-x^2}} \;=\; -\frac{\sqrt{4-x^2}}{4x} + C$     Let $x = 2/z$.

35. $\displaystyle\int \frac{dx}{x^2(4+x^2)} \;=\; -\frac{1}{4x} + \frac{1}{8}\arctan\frac{2}{x} + C$     36. $\displaystyle\int \sqrt{1+\sqrt{x}}\,dx \;=\; \frac{4}{5}(1+\sqrt{x})^{5/2} - \frac{4}{3}(1+\sqrt{x})^{3/2} + C$

37. $\displaystyle\int \frac{dx}{3(1-x^2) - (5+4x)\sqrt{1-x^2}} \;=\; \frac{2\sqrt{1+x}}{3\sqrt{1+x} - \sqrt{1-x}} + C$

# Chapter 31

## Integration of Hyperbolic Functions

### RULES OF INTEGRATION

$$\int \sinh u \, du \;=\; \cosh u + C \qquad\qquad \int \operatorname{sech}^2 u \, du \;=\; \tanh u + C$$

$$\int \cosh u \, du \;=\; \sinh u + C \qquad\qquad \int \operatorname{csch}^2 u \, du \;=\; -\coth u + C$$

$$\int \tanh u \, du \;=\; \ln \cosh u + C \qquad\qquad \int \operatorname{sech} u \tanh u \, du \;=\; -\operatorname{sech} u + C$$

$$\int \coth u \, du \;=\; \ln |\sinh u| + C \qquad\qquad \int \operatorname{csch} u \coth u \, du \;=\; -\operatorname{csch} u + C$$

$$\int \frac{du}{\sqrt{u^2 + a^2}} \;=\; \sinh^{-1}\frac{u}{a} + C \qquad\qquad \int \frac{du}{a^2 - u^2} \;=\; \frac{1}{a}\tanh^{-1}\frac{u}{a} + C, \quad u^2 < a^2$$

$$\int \frac{du}{\sqrt{u^2 - a^2}} \;=\; \cosh^{-1}\frac{u}{a} + C, \quad u > a > 0 \qquad\qquad \int \frac{du}{u^2 - a^2} \;=\; -\frac{1}{a}\coth^{-1}\frac{u}{a} + C, \quad u^2 > a^2$$

## Solved Problems

1. $\displaystyle \int \sinh \tfrac{1}{2}x \, dx \;=\; 2\cosh \tfrac{1}{2}x + C$

3. $\displaystyle \int \operatorname{sech}^2 (2x-1) \, dx \;=\; \tfrac{1}{2}\tanh (2x-1) + C$

2. $\displaystyle \int \cosh 2x \, dx \;=\; \tfrac{1}{2}\sinh 2x + C$

4. $\displaystyle \int \operatorname{csch} 3x \coth 3x \, dx \;=\; -\tfrac{1}{3}\operatorname{csch} 3x + C$

5. $\displaystyle \int \operatorname{sech} x \, dx \;=\; \int \frac{1}{\cosh x} dx \;=\; \int \frac{\cosh x}{\cosh^2 x} \;=\; \int \frac{\cosh x}{1 + \sinh^2 x} dx \;=\; \arctan (\sinh x) + C$

6. $\displaystyle \int \sinh^2 x \, dx \;=\; \frac{1}{2}\int (\cosh 2x - 1) \, dx \;=\; \tfrac{1}{4}\sinh 2x - \tfrac{1}{2}x + C$

7. $\displaystyle \int \tanh^2 2x \, dx \;=\; \int (1 - \operatorname{sech}^2 2x) \, dx \;=\; x - \tfrac{1}{2}\tanh 2x + C$

8. $\displaystyle \int \cosh^3 \tfrac{1}{2}x \, dx \;=\; \int (1 + \sinh^2 \tfrac{1}{2}x) \cosh \tfrac{1}{2}x \, dx \;=\; 2\sinh \tfrac{1}{2}x + \tfrac{2}{3}\sinh^3 \tfrac{1}{2}x + C$

9. $\displaystyle \int \operatorname{sech}^4 x \, dx \;=\; \int (1 - \tanh^2 x) \operatorname{sech}^2 x \, dx \;=\; \tanh x - \tfrac{1}{3}\tanh^3 x + C$

10. $\displaystyle \int e^x \cosh x \, dx \;=\; \int e^x \left(\frac{e^x + e^{-x}}{2}\right) dx \;=\; \frac{1}{2}\int (e^{2x} + 1) \, dx \;=\; \frac{1}{4}e^{2x} + \frac{1}{2}x + C$

11. $\displaystyle \int x \sinh x \, dx \;=\; \int x\left(\frac{e^x - e^{-x}}{2}\right) dx \;=\; \frac{1}{2}\int xe^x \, dx - \frac{1}{2}\int xe^{-x} \, dx$

$$\qquad\qquad =\; \frac{1}{2}(xe^x - e^x) - \frac{1}{2}(-xe^{-x} - e^{-x}) + C \;=\; x\left(\frac{e^x + e^{-x}}{2}\right) - \frac{e^x - e^{-x}}{2} + C$$

$$\qquad\qquad =\; x\cosh x - \sinh x + C$$

**12.** $\displaystyle\int \frac{dx}{\sqrt{4x^2-9}} = \frac{1}{2}\cosh^{-1}\frac{2x}{3} + C$      **13.** $\displaystyle\int \frac{dx}{9x^2-25} = -\frac{1}{15}\coth^{-1}\frac{3x}{5} + C$

**14.** Find $\displaystyle\int \sqrt{x^2+4}\,dx$.    Let $x = 2\sinh z$. Then $dx = 2\cosh z\,dz$, $\sqrt{x^2+4} = 2\cosh z$,   and

$$\int \sqrt{x^2+4}\,dx = 4\int \cosh^2 z\,dz = 2\int (\cosh 2z + 1)\,dz = \sinh 2z + 2z + C$$

$$= 2\sinh z\cosh z + 2z + C = \tfrac{1}{2}x\sqrt{x^2+4} + 2\sinh^{-1}\tfrac{1}{2}x + C$$

**15.** Find $\displaystyle\int \frac{dx}{x\sqrt{1-x^2}}$.    Let $x = \operatorname{sech} z$. Then $dx = -\operatorname{sech} z\tanh z\,dz$, $1-x^2 = \tanh z$,   and

$$\int \frac{dx}{x\sqrt{1-x^2}} = -\int \frac{\operatorname{sech} z\tanh z}{\operatorname{sech} z\tanh z}\,dz = -\int dz = -z + C = -\operatorname{sech}^{-1}x + C$$

## Supplementary Problems

**16.** $\displaystyle\int \sinh 3x\,dx = \tfrac{1}{3}\cosh 3x + C$      **22.** $\displaystyle\int \cosh^2 \tfrac{1}{2}x\,dx = \tfrac{1}{2}(\sinh x + x) + C$

**17.** $\displaystyle\int \cosh \tfrac{1}{4}x\,dx = 4\sinh \tfrac{1}{4}x + C$      **23.** $\displaystyle\int \coth^2 3x\,dx = x - \tfrac{1}{3}\coth 3x + C$

**18.** $\displaystyle\int \coth \tfrac{3}{2}x\,dx = \tfrac{2}{3}\ln|\sinh \tfrac{3}{2}x| + C$      **24.** $\displaystyle\int \sinh^3 x\,dx = \tfrac{1}{3}\cosh^3 x - \cosh x + C$

**19.** $\displaystyle\int \operatorname{csch}^2 (1+3x)\,dx = -\tfrac{1}{3}\coth (1+3x) + C$      **25.** $\displaystyle\int e^x \sinh x\,dx = \tfrac{1}{4}e^{2x} - \tfrac{1}{2}x + C$

**20.** $\displaystyle\int \operatorname{sech} 2x\tanh 2x\,dx = -\tfrac{1}{2}\operatorname{sech} 2x + C$      **26.** $\displaystyle\int e^{2x}\cosh x\,dx = \tfrac{1}{6}e^{3x} + \tfrac{1}{2}e^x + C$

**21.** $\displaystyle\int \operatorname{csch} x\,dx = \ln \sqrt{\frac{\cosh x - 1}{\cosh x + 1}} + C$      **27.** $\displaystyle\int x\cosh x\,dx = x\sinh x - \cosh x + C$

**28.** $\displaystyle\int x^2 \sinh x\,dx = (x^2+2)\cosh x - 2x\sinh x + C$

**29.** $\displaystyle\int \sinh^3 x\cosh^2 x\,dx = \tfrac{1}{5}\cosh^5 x - \tfrac{1}{3}\cosh^3 x + C$

**30.** $\displaystyle\int \sinh x\ln\cosh^2 x\,dx = \cosh x\,(\ln\cosh^2 x - 2) + C$

**31.** $\displaystyle\int \frac{dx}{\sqrt{x^2+9}} = \sinh^{-1}\frac{x}{3} + C$      **36.** $\displaystyle\int \frac{dx}{\sqrt{x^2-2x+17}} = \sinh^{-1}\frac{x-1}{4} + C$

**32.** $\displaystyle\int \frac{dx}{\sqrt{x^2-25}} = \cosh^{-1}\frac{x}{5} + C$      **37.** $\displaystyle\int \frac{dx}{4x^2+12x+5} = -\frac{1}{4}\coth^{-1}\left(x+\frac{3}{2}\right) + C$

**33.** $\displaystyle\int \frac{dx}{4-9x^2} = \frac{1}{6}\tanh^{-1}\frac{3}{2}x + C$      **38.** $\displaystyle\int \frac{x^2}{(x^2+4)^{3/2}}\,dx = \sinh^{-1}\frac{1}{2}x - \frac{x}{\sqrt{x^2+4}} + C$

**34.** $\displaystyle\int \frac{dx}{16x^2-9} = -\frac{1}{12}\coth^{-1}\frac{4}{3}x + C$      **39.** $\displaystyle\int \frac{\sqrt{x^2+1}}{x^2}\,dx = \sinh^{-1}x - \frac{\sqrt{1+x^2}}{x} + C$

**35.** $\displaystyle\int \sqrt{x^2-9}\,dx = \frac{1}{2}x\sqrt{x^2-9} - \frac{9}{2}\cosh^{-1}\frac{x}{3} + C$

# Chapter 32

## Applications of Indefinite Integrals

**WHEN THE EQUATION** $y = f(x)$ of a curve is known, the slope $m$ at any point $P(x, y)$ on it is given by $m = f'(x)$. Conversely, when the slope of a curve at a point $P(x, y)$ on it is given by $m = dy/dx = f'(x)$, a family of curves, $y = f(x) + C$, is found by integration. In order to single out a particular curve of the family, it is necessary to assign or to determine a value of $C$. This may be done by prescribing that the curve pass through a given point.

See Problems 1-4.

**AN EQUATION** $s = f(t)$, where $s$ is the distance at time $t$ of a body from a fixed point in its (straight line) path, completely defines the motion of the body. The velocity and acceleration at time $t$ are given by

$$v = \frac{ds}{dt} = f'(t) \quad \text{and} \quad a = \frac{dv}{dt} = \frac{d^2s}{dt^2} = f''(t)$$

Conversely if the velocity (acceleration) is known at time $t$, together with the position (position and velocity) at some given instant, usually at $t = 0$, the equation of motion may be obtained.

See Problems 7-10.

## Solved Problems

1. Find the equation of the family of curves whose slope at any point is equal to the negative of twice the abscissa of the point. Find the curve of the family which passes through the point $(1, 1)$.

    It is given that $dy/dx = -2x$. Then, $dy = -2x\, dx$, $\int dy = \int -2x\, dx$, and $y = -x^2 + C$. This is the equation of a family of parabolas.

    Setting $x = 1$, $y = 1$ in the equation of the family, $1 = -1 + C$ and $C = 2$.

    The equation of the curve of the family passing through the point $(1, 1)$ is $y = -x^2 + 2$.

2. Find the equation of the family of curves whose slope at any point $P(x, y)$ is $m = 3x^2y$ and the equation of the curve of the family which passes through the point $(0, 8)$.

    $m = \dfrac{dy}{dx} = 3x^2y$ or $\dfrac{dy}{y} = 3x^2\, dx$. Then $\ln y = x^3 + C = x^3 + \ln c$ and $y = ce^{x^3}$.

    When $x = 0$ and $y = 8$, $8 = ce^0 = c$. The equation of the required curve is $y = 8e^{x^3}$.

3. At every point of a certain curve, $y'' = x^2 - 1$. Find the equation of the curve if it passes through the point $(1, 1)$ and is there tangent to the line $x + 12y = 13$.

    $\dfrac{d^2y}{dx^2} = \dfrac{d}{dx}(y') = x^2 - 1$. Then $\int \dfrac{d}{dx}(y')\, dx = \int (x^2 - 1)\, dx$ and $y' = \dfrac{x^3}{3} - x + C_1$.

    At $(1, 1)$, the slope $y'$ of the curve equals the slope $-\frac{1}{12}$ of the line. Then $-\frac{1}{12} = \frac{1}{3} - 1 + C_1$, $C_1 = \frac{7}{12}$, and

    $$y' = \frac{dy}{dx} = \tfrac{1}{3}x^3 - x + \tfrac{7}{12}, \quad \int dy = \int (\tfrac{1}{3}x^3 - x + \tfrac{7}{12})\, dx, \quad y = \tfrac{1}{12}x^4 - \tfrac{1}{2}x^2 + \tfrac{7}{12}x + C_2$$

    At $(1, 1)$, $1 = \frac{1}{12} - \frac{1}{2} + \frac{7}{12} + C_2$ and $C_2 = \frac{5}{6}$. The required equation is $y = \tfrac{1}{12}x^4 - \tfrac{1}{2}x^2 + \tfrac{7}{12}x + \tfrac{5}{6}$.

4. The family of *orthogonal trajectories* of a given system of curves is another system of curves each of which cuts every curve of the given system at right angles. Find the equations of the orthogonal trajectories of the family of hyperbolas $x^2 - y^2 = c$.

159

At any point $P(x, y)$, the slope of the hyperbola through the point is given by $m_1 = x/y$, and the slope of the orthogonal trajectory through $P$ is given by $m_2 = dy/dx = -y/x$. Then

$$\frac{dy}{y} = -\frac{dx}{x}, \quad \ln|y| = -\ln|x| + \ln C' \quad \text{and} \quad |xy| = C'$$

Now the required equation is $xy = \pm C'$ or, simply, $xy = C$.

5. A certain quantity $q$ increases at a rate proportional to itself. If $q = 25$ when $t = 0$ and $q = 75$ when $t = 2$, find $q$ when $t = 6$.

Since $\frac{dq}{dt} = kq$, we have $\frac{dq}{q} = k \, dt$. Then $\ln q = kt + \ln c$ or $q = ce^{kt}$.

When $t = 0$, $q = 25 = ce^0 = c$; hence, $q = 25e^{kt}$.

When $t = 2$, $q = 25e^{2k} = 75$; then $e^{2k} = 3 = e^{1.10}$ and $k = .55$.

When $t = 6$, $q = 25e^{.55t} = 25e^{3.3} = 25(e^{1.1})^3 = 25(27) = 675$.

6. A substance is being transformed into another at a rate proportional to the amount untransformed. If the original amount is 50 and is 25 when $t = 3$, when will $\frac{1}{10}$ of the substance remain untransformed?

Let $q$ represent the amount transformed in time $t$. Then

$$\frac{dq}{dt} = k(50 - q), \quad \frac{dq}{50 - q} = k \, dt, \quad \ln(50 - q) = -kt + \ln c, \quad \text{and} \quad 50 - q = ce^{-kt}$$

When $t = 0$, $q = 0$ and $c = 50$; thus $50 - q = 50e^{-kt}$.

When $t = 3$, $50 - q = 25 = 50e^{-3k}$; then $e^{-3k} = .5 = e^{-.69}$, $k = .23$, and $50 - q = 50e^{-.23t}$.

When the amount untransformed is 5, $50e^{-.23t} = 5$; then $e^{-.23t} = .1 = e^{-2.30}$ and $t = 10$.

7. A ball is rolled over a level lawn with initial velocity 25 ft/sec. Due to friction, the velocity decreases at the rate of 6 ft/sec². How far will the ball roll?

$\frac{dv}{dt} = -6$ and $v = -6t + C_1$. When $t = 0$, $v = 25$; hence $C_1 = 25$ and $v = -6t + 25$.

$v = ds/dt = -6t + 25$ and $s = -3t^2 + 25t + C_2$. When $t = 0$, $s = 0$; hence $C_2 = 0$ and $s = -3t^2 + 25t$.

When $v = 0$, $t = 25/6$, that is, the ball rolls for 25/6 sec before coming to rest.

When $t = 25/6$, $s = -3(25/6)^2 + 25(25/6) = -625/12 + 625/6 = 625/12$ ft.

8. A stone was thrown straight down from a stationary balloon, 10,000 ft above the ground, with a speed of 48 ft/sec. Locate the stone and find its speed 20 seconds later.

Take the upward direction as positive. When the stone leaves the balloon,

$$a = dv/dt = -32 \text{ ft/sec}^2 \quad \text{and} \quad v = -32t + C_1$$

When $t = 0$, $v = -48$; hence $C_1 = -48$. Then $v = ds/dt = -32t - 48$ and $s = -16t^2 - 48t + C_2$.

When $t = 0$, $s = 10,000$; hence $C_2 = 10,000$ and $s = -16t^2 - 48t + 10,000$.

When $t = 20$, $s = -16(20)^2 - 48(20) + 10,000 = 2640$ and $v = -32(20) - 48 = -688$.

After 20 sec the stone is 2640 ft above the ground and its speed is 688 ft/sec.

9. A ball was dropped from a balloon 640 ft above the ground. If the balloon was rising at the rate of 48 ft/sec, find
   (a) the greatest distance above the ground attained by the ball,
   (b) the time the ball was in the air,
   (c) the speed of the ball when it struck the ground.

Take the upward direction as positive. Then

$$a = dv/dt = -32 \text{ ft/sec}^2 \quad \text{and} \quad v = -32t + C_1$$

When $t = 0$, $v = 48$; hence $C_1 = 48$. Then $v = ds/dt = -32t + 48$ and $s = -16t^2 + 48t + C_2$.

When $t = 0$, $s = 640$; hence $C_2 = 640$ and $s = -16t^2 + 48t + 640$.

   (a) When $v = 0$, $t = 3/2$ and $s = -16(3/2)^2 + 48(3/2) + 640 = 676$. The greatest height attained by the ball was 676 ft.
   (b) When $s = 0$, $-16t^2 + 48t + 640 = 0$ and $t = -5, 8$. The ball was in the air for 8 sec.
   (c) When $t = 8$, $v = -32(8) + 48 = -208$. The ball struck the ground with speed 208 ft/sec.

10. The velocity with which water will flow from a small orifice at a depth $h$ ft below the surface is $0.6\sqrt{2gh}$ ft/sec, where $g = 32$ ft/sec². Find the time required to empty an upright cylindrical tank, height 5 ft and radius 1 ft, through a 1 inch hole in the bottom.

Let $h$ be the depth of the water at time $t$. The water which flows out in time $dt$ generates a cylinder of height $v\,dt$ ft, radius 1/24 ft, and volume $\pi(1/24)^2\,v\,dt = 0.6\pi(1/24)^2\sqrt{2gh}\,dt$ ft³.

Let $-dh$ ft represent the corresponding drop in the surface level. The loss in volume is $-\pi(1)^2\,dh$ ft³. Then $0.6\pi(1/24)^2 \cdot 8\sqrt{h}\,dt = -\pi\,dh$, or $dt = -(120\,dh)/\sqrt{h}$ and $t = -240\sqrt{h} + C$.

At $t = 0$, $h = 5$ and $C = 240\sqrt{5}$; thus $t = -240\sqrt{h} + 240\sqrt{5}$.

When the tank is empty, $h = 0$ and $t = 240\sqrt{5}$ sec $= 9$ min, approximately.

# Supplementary Problems

11. Find the equation of the family of curves having the given slope, and the equation of the curve of the family which passes through the given point.

(a) $m = 4x$; $(1, 5)$     (c) $m = (x-1)^3$; $(3, 0)$     (e) $m = x/y$; $(4, 2)$     (g) $m = 2y/x$; $(2, 8)$

(b) $m = \sqrt{x}$; $(9, 18)$     (d) $m = 1/x^2$; $(1, 2)$     (f) $m = x^2/y^3$; $(3, 2)$     (h) $m = xy/(1+x^2)$; $(3, 5)$

*Ans.*   (a) $y = 2x^2 + C$; $y = 2x^2 + 3$        (e) $x^2 - y^2 = C$; $x^2 - y^2 = 12$

      (b) $3y = 2x^{3/2} + C$; $3y = 2x^{3/2}$      (f) $3y^4 = 4x^3 + C$; $3y^4 = 4x^3 - 60$

      (c) $4y = (x-1)^4 + C$; $4y = (x-1)^4 - 16$    (g) $y = Cx^2$; $y = 2x^2$

      (d) $xy = Cx - 1$; $xy = 3x - 1$          (h) $y^2 = C(1+x^2)$; $2y^2 = 5(1+x^2)$

12. (a) For a certain curve $y'' = 2$. Find its equation given that it passes through $P(2, 6)$ with slope 10.
     *Ans.* $y = x^2 + 6x - 10$

   (b) For a certain curve $y'' = 6x - 8$. Find its equation given that it passes through $P(1, 0)$ with slope 4.      *Ans.* $y = x^3 - 4x^2 + 9x - 6$

13. A particle moves along a straight line from the origin $O$, at $t = 0$, with the given velocity $v$. Find the distance the particle moves during the interval $t = t_1$ to $t = t_2$:

(a) $v = 4t + 1$; $0, 4$        (c) $v = 3t^2 + 2t$; $2, 4$        (e) $v = 2t - 2$; $0, 5$

(b) $v = 6t + 3$; $1, 3$        (d) $v = \sqrt{t} + 5$; $4, 9$       (f) $v = t^2 - 3t + 2$; $0, 4$

*Ans.* (a) 36, (b) 30, (c) 68, (d) $37\frac{2}{3}$, (e) 17, (f) 17/3

14. Find the equation of the family of curves whose subtangent at any point is equal to twice the abscissa of the point.     *Ans.* $y^2 = Cx$

15. Find the equation of the family of orthogonal trajectories of the system of parabolas $y^2 = 2x + C$.
    *Ans.* $y = Ce^{-x}$

16. A particle moves in a straight line from the origin (at $t = 0$) with given initial velocity $v_0$ and acceleration $a$. Find $s$ at time $t$.

(a) $a = 32$; $v_0 = 2$     (b) $a = -32$; $v_0 = 96$     (c) $a = 12t^2 + 6t$; $v_0 = -3$     (d) $a = 1/\sqrt{t}$; $v_0 = 4$

*Ans.* (a) $s = 16t^2 + 2t$    (b) $s = -16t^2 + 96t$    (c) $s = t^4 + t^3 - 3t$    (d) $s = \frac{4}{3}(t^{3/2} + 3t)$

17. A car is slowing down at the rate 0.8 ft/sec². How far will the car move before it stops if its speed is initially 15 mi/hr?     *Ans.* $302\frac{1}{2}$ ft

18. A particle is projected vertically upward from a point 112 ft above the ground with initial velocity 96 ft/sec. (a) How fast is it moving when it is 240 ft above the ground? (b) When will it reach the highest point in its path? (c) At what speed will it strike the ground?
*Ans.* (a) 32 ft/sec, (b) after 3 sec, (c) 128 ft/sec

19. A block of ice slides down a chute with acceleration 4 ft/sec². The chute is 60 ft long and the ice reaches the bottom in 5 sec. What was the initial velocity of the ice and the velocity when it is 20 ft from the bottom of the chute?     *Ans.* 2 ft/sec, 18 ft/sec

20. What constant acceleration is required (a) to move a particle 50 ft in 5 sec, (b) to slow a particle from a velocity of 45 ft/sec to a dead stop in 15 ft?     *Ans.* (a) 4 ft/sec², (b) $-67\frac{1}{2}$ ft/sec²

21. The bacteria in a certain culture increase according to the law $dN/dt = 0.25N$. If originally $N = 200$, find $N$ when $t = 8$.     *Ans.* 1478

# Chapter 33

## The Definite Integral

**THE DEFINITE INTEGRAL.** Let $a \leqq x \leqq b$ be an interval on which a given function $f(x)$ is continuous. Divide the interval into $n$ subintervals $h_1, h_2, \ldots, h_n$ by the insertion of $n-1$ points $\xi_1, \xi_2, \ldots, \xi_{n-1}$, where $a < \xi_1 < \xi_2 < \cdots < \xi_{n-1} < b$, and relabel $a$ as $\xi_0$ and $b$ as $\xi_n$. Denote the length of the subinterval $h_1$ by $\Delta_1 x = \xi_1 - \xi_0$, of $h_2$ by

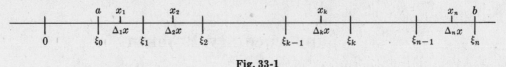

**Fig. 33-1**

$\Delta_2 x = \xi_2 - \xi_1$, ..., of $h_n$ by $\Delta_n x = \xi_n - \xi_{n-1}$. (These are directed distances, each being positive in view of the above inequality.) On each subinterval select a point — $x_1$ on the subinterval $h_1$, $x_2$ on $h_2$, ..., $x_n$ on $h_n$ — and form the sum

**(i)** $$S_n = \sum_{k=1}^{n} f(x_k)\, \Delta_k x = f(x_1)\, \Delta_1 x + f(x_2)\, \Delta_2 x + \cdots + f(x_n)\, \Delta_n x$$

each term being the product of the length of a subinterval and the value of the function at the selected point on that subinterval. Denote by $\lambda_n$ the length of the longest subinterval appearing in (i). Now let the number of subintervals increase indefinitely in such a manner that $\lambda_n \to 0$. (One way of doing this would be to bisect each of the original subintervals, in turn bisect each of these, and so on.) Then

**(ii)** $$\lim_{n \to +\infty} S_n = \lim_{n \to +\infty} \sum_{k=1}^{n} f(x_k)\, \Delta_k x$$

exists and is the same for all methods of subdividing the interval $a \leqq x \leqq b$, so long as the condition $\lambda_n \to 0$ is met, and for all choices of the points $x_k$ in the resulting subintervals.

A proof of the theorem is beyond the scope of this book. In Problems 1-3 the limit is evaluated for selected functions $f(x)$. It must be understood, however, that for an arbitrary function this procedure is too difficult to attempt. Moreover, in order to succeed in the evaluations made here, it is necessary to prescribe some relation among the lengths of the subintervals (we take them all of equal length) and to follow some pattern in choosing a point on each subinterval (for example, choose the left hand endpoint or the right hand endpoint or the midpoint of each subinterval).

By agreement, we write

$$\int_a^b f(x)\, dx = \lim_{n \to +\infty} S_n = \lim_{n \to +\infty} \sum_{k=1}^{n} f(x_k)\, \Delta_k x$$

The symbol $\int_a^b f(x)\, dx$ is read "the *definite integral* of $f(x)$, with respect to $x$, from $x = a$ to $x = b$". The function $f(x)$ is called the *integrand* while $a$ and $b$ are called respectively the *lower* and *upper limits* (boundaries) *of integration*.

See Problems 1-3.

**PROPERTIES OF DEFINITE INTEGRALS.** If $f(x)$ and $g(x)$ are continuous on the interval of integration $a \leqq x \leqq b$:

1. $\displaystyle\int_a^a f(x)\,dx = 0$

2. $\displaystyle\int_a^b f(x)\,dx = -\int_b^a f(x)\,dx$

3. $\displaystyle\int_a^b c\,f(x)\,dx = c\int_a^b f(x)\,dx$, for any constant $c$.

   For proofs of the above, see Problem 4.

4. $\displaystyle\int_a^b \{f(x) \pm g(x)\}\,dx = \int_a^b f(x)\,dx \pm \int_a^b g(x)\,dx$

5. $\displaystyle\int_a^c f(x)\,dx + \int_c^b f(x)\,dx = \int_a^b f(x)\,dx$, when $a < c < b$

6. The First Mean Value Theorem:

   $$\int_a^b f(x)\,dx = (b-a)\,f(x_0) \quad \text{for at least one value } x = x_0 \text{ between } a \text{ and } b.$$

   For a proof, see Problem 5.

7. If $\displaystyle F(u) = \int_a^u f(x)\,dx$, then $\dfrac{d}{du}F(u) = f(u)$.　　For a proof, see Problem 6.

**FUNDAMENTAL THEOREM OF INTEGRAL CALCULUS.** If $f(x)$ is continuous on the interval $a \leqq x \leqq b$ and if $F(x)$ is any indefinite integral of $f(x)$, then

$$\int_a^b f(x)\,dx = F(x)\Big|_a^b = F(b) - F(a)$$

For a proof, see Problem 7.

**Example 1:**

(a) Take $f(x) = c$, a constant, and $F(x) = cx$; then $\displaystyle\int_a^b c\,dx = cx\Big|_a^b = c(b-a)$.

(b) Take $f(x) = x$ and $F(x) = \frac{1}{2}x^2$; then $\displaystyle\int_0^5 x\,dx = \frac{1}{2}x^2\Big|_0^5 = \frac{25}{2} - 0 = \frac{25}{2}$.

(c) Take $f(x) = x^3$ and $F(x) = \frac{1}{4}x^4$; then $\displaystyle\int_1^3 x^3\,dx = \frac{1}{4}x^4\Big|_1^3 = \frac{81}{4} - \frac{1}{4} = 20$.

These results are to be compared with those of Problems 1-3. The reader will show that *any* indefinite integral of $f(x)$ may be used by resolving (c) with $F(x) = \frac{1}{4}x^4 + C$.

See Problems 8-20.

**THE THEOREM OF BLISS.** If $f(x)$ and $g(x)$ are continuous on the interval $a \leqq x \leqq b$, if the interval is divided into subintervals as before, and if two points are selected in each subinterval (i.e., $x_k$ and $x'_k$ in the $k$th subinterval), then

$$\lim_{n \to +\infty} \sum_{k=1}^n f(x_k) \cdot g(x'_k)\,\Delta_k x = \int_a^b f(x) \cdot g(x)\,dx$$

We note first that the theorem is true if the points $x_k$ and $x'_k$ are identical. The force of the theorem is that when the points of each pair are distinct, the result is the same as if they were coincident. An intuitive feeling for the validity of the theorem follows from writing

$$\sum_{k=1}^n f(x_k) \cdot g(x'_k)\,\Delta_k x = \sum_{k=1}^n f(x_k) \cdot g(x_k)\,\Delta_k x + \sum_{k=1}^n f(x_k)\,\{g(x'_k) - g(x_k)\}\,\Delta_k x$$

and noting that as $n \to +\infty$ (that is, $\Delta_k x \to 0$) $x_k$ and $x'_k$ must become more nearly identical and, since $g(x)$ is continuous, $g(x'_k) - g(x_k)$ must then $\to 0$.

# Solved Problems

In Problems 1-3 evaluate the definite integral by setting up $S_n$ and obtaining the limit as $n \to +\infty$.

1. $\displaystyle\int_a^b c\,dx = c(b-a)$, $c$ being a constant.

   Let the interval $a \leq x \leq b$ be divided into $n$ equal subintervals of length $\Delta x = (b-a)/n$. Since the integrand is $f(x) = c$, then $f(x_k) = c$ for any choice of the point $x_k$ on the $k$th subinterval, and

   $$S_n = \sum_{k=1}^{n} f(x_k)\,\Delta_k x = \sum_{k=1}^{n} c(\Delta x) = (c+c+\cdots+c)(\Delta x) = nc \cdot \Delta x = nc\frac{b-a}{n} = c(b-a)$$

   Hence $\displaystyle\int_a^b c\,dx = \lim_{n \to +\infty} S_n = \lim_{n \to +\infty} c(b-a) = c(b-a)$

2. $\displaystyle\int_0^5 x\,dx = 25/2.$

**Fig. 33-2**

   Let the interval $0 \leq x \leq 5$ be divided into $n$ equal subintervals of length $\Delta x = 5/n$. Take the points $x_k$ as the right hand endpoints of the subintervals, that is, $x_1 = \Delta x$, $x_2 = 2\,\Delta x$, ..., $x_n = n\,\Delta x$. Then

   $$S_n = \sum_{k=1}^{n} f(x_k)\,\Delta_k x = \sum_{k=1}^{n} (k \cdot \Delta x)\,\Delta x = (1+2+\cdots+n)(\Delta x)^2 = \frac{n(n+1)}{2}\left(\frac{5}{n}\right)^2 = \frac{25}{2}\left(1+\frac{1}{n}\right)$$

   and $\displaystyle\int_0^5 x\,dx = \lim_{n \to +\infty} S_n = \lim_{n \to +\infty} \frac{25}{2}\left(1+\frac{1}{n}\right) = \frac{25}{2}$

3. $\displaystyle\int_1^3 x^3\,dx = 20.$

   Let the interval $1 \leq x \leq 3$ be divided into $n$ subintervals of length $\Delta x = 2/n$.

   I.  Take the points $x_k$ as the left hand endpoints of the subintervals as in Fig. 33-3 below, that is, $x_1 = 1$, $x_2 = 1 + \Delta x$, ..., $x_n = 1 + (n-1)\,\Delta x$. Then

   $$S_n = \sum_{k=1}^{n} f(x_k)\,\Delta_k n = x_1^3 \cdot \Delta x + x_2^3 \cdot \Delta x + \cdots + x_n^3 \cdot \Delta x$$

   $$= \left[ 1 + (1+\Delta x)^3 + (1+2\cdot\Delta x)^3 + \cdots + \{1 + (n-1)\,\Delta x\}^3 \right]\Delta x$$

   $$= \left[ n + 3\{1+2+\cdots+(n-1)\}\,\Delta x + 3\{1^2+2^2+\cdots+(n-1)^2\}\,(\Delta x)^2 \right.$$
   $$\left. + \{1^3+2^3+\cdots+(n-1)^3\}\,(\Delta x)^3 \right]\Delta x$$

   $$= \left[ n + 3\frac{(n-1)n}{1\cdot 2}\left(\frac{2}{n}\right) + 3\frac{(n-1)n(2n-1)}{1\cdot 2\cdot 3}\left(\frac{2}{n}\right)^2 + \frac{(n-1)^2 n^2}{(1\cdot 2)^2}\left(\frac{2}{n}\right)^3 \right]\frac{2}{n}$$

   $$= 2 + \left(6 - \frac{6}{n}\right) + \left(8 - \frac{12}{n} + \frac{4}{n^2}\right) + \left(4 - \frac{8}{n} + \frac{4}{n^2}\right) = 20 - \frac{26}{n} + \frac{8}{n^2}$$

   and $\displaystyle\int_1^3 x^3\,dx = \lim_{n \to +\infty} \left(20 - \frac{26}{n} + \frac{8}{n^2}\right) = 20$

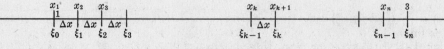

**Fig. 33-3**

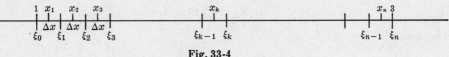

**Fig. 33-4**

**II.** Take the points $x_k$ as the midpoints of the subintervals as in Fig. 33-4 above, that is, $x_1 = 1 + \frac{1}{2}\Delta x,\ x_2 = 1 + \frac{3}{2}\Delta x,\ \ldots,\ x_n = 1 + \frac{2n-1}{2}\Delta x.$ Then

$$S_n = \left[(1+\tfrac{1}{2}\Delta x)^3 + (1+\tfrac{3}{2}\Delta x)^3 + \cdots + \left(1 + \frac{2n-1}{2}\Delta x\right)^3\right]\Delta x$$

$$= \left[\{1 + 3(\tfrac{1}{2})\Delta x + 3(\tfrac{1}{2})^2(\Delta x)^2 + (\tfrac{1}{2})^3(\Delta x)^3\} + \{1 + 3(\tfrac{3}{2})(\Delta x) + 3(\tfrac{3}{2})^2(\Delta x)^2 + (\tfrac{3}{2})^3(\Delta x)^3\} + \cdots\right.$$

$$\left. + \left\{1 + 3\left(\frac{2n-1}{2}\right)(\Delta x) + 3\left(\frac{2n-1}{2}\right)^2(\Delta x)^2 + \left(\frac{2n-1}{2}\right)^3(\Delta x)^3\right\}\right]\Delta x$$

$$= n\left(\frac{2}{n}\right) + \frac{3}{2}n^2\left(\frac{2}{n}\right)^2 + \frac{1}{4}(4n^3 - n)\left(\frac{2}{n}\right)^3 + \frac{1}{8}(2n^4 - n^2)\left(\frac{2}{n}\right)^4$$

$$= 2 + 6 + \left(8 - \frac{2}{n^2}\right) + \left(4 - \frac{2}{n^2}\right) = 20 - \frac{4}{n^2}$$

and

$$\int_1^3 x^3\,dx = \lim_{n \to +\infty}\left(20 - \frac{4}{n^2}\right) = 20$$

**4.** Prove:

(a) $\displaystyle\int_a^a f(x)\,dx = 0.$ Here the interval of integration is of length 0; hence $\Delta x = 0,\ S_n = 0,$ and

$$\int_a^a f(x)\,dx = \lim_{n \to +\infty} S_n = 0.$$

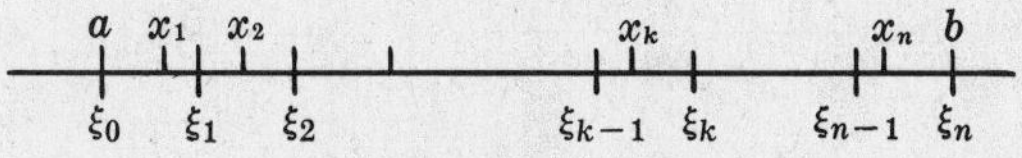

**Fig. 33-5**    **Fig. 33-6**

(b) $\displaystyle\int_a^b f(x)\,dx = -\int_b^a f(x)\,dx.$ Let the interval $a \le x \le b$ be subdivided and the points $x_k$ be selected as in Fig. 33-5 for $\displaystyle\int_a^b f(x)\,dx.$ For $\displaystyle\int_b^a f(x)\,dx$ let the interval (Fig. 33-6) be exactly as before except that the points $\xi_k$ and $x_k$ are numbered from right to left instead of from left to right. Now $S_n$ when computed from Fig. 33-5 and when computed from Fig. 33-6 are identical except for the signs of $\Delta_k x$ which are positive in the first and negative in the second. Thus

$$\int_a^b f(x)\,dx = -\int_b^a f(x)\,dx.$$

(c) $\displaystyle\int_a^b c \cdot f(x)\,dx = c\int_a^b f(x)\,dx.$ For a proper subdivision of the interval and any choice of points on the subintervals,

$$S_n = \sum_{k=1}^n c\,f(x_k)\,\Delta_k x = c\sum_{k=1}^n f(x_k)\,\Delta_k x$$

Then

$$\int_a^b c\,f(x)\,dx = c\lim_{n \to +\infty}\sum_{k=1}^n f(x_k)\,\Delta_k x = c\int_a^b f(x)\,dx$$

**5.** Prove the First Mean Value Theorem of Integral Calculus: If $f(x)$ is continuous on the interval $a \le x \le b,$ then $\displaystyle\int_a^b f(x)\,dx = (b - a)\cdot f(x_0)$ for at least one value $x = x_0$ between $a$ and $b.$

The theorem is true, by Example 1(a), when $f(x) = c,$ a constant. Otherwise, let $m$ be the absolute minimum value and $M$ be the absolute maximum value of $f(x)$ on the interval $a \le x \le b.$ For any proper subdivision of the interval and any choice of the points $x_k$ on the subintervals,

$$\sum_{k=1}^{n} m \, \Delta_k x \;\; < \;\; \sum_{k=1}^{n} f(x_k) \, \Delta_k x \;\; < \;\; \sum_{k=1}^{n} M \, \Delta_k x$$

Now when $n \to +\infty$, we have

$$\int_a^b m \, dx \;\; < \;\; \int_a^b f(x) \, dx \;\; < \;\; \int_a^b M \, dx$$

which, by Problem 1, becomes

$$m(b-a) \;\; < \;\; \int_a^b f(x) \, dx \;\; < \;\; M(b-a)$$

Then

$$m \;\; < \;\; \frac{1}{b-a} \int_a^b f(x) \, dx \;\; < \;\; M$$

so that $\dfrac{1}{b-a} \displaystyle\int_a^b f(x) \, dx = N$, where $N$ is some number between $m$ and $M$. Now since $f(x)$ is continuous on the interval $a \leqq x \leqq b$, it must by Theorem I, Chapter 3, take on at least once every value from $m$ to $M$. Hence, there must be a value of $x$, say $x = x_0$, such that $f(x_0) = N$. Then

$$\frac{1}{b-a} \int_a^b f(x) \, dx = N = f(x_0) \quad \text{and} \quad \int_a^b f(x) \, dx = (b-a) \, f(x_0)$$

6. Prove: If $F(u) = \displaystyle\int_a^u f(x) \, dx$, then $\dfrac{d}{du} F(u) = f(u)$.

By the step rule for finding derivatives,

$$F(u + \Delta u) - F(u) = \int_a^{u+\Delta u} f(x) \, dx - \int_a^u f(x) \, dx$$

which by using Properties 2, 5, and 6 in turn becomes

$$F(u + \Delta u) - F(u) = \int_u^a f(x) \, dx + \int_a^{u+\Delta u} f(x) \, dx = \int_u^{u+\Delta u} f(x) \, dx$$

$$= f(u_0) \cdot \Delta u, \quad \text{where} \;\; u < u_0 < u + \Delta u$$

Then

$$\frac{F(u + \Delta u) - F(u)}{\Delta u} = f(u_0) \quad \text{and} \quad \frac{dF}{du} = \lim_{\Delta u \to 0} \frac{F(u + \Delta u) - F(u)}{\Delta u} = \lim_{\Delta u \to 0} f(u_0) = f(u)$$

since as $\Delta u \to 0$, $u_0 \to u$.

This property is most frequently stated as:

(i) $\qquad\qquad\qquad$ If $F(x) = \displaystyle\int_a^x f(x) \, dx$, then $F'(x) = f(x)$.

The use of the letter $u$ above was merely an attempt to avoid the possibility of confusing the roles of the several $x$'s. Note carefully in (i) that $F(x)$ is a function of the upper limit $x$ of integration and not of the *dummy* letter $x$ in $f(x) \, dx$. In other words, the property might also be stated:

$$\text{If} \;\; F(x) = \int_a^x f(t) \, dt, \;\; \text{then} \;\; F'(x) = f(x).$$

It follows from (i) that $F(x)$ is simply an indefinite integral of $f(x)$.

7. Prove: If $f(x)$ is continuous on the interval $a \leqq x \leqq b$ and if $F(x)$ is any indefinite integral of $f(x)$, then

$$\int_a^b f(x) \, dx = F(b) - F(a)$$

Use the last statement in Problem 6 to write

$$\int_a^x f(x) \, dx = F(x) + C$$

When the upper limit of integration is $x = a$, we have

$$\int_a^a f(x) \, dx = 0 = F(a) + C \quad \text{and} \quad C = -F(a)$$

Then $\displaystyle\int_a^x f(x) \, dx = F(x) - F(a)$ and when the upper limit of integration is $x = b$, we have, as required,

$$\int_a^b f(x) \, dx = F(b) - F(a)$$

Use the Fundamental Theorem of Integral Calculus to evaluate each of the following.

8. $\displaystyle\int_{-1}^{1} (2x^2 - x^3)\, dx \;=\; \left[\frac{2x^3}{3} - \frac{x^4}{4}\right]_{-1}^{1} \;=\; \left(\frac{2}{3} - \frac{1}{4}\right) - \left(-\frac{2}{3} - \frac{1}{4}\right) \;=\; \frac{4}{3}$

9. $\displaystyle\int_{-3}^{-1} \left(\frac{1}{x^2} - \frac{1}{x^3}\right) dx \;=\; \left[-\frac{1}{x} + \frac{1}{2x^2}\right]_{-3}^{-1} \;=\; \left(1 + \frac{1}{2}\right) - \left(\frac{1}{3} + \frac{1}{18}\right) \;=\; \frac{10}{9}$

10. $\displaystyle\int_{1}^{4} \frac{dx}{\sqrt{x}} \;=\; 2\sqrt{x}\,\Big]_{1}^{4} \;=\; 2(\sqrt{4} - \sqrt{1}) \;=\; 2$

11. $\displaystyle\int_{-2}^{3} e^{-x/2}\, dx \;=\; -2e^{-x/2}\,\Big]_{-2}^{3} \;=\; -2(e^{-3/2} - e) \;=\; 4.9904$

12. $\displaystyle\int_{-6}^{-10} \frac{dx}{x+2} \;=\; \ln|x+2|\,\Big]_{-6}^{-10} \;=\; \ln 8 - \ln 4 \;=\; \ln 2$

13. $\displaystyle\int_{\pi/2}^{3\pi/4} \sin x\, dx \;=\; -\cos x\,\Big]_{\pi/2}^{3\pi/4} \;=\; -\left(-\frac{1}{2}\sqrt{2} - 0\right) \;=\; \frac{1}{2}\sqrt{2}$

14. $\displaystyle\int_{-2}^{2} \frac{dx}{x^2+4} \;=\; \frac{1}{2}\arctan\frac{1}{2}x\,\Big]_{-2}^{2} \;=\; \frac{1}{2}\left[\frac{1}{4}\pi - \left(-\frac{1}{4}\pi\right)\right] \;=\; \frac{1}{4}\pi$

15. $\displaystyle\int_{-5}^{-3} \sqrt{x^2 - 4}\, dx \;=\; \left[\frac{1}{2}x\sqrt{x^2-4} - 2\ln|x + \sqrt{x^2-4}\,|\right]_{-5}^{-3} \;=\; \frac{5}{2}\sqrt{21} - \frac{3}{2}\sqrt{5} - 2\ln\frac{3-\sqrt{5}}{5-\sqrt{21}}$

16. $\displaystyle\int_{-1}^{2} \frac{dx}{x^2-9} \;=\; \frac{1}{6}\ln\left|\frac{x-3}{x+3}\right|\,\Big]_{-1}^{2} \;=\; \frac{1}{6}\left(\ln\frac{1}{5} - \ln 2\right) \;=\; \frac{1}{6}\ln 0.1$

17. $\displaystyle\int_{1}^{e} \ln x\, dx \;=\; \left[x\ln x - x\right]_{1}^{e} \;=\; (e\ln e - e) - (\ln 1 - 1) \;=\; 1$

18. Find $\displaystyle\int_{3}^{6} xy\, dx$ when $x = 6\cos\theta$, $y = 2\sin\theta$.

Here we express $x$, $y$, and $dx$ in terms of the parameter $\theta$ and $d\theta$, change the limits of integration to corresponding values of the parameter, and evaluate the resulting integral.

$dx = -6\sin\theta\, d\theta$.  When $x = 6\cos\theta = 6$, $\theta = 0$; and when $x = 6\cos\theta = 3$, $\theta = \pi/3$.  Hence

$$\int_{3}^{6} xy\, dx \;=\; \int_{\pi/3}^{0} (6\cos\theta)(2\sin\theta)(-6\sin\theta)\, d\theta$$

$$=\; -72\int_{\pi/3}^{0} \sin^2\theta\cos\theta\, d\theta \;=\; -24\sin^3\theta\,\Big]_{\pi/3}^{0} \;=\; -24\{0 - (\sqrt{3}/2)^3\} \;=\; 9\sqrt{3}$$

19. Find $\displaystyle\int_{0}^{2\pi/3} \frac{d\theta}{5 + 4\cos\theta}\,.$    $\displaystyle\int \frac{d\theta}{5 + 4\cos\theta} \;=\; \int \frac{\dfrac{2\,dz}{1+z^2}}{5 + 4\dfrac{1-z^2}{1+z^2}} \;=\; \int \frac{2\,dz}{9 + z^2}\,.$

To determine the $z$ limits of integration ($\theta = 2\arctan z$): When $\theta = 0$, $z = 0$; when $\theta = 2\pi/3$, $\arctan z = \pi/3$ and $z = \sqrt{3}$.  Then

$$\int_{0}^{2\pi/3} \frac{d\theta}{5 + 4\cos\theta} \;=\; 2\int_{0}^{\sqrt{3}} \frac{dz}{9 + z^2} \;=\; \frac{2}{3}\arctan\frac{z}{3}\,\Big]_{0}^{\sqrt{3}} \;=\; \frac{\pi}{9}$$

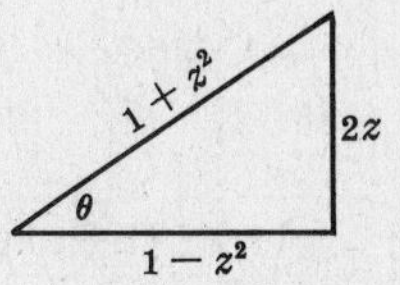

Fig. 33-7

20. Find $\displaystyle\int_0^{\pi/3} \frac{dx}{1 - \sin x}$.  $\displaystyle\int \frac{dx}{1 - \sin x} = \int \frac{\dfrac{2\,dz}{1+z^2}}{1 - \dfrac{2z}{1+z^2}} = \int \frac{2\,dz}{(1-z)^2}$.

When $x = 0$, $\arctan z = 0$ and $z = 0$; when $x = \pi/3$, $\arctan z = \pi/6$ and $z = \sqrt{3}/3$.  Then

$$\int_0^{\pi/3} \frac{dx}{1 - \sin x} = 2\int_0^{\sqrt{3}/3} \frac{dz}{(1-z)^2} = \frac{2}{1-z}\bigg]_0^{\sqrt{3}/3} = \frac{2}{1 - \sqrt{3}/3} - 2 = \sqrt{3} + 1$$

# Supplementary Problems

21. Evaluate $\displaystyle\int_a^b c\,dx$ of Problem 1 by dividing the interval $a \leqq x \leqq b$ into $n$ subintervals of lengths $\Delta_1 x, \Delta_2 x, \ldots, \Delta_n x$.  Note that $\displaystyle\sum_{k=1}^n \Delta_k x = b - a$.

22. Evaluate $\displaystyle\int_0^5 x\,dx$ of Problem 2 using subintervals of equal length and (a) choosing the points $x_k$ as the left hand endpoints of the subintervals, (b) choosing the points $x_k$ as the midpoints of the subintervals, (c) choosing the points $x_k$ one third the way into each subinterval, that is, taking $x_1 = \frac{1}{3}\Delta x$, $x_2 = \frac{4}{3}\Delta x$, $\ldots$

23. Evaluate $\displaystyle\int_1^4 x^2\,dx = 21$ using subintervals of equal length and choosing the points $x_k$ as (a) the right hand endpoints of the subintervals, (b) the left hand endpoints of the subintervals, (c) the midpoints of the subintervals.

24. Using the same choice of subintervals and points as in Problem 23(a), evaluate $\displaystyle\int_1^4 x\,dx$ and $\displaystyle\int_1^4 (x^2 + x)\,dx$, and verify that $\displaystyle\int_a^b \{f(x) + g(x)\}\,dx = \int_a^b f(x)\,dx + \int_a^b g(x)\,dx$.

25. Evaluate $\displaystyle\int_1^2 x^2\,dx$ and $\displaystyle\int_2^4 x^2\,dx$.  Compare the sum with the result of Problem 23 to verify

$$\int_a^c f(x)\,dx + \int_c^b f(x)\,dx = \int_a^b f(x)\,dx \quad \text{when } a < c < b.$$

26. Evaluate $\displaystyle\int_0^1 e^x\,dx = e - 1$.

    *Hint.*  $S_n = \displaystyle\sum_{k=1}^n e^{k\cdot\Delta x}\Delta x = e^{\Delta x}(e-1)\frac{\Delta x}{e^{\Delta x} - 1}$ and $\displaystyle\lim_{n\to+\infty} \frac{\Delta x}{e^{\Delta x} - 1} = \lim_{\Delta x \to 0} \frac{\Delta x}{e^{\Delta x} - 1}$ is indeterminate of the type 0/0.

27. Prove the properties 4 and 5 of definite integrals.

28. Use the Fundamental Theorem to evaluate:

(a) $\displaystyle\int_0^2 (2 + x)\,dx = 6$

(b) $\displaystyle\int_0^2 (2 - x)^2\,dx = 8/3$

(c) $\displaystyle\int_0^3 (3 - 2x + x^2)\,dx = 9$

(d) $\displaystyle\int_{-1}^2 (1 - t^2)t\,dt = -9/4$

(e) $\displaystyle\int_1^4 (1 - u)\sqrt{u}\,du = -116/15$

(f) $\displaystyle\int_1^8 \sqrt{1 + 3x}\,dx = 26$

(g) $\displaystyle\int_0^2 x^2(x^3 + 1)\,dx = 40/3$

(h) $\displaystyle\int_0^3 \frac{dx}{\sqrt{1 + x}} = 2$

(i) $\displaystyle\int_0^1 x(1 - \sqrt{x})^2\,dx = 1/30$

(j) $\displaystyle\int_4^8 \frac{x\,dx}{\sqrt{x^2 - 15}} = 6$

(k) $\displaystyle\int_0^a \sqrt{a^2 - x^2}\,dx = \frac{1}{4}a^2\pi$

(l) $\displaystyle\int_{-1}^1 x^2\sqrt{4 - x^2}\,dx = \frac{2}{3}\pi - \frac{1}{2}\sqrt{3}$

$(m)$ $\displaystyle\int_3^4 \frac{dx}{25-x^2} = \frac{1}{5}\ln\frac{3}{2}$

$(q)$ $\displaystyle\int_0^1 \ln(x^2+1)\,dx = \ln 2 + \frac{1}{2}\pi - 2$

$(n)$ $\displaystyle\int_{-\frac{1}{2}}^0 \frac{x^3\,dx}{x^2+x+1} = \frac{\sqrt{3}\,\pi}{9} - \frac{5}{8}$

$(r)$ $\displaystyle\int_0^{2\pi} \sin\frac{1}{2}t\,dt = 4$

$(o)$ $\displaystyle\int_2^4 \frac{\sqrt{16-x^2}}{x}\,dx = 4\ln(2+\sqrt{3}) - 2\sqrt{3}$

$(s)$ $\displaystyle\int_0^{\pi/3} x^2 \sin 3x\,dx = \frac{1}{27}(\pi^2 - 4)$

$(p)$ $\displaystyle\int_8^{27} \frac{dx}{x - x^{1/3}} = \frac{3}{2}\ln\frac{8}{3}$

$(t)$ $\displaystyle\int_0^{\pi/2} \frac{dx}{3 + \cos 2x} = \frac{\sqrt{2}\,\pi}{8}$

29. Show that $\displaystyle\int_3^5 \frac{dx}{\sqrt{x^2+16}} = \int_{-5}^{-3} \frac{dx}{\sqrt{x^2+16}}$.

30. Evaluate $\displaystyle\int_{\theta=0}^{\theta=2\pi} y\,dx = 3\pi$, given $x = \theta - \sin\theta$, $y = 1 - \cos\theta$.

31. Evaluate $\displaystyle\int_1^4 \sqrt{1+(y')^2}\,dx = \frac{15}{2} + \frac{1}{2}\ln 2$, given $y = \frac{1}{2}x^2 - \frac{1}{4}\ln x$.

32. Evaluate $\displaystyle\int_2^3 \sqrt{\left(\frac{dx}{dt}\right)^2 + \left(\frac{dy}{dt}\right)^2}\,dt = \sqrt{2}\,e^2(e-1)$, given $x = e^t\cos t$, $y = e^t\sin t$.

33. Use the appropriate reduction formulas (Chapter 26) to establish Wallis' Formulas:

$$\int_0^{\pi/2} \sin^n x\,dx = \int_0^{\pi/2} \cos^n x\,dx = \frac{1\cdot 3\ldots(n-3)(n-1)}{2\cdot 4\ldots(n-2)n}\cdot\frac{\pi}{2} \quad \text{if } n \text{ is even and} > 0$$

$$= \frac{2\cdot 4\ldots(n-3)(n-1)}{1\cdot 3\ldots(n-2)n} \quad \text{if } n \text{ is odd and} > 1$$

$$\int_0^{\pi/2} \sin^m x \cos^n x\,dx = \frac{1\cdot 3\ldots(m-1)\cdot 1\cdot 3\ldots(n-1)}{2\cdot 4\ldots(m+n-2)(m+n)}\cdot\frac{\pi}{2} \quad \text{if } m \text{ and } n \text{ are even and} > 0$$

$$= \frac{2\cdot 4\ldots(m-3)(m-1)}{(n+1)(n+3)\ldots(n+m)} \quad \text{if } m \text{ is odd and} > 1$$

$$= \frac{2\cdot 4\ldots(n-3)(n-1)}{(m+1)(m+3)\ldots(m+n)} \quad \text{if } n \text{ is odd and} > 1$$

34. Evaluate:

$(a)$ $\displaystyle\int_3^{11} \sqrt{2x+3}\,dx = 98/3$

$(c)$ $\displaystyle\int_4^9 \frac{1-\sqrt{x}}{1+\sqrt{x}}\,dx = 4\ln\frac{3}{4} - 1$

$(b)$ $\displaystyle\int_0^{\pi/4} \frac{\cos 2x - 1}{\cos 2x + 1}\,dx = \frac{1}{4}\pi - 1$

$(d)$ $\displaystyle\int_0^{\sqrt{2}} x^3 e^{x^2}\,dx = \frac{1}{2}(e^2 + 1)$

$(e)$ $\displaystyle\int_{\pi/4}^{3\pi/4} \frac{\sin x\,dx}{\cos^2 x - 5\cos x + 4} = \frac{1}{3}\ln\frac{7+3\sqrt{2}}{7-3\sqrt{2}}$

$(f)$ $\displaystyle\int_{-2}^{-1} \frac{x-1}{\sqrt{x^2-4x+3}}\,dx = \ln\frac{3-2\sqrt{2}}{4-\sqrt{15}} + 2\sqrt{2} - \sqrt{15}$

$(g)$ $\displaystyle\int_{\pi/6}^{\pi/3} \frac{dx}{\sin 2x} = \ln\sqrt{3}$

$(h)$ $\displaystyle\int_1^3 \ln(x + \sqrt{x^2-1})\,dx = 3\ln(3+2\sqrt{2}) - 2\sqrt{2}$

$(i)$ $\displaystyle\int_{-1}^{-2} \frac{dx}{\sqrt{x^2+2x+2}} = \ln(\sqrt{2}-1)$

$(k)$ $\displaystyle\int_{-8}^{-3} \frac{(x+2)\,dx}{x(x-2)^2} = \frac{1}{2}\ln\frac{3}{4} + \frac{1}{5}$

$(j)$ $\displaystyle\int_{1/4}^{3/4} \frac{(x+1)\,dx}{x^2(x-1)} = 4\ln\frac{1}{3} - \frac{8}{3}$

$(l)$ $\displaystyle\int_0^{\pi/4} \frac{dx}{2 + \tan x} = \frac{1}{5}\ln\frac{3\sqrt{2}}{4} + \frac{\pi}{10}$

# Chapter 34

## Plane Areas by Integration

**AREA AS THE LIMIT OF A SUM.** If $f(x)$ is continuous and non-negative on the interval $a \le x \le b$, the definite integral $\int_a^b f(x)\,dx = \lim\limits_{n \to +\infty} \sum\limits_{k=1}^n f(x_k)\,\Delta_k x$ can be given a geometric interpretation. Let the interval $a \le x \le b$ be subdivided and points $x_k$ be selected as in the preceding chapter. Through each of the endpoints $\xi_0 = a, \xi_1, \xi_2, \ldots, \xi_n = b$ erect perpendiculars to the $x$-axis, thus dividing the portion of the plane bounded above by the curve $y = f(x)$, below by the $x$-axis, and laterally by the ordinates $x = a$ and $x = b$ into $n$ strips. Approximate each strip by a rectangle whose base is the lower base of the strip and whose altitude is the ordinate erected at the point $x_k$ of the subinterval. The area of the representative approximating rectangle shown in Fig. 34-1 is $f(x_k)\,\Delta_k x$. Hence $\sum\limits_{k=1}^n f(x_k)\,\Delta_k x$ is simply the sum of the areas of the $n$ approximating rectangles.

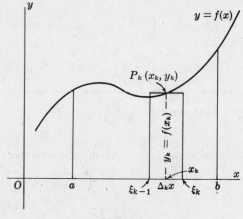

**Fig. 34-1**

The limit of this sum, $\int_a^b f(x)\,dx$, as the number of strips is indefinitely increased in the prescribed manner of Chapter 33, is by definition the area of the portion of the plane described above or, briefly, the area under the curve from $x = a$ to $x = b$.

See Problems 1-2.

Similarly, if $x = g(y)$ is continuous and non-negative on the interval $c \le y \le d$, the definite integral $\int_c^d g(y)\,dy$ is by definition the area bounded by the curve $x = g(y)$, the $y$-axis, and the abscissas $y = c$ and $y = d$. See Problem 3.

If $y = f(x)$ is continuous and non-positive on the interval $a \le x \le b$, then $\int_a^b f(x)\,dx$ is negative, indicating that the area lies below the $x$-axis. Similarly, if $x = g(y)$ is continuous and non-positive on the interval $c \le y \le d$, $\int_c^d g(y)\,dy$ is negative, indicating that the area lies to the left of the $y$-axis. See Problem 4.

If $y = f(x)$ changes sign on the interval $a \le x \le b$ or if $x = g(y)$ changes sign on the interval $c \le y \le d$, then the area "under the curve" is given by the sum of two or more definite integrals. See Problem 5.

**AREAS BY INTEGRATION.** The steps necessary for setting up the definite integral which yields a required area are:

170

(1) Make a sketch showing (a) the area sought, (b) a representative strip, and (c) the approximating rectangle. As a matter of policy, we shall show the representative subinterval of length $\Delta x$ (or $\Delta y$) and the point $x_k$ (or $y_k$) on this subinterval as its midpoint.

(2) Write the area of the approximating rectangle and the sum for the $n$ rectangles.

(3) Assume the number of rectangles to be indefinitely increased and apply the Fundamental Theorem of the preceding chapter.

See Problems 6-14.

# Solved Problems

1. Find the area bounded by the curve $y = x^2$, the $x$-axis, and the ordinates $x = 1$ and $x = 3$.

Fig. 34-2 shows the area $KLMN$ sought, a representative strip $RSTU$, and its approximating rectangle $RVWU$. For this rectangle, the base is $\Delta_k x$, the altitude is $y_k = f(x_k) = x_k^2$, and the area is $x_k^2 \cdot \Delta_k x$. Then

$$A = \lim_{n \to +\infty} \sum_{k=1}^{n} x_k^2 \Delta_k x = \int_1^3 x^2 \, dx$$

$$= \frac{x^3}{3} \Big|_1^3 = 9 - \frac{1}{3} = \frac{26}{3} \text{ square units}$$

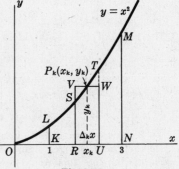

**Fig. 34-2**

2. Find the area lying above the $x$-axis and under the parabola $y = 4x - x^2$.

The given curve crosses the $x$-axis at $x = 0$ and $x = 4$. When vertical slicing is used, these values become the limits of integration. For the approximating rectangle shown in Fig. 34-3, the width is $\Delta_k x$, the height is $y_k = 4x_k - x_k^2$, and the area is $(4x_k - x_k^2) \cdot \Delta_k x$. Then

$$A = \lim_{n \to +\infty} \sum_{k=1}^{n} (4x_k - x_k^2) \Delta_k x = \int_0^4 (4x - x^2) \, dx$$

$$= \left[ 2x^2 - \frac{1}{3} x^3 \right]_0^4 = 32/3 \text{ square units}$$

With the complete procedure, as given above, always in mind, an abbreviation of the work is possible. It will be seen that, aside from the limits of integration, the definite integral can be formulated once the area of the approximating rectangle has been set down.

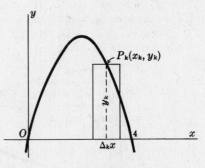

**Fig. 34-3**

3. Find the area bounded by the parabola $x = 8 + 2y - y^2$, the $y$-axis, and the lines $y = -1$ and $y = 3$.

Here we slice the area into horizontal strips. For the approximating rectangle shown in Fig. 34-4, the width is $\Delta y$, the length is $x = 8 + 2y - y^2$, and the area is $(8 + 2y - y^2)\Delta y$. The required area is

$$\int_{-1}^{3} (8 + 2y - y^2) \, dy = \left[ 8y + y^2 - \frac{y^3}{3} \right]_{-1}^{3} = \frac{92}{3} \text{ sq. un.}$$

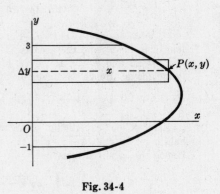

**Fig. 34-4**

**4.** Find the area bounded by the parabola $y = x^2 - 7x + 6$, the $x$-axis, and the lines $x = 2$ and $x = 6$.

For the approximating rectangle shown in Fig. 34-5, the width is $\Delta x$, the height is $-y = -(x^2 - 7x + 6)$, and the area is $-(x^2 - 7x + 6)\,\Delta x$. The required area is then

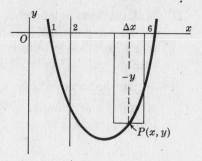

$$A = \int_2^6 -(x^2 - 7x + 6)\,dx = -\left(\frac{x^3}{3} - \frac{7x^2}{2} + 6x\right)\Big]_2^6$$

$$= \frac{56}{3} \text{ square units}$$

**Fig. 34-5**

**5.** Find the area between the curve $y = x^3 - 6x^2 + 8x$ and the $x$-axis.

The curve crosses the $x$-axis at $x = 0$, $x = 2$, and $x = 4$ as in Fig. 34-6. Using vertical slices, the area of the approximating rectangle with base on the interval $0 < x < 2$ is $(x^3 - 6x^2 + 8x)\,\Delta x$, and the area of the portion lying above the $x$-axis is given by $\int_0^2 (x^3 - 6x^2 + 8x)\,dx$. The area of the approximating rectangle with base on the interval $2 < x < 4$ is $-(x^3 - 6x^2 + 8x)\,\Delta x$, and the area of the portion lying below the $x$-axis is given by $\int_2^4 -(x^3 - 6x^2 + 8x)\,dx$. The required area is, therefore,

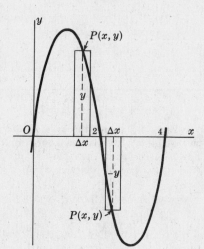

$$A = \int_0^2 (x^3 - 6x^2 + 8x)\,dx + \int_2^4 -(x^3 - 6x^2 + 8x)\,dx$$

$$= \left[\frac{x^4}{4} - 2x^3 + 4x^2\right]_0^2 - \left[\frac{x^4}{4} - 2x^3 + 4x^2\right]_2^4$$

$$= 4 + 4 = 8 \text{ square units}$$

The use of two definite integrals is necessary here since the integrand changes sign on the interval of integration. Failure to note this would have resulted in the incorrect integral $\int_0^4 (x^3 - 6x^2 + 8x)\,dx = 0$.

**Fig. 34-6**

**6.** Find the area bounded by the parabola $x = 4 - y^2$ and the $y$-axis.

The parabola crosses the $x$-axis at the point $(4, 0)$ and the $y$-axis at the points $(0, 2)$ and $(0, -2)$. We shall give two solutions.

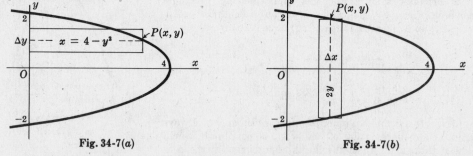

**Fig. 34-7(a)**                                      **Fig. 34-7(b)**

*Using horizontal slicing.* For the approximating rectangle of Fig. 34-7(a), the width is $\Delta y$, the length is $4 - y^2$, and the area is $(4 - y^2)\Delta y$. The limits of integration of the resulting definite integral are $y = -2$ and $y = 2$. However, it will be noted that the area lying below the $x$-axis is equal to that lying above. Hence we have for the required area

$$\int_{-2}^2 (4 - y^2)\,dy = 2\int_0^2 (4 - y^2)\,dy = 2\left(4y - \frac{y^3}{3}\right)\Big]_0^2 = \frac{32}{3} \text{ square units}$$

*Using vertical slicing.* For the approximating rectangle of Fig. 34-7(b), the width is $\Delta x$, the height is $2y = 2\sqrt{4 - x}$, and the area is $2\sqrt{4 - x}\,\Delta x$. The limits of integration are $x = 0$ and $x = 4$. Hence the required area is

$$\int_0^4 2\sqrt{4 - x}\,dx = -\frac{4}{3}(4 - x)^{3/2}\Big]_0^4 = \frac{32}{3} \text{ square units}$$

7.   Find the area bounded by the parabola $y^2 = 4x$ and the line $y = 2x - 4$.

The line intersects the parabola at the points $(1, -2)$ and $(4, 4)$. It will be seen from the two figures below that when vertical slicing is used certain strips run from the line to the parabola and others from one branch of the parabola to the other branch, while when horizontal slicing is used each strip runs from the parabola to the line. We give both solutions here to show the superiority of one over the other and to indicate that both methods of slicing should be considered before beginning to set up a definite integral.

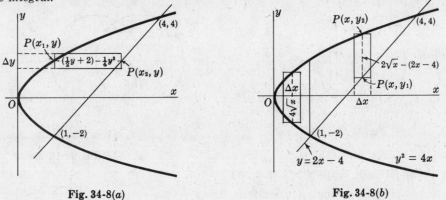

Fig. 34-8(a)                                              Fig. 34-8(b)

*Using horizontal slicing.*  See Fig. 34-8(a).  For the approximating rectangle of Fig. 34-8(a), the width is $\Delta y$, the length is   {(value of $x$ of the line) $-$ (value of $x$ of the parabola)} $= (\frac{1}{2}y + 2) - \frac{1}{4}y^2 = 2 + \frac{1}{2}y - \frac{1}{4}y^2$,  and the area is  $(2 + \frac{1}{2}y - \frac{1}{4}y^2)\Delta y$.   The required area is

$$\int_{-2}^{4} (2 + \tfrac{1}{2}y - \tfrac{1}{4}y^2)\, dy \;=\; \left[ 2y + \frac{y^2}{4} - \frac{y^3}{12} \right]_{-2}^{4} \;=\; 9 \text{ square units}$$

*Using vertical slicing.*  See Fig. 34-8(b).  Divide the area by the line $x = 1$.  For the approximating rectangle to the left of this line, the width is $\Delta x$, the height (making use of symmetry) is $2y = 4\sqrt{x}$, and the area is $4\sqrt{x}\, \Delta x$.  For the approximating rectangle to the right, the width is $\Delta x$, the height is $2\sqrt{x} - (2x - 4) = 2\sqrt{x} - 2x + 4$,  and the area is  $(2\sqrt{x} - 2x + 4)\Delta x$.  The required area is

$$\int_{0}^{1} 4\sqrt{x}\, dx + \int_{1}^{4} (2\sqrt{x} - 2x + 4)\, dx \;=\; \left[ \frac{8}{3}x^{3/2} \right]_{0}^{1} + \left[ \frac{4}{3}x^{3/2} - x^2 + 4x \right]_{1}^{4} \;=\; \frac{8}{3} + \frac{19}{3} \;=\; 9 \text{ sq. un.}$$

8.   Find the area bounded by the parabolas $y = 6x - x^2$ and $y = x^2 - 2x$.

The parabolas intersect at the points $(0, 0)$ and $(4, 8)$. It is readily seen that vertical slicing will yield the simpler solution.

For the approximating rectangle, the width is $\Delta x$, the height is {(value of $y$ of the upper boundary) $-$ (value of $y$ of the lower boundary)} $= (6x - x^2) - (x^2 - 2x) = 8x - 2x^2$,  and the area is $(8x - 2x^2)\, \Delta x$.  The required area is

$$\int_{0}^{4} (8x - 2x^2)\, dx \;=\; \left[ 4x^2 - \frac{2}{3}x^3 \right]_{0}^{4} \;=\; \frac{64}{3} \text{ square units}$$

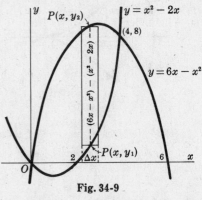

Fig. 34-9

9.   Find the area enclosed by the curve $y^2 = x^2 - x^4$.

The curve is symmetric with respect to the coordinate axes.  Hence the required area is 4 times the portion lying in the first quadrant.

For the approximating rectangle, the width is $\Delta x$, the height is  $y = \sqrt{x^2 - x^4} = x\sqrt{1 - x^2}$,  and the area is $x\sqrt{1 - x^2}\, \Delta x$.  Hence the required area is

$$4 \int_{0}^{1} x\sqrt{1 - x^2}\, dx \;=\; \left[ -\frac{4}{3}(1 - x^2)^{3/2} \right]_{0}^{1}$$

$$=\; \frac{4}{3} \text{ square units}$$

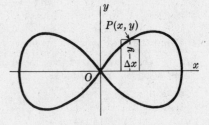

Fig. 34-10

10. Find the smaller area cut from the circle $x^2 + y^2 = 25$ by the line $x = 3$. See Fig. 34-11 below.

$$A = \int_3^5 2y\,dx = 2\int_3^5 \sqrt{25 - x^2}\,dx = 2\left[\frac{x}{2}\sqrt{25 - x^2} + \frac{25}{2}\arcsin\frac{x}{5}\right]_3^5$$

$$= \left(\frac{25}{2}\pi - 12 - 25\arcsin\frac{3}{5}\right)\text{ square units}$$

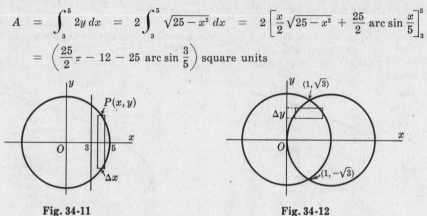

Fig. 34-11                Fig. 34-12

11. Find the area common to the circles $x^2 + y^2 = 4$ and $x^2 + y^2 = 4x$. See Fig. 34-12 above.

The circles intersect in the points $(1, \pm\sqrt{3})$.

The approximating rectangle extends from $x = 2 - \sqrt{4 - y^2}$ to $x = \sqrt{4 - y^2}$.

$$A = 2\int_0^{\sqrt{3}} \{\sqrt{4 - y^2} - (2 - \sqrt{4 - y^2})\}\,dy = 4\int_0^{\sqrt{3}} (\sqrt{4 - y^2} - 1)\,dy$$

$$= 4\left[\frac{y}{2}\sqrt{4 - y^2} + 2\arcsin\frac{1}{2}y - y\right]_0^{\sqrt{3}} = \left(\frac{8\pi}{3} - 2\sqrt{3}\right)\text{ square units}$$

12. Find the area of the loop of the curve $y^2 = x^4(4 + x)$. See Fig. 34-13 below.

$$A = \int_{-4}^0 2y\,dx = 2\int_{-4}^0 x^2\sqrt{4 + x}\,dx. \quad \text{Let } 4 + x = z^2; \quad \text{then}$$

$$A = 4\int_0^2 (z^2 - 4)^2 z^2\,dz = 4\left[\frac{z^7}{7} - \frac{8z^5}{5} + \frac{16z^3}{3}\right]_0^2 = \frac{4096}{105}\text{ square units}$$

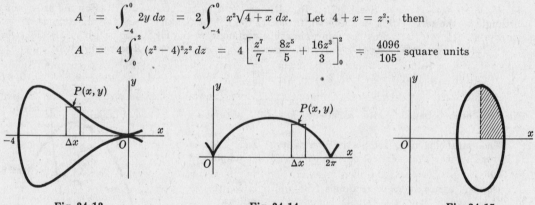

Fig. 34-13                Fig. 34-14                Fig. 34-15

13. Find the area of an arch of the cycloid $x = \theta - \sin\theta$, $y = 1 - \cos\theta$. See Fig. 34-14 above.

An arch is described as $\theta$ varies from 0 to $2\pi$. Then $dx = (1 - \cos\theta)\,d\theta$ and

$$A = \int_{\theta=0}^{\theta=2\pi} y\,dx = \int_0^{2\pi} (1 - \cos\theta)(1 - \cos\theta)\,d\theta = \int_0^{2\pi}\left(\frac{3}{2} - 2\cos\theta + \frac{1}{2}\cos 2\theta\right)d\theta$$

$$= \left[\frac{3}{2}\theta - 2\sin\theta + \frac{1}{4}\sin 2\theta\right]_0^{2\pi} = 3\pi\text{ square units}$$

14. Find the area bounded by the curve $x = 3 + \cos\theta$, $y = 4\sin\theta$. See Fig. 34-15 above.

The boundary of the shaded area in the figure, ($\frac{1}{4}$ the required area), is described from *right* to *left* as $\theta$ varies from 0 to $\frac{1}{2}\pi$.

$$A = -4\int_0^{\pi/2} (4\sin\theta)(-\sin\theta)\,d\theta = 16\int_0^{\pi/2}\sin^2\theta\,d\theta = 8\int_0^{\pi/2}(1 - \cos 2\theta)\,d\theta$$

$$= 8\left[\theta - \frac{1}{2}\sin 2\theta\right]_0^{\pi/2} = 4\pi\text{ square units}$$

# Supplementary Problems

15. Find the area bounded as follows:

(a) $y = x^2$, $y = 0$, $x = 2$, $x = 5$

(b) $y = x^3$, $y = 0$, $x = 1$, $x = 3$

(c) $y = 4x - x^2$, $y = 0$, $x = 1$, $x = 3$

(d) $x = 1 + y^2$, $x = 10$

(e) $x = 3y^2 - 9$, $x = 0$, $y = 0$, $y = 1$

(f) $x = y^2 + 4y$, $x = 0$

(g) $y = 9 - x^2$, $y = x + 3$

(h) $y = 2 - x^2$, $y = -x$

(i) $y = x^2 - 4$, $y = 8 - 2x^2$

(j) $y = x^4 - 4x^2$, $y = 4x^2$

(k) A loop of $y^2 = x^2(a^2 - x^2)$

(l) The loop of $9ay^2 = x(3a - x)^2$

(m) $y = e^x$, $y = e^{-x}$, $x = 0$, $x = 2$

(n) $y = e^{x/a} + e^{-x/a}$, $y = 0$, $x = \pm a$

(o) $xy = 12$, $y = 0$, $x = 1$, $x = e^2$

(p) $y = 1/(1 + x^2)$, $y = 0$, $x = \pm 1$

(q) $y = \tan x$, $x = 0$, $x = \frac{1}{4}\pi$

(r) A circular sector of radius $r$ and angle $\alpha$.

(s) The ellipse $x = a \cos t$, $y = b \sin t$.

(t) $x = 2 \cos \theta - \cos 2\theta - 1$, $y = 2 \sin \theta - \sin 2\theta$.

(u) $x = a \cos^3 t$, $y = a \sin^3 t$.

(v) First arch of $y = e^{-ax} \sin ax$.

(w) $y = xe^{-x^2}$, $y = 0$, and the maximum ordinate.

(x) The two branches of $(2x - y)^2 = x^3$ and $x = 4$.

(y) Within $y = 25 - x^2$, $256x = 3y^2$, $16y = 9x^2$.

*Answers:* (a) 39 square units, (b) 20, (c) 22/3, (d) 36, (e) 8, (f) 32/3, (g) 125/6, (h) 9/2, (i) 32, (j) $512\sqrt{2}/15$, (k) $2a^3/3$, (l) $8\sqrt{3}a^2/5$, (m) $(e^2 + 1/e^2 - 2)$, (n) $2a(e - 1/e)$, (o) 24, (p) $\frac{1}{2}\pi$, (q) $\frac{1}{2}\ln 2$, (r) $\frac{1}{2}r^2\alpha$, (s) $\pi ab$, (t) $6\pi$, (u) $3\pi a^2/8$, (v) $(1 + 1/e^\pi)/2a$, (w) $\frac{1}{2}(1 - 1/\sqrt{e})$, (x) 128/5, (y) 98/3 square units

By the *average ordinate* of the curve $y = f(x)$ over the interval $a \leqq x \leqq b$ is meant the quantity

$$\frac{\text{Area}}{\text{Base}} = \frac{\int_a^b f(x)\, dx}{b - a}$$

16. Find the average ordinate (a) of a semicircle, (b) of the parabola $y = 4 - x^2$ from $x = -2$ to $x = 2$.
*Ans.* (a) $\pi r/4$, (b) 8/3

17. (a) Find the average ordinate of an arch of the cycloid $x = a(\theta - \sin \theta)$, $y = a(1 - \cos \theta)$ with respect to $x$.

(b) Same, with respect to $\theta$.

*Ans.* (a) $\dfrac{1}{2\pi a} \displaystyle\int_0^{2\pi} a^2(1 - \cos \theta)^2\, d\theta = \dfrac{3a}{2}$,   (b) $\dfrac{1}{2\pi} \displaystyle\int_0^{2\pi} a(1 - \cos \theta)\, d\theta = a$

18. For a freely falling body, $s = \frac{1}{2}gt^2$ and $v = gt = \sqrt{2gs}$.

(a) Show that the average value of $v$ with respect to $t$ for the interval $0 \leqq t \leqq t_1$ is one-half the final velocity.

(b) Show that the average value of $v$ with respect to $s$ for the interval $0 \leqq s \leqq s_1$ is two-thirds the final velocity.

# Chapter 35

## Volumes of Solids of Revolution

**A SOLID OF REVOLUTION** is generated by revolving a plane area about a line, called the axis of revolution, in the plane. The *volume* of a solid of revolution may be found by using one of the following procedures.

### DISC METHOD

A.  The axis of rotation is a part of the boundary of the plane area.
    (1) Make a sketch showing the area involved, a representative strip perpendicular to the axis of rotation, and the approximating rectangle as in the preceding chapter.
    (2) Write the volume of the disc (cylinder) generated when the approximating rectangle is revolved about the axis of rotation and sum for the $n$ rectangles.
    (3) Assume the number of rectangles to be indefinitely increased and apply the Fundamental Theorem.

    See Problems 1-2.

B.  The axis of rotation is not a part of the boundary of the plane area.
    (1) As in (1) above.
    (2) Extend the sides of the approximating rectangle $ABCD$ to meet the axis of rotation in $E$ and $F$ as in Fig. 35-3 of Problem 3 below. When the approximating rectangle is revolved about the axis of rotation, a washer is formed whose volume is the difference between the volumes generated by revolving the rectangles $EABF$ and $ECDF$ about the axis. Write the difference of the two volumes and proceed as in (2) above.
    (3) As in (3) above.

    See Problems 3-4.

### SHELL METHOD

(1) Make a sketch showing the area involved, a representative strip parallel to the axis of rotation, and the approximating rectangle.
(2) Write the volume (= mean circumference × height × thickness) of the cylindrical shell generated when the approximating rectangle is revolved about the axis of rotation and sum for the $n$ rectangles.
(3) Assume the number of rectangles to be indefinitely increased and apply the Fundamental Theorem.

See Problems 5-8.

## Solved Problems

1.  Find the volume generated by revolving the first quadrant area bounded by the parabola $y^2 = 8x$ and its latus rectum ($x = 2$) about the $x$-axis.

Refer to Fig. 35-1 below. Divide the area by vertical slicing. When the approximating rectangle of Fig. 35-1 is revolved about the x-axis, a disc whose radius is $y$, whose height is $\Delta x$, and whose volume is $\pi y^2 \, \Delta x$ is generated. The sum of the volumes of the $n$ discs, corresponding to the $n$ approximating rectangles, is $\Sigma \pi y^2 \, \Delta x$ and the required volume is given by

$$V = \int_a^b dV = \int_0^2 \pi y^2 \, dx = \pi \int_0^2 8x \, dx = 4\pi x^2 \bigg]_0^2 = 16\pi \text{ cubic units}$$

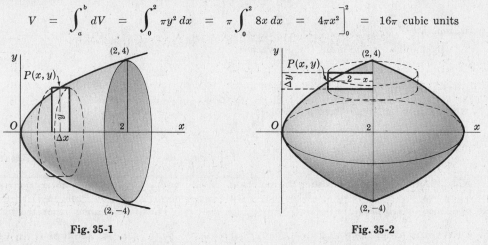

Fig. 35-1                           Fig. 35-2

2.  Find the volume generated by revolving the area bounded by the parabola $y^2 = 8x$ and its latus rectum $(x = 2)$ about the latus rectum.

Refer to Fig. 35-2 above. Divide the area by horizontal slicing. When the approximating rectangle of Fig. 35-2 is revolved about the latus rectum, it generates a disc whose radius is $2 - x$, whose height is $\Delta y$, and whose volume is $\pi(2 - x)^2 \, \Delta y$. The required volume is then

$$V = \int_{-4}^4 \pi(2-x)^2 \, dy = 2\pi \int_0^4 (2-x)^2 \, dy = 2\pi \int_0^4 \left(2 - \frac{y^2}{8}\right)^2 dy = \frac{256}{15}\pi \text{ cubic units}$$

3.  Find the volume generated by revolving the area bounded by the parabola $y^2 = 8x$ and its latus rectum $(x = 2)$ about the y-axis.

Refer to Fig. 35-3 below. Divide the area by horizontal slicing. When the approximating rectangle of Fig. 35-3 is revolved about the y-axis, it generates a washer whose volume is the difference between the volumes generated by revolving the rectangle $ECDF$ (of dimensions 2 by $\Delta y$) and the rectangle $EABF$ (of dimensions $x$ by $\Delta y$) about the y-axis, i.e., $\pi(2)^2 \, \Delta y - \pi(x)^2 \, \Delta y$. The required volume is then

$$V = \int_{-4}^4 4\pi \, dy - \int_{-4}^4 \pi x^2 \, dy = 2\pi \int_0^4 (4 - x^2) \, dy = 2\pi \int_0^4 \left(4 - \frac{y^4}{64}\right) dy = \frac{128}{5}\pi \text{ cubic units}$$

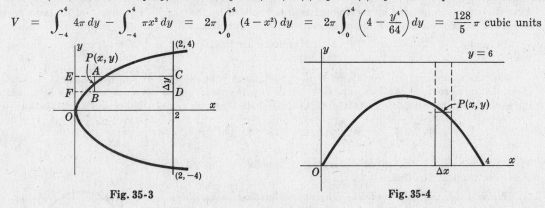

Fig. 35-3                           Fig. 35-4

4.  Find the volume generated by revolving the area cut off from the parabola $y = 4x - x^2$ by the x-axis about the line $y = 6$.

Refer to Fig. 35-4 above. Divide the area by vertical slicing. The solid generated by revolving the approximating rectangle about the line $y = 6$ is a washer, whose volume is $\pi(6)^2 \, \Delta x - \pi(6-y)^2 \, \Delta x$. The required volume is then given by

$$V = \pi \int_0^4 \{(6)^2 - (6-y)^2\} \, dx = \pi \int_0^4 (12y - y^2) \, dx$$

$$= \pi \int_0^4 (48x - 28x^2 + 8x^3 - x^4) \, dx = \frac{1408\pi}{15} \text{ cubic units}$$

5.  Refer to Fig. 35-5 below. Suppose the volume in question is generated by revolving about the $y$-axis
the first quadrant area under the curve $y = f(x)$ from $x = a$ to $x = b$. Let this area be divided into $n$
strips and each strip approximated by a rectangle. When the representative rectangle is revolved
about the $y$-axis a cylindrical shell of height $y_k$, inner radius $\xi_{k-1}$, outer radius $\xi_k$, and volume

(i)
$$\Delta_k V = \pi(\xi_k^2 - \xi_{k-1}^2)y_k$$

is generated.

By the law of the mean for derivatives,

(ii)
$$\xi_k^2 - \xi_{k-1}^2 = \left.\frac{d}{dx}(x^2)\right|_{x=x_k'} \cdot (\xi_k - \xi_{k-1}) = 2x_k'\Delta_k x$$

where $\xi_{k-1} < x_k' < \xi_k$. Then (i) becomes
$$\Delta_k V = 2\pi\, x_k'\, y_k\, \Delta_k x = 2\pi\, x_k'\, f(x_k)\, \Delta_k x$$

and
$$V = 2\pi \lim_{n \to +\infty} \sum_{k=1}^n x_k'\, f(x_k)\, \Delta_k x = 2\pi \int_a^b x\, f(x)\, dx \qquad \text{by the Theorem of Bliss}$$

*Note.* If the policy of choosing the points $x_k$ as the midpoints of the subintervals, used in the
preceding chapter, is followed, the Theorem of Bliss is not needed. For, by Problem 17($b$), Chapter 21,
the $x_k'$ defined by (ii) above is then $x_k' = \frac{1}{2}(\xi_k + \xi_{k-1}) = x_k$. Thus, the volume generated by revolving

the $n$ rectangles about the $y$-axis is $\displaystyle\sum_{k=1}^n 2\pi x_k\, f(x_k)\, \Delta_k x = \sum_{k=1}^n g(x_k)\, \Delta_k x$ of the type (i) in Chapter 33.

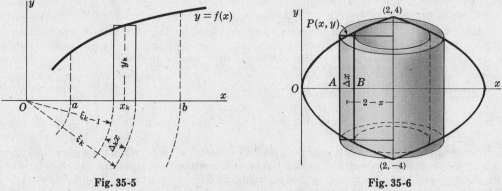

Fig. 35-5                                             Fig. 35-6

6.  Find the volume generated by revolving the area bounded by the parabola $y^2 = 8x$ and its latus rectum
about the latus rectum. Use the shell method. (See Problem 2.)

Refer to Fig. 35-6 above. Divide the area by vertical slicing and, for convenience, choose the
point $P$ so that $x$ is the midpoint of the segment $AB$.

For the approximating rectangle of Fig. 35-6 the height is $2y = 4\sqrt{2x}$, the width is $\Delta x$ and its
mean distance from the latus rectum is $2 - x$. When the rectangle is revolved about the latus rectum
the volume of the cylindrical shell generated is $2\pi(2 - x) \cdot 4\sqrt{2x}\,\Delta x$. The required volume is then

$$V = 8\sqrt{2}\,\pi \int_0^2 (2 - x)\sqrt{x}\, dx = 8\sqrt{2}\,\pi \int_0^2 (2x^{1/2} - x^{3/2})\, dx = \frac{256\pi}{15} \text{ cubic units}$$

7.  Find the volume of the torus generated by revolving the circle
$x^2 + y^2 = 4$ about the line $x = 3$.

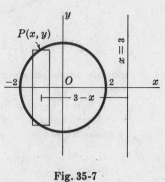

We will use the shell method. The approximating rectangle is
of height $2y$, thickness $\Delta x$, and mean distance from the axis of
revolution $3 - x$. The required volume is then

$$V = 2\pi \int_{-2}^2 2y(3 - x)\, dx = 4\pi \int_{-2}^2 (3 - x)\sqrt{4 - x^2}\, dx$$

$$= 12\pi \int_{-2}^2 \sqrt{4 - x^2}\, dx - 4\pi \int_{-2}^2 x\sqrt{4 - x^2}\, dx$$

$$= \left[12\pi\left(\frac{x}{2}\sqrt{4 - x^2} + 2\arcsin\frac{x}{2}\right) + \frac{4\pi}{3}(4 - x^2)^{3/2}\right]_{-2}^2$$

Fig. 35-7

$$= 24\pi^2 \text{ cubic units}$$

8.　Find the volume of the solid generated by revolving about the $y$-axis the area between the first arch of the cycloid $x = \theta - \sin\theta$, $y = 1 - \cos\theta$ and the $x$-axis. Use the shell method.

$$V = 2\pi \int_{\theta=0}^{\theta=2\pi} xy\,dx = 2\pi \int_{0}^{2\pi} (\theta - \sin\theta)(1 - \cos\theta)(1 - \cos\theta)\,d\theta$$

$$= 2\pi \int_{0}^{2\pi} (\theta - 2\theta\cos\theta + \theta\cos^2\theta - \sin\theta + 2\sin\theta\cos\theta$$
$$- \cos^2\theta\sin\theta)\,d\theta$$

$$= 2\pi \left[ \tfrac{3}{4}\theta^2 - 2(\theta\sin\theta + \cos\theta) + \tfrac{1}{2}(\tfrac{1}{2}\theta\sin 2\theta + \tfrac{1}{4}\cos 2\theta) \right.$$
$$\left. + \cos\theta + \sin^2\theta + \tfrac{1}{3}\cos^3\theta \right]_{0}^{2\pi} = 6\pi^3 \text{ cu. un.}$$

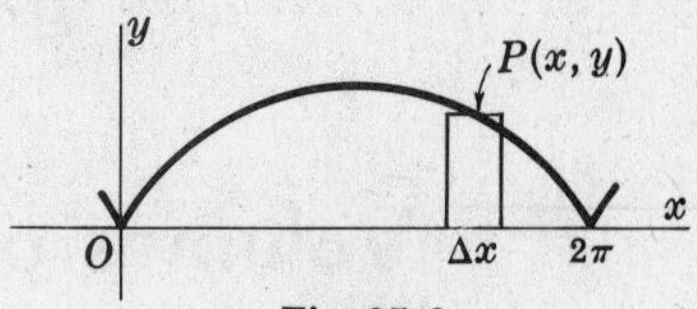

**Fig. 35-8**

9.　Find the volume when the plane area bounded by $y = -x^2 - 3x + 6$ and $x + y - 3 = 0$ is revolved (a) about $x = 3$, (b) about $y = 0$.

(a)　$V = 2\pi \int_{-3}^{1} (y_C - y_L)(3 - x)\,dx$

$$= 2\pi \int_{-3}^{1} (x^3 - x^2 - 9x + 9)\,dx = 256\pi/3 \text{ cubic units}$$

(b)　$V = \pi \int_{-3}^{1} \{(y_C)^2 - (y_L)^2\}\,dx$

$$= \pi \int_{-3}^{1} (x^4 + 6x^3 - 4x^2 - 30x + 27)\,dx = 1792\pi/15 \text{ cu. un.}$$

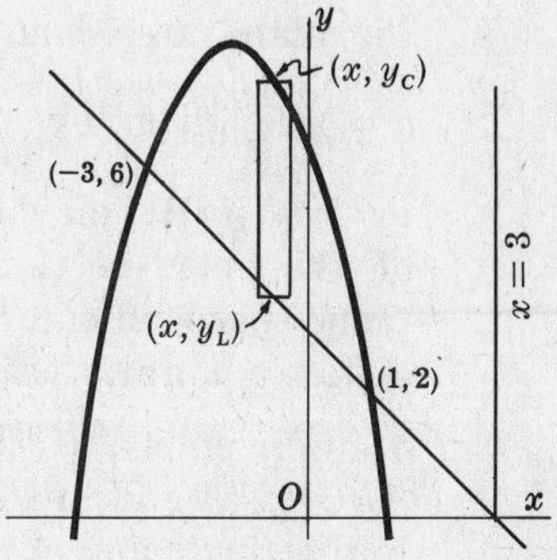

**Fig. 35-9**

# Supplementary Problems

In Problems 10-19 find the volume generated by revolving the given plane area about the given line, using the disc method $A$. (Answers are in cubic units.)

10.　$y = 2x^2$, $y = 0$, $x = 0$, $x = 5$; $x$-axis　*Ans.* $2500\pi$
11.　$x^2 - y^2 = 16$, $y = 0$, $x = 8$; $x$-axis　*Ans.* $256\pi/3$
12.　$y = 4x^2$, $x = 0$, $y = 16$; $y$-axis　*Ans.* $32\pi$
13.　$y = 4x^2$, $x = 0$, $y = 16$; $y = 16$　*Ans.* $4096\pi/15$
14.　$y^2 = x^3$, $y = 0$, $x = 2$; $x$-axis　*Ans.* $4\pi$

15.　$y = x^3$, $y = 0$, $x = 2$; $x = 2$　*Ans.* $16\pi/5$
16.　$y^2 = x^4(1 - x^2)$; $x$-axis　*Ans.* $4\pi/35$
17.　$4x^2 + 9y^2 = 36$; $x$-axis　*Ans.* $16\pi$
18.　$4x^2 + 9y^2 = 36$, $y$-axis　*Ans.* $24\pi$
19.　Within $x = 9 - y^2$, between
　　$x - y - 7 = 0$, $x = 0$; $y$-axis　*Ans.* $963\pi/5$

In Problems 20-26 find the volume generated by revolving the given plane area about the given line, using the disc method $B$. (Answers are in cubic units.)

20.　$y = 2x^2$, $y = 0$, $x = 0$, $x = 5$; $y$-axis　*Ans.* $625\pi$
21.　$x^2 - y^2 = 16$, $y = 0$, $x = 8$; $y$-axis　*Ans.* $128\sqrt{3}\,\pi$
22.　$y = 4x^2$, $x = 0$, $y = 16$; $x$-axis　*Ans.* $2048\pi/5$
23.　$y = x^3$, $x = 0$, $y = 8$; $x = 2$　*Ans.* $144\pi/5$

24.　$y = x^2$, $y = 4x - x^2$; $x$-axis　*Ans.* $32\pi/3$
25.　$y = x^2$, $y = 4x - x^2$; $y = 6$　*Ans.* $64\pi/3$
26.　$x = 9 - y^2$, $x - y - 7 = 0$; $x = 4$　*Ans.* $153\pi/5$

In Problems 27-32 find the volume generated by revolving the given plane area about the given line, using the shell method. (Answers are in cubic units.)

27.　$y = 2x^2$, $y = 0$, $x = 0$, $x = 5$; $y$-axis
28.　$y = 2x^2$, $y = 0$, $x = 0$, $x = 5$; $x = 6$
29.　$y = x^3$, $y = 0$, $x = 2$; $y = 8$

30.　$y = x^2$, $y = 4x - x^2$; $x = 5$
31.　$y = x^2 - 5x + 6$, $y = 0$; $y$-axis
32.　Within $x = 9 - y^2$, between $x - y - 7 = 0$, $x = 0$; $y = 3$

*Ans.* (27) $625\pi$, (28) $375\pi$, (29) $320\pi/7$, (30) $64\pi/3$, (31) $5\pi/6$, (32) $369\pi/2$

In Problems 33-39 find the volume generated by revolving the given plane area about the given line, using any appropriate method.

33.　$y = e^{-x^2}$, $y = 0$, $x = 0$, $x = 1$; $y$-axis　　　*Ans.* $\pi(1 - 1/e)$ cubic units
34.　An arch of $y = \sin 2x$; $x$-axis　　　*Ans.* $\tfrac{1}{4}\pi^2$ cubic units
35.　First arch of $y = e^x \sin x$; $x$-axis　　　*Ans.* $\pi(e^{2\pi} - 1)/8$ cubic units
36.　First arch of $y = e^x \sin x$; $y$-axis　　　*Ans.* $\pi[(\pi - 1)e^\pi - 1]$ cubic units
37.　First arch of $x = \theta - \sin\theta$, $y = 1 - \cos\theta$; $x$-axis　　　*Ans.* $5\pi^2$ cubic units
38.　The cardioid $x = 2\cos\theta - \cos 2\theta - 1$, $y = 2\sin\theta - \sin 2\theta$; $x$-axis　　　*Ans.* $64\pi/3$ cubic units
39.　$y = 2x^2$, $2x - y + 4 = 0$; $x = 2$　　　*Ans.* $27\pi$ cubic units
40.　Obtain the volume of the frustum of a cone whose lower base is of radius $R$, upper base is of radius $r$, and altitude is $h$.　　　*Ans.* $\tfrac{1}{3}\pi h(r^2 + rR + R^2)$ cubic units

# Chapter 36

## Volumes of Solids with Known Cross Sections

**THE VOLUME OF A SOLID OF REVOLUTION** generated by revolving about the $x$-axis the plane area bounded by the curve $y = f(x)$, the $x$-axis, and the lines $x = a$ and

$x = b$ is given by $\displaystyle\int_a^b \pi y^2\, dx$. The integrand $\pi y^2 = \pi \{f(x)\}^2$ may be interpreted as the area of the cross section of the solid made by a plane perpendicular to the $x$-axis and at a distance $x$ units from the origin.

Conversely, if the area of a cross section $ABC$ of a solid, made by a plane perpendicular to the $x$-axis and at a distance $x$ from the origin can be expressed as a function $A(x)$ of $x$, the

volume of the solid is given by $\displaystyle V = \int_\alpha^\beta A(x)\, dx$.

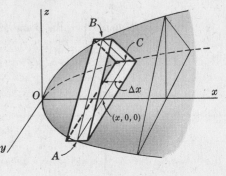

Fig. 36-1

## Solved Problems

1. A solid has a circular base of radius 4 units. Find the volume of the solid if every plane section perpendicular to a fixed diameter is an equilateral triangle.

Take the circle as in Fig. 36-2, with the $x$-axis as the fixed diameter. The equation of the circle is $x^2 + y^2 = 16$. The cross section $ABC$ of the solid is an equilateral triangle of side $2y$ and area $A(x) = \sqrt{3}\, y^2 = \sqrt{3}\,(16 - x^2)$.

$$V = \int_\alpha^\beta A(x)\, dx = \sqrt{3} \int_{-4}^4 (16 - x^2)\, dx$$

$$= \sqrt{3} \left[ 16x - \frac{x^3}{3} \right]_{-4}^4 = \frac{256}{3}\sqrt{3} \text{ cubic units}$$

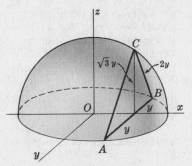

Fig. 36-2

2. A solid has a base in the form of an ellipse with major axis 10 and minor axis 8. Find the volume if every section perpendicular to the major axis is an isosceles triangle with altitude 6.

Take the ellipse as in Fig. 36-3, with equation $\dfrac{x^2}{25} + \dfrac{y^2}{16} = 1$. The section $ABC$ is an isosceles triangle of base $2y$, altitude 6, and area $A(x) = 6y = 6 \cdot \frac{4}{5}\sqrt{25 - x^2}$.

$$V = \frac{24}{5} \int_{-5}^5 \sqrt{25 - x^2}\, dx = 60\pi \text{ cubic units}$$

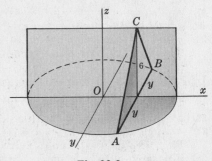

Fig. 36-3

180

3.   Find the volume of the solid cut from the paraboloid $\dfrac{x^2}{16} + \dfrac{y^2}{25} = z$ by the plane $z = 10$.

Refer to Fig. 36-4 below.  The section of the solid by a plane parallel to the plane $xOy$ and at a distance $z$ from the origin, is an ellipse of area $\pi xy = \pi(4\sqrt{z})(5\sqrt{z}) = 20\pi z$.  Hence

$$V = 20\pi \int_0^{10} z\,dz = 1000\pi \text{ cubic units}$$

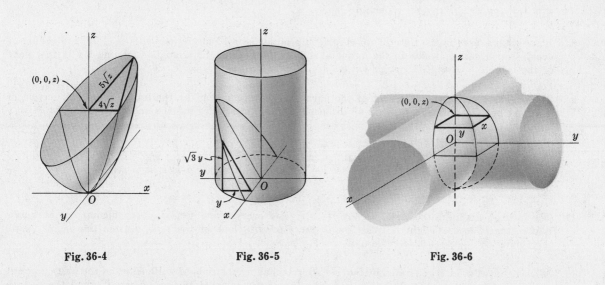

Fig. 36-4                    Fig. 36-5                    Fig. 36-6

4.   Two cuts are made on a circular log of radius 8 inches, the first perpendicular to the axis of the log and the second inclined at an angle of $60°$ with the first.  If the two cuts meet on a line through the center, find the volume of the wood cut out.

Refer to Fig. 36-5 above.  Take the origin at the center of the log, the $x$-axis along the intersection of the two cuts, and the positive side of the $y$-axis in the face of the first cut.  A section of the cut made by a plane perpendicular to the $x$-axis is a right triangle having one angle of $60°$ and the adjacent leg of length $y$.  The other leg is of length $\sqrt{3}\,y$ and the area of the section is $\frac{1}{2}\sqrt{3}\,y^2 = \frac{1}{2}\sqrt{3}\,(64 - x^2)$.  Then

$$V = \frac{1}{2}\sqrt{3}\int_{-8}^{8} (64 - x^2)\,dx = \frac{1024}{3}\sqrt{3} \text{ in}^3$$

5.   The axes of two circular cylinders of equal radii $r$ intersect at right angles.  Find their common volume.

Refer to Fig. 36-6 above.  Let the cylinders have equations $x^2 + z^2 = r^2$ and $y^2 + z^2 = r^2$.  A section of the solid whose volume is required by a plane perpendicular to the $z$-axis is a square of side $2x = 2y = 2\sqrt{r^2 - z^2}$ and area $4(r^2 - z^2)$.  Hence

$$V = 4\int_{-r}^{r} (r^2 - z^2)\,dz = \frac{16r^3}{3} \text{ cubic units}$$

6.   Find the volume of the right cone of height $h$ whose base is an ellipse of major axis $2a$ and minor axis $2b$.

Refer to Fig. 36-7.  A section of the cone by a plane parallel to the base is an ellipse whose major axis is $2x$ and minor axis $2y$.

From similar triangles of the adjacent Fig. 36-7,

$$\dfrac{PC}{OA} = \dfrac{PM}{OM} \text{ or } \dfrac{x}{a} = \dfrac{h-z}{h}; \text{ also, } \dfrac{PD}{OB} = \dfrac{PM}{OM} \text{ or } \dfrac{y}{b} = \dfrac{h-z}{h}$$

The area of the section is $\pi xy = \dfrac{\pi ab(h-z)^2}{h^2}$.  Hence

$$V = \dfrac{\pi ab}{h^2}\int_0^h (h-z)^2\,dz = \dfrac{1}{3}\pi abh \text{ cubic units}$$

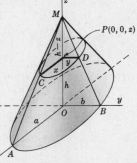

Fig. 36-7

# Supplementary Problems

7. A solid has a circular base of radius 4 units. Find the volume of the solid if every plane perpendicular to a fixed diameter ($x$-axis of the figure of Problem 1) is (a) a semicircle, (b) a square, (c) an isosceles right triangle with the hypotenuse in the plane of the base.
   *Ans.* (a) $128\pi/3$, (b) 1024/3, (c) 256/3 cubic units

8. A solid has a base in the form of an ellipse with major axis 10 and minor axis 8. Find the volume if every section perpendicular to the major axis is an isosceles right triangle with one leg in the plane of the base.    *Ans.* 640/3 cubic units

9. The base of a solid is the segment of the parabola $y^2 = 12x$ cut off by the latus rectum. A section of the solid perpendicular to the axis of the parabola is a square. Find the volume.
   *Ans.* 216 cubic units

10. The base of a solid is the first quadrant area bounded by the line $4x + 5y = 20$ and the coordinate axes. Find the volume if every plane section perpendicular to the $x$-axis is a semicircle.
    *Ans.* $10\pi/3$ cubic units

11. The base of a solid is the circle $x^2 + y^2 = 16x$, and every plane section perpendicular to the $x$-axis is a rectangle whose height is twice the distance of the plane of the section from the origin. Find the volume.    *Ans.* $1024\pi$ cubic units

12. A horn shaped solid is generated by moving a circle, having the ends of a diameter on the first quadrant arcs of the parabolas $y^2 + 8x = 64$ and $y^2 + 16x = 64$, parallel to the $xz$ plane. Find the volume generated.    *Ans.* $256\pi/15$ cubic units

13. The vertex of a cone is at $(a, 0, 0)$ and its base is the circle $y^2 + z^2 - 2by = 0$, $x = 0$. Find its volume.
    *Ans.* $\frac{1}{3}\pi ab^2$ cubic units

14. Find the volume of the solid bounded by the paraboloid $y^2 + 4z^2 = x$ and the plane $x = 4$.
    *Ans.* $4\pi$ cubic units

15. A barrel has the shape of an ellipsoid of revolution with equal pieces cut from the ends. Find the volume if the height is 6 ft, the midsection of radius 3 ft and the ends of radius 2 ft.
    *Ans.* $44\pi$ ft³

16. The section of a certain solid by any plane perpendicular to the $x$-axis is a circle with the ends of a diameter lying on the parabolas $y^2 = 9x$ and $x^2 = 9y$. Find its volume.    *Ans.* $6561\pi/280$ cubic units

17. The section of a certain solid by any plane perpendicular to the $x$-axis is a square with the ends of a diagonal lying on the parabolas $y^2 = 4x$ and $x^2 = 4y$. Find its volume.    *Ans.* 144/35 cubic units

18. A hole of radius 1 in. is bored through a sphere of radius 3 in., the axis of the hole being a diameter of the sphere. Find the volume of the sphere which remains.    *Ans.* $64\pi\sqrt{2}/3$ in³

# Chapter 37

## Centroids

### *Plane Areas and Solids of Revolution*

**THE MASS OF A PHYSICAL BODY** is a measure of the quantity of matter in it while the volume of the body is a measure of the space it occupies. If the mass per unit volume is the same throughout, the body is said to be *homogeneous* or to have *constant density*.

It is highly desirable in physics and mechanics to consider a given mass as concentrated at a point, called its center of mass (also, its center of gravity). For a homogeneous body, this point coincides with its geometric center or *centroid*. For example, the center of mass of a homogeneous rubber ball coincides with the centroid (center) of the ball considered as a geometric solid (a sphere).

The centroid of a rectangular sheet of paper lies midway between the two surfaces but it may well be considered as located on one of the surfaces at the intersection of the diagonals. Then the center of mass of a thin sheet coincides with the centroid of the sheet considered as a plane area.

The discussion in this and the next chapter will be limited to plane areas and solids of revolution. Other solids, arcs of curves (a piece of fine homogeneous wire), and non-homogeneous masses will be treated in later chapters.

**THE (FIRST) MOMENT $M_L$ OF A PLANE AREA** with respect to a line $L$ is the product of the area and the directed distance of its centroid from the line. The moment of a composite area with respect to a line is the sum of the moments of the individual areas with respect to the line.

The moment of a plane area with respect to a coordinate axis may be found as follows:

(1) Sketch the area, showing a representative strip and the approximating rectangle.

(2) Form the product of the area of the rectangle and the distance of its centroid from the axis, and sum for all the rectangles.

(3) Assume the number of rectangles to be indefinitely increased and apply the Fundamental Theorem. (See Problem 2.)

For a plane area $A$ having centroid $(\bar{x}, \bar{y})$ and moments $M_x$ and $M_y$ with respect to the $x$- and $y$-axes,

$$A\bar{x} = M_y \quad \text{and} \quad A\bar{y} = M_x$$

See Problems 1-8.

**THE (FIRST) MOMENT OF A SOLID** of volume $V$, generated by revolving a plane area about a coordinate axis, with respect to the plane through the origin and perpendicular to the axis may be found as follows:

(1) Sketch the area, showing a representative strip and the approximating rectangle.

(2) Form the product of the volume, disc or shell, generated by revolving the rectangle about the axis and the distance of the centroid of the rectangle from the plane, and sum for all the rectangles.

(3)   Assume the number of rectangles to be indefinitely increased and apply the Fundamental Theorem.

When the area is revolved about the $x$-axis, the centroid $(\bar{x}, \bar{y})$ is on that axis. If $M_{yz}$ is the moment of the solid with respect to the plane through the origin and perpendicular to the $x$-axis,

$$V\bar{x} = M_{yz}, \quad \bar{y} = 0$$

Similarly, when the area is revolved about the $y$-axis, the centroid $(\bar{x}, \bar{y})$ is on that axis. If $M_{xz}$ is the moment of the solid with respect to the plane through the origin and perpendicular to the $y$-axis,

$$V\bar{y} = M_{xz}, \quad \bar{x} = 0$$

See Problems 9-12.

**FIRST THEOREM OF PAPPUS.** If a plane area is revolved about an axis in its plane and not crossing the area, the volume of the solid generated is equal to the product of the area and the length of the path described by the centroid of the area.

See Problems 13-15.

# Solved Problems

1.   For the plane area shown, find   (a) the moments with respect to the coordinate axes and   (b) the coordinates of the centroid $(\bar{x}, \bar{y})$.

(a)   The upper rectangle has area $5 \times 2 = 10$ units and centroid $A(2.5, 9)$. Similarly, the areas and centroids of the other rectangles are: 12 units, $B(1,5)$; 2 units, $C(2.5, 5)$; 10 units, $D(2.5, 1)$.

The moments of the rectangles with respect to the $x$-axis are $10(9)$, $12(5)$, $2(5)$, and $10(1)$. Hence the moment of area of the figure with respect to the $x$-axis is

$$M_x = 10(9) + 12(5) + 2(5) + 10(1) = 170$$

Similarly, the moment of area of the figure with respect to the $y$-axis is

$$M_y = 10(2.5) + 12(1) + 2(2.5) + 10(2.5) = 67$$

(b)   The area of the figure is $A = 10 + 12 + 2 + 10 = 34$.

Since $A\bar{x} = M_y$, $34\bar{x} = 67$ and $\bar{x} = 67/34$.

Since $A\bar{y} = M_x$, $34\bar{y} = 170$ and $\bar{y} = 5$.

The point $(67/34, 5)$ is the centroid.

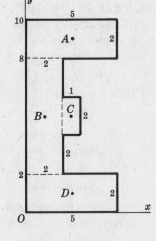

Fig. 37-1

2.   Find the moments with respect to the coordinate axes of the plane area in the second quadrant bounded by the curve $x = y^2 - 9$.

Use the approximating rectangle of the figure. Its area is $-x \cdot \Delta y$, its centroid is $(\frac{1}{2}x, y)$, and its moment with respect to the $x$-axis is $y(-x \cdot \Delta y)$. Then

$$M_x = -\int_0^3 y \cdot x \, dy = -\int_0^3 y(y^2 - 9) \, dy = \frac{81}{4}$$

Similarly, the moment of the approximating rectangle with respect to the $y$-axis is $\frac{1}{2}x(-x \cdot \Delta y)$. Then

$$M_y = -\frac{1}{2}\int_0^3 x^2 \, dy = -\frac{1}{2}\int_0^3 (y^2 - 9)^2 \, dy = -\frac{324}{5}$$

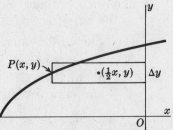

Fig. 37-2

3.  Determine the centroid of the first quadrant area bounded by the
    parabola $y = 4 - x^2$.

    The centroid of the approximating rectangle is $(x, \frac{1}{2}y)$.

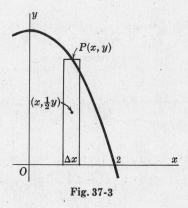
Fig. 37-3

$$A = \int_0^2 y \, dx = \int_0^2 (4 - x^2) \, dx = 16/3$$

$$M_x = \int_0^2 \frac{1}{2}y \cdot y \, dx = \frac{1}{2} \int_0^2 (4 - x^2)^2 \, dx = 128/15$$

$$M_y = \int_0^2 x \cdot y \, dx = \int_0^2 x(4 - x^2) \, dx = 4$$

Then $\bar{x} = M_y/A = 3/4$, $\bar{y} = M_x/A = 8/5$, and the centroid has
coordinates $(3/4, 8/5)$.

4.  Find the centroid of the first quadrant area bounded by the
    parabola $y = x^2$ and the line $y = x$.

    The centroid of the approximating rectangle is $[x, \frac{1}{2}(x + x^2)]$.

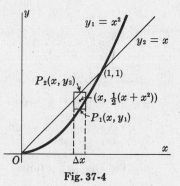
Fig. 37-4

$$A = \int_0^1 (x - x^2) \, dx = 1/6$$

$$M_x = \int_0^1 \frac{1}{2}(x + x^2)(x - x^2) \, dx = 1/15$$

$$M_y = \int_0^1 x(x - x^2) \, dx = 1/12$$

Then $\bar{x} = M_y/A = 1/2$, $\bar{y} = M_x/A = 2/5$, and the coordinates
of the centroid are $(1/2, 2/5)$.

5.  Find the centroid of the area bounded by the parabolas
    $x = y^2$ and $x^2 = -8y$.

    The centroid of the approximating rectangle is
    $[x, \frac{1}{2}(-x^2/8 - \sqrt{x}\,)]$.

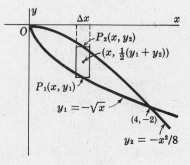
Fig. 37-5

$$A = \int_0^4 \left(-\frac{x^2}{8} + \sqrt{x}\,\right) dx = \frac{8}{3}$$

$$M_x = \int_0^4 \frac{1}{2}\left(-\frac{x^2}{8} - \sqrt{x}\,\right)\left(-\frac{x^2}{8} + \sqrt{x}\,\right) dx = -\frac{12}{5}$$

$$M_y = \int_0^4 x\left(-\frac{x^2}{8} + \sqrt{x}\,\right) dx = \frac{24}{5}$$

and the centroid is $(\bar{x}, \bar{y}) = (9/5, -9/10)$.

6.  Find the centroid of the area under the curve $y = 2\sin 3x$ from $x = 0$ to $x = \pi/3$. See Fig. 37-6.

    Use the approximating rectangle of Fig. 37-6 below, whose centroid is $(x, \frac{1}{2}y)$.

$$A = \int_0^{\pi/3} y \, dx = \int_0^{\pi/3} 2\sin 3x \, dx = -\frac{2}{3}\cos 3x \Big]_0^{\pi/3} = \frac{4}{3}$$

$$M_x = \int_0^{\pi/3} \frac{1}{2}y \cdot y \, dx = 2\int_0^{\pi/3} \sin^2 3x \, dx$$

$$= 2\left[\frac{1}{2}x - \frac{1}{12}\sin 6x\right]_0^{\pi/3} = \frac{\pi}{3}$$

$$M_y = \int_0^{\pi/3} x \cdot y \, dx = 2\int_0^{\pi/3} x \sin 3x \, dx$$

$$= \frac{2}{9}\left[\sin 3x - 3x\cos 3x\right]_0^{\pi/3} = \frac{2}{9}\pi$$

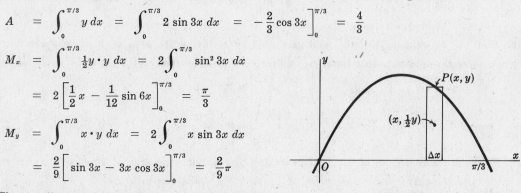
Fig. 37-6

The coordinates of the centroid are $(M_y/A, M_x/A) =$
$(\pi/6, \pi/4)$.

7. Determine the centroid of the first quadrant area of the hypocycloid $x = a\cos^3\theta$, $y = a\sin^3\theta$. See Fig. 37-7.

By symmetry, $\bar{x} = \bar{y}$.

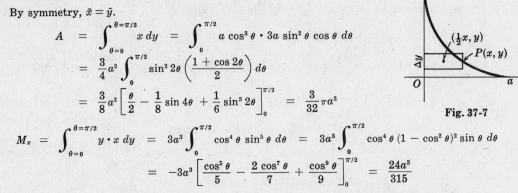

$$A = \int_{\theta=0}^{\theta=\pi/2} x\,dy = \int_0^{\pi/2} a\cos^3\theta \cdot 3a\sin^2\theta\cos\theta\,d\theta$$

$$= \frac{3}{4}a^2 \int_0^{\pi/2} \sin^2 2\theta \left(\frac{1+\cos 2\theta}{2}\right)d\theta$$

$$= \frac{3}{8}a^2\left[\frac{\theta}{2} - \frac{1}{8}\sin 4\theta + \frac{1}{6}\sin^3 2\theta\right]_0^{\pi/2} = \frac{3}{32}\pi a^2$$

Fig. 37-7

$$M_x = \int_{\theta=0}^{\theta=\pi/2} y \cdot x\,dy = 3a^3 \int_0^{\pi/2}\cos^4\theta\sin^5\theta\,d\theta = 3a^3\int_0^{\pi/2}\cos^4\theta(1-\cos^2\theta)^2\sin\theta\,d\theta$$

$$= -3a^3\left[\frac{\cos^5\theta}{5} - \frac{2\cos^7\theta}{7} + \frac{\cos^9\theta}{9}\right]_0^{\pi/2} = \frac{24a^3}{315}$$

Then $\bar{y} = M_x/A = 256a/315\pi$ and the centroid has coordinates $(256a/315\pi, 256a/315\pi)$.

8. Show that the centroid of the area of a circular sector of radius $r$ and angle $2\theta$ is at a distance $\dfrac{2r\sin\theta}{3\theta}$ from the center of the circle.

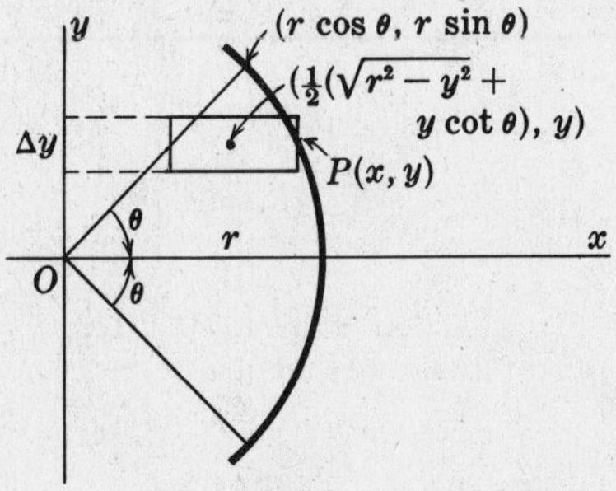

Take the sector so that the centroid lies on the $x$-axis. By symmetry the abscissa of the required centroid is that of the centroid of the area lying above the $x$-axis bounded by the circle and the line $y = x\tan\theta$. For this latter sector:

$$A = \int_0^{r\sin\theta} (\sqrt{r^2-y^2} - y\cot\theta)\,dy$$

$$= \left[\frac{1}{2}y\sqrt{r^2-y^2} + \frac{1}{2}r^2\arcsin\frac{y}{r} - \frac{1}{2}y^2\cot\theta\right]_0^{r\sin\theta} = \frac{1}{2}r^2\theta$$

Fig. 37-8

$$M_y = \int_0^{r\sin\theta}\frac{1}{2}(\sqrt{r^2-y^2} + y\cot\theta)(\sqrt{r^2-y^2} - y\cot\theta)\,dy = \frac{1}{2}\int_0^{r\sin\theta}(r^2 - y^2 - y^2\cot^2\theta)\,dy$$

$$= \frac{1}{2}\left[r^2y - \frac{1}{3}y^3 - \frac{1}{3}y^3\cot^2\theta\right]_0^{r\sin\theta} = \frac{1}{3}r^3\sin\theta, \quad \text{and} \quad \bar{x} = \frac{M_y}{A} = \frac{2r\sin\theta}{3\theta}$$

9. Find the centroid $(\bar{x}, 0)$ of the solid generated by revolving the area of Problem 3 about the $x$-axis. Use the approximating rectangle of Problem 3 and the disc method.

$$V = \pi\int_0^2 y^2\,dx = \pi\int_0^2 (4-x^2)^2\,dx = 256\pi/15,$$

$$M_{yz} = \pi\int_0^2 x \cdot y^2\,dx = \pi\int_0^2 x(4-x^2)^2\,dx = 32\pi/3, \quad \text{and} \quad \bar{x} = M_{yz}/V = 5/8$$

10. Find the centroid $(0, \bar{y})$ of the solid generated by revolving the area of Problem 3 about the $y$-axis. Use the approximating rectangle of Problem 3 and the shell method.

$$V = 2\pi\int_0^2 xy\,dx = 2\pi\int_0^2 x(4-x^2)\,dx = 8\pi,$$

$$M_{xz} = 2\pi\int_0^2 \frac{1}{2}y \cdot xy\,dx = \pi\int_0^2 x(4-x^2)^2\,dx = 32\pi/3, \quad \text{and} \quad \bar{y} = M_{xz}/V = 4/3$$

11. Find the centroid $(\bar{x}, 0)$ of the solid generated by revolving the area of Problem 4 about the $x$-axis. Use the approximating rectangle of Problem 4 and the disc method.

$$V = \pi\int_0^1 (x^2 - x^4)\,dx = 2\pi/15, \quad M_{yz} = \pi\int_0^1 x(x^2 - x^4)\,dx = \pi/12, \quad \text{and} \quad \bar{x} = M_{yz}/V = 5/8.$$

12. Find the centroid $(0, \bar{y})$ of the solid generated by revolving the area of Problem 4 about the $y$-axis.

Use the approximating rectangle of Problem 4 and the shell method.

$$V = 2\pi \int_0^1 x(x - x^2)\, dx = \pi/6,$$

$$M_{xz} = 2\pi \int_0^1 \tfrac{1}{2}(x + x^2) \cdot x(x - x^2)\, dx = \pi/12, \quad \text{and} \quad \bar{y} = M_{xz}/V = 1/2$$

13. Find the centroid of the area of a semi-circle of radius $r$.

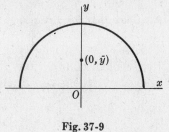

Fig. 37-9

Take the semi-circle as in the figure, so that $\bar{x} = 0$.

The area of the semi-circle is $\tfrac{1}{2}\pi r^2$, the solid generated by revolving it about the $x$-axis is a sphere of volume $\tfrac{4}{3}\pi r^3$, and the centroid $(0, \bar{y})$ of the area describes a circle of radius $\bar{y}$. Then, by the Theorem of Pappus, $\tfrac{1}{2}\pi r^2 \cdot 2\pi \bar{y} = \tfrac{4}{3}\pi r^3$ and $\bar{y} = 4r/3\pi$. The centroid is at the point $(0, 4r/3\pi)$.

14. Find the volume of the torus generated by revolving the circle $x^2 + y^2 = 4$ about the line $x = 3$. See Fig. 37-10 below.

The centroid of the area of the circle describes a circle of radius 3.

Hence, $V = 4\pi(6\pi) = 24\pi^2$ cubic units.

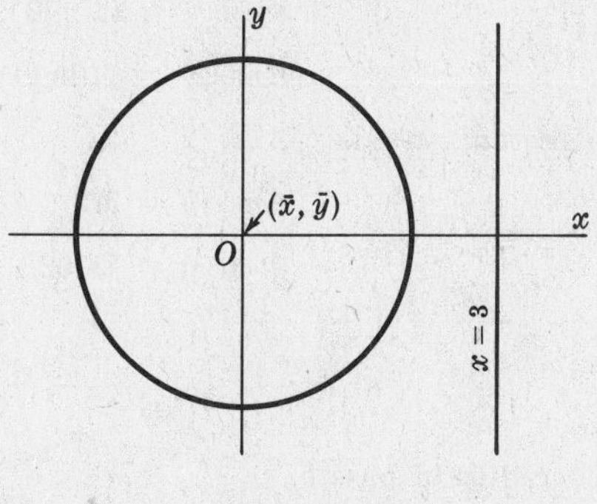

Fig. 37-10 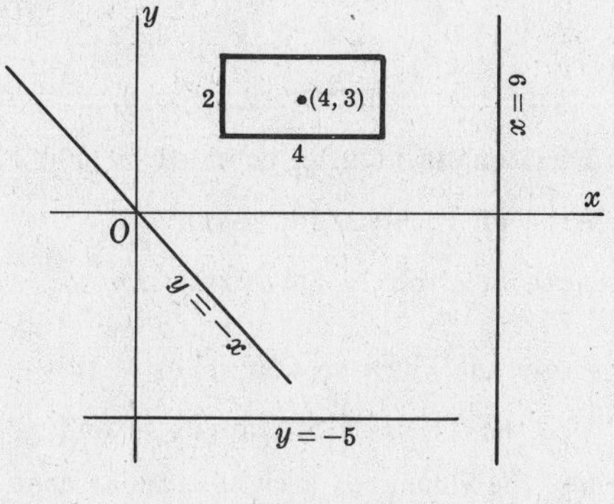                    Fig. 37-11

15. The rectangle of Fig. 37-11 above, is revolved about (1) the line $x = 9$, (2) the line $y = -5$, and (3) the line $y = -x$. Find the volume generated in each case.

(1) The centroid $(4, 3)$ of the rectangle describes a circle of radius 5. Hence, $V = 8(10\pi) = 80\pi$ cu. un.

(2) The centroid describes a circle of radius 8. Hence, $V = 8(16\pi) = 128\pi$ cubic units.

(3) The centroid describes a circle of radius $(4 + 3)/\sqrt{2}$. Hence, $V = 56\sqrt{2}\,\pi$ cubic units.

# Supplementary Problems

In Problems 16-26 find the centroid of the given area.

16. $y = x^2,\ y = 9$            Ans. $(0, 27/5)$

17. $y = 4x - x^2,\ y = 0$           Ans. $(2, 8/5)$

18. $y = 4x - x^2,\ y = x$          Ans. $(3/2, 12/5)$

19. $3y^2 = 4(3 - x),\ x = 0$       Ans. $(6/5, 0)$

20. $x^2 = 8y,\ y = 0,\ x = 4$        Ans. $(3, 3/5)$

21. $y = x^2,\ 4y = x^3$            Ans. $(12/5, 192/35)$

22. $x^2 - 8y + 4 = 0$, $x^2 = 4y$, first quadrant.        *Ans.* (3/4, 2/5)

23. First quadrant area of $x^2 + y^2 = a^2$.        *Ans.* ($4a/3\pi$, $4a/3\pi$)

24. First quadrant area of $9x^2 + 16y^2 = 144$.        *Ans.* ($16/3\pi$, $4/\pi$)

25. Right loop of $y^2 = x^4(1 - x^2)$.        *Ans.* ($32/15\pi$, 0)

26. First arch of $x = \theta - \sin\theta$, $y = 1 - \cos\theta$.        *Ans.* ($\pi$, 5/6)

27. Show that the distance of the centroid of a triangle from the base is 1/3 the altitude.


In Problems 28-38, find the centroid of the solid generated by revolving the given plane area about the given line.

28. $y = x^2$, $y = 9$, $x = 0$; $y$-axis        *Ans.* $\bar{y} = 6$

29. $y = x^2$, $y = 9$, $x = 0$; $x$-axis        *Ans.* $\bar{x} = 5/4$

30. $y = 4x - x^2$, $y = x$; $x$-axis        *Ans.* $\bar{x} = 27/16$

31. $y = 4x - x^2$, $y = x$; $y$-axis        *Ans.* $\bar{y} = 27/10$

32. $x^2 - y^2 = 16$, $y = 0$, $x = 8$; $x$-axis        *Ans.* $\bar{x} = 27/4$

33. $x^2 - y^2 = 16$, $y = 0$, $x = 8$; $y$-axis        *Ans.* $\bar{y} = 3\sqrt{3}/2$

34. $(x - 2)y^2 = 4$, $y = 0$, $x = 3$, $x = 5$; $x$-axis        *Ans.* $\bar{x} = (2 + 2\ln 3)/(\ln 3)$

35. $x^2 y = 16(4 - y)$, $x = 0$, $y = 0$, $x = 4$; $y$-axis        *Ans.* $\bar{y} = 1/(\ln 2)$

36. First quadrant area bounded by $y^2 = 12x$ and its latus rectum; $x$-axis        *Ans.* $\bar{x} = 2$

37. Area of Problem 36; $y$-axis.        *Ans.* $\bar{y} = 5/2$

38. Area of Problem 36; directrix.        *Ans.* $\bar{y} = 75/32$


39. Prove the Theorem of Pappus of this chapter.

40. Use the Theorem of Pappus to find:

  (a) the volume of a right circular cone of altitude $a$ and radius of base $b$.

  (b) the ring obtained by revolving the ellipse $4(x - 6)^2 + 9(y - 5)^2 = 36$ about the $x$-axis.

  *Ans.* (a) $\frac{1}{3}\pi ab^2$ c.u. (b) $60\pi^2$ c.u.

41. For the area $A$ bounded by $y = -x^2 - 3x + 6$ and $x + y - 3 = 0$, find (a) its centroid, (b) the volume generated when $A$ is revolved about the bounding line.

  *Ans.* (a) $(-1, 28/5)$,  (b) $2\pi\left(\dfrac{\bar{x} + \bar{y} - 3}{\sqrt{2}}\right) \cdot A = \dfrac{256\sqrt{2}}{15}\pi$ cu. un.

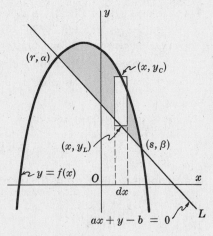

42. For the volume generated by revolving the area $A$ (shaded in Fig. 37-12) about the bounding line $L$, obtain

$$V = 2\pi\left(\frac{a\bar{x} + \bar{y} - b}{\sqrt{a^2 + 1}}\right) \cdot A = \frac{2\pi}{\sqrt{a^2 + 1}}(aM_y + M_x - bA)$$

$$= \frac{\pi}{\sqrt{a^2 + 1}} \int_r^s (y_C - y_L)^2\, dx$$

43. Use the formula of Problem 42 to obtain the volume generated by revolving the given area about the bounding line:

  (a) $y = -x^2 - 3x + 6$, $x + y - 3 = 0$

  (b) $y = 2x^2$, $2x - y + 4 = 0$

  *Ans.* (a) See Problem 41, (b) $162\sqrt{5}\,\pi/25$ cu. un.

Fig. 37-12

# Chapter 38

## Moments of Inertia
### *Plane Areas and Solids of Revolution*

**THE MOMENT OF INERTIA $I_L$ OF A PLANE AREA $A$** with respect to a line $L$ in its plane may be found as follows:

(1) Make a sketch of the area, showing a representative strip parallel to the line and the approximating rectangle.

(2) Form the product of the area of the rectangle and the square of the distance of its centroid from the line, and sum for all the rectangles.

(3) Assume the number of rectangles to be indefinitely increased and apply the Fundamental Theorem.

See Problems 1-4.

**THE MOMENT OF INERTIA $I_L$ OF A SOLID** of volume $V$, generated by revolving a plane area about a line $L$ in its plane, with respect to that line (the axis of the solid) may be found as follows:

(1) Make a sketch showing a representative strip parallel to the axis and the approximating rectangle.

(2) Form the product of the volume generated by revolving the rectangle about the axis (a shell) and the square of the distance of the centroid of the rectangle from the axis, and sum for all the rectangles.

(3) Assume the number of rectangles to be indefinitely increased and apply the Fundamental Theorem.

See Problems 5-8.

**RADIUS OF GYRATION.** The positive number $R$ defined by the relation $I_L = AR^2$ in the case of a plane area $A$ and by $I_L = VR^2$ in the case of a solid of revolution is called the radius of gyration of the area or volume with respect to $L$.

**PARALLEL AXIS THEOREM.** The moment of inertia of an area, arc length, or volume with respect to any axis is equal to the moment of inertia with respect to a parallel axis through the centroid plus the product of the area, arc length, or volume and the square of the distance between the parallel axes.

See Problems 9-10.

## Solved Problems

1. Find the moment of inertia of a rectangular area $A$ of dimensions $a$ and $b$ with respect to a side.

   Take the rectangular area as in the figure and let the side in question be that along the $y$-axis.

   The approximating rectangle has area $= b \cdot \Delta x$ and centroid $(x, \frac{1}{2}b)$. Hence its moment element is $x^2 b\, \Delta x$. Then

   $$I_y = \int_0^a x^2 b\, dx = b\frac{x^3}{3}\bigg]_0^a = \frac{ba^3}{3} = \frac{1}{3}Aa^2$$

   Thus the moment of inertia of a rectangular area with respect to a side is $\frac{1}{3}$ the product of the area and the square of the length of the other side.

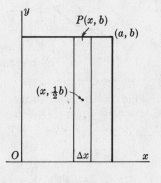

**Fig. 38-1**

**2.** Find the moment of inertia with respect to the $y$-axis of the plane area between the parabola $y = 9 - x^2$ and the $x$-axis.

*First Solution.* For the approximating rectangle of Fig. 38-2 below, $A = y \cdot \Delta x$ and the centroid is $(x, \frac{1}{2}y)$. Then

$$I_y = \int_{-3}^{3} x^2 y \, dx = 2 \int_{0}^{3} (9x^2 - x^4) \, dx = \frac{324}{5}$$

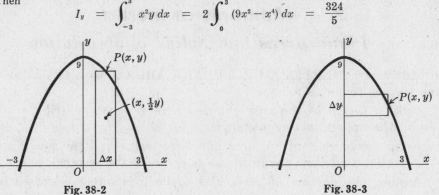

**Fig. 38-2**                   **Fig. 38-3**

*Second Solution.* For the approximating rectangle of Fig. 38-3 above, the area is $x \cdot \Delta y$ and the dimension perpendicular to the $y$-axis is $x$. Hence, from Problem 1, the moment element is $\frac{1}{3}(x \, \Delta y)x^2$. Thus, due to symmetry,

$$I_y = 2 \cdot \frac{1}{3} \int_{0}^{9} x^3 \, dy = \frac{2}{3} \int_{0}^{9} (9 - y)^{3/2} \, dy = \frac{324}{5}$$

Since $A = 2 \int_{0}^{9} x \, dy = 2 \int_{0}^{9} \sqrt{9 - y} \, dy = 36$, $I_y = 324/5 = AR^2$ and the radius of gyration is $R = 3/\sqrt{5}$.

**3.** Find the moment of inertia with respect to the $y$-axis of the first quadrant area bounded by the parabola $x^2 = 4y$ and the line $y = x$. See Fig. 38-4 below.

Use the approximating rectangle of Fig. 38-4 below, whose area is $(x - \frac{1}{4}x^2) \Delta x$ and whose centroid is $[x, \frac{1}{2}(x + \frac{1}{4}x^2)]$. Then

$$A = \int_{0}^{4} (x - \tfrac{1}{4}x^2) \, dx = \frac{8}{3} \quad \text{and} \quad I_y = \int_{0}^{4} x^2(x - \tfrac{1}{4}x^2) \, dx = \frac{64}{5} = \frac{24}{5}A$$

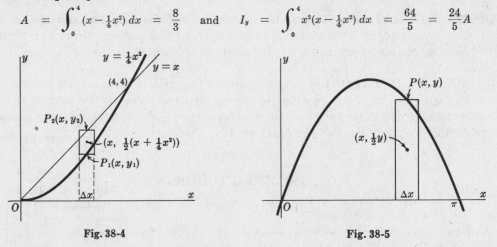

**Fig. 38-4**                   **Fig. 38-5**

**4.** Find the moment of inertia with respect to each coordinate axis of the area between the curve $y = \sin x$ from $x = 0$ to $x = \pi$ and the $x$-axis. See Fig. 38-5 above.

$$A = \int_{0}^{\pi} \sin x \, dx = -\cos x \Big]_{0}^{\pi} = 2$$

$$I_x = \int_{0}^{\pi} y^2 \cdot \frac{1}{3} \sin x \, dx = \frac{1}{3} \int_{0}^{\pi} \sin^3 x \, dx = \frac{1}{3} \Big[ -\cos x + \frac{1}{3} \cos^3 x \Big]_{0}^{\pi} = \frac{4}{9} = \frac{2}{9}A$$

$$I_y = \int_{0}^{\pi} x^2 \sin x \, dx = \Big[ 2 \cos x + 2x \sin x - x^2 \cos x \Big]_{0}^{\pi} = (\pi^2 - 4) = \frac{1}{2}(\pi^2 - 4)A$$

5.   Find the moment of inertia with respect to its axis of a right circular cylinder of height $b$ and radius of base $a$.  See Fig. 38-6 below.

Let the cylinder be generated by revolving the rectangle of dimensions $a$ and $b$ about the $y$-axis as in Fig. 38-6 below.  For the approximating rectangle of the figure, the centroid is $(x, \frac{1}{2}b)$ and the volume of the shell generated by revolving the rectangle about the $y$-axis is  $\Delta V = 2\pi bx \cdot \Delta x$.  Then, since  $V = \pi ba^2$,

$$I_y \;=\; 2\pi \int_0^a x^2 \cdot bx \, dx \;=\; \tfrac{1}{2}\pi ba^4 \;=\; \tfrac{1}{2}\pi ba^2 \cdot a^2 \;=\; \tfrac{1}{2}Va^2$$

Thus the moment of inertia with respect to its axis of a right circular cylinder is equal to one-half the product of its volume and the square of its radius.

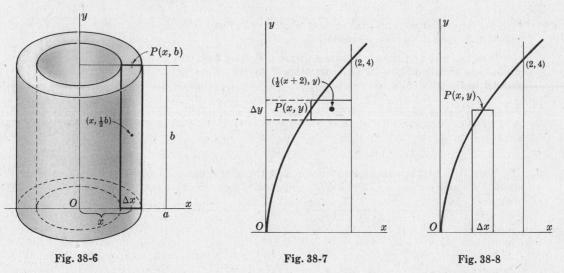

Fig. 38-6                          Fig. 38-7                          Fig. 38-8

6.   Find the moment of inertia with respect to its axis of the solid generated by revolving about the $x$-axis the area in the first quadrant bounded by the parabola $y^2 = 8x$, the $x$-axis, and the line $x = 2$.

*First Solution.*  Refer to Fig. 38-7 above.  The centroid of the approximating rectangle of Fig. 38-7 is $[\frac{1}{2}(x + 2), \, y]$ and the volume generated by revolving the rectangle about the $x$-axis is $2\pi y(2 - x) \, \Delta y = 2\pi y(2 - y^2/8) \, \Delta y$.  Then

$$V \;=\; 2\pi \int_0^4 y(2 - y^2/8) \, dy \;=\; 16\pi \quad \text{and} \quad I_x \;=\; 2\pi \int_0^4 y^2 \cdot y(2 - y^2/8) \, dy \;=\; \frac{256}{3}\pi \;=\; \frac{16}{3}V$$

*Second Solution.*  Refer to Fig. 38-8 above.  The volume generated by revolving the approximating rectangle of Fig. 38-8 about the $x$-axis is $\pi y^2 \Delta x$ and, using the result of Problem 5, its moment of inertia with respect to the $x$-axis is $\frac{1}{2}y^2(\pi y^2 \Delta x) = \frac{1}{2}\pi y^4 \Delta x$.  Then

$$V \;=\; \pi \int_0^2 y^2 \, dx \;=\; 8\pi \int_0^2 x \, dx \;=\; 16\pi$$

and

$$I_x \;=\; \tfrac{1}{2}\pi \int_0^2 y^4 \, dx \;=\; 32\pi \int_0^2 x^2 \, dx \;=\; \frac{256}{3}\pi \;=\; \frac{16}{3}V$$

7.   Find the moment of inertia with respect to its axis of the solid generated by revolving the area of Problem 6 about the $y$-axis.  Refer to Fig. 38-8 above.

The volume generated by revolving the rectangle of Fig. 38-8 about the $y$-axis is $2\pi xy \, \Delta x$.  Then

$$V \;=\; 2\pi \int_0^2 xy \, dx \;=\; 4\sqrt{2}\,\pi \int_0^2 x^{3/2} \, dx \;=\; \frac{64}{5}\pi$$

and

$$I_y \;=\; 2\pi \int_0^2 x^2 \cdot xy \, dx \;=\; 4\sqrt{2}\,\pi \int_0^2 x^{7/2} \, dx \;=\; \frac{256}{9}\pi \;=\; \frac{20}{9}V$$

8.   Find the moment of inertia with respect to its axis of the volume of the sphere generated by revolving a circle of radius $r$ about a fixed diameter.

Take the circle as in the figure, with the fixed diameter along the $x$-axis.  Using the shell method,

and
$$V = 2\pi \int_0^r 2x \cdot y \, dy = \tfrac{4}{3}\pi r^3$$

$$I_x = 4\pi \int_0^r y^2 \cdot xy \, dy = 4\pi \int_0^r y^3 \sqrt{r^2 - y^2} \, dy$$

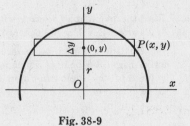

Let $y = r \sin z$; then $\sqrt{r^2 - y^2} = r \cos z$, $dy = r \cos z \, dz$.

To change the $y$ limits of integration to the $z$ limits, consider that when $y = 0$ then $0 = r \sin z$, $0 = \sin z$, $z = 0$; and when $y = r$ then $r = r \sin z$, $1 = \sin z$, $z = \tfrac{1}{2}\pi$. Now

**Fig. 38-9**

$$I_x = 4\pi r^5 \int_0^{\pi/2} \sin^3 z \cos^2 z \, dz = 4\pi r^5 \int_0^{\pi/2} (1 - \cos^2 z) \cos^2 z \sin z \, dz = \tfrac{8}{15}\pi r^5 = \tfrac{2}{5}r^2 V$$

9. Find the moment of inertia of the area of a circle of radius $r$ with respect to a line $s$ units from its center.

Take the center of the circle at the origin. We find first the moment of inertia of the circle with respect to the diameter parallel to the given line as

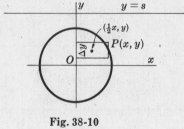

$$I_x = 4 \int_0^r y^2 \cdot x \, dy = 4 \int_0^r y^2 \sqrt{r^2 - y^2} \, dy = \tfrac{1}{4}r^4\pi = \tfrac{1}{4}r^2 A$$

Then
$$I_s = I_x + A \cdot s^2 = (\tfrac{1}{4}r^2 + s^2)A$$

**Fig. 38-10**

10. The moment of inertia with respect to its axis of the solid generated by revolving an arch of $y = \sin 3x$ about the $x$-axis is $I_x = \pi^2/16 = 3V/8$. Find the moment of inertia of the solid with respect to the line $y = 2$.

$$I_{y=2} = I_x + 2^2 V = 3V/8 + 4V = 35V/8$$

# Supplementary Problems

11. Find the moment of inertia of the given plane area with respect to the given line:
    - (a) $y = 4 - x^2$, $x = 0$, $y = 0$; $x$-axis, $y$-axis      *Ans.* $128A/35$, $4A/5$
    - (b) $y = 8x^3$, $y = 0$, $x = 1$; $x$-axis, $y$-axis      *Ans.* $128A/15$, $2A/3$
    - (c) $x^2 + y^2 = a^2$; a diameter      *Ans.* $a^2 A/4$
    - (d) $y^2 = 4x$, $x = 1$; $x$-axis, $y$-axis      *Ans.* $4A/5$, $3A/7$
    - (e) $4x^2 + 9y^2 = 36$; $x$-axis, $y$-axis      *Ans.* $A$, $9A/4$

12. Use the results of Problem 11 and the parallel axis theorem to obtain the moment of inertia of the given area with respect to the given line:
    - (a) $y = 4 - x^2$, $y = 0$; $x = 4$    (b) $x^2 + y^2 = a^2$; a tangent    (c) $y^2 = 4x$, $x = 1$; $x = 1$
    *Ans.* (a) $84A/5$    (b) $5a^2 A/4$    (c) $10A/7$

13. Find the moment of inertia with respect to its axis of the solid generated by revolving the given plane area about the given line:
    - (a) $y = 4x - x^2$, $y = 0$; $x$-axis, $y$-axis      (c) $4x^2 + 9y^2 = 36$; $x$-axis, $y$-axis
    - (b) $y^2 = 8x$, $x = 2$; $x$-axis, $y$-axis      (d) $x^2 + y^2 = a^2$; $y = b$, $b > a$
    *Ans.* (a) $128V/21$, $32V/5$    (b) $16V/3$, $20V/9$    (c) $8V/5$, $18V/5$    (d) $(b^2 + \tfrac{3}{4}a^2)V$

14. Use the parallel axis theorem to obtain the moment of inertia of: (a) a sphere of radius $r$ about a line tangent to it; (b) a right circular cylinder about one of its elements.    *Ans.* (a) $7r^2 V/5$, (b) $3r^2 V/2$

15. Prove: The moment of inertia of a plane area with respect to a line $L$ perpendicular to its plane (or with respect to the foot of that perpendicular) is equal to the sum of the moments of inertia with respect to any two mutually perpendicular lines in the plane and through the foot of $L$.

16. Find the *polar moment of inertia* $I_0$ (the moment of inertia with respect to the origin) of: (a) the triangle bounded by $y = 2x$, $y = 0$, $x = 4$; (b) the circle of radius $r$ and center at the origin; (c) the circle $x^2 - 2rx + y^2 = 0$; (d) the area bounded by the line $y = x$ and the parabola $y^2 = 2x$.
    *Ans.* (a) $I_0 = I_x + I_y = 56A/3$, (b) $\tfrac{1}{2}r^2 A$, (c) $3r^2 A/2$, (d) $72A/35$

# Chapter 39

## Fluid Pressure

**PRESSURE** = force per unit area = $\dfrac{\text{force acting perpendicular to an area}}{\text{area over which the force is distributed}}$.

The pressure $p$ on a horizontal surface of area $A$ due to a column of fluid of height $h$ resting on it is $p = wh$, where $w$ = weight of fluid per unit of volume. The force on this surface = pressure × surface area = $whA$.

The pressure exerted by a fluid at any point within it is equal in all directions.

### FORCE ON A SUBMERGED PLANE AREA

Fig. 39-1 shows a plane area submerged vertically in a liquid of weight $w$ lb per unit of volume. Take the area in the $xy$-plane with the $x$-axis in the surface of the liquid and the positive $y$-axis directed downward. Divide the area into strips (always parallel to the surface of the liquid) and approximate each by a rectangle (as in Chapter 34).

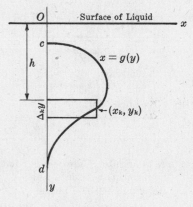

**Fig. 39-1**

Denote by $h$ the depth of the upper edge of the representative rectangle of the figure. The force exerted on this rectangle of width $\Delta_k y$ and length $x_k = g(y_k)$ is $w \cdot y_k' \cdot g(y_k)\,\Delta_k y$, where $y_k'$ is some value of $y$ between $h$ and $h + \Delta_k y$. The total force on the plane area is, by the theorem of Bliss,

$$F = \lim_{n \to +\infty} \sum_{k=1}^{n} w \cdot y_k' \cdot g(y_k)\,\Delta_k y = w \int_c^d y \cdot g(y)\,dy = w \int_c^d yx\,dy$$

The force exerted on a plane area submerged vertically in a liquid is equal to the product of the weight of a unit volume of the liquid, the area submerged, and the depth of the centroid of the area below the surface of the liquid. This, rather than a formula, should be used as the working principle in setting up all integrals.

## Solved Problems

1. Find the force on one face of the rectangle submerged in water as shown in Fig. 39-2. Water weighs 62.5 pounds per cubic foot.

   The submerged area is $2 \times 8 = 16$ ft² and its centroid is 1 ft below the water level. Hence,

   $\qquad F$ = specific weight × area × depth of centroid
   $\qquad\quad$ = 62.5 lb/ft³ × 16 ft² × 1 ft = 1000 lb

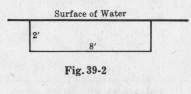

**Fig. 39-2**

2. Find the force on one face of the rectangle submerged in water as shown in Fig. 39-3.

   The submerged area is 90 ft² and its centroid is 5 ft below the water level.

   $\qquad F$ = 62.5 lb/ft³ × 90 ft² × 5 ft = 28,125 lb

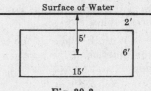

**Fig. 39-3**

193

3.  Find the force on one face of the triangle shown in Fig. 39-4 below, the units being feet and the liquid weighing 50 lb/ft³.

*First Solution.* The submerged area is bounded by the lines $x = 0$, $y = 2$, and $3x + 2y = 10$. The force exerted on the approximating rectangle of area $x \cdot \Delta y$ and depth $y$ is $w \cdot y \cdot x \cdot \Delta y = wy\left(\dfrac{10 - 2y}{3}\right)\Delta y$. Then

$$F = w \int_2^5 y\left(\frac{10 - 2y}{3}\right) dy = 9w = 450 \text{ lb}$$

*Second Solution.* The submerged area is 3 ft² and its centroid is $2 + \frac{1}{3}(3) = 3$ ft below the surface of the liquid. Hence, $F = 50 \cdot 3 \cdot 3 = 450$ lb.

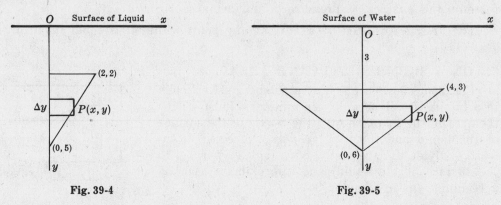

Fig. 39-4                                              Fig. 39-5

4.  A triangular plate whose edges are 5, 5, and 8 ft is placed vertically in water with its longest edge uppermost, horizontal, and 3 ft below the water level. Calculate the force on a side of the plate. Refer to Fig. 39-5 above.

*First Solution.* Choosing the axes as in Fig. 39-5 above, it is seen that the required force is twice that on the area bounded by the lines $y = 3$, $x = 0$, and $3x + 4y = 24$. The area of the approximating rectangle is $x \cdot \Delta y$ and its mean depth is $y$. Hence $\Delta F = wyx \cdot \Delta y = wy(8 - 4y/3)\Delta y$ and

$$F = 2w \int_3^6 y(8 - \tfrac{4}{3}y) \, dy = 48w = 3000 \text{ lb}$$

*Second Solution.* The submerged area is 12 ft² and its centroid is $3 + \frac{1}{3}(3) = 4$ ft below the water level. Hence $F = 62.5(12)(4) = 3000$ lb.

5.  Find the force on the end of a trough in the form of a semi-circle of radius 2 ft, when filled with a liquid weighing 60 lb/ft³.

With the choice of the coordinate system of Fig. 39-6, the force on the approximating rectangle is $wyx \cdot \Delta y = wy\sqrt{4 - y^2}\,\Delta y$. Hence

$$F = 2w \int_0^2 y\sqrt{4 - y^2}\, dy = \frac{16}{3}w = 320 \text{ lb}$$

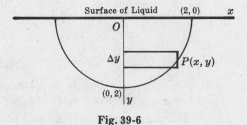

Fig. 39-6

6.  A plate in the form of a parabolic segment of base 12 ft and height 4 ft is submerged in water so that its base is in the surface of the liquid. Find the force on a face of the plate.

With the choice of the coordinate system of Fig. 39-7, the equation of the parabola is $x^2 = 9y$. The area of the approximating rectangle is $2x \cdot \Delta y$ and the mean depth is $4 - y$. Then

$$\Delta F = 2w(4 - y)x \cdot \Delta y = 2w(4 - y) \cdot 3\sqrt{y}\,\Delta y$$

and  $F = 6w \int_0^4 (4 - y)\sqrt{y}\, dy = \dfrac{256}{5}w = 3200$ lb

Fig. 39-7

7.  Find the force on the plate of Problem 6 if it is
    partly submerged in a liquid weighing 48 lb/ft³
    so that its axis is parallel to and 3 ft below the
    surface of the liquid.

    Choosing the coordinate system as in the
    adjacent Fig. 39-8, the equation of the parabola
    is $y^2 = 9x$.

    The area of the approximating rectangle is
    $(4 - x)\Delta y$, its mean depth is $3 - y$, and the force
    on it is

    $$\Delta F = w(3 - y)(4 - x)\Delta y = w(3 - y)(4 - y^2/9)\Delta y$$

    Then  $F = w\displaystyle\int_{-6}^{3} (3 - y)\left(4 - \frac{y^2}{9}\right) dy$

    $\qquad = \dfrac{405}{4}w = 4860$ lb

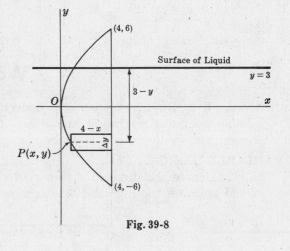

**Fig. 39-8**

# Supplementary Problems

8.  A rectangular plate  $6' \times 8'$  is submerged vertically in a liquid weighing $w$ lb/ft³.  Find the force
    on one face

    (a)  if the shorter side is upppermost and lies in the surface of the liquid,

    (b)  if the shorter side is uppermost and lies 2 ft below the surface of the liquid,

    (c)  if the longer side is uppermost and lies in the surface of the liquid,

    (d)  if the plate is held by a rope attached to a corner 2 ft below the liquid surface.

    *Ans.*  (a) 192$w$ lb,  (b) 288$w$ lb,  (c) 144$w$ lb,  (d) 336$w$ lb

9.  Assuming the $x$-axis horizontal and the positive $y$-axis directed downward, find the force on a side of
    each of the following areas.  The dimensions are in feet and the fluid weighs $w$ lb/ft³.

    (a)  $y = x^2$, $y = 4$; fluid surface at $y = 0$.        *Ans.*  128$w$/5 lb

    (b)  $y = x^2$, $y = 4$; fluid surface at $y = -2$.       *Ans.*  704$w$/15 lb

    (c)  $y = 4 - x^2$, $y = 0$; fluid surface at $y = 0$.    *Ans.*  256$w$/15 lb

    (d)  $y = 4 - x^2$, $y = 0$; fluid surface at $y = -3$.   *Ans.*  736$w$/15 lb

    (e)  $y = 4 - x^2$, $y = 2$; fluid surface at $y = -1$.   *Ans.*  $152\sqrt{2}\,w$/15 lb

10. A trough of trapezoidal cross section is 2 ft wide at the bottom, 4 ft wide at the top and 3 ft deep.
    Find the force on an end  (a) if it is full of water,  (b) if it contains 2 ft of water.
    *Ans.*  (a) 750 lb,  (b) 305.6 lb

11. A circular plate of radius 2 ft is lowered into a liquid ($w$ lb/ft³) so that its center is 4 ft below the
    surface.  Find the force on the lower half and on the upper half of the plate.
    *Ans.*  $(8\pi + 16/3)w$ lb,  $(8\pi - 16/3)w$ lb

12. A cylindrical tank 6 ft in radius is lying on its side.  If it contains oil weighing $w$ lb/ft³ to a depth of
    9 ft, find the force on an end.      *Ans.*  $(72\pi + 81\sqrt{3})w$ lb

13. The center of pressure of the area of Fig. 39-1 is that point $(\bar{x}, \bar{y})$ where a concentrated force of
    magnitude $F$ would yield the same moment with respect to any horizontal (vertical) line as the
    distributed forces.

    (a)  Show that        $F\bar{x} = \frac{1}{2}w\displaystyle\int_{c}^{d} yx^2\,dy$   and   $F\bar{y} = w\displaystyle\int_{c}^{d} y^2x\,dy$

    (b)  Show that the depth of the center of pressure below the surface of the liquid is equal to the
         moment of inertia of the area divided by the first moment of the area, each with respect to a line
         in the surface of the liquid.

14. Use (b) of Problem 13 to find the depth of the center of pressure below the surface of the liquid in
    (a) Problem 5,  (b) Problem 6,  (c) Problem 7,  (d) Problem 9(a),  (e) Problem 9(b).
    *Ans.*  (a) 3$\pi$/8,  (b) 16/7,  (c) 126/25,  (d) 20/7,  (e) 358/77

# Chapter 40

## Work

**CONSTANT FORCE.** The work $W$ done by a constant force $F$ acting over a directed distance $s$ along a straight line is $F \cdot s$ units.

**VARIABLE FORCE.** Consider a continuously varying force acting along a straight line. Let $x$ denote the directed distance of the point of application of the force from a fixed point on the line, and let the force be given as some function $F(x)$ of $x$.

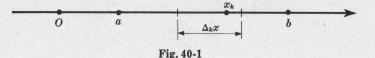

**Fig. 40-1**

To find the work done as the point of application moves from $x = a$ to $x = b$:

(a)  Divide the interval $a \leq x \leq b$ into $n$ subintervals of length $\Delta_k x$ and let $x_k$ be any point in the $k$-th subinterval.

(b)  Assume that during the displacement over the $k$-th subinterval the force is constant and equal to $F(x_k)$. The work done during this displacement is then $F(x_k) \Delta_k x$, and the total work done by the set of $n$ assumed forces is given by

$$\sum_{k=1}^{n} F(x_k) \Delta_k x.$$

(c)  Increase the number of subintervals indefinitely in such a manner that each $\Delta_k x \to 0$ and apply the Fundamental Theorem to obtain

$$W = \lim_{n \to \infty} \sum_{k=1}^{n} F(x_k) \Delta_k x = \int_a^b F(x)\,dx$$

## Solved Problems

1.  Within certain limits, the force required to stretch a spring is proportional to the stretch, the constant of proportionality being called the *modulus* of the spring. If a given spring of normal length 10 in. requires a force of 25 lb to stretch it $\frac{1}{4}$ in., calculate the work done in stretching it from 11 to 12 inches.

    Let $x$ denote the stretch; then $F(x) = kx$.

    When $x = \frac{1}{4}$, $F(x) = 25$; hence $25 = \frac{1}{4}k$, $k = 100$, and $F(x) = 100x$.

    The work corresponding to a stretch $\Delta x$ is $100x \cdot \Delta x$, and the required work is given by

$$W = \int_1^2 100x\,dx = 150 \text{ in. lb}$$

2.  The modulus of the spring on a bumping post in a freight yard is 270,000 lb/ft. Find the work done in compressing the spring 1 inch.

    Let $x$ be the displacement of the free end of the spring in feet. Then $F(x) = 270{,}000x$ and the work corresponding to a displacement $\Delta x$ is $270{,}000x \cdot \Delta x$. Hence

$$W = \int_0^{1/12} 270{,}000x\,dx = 937.5 \text{ ft lb}$$

3.  A cable weighing 3 lb/ft is unwinding from a cylindrical drum. If 50 ft are already unwound, find the work done by the force of gravity as an additional 250 ft are unwound.

Let $x$ = length of cable unwound at any time. Then $F(x) = 3x$ and

$$W = \int_{50}^{300} 3x\, dx = 131{,}250 \text{ ft lb}$$

4.  A 100 ft cable weighing 5 lb/ft supports a safe weighing 500 lb. Find the work done in winding 80 ft of the cable on a drum.

Let $x$ denote the length of cable which has been wound on the drum.

The total weight (unwound cable and safe) is $500 + 5(100 - x) = 1000 - 5x$, and the work done in raising the safe a distance $\Delta x$ is $(1000 - 5x)\,\Delta x$. Thus the required work

$$W = \int_{0}^{80} (1000 - 5x)\, dx = 64{,}000 \text{ ft lb}$$

5.  A right circular cylindrical tank of radius 2 ft and height 8 ft is full of water. Find the work done in pumping the water to the top of the tank. Assume that the water weighs 62.5 lb per cubic foot.

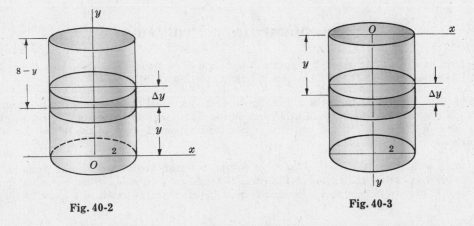

**Fig. 40-2**                                    **Fig. 40-3**

*First Solution.* Refer to Fig. 40-2. Imagine the water being pushed out by means of a piston which is forced upward from the bottom of the tank. Fig. 40-2 shows the piston when it is $y$ ft from the bottom. The lifting force, being equal to the weight of the water on the piston, is approximately $F(y) = \pi r^2 w(8 - y) = 4\pi w(8 - y)$, and the work corresponding to a displacement $\Delta y$ of the piston is approximately $4\pi w(8 - y) \cdot \Delta y$. The work done in emptying the tank is

$$W = 4\pi w \int_{0}^{8} (8 - y)\, dy = 128\pi w = 128\pi(62.5) = 8000\pi \text{ ft lb}$$

*Second Solution.* Refer to Fig. 40-3. Imagine the water in the tank to be sliced into $n$ disks of thickness $\Delta y$ and that the tank is to be emptied by lifting each disk to the top. For the representative disk of Fig. 40-3, whose mean distance from the top is $y$, the weight is $4\pi w \cdot \Delta y$ and the work done in moving it to the top of the tank is $4\pi wy \cdot \Delta y$. Summing for the $n$ disks and applying the Fundamental Theorem, we have

$$W = 4\pi w \int_{0}^{8} y\, dy = 128\pi w = 8000\pi \text{ ft lb}$$

6.  The expansion of a gas in a cylinder causes a piston to move so that the volume of the enclosed gas increases from 15 to 25 cubic inches. Assuming the relation between the pressure ($p$ lb/in²) and the volume ($v$ in³) to be $pv^{1.4} = 60$, find the work done.

If $A$ denotes the area of a cross section of the cylinder, $pA$ is the force exerted by the gas. A volume increase $\Delta v$ causes the piston to move a distance $\Delta v/A$, and the work corresponding to this displacement is $pA \cdot \dfrac{\Delta v}{A} = \dfrac{60}{v^{1.4}} \Delta v$. Then,

$$W = 60 \int_{15}^{25} \frac{dv}{v^{1.4}} = -\frac{60}{.4} v^{-.4} \Big]_{15}^{25} = -150\left(\frac{1}{25^{.4}} - \frac{1}{15^{.4}}\right) = 9.39 \text{ in. lb}$$

7.  A conical vessel is 12 ft across the top and 15 ft deep.
    If it contains a liquid weighing $w$ lb/ft³ to a depth of
    10 ft, find the work done in pumping the liquid to a
    height 3 ft above the top of the vessel.

    Consider the representative disk of the adjacent
    Fig. 40-4 whose radius is $x$, thickness is $\Delta y$, and mean
    distance from the bottom of the vessel is $y$. Its weight
    is $\pi w x^2 \Delta y$, and the work done in lifting it to the re-
    quired height is $\pi w x^2 (18 - y) \Delta y$.

    From similar triangles, $x/y = 6/15$ or $x = \frac{2}{5}y$.
    Then

$$W = \frac{4}{25}\pi w \int_0^{10} y^2(18 - y)\, dy = 560\pi w \text{ ft lb}$$

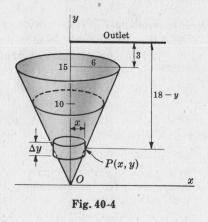

**Fig. 40-4**

# Supplementary Problems

8.  If a force of 80 lb stretches a 12 ft spring 1 foot, find the work done in stretching it  (a) from 12 to
    15 ft,  (b) from 15 to 16 ft.      *Ans.*  (a) 360 ft lb,  (b) 280 ft lb

9.  Two particles repel each other with a force inversely proportional to the square of the distance between
    them.  If one particle remains fixed at a point on the $x$-axis 2 units to the right of the origin, find the
    work done in moving the second along the $x$-axis from a point 3 units to the left of the origin up to
    the origin.      *Ans.*  $3k/10$

10. The force with which the earth attracts a weight $w$ lb at a distance $s$ miles from its center is
    $F = (4000)^2 w/s^2$, where the radius of the earth is taken as 4000 miles.  Find the work done against
    the force of gravity in moving a 1 lb mass from the surface of the earth to a point 1000 miles above
    the surface.      *Ans.*  800 mi lb

11. Find the work done against the force of gravity in moving a rocket weighing 8 tons to a height
    200 miles above the surface of the earth.      *Ans.*   32,000/21 mi ton

12. Find the work done in lifting 1000 lb of coal from a mine 1500 ft deep by means of a cable weighing
    2 lb/ft.      *Ans.*  1875 ft ton

13. A cistern is 10 ft square and 8 ft deep.  Find the work done in emptying it over the top if  (a) it is
    full of water,  (b) it is $\frac{3}{4}$ full of water.      *Ans.*  (a) 200,000 ft lb,  (b) 187,500 ft lb

14. A hemispherical tank of radius 3 ft is full of water.  (a) Find the work done in pumping the water
    over the edge of the tank.  (b) Find the work done in emptying the tank by an outlet pipe 2 ft above
    the top of the tank.      *Ans.*  (a) 3976 ft lb,  (b) 11,045 ft lb

15. How much work is done in filling an upright cylindrical tank of radius 3 ft and height 10 ft with
    liquid weighing $w$ lb/ft³ through a hole in the bottom?  How much if the tank is horizontal?
    *Ans.*  $450\pi w$ ft lb, $270\pi w$ ft lb

16. Show that the work done in pumping out a tank is equal to the work that would be done by lifting
    the contents from the center of gravity of the liquid to the outlet.

17. A 200 lb weight is to be dragged 60 ft up a 30° ramp.  If the force of friction opposing the motion is
    $N\mu$, where $\mu = 1/\sqrt{3}$ is the coefficient of friction and $N = 200 \cos 30°$  is the normal force between
    weight and ramp, find the work done.      *Ans.*  12,000 ft lb

18. Solve Problem 17 for a 45° ramp with the coefficient of friction $\mu = 1/\sqrt{2}$.      *Ans.*  $6000(1 + \sqrt{2})$ ft lb

19. Air is confined in a cylinder fitted with a piston.  At a pressure of 20 lb/ft², the volume is 100 ft³.
    Find the work done on the piston when the air is compressed to 2 ft³  (a) assuming $pv = $ constant,
    (b) assuming $pv^{1.4} = $ constant.      *Ans.*  (a) 7824 ft lb,  (b) 18,910 ft lb

# Chapter 41

## Length of Arc

**THE LENGTH OF AN ARC** $AB$ of a curve is by definition the limit of the sum of the lengths of a set of consecutive chords $AP_1$, $P_1P_2$, ..., $P_{n-1}B$, joining points on the arc, when the number of points is indefinitely increased in such a manner that the length of each chord approaches zero.

If $A(a,c)$ and $B(b,d)$ are two points on the curve $y = f(x)$, where $f(x)$ and its derivative $f'(x)$ are continuous on the interval $a \leqq x \leqq b$, the length of arc $AB$ is given by

$$s = \int_{AB} ds = \int_a^b \sqrt{1 + \left(\frac{dy}{dx}\right)^2}\, dx$$

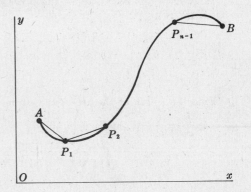

**Fig. 41-1**

Similarly, if $A(a,c)$ and $B(b,d)$ are two points on the curve $x = g(y)$, where $g(y)$ and its derivative with respect to $y$ are continuous on the interval $c \leqq y \leqq d$, the length of arc $AB$ is given by

$$s = \int_{AB} ds = \int_c^d \sqrt{1 + \left(\frac{dx}{dy}\right)^2}\, dy$$

If $A(u = u_1)$ and $B(u = u_2)$ are two points on a curve defined by the parametric equations $x = f(u)$, $y = g(u)$, and if conditions of continuity are satisfied, the length of arc $AB$ is given by

$$s = \int_{AB} ds = \int_{u_1}^{u_2} \sqrt{\left(\frac{dx}{du}\right)^2 + \left(\frac{dy}{du}\right)^2}\, du$$

For a derivation, see Problem 1.

## Solved Problems

1. Derive the arc length formula

$$s = \int_a^b \sqrt{1 + \left(\frac{dy}{dx}\right)^2}\, dx$$

Let the interval $a \leqq x \leqq b$ be divided into subintervals by the insertion of points $\xi_0 = a$, $\xi_1$, $\xi_2$, ..., $\xi_{n-1}$, $\xi_n = b$ and erect perpendiculars to determine the points $P_0 = A$, $P_1$, $P_2$, ..., $P_{n-1}$, $P_n = B$ on the arc as in Fig. 41-2. For the representative chord of the figure,

$$P_{k-1}P_k = \sqrt{(\Delta_k x)^2 + (\Delta_k y)^2} = \sqrt{1 + \left(\frac{\Delta_k y}{\Delta_k x}\right)^2}\, \Delta_k x$$

By the Law of the Mean (Chapter 21) there is at least one point, say $x = x_k$, on the arc $P_{k-1}P_k$, where the slope of the tangent $f'(x_k)$ is equal to the slope $\frac{\Delta_k y}{\Delta_k x}$ of the chord $P_{k-1}P_k$. Thus

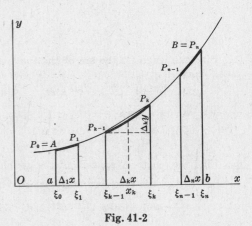

**Fig. 41-2**

$$P_{k-1}P_k = \sqrt{1 + \{f'(x_k)\}^2}\, \Delta_k x, \qquad \xi_{k-1} < x_k < \xi_k$$

and, using the Fundamental Theorem,

$$AB = \lim_{n \to +\infty} \sum_{k=1}^{n} \sqrt{1 + \{f'(x_k)\}^2}\, \Delta_k x = \int_a^b \sqrt{1 + \left(\frac{dy}{dx}\right)^2}\, dx$$

2.  Find the length of the arc of the curve $y = x^{3/2}$ from $x = 0$ to $x = 5$.

$\dfrac{dy}{dx} = \dfrac{3}{2} x^{1/2}$   and

$$s = \int_a^b \sqrt{1 + \left(\frac{dy}{dx}\right)^2}\, dx = \int_0^5 \sqrt{1 + \frac{9}{4}x}\, dx = \frac{8}{27}\left(1 + \frac{9}{4}x\right)^{3/2}\Big]_0^5 = \frac{335}{27}\ \text{units}$$

3.  Find the length of the arc of the curve $x = 3y^{3/2} - 1$ from $y = 0$ to $y = 4$.

$\dfrac{dx}{dy} = \dfrac{9}{2} y^{1/2}$   and

$$s = \int_c^d \sqrt{1 + \left(\frac{dx}{dy}\right)^2}\, dy = \int_0^4 \sqrt{1 + \frac{81}{4}y}\, dy = \frac{8}{243}(82\sqrt{82} - 1)\ \text{units}$$

4.  Find the length of the arc of $24xy = x^4 + 48$ from $x = 2$ to $x = 4$.

$\dfrac{dy}{dx} = \dfrac{x^4 - 16}{8x^2}$   and   $1 + \left(\dfrac{dy}{dx}\right)^2 = \dfrac{1}{64}\left(\dfrac{x^4 + 16}{x^2}\right)^2$.   Then   $s = \dfrac{1}{8}\displaystyle\int_2^4 \left(x^2 + \dfrac{16}{x^2}\right) dx = \dfrac{17}{6}\ \text{units}$

5.  Find the length of the arc of the catenary $y = \tfrac{1}{2}a(e^{x/a} + e^{-x/a})$ from $x = 0$ to $x = a$.

$\dfrac{dy}{dx} = \tfrac{1}{2}(e^{x/a} - e^{-x/a})$   and   $1 + \left(\dfrac{dy}{dx}\right)^2 = 1 + \tfrac{1}{4}(e^{2x/a} - 2 + e^{-2x/a}) = \tfrac{1}{4}(e^{x/a} + e^{-x/a})^2$.   Then

$$s = \frac{1}{2}\int_0^a (e^{x/a} + e^{-x/a})\, dx = \frac{1}{2}a\left[e^{x/a} - e^{-x/a}\right]_0^a = \frac{1}{2}a\left(e - \frac{1}{e}\right)\ \text{units}$$

6.  Find the length of the arc of the parabola $y^2 = 12x$ cut off by its latus rectum.

The required length is twice that from the point $(0, 0)$ to the point $(3, 6)$.

$\dfrac{dx}{dy} = \dfrac{y}{6}$   and   $1 + \left(\dfrac{dx}{dy}\right)^2 = \dfrac{36 + y^2}{36}$.   Then

$$s = 2\left(\frac{1}{6}\right)\int_0^6 \sqrt{36 + y^2}\, dy = \frac{1}{3}\left[\frac{1}{2}y\sqrt{36 + y^2} + 18\ln(y + \sqrt{36 + y^2})\right]_0^6$$
$$= 6\{\sqrt{2} + \ln(1 + \sqrt{2})\}\ \text{units}$$

7.  Find the length of the arc of the curve $x = t^2$, $y = t^3$ from $t = 0$ to $t = 4$.

$\dfrac{dx}{dt} = 2t,\ \dfrac{dy}{dt} = 3t^2$,   and   $\left(\dfrac{dx}{dt}\right)^2 + \left(\dfrac{dy}{dt}\right)^2 = 4t^2 + 9t^4 = 4t^2\left(1 + \dfrac{9}{4}t^2\right)$.   Then

$$s = \int_0^4 \sqrt{1 + \frac{9}{4}t^2}\cdot 2t\, dt = \frac{8}{27}(37\sqrt{37} - 1)\ \text{units}$$

8.  Find the length of an arch of the cycloid $x = \theta - \sin\theta$, $y = 1 - \cos\theta$.

An arch is described as $\theta$ varies from $\theta = 0$ to $\theta = 2\pi$.

$\dfrac{dx}{d\theta} = 1 - \cos\theta,\ \dfrac{dy}{d\theta} = \sin\theta$,   and   $\left(\dfrac{dx}{d\theta}\right)^2 + \left(\dfrac{dy}{d\theta}\right)^2 = 2(1 - \cos\theta) = 4\sin^2\tfrac{1}{2}\theta$.   Then

$$s = 2\int_0^{2\pi} \sin\frac{1}{2}\theta\, d\theta = -4\cos\frac{1}{2}\theta\Big]_0^{2\pi} = 8\ \text{units}$$

# Supplementary Problems

In Problems 9-20, find the length of the entire curve or indicated arc.

9.  $y^3 = 8x^2$ from $x = 1$ to $x = 8$.           *Ans.* $(104\sqrt{13} - 125)/27$ units

10. $6xy = x^4 + 3$ from $x = 1$ to $x = 2$.        *Ans.* 17/12 units

11. $y = \ln x$ from $x = 1$ to $x = 2\sqrt{2}$.      *Ans.* $3 - \sqrt{2} + \ln \frac{1}{2}(2 + \sqrt{2})$ units

12. $27y^2 = 4(x - 2)^3$ from $(2, 0)$ to $(11, 6\sqrt{3})$.   *Ans.* 14 units

13. $y = \ln (e^x - 1)/(e^x + 1)$ from $x = 2$ to $x = 4$.   *Ans.* $\ln (e^4 + 1) - 2$ units

14. $y = \ln (1 - x^2)$ from $x = 1/4$ to $x = 3/4$.   *Ans.* $\ln 21/5 - 1/2$ units

15. $y = \frac{1}{2}x^2 - \frac{1}{4}\ln x$ from $x = 1$ to $x = e$.   *Ans.* $\frac{1}{2}e^2 - \frac{1}{4}$ units

16. $y = \ln \cos x$ from $x = \pi/6$ to $x = \frac{1}{4}\pi$.   *Ans.* $\ln (1 + \sqrt{2})/\sqrt{3}$ units

17. $x = a \cos \theta$, $y = a \sin \theta$.         *Ans.* $2\pi a$ units

18. $x = e^t \cos t$, $y = e^t \sin t$ from $t = 0$ to $t = 4$.   *Ans.* $\sqrt{2}(e^4 - 1)$ units

19. $x = \ln\sqrt{1 + t^2}$, $y = \arctan t$ from $t = 0$ to $t = 1$.   *Ans.* $\ln (1 + \sqrt{2})$ units

20. $x = 2 \cos \theta + \cos 2\theta + 1$, $y = 2 \sin \theta + \sin 2\theta$.   *Ans.* 16 units

21. The position of a point at time $t$ is given as $x = \frac{1}{2}t^2$, $y = \frac{1}{9}(6t + 9)^{3/2}$. Find the distance the point travels from $t = 0$ to $t = 4$.    *Ans.* 20 units

22. Let $P(x, y)$ be a fixed point and $Q(x + \Delta x, y + \Delta y)$ be a variable point on the curve $y = f(x)$. See Fig. 17-1, Chapter 17. Show that

$$\lim_{Q \to P} \frac{\text{arc } PQ}{\text{chord } PQ} = \lim_{Q \to P} \frac{\Delta s}{\sqrt{(\Delta x)^2 + (\Delta y)^2}} = \frac{ds/dx}{\sqrt{1 + (dy/dx)^2}} = 1$$

23. (a) Show that the length of the first quadrant arc of $x = a \cos^3 \theta$, $y = a \sin^3 \theta$ is $3a/2$.

(b) Show that when the arc length of (a) is computed from $x^{2/3} + y^{2/3} = a^{2/3}$ we obtain $a^{1/3} \int_0^a \dfrac{dx}{x^{1/3}}$

in which the integrand is infinite at the lower limit of integration. Definite integrals of this type will be considered in Chapter 46.

24. A problem leading to the so-called *curve of pursuit* may be formulated: A dog at $A(1, 0)$ sees his master at $0(0, 0)$ walking along the $y$-axis and runs (in the first quadrant) to meet him. Find the path of the dog assuming that it is always headed toward its master and that each moves at a constant rate, $p$ for the master and $q > p$ for the dog. This problem can be solved in Chapter 70. Verify here that the equation $y = f(x)$ of the path may be found by integrating

$$y' = \tfrac{1}{2}(x^{p/q} - x^{-p/q})$$

*Hint.* Let $P(a, b)$, $0 < a < 1$, be a position of the dog and denote by $Q$ the intersection of the $y$-axis and the tangent to $y = f(x)$ at $P$. Find the time required for the dog to reach $P$ and show that the master is then at $Q$.

# Chapter 42

## Area of Surface of Revolution

**THE AREA OF THE SURFACE** generated by revolving the arc $AB$ of a continuous curve about a line in its plane is by definition the limit of the sum of the areas generated by the $n$ chords $AP_1, P_1P_2, \ldots, P_{n-1}B$, when revolved about the line, as the number of chords is indefinitely increased in such a manner that the length of each chord approaches zero.

If $A(a, c)$ and $B(b, d)$ are two points of the curve $y = f(x)$, where $f(x)$ and $f'(x)$ are continuous and $f(x)$ does not change sign on the interval $a \le x \le b$, the area of the surface generated by revolving the arc $AB$ about the $x$-axis is given by

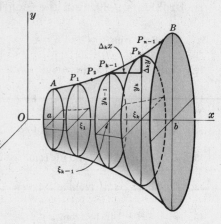

**Fig. 42-1**

$$S_x = 2\pi \int_{AB} y \, ds = 2\pi \int_a^b y \sqrt{1 + \left(\frac{dy}{dx}\right)^2} \, dx$$

When, in addition, $f'(x) \ne 0$ on the interval an alternate form is

$$S_x = 2\pi \int_{AB} y \, ds = 2\pi \int_c^d y \sqrt{1 + \left(\frac{dx}{dy}\right)^2} \, dy$$

If $A(a, c)$ and $B(b, d)$ are two points of the curve $x = g(y)$, where $g(y)$ and its derivative with respect to $y$ satisfy conditions similar to those listed in the above paragraph, the area of the surface generated by revolving the arc $AB$ about the $y$-axis is given by

$$S_y = 2\pi \int_{AB} x \, ds = 2\pi \int_a^b x \sqrt{1 + \left(\frac{dy}{dx}\right)^2} \, dx = 2\pi \int_c^d x \sqrt{1 + \left(\frac{dx}{dy}\right)^2} \, dy$$

If $A(u = u_1)$ and $B(u = u_2)$ are two points on the curve defined by the parametric equations $x = f(u)$, $y = g(u)$ and if conditions of continuity are satisfied, the area of the surface generated by revolving the arc $AB$ about the $x$-axis is given by

$$S_x = 2\pi \int_{AB} y \, ds = 2\pi \int_{u_1}^{u_2} y \sqrt{\left(\frac{dx}{du}\right)^2 + \left(\frac{dy}{du}\right)^2} \, du$$

and the area generated by revolving the arc $AB$ about the $y$-axis is given by

$$S_y = 2\pi \int_{AB} x \, ds = 2\pi \int_{u_1}^{u_2} x \sqrt{\left(\frac{dx}{du}\right)^2 + \left(\frac{dy}{du}\right)^2} \, du$$

## Solved Problems

1.  Find the area of the surface of revolution generated by revolving about the $x$-axis the arc of the parabola $y^2 = 12x$ from $x = 0$ to $x = 3$.

    (a) Solution using $S_x = 2\pi \int_a^b y \sqrt{1 + \left(\frac{dy}{dx}\right)^2} \, dx$.

    $$1 + \left(\frac{dy}{dx}\right)^2 = \frac{y^2 + 36}{y^2}$$

**Fig. 42-2**

202

and    $S_x = 2\pi \int_0^3 y \dfrac{\sqrt{y^2+36}}{y} \, dx = 2\pi \int_0^3 \sqrt{12x+36} \, dx = 24(2\sqrt{2}-1)\pi$ square units

(b) Solution using $S_x = 2\pi \int_c^d y\sqrt{1+\left(\dfrac{dx}{dy}\right)^2} \, dy$.   $\dfrac{dx}{dy} = \dfrac{y}{6}$, $1+\left(\dfrac{dx}{dy}\right)^2 = \dfrac{36+y^2}{36}$,  and

$$S_x = 2\pi \int_0^6 y\dfrac{\sqrt{36+y^2}}{6} \, dy = \dfrac{\pi}{9}(36+y^2)^{3/2}\Big]_0^6 = 24(2\sqrt{2}-1)\pi \text{ square units}$$

2.  Find the area of the surface of revolution generated by revolving about the $y$-axis the arc of $x = y^3$ from $y=0$ to $y=1$.

$$S_y = 2\pi \int_c^d x\sqrt{1+\left(\dfrac{dx}{dy}\right)^2} \, dy = 2\pi \int_0^1 y^3\sqrt{1+9y^4} \, dy$$

$$= \dfrac{\pi}{27}(1+9y^4)^{3/2}\Big]_0^1 = \dfrac{\pi}{27}(10\sqrt{10}-1) \text{ square units}$$

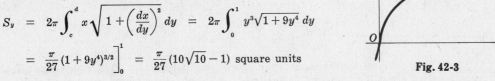

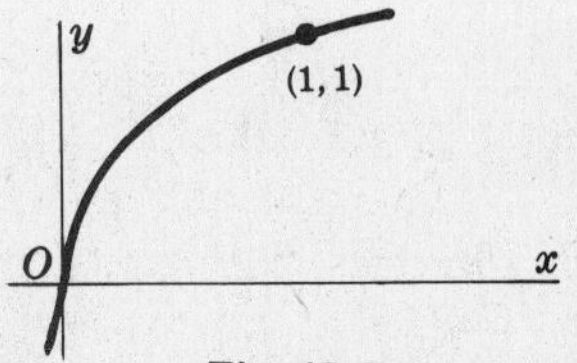

(1,1)

**Fig. 42-3**

3.  Find the area of the surface of revolution generated by revolving about the $x$-axis the arc of $y^2 + 4x = 2 \ln y$  from  $y=1$ to $y=3$.

$$S_x = 2\pi \int_c^d y\sqrt{1+\left(\dfrac{dx}{dy}\right)^2} \, dy = 2\pi \int_1^3 y\dfrac{1+y^2}{2y} \, dy = \pi \int_1^3 (1+y^2) \, dy = \dfrac{32}{3}\pi \text{ square units}$$

4.  Find the area of the surface of revolution generated by revolving a loop of the curve $8a^2y^2 = a^2x^2 - x^4$ about the $x$-axis.

$$\dfrac{dy}{dx} = \dfrac{a^2x - 2x^3}{8a^2y} \quad \text{and} \quad 1+\left(\dfrac{dy}{dx}\right)^2 = 1 + \dfrac{(a^2-2x^2)^2}{8a^2(a^2-x^2)} = \dfrac{(3a^2-2x^2)^2}{8a^2(a^2-x^2)}$$

$$S_x = 2\pi \int_a^b y\sqrt{1+\left(\dfrac{dy}{dx}\right)^2} \, dx = 2\pi \int_0^a \dfrac{x\sqrt{a^2-x^2}}{2a\sqrt{2}} \cdot \dfrac{3a^2-2x^2}{2a\sqrt{2}\sqrt{a^2-x^2}} \, dx$$

$$= \dfrac{\pi}{4a^2} \int_0^a (3a^2-2x^2)x \, dx = \tfrac{1}{4}\pi a^2 \text{ square units}$$

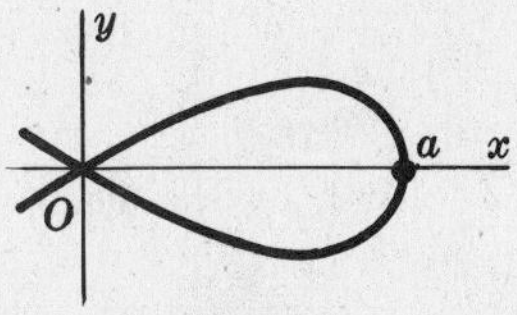

**Fig. 42-4**

5.  Find the area of the surface of revolution generated by revolving about the $x$-axis the ellipse $\dfrac{x^2}{16} + \dfrac{y^2}{4} = 1$.

$$S_x = 2\pi \int_a^b y\sqrt{1+\left(\dfrac{dy}{dx}\right)^2} \, dx = 2\pi \int_{-4}^4 y\dfrac{\sqrt{16y^2+x^2}}{4y} \, dx = \dfrac{1}{2}\pi \int_{-4}^4 \sqrt{64-3x^2} \, dx$$

$$= \dfrac{\pi}{2\sqrt{3}}\left[\dfrac{x\sqrt{3}}{2}\sqrt{64-3x^2} + 32 \arcsin\dfrac{x\sqrt{3}}{8}\right]_{-4}^4 = 8\pi\left(1+\dfrac{4\sqrt{3}}{9}\pi\right) \text{ square units}$$

6.  Find the area of the surface of revolution generated by revolving about the $x$-axis the hypocycloid $x = a\cos^3\theta$, $y = a\sin^3\theta$.

The required surface is generated by revolving the arc from $\theta=0$ to $\theta=\pi$.

$$\dfrac{dx}{d\theta} = -3a\cos^2\theta\sin\theta, \quad \dfrac{dy}{d\theta} = 3a\sin^2\theta\cos\theta, \quad \text{and} \quad \left(\dfrac{dx}{d\theta}\right)^2 + \left(\dfrac{dy}{d\theta}\right)^2 = 9a^2\cos^2\theta\sin^2\theta.$$

$$S_x = 2 \cdot 2\pi \int_0^{\pi/2} y\sqrt{\left(\dfrac{dx}{d\theta}\right)^2 + \left(\dfrac{dy}{d\theta}\right)^2} \, d\theta = 2 \cdot 2\pi \int_0^{\pi/2} (a\sin^3\theta)3a\cos\theta\sin\theta \, d\theta = \dfrac{12a^2\pi}{5} \text{ sq. un.}$$

*Note.* It would seem natural to write $2\pi \int_0^\pi (a\sin^3\theta)3a\cos\theta\sin\theta \, d\theta$,  but the value is then 0.

It must be remembered that while areas, volumes, etc., are given by definite integrals, not every definite integral can be interpreted as an area, etc.

7. Find the area of the surface of revolution generated by revolving about the $x$-axis the cardioid $x = 2 \cos \theta - \cos 2\theta$, $y = 2 \sin \theta - \sin 2\theta$.

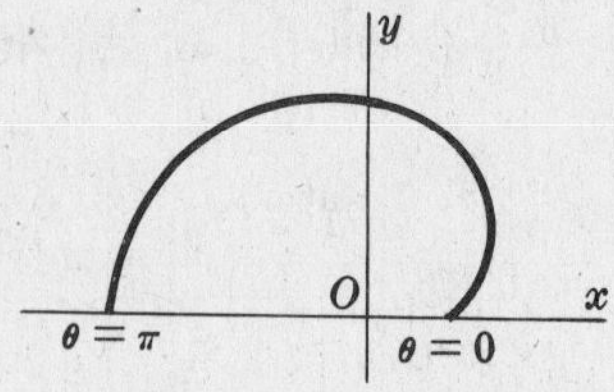

The required surface is generated by revolving the arc from $\theta = 0$ to $\theta = \pi$.

$dx/d\theta = -2 \sin \theta + 2 \sin 2\theta$, $dy/d\theta = 2 \cos \theta - 2 \cos 2\theta$, and
$(dx/d\theta)^2 + (dy/d\theta)^2 = 8(1 - \sin \theta \sin 2\theta - \cos \theta \cos 2\theta) = 8(1 - \cos \theta)$.

$$S_x = 2\pi \int_0^\pi (2 \sin \theta - \sin 2\theta) \cdot 2\sqrt{2} \sqrt{1 - \cos \theta} \, d\theta$$

**Fig. 42-5**

$$= 8\sqrt{2}\,\pi \int_0^\pi \sin \theta (1 - \cos \theta)^{3/2} \, d\theta = \frac{16\sqrt{2}}{5} \pi \, (1 - \cos \theta)^{5/2} \Big]_0^\pi = \frac{128\pi}{5} \text{ square units}$$

8. Derive: $S_x = 2\pi \int_a^b y \sqrt{1 + \left(\dfrac{dy}{dx}\right)^2} \, dx$.

Let the arc $AB$ be approximated by $n$ chords, as in Fig. 42-1. The representative chord $P_{k-1}P_k$ when revolved about the $x$-axis generates the frustum of a cone whose bases are of radii $y_{k-1}$ and $y_k$, whose slant height is

$$P_{k-1}P_k = \sqrt{(\Delta_k x)^2 + (\Delta_k y)^2} = \sqrt{1 + \left(\frac{\Delta_k y}{\Delta_k x}\right)^2} \, \Delta_k x = \sqrt{1 + \{f'(x_k)\}^2} \, \Delta_k x$$

(see Problem 1, Chapter 41), and whose lateral area (circumference of midsection × slant height) is

$$S_k = 2\pi \left(\frac{y_{k-1} + y_k}{2}\right) \sqrt{1 + \{f'(x_k)\}^2} \, \Delta_k x$$

Since $f(x)$ is continuous, there exists at least one point $x_k'$ on the arc $P_{k-1}P_k$ such that
$$f(x_k') = \tfrac{1}{2}(y_{k-1} + y_k) = \tfrac{1}{2}\{f(\xi_{k-1}) + f(\xi_k)\}$$

Hence
$$S_k = 2\pi f(x_k') \sqrt{1 + \{f'(x_k)\}^2} \, \Delta_k x$$

and, by the Theorem of Bliss,

$$S_x = \lim_{n \to +\infty} \sum_{k=1}^n S_k = \lim_{n \to +\infty} \sum_{k=1}^n 2\pi f(x_k') \sqrt{1 + \{f'(x_k)\}^2} \, \Delta_k x$$

$$= 2\pi \int_a^b f(x) \sqrt{1 + \{f'(x)\}^2} \, dx = 2\pi \int_a^b y \sqrt{1 + \left(\frac{dy}{dx}\right)^2} \, dx$$

# Supplementary Problems

In Problems 9-18, find the area of the surface generated by revolving the given arc about the given axis.

9. $y = mx$ from $x = 0$ to $x = 2$; $x$-axis.      *Ans.* $4m\pi\sqrt{1 + m^2}$ square units

10. $y = \tfrac{1}{3}x^3$ from $x = 0$ to $x = 3$; $x$-axis.      *Ans.* $\pi(82\sqrt{82} - 1)/9$ square units

11. $y = \tfrac{1}{3}x^3$ from $x = 0$ to $x = 3$; $y$-axis.      *Ans.* $\tfrac{1}{2}\pi[9\sqrt{82} + \ln(9 + \sqrt{82})]$ square units

12. $8y^2 = x^2(1 - x^2)$, loop; $x$-axis.      *Ans.* $\tfrac{1}{4}\pi$ square units

13. $y = x^3/6 + 1/2x$ from $x = 1$ to $x = 2$; $y$-axis.      *Ans.* $(15/4 + \ln 2)\pi$ square units

14. $y = \ln x$ from $x = 1$ to $x = 7$; $y$-axis.      *Ans.* $[34\sqrt{2} + \ln(3 + 2\sqrt{2})]\pi$ square units

15. $9y^2 = x(3 - x)^2$ loop; $y$-axis.      *Ans.* $28\pi\sqrt{3}/5$ square units

16. $y = a \cosh x/a$ from $x = -a$ to $x = a$; $x$-axis.      *Ans.* $\tfrac{1}{2}\pi a^2(e^2 - e^{-2} + 4)$ square units

17. An arch of $x = a(\theta - \sin \theta)$, $y = a(1 - \cos \theta)$; $x$-axis.      *Ans.* $64\pi a^2/3$ square units

18. $x = e^t \cos t$, $y = e^t \sin t$ from $t = 0$ to $t = \tfrac{1}{2}\pi$; $x$-axis.      *Ans.* $2\pi\sqrt{2}\,(2e^\pi + 1)/5$ square units

19. Find the surface of a zone cut from a sphere of radius $r$ by two parallel planes, each at a distance $\tfrac{1}{2}a$ from the center.      *Ans.* $2\pi ar$.

20. Find the surface cut from a sphere of radius $r$ by a circular cone of half angle $\alpha$ with its vertex at the center of the sphere.      *Ans.* $2\pi r^2(1 - \cos \alpha)$ square units

# Chapter 43

## Centroid and Moment of Inertia
### *Arcs and Surfaces of Revolution*

**CENTROID OF AN ARC.** The coordinates $(\bar{x}, \bar{y})$ of the centroid of an arc $AB$ of a plane curve of equation $F(x, y) = 0$ or $x = f(u)$, $y = g(u)$ satisfy the relations

$$\bar{x} \cdot s = \bar{x}\int_{AB} ds = \int_{AB} x\, ds \quad \text{and} \quad \bar{y} \cdot s = \bar{y}\int_{AB} ds = \int_{AB} y\, ds$$

See Problems 1-2.

**SECOND THEOREM OF PAPPUS.** If an arc of a curve is revolved about an axis in its plane but not crossing the arc, the area of the surface generated is equal to the product of the length of the arc and the length of the path described by the centroid of the arc.

See Problem 3.

**MOMENTS OF INERTIA OF AN ARC.** The moments of inertia with respect to the coordinate axes of an arc $AB$ of a curve (a piece of homogeneous fine wire, for example) are given respectively by

$$I_x = \int_{AB} y^2\, ds \quad \text{and} \quad I_y = \int_{AB} x^2\, ds$$

See Problems 4-5.

**CENTROID OF A SURFACE OF REVOLUTION.** The coordinate $\bar{x}$ of the centroid of the surface generated by revolving an arc $AB$ of a curve about the $x$-axis satisfies the relation

$$\bar{x} \cdot S_x = 2\pi \int_{AB} x \cdot y\, ds$$

**MOMENT OF INERTIA OF A SURFACE OF REVOLUTION.** The moment of inertia with respect to the axis of rotation of the surface generated by revolving an arc $AB$ of a curve about the $x$-axis is given by

$$I_x = 2\pi \int_{AB} y^2 \cdot y\, ds$$

## Solved Problems

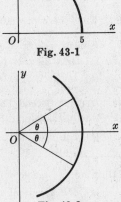

**Fig. 43-1**

**Fig. 43-2**

1. Find the centroid of the first quadrant arc of the circle $x^2 + y^2 = 25$.

   $\dfrac{dy}{dx} = -\dfrac{x}{y}$ and $1 + \left(\dfrac{dy}{dx}\right)^2 = 1 + \dfrac{x^2}{y^2} = \dfrac{25}{y^2}$. Since $s = \dfrac{5}{2}\pi$,

   $$\frac{5}{2}\pi\bar{y} = \int_0^5 y\sqrt{1 + \left(\frac{dy}{dx}\right)^2}\, dx = \int_0^5 5\, dx = 25 \quad \text{or} \quad \bar{y} = 10/\pi$$

   By symmetry, $\bar{x} = \bar{y}$ and the coordinates of the centroid are $(10/\pi, 10/\pi)$.

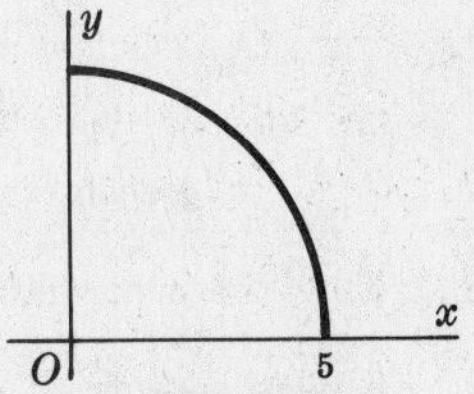

2. Find the centroid of a circular arc of radius $r$ and central angle $2\theta$.

   Take the arc as in Fig. 43-2, so that $\bar{x}$ is identical with the abscissa of the centroid of the upper half of the arc and $\bar{y} = 0$.

   $\dfrac{dx}{dy} = -\dfrac{y}{x}$ and $1 + \left(\dfrac{dx}{dy}\right)^2 = \dfrac{r^2}{x^2}$. For the upper half of the arc, $s = r\theta$,

   $$r\theta \cdot \bar{x} = \int_0^{r\sin\theta} x\sqrt{1 + \left(\frac{dx}{dy}\right)^2}\, dy = r\int_0^{r\sin\theta} dy = r^2\sin\theta$$

   and $\bar{x} = (r\sin\theta)/\theta$. Thus, the centroid is on the bisecting radius at a distance $(r\sin\theta)/\theta$ from the center of the circle.

205

3. Find the area of the surface generated by revolving the rectangle of dimensions $a$ by $b$ about an axis $c$ ($> a, b$) units from the centroid.

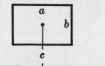

    The perimeter of the rectangle is $2(a + b)$ and the centroid describes a circle of radius $c$.  Then

$$S = 2(a+b) \cdot 2\pi c = 4\pi(a+b)c \text{ square units}$$

**Fig. 43-3**

4. Find the moment of inertia of the arc of a circle with respect to a fixed diameter.

    Take the circle as in Fig. 43-4, with the fixed diameter along the $x$-axis.  The required moment is 4 times that of the first quadrant arc.

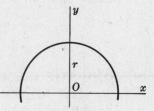

$$\frac{dy}{dx} = -\frac{x}{y} \text{ and } \sqrt{1+\left(\frac{dy}{dx}\right)^2} = \frac{r}{y}. \text{ Also, } s = 2\pi r. \text{ Then}$$

$$I_x = 4\int_0^r y^2\, ds = 4\int_0^r y^2 \cdot \frac{r}{y}\, dx = 4r\int_0^r \sqrt{r^2 - x^2}\, dx$$

$$= 4r\left[\frac{1}{2}x\sqrt{r^2 - x^2} + \frac{1}{2}r^2 \arcsin\frac{x}{r}\right]_0^r = \pi r^3 = \frac{1}{2}r^2 s$$

**Fig. 43-4**

5. Find the moment of inertia with respect to the $x$-axis of the arc of the hypocycloid $x = a\sin^3\theta$, $y = a\cos^3\theta$.

    The required moment is 4 times that of the first quadrant arc.

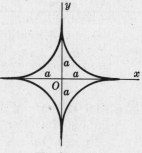

$$\frac{dx}{d\theta} = 3a\sin^2\theta\cos\theta, \quad \frac{dy}{d\theta} = -3a\cos^2\theta\sin\theta, \quad \text{and}$$

$$s = 4\int ds = 4\int_0^{\pi/2} 3a\sin\theta\cos\theta\, d\theta = 6a$$

$$I_x = 4\int y^2\, ds = 12a^3\int_0^{\pi/2} \cos^6\theta\sin\theta\cos\theta\, d\theta = \frac{3}{2}a^3 = \frac{1}{4}a^2 s$$

**Fig. 43-5**

# Supplementary Problems

6. Find the centroid of
   (a) the first quadrant arc of $x^{2/3} + y^{2/3} = a^{2/3}$.  Use $s = 3a/2$.    Ans. $(2a/5, 2a/5)$
   (b) the first quadrant arc of the loop of $9y^2 = x(3-x)^2$.  Use $s = 2\sqrt{3}$.    Ans. $(7/5, \sqrt{3}/4)$
   (c) the first arch of $x = a(\theta - \sin\theta)$, $y = a(1 - \cos\theta)$.    Ans. $(\pi a, 4a/3)$
   (d) the first quadrant arc of $x = a\cos^3\theta$, $y = a\sin^3\theta$.    Ans. See (a).

7. Find the moment of inertia of the given arc with respect to the given line:
   (a) loop of $9y^2 = x(3-x)^2$; $x$-axis, $y$-axis.  Use $s = 4\sqrt{3}$.    Ans. $I_x = 8s/35$, $I_y = 99s/35$
   (b) $y = a\cosh x/a$ from $x = 0$ to $x = a$; $x$-axis.    Ans. $(a^2 + \frac{1}{3}s^2)s$

8. Find the centroid of a hemispherical surface.    Ans. $\bar{y} = \frac{1}{2}r$

9. Find the centroid of the surface generated by revolving
   (a) $4y + 3x = 8$ from $x = 0$ to $x = 2$ about the $x$-axis.    Ans. $\bar{x} = 4/5$
   (b) an arch of $x = a(\theta - \sin\theta)$, $y = a(1 - \cos\theta)$ about the $y$-axis.    Ans. $\bar{y} = 4a/3$

10. Use the Second Theorem of Pappus to obtain
    (a) the centroid of the first quadrant arc of a circle of radius $r$.    Ans. $(2r/\pi, 2r/\pi)$
    (b) the area of the surface generated by revolving an equilateral triangle of side $a$ about an axis $c$ units from its centroid.    Ans. $6\pi ac$ square units

11. Find the moment of inertia with respect to the axis of rotation of
    (a) the spherical surface of radius $r$.    Ans. $\frac{2}{3}Sr^2$
    (b) the lateral surface of a cone generated by revolving the line $y = 2x$ from $x = 0$ to $x = 2$ about the $x$-axis.    Ans. $8S$

12. Derive each of the formulas of this chapter.

# Chapter 44

## Plane Area and Centroid of Area
### *Polar Coordinates*

**THE PLANE AREA** bounded by the curve $\rho = f(\theta)$ and the radius vectors $\theta = \theta_1$ and $\theta = \theta_2$ is given by

$$A = \frac{1}{2} \int_{\theta_1}^{\theta_2} \rho^2 \, d\theta$$

When using polar coordinates, considerable care must be taken to determine the proper limits of integration. This requires that, by taking advantage of any symmetry, the limits be as narrow as possible.

See Problems 1-7.

**CENTROID OF PLANE AREA.** The coordinates $(\bar{x}, \bar{y})$ of the centroid of a plane area bounded by the curve $\rho = f(\theta)$ and the radius vectors $\theta = \theta_1$ and $\theta = \theta_2$ are given by

$$A\bar{x} = \bar{x} \cdot \frac{1}{2} \int_{\theta_1}^{\theta_2} \rho^2 \, d\theta = \frac{1}{3} \int_{\theta_1}^{\theta_2} \rho^3 \cos \theta \, d\theta$$

$$= \frac{1}{2} \int_{\theta_1}^{\theta_2} \frac{2}{3} x \cdot \rho^2 \, d\theta$$

and

$$A\bar{y} = \bar{y} \cdot \frac{1}{2} \int_{\theta_1}^{\theta_2} \rho^2 \, d\theta = \frac{1}{3} \int_{\theta_1}^{\theta_2} \rho^3 \sin \theta \, d\theta$$

$$= \frac{1}{2} \int_{\theta_1}^{\theta_2} \frac{2}{3} y \cdot \rho^2 \, d\theta$$

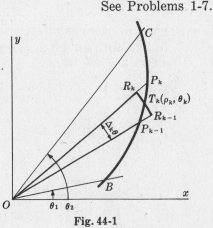

Fig. 44-1

See Problems 8-10.

## Solved Problems

1. Derive $A = \frac{1}{2} \int_{\theta_1}^{\theta_2} \rho^2 \, d\theta$.

   Let the angle $BOC$ of the figure above be divided into $n$ parts by rays $OP_0 = OB, OP_1, OP_2, \ldots, OP_{n-1}, OP_n = OC$. The figure shows a representative slice $P_{k-1}OP_k$ of central angle $\Delta_k\theta$ and its approximating circular sector $R_{k-1}OR_k$ of radius $\rho_k$, central angle $\Delta_k\theta$, and (see Problem 15($r$), Chap. 34) of area $\frac{1}{2}\rho_k^2 \Delta_k\theta = \frac{1}{2}\{f(\theta_k)\}^2 \Delta_k\theta$. Hence, by the Fundamental Theorem,

$$A = \lim_{n \to +\infty} \sum_{k=1}^{n} \frac{1}{2} \{f(\theta_k)\}^2 \Delta_k\theta = \frac{1}{2} \int_{\theta_1}^{\theta_2} \{f(\theta)\}^2 \, d\theta = \frac{1}{2} \int_{\theta_1}^{\theta_2} \rho^2 \, d\theta$$

2. Find the area bounded by the curve $\rho^2 = a^2 \cos 2\theta$.

   From Fig. 44-2 below it is seen that the required area consists of four equal pieces, one of which is swept over as $\theta$ varies from $\theta = 0$ to $\theta = \frac{1}{4}\pi$. Thus,

$$A = 4 \cdot \frac{1}{2} \int_{0}^{\pi/4} \rho^2 \, d\theta = 2a^2 \int_{0}^{\pi/4} \cos 2\theta \, d\theta = a^2 \sin 2\theta \Big]_{0}^{\pi/4} = a^2 \text{ square units}$$

Since portions of the required area lie in each of the quadrants, it might appear reasonable to use

$$\frac{1}{2}\int_0^{2\pi} \rho^2\, d\theta = \frac{1}{2}a^2\int_0^{2\pi} \cos 2\theta\, d\theta = \frac{1}{4}a^2\sin 2\theta\Big]_0^{2\pi} = 0$$

or

$$2\cdot\frac{1}{2}\int_0^{\pi}\rho^2\, d\theta = a^2\int_0^{\pi}\cos 2\theta\, d\theta = 0$$

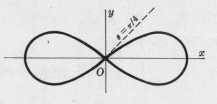

**Fig. 44-2**

The reason for these incorrect results may be found by considering

$$\frac{1}{2}\int_0^{\pi}\rho^2\, d\theta = \frac{1}{2}\int_0^{\pi/4}\rho^2\, d\theta + \frac{1}{2}\int_{\pi/4}^{3\pi/4}\rho^2\, d\theta + \frac{1}{2}\int_{3\pi/4}^{\pi}\rho^2\, d\theta = \frac{1}{4}a^2 - \frac{1}{2}a^2 + \frac{1}{4}a^2$$

On the intervals $[0,\pi/4]$ and $[3\pi/4,\pi]$, $\rho = a\sqrt{\cos 2\theta}$ is real; thus the first and third integrals give the areas swept over as $\theta$ ranges over these intervals. But on the interval $[\pi/4, 3\pi/4]$, $\rho^2 < 0$ and $\rho$ is imaginary. Thus, while $\dfrac{1}{2}\displaystyle\int_{\pi/4}^{3\pi/4} a^2\cos 2\theta\, d\theta$ is a perfectly valid integral, it cannot be interpreted here as an area.

3.  Find the area bounded by the three-leaved rose $\rho = a\cos 3\theta$.

The required area is 6 times the area shaded in Fig. 44-3 below, that is, the area swept over as $\theta$ varies from 0 to $\pi/6$.  Hence,

$$A = 6\cdot\frac{1}{2}\int_{\theta_1}^{\theta_2}\rho^2\, d\theta = 6\cdot\frac{1}{2}\int_0^{\pi/6} a^2\cos^2 3\theta\, d\theta = 3a^2\int_0^{\pi/6}(\tfrac{1}{2}+\tfrac{1}{2}\cos 6\theta)\, d\theta = \tfrac{1}{4}\pi a^2 \text{ sq. un.}$$

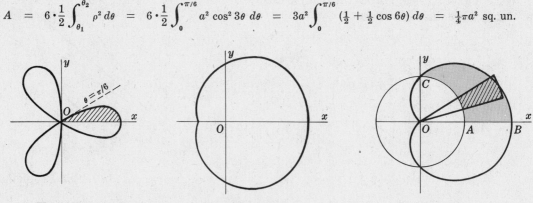

**Fig. 44-3**            **Fig. 44-4**            **Fig. 44-5**

4.  Find the area bounded by the limacon $\rho = 2 + \cos\theta$ shown in Fig. 44-4 above.

The required area is twice that swept over as $\theta$ varies from 0 to $\pi$.

$$A = 2\cdot\frac{1}{2}\int_0^{\pi}(2+\cos\theta)^2\, d\theta = \int_0^{\pi}(4 + 4\cos\theta + \cos^2\theta)\, d\theta$$

$$= \left[4\theta + 4\sin\theta + \frac{1}{2}\theta + \frac{1}{4}\sin 2\theta\right]_0^{\pi} = 9\pi/2 \text{ square units}$$

5.  Find the area inside the cardioid $\rho = 1 + \cos\theta$ and outside the circle $\rho = 1$.

In Fig. 44-5 above, area $ABC$ = area $OBC$ − area $OAC$ is one half the required area.  Thus,

$$A = 2\cdot\frac{1}{2}\int_0^{\pi/2}(1+\cos\theta)^2\, d\theta - 2\cdot\frac{1}{2}\int_0^{\pi/2}(1)^2\, d\theta = \int_0^{\pi/2}(2\cos\theta + \cos^2\theta)\, d\theta = 2 + \tfrac{1}{4}\pi \text{ sq. un.}$$

6.  Find the area of each loop of $\rho = \frac{1}{2} + \cos\theta$.

*Larger loop.*  The required area is twice that swept over as $\theta$ varies from 0 to $2\pi/3$.  Hence,

$$A = 2\cdot\frac{1}{2}\int_0^{2\pi/3}(\tfrac{1}{2}+\cos\theta)^2\, d\theta = \int_0^{2\pi/3}(\tfrac{1}{4}+\cos\theta+\cos^2\theta)\, d\theta$$

$$= \frac{\pi}{2} + \frac{3\sqrt{3}}{8} \text{ square units}$$

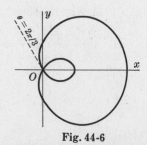

**Fig. 44-6**

*Smaller loop.*  The required area is twice that swept over as $\theta$ varies from $2\pi/3$ to $\pi$.  Hence,

$$A \;=\; 2\cdot\frac{1}{2}\int_{2\pi/3}^{\pi}(\tfrac{1}{2}+\cos\theta)^2\,d\theta \;=\; \frac{\pi}{4}-\frac{3\sqrt{3}}{8}\ \text{square units}$$

7.  Find the area common to the circle $\rho = 3\cos\theta$ and the cardioid $\rho = 1+\cos\theta$.

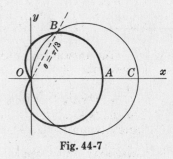

Fig. 44-7

Area $OAB$ consists of two portions, one swept over by the radius vector $\rho = 1+\cos\theta$ as $\theta$ varies from 0 to $\pi/3$ and the other swept over by $\rho = 3\cos\theta$ as $\theta$ varies from $\pi/3$ to $\pi/2$.

$$A \;=\; 2\cdot\frac{1}{2}\int_0^{\pi/3}(1+\cos\theta)^2\,d\theta \;+\; 2\cdot\frac{1}{2}\int_{\pi/3}^{\pi/2}9\cos^2\theta\,d\theta$$

$$=\; 5\pi/4\ \text{square units}$$

8.  Derive the formulas  $A\bar{x} = \dfrac{1}{3}\displaystyle\int_{\theta_1}^{\theta_2}\rho^3\cos\theta\,d\theta$,  $A\bar{y} = \dfrac{1}{3}\displaystyle\int_{\theta_1}^{\theta_2}\rho^3\sin\theta\,d\theta$,  where $(\bar{x},\bar{y})$ are the coordinates of the centroid of the plane area $BOC$ of Fig. 44-1.

Consider the representative approximating circular sector $R_{k-1}OR_k$ and suppose, for convenience that $OT_k$ bisects the angle $P_{k-1}OP_k$.  To approximate the centroid $C_k(\bar{x}_k,\bar{y}_k)$ of this sector, consider it to be a true triangle.  Then its centroid will lie on $OT_k$ at a distance $\frac{2}{3}\rho_k$ from $O$; thus, approximately,

$$\bar{x}_k = \tfrac{2}{3}\rho_k\cos\theta_k = \tfrac{2}{3}f(\theta_k)\cos\theta_k \quad\text{and}\quad \bar{y}_k = \tfrac{2}{3}f(\theta_k)\sin\theta_k$$

Now the first moment of the sector about the $y$-axis is

$$\bar{x}_k\cdot\tfrac{1}{2}\rho_k^2\,\Delta_k\theta \;=\; \tfrac{2}{3}\rho_k\cos\theta_k\cdot\tfrac{1}{2}\rho_k^2\,\Delta_k\theta \;=\; \tfrac{1}{3}\{f(\theta_k)\}^3\cos\theta_k\,\Delta_k\theta$$

and, by the Fundamental Theorem,

$$A\bar{x} \;=\; \lim_{n\to+\infty}\sum_{k=1}^{n}\frac{1}{3}\{f(\theta_k)\}^3\cos\theta_k\,\Delta_k\theta \;=\; \frac{1}{3}\int_{\theta_1}^{\theta_2}\rho^3\cos\theta\,d\theta$$

It is left as an exercise to obtain the formula for $A\bar{y}$.

*Note.*  From Problem 8, Chapter 37, the centroid of the sector $R_{k-1}OR_k$ lies on $OT_k$ at a distance $\dfrac{2\rho_k\sin\frac{1}{2}\Delta_k\theta}{3\cdot\frac{1}{2}\Delta_k\theta}$ from $O$.  The reader may wish to use this to derive the formulas.

9.  Find the centroid of the area of the first quadrant loop of the rose $\rho = \sin 2\theta$.

$$A \;=\; \frac{1}{2}\int_0^{\pi/2}\sin^2 2\theta\,d\theta \;=\; \frac{1}{4}\left[\theta-\frac{1}{4}\sin 4\theta\right]_0^{\pi/2} \;=\; \frac{\pi}{8}$$

Fig. 44-8

$$\frac{\pi}{8}\bar{x} \;=\; \frac{1}{3}\int_0^{\pi/2}\rho^3\cos\theta\,d\theta \;=\; \frac{1}{3}\int_0^{\pi/2}\sin^3 2\theta\cos\theta\,d\theta$$

$$=\; \frac{8}{3}\int_0^{\pi/2}\sin^3\theta\cos^4\theta\,d\theta \;=\; \frac{8}{3}\int_0^{\pi/2}(1-\cos^2\theta)\cos^4\theta\sin\theta\,d\theta \;=\; \frac{16}{105},\ \text{and}\ \bar{x} = \frac{128}{105\pi}.$$

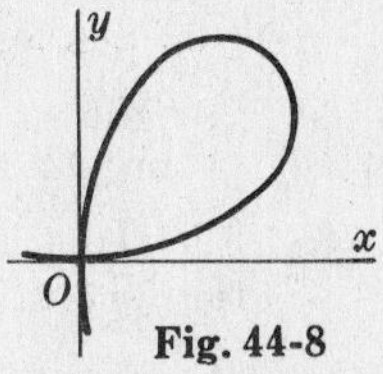

By symmetry, $\bar{y} = 128/105\pi$.  The coordinates of the centroid are $(128/105\pi,\,128/105\pi)$.

10.  Find the centroid of the first quadrant area bounded by the parabola $\rho = \dfrac{6}{1+\cos\theta}$ shown in Fig. 44-9 below.

$$A \;=\; \frac{1}{2}\int_0^{\pi/2}\frac{36}{(1+\cos\theta)^2}\,d\theta \;=\; \frac{9}{2}\int_0^{\pi/2}\sec^4\tfrac{1}{2}\theta\,d\theta$$

$$=\; \frac{9}{2}\int_0^{\pi/2}(1+\tan^2\tfrac{1}{2}\theta)\sec^2\tfrac{1}{2}\theta\,d\theta \;=\; 9\left[\tan\tfrac{1}{2}\theta+\frac{1}{3}\tan^3\tfrac{1}{2}\theta\right]_0^{\pi/2} \;=\; 12$$

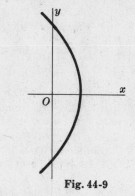

$$12\bar{x} = \frac{1}{3}\int_0^{\pi/2} \frac{216\cos\theta}{(1+\cos\theta)^3}\,d\theta = 9\int_0^{\pi/2} \frac{2\cos^2\frac{1}{2}\theta - 1}{\cos^6\frac{1}{2}\theta}\,d\theta$$

$$= 9\int_0^{\pi/2} (2\sec^4\tfrac{1}{2}\theta - \sec^6\tfrac{1}{2}\theta)\,d\theta = 18\left[\tan\tfrac{1}{2}\theta - \tfrac{1}{5}\tan^5\tfrac{1}{2}\theta\right]_0^{\pi/2}$$

$$= 72/5, \text{ and } \bar{x} = 6/5.$$

$$12\bar{y} = \frac{1}{3}\int_0^{\pi/2} \frac{216\sin\theta}{(1+\cos\theta)^3}\,d\theta = 27, \text{ and } \bar{y} = 9/4.$$

The centroid is $(6/5,\ 9/4)$.

**Fig. 44-9**

# Supplementary Problems

**11.** Find the area bounded by each of the following curves.

    (a) $\rho^2 = 1 + \cos 2\theta$          *Ans.* $\pi$ sq. un.

    (b) $\rho^2 = a^2\sin\theta\,(1-\cos\theta)$      *Ans.* $a^2$ sq. un.

    (c) $\rho = 4\cos\theta$          *Ans.* $4\pi$ sq. un.

    (d) $\rho = a\cos 2\theta$          *Ans.* $\frac{1}{2}\pi a^2$ sq. un.

    (e) $\rho = 4\sin^2\theta$          *Ans.* $6\pi$ sq. un.

    (f) $\rho = 4(1-\sin\theta)$          *Ans.* $24\pi$ sq. un.

**12.** Find the area

    (a) inside $\rho = \cos\theta$ and outside $\rho = 1 - \cos\theta$.      *Ans.* $(\sqrt{3} - \pi/3)$ sq. un.

    (b) inside $\rho = \sin\theta$ and outside $\rho = 1 - \cos\theta$.      *Ans.* $(1 - \pi/4)$ sq. un.

    (c) between the inner and outer ovals of $\rho^2 = a^2(1 + \sin\theta)$.      *Ans.* $4a^2$ sq. un.

    (d) between the loops of $\rho = 2 - 4\sin\theta$.      *Ans.* $4(\pi + 3\sqrt{3})$ sq. un.

**13.** (a) For the spiral of Archimedes, $\rho = a\theta$, show that the additional area swept over by the $n$th revolution, $n > 2$, is $(n - 1)$ times that added by the second revolution.

    (b) For the equiangular spiral $\rho = ae^\theta$, show that the additional area swept over by the $n$th revolution, $n > 2$, is $e^{4\pi}$ times that added by the previous sweep.

**14.** Find the centroid of the following areas:

    (a) Right half of $\rho = a(1 - \sin\theta)$.      *Ans.* $(16a/9\pi,\ -5a/6)$

    (b) First quadrant area of $\rho = 4\sin^2\theta$.      *Ans.* $(128/63\pi,\ 2048/315\pi)$

    (c) Upper half of $\rho = 2 + \cos\theta$.      *Ans.* $(17/18,\ 80/27\pi)$

    (d) First quadrant area of $\rho = 1 + \cos\theta$.      *Ans.* $\left(\dfrac{16 + 5\pi}{16 + 6\pi},\ \dfrac{10}{8 + 3\pi}\right)$

    (e) First quadrant area of Problem 5.      *Ans.* $\left(\dfrac{32 + 15\pi}{48 + 6\pi},\ \dfrac{22}{24 + 3\pi}\right)$

**15.** Use the First Theorem of Pappus to obtain the volume generated by revolving

    (a) $\rho = a(1 - \sin\theta)$ about the 90° line.      *Ans.* $8\pi a^3/3$ cu. un.

    (b) $\rho = 2 + \cos\theta$ about the polar axis.      *Ans.* $40\pi/3$ cu. un.

# Chapter 45

## Length and Centroid of Arc.  Area of Surface Revolution
### *Polar Coordinates*

**THE LENGTH OF THE ARC** of the curve $\rho = f(\theta)$ from $\theta = \theta_1$ to $\theta = \theta_2$ is given by

$$ s \;=\; \int_{\theta_1}^{\theta_2} ds \;=\; \int_{\theta_1}^{\theta_2} \sqrt{\rho^2 + \left(\frac{d\rho}{d\theta}\right)^2}\, d\theta $$

See Problems 1-4.

**CENTROID OF ARC.**  The coordinates $(\bar{x}, \bar{y})$ of the centroid of the arc of the curve $\rho = f(\theta)$ from $\theta = \theta_1$ to $\theta = \theta_2$ satisfy the relations

$$ \bar{x} \cdot s \;=\; \bar{x} \int_{\theta_1}^{\theta_2} ds \;=\; \int_{\theta_1}^{\theta_2} \rho \cos\theta\, ds \;=\; \int_{\theta_1}^{\theta_2} x\, ds $$

$$ \bar{y} \cdot s \;=\; \bar{y} \int_{\theta_1}^{\theta_2} ds \;=\; \int_{\theta_1}^{\theta_2} \rho \sin\theta\, ds \;=\; \int_{\theta_1}^{\theta_2} y\, ds $$

See Problems 5-6.

**THE AREA OF THE SURFACE** generated by revolving the arc of the curve $\rho = f(\theta)$ from $\theta = \theta_1$ to $\theta = \theta_2$ about

the polar axis is $\qquad S_x \;=\; 2\pi \int_{\theta_1}^{\theta_2} y\, ds \;=\; 2\pi \int_{\theta_1}^{\theta_2} \rho \sin\theta\, ds$

the 90°-line is $\qquad S_y \;=\; 2\pi \int_{\theta_1}^{\theta_2} x\, ds \;=\; 2\pi \int_{\theta_1}^{\theta_2} \rho \cos\theta\, ds$

The limits of integration should be taken as narrow as possible.

See Problems 7-10.

## Solved Problems

1. Find the length of the spiral $\rho = e^{2\theta}$ from $\theta = 0$ to $\theta = 2\pi$.

    $d\rho/d\theta = 2e^{2\theta}$ and $\rho^2 + (d\rho/d\theta)^2 = 5e^{4\theta}$.

    $$ s \;=\; \int_0^{2\pi} \sqrt{\rho^2 + (d\rho/d\theta)^2}\, d\theta \;=\; \sqrt{5} \int_0^{2\pi} e^{2\theta}\, d\theta \;=\; \tfrac{1}{2}\sqrt{5}\,(e^{4\pi} - 1) \text{ units.} $$

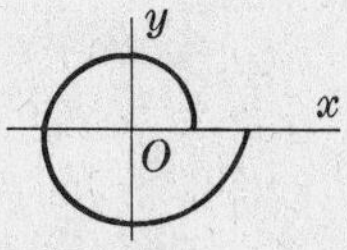

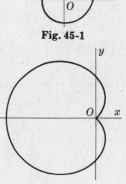

**Fig. 45-1**

2. Find the length of the cardioid $\rho = a(1 - \cos\theta)$.

    The cardioid is described as $\theta$ varies from 0 to $2\pi$.

    $$ \rho^2 + (d\rho/d\theta)^2 \;=\; a^2(1 - \cos\theta)^2 + (a \sin\theta)^2 \;=\; 4a^2 \sin^2 \tfrac{1}{2}\theta $$

    $$ s \;=\; \int_0^{2\pi} \sqrt{\rho^2 + (d\rho/d\theta)^2}\, d\theta \;=\; 2a \int_0^{2\pi} \sin\tfrac{1}{2}\theta\, d\theta \;=\; 8a \text{ units} $$

    In this solution the instruction to take the limits of integration as narrow as possible has not been followed, since the required length is twice that described as $\theta$ varies from 0 to $\pi$.  However, see Problem 3 below.

**Fig. 45-2**

211

**3.** Find the length of the cardioid $\rho = a(1 - \sin\theta)$.

$$\rho^2 + \left(\frac{d\rho}{d\theta}\right)^2 = a^2(1 - \sin\theta)^2 + (-a\cos\theta)^2 = 2a^2(\sin\tfrac{1}{2}\theta - \cos\tfrac{1}{2}\theta)^2$$

Following Problem 2, we write

$$s = \int_0^{2\pi} \sqrt{\rho^2 + (d\rho/d\theta)^2}\, d\theta = \sqrt{2}\,a\int_0^{2\pi}(\sin\tfrac{1}{2}\theta - \cos\tfrac{1}{2}\theta)\, d\theta$$

$$= 2\sqrt{2}\,a(-\cos\tfrac{1}{2}\theta - \sin\tfrac{1}{2}\theta)\Big]_0^{2\pi} = 4\sqrt{2}\,a \text{ units}$$

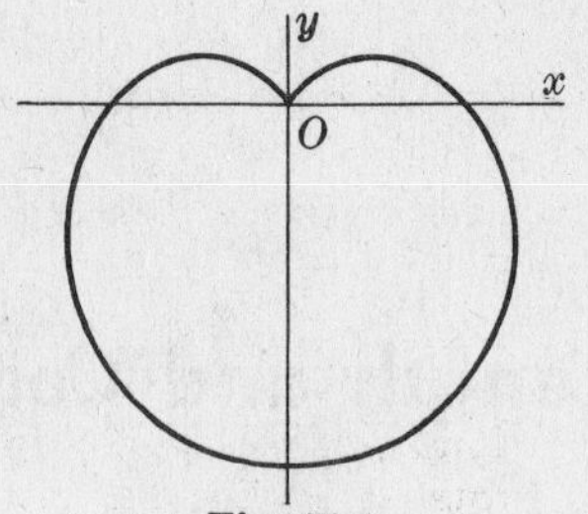

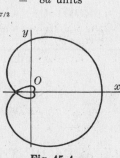

**Fig. 45-3**

Now the cardioids of the two problems differ only in their positions in the plane; hence their lengths should agree. An explanation for the disagreement is to be found in a study of the two integrands $\sin\tfrac{1}{2}\theta$ and $\sin\tfrac{1}{2}\theta - \cos\tfrac{1}{2}\theta$. The first is never negative, while the second is negative as $\theta$ varies from 0 to $\tfrac{1}{2}\pi$ and positive otherwise. By symmetry, the required length is twice that described as $\theta$ varies from $\pi/2$ to $3\pi/2$. Hence,

$$s = 2\sqrt{2}\,a\int_{\pi/2}^{3\pi/2}(\sin\tfrac{1}{2}\theta - \cos\tfrac{1}{2}\theta)\, d\theta = 4\sqrt{2}\,a(-\cos\tfrac{1}{2}\theta - \sin\tfrac{1}{2}\theta)\Big]_{\pi/2}^{3\pi/2} = 8a \text{ units}$$

**4.** Find the length of the curve $\rho = a\cos^4\tfrac{1}{4}\theta$.

The required length is twice that described as $\theta$ varies from 0 to $2\pi$.

$$d\rho/d\theta = -a\cos^3\tfrac{1}{4}\theta \sin\tfrac{1}{4}\theta \quad \text{and} \quad \rho^2 + (d\rho/d\theta)^2 = a^2\cos^6\tfrac{1}{4}\theta.$$

$$s = 2\cdot a\int_0^{2\pi}\cos^3\tfrac{1}{4}\theta\, d\theta = 8a\left[\sin\tfrac{1}{4}\theta - \tfrac{1}{3}\sin^3\tfrac{1}{4}\theta\right]_0^{2\pi}$$

$$= 16a/3 \text{ units}$$

**Fig. 45-4**

**5.** Find the centroid of the arc of the cardioid $\rho = a(1 - \cos\theta)$. Refer to Problem 2.

By symmetry, $\bar{y} = 0$ and the abscissa of the centroid of the entire arc is the same as that for the upper half. From Problem 2, half the length of the cardioid is $4a$; hence,

$$4a\cdot\bar{x} = \int_0^{\pi}\rho\cos\theta\sqrt{\rho^2 + (d\rho/d\theta)^2}\, d\theta = 2a^2\int_0^{\pi}(1 - \cos\theta)\cos\theta\sin\tfrac{1}{2}\theta\, d\theta$$

$$= 4a^2\int_0^{\pi}(-2\cos^4\tfrac{1}{2}\theta + 3\cos^2\tfrac{1}{2}\theta - 1)\sin\tfrac{1}{2}\theta\, d\theta = 4a^2\left[\tfrac{4}{5}\cos^5\tfrac{1}{2}\theta - 2\cos^3\tfrac{1}{2}\theta + 2\cos\tfrac{1}{2}\theta\right]_0^{\pi}$$

$$= -16a^2/5, \text{ and } \bar{x} = -4a/5. \text{ The coordinates of the centroid are } (-4a/5, 0).$$

**6.** Find the centroid of the arc of the circle $\rho = 2\sin\theta + 4\cos\theta$ from $\theta = 0$ to $\theta = \tfrac{1}{2}\pi$.

$$d\rho/d\theta = 2\cos\theta - 4\sin\theta \quad \text{and} \quad \rho^2 + (d\rho/d\theta)^2 = 20. \text{ Since the radius is}$$
$\sqrt{5}$, $s = \sqrt{5}\,\pi$.

$$\sqrt{5}\,\pi\cdot\bar{x} = \int_0^{\pi/2}\rho\cos\theta\sqrt{\rho^2 + (d\rho/d\theta)^2}\, d\theta$$

$$= 4\sqrt{5}\int_0^{\pi/2}(\sin\theta\cos\theta + 2\cos^2\theta)\, d\theta$$

$$= 4\sqrt{5}\left[\tfrac{1}{2}\sin^2\theta + \theta + \tfrac{1}{2}\sin 2\theta\right]_0^{\pi/2}$$

$$= 2\sqrt{5}\,(\pi + 1), \text{ and } \bar{x} = \frac{2(\pi + 1)}{\pi}.$$

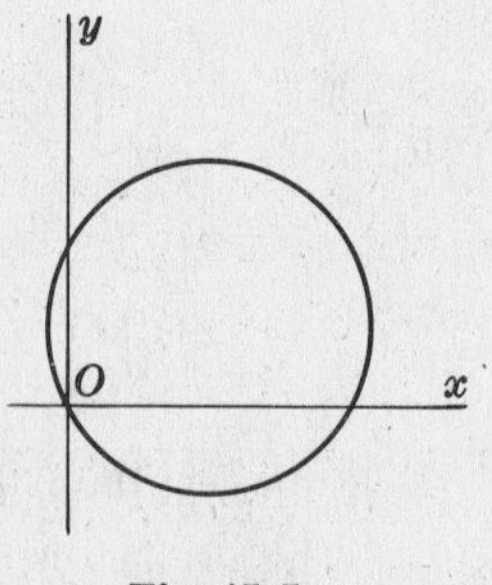

**Fig. 45-5**

$$\sqrt{5}\,\pi\cdot\bar{y} = \int_0^{\pi/2}\rho\sin\theta\sqrt{\rho^2 + (d\rho/d\theta)^2}\, d\theta = 4\sqrt{5}\int_0^{\pi/2}(\sin^2\theta + 2\sin\theta\cos\theta)\, d\theta$$

$$= 4\sqrt{5}\left[\tfrac{1}{2}\theta - \tfrac{1}{4}\sin 2\theta + \sin^2\theta\right]_0^{\pi/2} = 4\sqrt{5}\,(\tfrac{1}{4}\pi + 1), \text{ and } \bar{y} = \frac{\pi + 4}{\pi}.$$

**7.** Find the area of the surface generated by revolving the upper half of the cardioid $\rho = a(1 - \cos\theta)$ about the polar axis.

From Problem 2,   $\rho^2 + (d\rho/d\theta)^2 \;=\; 4a^2 \sin^2 \tfrac{1}{2}\theta.$

$$S \;=\; 2\pi \int_0^\pi \rho \sin\theta \sqrt{\rho^2 + (d\rho/d\theta)^2}\; d\theta \;=\; 4a^2\pi \int_0^\pi (1 - \cos\theta)\sin\theta \sin\tfrac{1}{2}\theta\; d\theta$$

$$=\; 16a^2\pi \int_0^\pi \sin^4 \tfrac{1}{2}\theta \cos\tfrac{1}{2}\theta\; d\theta \;=\; \tfrac{32}{5}a^2\pi \text{ square units}$$

8. Find the area of the surface generated by revolving the lemniscate $\rho^2 = a^2 \cos 2\theta$ about the polar axis.

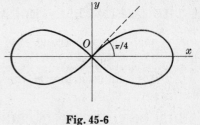

Fig. 45-6

The required area is twice that generated by revolving the first quadrant arc.

$$\rho^2 + \left(\frac{d\rho}{d\theta}\right)^2 \;=\; a^2 \cos 2\theta + \left(-\frac{a^2 \sin 2\theta}{\rho}\right)^2 \;=\; \frac{a^4}{\rho^2}$$

$$S \;=\; 2 \cdot 2\pi \int_0^{\pi/4} \rho \sin\theta \,\frac{a^2}{\rho}\, d\theta \;=\; 4a^2\pi \int_0^{\pi/4} \sin\theta\; d\theta$$

$$=\; 2a^2\pi(2 - \sqrt{2}) \text{ square units}$$

9. Find the area of the surface generated by revolving about the 90°-line a loop of the lemniscate $\rho^2 = a^2 \cos 2\theta$.

The required area is twice that generated by revolving the first quadrant arc.

$$S \;=\; 2 \cdot 2\pi \int_0^{\pi/4} \rho \cos\theta \,\frac{a^2}{\rho}\, d\theta \;=\; 4a^2\pi \int_0^{\pi/4} \cos\theta\; d\theta \;=\; 2\sqrt{2}\,a^2\pi \text{ square units}$$

10. Use the Theorem of Pappus to find the centroid of the arc of the cardioid $\rho = a(1 - \cos\theta)$ from $\theta = 0$ to $\theta = \pi$.

If the arc is revolved about the polar axis, $S = 2\pi\bar{y}s$ or, from Problems 2 and 7, $32a^2\pi/5 = 2\pi\bar{y} \cdot 4a$ and $\bar{y} = 4a/5$.

By Problem 5, the centroid has coordinates $(-4a/5,\, 4a/5)$.

# Supplementary Problems

11. Find the length of

 (a) $\rho = \theta^2$ from $\theta = 0$ to $\theta = 2\sqrt{3}$   *Ans.* 56/3 units   (d) $\rho = \sin^3 \theta/3$   *Ans.* $3\pi/2$ units

 (b) $\rho = e^{\theta/2}$ from $\theta = 0$ to $\theta = 8$   *Ans.* $\sqrt{5}(e^4 - 1)$ units   (e) $\rho = \cos^4 \theta/4$   *Ans.* 16/3 units

 (c) $\rho = \cos^2 \tfrac{1}{2}\theta$   *Ans.* 4 units

 (f) $\rho = a/\theta$ from $(\rho_1, \theta_1)$ to $(\rho_2, \theta_2)$   *Ans.* $\sqrt{a^2 + \rho_1^2} - \sqrt{a^2 + \rho_2^2} + a \ln \dfrac{\rho_1(a + \sqrt{a^2 + \rho_2^2})}{\rho_2(a + \sqrt{a^2 + \rho_1^2})}$ units

 (g) $\rho = 2a \tan\theta \sin\theta$ from $\theta = 0$ to $\theta = \pi/3$   *Ans.* $2a\sqrt{3}\left\{\dfrac{\sqrt{7} - 2}{\sqrt{3}} + \ln \dfrac{2(2 + \sqrt{3})}{\sqrt{7} + \sqrt{3}}\right\}$ units

12. Find the centroid of the upper half of $\rho = 8\cos\theta$.   *Ans.* $(4, 8/\pi)$

13. For $\rho = a\sin\theta + b\cos\theta$, show that $s = \pi\sqrt{a^2 + b^2}$, $S_x = a\pi s$, and $S_y = b\pi s$.

14. Find the area of the surface generated by revolving $\rho = 4\cos\theta$ about the polar axis. *Ans.* $16\pi$ sq. un.

15. Find the area of the surface generated by revolving each loop of $\rho = \sin^3 \theta/3$ about the 90°-axis. *Ans.* $\pi/256$ sq. un.; $513\pi/256$ sq. un.

16. Find the area of the surface generated by revolving one loop of $\rho^2 = \cos 2\theta$ about the 90°-line. *Ans.* $2\sqrt{2}\,\pi$ sq. un.

17. Show that when the two loops of $\rho = \cos^4 \theta/4$ are revolved about the polar axis, they generate equal surface areas.

18. Find the centroid of the surface generated by revolving the right hand loop of $\rho^2 = a^2 \cos 2\theta$ about the polar axis.   *Ans.* $\bar{x} = \sqrt{2}a(\sqrt{2} + 1)/6$

19. Find the area of the surface generated by revolving $\rho = \sin^2 \theta/2$ about the line $\rho = \csc\theta$. *Ans.* $8\pi$ sq. un.

20. Derive the formulas of this chapter.

# Chapter 46

# Improper Integrals

**THE DEFINITE INTEGRAL** $\int_a^b f(x)\,dx$ is called an *improper integral* if

   (a)  the integrand $f(x)$ has one or more points of discontinuity on the interval $a \leqq x \leqq b$, or

   (b)  at least one of the limits of integration is infinite.

**DISCONTINUOUS INTEGRAND.** If $f(x)$ is continuous on the interval $a \leqq x < b$ but is discontinuous at $x = b$, we define

$$\int_a^b f(x)\,dx = \lim_{\epsilon \to 0^+} \int_a^{b-\epsilon} f(x)\,dx \quad \text{provided the limit exists}$$

If $f(x)$ is continuous on the interval $a < x \leqq b$ but is discontinuous at $x = a$, we define

$$\int_a^b f(x)\,dx = \lim_{\epsilon \to 0^+} \int_{a+\epsilon}^b f(x)\,dx \quad \text{provided the limit exists}$$

If $f(x)$ is continuous for all values of $x$ on the interval $a \leqq x \leqq b$ except $x = c$, where $a < c < b$, we define

$$\int_a^b f(x)\,dx = \lim_{\epsilon \to 0^+} \int_a^{c-\epsilon} f(x)\,dx + \lim_{\epsilon' \to 0^+} \int_{c+\epsilon'}^b f(x)\,dx$$

provided *both* limits exist. See Problems 1-6.

**INFINITE LIMITS OF INTEGRATION.** If $f(x)$ is continuous on the interval $a \leqq x \leqq u$, we define

$$\int_a^{+\infty} f(x)\,dx = \lim_{u \to +\infty} \int_a^u f(x)\,dx \quad \text{provided the limit exists}$$

If $f(x)$ is continuous on the interval $u' \leqq x \leqq b$, we define

$$\int_{-\infty}^b f(x)\,dx = \lim_{u' \to -\infty} \int_{u'}^b f(x)\,dx \quad \text{provided the limit exists}$$

If $f(x)$ is continuous on the interval $u' \leqq x \leqq u$, we define

$$\int_{-\infty}^{+\infty} f(x)\,dx = \lim_{u \to +\infty} \int_a^u f(x)\,dx + \lim_{u' \to -\infty} \int_{u'}^a f(x)\,dx$$

provided *both* limits exist. See Problems 7-13.

# Solved Problems

1.  Evaluate $\int_0^3 \dfrac{dx}{\sqrt{9 - x^2}}$.   The integrand is discontinuous at $x = 3$. We consider

$$\lim_{\epsilon \to 0^+} \int_0^{3-\epsilon} \frac{dx}{\sqrt{9-x^2}} = \lim_{\epsilon \to 0^+} \arcsin \frac{x}{3}\bigg]_0^{3-\epsilon} = \lim_{\epsilon \to 0^+} \arcsin \frac{3-\epsilon}{3} = \arcsin 1 = \frac{1}{2}\pi$$

Hence, $\qquad\qquad\qquad\qquad\qquad\qquad\qquad \displaystyle\int_0^3 \frac{dx}{\sqrt{9-x^2}} = \frac{1}{2}\pi$

2.  Show that $\int_0^2 \dfrac{dx}{2-x}$ is meaningless.　　　The integrand is discontinuous at $x=2$. We consider

$$\lim_{\epsilon \to 0^+} \int_0^{2-\epsilon} \frac{dx}{2-x} \;=\; \lim_{\epsilon \to 0^+} \ln \frac{1}{2-x}\Big]_0^{2-\epsilon} \;=\; \lim_{\epsilon \to 0^+} \left( \ln \frac{1}{\epsilon} - \ln \frac{1}{2} \right)$$

The limit does not exist and the integral is meaningless.

3.  Show that $\int_0^4 \dfrac{dx}{(x-1)^2}$ is meaningless.

　　　The integrand is discontinuous at $x=1$, a value between the limits of integration 0 and 4. We consider

$$\lim_{\epsilon \to 0^+} \int_0^{1-\epsilon} \frac{dx}{(x-1)^2} + \lim_{\epsilon' \to 0^+} \int_{1+\epsilon'}^4 \frac{dx}{(x-1)^2}$$

$$= \lim_{\epsilon \to 0^+} \frac{-1}{x-1}\Big]_0^{1-\epsilon} + \lim_{\epsilon' \to 0^+} \frac{-1}{x-1}\Big]_{1+\epsilon'}^4 = \lim_{\epsilon \to 0^+} \left( \frac{1}{\epsilon} - 1 \right) + \lim_{\epsilon' \to 0^+} \left( -\frac{1}{3} + \frac{1}{\epsilon'} \right)$$

The limits do not exist.

　　　If the point of discontinuity is overlooked, $\int_0^4 \dfrac{dx}{(x-1)^2} = -\dfrac{1}{x-1}\Big]_0^4 =$ $-\dfrac{4}{3}$. This is an absurd result.

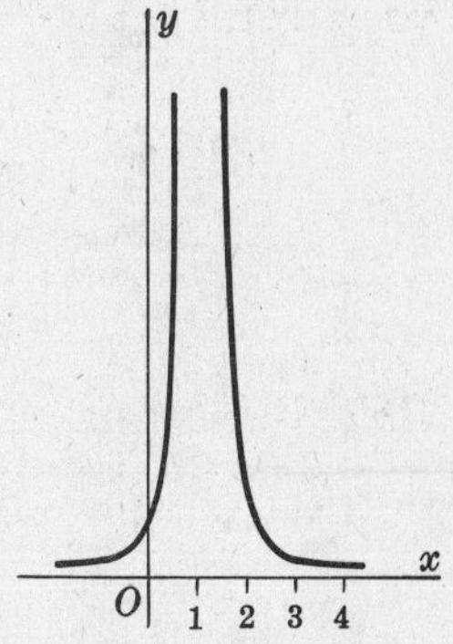

**Fig. 46-1**

4.  Evaluate $\int_0^4 \dfrac{dx}{\sqrt[3]{x-1}}$.　　　The integrand is discontinuous at $x=1$. We consider

$$\lim_{\epsilon \to 0^+} \int_0^{1-\epsilon} \frac{dx}{\sqrt[3]{x-1}} + \lim_{\epsilon' \to 0^+} \int_{1+\epsilon'}^4 \frac{dx}{\sqrt[3]{x-1}} \;=\; \lim_{\epsilon \to 0^+} \frac{3}{2}(x-1)^{2/3}\Big]_0^{1-\epsilon} + \lim_{\epsilon' \to 0^+} \frac{3}{2}(x-1)^{2/3}\Big]_{1+\epsilon'}^4$$

$$= \lim_{\epsilon \to 0^+} \frac{3}{2}\left( (-\epsilon)^{2/3} - 1 \right) + \lim_{\epsilon' \to 0^+} \frac{3}{2}\left( \sqrt[3]{9} - \epsilon'^{2/3} \right) \;=\; \frac{3}{2}(\sqrt[3]{9} - 1)$$

Hence, $\int_0^4 \dfrac{dx}{\sqrt[3]{x-1}} = \dfrac{3}{2}(\sqrt[3]{9} - 1)$.

5.  Show that $\int_0^{\pi/2} \sec x \, dx$ is meaningless.　　　The integrand is discontinuous at $x = \frac{1}{2}\pi$. We consider

$$\lim_{\epsilon \to 0^+} \int_0^{\frac{1}{2}\pi-\epsilon} \sec x \, dx \;=\; \lim_{\epsilon \to 0^+} \ln (\sec x + \tan x)\Big]_0^{\frac{1}{2}\pi-\epsilon} \;=\; \lim_{\epsilon \to 0^+} \ln \{ \sec (\tfrac{1}{2}\pi - \epsilon) + \tan (\tfrac{1}{2}\pi - \epsilon) \}$$

The limit does not exist and the integral is meaningless.

6.  Evaluate $\int_0^{\pi/2} \dfrac{\cos x}{\sqrt{1 - \sin x}} \, dx$.　　　The integrand is discontinuous at $x = \frac{1}{2}\pi$. We consider

$$\lim_{\epsilon \to 0^+} \int_0^{\frac{1}{2}\pi-\epsilon} \frac{\cos x}{\sqrt{1 - \sin x}} \, dx \;=\; \lim_{\epsilon \to 0^+} -2(1 - \sin x)^{1/2}\Big]_0^{\frac{1}{2}\pi-\epsilon} \;=\; 2 \lim_{\epsilon \to 0^+} \{ -[1 - \sin (\tfrac{1}{2}\pi - \epsilon)] + 1 \}$$

$$= 2(0+1) \;=\; 2. \qquad \text{Hence,} \quad \int_0^{\pi/2} \frac{\cos x}{\sqrt{1 - \sin x}} \, dx = 2.$$

7.  Evaluate $\int_0^{+\infty} \dfrac{dx}{x^2 + 4}$.　　　The upper limit of integration is infinite. We consider

$$\lim_{u \to +\infty} \int_0^u \frac{dx}{x^2 + 4} \;=\; \lim_{u \to +\infty} \frac{1}{2} \arctan \frac{1}{2}x \Big]_0^u \;=\; \frac{1}{4}\pi. \qquad \text{Hence,} \quad \int_0^{+\infty} \frac{dx}{x^2 + 4} = \frac{1}{4}\pi.$$

8.  Evaluate $\int_{-\infty}^0 e^{2x} \, dx$.　　　The lower limit of integration is infinite. We consider

$$\lim_{u \to -\infty} \int_u^0 e^{2x} \, dx \;=\; \lim_{u \to -\infty} \frac{1}{2} e^{2x} \Big]_u^0 \;=\; \frac{1}{2}(1) - \lim_{u \to -\infty} \frac{1}{2} e^{2u} \;=\; \frac{1}{2} - 0. \qquad \text{Hence,} \quad \int_{-\infty}^0 e^{2x} \, dx = \frac{1}{2}.$$

9.   Show that $\int_1^{+\infty} \dfrac{dx}{\sqrt{x}}$ is meaningless.    The upper limit of integration is infinite.   We consider

$$\lim_{u \to +\infty} \int_1^u \frac{dx}{\sqrt{x}} \;=\; \lim_{u \to +\infty} 2\sqrt{x}\Big]_1^u \;=\; \lim_{u \to +\infty} (2\sqrt{u} - 2). \quad \text{The limit does not exist.}$$

10.   Evaluate   $\int_{-\infty}^{+\infty} \dfrac{dx}{e^x + e^{-x}} = \int_{-\infty}^{+\infty} \dfrac{e^x\, dx}{e^{2x} + 1}$.     Both limits of integration are infinite.   We consider

$$\lim_{u \to +\infty} \int_0^u \frac{e^x\, dx}{e^{2x}+1} + \lim_{u' \to -\infty} \int_{u'}^0 \frac{e^x\, dx}{e^{2x}+1} \;=\; \lim_{u \to +\infty} \arctan e^x\Big]_0^u + \lim_{u' \to -\infty} \arctan e^x\Big]_{u'}^0$$

$$= \lim_{u \to +\infty} (\arctan e^u - \tfrac{1}{4}\pi) + \lim_{u' \to -\infty} (\tfrac{1}{4}\pi - \arctan e^{u'})$$

$$= \tfrac{1}{2}\pi - \tfrac{1}{4}\pi + \tfrac{1}{4}\pi - 0 \;=\; \tfrac{1}{2}\pi$$

11.   Evaluate $\int_0^{+\infty} e^{-x} \sin x \, dx$.    The upper limit of integration is infinite.   We consider

$$\lim_{u \to +\infty} \int_0^u e^{-x} \sin x \, dx \;=\; \lim_{u \to +\infty} -\tfrac{1}{2} e^{-x}(\sin x + \cos x)\Big]_0^u \;=\; \lim_{u \to +\infty} \{-\tfrac{1}{2}e^{-u}(\sin u + \cos u)\} + \tfrac{1}{2}$$

As $u \to +\infty$, $e^{-u} \to 0$ while $\sin u$ and $\cos u$ vary from 1 to $-1$. Hence, $\int_0^{+\infty} e^{-x} \sin x \, dx = \tfrac{1}{2}$.

12.   Find the area between the curve $y^2 = \dfrac{x^2}{1 - x^2}$ and its asymptotes.   See Fig. 46-2 below.

$A = 4 \int_0^1 \dfrac{x\, dx}{\sqrt{1 - x^2}}$,   the integrand being discontinuous at $x = 1$.   We consider

$$\lim_{\epsilon \to 0^+} \int_0^{1-\epsilon} \frac{x\, dx}{\sqrt{1 - x^2}} \;=\; \lim_{\epsilon \to 0^+} -(1 - x^2)^{1/2}\Big]_0^{1-\epsilon} \;=\; \lim_{\epsilon \to 0^+} (1 - \sqrt{2\epsilon - \epsilon^2}) \;=\; 1$$

The required area is $4(1) = 4$ square units.

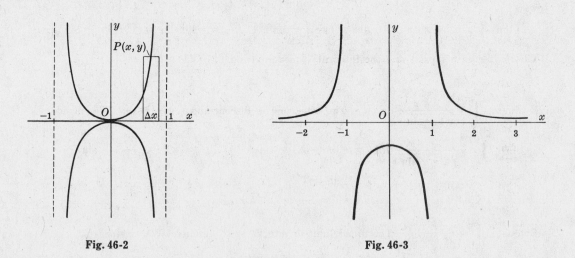

Fig. 46-2            Fig. 46-3

13.   Find the area lying to the right of $x = 3$ and between the curve $y = \dfrac{1}{x^2 - 1}$ and the $x$-axis. See Fig. 46-3 above.

$$A \;=\; \int_3^{+\infty} \frac{dx}{x^2 - 1} \;=\; \lim_{u \to +\infty} \int_3^u \frac{dx}{x^2 - 1} \;=\; \frac{1}{2} \lim_{u \to +\infty} \ln \frac{x-1}{x+1}\Big]_3^u$$

$$= \frac{1}{2} \lim_{u \to +\infty} \ln \frac{u-1}{u+1} - \frac{1}{2}\ln\frac{1}{2} \;=\; \frac{1}{2} \lim_{u \to +\infty} \ln \frac{1 - 1/u}{1 + 1/u} + \frac{1}{2}\ln 2$$

$$= \frac{1}{2}(\ln 2) \text{ square units}$$

# Supplementary Problems

**14.** Evaluate:

(a) $\int_0^1 \dfrac{dx}{\sqrt{x}} = 2$     (d) $\int_0^4 \dfrac{dx}{(4-x)^{3/2}}$ (Meaningless)     (g) $\int_0^4 \dfrac{dx}{(x-2)^{2/3}} = 6\sqrt[3]{2}$

(b) $\int_0^4 \dfrac{dx}{4-x}$ (Meaningless)     (e) $\int_{-2}^2 \dfrac{dx}{\sqrt{4-x^2}} = \pi$     (h) $\int_{-1}^1 \dfrac{dx}{x^4}$ (Meaningless)

(c) $\int_0^4 \dfrac{dx}{\sqrt{4-x}} = 4$     (f) $\int_{-1}^8 \dfrac{dx}{x^{1/3}} = 9/2$     (i) $\int_0^1 \ln x\, dx = -1$

(j) $\int_0^1 x \ln x\, dx = -1/4$

**15.** Find the area between the given curve and its asymptotes.

(a) $y^2 = \dfrac{x^4}{4-x^2}$,   (b) $y^2 = \dfrac{4-x}{x}$,   (c) $y^2 = \dfrac{1}{x(1-x)}$

*Ans.* (a) $4\pi$ sq. un., (b) $4\pi$ sq. un., (c) $2\pi$ sq. un.

**16.** Evaluate:

(a) $\int_1^{+\infty} \dfrac{dx}{x^2} = 1$     (f) $\int_1^{+\infty} \dfrac{e^{-\sqrt{x}}}{\sqrt{x}}\, dx = 2/e$

(b) $\int_{-\infty}^0 \dfrac{dx}{(4-x)^2} = \dfrac{1}{4}$     (g) $\int_{-\infty}^{+\infty} xe^{-x^2}\, dx = 0$

(c) $\int_0^{+\infty} e^{-x}\, dx = 1$     (h) $\int_{-\infty}^{+\infty} \dfrac{dx}{1+4x^2} = \pi/2$

(d) $\int_{-\infty}^6 \dfrac{dx}{(4-x)^2}$ (Meaningless)     (i) $\int_{-\infty}^0 xe^x\, dx = -1$

(e) $\int_2^{+\infty} \dfrac{dx}{x \ln^2 x} = \dfrac{1}{\ln 2}$     (j) $\int_0^{+\infty} x^3 e^{-x}\, dx = 6$

**17.** Find the area between the given curve and its asymptote:

(a) $y = \dfrac{8}{x^2+4}$,   (b) $y = \dfrac{x}{(4+x^2)^2}$,   (c) $y = xe^{-x^2/2}$

*Ans.* (a) $4\pi$ sq. un., (b) $\frac{1}{4}$ sq. un., (c) $2$ sq. un.

**18.** Find the area:

(a) under $y = \dfrac{1}{x^2-4}$ and to the right of $x = 3$.     *Ans.* $\frac{1}{4} \ln 5$ sq. un.

(b) under $y = \dfrac{1}{x(x-1)^2}$ and to the right of $x = 2$.     *Ans.* $1 - \ln 2$ sq. un.

**19.** Show that the following are meaningless:

(a) area under $y = \dfrac{1}{4-x^2}$ from $x = 2$ to $x = -2$.

(b) area under $xy = 9$ to the right of $x = 1$.

**20.** Show that the first quadrant area under $y = e^{-2x}$ is $\frac{1}{2}$ sq. un. and the volume generated by revolving the area about the $x$-axis is $\frac{1}{4}\pi$ cubic units.

**21.** Show that when the portion $R$ of the plane under $xy = 9$ and to the right of $x = 1$ is revolved about the $x$-axis the volume generated is $81\pi$ cu. un. but the area of the surface is infinite.

**22.** Find the length of the indicated arc:

(a) $9y^2 = x(3-x)^2$, loop.    (b) $x^{2/3} + y^{2/3} = a^{2/3}$, entire length.    (c) $9y^2 = x^2(2x+3)$, loop.

*Ans.*   (a) $4\sqrt{3}$ units   (b) $6a$ units   (c) $2\sqrt{3}$ units

**23.** Find the moment of inertia of a circle of radius $r$ with respect to a tangent.    *Ans.* $3r^2s/2$

**24.** Show that $\displaystyle\int_0^{+\infty} \frac{dx}{x^p}$ diverges for all values of $p$.

**25.** (a) Show that $\displaystyle\int_a^b \frac{N\,dx}{(x-b)^p}$ exists for $p < 1$ and is meaningless for $p \geqq 1$.

(b) Show that $\displaystyle\int_a^{+\infty} \frac{N\,dx}{x^p}$ exists for $p > 1$ and is meaningless for $p \leqq 1$.

**26.** Let $f(x) \leqq g(x)$ be defined and non-negative everywhere on the interval $a \leqq x < b$ while $\lim\limits_{x \to b^-} f(x) = +\infty$ and $\lim\limits_{x \to b^-} g(x) = +\infty$. From Fig. 46-4 below, it appears reasonable to assume:

(1) If $\displaystyle\int_a^b g(x)\,dx$ exists so also does $\displaystyle\int_a^b f(x)\,dx$.

(2) If $\displaystyle\int_a^b f(x)\,dx$ does not exist neither does $\displaystyle\int_a^b g(x)\,dx$.

Determine whether or not each of the following exists:

(a) $\displaystyle\int_0^1 \frac{dx}{1-x^4}$.   For $0 \leqq x < 1$,  $1 - x^4 = (1-x)(1+x)(1+x^2) < 4(1-x)$   and   $\dfrac{\frac{1}{4}}{1-x} < \dfrac{1}{1-x^4}$.

Since $\dfrac{1}{4}\displaystyle\int_0^1 \frac{dx}{1-x}$ does not exist, neither does the given integral.

(b) $\displaystyle\int_0^1 \frac{dx}{x^2 + \sqrt{x}}$.   For $0 < x \leqq 1$,  $\dfrac{1}{x^2 + \sqrt{x}} < \dfrac{1}{\sqrt{x}}$.   Since $\displaystyle\int_0^1 \frac{dx}{\sqrt{x}}$ exists so also does the given integral.

(c) $\displaystyle\int_0^1 \frac{e^x\,dx}{x^{1/3}}$ exists.   (d) $\displaystyle\int_0^{\pi/4} \frac{\cos x}{x}\,dx$ does not exist.   (e) $\displaystyle\int_0^{\pi/4} \frac{\cos x}{\sqrt{x}}\,dx$ exists.

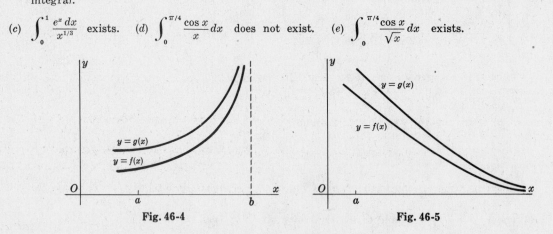

**Fig. 46-4**         **Fig. 46-5**

**27.** Let $f(x) \leqq g(x)$ be defined and non-negative everywhere on the interval $x \geqq a$ while $\lim\limits_{x \to +\infty} f(x) = \lim\limits_{x \to +\infty} g(x) = 0$. From Fig. 46-5 above, it appears reasonable to assume:

(3) If $\displaystyle\int_a^{+\infty} g(x)\,dx$ exists so also does $\displaystyle\int_a^{+\infty} f(x)\,dx$.

(4) If $\displaystyle\int_a^{+\infty} f(x)\,dx$ does not exist neither does $\displaystyle\int_a^{+\infty} g(x)\,dx$.

Determine whether or not each of the following exists.

(a) $\displaystyle\int_1^{+\infty} \frac{dx}{\sqrt{x^4 + 2x + 6}}$.   For $x \geqq 1$,  $\dfrac{1}{\sqrt{x^4 + 2x + 6}} < \dfrac{1}{x^2}$.   Since $\displaystyle\int_1^{+\infty} \frac{dx}{x^2}$ exists so also does the given integral.

(b) $\displaystyle\int_2^{+\infty} \frac{dx}{\sqrt{x^3 + 2x}}$ exists.   (c) $\displaystyle\int_1^{+\infty} e^{-x^2}\,dx$ exists.   (d) $\displaystyle\int_0^{+\infty} \frac{dx}{\sqrt{x + x^4}}$ exists.

# Chapter 47

## Infinite Sequences and Series

**AN INFINITE SEQUENCE** $\{s_n\} = s_1, s_2, s_3, \ldots, s_n, \ldots$ is a function of $n$ whose domain of definition is the set of positive integers. (See Chapter 1.)

A sequence $\{s_n\}$ is said to be *bounded* if there exist numbers $P$ and $Q$ such that $P \le s_n \le Q$ for all values of $n$. For example, $3/2, 5/4, 7/6, \ldots, \dfrac{2n+1}{2n}, \ldots$ is bounded since, for all $n$, $1 \le s_n \le 2$; but $2, 4, 6, \ldots, 2n, \cdots$ is not bounded.

A sequence $\{s_n\}$ is called *nondecreasing* if $s_1 \le s_2 \le s_3 \le \cdots \le s_n \le \cdots$, and is called *nonincreasing* if $s_1 \ge s_2 \ge s_3 \ge \cdots \ge s_n \ge \cdots$. For example, the sequences $\left\{\dfrac{n^2}{n+1}\right\} = 1/2, 4/3, 9/4, 16/5, \ldots$ and $\{2n-(-1)^n\} = 3, 3, 7, 7, \ldots$ are nondecreasing; and the sequences $\{1/n\} = 1, 1/2, 1/3, 1/4, \ldots$ and $\{-n\} = -1, -2, -3, -4, \ldots$ are nonincreasing.

A sequence $\{s_n\}$ is said to converge to the finite number $s$ as limit, $\left[\lim\limits_{n \to +\infty} s_n = s\right]$, if for any positive number $\epsilon$, however small, there exists a positive integer $m$ such that whenever $n > m$, then $|s - s_n| < \epsilon$. If a sequence has a limit, it is called a *convergent sequence*; otherwise, a *divergent sequence*. See Problems 1-2.

A sequence $\{s_n\}$ is said to diverge to $\infty$, $\left[\lim\limits_{n \to +\infty} s_n = \infty\right]$, if for any positive number $M$, however large, there exists a positive integer $m$ such that whenever $n > m$, then $|s_n| > M$. If $s_n > M$, $\lim\limits_{n \to +\infty} s_n = +\infty$; if $s_n < -M$, $\lim\limits_{n \to +\infty} s_n = -\infty$.

**THEOREMS ON SEQUENCES.** A proof of the basic theorem

I.      Every bounded nondecreasing (nonincreasing) sequence is convergent.

is beyond the scope of this book.

II.     Every unbounded sequence is divergent.      For a proof, see Problem 3.

A number of the remaining theorems are merely restatements of those given in Chapter 2.

III.    A convergent (divergent) sequence remains convergent (divergent) after any or all of its first $n$ terms are altered.

IV.     The limit of a convergent sequence is unique.      For a proof, see Problem 4.

If $\lim\limits_{n \to +\infty} s_n = s$ and $\lim\limits_{n \to +\infty} t_n = t$:

V.      $\lim\limits_{n \to +\infty} (k \cdot s_n) = k \lim\limits_{n \to +\infty} s_n = ks$, $k$ being any constant.

VI.     $\lim\limits_{n \to +\infty} (s_n \pm t_n) = \lim\limits_{n \to +\infty} s_n \pm \lim\limits_{n \to +\infty} t_n = s \pm t$.

VII.    $\lim\limits_{n \to +\infty} (s_n \cdot t_n) = \lim\limits_{n \to +\infty} s_n \cdot \lim\limits_{n \to +\infty} t_n = s \cdot t$.

**VIII.**   $\lim\limits_{n \to +\infty} (s_n/t_n) = \lim\limits_{n \to +\infty} s_n \Big/ \lim\limits_{n \to +\infty} t_n = s/t$, provided $t \neq 0$ and $t_n \neq 0$ for all $n$.

**IX.**   If $\{s_n\}$ is a sequence of non-zero terms and if $\lim\limits_{n \to +\infty} s_n = \infty$, then $\lim\limits_{n \to +\infty} 1/s_n = 0$.
For a proof, see Problem 5.

**X.**   If $a > 1$, then $\lim\limits_{n \to +\infty} a^n = +\infty$.
For a proof, see Problem 6.

**XI.**   If $|r| < 1$, then $\lim\limits_{n \to +\infty} r^n = 0$.

## THE INDICATED SUM

$$\sum s_n = \sum_{n=1}^{+\infty} s_n = s_1 + s_2 + s_3 + \cdots + s_n + \cdots \qquad (1)$$

of an infinite sequence $\{s_n\}$ is called an *infinite series*. With each series there is associated a sequence of *partial sums*: $S_1 = s_1$, $S_2 = s_1 + s_2$, $S_3 = s_1 + s_2 + s_3$, ..., $S_n = s_1 + s_2 + s_3 + \cdots + s_n$, $\cdots$.

If $\lim\limits_{n \to +\infty} S_n = S$, a finite number, the series (1) is said to *converge* and $S$ is called its *sum*. If $\lim\limits_{n \to +\infty} S_n$ does not exist, series (1) is said to *diverge*. A series diverges either because $\lim\limits_{n \to +\infty} S_n = \infty$ or because, as $n$ increases, $S_n$ increases and decreases without approaching a limit. An example of the latter is the *oscillating* series $1 - 1 + 1 - 1 + \cdots$. Here, $S_1 = 1$, $S_2 = 0$, $S_3 = 1$, $S_4 = 0$, $\ldots$.

See Problems 7-8.

From the theorems above, follow:

**XII.**   A convergent (divergent) series remains convergent (divergent) after any or all of its first $n$ terms are altered.      See Problem 9.

**XIII.**   The sum of a convergent series is unique.

**XIV.**   If $\Sigma s_n$ converges to $S$, then $\Sigma k s_n$, $k$ being any constant, converges to $kS$. If $\Sigma s_n$ diverges, so also does $\Sigma k s_n$.

**XV.**   If $\Sigma s_n$ converges, then $\lim\limits_{n \to +\infty} s_n = 0$.      For a proof, see Problem 10.
The converse is not true. For the harmonic series [Problem 7(c)] $\lim\limits_{n \to +\infty} s_n = 0$ but the series diverges.

**XVI.**   If $\lim\limits_{n \to +\infty} s_n \neq 0$, then $\Sigma s_n$ diverges.
The converse is not true; see Problem 7(c).      See Problem 11.

# Solved Problems

1. Let the sequence $\{s_n\}$ converge to $s$. Lay off on a number scale (Fig. 47-1) the points $s$, $s - \epsilon$, $s + \epsilon$, where $\epsilon$ is any small positive number. Now locate in order the points $s_1, s_2, s_3, \ldots$. The definition

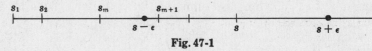

**Fig. 47-1**

of convergence assures us that while the first $m$ points lie outside the $\epsilon$-neighborhood of $s$, the point $s_{m+1}$ and all subsequent points will lie within the neighborhood.

In Fig. 47-2 an ordinary rectangular coordinate system is used. First draw in the lines $y = s$, $y = s - \epsilon$, and $y = s + \epsilon$, determining a band (shaded) of width $2\epsilon$. Now locate in turn the points $(1, s_1)$, $(2, s_2)$, $(3, s_3)$, .... As before, the point $(m + 1, s_{m+1})$ and all subsequent points lie within the band.

It is important to note that only a finite number of points of a convergent sequence lie outside an $\epsilon$-interval or $\epsilon$-band.

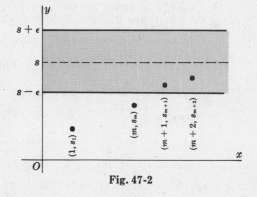

Fig. 47-2

2.  Use Theorem 1 to show that the sequences  $(a)$ $\left\{1 - \dfrac{1}{n}\right\}$ and $(b)$ $\left\{\dfrac{1 \cdot 3 \cdot 5 \cdot 7 \ldots (2n - 1)}{2 \cdot 4 \cdot 6 \cdot 8 \ldots (2n)}\right\}$ are convergent.

$(a)$  The sequence is bounded since $0 \leqq s_n \leqq 1$, for all $n$.

Since $s_{n+1} = 1 - \dfrac{1}{n+1} = 1 - \dfrac{1}{n} + \dfrac{1}{n(n+1)} = s_n + \dfrac{1}{n(n+1)}$, that is $s_{n+1} \geqq s_n$, the sequence is nondecreasing.  Thus the sequence converges to $s \leqq 1$.

$(b)$  The sequence is bounded since $0 \leqq s_n \leqq 1$, for every $n$.  Since $s_{n+1} = \dfrac{1 \cdot 3 \cdot 5 \cdot 7 \ldots (2n + 1)}{2 \cdot 4 \cdot 6 \cdot 8 \ldots (2n + 2)} = \dfrac{2n+1}{2n+2} s_n$, the sequence is nonincreasing.  Thus the sequence converges to $s \geqq 0$.

3.  Prove:  Every unbounded sequence $\{s_n\}$ is divergent.

Suppose $\{s_n\}$ were convergent.  Then for any positive $\epsilon$, however small, there would exist a positive integer $m$ such that whenever $n > m$, then $|s_n - s| < \epsilon$.  Since all but a finite number of the terms of the sequence lie within this interval, the sequence must be bounded.  But this is contrary to the hypothesis; hence the sequence is divergent.

4.  Prove:  The limit of a convergent sequence is unique.

Suppose the contrary so that $\lim\limits_{n \to +\infty} s_n = s$ and $\lim\limits_{n \to +\infty} s_n = t$, where $|s - t| > 2\epsilon > 0$.  Now the $\epsilon$-neighborhoods of $s$ and $t$ have the contradictory properties:  (i) have no points in common, and (ii) each contains all but a finite number of terms of the sequence.  Thus $s = t$ and the limit is unique.

5.  Prove:  If $\{s_n\}$ is a sequence of non-zero terms and if $\lim\limits_{n \to +\infty} s_n = \infty$ then $\lim\limits_{n \to +\infty} 1/s_n = 0$.

Let $\epsilon > 0$ be chosen.  From $\lim\limits_{n \to +\infty} s_n = \infty$ it follows that for any $M > 1/\epsilon$, there exists a positive integer $m$ such that whenever $n > m$ then $|s_n| > M > 1/\epsilon$.  For this $m$, $|1/s_n| < 1/M < \epsilon$ whenever $n > m$; hence, $\lim\limits_{n \to +\infty} 1/s_n = 0$.

6.  Prove:  If $a > 1$, then $\lim\limits_{n \to +\infty} a^n = +\infty$.

Let $M > 0$ be chosen.  Suppose $a = 1 + b$, $b > 0$;  then

$$a^n = (1 + b)^n = 1 + nb + \frac{n(n-1)}{1 \cdot 2} b^2 + \cdots > 1 + nb > M$$

when $n > M/b$.  Thus an effective $m$ is the largest integer in $M/b$.

7.  Prove:

$(a)$  The infinite arithmetic series $a + (a + d) + (a + 2d) + \cdots + [a + (n - 1)d] + \cdots$ diverges when $a^2 + d^2 > 0$.

(b) The infinite geometric series $a + ar + ar^2 + \cdots + ar^{n-1} + \cdots$, where $a \neq 0$, converges to $\dfrac{a}{1-r}$ if $|r| < 1$ and diverges if $|r| \geqq 1$.

(c) The *harmonic* series $1 + 1/2 + 1/3 + 1/4 + \cdots + 1/n + \cdots$ diverges.

(a) Here $S_n = \frac{1}{2}n[2a + (n-1)d]$ and $\lim\limits_{n \to +\infty} S_n = \infty$ unless $a = d = 0$.

Thus the series diverges when $a^2 + d^2 > 0$.

(b) Here $S_n = \dfrac{a - ar^n}{1-r} = \dfrac{a}{1-r} - \dfrac{a}{1-r}r^n$, $r \neq 1$.

If $|r| < 1$, $\lim\limits_{n \to +\infty} r^n = 0$, and $\lim\limits_{n \to +\infty} S_n = \dfrac{a}{1-r}$.

If $|r| > 1$, $\lim\limits_{n \to +\infty} r^n = \infty$, and $S_n$ diverges.

If $|r| = 1$, the series is either $a + a + a + \cdots$ or $a - a + a - a + \cdots$ and diverges.

(c) When the partial sums are formed, it is found that

$$S_4 > 2, \quad S_8 > 2.5, \quad S_{16} > 3, \quad S_{32} > 3.5, \quad S_{64} > 4, \quad \ldots$$

Thus the sequence of partial sums is unbounded and diverges; hence the series diverges.

8. (a) For the series $\dfrac{1}{5} + \dfrac{1}{5^2} + \dfrac{1}{5^3} + \cdots$:

$$S_1 = \frac{1}{5} = \frac{1}{4}\left(1 - \frac{1}{5}\right), \quad S_2 = \frac{1}{5} + \frac{1}{5^2} = \frac{1}{4}\left(1 - \frac{1}{5^2}\right), \quad S_3 = \frac{1}{5} + \frac{1}{5^2} + \frac{1}{5^3} = \frac{1}{4}\left(1 - \frac{1}{5^3}\right), \quad \ldots,$$

$$S_n = \frac{1}{4}\left(1 - \frac{1}{5^n}\right), \quad \text{and} \quad S = \lim_{n \to +\infty} \frac{1}{4}\left(1 - \frac{1}{5^n}\right) = \frac{1}{4}.$$

(b) For the series $\dfrac{1}{1 \cdot 2} + \dfrac{1}{2 \cdot 3} + \dfrac{1}{3 \cdot 4} + \dfrac{1}{4 \cdot 5} + \cdots$:

$$S_1 = \frac{1}{1 \cdot 2} = 1 - \frac{1}{2}, \quad S_2 = \frac{1}{1 \cdot 2} + \frac{1}{2 \cdot 3} = 1 - \frac{1}{2} + \frac{1}{2} - \frac{1}{3} = 1 - \frac{1}{3},$$

$$S_3 = S_2 + \frac{1}{3 \cdot 4} = 1 - \frac{1}{3} + \frac{1}{3} - \frac{1}{4} = 1 - \frac{1}{4}, \quad \ldots, \quad S_n = 1 - \frac{1}{n+1}$$

and $S = \lim\limits_{n \to +\infty} \left(1 - \dfrac{1}{n+1}\right) = 1$.

9. The series $1 + \frac{1}{2} + \frac{1}{4} + \frac{1}{8} + \frac{1}{16} + \cdots$ converges to 2. Examine the series which results when (a) its first four terms are dropped, (b) when the terms $8 + 4 + 2$ are adjoined to the series.

(a) The series $\frac{1}{16} + \frac{1}{32} + \cdots$ is an infinite geometric series with $r = \frac{1}{2}$. It converges to

$$S = 2 - (1 + \tfrac{1}{2} + \tfrac{1}{4} + \tfrac{1}{8}) = \tfrac{1}{8}$$

(b) The series $8 + 4 + 2 + 1 + \frac{1}{2} + \frac{1}{4} + \cdots$ is an infinite geometric series with $r = \frac{1}{2}$. It converges to

$$2 + (8 + 4 + 2) = 16$$

10. Prove: If $\Sigma s_n = S$, then $\lim\limits_{n \to +\infty} s_n = 0$.

Since $\Sigma s_n = S$, $\lim\limits_{n \to +\infty} S_n = S$ and $\lim\limits_{n \to +\infty} S_{n-1} = S$. Now $s_n = S_n - S_{n-1}$; hence,

$$\lim_{n \to +\infty} s_n = \lim_{n \to +\infty} (S_n - S_{n-1}) = \lim_{n \to +\infty} S_n - \lim_{n \to +\infty} S_{n-1} = S - S = 0$$

11. Show that the series (a) $1/3 + 2/5 + 3/7 + 4/9 + \cdots$ and (b) $1/2 + 3/4 + 7/8 + 15/16 + \cdots$ diverge.

(a) $s_n = \dfrac{n}{2n+1}$ and $\lim\limits_{n \to +\infty} s_n = \lim\limits_{n \to +\infty} \dfrac{n}{2n+1} = \lim\limits_{n \to +\infty} \dfrac{1}{2 + 1/n} = \dfrac{1}{2} \neq 0$.

(b) $s_n = \dfrac{2^n - 1}{2^n}$ and $\lim\limits_{n \to +\infty} \dfrac{2^n - 1}{2^n} = \lim\limits_{n \to +\infty} \left(1 - \dfrac{1}{2^n}\right) = 1 \neq 0$.

12. A series $\Sigma s_n$ converges to $S$ as limit if the sequence $\{S_n\}$ of partial sums converges to $S$, that is, if for any $\epsilon > 0$, however small, there exists an integer $m$ such that whenever $n > m$ then $|S - S_n| < \epsilon$. Show that the series of Problem 8 converge by producing for each an effective $m$ for any given $\epsilon$.

(a) If $|S - S_n| = \left| \dfrac{1}{4} - \dfrac{1}{4}\left(1 - \dfrac{1}{5^n}\right) \right| = \dfrac{1}{4 \cdot 5^n} < \epsilon$, then $5^n > \dfrac{1}{4\epsilon}$, $n \ln 5 > -\ln(4\epsilon)$, and $n > -\dfrac{\ln 4\epsilon}{\ln 5}$.

Thus, $m = $ greatest integer in $-\dfrac{\ln 4\epsilon}{\ln 5}$ is effective.

(b) If $|S - S_n| = \left| 1 - \left(1 - \dfrac{1}{n+1}\right) \right| = \dfrac{1}{n+1} < \epsilon$, then $n + 1 > \dfrac{1}{\epsilon}$ and $n > \dfrac{1}{\epsilon} - 1$. Thus, $m = $ greatest integer in $\dfrac{1}{\epsilon} - 1$ is effective.

# Supplementary Problems

13. Determine for each sequence whether or not it is bounded, whether nonincreasing or nondecreasing, whether convergent or divergent, and whether oscillating.

(a) $\left\{ n + \dfrac{2}{n} \right\}$    (b) $\left\{ \dfrac{(-1)^n}{n} \right\}$    (c) $\{\sin \frac{1}{4} n\pi\}$    (d) $\{\sqrt[3]{n^2}\}$    (e) $\left\{ \dfrac{n!}{10^n} \right\}$    (f) $\left\{ \dfrac{\ln n}{n} \right\}$

14. Show that $\lim\limits_{n \to +\infty} \sqrt[n]{1/n^p} = 1$, $p > 0$. *Hint.* $n^{p/n} = e^{(p \ln n)/n}$.

15. For the sequence $\left\{ \dfrac{n}{n+1} \right\}$, verify: (a) the neighborhood $|1 - s_n| < 0.01$ contains all but the first 99 terms of the sequence, (b) the sequence is bounded, (c) $\lim\limits_{n \to +\infty} s_n = 1$.

16. Prove: If $|r| < 1$, then $\lim\limits_{n \to +\infty} r^n = 0$.

17. Examine each of the following geometric series for convergence. If the series converges, find its sum.

(a) $1 + 1/2 + 1/4 + 1/8 + \cdots$    (b) $4 - 1 + 1/4 - 1/16 + \cdots$    (c) $1 + 3/2 + 9/4 + 27/8 + \cdots$

*Ans.* (a) $S = 2$    (b) $S = 16/5$    (c) Diverges

18. Find the sum of each of the following series.

(a) $\sum 3^{-n}$

(b) $\sum \dfrac{1}{(2n-1)(2n+1)}$

(c) $\sum \left( \dfrac{1}{n^p} - \dfrac{1}{(n+1)^p} \right)$, $p > 0$

(d) $\sum \dfrac{1}{n(n+2)}$

(e) $\sum \dfrac{1}{n(n+3)}$

(f) $\sum \dfrac{n}{(n+1)!}$

(g) $\sum \dfrac{1}{(4n-3)(4n+1)}$

(h) $\sum \dfrac{1}{n(n+1)(n+2)}$

*Ans.* (a) 1/2, (b) 1/2, (c) 1, (d) 3/4, (e) 11/18, (f) 1, (g) 1/4, (h) 1/4

19. Show that each of the following diverges.

(a) $3 + 5/2 + 7/3 + 9/4 + \cdots$

(b) $2 + \sqrt{2} + \sqrt[3]{2} + \sqrt[4]{2} + \cdots$

(c) $e + e^2/8 + e^3/27 + e^4/64 + \cdots$

(d) $\sum \dfrac{1}{\sqrt{n} + \sqrt{n-1}}$

20. Prove: If $\lim\limits_{n \to +\infty} s_n \neq 0$, then $\Sigma s_n$ diverges.

21. Prove the series of Problem 18(a)-(d) convergent by producing an effective positive integer $m$ such that for $\epsilon > 0$, $|S - S_n| < \epsilon$ whenever $n > m$.

*Ans.* $m = $ greatest integer in (a) $-\dfrac{\ln 2\epsilon}{\ln 3}$, (b) $\dfrac{1}{4\epsilon} - \dfrac{1}{2}$, (c) $\sqrt[p]{1/\epsilon} - 1$, (d) the positive root of $2\epsilon m^2 - 2(1 - 3\epsilon)m - (3 - 4\epsilon) = 0$.

# Chapter 48

# Tests for Convergence and Divergence
## of Positive Series

**SERIES OF POSITIVE TERMS.** A series $\Sigma s_n$, all of whose terms are positive, is called a *positive* series.

**I.** A positive series $\Sigma s_n$ is convergent if the sequence of partial sums $\{S_n\}$ is bounded.

This theorem follows from the fact that the sequence of partial sums of a positive series is always nondecreasing.

**II.** **THE INTEGRAL TEST.** Let $f(n)$ denote the general term $s_n$ of the series $\Sigma s_n$ of positive terms. If $f(x) > 0$ and never increases on the interval $x > \xi$, where $\xi$ is some positive integer, then the series $\Sigma s_n$ converges or diverges according as $\displaystyle\int_{\xi}^{+\infty} f(x)\,dx$ exists or does not exist. $\hspace{1em}$ See Problems 1-5.

**III.** **THE COMPARISON TEST FOR CONVERGENCE.** A positive series $\Sigma s_n$ is convergent if each term (perhaps, after a finite number) is less than or equal to the corresponding term of a known convergent positive series $\Sigma c_n$.

**IV.** **THE COMPARISON TEST FOR DIVERGENCE.** A positive series $\Sigma s_n$ is divergent if each term (perhaps, after a finite number) is equal to or greater than the corresponding term of a known divergent positive series $\Sigma d_n$.

See Problems 6-11.

**V.** **THE RATIO TEST.** A positive series $\Sigma s_n$ converges if $\displaystyle\lim_{n \to +\infty} \frac{s_{n+1}}{s_n} < 1$, and diverges if $\displaystyle\lim_{n \to +\infty} \frac{s_{n+1}}{s_n} > 1$. If $\displaystyle\lim_{n \to +\infty} \frac{s_{n+1}}{s_n} = 1$, the test fails to indicate convergence or divergence. $\hspace{1em}$ See Problems 12-18.

# Solved Problems

### THE INTEGRAL TEST

1. Prove the Integral Test: Let $f(n)$ denote the general term $s_n$ of the positive series $\Sigma s_n$. If $f(x) > 0$ and never increases on the interval $x > \xi$, where $\xi$ is a positive integer, then the series $\Sigma s_n$ converges or diverges according as $\displaystyle\int_{\xi}^{+\infty} f(x)\,dx$ exists or does not exist.

In the figure the area under the curve $y = f(x)$ from $x = \xi$ to $x = n$ has been approximated by two sets of rectangles having unit bases. Expressing the fact that the area under the curve lies between the sum of the areas of the small rectangles and the sum of the areas of the large rectangles, we have

$$s_{\xi+1} + s_{\xi+2} + \cdots + s_n \; < \; \int_{\xi}^{n} f(x)\,dx \; < \; s_{\xi} + s_{\xi+1} + \cdots + s_{n-1}$$

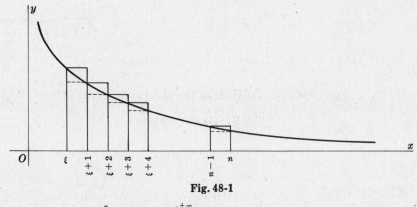

**Fig. 48-1**

(1) Suppose $\lim\limits_{n \to +\infty} \int_\xi^n f(x)\, dx = \int_\xi^{+\infty} f(x)\, dx = A$.   Then

$$s_{\xi+1} + s_{\xi+2} + \cdots + s_n < A$$

and $\qquad\qquad S_n = s_\xi + s_{\xi+1} + s_{\xi+2} + \cdots + s_n$

is bounded and nondecreasing, as $n$ increases.   Thus, by Theorem I, $\Sigma s_n$ converges.

(2) Suppose $\lim\limits_{n \to +\infty} \int_\xi^n f(x)\, dx = \int_\xi^{+\infty} f(x)\, dx$   does not exist.   Then

$$S_n = s_\xi + s_{\xi+1} + \cdots + s_n \quad \text{is unbounded and } \Sigma s_n \text{ diverges.}$$

**Examine Problems 2-5 for convergence, using the integral test.**

2.   $\dfrac{1}{\sqrt{3}} + \dfrac{1}{\sqrt{5}} + \dfrac{1}{\sqrt{7}} + \dfrac{1}{\sqrt{9}} + \cdots$   $\qquad\qquad f(n) = s_n = \dfrac{1}{\sqrt{2n+1}}$; take, $f(x) = \dfrac{1}{\sqrt{2x+1}}$.

On the interval $x > 1$, $f(x) > 0$ and decreases as $x$ increases.   Take $\xi = 1$ and consider

$$\int_1^{+\infty} f(x)\, dx = \lim_{u \to +\infty} \int_1^u \frac{dx}{\sqrt{2x+1}} = \lim_{u \to +\infty} \sqrt{2x+1}\,\Big]_1^u = \lim_{u \to +\infty} \sqrt{2u+1} - \sqrt{3} = \infty$$

The integral does not exist and the series is divergent.

3.   $\dfrac{1}{4} + \dfrac{1}{16} + \dfrac{1}{36} + \dfrac{1}{64} + \cdots$   $\qquad\qquad f(n) = s_n = \dfrac{1}{4n^2}$; take $f(x) = \dfrac{1}{4x^2}$.

On the interval $x > 1$, $f(x) > 0$ and decreases as $x$ increases.   Take $\xi = 1$ and consider

$$\int_1^{+\infty} f(x)\, dx = \frac{1}{4} \lim_{u \to +\infty} \int_1^u \frac{dx}{x^2} = \frac{1}{4} \lim_{u \to +\infty} \left(-\frac{1}{x}\right)\Big]_1^u = \frac{1}{4} \lim_{u \to +\infty} \left(-\frac{1}{u} + 1\right) = \frac{1}{4}$$

The integral exists and the series is convergent.

4.   $\sin \pi + \dfrac{1}{4} \sin \tfrac{1}{2}\pi + \dfrac{1}{9} \sin \tfrac{1}{3}\pi + \dfrac{1}{16} \sin \tfrac{1}{4}\pi + \cdots$   $\qquad f(n) = s_n = \dfrac{1}{n^2} \sin \dfrac{1}{n}\pi$; take $f(x) = \dfrac{1}{x^2} \sin \dfrac{1}{x}\pi$.

On the interval $x > 2$, $f(x) > 0$ and decreases as $x$ increases.   Take $\xi = 2$ and consider

$$\int_2^{+\infty} f(x)\, dx = \lim_{u \to +\infty} \int_2^u \frac{1}{x^2} \sin \frac{1}{x}\pi\, dx = \frac{1}{\pi} \lim_{u \to +\infty} \cos \frac{1}{x}\pi\,\Big]_2^u = \frac{1}{\pi}$$

The series converges.

5.   $1 + \dfrac{1}{2^p} + \dfrac{1}{3^p} + \dfrac{1}{4^p} + \cdots$ $(p > 0)$.   (The $p$-series.)   $\qquad f(n) = s_n = \dfrac{1}{n^p}$; take $f(x) = \dfrac{1}{x^p}$.

On the interval $x > 1$, $f(x) > 0$ and decreases as $x$ increases.   Take $\xi = 1$ and consider

$$\int_1^{+\infty} f(x)\, dx = \lim_{u \to +\infty} \int_1^u \frac{dx}{x^p} = \lim_{u \to +\infty} \left(\frac{x^{1-p}}{1-p}\right)\Big]_1^u = \frac{1}{1-p} \left\{ \lim_{u \to +\infty} u^{1-p} - 1 \right\}, \quad (p \neq 1)$$

If $p > 1$, $\dfrac{1}{1-p}\left\{\lim\limits_{u \to +\infty} u^{1-p} - 1\right\} = \dfrac{1}{1-p}\left\{\lim\limits_{u \to +\infty} \dfrac{1}{u^{p-1}} - 1\right\} = \dfrac{1}{p-1}$  and the series converges.

If $p = 1$, $\displaystyle\int_1^{+\infty} f(x)\,dx = \lim\limits_{u \to +\infty} \ln u = +\infty$  and the series diverges.

If $p < 1$, $\dfrac{1}{1-p}\left\{\lim\limits_{u \to +\infty} u^{1-p} - 1\right\} = +\infty$  and the series diverges.

Note the second proof that the harmonic series diverges.

## THE COMPARISON TEST

The general term of a given series to be tested for convergence is to be compared with general terms of known convergent and divergent series. The following series will be found useful as test series:

(a)  The geometric series $a + ar + ar^2 + \cdots + ar^n + \cdots$, $a \neq 0$, which converges for $0 < r < 1$ and diverges for $r \geqq 1$.

(b)  The $p$-series $1 + \dfrac{1}{2^p} + \dfrac{1}{3^p} + \dfrac{1}{4^p} + \cdots + \dfrac{1}{n^p} + \cdots$ which converges for $p > 1$ and diverges for $p \leqq 1$.

(c)  Each new series tested.

Examine Problems 6-11 for convergence, using the comparison tests.

6.  $\dfrac{1}{2} + \dfrac{1}{5} + \dfrac{1}{10} + \dfrac{1}{17} + \cdots + \dfrac{1}{n^2+1} + \cdots$

   The general term of the series is $s_n = \dfrac{1}{n^2+1} < \dfrac{1}{n^2}$; hence the given series is term by term less than the $p$-series $1 + \dfrac{1}{4} + \dfrac{1}{9} + \cdots + \dfrac{1}{n^2} + \cdots$.

   The test series is convergent since $p = 2$, and so also is the given series. (The integral test may be used here.)

7.  $\dfrac{1}{\sqrt{1}} + \dfrac{1}{\sqrt{2}} + \dfrac{1}{\sqrt{3}} + \dfrac{1}{\sqrt{4}} + \cdots$

   The general term of the series is $\dfrac{1}{\sqrt{n}}$. Since $\dfrac{1}{\sqrt{n}} \geqq \dfrac{1}{n}$, the given series is term by term greater than or equal to the harmonic series and is divergent. (The integral test may be used here.)

8.  $1 + \dfrac{1}{2!} + \dfrac{1}{3!} + \dfrac{1}{4!} + \cdots$

   The general term of the series is $\dfrac{1}{n!}$. Since $n! \geqq 2^{n-1}$, $\dfrac{1}{n!} \leqq \dfrac{1}{2^{n-1}}$.

   The given series is term by term less than or equal to the convergent geometric series $1 + \dfrac{1}{2} + \dfrac{1}{4} + \dfrac{1}{8} + \cdots$ and is convergent. (The integral test cannot be used here.)

9.  $2 + \dfrac{3}{2^3} + \dfrac{4}{3^3} + \dfrac{5}{4^3} + \cdots$     The general term of the series is $\dfrac{n+1}{n^3}$.

   Since $\dfrac{n+1}{n^3} \leqq \dfrac{2n}{n^3} = \dfrac{2}{n^2}$, the given series is term by term less than or equal to twice the convergent $p$-series $1 + \dfrac{1}{2^2} + \dfrac{1}{3^2} + \dfrac{1}{4^2} + \cdots$ and is convergent.

**10.** $1 + \dfrac{1}{2^2} + \dfrac{1}{3^3} + \dfrac{1}{4^4} + \cdots$

The general term of the series is $\dfrac{1}{n^n}$. Since $\dfrac{1}{n^n} \leqq \dfrac{1}{2^{n-1}}$, the given series is term by term less than or equal to the convergent geometric series $1 + \dfrac{1}{2} + \dfrac{1}{4} + \dfrac{1}{8} + \cdots$.

Also, the given series is term by term less than or equal to the convergent $p$-series with $p = 2$.

**11.** $1 + \dfrac{2^2 + 1}{2^3 + 1} + \dfrac{3^2 + 1}{3^3 + 1} + \dfrac{4^2 + 1}{4^3 + 1} + \cdots$

The general term is $\dfrac{n^2 + 1}{n^3 + 1} \geqq \dfrac{1}{n}$. Hence the given series is term by term greater than or equal to the harmonic series and is divergent.

## THE RATIO TEST

**12.** Prove the Ratio Test:

A positive series $\Sigma s_n$ converges if $\displaystyle\lim_{n \to +\infty} \dfrac{s_{n+1}}{s_n} < 1$ and diverges if $\displaystyle\lim_{n \to +\infty} \dfrac{s_{n+1}}{s_n} > 1$.

Suppose $\displaystyle\lim_{n \to +\infty} \dfrac{s_{n+1}}{s_n} = L < 1$. Then for any $r$, where $L < r < 1$, there exists a positive integer $m$ such that whenever $n > m$ then $\dfrac{s_{n+1}}{s_n} < r$, that is,

$$\dfrac{s_{m+2}}{s_{m+1}} < r \quad \text{or} \quad s_{m+2} < r \cdot s_{m+1}$$

$$\dfrac{s_{m+3}}{s_{m+2}} < r \quad \text{or} \quad s_{m+3} < r \cdot s_{m+2} < r^2 \cdot s_{m+1}$$

$$\dfrac{s_{m+4}}{s_{m+3}} < r \quad \text{or} \quad s_{m+4} < r \cdot s_{m+3} < r^3 \cdot s_{m+1}$$

$$\cdots\cdots\cdots\cdots\cdots\cdots\cdots\cdots\cdots\cdots\cdots\cdots$$

Thus each term of the series $s_{m+1} + s_{m+2} + s_{m+3} + \cdots$ is $\leqq$ the corresponding term of the geometric series $s_{m+1} + r \cdot s_{m+1} + r^2 \cdot s_{m+1} + \cdots$ which converges since $r < 1$. Hence $\Sigma s_n$ is convergent by Theorem III.

Suppose $\displaystyle\lim_{n \to +\infty} \dfrac{s_{n+1}}{s_n} = L > 1$ (or $= +\infty$). Then there exists a positive integer $m$ such that whenever $n > m$, $\dfrac{s_{n+1}}{s_n} > 1$. Now $s_{n+1} > s_n$ and $\{s_n\}$ does not converge to 0. Hence $\Sigma s_n$ diverges by Theorem XVI (Chapter 47).

Suppose $\displaystyle\lim_{n \to +\infty} \dfrac{s_{n+1}}{s_n} = 1$. An example is the $p$-series $\displaystyle\Sigma \dfrac{1}{n^p}$, $p > 0$, for which

$$\lim_{n \to +\infty} \dfrac{s_{n+1}}{s_n} = \lim_{n \to +\infty} \dfrac{n^p}{(n+1)^p} = \lim_{n \to +\infty} \left( \dfrac{1}{1 + 1/n} \right)^p = 1$$

Since the series converges when $p > 1$ and diverges when $p \leqq 1$, the test fails to indicate convergence or divergence.

Investigate the series of Problems 13-23 for convergence, using the ratio test.

**13.** $\dfrac{1}{3} + \dfrac{2}{3^2} + \dfrac{3}{3^3} + \dfrac{4}{3^4} + \cdots$ $\qquad s_n = \dfrac{n}{3^n}, \quad s_{n+1} = \dfrac{n+1}{3^{n+1}}, \quad \dfrac{s_{n+1}}{s_n} = \dfrac{n+1}{3^{n+1}} \cdot \dfrac{3^n}{n} = \dfrac{n+1}{3n}.$

Then $\displaystyle\lim_{n \to +\infty} \dfrac{s_{n+1}}{s_n} = \lim_{n \to +\infty} \dfrac{n+1}{3n} = \dfrac{1}{3}$ and the series converges.

14. $\dfrac{1}{3} + \dfrac{2!}{3^2} + \dfrac{3!}{3^3} + \dfrac{4!}{3^4} + \cdots$                                      $s_n = \dfrac{n!}{3^n},\ \ s_{n+1} = \dfrac{(n+1)!}{3^{n+1}},\ \ \dfrac{s_{n+1}}{s_n} = \dfrac{n+1}{3}.$

Then $\displaystyle\lim_{n \to +\infty} \frac{s_{n+1}}{s_n} = \lim_{n \to +\infty} \frac{n+1}{3} = \infty$ and the series diverges.

15. $1 + \dfrac{1 \cdot 2}{1 \cdot 3} + \dfrac{1 \cdot 2 \cdot 3}{1 \cdot 3 \cdot 5} + \dfrac{1 \cdot 2 \cdot 3 \cdot 4}{1 \cdot 3 \cdot 5 \cdot 7} + \cdots$

$$s_n = \frac{n!}{1 \cdot 3 \cdot 5 \ldots (2n-1)},\ \ s_{n+1} = \frac{(n+1)!}{1 \cdot 3 \cdot 5 \ldots (2n+1)},\ \ \frac{s_{n+1}}{s_n} = \frac{n+1}{2n+1}.$$

Then $\displaystyle\lim_{n \to +\infty} \frac{n+1}{2n+1} = \frac{1}{2}$ and the series converges.

16. $\dfrac{1}{1 \cdot 2} + \dfrac{1}{2 \cdot 2^2} + \dfrac{1}{3 \cdot 2^3} + \dfrac{1}{4 \cdot 2^4} + \cdots$               $s_n = \dfrac{1}{n \cdot 2^n},\ \ s_{n+1} = \dfrac{1}{(n+1)2^{n+1}},\ \ \dfrac{s_{n+1}}{s_n} = \dfrac{n}{2(n+1)}.$

Then $\displaystyle\lim_{n \to +\infty} \frac{n}{2(n+1)} = \frac{1}{2}$ and the series converges.

17. $2 + \dfrac{3}{2} \cdot \dfrac{1}{4} + \dfrac{4}{3} \cdot \dfrac{1}{4^2} + \dfrac{5}{4} \cdot \dfrac{1}{4^3} + \cdots$         $s_n = \dfrac{n+1}{n} \cdot \dfrac{1}{4^{n-1}},\ \ s_{n+1} = \dfrac{n+2}{n+1} \cdot \dfrac{1}{4^n},\ \ \dfrac{s_{n+1}}{s_n} = \dfrac{n(n+2)}{4(n+1)^2}.$

Then $\displaystyle\lim_{n \to +\infty} \frac{n(n+2)}{4(n+1)^2} = \frac{1}{4}$ and the series converges.

18. $1 + \dfrac{2^2+1}{2^3+1} + \dfrac{3^2+1}{3^3+1} + \dfrac{4^2+1}{4^3+1} + \cdots$

$$s_n = \frac{n^2+1}{n^3+1},\ \ s_{n+1} = \frac{(n+1)^2+1}{(n+1)^3+1},\ \ \frac{s_{n+1}}{s_n} = \frac{(n+1)^2+1}{(n+1)^3+1} \cdot \frac{n^3+1}{n^2+1}.$$

Then $\displaystyle\lim_{n \to +\infty} \frac{s_{n+1}}{s_n} = 1$ and the test fails.  See Problem 11 above.

# Supplementary Problems

19. Verify that the integral test may be applied and use the test to determine convergence or divergence.

(a) $\displaystyle\sum \frac{1}{n}$             (c) $\displaystyle\sum \frac{1}{n \ln n}$          (e) $\displaystyle\sum \frac{n}{n^2+1}$        (g) $\displaystyle\sum \frac{2n}{(n+1)(n+2)(n+3)}$

(b) $\displaystyle\sum \frac{50}{n(n+1)}$      (d) $\displaystyle\sum \frac{n}{(n+1)(n+2)}$     (f) $\displaystyle\sum \frac{n}{e^n}$          (h) $\displaystyle\sum \frac{1}{(2n+1)^2}$

*Ans.* (a), (c), (d), (e) divergent.

20. Determine convergence or divergence using the comparison test.

(a) $\displaystyle\sum \frac{1}{n^3-1}$         (e) $\displaystyle\sum \frac{n+2}{n(n+1)}$       (i) $\displaystyle\sum \frac{1}{3^n+1}$        (m) $\displaystyle\sum \frac{n}{3n^2-4}$

(b) $\displaystyle\sum \frac{n-2}{n^3}$          (f) $\displaystyle\sum \frac{1}{n^{n-1}}$            (j) $\displaystyle\sum \frac{\ln n}{\sqrt{n}}$          (n) $\displaystyle\sum \frac{1}{1 + \ln n}$

(c) $\displaystyle\sum \frac{1}{\sqrt[3]{n}}$           (g) $\displaystyle\sum \frac{1}{3n+1}$          (k) $\displaystyle\sum \frac{1}{3^n-1}$        (o) $\displaystyle\sum \frac{n^4+5}{n^5}$

(d) $\displaystyle\sum \frac{1}{n^2+5}$        (h) $\displaystyle\sum \frac{\ln n}{n}$            (l) $\displaystyle\sum \frac{\ln n}{n^p}$         (p) $\displaystyle\sum \frac{n+1}{n\sqrt{3n-2}}$

*Ans.* (a), (b), (d), (f), (i), (k), (l) for $p > 2$ convergent.

**21.** Determine convergence or divergence, using the ratio test.

(a) $\sum \dfrac{(n+1)(n+2)}{n!}$

(b) $\sum \dfrac{5^n}{n!}$

(c) $\sum \dfrac{n}{2^{2n}}$

(d) $\sum \dfrac{3^{2n-1}}{n^2+n}$

(e) $\sum \dfrac{(n+1)2^n}{n!}$

(f) $\sum n\left(\dfrac{3}{4}\right)^n$

(g) $\sum \dfrac{n^3}{(\ln 2)^n}$

(h) $\sum \dfrac{n^3}{(\ln 3)^n}$

(i) $\sum \dfrac{2^n}{n(n+2)}$

(j) $\sum \dfrac{n^n}{n!}$

(k) $\sum \dfrac{2^n}{2n-1}$

(l) $\sum \dfrac{n^3}{3^n}$

*Ans.* $(a), (b), (c), (e), (f), (h), (l)$ convergent.

**22.** Determine convergence or divergence.

(a) $\dfrac{1}{4^2} + \dfrac{1}{7^2} + \dfrac{1}{10^2} + \dfrac{1}{13^2} + \cdots$

(b) $3 + \dfrac{3}{\sqrt[3]{2}} + \dfrac{3}{\sqrt[3]{3}} + \dfrac{3}{\sqrt[3]{4}} + \cdots$

(c) $1 + \dfrac{1}{5} + \dfrac{1}{9} + \dfrac{1}{13} + \cdots$

(d) $\dfrac{1}{2} + \dfrac{1}{3\cdot 4} + \dfrac{1}{4\cdot 5\cdot 6} + \dfrac{1}{5\cdot 6\cdot 7\cdot 8} + \cdots$

(e) $3 + \dfrac{3}{4} + \dfrac{11}{27} + \dfrac{9}{32} + \cdots$

(f) $\dfrac{2}{3} + \dfrac{3}{2\cdot 3^2} + \dfrac{4}{3\cdot 3^3} + \dfrac{5}{4\cdot 3^4} + \cdots$

(g) $\dfrac{1}{2} + \dfrac{1}{2\cdot 2^2} + \dfrac{1}{3\cdot 2^3} + \dfrac{1}{4\cdot 2^4} + \cdots$

(h) $\dfrac{2}{1\cdot 3} + \dfrac{3}{2\cdot 4} + \dfrac{4}{3\cdot 5} + \dfrac{5}{4\cdot 6} + \cdots$

(i) $\dfrac{1}{2} + \dfrac{2}{3^2} + \dfrac{3}{4^3} + \dfrac{4}{5^4} + \cdots$

(j) $1 + \dfrac{1}{2^2} + \dfrac{1}{3^{5/2}} + \dfrac{1}{4^3} + \cdots$

(k) $2 + \dfrac{3}{5} + \dfrac{4}{10} + \dfrac{5}{17} + \cdots$

(l) $\dfrac{2}{5} + \dfrac{2\cdot 4}{5\cdot 8} + \dfrac{2\cdot 4\cdot 6}{5\cdot 8\cdot 11} + \dfrac{2\cdot 4\cdot 6\cdot 8}{5\cdot 8\cdot 11\cdot 14} + \cdots$

*Ans.* $(a), (d), (f), (g), (i), (j), (l)$ convergent.

**23.** Prove the comparison test for convergence. *Hint.* If $\Sigma c_n = C$, then $\{S_n\}$ is bounded.

**24.** Prove the comparison test for divergence. *Hint.* $\displaystyle\sum_1^n s_i \geqq \sum_1^n d_i > M$ for $n > m$.

**25.** Prove the Polynomial Test: If $P(n)$ and $Q(n)$ are polynomials of degree $p$ and $q$ respectively, the series $\sum \dfrac{P(n)}{Q(n)}$ converges if $q > p+1$ and diverges if $q \leqq p+1$. *Hint.* Compare with $1/n^{q-p}$.

**26.** Use the polynomial test to determine convergence or divergence.

(a) $\dfrac{1}{1\cdot 2} + \dfrac{1}{2\cdot 3} + \dfrac{1}{3\cdot 4} + \dfrac{1}{4\cdot 5} + \cdots$

(b) $\dfrac{1}{2} + \dfrac{1}{7} + \dfrac{1}{12} + \dfrac{1}{17} + \cdots$

(c) $\dfrac{3}{2} + \dfrac{5}{10} + \dfrac{7}{30} + \dfrac{9}{68} + \cdots$

(d) $\dfrac{3}{2} + \dfrac{5}{24} + \dfrac{7}{108} + \dfrac{9}{320} + \cdots$

(e) $\dfrac{1}{2^2-1} + \dfrac{2}{3^2-2} + \dfrac{3}{4^2-3} + \dfrac{4}{5^2-4} + \cdots$

(f) $\dfrac{1}{2^3-1^2} + \dfrac{1}{3^3-2^2} + \dfrac{1}{4^3-3^2} + \dfrac{1}{5^3-4^2} + \cdots$

(g) $\dfrac{2}{1\cdot 3} + \dfrac{3}{2\cdot 4} + \dfrac{4}{3\cdot 5} + \dfrac{5}{4\cdot 6} + \cdots$

*Ans.* $(a), (c), (d), (f)$ convergent.

**27.** Prove the Root Test: A positive series $\Sigma s_n$ converges if $\displaystyle\lim_{n\to+\infty} \sqrt[n]{s_n} < 1$ and diverges if $\displaystyle\lim_{n\to+\infty} \sqrt[n]{s_n} > 1$. The test fails if $\displaystyle\lim_{n\to+\infty} \sqrt[n]{s_n} = 1$. *Hint.* If $\displaystyle\lim_{n\to+\infty} \sqrt[n]{s_n} < 1$, then $\sqrt[n]{s_n} < r < 1$, for $n > m$, and $s_n < r^n$.

**28.** Determine convergence or divergence, using the Root Test.

(a) $\sum \dfrac{1}{n^n}$,  (b) $\sum \dfrac{1}{(\ln n)^n}$,  (c) $\sum \dfrac{2^{n-1}}{n^n}$,  (d) $\sum \left(\dfrac{n}{n^2+2}\right)^n$      *Ans.* All convergent.

# Chapter 49

## Series with Negative Terms

**A SERIES,** all of whose terms are negative, may be treated as the negative of a positive series.

**ALTERNATING SERIES.** A series whose terms are alternately positive and negative, as

$$\sum (-1)^{n-1} s_n = s_1 - s_2 + s_3 - s_4 + \cdots + (-1)^{n-1} s_n + \cdots \qquad (1)$$

in which each $s_i$ is *positive*, is called an *alternating series*.

**I.** An alternating series (1) converges if (i) $s_n > s_{n+1}$, for every value of $n$, and (ii) $\lim_{n \to +\infty} s_n = 0$.

See Problems 1-2.

**ABSOLUTE CONVERGENCE.** A series $\Sigma s_n = s_1 + s_2 + \cdots + s_n + \cdots$, with mixed (positive and negative) terms, is called *absolutely convergent* if $\Sigma |s_n| = |s_1| + |s_2| + |s_3| + \cdots + |s_n| + \cdots$ converges.

Every convergent positive series is absolutely convergent.

Every absolutely convergent series is convergent. For a proof, see Problem 3.

**CONDITIONAL CONVERGENCE.** If $\Sigma s_n$ converges while $\Sigma |s_n|$ diverges, $\Sigma s_n$ is called *conditionally convergent*.

For example, the series $1 - \frac{1}{2} + \frac{1}{3} - \frac{1}{4} + \cdots$ is conditionally convergent since it converges but $1 + \frac{1}{2} + \frac{1}{3} + \frac{1}{4} + \cdots$ diverges.

**THE RATIO TEST FOR ABSOLUTE CONVERGENCE.** A series $\Sigma s_n$ with mixed terms is absolutely convergent if $\lim_{n \to +\infty} \left| \frac{s_{n+1}}{s_n} \right| < 1$ and is divergent if $\lim_{n \to +\infty} \left| \frac{s_{n+1}}{s_n} \right| > 1$.

If the limit is 1, the test gives no information.

See Problems 4-12.

## Solved Problems

1. Prove: An alternating series $s_1 - s_2 + s_3 - s_4 + \cdots$ converges if (i) $s_n > s_{n+1}$, for every value of $n$, and (ii) $\lim_{n \to +\infty} s_n = 0$.

Consider the partial sum

$$S_{2m} = s_1 - s_2 + s_3 - s_4 + \cdots + s_{2m-1} - s_{2m}$$

which may be grouped as follows:

(a) $\quad S_{2m} = (s_1 - s_2) + (s_3 - s_4) + \cdots + (s_{2m-1} - s_{2m})$

(b) $\quad S_{2m} = s_1 - (s_2 - s_3) - \cdots - (s_{2m-2} - s_{2m-1}) - s_{2m}$

By hypothesis, $s_n > s_{n+1}$ and $s_n - s_{n+1} > 0$; hence, by $(a)$, $S_{2m} > 0$, and by $(b)$, $S_{2m} < s_1$. Thus the sequence $\{S_{2m}\}$ is bounded and converges to a limit $L < s_1$.

Consider next the partial sum $S_{2m+1} = S_{2m} + s_{2m+1}$; then

$$\lim_{m \to +\infty} S_{2m+1} = \lim_{m \to +\infty} S_{2m} + \lim_{m \to +\infty} s_{2m+1} = L + 0 = L$$

Thus $\lim\limits_{n \to +\infty} S_n = L$ and the series converges.

2. Show that the following alternating series converge:

$(a)$  $1 - \dfrac{1}{2^2} + \dfrac{1}{3^2} - \dfrac{1}{4^2} + \cdots$

$\qquad s_n = \dfrac{1}{n^2}$ and $s_{n+1} = \dfrac{1}{(n+1)^2}$; then $s_n > s_{n+1}$, $\lim\limits_{n \to +\infty} s_n = 0$, and the series converges.

$(b)$  $1/2 - 1/5 + 1/10 - 1/17 + \cdots$

$\qquad s_n = \dfrac{1}{n^2+1}$ and $s_{n+1} = \dfrac{1}{(n+1)^2+1}$; then $s_n \geqq s_{n+1}$, $\lim\limits_{n \to +\infty} \dfrac{1}{n^2+1} = 0$, and the series converges.

$(c)$  $\dfrac{1}{e} - \dfrac{2}{e^2} + \dfrac{3}{e^3} - \dfrac{4}{e^4} + \cdots$

$\qquad$ The series converges since $s_n \geqq s_{n+1}$ and $\lim\limits_{n \to +\infty} \dfrac{n}{e^n} = \lim\limits_{n \to +\infty} \dfrac{1}{e^n} = 0$.

3. Prove:  Every absolutely convergent series is convergent.

$\qquad$ Let $\qquad\qquad (a)$  $\Sigma s_n = s_1 + s_2 + s_3 + s_4 + \cdots + s_n + \cdots,$

having both positive and negative terms, be the given series whose corresponding positive series

$\qquad\qquad (b)$  $\Sigma |s_n| = |s_1| + |s_2| + |s_3| + \cdots + |s_n| + \cdots$

converges to $S'$.

$\qquad$ Suppose the $n$th partial sum $S_n = s_1 + s_2 + s_3 + \cdots + s_n$ of $(a)$ consists of $r$ positive terms of sum $P_r$ and $t = n - r$ negative terms of sum $-Q_t$. Then $S_n = P_r - Q_t$ while the corresponding partial sum of $(b)$ is $S'_n = P_r + Q_t$. Since $\lim\limits_{n \to +\infty} S'_n = S'$, the partial sums $S'_n$ are bounded. Then, as $n$ increases, the sequences $\{P_r\}$ and $\{Q_t\}$ are bounded and nondecreasing. Let $\lim\limits_{n \to +\infty} P_r = P$ and $\lim\limits_{n \to +\infty} Q_t = Q$. Now $\lim\limits_{n \to +\infty} S_n = \lim\limits_{n \to +\infty} P_r - \lim\limits_{n \to +\infty} Q_t = P - Q$ and $\Sigma s_n$ converges.

## ABSOLUTE AND CONDITIONAL CONVERGENCE

Examine each of the following convergent series for absolute or conditional convergence.

4.  $1 - \dfrac{1}{2} + \dfrac{1}{4} - \dfrac{1}{8} + \cdots$

$\qquad$ The series $1 + \dfrac{1}{2} + \dfrac{1}{4} + \dfrac{1}{8} + \cdots$, obtained by making all of the terms positive, is convergent, being a geometric series with $r = \frac{1}{2}$. Thus the given series is absolutely convergent.

5.  $1 - \dfrac{2}{3} + \dfrac{3}{3^2} - \dfrac{4}{3^3} + \cdots$

$\qquad$ The series $1 + \dfrac{2}{3} + \dfrac{3}{3^2} + \dfrac{4}{3^3} + \cdots$, obtained by making all the terms positive is convergent by the ratio test. Thus the given series is absolutely convergent.

6.  $1 - \dfrac{1}{\sqrt{2}} + \dfrac{1}{\sqrt{3}} - \dfrac{1}{\sqrt{4}} + \cdots$

$\qquad$ The series $1 + \dfrac{1}{\sqrt{2}} + \dfrac{1}{\sqrt{3}} + \dfrac{1}{\sqrt{4}} + \cdots$ diverges, being a $p$-series with $p = \frac{1}{2} < 1$. Thus the given series is conditionally convergent.

7.  $\dfrac{1}{2} - \dfrac{2}{3} \cdot \dfrac{1}{2^3} + \dfrac{3}{4} \cdot \dfrac{1}{3^3} - \dfrac{4}{5} \cdot \dfrac{1}{4^3} + \cdots$

   The series $1 + \dfrac{2}{3} \cdot \dfrac{1}{2^3} + \dfrac{3}{4} \cdot \dfrac{1}{3^3} + \dfrac{4}{5} \cdot \dfrac{1}{4^3} + \cdots$ converges, since it is term by term less than or equal to the $p$-series with $p = 3$. Thus the given series is absolutely convergent.

8.  $\dfrac{2}{3} - \dfrac{3}{4} \cdot \dfrac{1}{2} + \dfrac{4}{5} \cdot \dfrac{1}{3} - \dfrac{5}{6} \cdot \dfrac{1}{4} + \cdots$

   The series $\dfrac{2}{3} + \dfrac{3}{4} \cdot \dfrac{1}{2} + \dfrac{4}{5} \cdot \dfrac{1}{3} + \dfrac{5}{6} \cdot \dfrac{1}{4} + \cdots$ is divergent, being term by term greater than $\tfrac{1}{2}$(harmonic series). Thus the given series is conditionally convergent.

9.  $2 - \dfrac{2^3}{3!} + \dfrac{2^5}{5!} - \dfrac{2^7}{7!} + \cdots$

   The series $2 + \dfrac{2^3}{3!} + \dfrac{2^5}{5!} + \dfrac{2^7}{7!} + \cdots + \dfrac{2^{2n-1}}{(2n-1)!} + \cdots$ is convergent (ratio test) and the given series is absolutely convergent.

10. $\dfrac{1}{2} - \dfrac{4}{2^3 + 1} + \dfrac{9}{3^3 + 1} - \dfrac{16}{4^3 + 1} + \cdots$

   The series $\dfrac{1}{2} + \dfrac{4}{2^3 + 1} + \dfrac{9}{3^3 + 1} + \dfrac{16}{4^3 + 1} + \cdots + \dfrac{n^2}{n^3 + 1} + \cdots$ is divergent (integral test) and the given series is conditionally convergent.

11. $\dfrac{1}{2} - \dfrac{2}{2^3 + 1} + \dfrac{3}{3^3 + 1} - \dfrac{4}{4^3 + 1} + \cdots$

   The series $\dfrac{1}{2} + \dfrac{2}{2^3 + 1} + \dfrac{3}{3^3 + 1} + \dfrac{4}{4^3 + 1} + \cdots + \dfrac{n}{n^3 + 1} + \cdots$ is convergent, being term by term less than the $p$-series for $p = 2$. Thus the given series is absolutely convergent.

12. $\dfrac{1}{1 \cdot 2} - \dfrac{1}{2 \cdot 2^2} + \dfrac{1}{3 \cdot 2^3} - \dfrac{1}{4 \cdot 2^4} + \cdots$

   The series $\dfrac{1}{1 \cdot 2} + \dfrac{1}{2 \cdot 2^2} + \dfrac{1}{3 \cdot 2^3} + \dfrac{1}{4 \cdot 2^4} + \cdots$ is convergent, being term by term less than or equal to the convergent geometric series $\dfrac{1}{2} + \dfrac{1}{4} + \dfrac{1}{8} + \dfrac{1}{16} + \cdots$. Thus the given series is absolutely convergent.

# Supplementary Problems

13. Examine each of the following alternating series for convergence or divergence.

   (a) $\sum \dfrac{(-1)^{n-1}}{n!}$ $\qquad$ (c) $\sum (-1)^{n-1} \dfrac{n+1}{n}$ $\qquad$ (e) $\sum \dfrac{(-1)^{n-1}}{2n-1}$

   (b) $\sum \dfrac{(-1)^{n-1}}{\ln n}$ $\qquad$ (d) $\sum (-1)^{n-1} \dfrac{\ln n}{3n+2}$ $\qquad$ (f) $\sum (-1)^{n-1} \dfrac{1}{\sqrt[n]{3}}$

   *Ans.* (a), (b), (d), (e) convergent.

14. Examine each of the following for conditional or absolute convergence.

   (a) $\sum \dfrac{(-1)^{n-1}}{(2n-1)^3}$ $\quad$ (c) $\sum \dfrac{(-1)^{n-1}}{(n+1)^2}$ $\quad$ (e) $\sum \dfrac{(-1)^{n-1}}{3n-1}$ $\quad$ (g) $\sum (-1)^{n-1} \dfrac{n}{n^2 + 1}$

   (b) $\sum \dfrac{(-1)^{n-1}}{\sqrt{n(n+1)}}$ $\quad$ (d) $\sum \dfrac{(-1)^{n-1}}{n^2 + 2}$ $\quad$ (f) $\sum \dfrac{(-1)^{n-1}}{(n!)^3}$ $\quad$ (h) $\sum (-1)^{n-1} \dfrac{n^2}{n^4 + 2}$

   *Ans.* (a), (c), (d), (f), (h) absolutely convergent.

# Chapter 50

# Computations with Series

**OPERATIONS ON SERIES.** Let

$$\Sigma s_n = s_1 + s_2 + s_3 + \cdots + s_n + \cdots \tag{1}$$

be a given series and let $\Sigma t_n$ be obtained from it by the insertion of parentheses. For example,

$$\Sigma t_n = (s_1 + s_2) + (s_3 + s_4 + s_5) + (s_6 + s_7) + (s_8 + s_9 + s_{10} + s_{11}) + \cdots$$

**I.** Any series obtained from a convergent series by the insertion of parentheses converges to the same sum as the original series.

**II.** A series obtained from a divergent positive series by the insertion of parentheses diverges, but one obtained from a divergent series with mixed terms may or may not diverge. *See Problems 1-2.*

Let $\Sigma u_n$ be obtained from (1) by a reordering of the terms. For example,

$$\Sigma u_n = s_1 + s_3 + s_2 + s_4 + s_6 + s_5 + \cdots$$

**III.** Any series obtained from an absolutely convergent series by a reordering of the terms converges absolutely to the same sum as the original series.

**IV.** The terms of a conditionally convergent series can be rearranged to give either a divergent series or a convergent series whose sum is a preassigned number. *See Problem 3.*

**ADDITION, SUBTRACTION AND MULTIPLICATION.** If $\Sigma s_n$ and $\Sigma t_n$ are any two series, their *sum series* $\Sigma u_n$, their *difference series* $\Sigma v_n$, and their *product series* $\Sigma w_n$ are defined as

$$\Sigma u_n = \Sigma(s_n + t_n)$$
$$\Sigma v_n = \Sigma(s_n - t_n)$$
$$\Sigma w_n = s_1 t_1 + (s_1 t_2 + s_2 t_1) + (s_1 t_3 + s_2 t_2 + s_3 t_1) + \cdots$$

**V.** If $\Sigma s_n$ converges to $S$ and $\Sigma t_n$ converges to $T$, then $\Sigma(s_n + t_n)$ converges to $S + T$ and $\Sigma(s_n - t_n)$ converges to $S - T$. If $\Sigma s_n$ and $\Sigma t_n$ are both absolutely convergent, so also are $\Sigma(s_n \pm t_n)$.

**VI.** If $\Sigma s_n$ and $\Sigma t_n$ converge, their product series $\Sigma w_n$ may or may not converge. If $\Sigma s_n$ and $\Sigma t_n$ converge and at least one of them is absolutely convergent, then $\Sigma w_n$ converges to $ST$. If $\Sigma s_n$ and $\Sigma t_n$ are absolutely convergent, so also is $\Sigma w_n$. *See Problems 4-5.*

**COMPUTATIONS WITH SERIES.** The sum of a convergent series can be obtained readily provided the $n$th partial sum can be expressed as a function of $n$; for example, any convergent geometric series. On the other hand, any partial sum of a convergent series may be taken as an approximation of the sum of the series. If the approximation $S_n$ of $S$ is to be useful, information concerning the possible size of $|S_n - S|$ must be known.

For a convergent series $\Sigma s_n$ with sum $S$, we write

$$S = S_n + R_n$$

where $R_n$, called the "remainder after $n$ terms" is the error introduced by using $s_n$, the $n$th partial sum, instead of the true sum $S$. The theorems below give approximations of this error in the form $R_n < \alpha$ for positive series and $|R_n| \leqq \alpha$ for series with mixed terms.

For a convergent alternating series $s_1 - s_2 + s_3 - s_4 + \cdots$,

$$R_{2m} = s_{2m+1} - s_{2m+2} + s_{2m+3} - s_{2m+4} + \cdots < s_{2m+1}$$

and

$$R_{2m+1} = - s_{2m+2} + s_{2m+3} - s_{2m+4} + s_{2m+5} - \cdots > - s_{2m+2}$$

by Problem 1, Chapter 49. Thus,

**VII.**  For a convergent alternating series, $|R_n| < s_{n+1}$; moreover, $R_n$ is positive when $n$ is even and $R_n$ is negative when $n$ is odd.

See Problem 6.

**VIII.**  For the convergent geometric series $\Sigma ar^{n-1}$,

$$|R_n| = \left| \frac{ar^n}{1-r} \right|$$

**IX.**  If the positive series $\Sigma s_n$ converges by the Integral Test, then

$$R_n < \int_n^{+\infty} f(x)\,dx \quad \text{for } n > \xi \qquad \text{See Problems 7-9.}$$

**X.**  If $\Sigma c_n$ is a known convergent positive series and if for the positive series $\Sigma s_n$, $s_n \leqq c_n$ for every value of $n > n_1$, then

$$R_n \leqq \sum_{n+1}^{+\infty} c_j \quad \text{for } n > n_1 \qquad \text{See Problems 10-12.}$$

# Solved Problems

1.  Let $\Sigma s_n = s_1 + s_2 + s_3 + \cdots + s_n + \cdots$ be a given positive series and let $\Sigma t_n = (s_1 + s_2) + s_3 + (s_4 + s_5) + s_6 + \cdots$ be obtained from it by the insertion of parentheses according to the pattern $2, 1, 2, 1, 2, 1, \ldots$.

    For the partial sums of $\Sigma t_n$, we have $T_1 = S_2$, $T_2 = S_3$, $T_3 = S_5$, $T_4 = S_6$, $\cdots$. If $\Sigma s_n$ converges to $S$ so also does $\Sigma t_n$ since $\lim\limits_{n \to +\infty} T_n = \lim\limits_{n \to +\infty} S_n$. If $\Sigma s_n$ diverges, $\{S_n\}$ is unbounded and so also is $\{T_n\}$; hence $\Sigma t_n$ diverges.

2.  The series $\Sigma (-1)^{n-1} \left( \dfrac{2n+1}{n} \right)$ diverges. (Why?) When grouped as

    $$\left(3 - \frac{5}{2}\right) + \left(\frac{7}{3} - \frac{9}{4}\right) + \left(\frac{11}{5} - \frac{13}{6}\right) + \cdots + \left(\frac{4m-1}{2m-1} - \frac{4m+1}{2m}\right) + \cdots$$

    the series converges since the general term $\left(\dfrac{4m-1}{2m-1} - \dfrac{4m+1}{2m}\right) = \dfrac{1}{4m^2 - 2m} < \dfrac{1}{m^2}$.

3.  The series  (a) $1 - \dfrac{1}{2} + \dfrac{1}{3} - \dfrac{1}{4} + \cdots + \dfrac{1}{2n-1} - \dfrac{1}{2n} + \cdots$ is convergent and may be grouped as

    $\left(1 - \dfrac{1}{2}\right) + \left(\dfrac{1}{3} - \dfrac{1}{4}\right) + \cdots + \left(\dfrac{1}{2n-1} - \dfrac{1}{2n}\right) + \cdots$  to yield the convergent series  $\dfrac{1}{2} + \dfrac{1}{12} + \dfrac{1}{30} +$

    $\cdots = A$. When (a) is rearranged in the pattern $+ - - + - - \cdots$, we have $\left(1 - \dfrac{1}{2} - \dfrac{1}{4}\right) +$

    $\left(\dfrac{1}{3} - \dfrac{1}{6} - \dfrac{1}{8}\right) + \cdots + \left(\dfrac{1}{2n-1} - \dfrac{1}{4n-2} - \dfrac{1}{4n}\right) + \cdots$  or  $\dfrac{1}{4} + \dfrac{1}{24} + \dfrac{1}{60} + \cdots = \dfrac{1}{2}A$.

4.  Show that $\dfrac{3+1}{3\cdot 1} + \dfrac{3^2+2^3}{3^2\cdot 2^3} + \dfrac{3^3+3^3}{3^3\cdot 3^3} + \cdots + \dfrac{3^n+n^3}{3^n\cdot n^3} + \cdots$  converges.

Since $\dfrac{3^n+n^3}{3^n\cdot n^3} = \dfrac{1}{n^3} + \dfrac{1}{3^n}$, the given series is the sum of the two series $\Sigma\dfrac{1}{n^3}$ and $\Sigma\dfrac{1}{3^n}$.  Each is convergent; hence by Theorem V the given series converges.

5.  Show that the series $\dfrac{3^n+n}{n\cdot 3^n}$ diverges.

Suppose $\Sigma\dfrac{3^n+n}{n\cdot 3^n} = \Sigma\left(\dfrac{1}{n}+\dfrac{1}{3^n}\right)$ converges.  Then, since $\Sigma\dfrac{1}{3^n}$ converges so also (by Theorem V) does $\Sigma\dfrac{1}{n}$.  But this is false; hence the given series diverges.

6.  (a)  Estimate the error when  $\Sigma s_n = 1 - \frac{1}{4} + \frac{1}{9} - \frac{1}{16} + \cdots$  is approximated by its first 10 terms.
    (b)  How many terms must be used to compute the value of the series with allowable error 0.05?

(a)  This is a convergent alternating series.  The error $R_{10} < s_{11} = 1/11^2 = 0.0083$.

(b)  Since $|R_n| < s_{n+1}$, set $s_{n+1} = \dfrac{1}{(n+1)^2} = 0.05$.  Then $(n+1)^2 = 20$ and $n = 3.5$.  Hence four terms are required.

7.  Establish  $R_n < \displaystyle\int_n^{+\infty} f(x)\,dx$  as given in Theorem IX.

In the figure of Problem 1, Chapter 48, let the approximation (by the smaller rectangles) of the area under the curve be extended to the right of $x = n$.  Then

$$R_n = s_{n+1} + s_{n+2} + s_{n+3} + \cdots < \int_n^{+\infty} f(x)\,dx$$

8.  Estimate the error when $\Sigma\dfrac{1}{4n^2}$ is approximated by its first 10 terms.

This series converges by the integral test (Problem 3, Chapter 48).  Then

$$R_{10} < \frac{1}{4}\int_{10}^{+\infty}\frac{dx}{x^2} = \frac{1}{4}\lim_{u\to+\infty}\int_{10}^{u}\frac{dx}{x^2} = \frac{1}{4}\lim_{u\to+\infty}\left(-\frac{1}{u}+\frac{1}{10}\right) = \frac{1}{40} = 0.025$$

9.  Estimate the number of terms necessary to compute  $\Sigma\dfrac{1}{n^5+1}$  with allowable error 0.00001.

This series converges by comparison with $\Sigma\dfrac{1}{n^5}$ which, in turn, converges by the integral test.  Then $R_n < \displaystyle\int_n^{+\infty}\frac{dx}{x^5} = \frac{1}{4n^4}$.  Setting $\dfrac{1}{4n^4} = 0.00001$, we find $n^4 = 25,000$ and $n = 12.6$.  Thus 13 terms are necessary.

10.  Estimate the error when $\Sigma\dfrac{1}{n!}$ is approximated by its first 12 terms.

This series was found to converge (Problem 8, Chapter 48) by comparison with the geometric series $\Sigma\dfrac{1}{2^{n-1}}$.  Thus the error $R_{12}$ for the given series is less than the error $R'_{12}$ for the geometric series, that is, $R_{12} < R'_{12} = \dfrac{(\frac{1}{2})^{12}}{1-\frac{1}{2}} = \dfrac{1}{2^{11}} = 0.0005$.

We can do better!  For $n > 6$, $\dfrac{1}{n!} < \dfrac{1}{4^{n-1}}$; hence, $R_{12} < \dfrac{(\frac{1}{4})^{12}}{1-\frac{1}{4}} = \dfrac{1}{3\cdot 4^{11}} = 0.00000008$.

11.  Estimate the error when $\Sigma s_n = \frac{2}{3} + \frac{1}{2}(\frac{2}{3})^2 + \frac{1}{3}(\frac{2}{3})^3 + \frac{1}{4}(\frac{2}{3})^4 + \cdots$ is approximated by its first 10 terms.

The series converges by the ratio test since $\dfrac{s_{n+1}}{s_n} = \dfrac{2}{3}\left(\dfrac{n}{n+1}\right)$ and $r = \displaystyle\lim_{n\to+\infty}\frac{s_{n+1}}{s_n} = \frac{2}{3}$.  Now $\dfrac{s_{n+1}}{s_n} < \dfrac{2}{3}$, for every value of $n$, so that the given series is term by term less than or equal to the geometric series $\Sigma s_1 r^{n-1}$.  Hence $R_{10} < \left(\dfrac{2}{3}\right)^{11} + \left(\dfrac{2}{3}\right)^{12} + \left(\dfrac{2}{3}\right)^{13} + \cdots = \dfrac{(2/3)^{11}}{1-2/3} = \dfrac{2^{11}}{3^{10}} = 0.04$.

A better approximation may be obtained by noting that after the 10th term the given series is term by term less than $\sum s_{11}\left(\frac{2}{3}\right)^{n-1} = \sum\frac{1}{11}\left(\frac{2}{3}\right)^{11}\left(\frac{2}{3}\right)^{n-1} = \frac{2^{11}}{11\cdot 3^{10}} = 0.004$.

12. Estimate the error when $\sum s_n = \frac{1}{3} + \frac{2}{3^2} + \frac{3}{3^3} + \frac{4}{3^4} + \cdots$ is approximated by its first 10 terms.

The series converges by the ratio test since $\frac{s_{n+1}}{s_n} = \frac{1}{3}\left(\frac{n+1}{n}\right)$ and $r = \frac{1}{3}$. Here $\frac{s_{n+1}}{s_n} \geqq \frac{1}{3}$, for every value of $n$, and we cannot use the geometric series $\Sigma(1/3)^n$ as comparison series. However, $\left\{\frac{s_{n+1}}{s_n}\right\}$ is a nonincreasing sequence and $\frac{s_{12}}{s_{11}} = \frac{4}{11}$; hence after the first 10 terms the given series is term by term less than or equal to the geometric series $\sum s_{11}\left(\frac{4}{11}\right)^{n-1} = \frac{11}{3^{11}}\left(\frac{4}{11}\right)^{n-1}$. Then $R_{10} < \sum\frac{11}{3^{11}}\left(\frac{4}{11}\right)^{n-1} = \frac{121}{7\cdot 3^{11}} = 0.00009758 < 0.0001$.

# Supplementary Problems

13. Rearrange the terms of $1 - \frac{1}{2} + \frac{1}{3} - \frac{1}{4} + \cdots$ to produce a convergent series whose sum is (a) 1, (b) −2.

   *Hint.* (a) Set down the first $n_1$ positive terms until their sum first exceeds 1, then follow with the first $n_2$ negative terms until the sum first falls below 1, and repeat.

14. Can the sum of two divergent series converge? Give an example.

15. (a) Estimate the error when the series $\sum\frac{(-1)^{n-1}}{2n-1}$ is approximated by its first 50 terms.

   (b) Estimate the number of terms necessary if the allowable error is 0.000005.
   *Ans.* (a) 0.01, (b) 100,000

16. (a) Estimate the error when $\sum\frac{(-1)^{n-1}}{n^4}$ is approximated by its first 8 terms.

   (b) Estimate the number of terms necessary if the allowable error is 0.00005.
   *Ans.* (a) 0.0002, (b) 11

17. (a) Estimate the error when the geometric series $\sum\frac{3}{2^n}$ is approximated by its first 6 terms.

   (b) How many terms are necessary if the allowable error is 0.00005?   *Ans.* (a) 0.05, (b) 16

18. Prove: If the positive series $\Sigma s_n$ converges by comparison with the geometric series $\Sigma r^n$, $0 < r < 1$, then $R_n < \frac{r^{n+1}}{1-r}$.

19. Estimate the error when:

   (a) $\sum\frac{1}{3^n+1}\left(<\sum\frac{1}{3^n}\right)$ is approximated by its first 6 terms.

   (b) $\sum\frac{1}{3+4^n}\left(<\sum\frac{1}{4^n}\right)$ is approximated by its first 6 terms.
   *Ans.* (a) 0.0007, (b) 0.00009

20. The series (a) $\sum\frac{n+1}{n\cdot 3^n}$ and (b) $\sum\frac{n}{(n+1)3^n}$ are convergent by the ratio test. Estimate the error when each is approximated by its first 8 terms.   *Ans.* (a) 0.00009, (b) 0.00007

21. For the convergent $p$-series, show that $R_n < \frac{1}{(p-1)n^{p-1}}$.   *Hint.* See Problem 9.

22. The series (a) $\sum\frac{1}{n^3+2}$ and (b) $\sum\frac{n-1}{n^5}$ are convergent by comparison with appropriate $p$-series. Estimate the error when each is approximated by its first 6 terms and find the number of terms necessary if the allowable error is 0.005.   *Ans.* (a) 0.014; 10 terms   (b) 0.002; 5 terms

# Chapter 51

## Power Series

**AN INFINITE SERIES** of the form

$$\sum c_i x^i = \sum_{i=0}^{+\infty} c_i x^i = c_0 + c_1 x + c_2 x^2 + \cdots + c_n x^n + \cdots \qquad (1)$$

where the $c$'s are constants, is called a *power series in $x$*. Similarly, an infinite series of the form

$$\sum c_i (x-a)^i = \sum_{i=0}^{+\infty} c_i (x-a)^i = c_0 + c_1 (x-a) + c_2 (x-a)^2 + \cdots + c_n (x-a)^n + \cdots \qquad (2)$$

is called a *power series in $(x-a)$*.

For any given value of $x$, both $(1)$ and $(2)$ become infinite series of constant terms and (see Chapters 48 and 49) either converge or diverge.

**INTERVAL OF CONVERGENCE.** The totality of values of $x$ for which a power series converges is called its *interval of convergence*. Clearly, $(1)$ converges for $x=0$ and $(2)$ converges for $x=a$. If there are other values of $x$ for which a power series $(1)$ or $(2)$ converges, it converges either for all values of $x$ or for all values of $x$ on some finite interval (closed, open, or half-open) having as midpoint $x=0$ for $(1)$ and $x=a$ for $(2)$.

The interval of convergence will be found here by using the ratio test for absolute convergence supplemented by other tests of Chapters 48 and 49 at the endpoints.

See Problems 1-9.

**CONVERGENCE AND UNIFORM CONVERGENCE.** The discussion and theorems given below involve series of type $(1)$ but apply equally after only minor changes to series of type $(2)$.

Consider the power series $(1)$. Denote by

$$S_n(x) = \sum_{j=0}^{n-1} c_j x^j = c_0 + c_1 x + c_2 x^2 + \cdots + c_{n-1} x^{n-1}$$

the $n$th *partial sum* and by

$$R_n(x) = \sum_{k=n}^{+\infty} c_k x^k = c_n x^n + c_{n+1} x^{n+1} + c_{n+2} x^{n+2} + \cdots$$

the *remainder after $n$ terms*. Then

$$\sum c_i x^i = S_n(x) + R_n(x) \qquad (3)$$

If for $x=x_0$, $\Sigma c_i x^i$ converges to $S(x_0)$, a finite number, then $\lim_{n \to +\infty} S_n(x_0) = S(x_0)$.

Since $|S(x_0) - S_n(x_0)| = |R_n(x_0)|$, $\lim_{n \to +\infty} |S(x_0) - S_n(x_0)| = \lim_{n \to +\infty} |R_n(x_0)| = 0$.

Thus, $\Sigma c_i x^i$ converges for $x=x_0$ if for any positive $\epsilon$, however small, there exists a positive integer $m$ such that whenever $n > m$ then $|R_n(x_0)| < \epsilon$.

Note that here $m$ depends not only upon $\epsilon$ (see Problem 12, Chapter 47) but also upon the choice $x_0$ of $x$. See Problem 10.

In Problem 11, we prove

I.    If $\Sigma c_i x^i$ converges for $x=x_1$ and if $|x_2| < |x_1|$, then the series converges absolutely for $x=x_2$.

Suppose now that (1) converges absolutely, that is $\Sigma |c_i x^i|$ converges for all values of $x$ such that $|x| < P$. Choose a value of $x$, either $x = p$ or $x = -p$, so that $|x| = p < P$. Since (1) converges for $|x| = p$, it follows that for any $\epsilon > 0$, however small, there exists a positive integer $m$ such that whenever $n > m$, then $|R_n(p)| = \sum_{k=n}^{+\infty} |c_k p^k| < \epsilon$. Now let $x$ vary over the interval $|x| \leqq p$. Every term of $|R_n(x)| = \sum_{k=n}^{+\infty} |c_k x^k|$ has its maximum value at $|x| = p$; hence $|R_n(x)|$ has its maximum value on the interval $|x| \leqq p$ when $|x| = p$.

Let $\epsilon$ be chosen and $m$ be found when $|x| = p$. Then for this $\epsilon$ and $m$, $|R_n(x)| < \epsilon$ for *all* $x$ such that $|x| \leqq p$, that is, $m$ depends on $\epsilon$ and $p$ but not on the choice $x_0$ of $x$ on the interval $|x| \leqq p$ as in ordinary convergence. We say that (1) is *uniformly convergent* on the interval $|x| \leqq p$. We have proved

**II.**  If $\Sigma c_i x^i$ converges absolutely for $|x| < P$, then it converges uniformly for $|x| \leqq p < P$.

For example, the series $\Sigma (-1)^i x^i$ is convergent for $|x| < 1$. By Theorem I it is absolutely convergent for $|x| \leqq 0.99$ and by Theorem II it is uniformly convergent for $|x| \leqq 0.9$.

**III.**  A power series *represents* a continuous function $f(x)$ *within* the interval of convergence of the series.                    For a proof, see Problem 12.

**IV.**  If $\Sigma c_i x^i$ converges to the function $f(x)$ on an interval I and if $a$ and $b$ are *within* the interval, then

$$\int_a^b f(x)\, dx = \sum_{i=0}^{+\infty} \int_a^b c_i x^i\, dx = \int_a^b c_0\, dx + \int_a^b c_1 x\, dx + \int_a^b c_2 x^2\, dx + \cdots$$

$$+ \int_a^b c_{n-1} x^{n-1}\, dx + \cdots$$

For a proof, see Problem 13.

**V.**  If $\Sigma c_i x^i$ converges to $f(x)$ on an interval I, then the indefinite integral $\sum_{i=0}^{+\infty} \int_0^x c_i x^i\, dx$ converges to $g(x) = \int_0^x f(x)\, dx$ for all $x$ *within* the interval I.

**VI.**  If $\Sigma c_i x^i$ converges to the function $f(x)$ on the interval I, then the derivative of the series $\sum \dfrac{d}{dx}(c_i x^i)$ converges to $f'(x)$ for all $x$ *within* the interval I.

**VII.**  The representation of a function $f(x)$ in powers of $x$ is unique.

# Solved Problems

1.  Find the interval of convergence of $x - \frac{1}{2}x^2 + \frac{1}{3}x^3 - \frac{1}{4}x^4 + \cdots + (-1)^{n-1}\frac{1}{n}x^n + \cdots$.

Using the ratio test,

$$\lim_{n \to +\infty} \left| \frac{s_{n+1}}{s_n} \right| = \lim_{n \to +\infty} \left| \frac{x^{n+1}}{n+1} \cdot \frac{n}{x^n} \right| = |x| \lim_{n \to +\infty} \frac{n}{n+1} = |x|$$

The series converges absolutely for $|x| < 1$ and diverges for $|x| > 1$. Individual tests *must* be made at the endpoints $x = 1$ and $x = -1$.

For $x = 1$, the series becomes $1 - \frac{1}{2} + \frac{1}{3} - \frac{1}{4} + \cdots$ and is conditionally convergent.

For $x = -1$, the series becomes $-(1 + \frac{1}{2} + \frac{1}{3} + \frac{1}{4} + \cdots)$ and is divergent.

Thus the given series converges on the interval $-1 < x \leqq 1$.

2.  Find the interval of convergence of $1 + x + \dfrac{x^2}{2!} + \dfrac{x^3}{3!} + \cdots + \dfrac{x^n}{n!} + \cdots$.

$$\lim_{n \to +\infty} \left| \frac{s_{n+1}}{s_n} \right| = \lim_{n \to +\infty} \left| \frac{x^{n+1}}{(n+1)!} \cdot \frac{n!}{x^n} \right| = |x| \lim_{n \to +\infty} \frac{1}{n+1} = 0$$

The given series converges for all values of $x$.

3.  Find the interval of convergence of $\dfrac{x-2}{1} + \dfrac{(x-2)^2}{2} + \dfrac{(x-2)^3}{3} + \cdots + \dfrac{(x-2)^n}{n} + \cdots$.

$$\lim_{n \to +\infty} \left| \frac{(x-2)^{n+1}}{n+1} \cdot \frac{n}{(x-2)^n} \right| = |x-2| \lim_{n \to +\infty} \frac{n}{n+1} = |x-2|$$

The series converges absolutely for $|x-2| < 1$ or $1 < x < 3$ and diverges for $|x-2| > 1$ or for $x < 1$ and $x > 3$.

For $x = 1$, the series becomes $-1 + \frac{1}{2} - \frac{1}{3} + \frac{1}{4} - \cdots$ and for $x = 3$, it becomes $1 + \frac{1}{2} + \frac{1}{3} + \frac{1}{4} + \cdots$. The first converges and the second diverges. Thus the given series converges on the interval $1 \leqq x < 3$ and diverges elsewhere.

4.  Find the interval of convergence of $1 + \dfrac{x-3}{1^9} + \dfrac{(x-3)^2}{2^9} + \dfrac{(x-3)^3}{3^9} + \cdots + \dfrac{(x-3)^{n-1}}{(n-1)^9} + \cdots$.

$$\lim_{n \to +\infty} \left| \frac{(x-3)^n}{n^2} \cdot \frac{(n-1)^2}{(x-3)^{n-1}} \right| = |x-3| \lim_{n \to +\infty} \left( \frac{n-1}{n} \right)^2 = |x-3|$$

The series converges absolutely for $|x-3| < 1$ or $2 < x < 4$, and diverges for $|x-3| > 1$ or for $x < 2$ and $x > 4$.

For $x = 2$, the series becomes $1 - 1 + \frac{1}{4} - \frac{1}{9} + \cdots$; and for $x = 4$, it becomes $1 + 1 + \frac{1}{4} + \frac{1}{9} + \cdots$. Since both are absolutely convergent, the given series converges absolutely on the interval $2 \leqq x \leqq 4$ and diverges elsewhere. Note that the first term of the series is not given by the general term with $n = 0$.

5.  Find the interval of convergence of $\dfrac{x+1}{\sqrt{1}} + \dfrac{(x+1)^2}{\sqrt{2}} + \dfrac{(x+1)^3}{\sqrt{3}} + \cdots + \dfrac{(x+1)^n}{\sqrt{n}} + \cdots$.

$$\lim_{n \to +\infty} \left| \frac{(x+1)^{n+1}}{\sqrt{n+1}} \cdot \frac{\sqrt{n}}{(x+1)^n} \right| = |x+1| \lim_{n \to +\infty} \sqrt{\frac{n}{n+1}} = |x+1|$$

The series converges absolutely for $|x+1| < 1$ or $-2 < x < 0$ and diverges for $x < -2$ and $x > 0$.

For $x = -2$, the series becomes $-1 + \dfrac{1}{\sqrt{2}} - \dfrac{1}{\sqrt{3}} + \dfrac{1}{\sqrt{4}} - \cdots$ and for $x = 0$, it becomes $1 + \dfrac{1}{\sqrt{2}} + \dfrac{1}{\sqrt{3}} + \dfrac{1}{\sqrt{4}} + \cdots$. The first is convergent and the second is divergent (why?). Thus, the given series converges on the interval $-2 \leqq x < 0$ and diverges elsewhere.

6.  Find the interval of convergence of $1 + \dfrac{m}{1}x + \dfrac{m(m-1)}{1 \cdot 2}x^2 + \dfrac{m(m-1)(m-2)}{1 \cdot 2 \cdot 3}x^3 + \cdots$.

This is the binomial series. For positive, integral values of $m$, the series is finite; for all other values of $m$, it is an infinite series.

$$\lim_{n \to +\infty} \left| \frac{m(m-1)(m-2)\cdots(m-n+1)x^n}{n!} \cdot \frac{(n-1)!}{m(m-1)(m-2)\cdots(m-n+2)x^{n-1}} \right|$$
$$= |x| \lim_{n \to +\infty} \left| \frac{m-n+1}{n} \right| = |x|$$

The infinite series converges absolutely for $|x| < 1$ and diverges for $|x| > 1$.

At the endpoints $x = \pm 1$, the series converges when $m \geqq 0$ and diverges when $m \leqq -1$. When $-1 < m < 0$, the series converges when $x = 1$ and diverges when $x = -1$. To establish these facts, tests more delicate than those of Chapter 48 are needed.

7.  Find the interval of convergence of $x - \dfrac{x^3}{3} + \dfrac{x^5}{5} - \dfrac{x^7}{7} + \cdots + (-1)^{n-1}\dfrac{x^{2n-1}}{2n-1} + \cdots$.

$$\lim_{n \to +\infty} \left| \frac{x^{2n+1}}{2n+1} \cdot \frac{2n-1}{x^{2n-1}} \right| = x^2 \lim_{n \to +\infty} \frac{2n-1}{2n+1} = x^2$$

The series is absolutely convergent on the interval $x^2 < 1$ or $-1 < x < 1$.

For $x = -1$, the series becomes $-1 + \frac{1}{3} - \frac{1}{5} + \frac{1}{7} - \cdots$ and for $x = 1$, it becomes $1 - \frac{1}{3} + \frac{1}{5} - \frac{1}{7} + \cdots$. Both series converge; thus the given series converges for $-1 \leqq x \leqq 1$ and diverges elsewhere.

8.  Find the interval of convergence of    $(x-1) + 2!\,(x-1)^2 + 3!\,(x-1)^3 + \cdots + n!\,(x-1)^n + \cdots$.

$$\lim_{n \to +\infty} \left| \frac{(n+1)!\,(x-1)^{n+1}}{n!\,(x-1)^n} \right| = |x-1| \lim_{n \to +\infty} (n+1) = \infty$$

The series converges for $x = 1$ only.

9.  Find the interval of convergence of $\dfrac{1}{2x} + \dfrac{2}{4x^2} + \dfrac{3}{8x^3} + \cdots + \dfrac{n}{2^n x^n} + \cdots$.  This is a power series in $1/x$.

$$\lim_{x \to +\infty} \left| \frac{n+1}{2^{n+1} x^{n+1}} \cdot \frac{2^n x^n}{n} \right| = \frac{1}{2|x|} \lim_{n \to +\infty} \frac{n+1}{n} = \frac{1}{2|x|}$$

The series converges absolutely for $\dfrac{1}{2|x|} < 1$ or $|x| > \frac{1}{2}$.

For $x = \frac{1}{2}$ the series becomes $1 + 2 + 3 + 4 + \cdots$ and for $x = -\frac{1}{2}$ the series becomes $-1 + 2 - 3 + 4 - \cdots$  Both of these diverge.  Thus the given series converges on the intervals $x < -\frac{1}{2}$ and $x > \frac{1}{2}$ and diverges on the interval $-\frac{1}{2} \le x \le \frac{1}{2}$.

10.  The series $1 - x + x^2 - x^3 + \cdots + (-1)^n x^n + \cdots$ converges for $|x| < 1$.  Given $\epsilon = 0.000001$, find $m$ when  (a) $x = \frac{1}{2}$ and  (b) $x = \frac{1}{4}$ so that  $|R_n(x)| < \epsilon$ for $n > m$.

$$R_n(x) = \sum_{k=n}^{+\infty} (-1)^k x^k \quad \text{so that}$$

$$\left| R_n\left(\tfrac{1}{2}\right) \right| = \left| \sum_{k=n}^{+\infty} (-1)^k \left(\tfrac{1}{2}\right)^k \right| = \tfrac{1}{3}\left(\tfrac{1}{2}\right)^{n-1} \quad \text{and} \quad \left| R_n\left(\tfrac{1}{4}\right) \right| = \left| \sum_{k=n}^{+\infty} (-1)^k \left(\tfrac{1}{4}\right)^k \right| = \tfrac{1}{5}\left(\tfrac{1}{4}\right)^{n-1}$$

(a)  We seek $m$ such that for $n > m$ then $\frac{1}{3}\left(\frac{1}{2}\right)^{n-1} < 0.000001$ or $1/2^{n-1} < 0.000003$.  Since $1/2^{18} = 0.000004$ and $1/2^{19} = 0.000002$,  $m = 19$.

(b)  We seek $m$ such that for $n > m$ then $\frac{1}{5}\left(\frac{1}{4}\right)^{n-1} < 0.000001$ or $1/4^{n-1} < 0.000005$.  Here, $m = 9$.

11.  Prove:  If a power series $\Sigma c_i x^i$ converges for $x = x_1$ and if $|x_2| < |x_1|$, the series converges absolutely for $x = x_2$.

Since $\Sigma c_i x_1^i$ converges, $\lim\limits_{n \to +\infty} c_n x_1^n = 0$ by Theorem XV, Chapter 47, and $\{|c_i x_1^i|\}$ being convergent is bounded, say, $0 < |c_n x_1^n| < K$ for all values of $n$.  Suppose $|x_2/x_1| = r$, $0 < r < 1$; then

$$|c_n x_2^n| = |c_n x_1^n| \cdot |x_2^n/x_1^n| = |c_n x_1^n| \cdot |x_2/x_1|^n < Kr^n$$

and $\Sigma |c_n x_2^n|$, being term by term less than the convergent geometric series $\Sigma Kr^n$ is convergent.  Thus $\Sigma c_i x_2^i$ converges, in fact, converges absolutely.

12.  Prove:  A power series represents a continuous function $f(x)$ within the interval of convergence of the series.

Set $f(x) = \Sigma c_i x^i = S_n(x) + R_n(x)$.  For any $x = x_0$ within the interval of convergence of $\Sigma c_i x^i$ there is, by Theorem I, an interval $I$ about $x_0$ on which the series is uniformly convergent.  In order to prove $f(x)$ continuous at $x = x_0$, it is necessary to show that $\lim\limits_{\Delta x \to 0} |f(x_0 + \Delta x) - f(x_0)| = 0$ when $x_0 + \Delta x$ is on $I$; that is, it is necessary to show that for a given $\epsilon > 0$, however small, $\Delta x$ may be chosen so that $x_0 + \Delta x$ is on $I$ and  $|f(x_0 + \Delta x) - f(x_0)| < \epsilon$.

Now for any $\Delta x$ such that $x_0 + \Delta x$ is on the interval $I$,

(i)     $\begin{aligned} |f(x_0 + \Delta x) - f(x_0)| &= |S_n(x_0 + \Delta x) + R_n(x_0 + \Delta x) - S_n(x_0) - R_n(x_0)| \\ &\le |S_n(x_0 + \Delta x) - S_n(x_0)| + |R_n(x_0 + x)| + |R_n(x_0)| \end{aligned}$

Let $\epsilon$ be chosen.  Since $x_0 + \Delta x$ is on the interval of convergence of the series, an integer $m > 0$ can be found so that whenever $n > m$ then $|R_n(x_0 + \Delta x)| < \epsilon/3$ and $|R_n(x_0)| < \epsilon/3$.  Also, since $S_n(x)$ is a polynomial, a smaller $|\Delta x|$ can be chosen, if necessary, so that $|S_n(x_0 + \Delta x) - S_n(x_0)| < \epsilon/3$.  For this new choice of $\Delta x$, $|R_n(x_0 + \Delta x)|$ remains less than $\epsilon/3$ since the series is uniformly convergent on $I$ and $|R_n(x_0)|$ is unchanged.  Hence by (i)

$$|f(x_0 + \Delta x) - f(x_0)| < \epsilon/3 + \epsilon/3 + \epsilon/3 = \epsilon$$

Thus $f(x)$ is continuous for all $x$ within the interval of convergence of the series.

13.  Prove:  If $\Sigma c_i x^i$ converges to the function $f(x)$ on an interval and if $x = a$ and $x = b$ are within the interval, then

$$\int_a^b f(x)\,dx = \int_a^b c_0\,dx + \int_a^b c_1 x\,dx + \int_a^b c_2 x^2\,dx + \cdots + \int_a^b c_{n-1} x^{n-1}\,dx + \cdots$$

Suppose $b > a$ and write $f(x) = \Sigma c_i x^i = S_n(x) + R_n(x)$.　Then

$$\int_a^b f(x)\,dx = \int_a^b S_n(x)\,dx + \int_a^b R_n(x)\,dx$$

and
$$\left| \int_a^b f(x)\,dx - \int_a^b S_n(x)\,dx \right| = \left| \int_a^b R_n(x)\,dx \right|$$

Since $\Sigma c_i x^i$ is convergent on an interval, say $|x| < P$, the series is uniformly convergent on the interval $|x| \leq p < P$ which includes both $x = a$ and $x = b$. Then for any $\epsilon > 0$, however small, $n$ can be chosen sufficiently large that $|R_n(x)| < \dfrac{\epsilon}{b-a}$ for all $|x| \leq p$. Thus,

$$\left| \int_a^b f(x)\,dx - \int_a^b S_n(x)\,dx \right| < \int_a^b \frac{\epsilon}{b-a}\,dx = \frac{\epsilon}{b-a}(b-a) = \epsilon ,$$

$$\lim_{n \to +\infty} \left| \int_a^b f(x)\,dx - \int_a^b S_n(x)\,dx \right| = 0 , \quad \text{and} \quad \int_a^b f(x)\,dx = \Sigma \int_a^b c_i x^i\,dx$$

as was to be proved.

# Supplementary Problems

14. Find the interval of convergence of each of the following series.

(a) $x + 2x^2 + 3x^3 + 4x^4 + \cdots$

(d) $\dfrac{x}{5} - \dfrac{x^2}{2 \cdot 5^2} + \dfrac{x^3}{3 \cdot 5^3} - \dfrac{x^4}{4 \cdot 5^4} + \cdots$

(b) $\dfrac{x}{1 \cdot 2} + \dfrac{x^2}{2 \cdot 3} + \dfrac{x^3}{3 \cdot 4} + \dfrac{x^4}{4 \cdot 5} + \cdots$

(e) $\dfrac{1}{1 \cdot 2 \cdot 3} + \dfrac{x^2}{2 \cdot 3 \cdot 4} + \dfrac{x^4}{3 \cdot 4 \cdot 5} + \dfrac{x^6}{4 \cdot 5 \cdot 6} + \cdots$

(c) $x - \dfrac{x^2}{2^2} + \dfrac{x^3}{3^3} - \dfrac{x^4}{4^4} + \cdots$

(f) $\dfrac{x^2}{(\ln 2)^2} + \dfrac{x^3}{(\ln 3)^3} + \dfrac{x^4}{(\ln 4)^4} + \dfrac{x^5}{(\ln 5)^5} + \cdots$

(g) The series obtained by differentiating (a) term by term.
(h) The series obtained by differentiating (b) term by term.

(i) $x + \dfrac{x^2}{1 + 2^3} + \dfrac{x^3}{1 + 3^3} + \dfrac{x^4}{1 + 4^3} + \cdots$

(j) The series obtained by differentiating (i) term by term.
(k) The series obtained by differentiating (j) term by term.
(l) The series obtained by integrating (a) term by term.
(m) The series obtained by integrating (c) term by term.

(n) $(x - 2) + \dfrac{(x-2)^2}{4} + \dfrac{(x-2)^3}{9} + \dfrac{(x-2)^4}{16} + \cdots$

(o) $\dfrac{x-3}{1 \cdot 3} + \dfrac{(x-3)^2}{2 \cdot 3^2} + \dfrac{(x-3)^3}{3 \cdot 3^3} + \dfrac{(x-3)^4}{4 \cdot 3^4} + \cdots$ (p) $1 - \dfrac{3x-2}{5} + \dfrac{(3x-2)^2}{5^2} - \dfrac{(3x-2)^3}{5^3} + \cdots$

(q) The series obtained by differentiating (n) term by term.
(r) The series obtained by integrating (n) term by term.

(s) $1 + \dfrac{x}{1-x} + \left(\dfrac{x}{1-x}\right)^2 + \left(\dfrac{x}{1-x}\right)^3 + \cdots$ (t) $1 - \dfrac{2}{x} + \dfrac{3}{x^2} - \dfrac{4}{x^3} + \cdots$

(u) $\dfrac{1}{2} + \dfrac{x^2 + 6x + 7}{2^2} + \dfrac{(x^2 + 6x + 7)^2}{2^3} + \dfrac{(x^2 + 6x + 7)^3}{2^4} + \cdots$

Ans. (a) $-1 < x < 1$　　(f) all values $x$　　(k) $-1 \leq x < 1$　　(p) $-1 < x < 7/3$　(t) $x < -1$,
　　　(b) $-1 \leq x \leq 1$　　(g) $-1 < x < 1$　　(l) $-1 < x < 1$　　(q) $1 \leq x < 3$　　　$x > 1$
　　　(c) all values of $x$　(h) $-1 \leq x < 1$　(m) all values of $x$　(r) $1 \leq x \leq 3$　(u) $-5 < x < -3$,
　　　(d) $-5 < x \leq 5$　　(i) $-1 \leq x \leq 1$　(n) $1 \leq x \leq 3$　　(s) $x < \frac{1}{2}$　　　$-3 < x < -1$
　　　(e) $-1 \leq x \leq 1$　　(j) $-1 \leq x \leq 1$　(o) $0 \leq x < 6$

15. Prove: A power series can be differentiated term by term within its interval of convergence.

Hint. $f(x) = \displaystyle\sum_{i=0}^{+\infty} c_i x^i$ and $\displaystyle\sum_{i=0}^{+\infty} \dfrac{d}{dx}(c_i x^i) = \displaystyle\sum_{j=1}^{+\infty} j c_j x^{j-1}$ converge for $|x| < \displaystyle\lim_{n \to +\infty} \left| \dfrac{c_n}{c_{n+1}} \right|$. Use

Theorems I, II, and V to show $\displaystyle\int_0^x f'(x)\,dx = f(x)$.

16. Prove: The representation of a function $f(x)$ in powers of $x$ is unique.

Hint. Let $f(x) = \Sigma s_n x^n$ and $f(x) = \Sigma t_n x^n$ on $|x| < a \neq 0$. Put $x = 0$ in $\Sigma(s_n - t_n)x^n = 0$,

$\dfrac{d}{dx} \Sigma(s_n - t_n)x^n = 0$, $\dfrac{d^2}{dx^2} \Sigma(s_n - t_n)x^n = 0$, $\ldots$ to obtain $s_j = t_j$, $j = 0, 1, 2, 3, \ldots$.

# Chapter 52

## Series Expansion of Functions

**POWER SERIES** in $x$ may be generated in various ways; for example, imagining the division continued indefinitely, we find

$$\frac{1}{1-x} = 1 + x + x^2 + x^3 + \cdots + x^{n-1} + \cdots \tag{1}$$

(Note that for, say, $x = 5$ this is a perfectly absurd statement.) In Problem 1 it is shown that the series $(1)$ represents $\dfrac{1}{1-x}$ only on the interval $|x| < 1$, that is,

$$\frac{1}{1-x} = 1 + x + x^2 + x^3 + \cdots + x^{n-1} + \cdots, \quad -1 < x < 1$$

Other methods for generating power series are illustrated in Problems 2-3.

**A GENERAL METHOD** for expanding a function in a powers series in $x$ and in $(x-a)$ is given below. Note the requirement that the function and its derivatives of *all* orders must exist at $x = 0$ or at $x = a$. Thus $1/x$, $\ln x$, and $\cot x$ cannot be expanded in powers of $x$.

*Maclaurin's Series.* Assuming that a given function can be represented by a power series in $x$, that series is necessarily of the form of *Maclaurin's series*:

$$f(x) = f(0) + \frac{f'(0)}{1!}x + \frac{f''(0)}{2!}x^2 + \frac{f'''(0)}{3!}x^3 + \cdots + \frac{f^{(n-1)}(0)}{(n-1)!}x^{n-1} + \cdots \tag{2}$$

*Taylor's Series.* Assuming that a given function can be represented by a power series in $(x-a)$, that series is necessarily of the form of *Taylor's series*:

$$f(x) = f(a) + \frac{f'(a)}{1!}(x-a) + \frac{f''(a)}{2!}(x-a)^2 + \frac{f'''(a)}{3!}(x-a)^3$$

$$+ \cdots + \frac{f^{(n-1)}(a)}{(n-1)!}(x-a)^{n-1} + \cdots \tag{3}$$

See Problem 4.

The question of the interval on which $f(x)$ is represented by its Maclaurin's or Taylor's series will be considered in the next chapter. For the functions of this book, the interval on which a series represents the function coincides with the interval of convergence of the series.

See Problems 5-11.

Another and very useful form of Taylor's series

$$f(a+h) = f(a) + \frac{h}{1!}f'(a) + \frac{h^2}{2!}f''(a) + \frac{h^3}{3!}f'''(a) + \cdots + \frac{h^{n-1}}{(n-1)!}f^{(n-1)}(a) + \cdots \tag{4}$$

is obtained by replacing $x$ by $a+h$ in $(3)$.

# Solved Problems

1. The power series $1 + x + x^2 + x^3 + \cdots + x^{n-1} + \cdots$ is an infinite geometric series with $a = 1$ and $r = x$. For $|r| = |x| < 1$, the series converges to $\dfrac{a}{1-r} = \dfrac{1}{1-x}$; for $|r| = |x| \geqq 1$, the series diverges.

2. By repeated differentiation of the series of Problem 1, we obtain other power series

   (i) $\qquad 1 + 2x + 3x^2 + 4x^3 + \cdots + nx^{n-1} + \cdots$

   (ii) $\qquad 2 + 6x + 12x^2 + 20x^3 + \cdots + n(n+1)x^{n-1} + \cdots$

   By repeated integration between the limits 0 and $x$ of the series of Problem 1, we obtain

   (iii) $\qquad x + \dfrac{1}{2}x^2 + \dfrac{1}{3}x^3 + \dfrac{1}{4}x^4 + \cdots + \dfrac{1}{n}x^n + \cdots$

   (iv) $\qquad \dfrac{1}{2}x^2 + \dfrac{1}{6}x^3 + \dfrac{1}{12}x^4 + \dfrac{1}{20}x^5 + \cdots + \dfrac{1}{n(n+1)}x^{n+1} + \cdots$

3. Find the power series $y = \Sigma c_n x^n$ satisfying the conditions:
   (i) $y = 2$ when $x = 0$, (ii) $y' = 1$ when $x = 0$, and (iii) $y'' + 2y' = 0$.

   Consider

   (a) $\qquad y = c_0 + c_1 x + c_2 x^2 + c_3 x^3 + \cdots$

   (b) $\qquad y' = c_1 + 2c_2 x + 3c_3 x^2 + 4c_4 x^3 + \cdots$

   (c) $\qquad y'' = 2c_2 + 6c_3 x + 12c_4 x^2 + 20c_5 x^3 + \cdots$

   From (a) with $x = 0$, $y = 2$ we find $c_0 = 2$; from (b) with $x = 0$, $y' = 1$ we find $c_1 = 1$. Since $y'' = -2y'$, we have

   $$2c_2 + 6c_3 x + 12c_4 x^2 + 20c_5 x^3 + \cdots = -2c_1 - 4c_2 x - 6c_3 x^2 - 8c_4 x^3 - \cdots$$

   from which it follows that $c_2 = -c_1 = -1$, $c_3 = -\frac{2}{3}c_2 = \frac{2}{3}$, $c_4 = -\frac{1}{2}c_3 = -\frac{1}{3}$, $\cdots$. Thus, $y = 2 + x - x^2 + \frac{2}{3}x^3 - \frac{1}{3}x^4 + \cdots$ is the required series.

4. Assuming (i) $f(x)$ together with its derivatives of all orders exist at $x = a$ and (ii) $f(x)$ can be represented as a power series in $(x - a)$, show that this series is

   $$f(x) = f(a) + \frac{f'(a)}{1!}(x - a) + \frac{f''(a)}{2!}(x - a)^2 + \cdots + \frac{f^{(n-1)}(a)}{(n-1)!}(x - a)^{n-1} + \cdots$$

   Let the series be

   (a) $\qquad f(x) = c_0 + c_1(x - a) + c_2(x - a)^2 + c_3(x - a)^3 + \cdots + c_{n-1}(x - a)^{n-1} + \cdots$

   Differentiating successively, we have

   (b) $\quad f'(x) = c_1 + 2c_2(x - a) + 3c_3(x - a)^2 + 4c_4(x - a)^3 + \cdots + nc_n(x - a)^{n-1} + \cdots$

   (c) $\quad f''(x) = 2c_2 + 6c_3(x - a) + 12c_4(x - a)^2 + 20c_5(x - a)^3 + \cdots + (n+1)nc_{n+1}(x - a)^{n-1} + \cdots$

   (d) $\quad f'''(x) = 6c_3 + 24c_4(x - a) + 60c_5(x - a)^2 + \cdots + (n+2)(n+1)nc_{n+2}(x - a)^{n-1} + \cdots$

   ............................................................

   ............................................................

   Setting $x = a$ in $(a), (b), (c), \ldots$ we find in turn

   $$c_0 = f(a), \quad c_1 = f'(a), \quad c_2 = \frac{1}{2!}f''(a), \quad \ldots, \quad c_{n-1} = \frac{1}{(n-1)!}f^{(n-1)}(a), \quad \ldots$$

   When these replacements are made in $(a)$, we have the required Taylor's series.

In Problems 5-10 obtain the expansion of the function in powers of $x$ or $x-a$ as indicated, under the assumptions of this chapter, and determine the interval of convergence of the series.

**5.** $e^{-2x}$; powers of $x$.

$$f(x) = e^{-2x} \qquad\qquad f(0) = 1$$
$$f'(x) = -2e^{-2x} \qquad\qquad f'(0) = -2$$
$$f''(x) = 2^2 e^{-2x} \qquad\qquad f''(0) = 2^2$$
$$f'''(x) = -2^3 e^{-2x} \qquad\qquad f'''(0) = -2^3$$

$$\cdots\cdots\cdots\cdots\qquad\qquad\cdots\cdots\cdots\cdots$$

Then $\quad e^{-2x} = 1 - 2x + \dfrac{2^2}{2!}x^2 - \dfrac{2^3}{3!}x^3 + \dfrac{2^4}{4!}x^4 - \cdots + (-1)^n \dfrac{2^n}{n!}x^n + \cdots$

Since $\qquad\qquad \lim\limits_{n \to +\infty} \left| \dfrac{2^{n+1}x^{n+1}}{(n+1)!} \cdot \dfrac{n!}{2^n x^n} \right| = |x| \lim\limits_{n \to +\infty} \dfrac{2}{n+1} = 0$

the series converges for every value of $x$.

**6.** $\sin x$; powers of $x$.

$$f(x) = \sin x \qquad\qquad f(0) = 0$$
$$f'(x) = \cos x \qquad\qquad f'(0) = 1$$
$$f''(x) = -\sin x \qquad\qquad f''(0) = 0$$
$$f'''(x) = -\cos x \qquad\qquad f'''(0) = -1$$

$$\cdots\cdots\cdots\cdots\qquad\qquad\cdots\cdots\cdots\cdots$$

The values of the derivatives at $x = 0$ form cycles of $0, 1, 0, -1$; hence

$$\sin x = 0 + 1x + \frac{0}{2!}x^2 + \frac{-1}{3!}x^3 + \frac{0}{4!}x^4 + \frac{1}{5!}x^5 + \cdots$$

$$= x - \frac{x^3}{3!} + \frac{x^5}{5!} - \frac{x^7}{7!} + \cdots + (-1)^{n-1}\frac{x^{2n-1}}{(2n-1)!} + \cdots$$

Since $\qquad \lim\limits_{n \to +\infty} \left| \dfrac{x^{2n+1}}{(2n+1)!} \cdot \dfrac{(2n-1)!}{x^{2n-1}} \right| = x^2 \lim\limits_{n \to +\infty} \dfrac{1}{2n(2n+1)} = 0$

the series converges for every value of $x$.

**7.** $\ln(1+x)$; powers of $x$.

$$f(x) = \ln(1+x) \qquad\qquad f(0) = 0$$
$$f'(x) = \frac{1}{1+x} \qquad\qquad f'(0) = 1$$
$$f''(x) = -\frac{1}{(1+x)^2} \qquad\qquad f''(0) = -1$$
$$f'''(x) = \frac{1 \cdot 2}{(1+x)^3} \qquad\qquad f'''(0) = 2!$$
$$f^{iv}(x) = -\frac{1 \cdot 2 \cdot 3}{(1+x)^4} \qquad\qquad f^{iv}(0) = -3!$$

$$\cdots\cdots\cdots\cdots\cdots\qquad\qquad\cdots\cdots\cdots\cdots$$

Hence $\quad \ln(1+x) = x - \dfrac{x^2}{2!} + 2!\dfrac{x^3}{3!} - 3!\dfrac{x^4}{4!} + \cdots + (-1)^{n-1}(n-1)!\dfrac{x^n}{n!} + \cdots$

$$= x - \frac{1}{2}x^2 + \frac{1}{3}x^3 - \frac{1}{4}x^4 + \cdots + (-1)^{n-1}\frac{1}{n}x^n + \cdots$$

By Problem 1, Chapter 51, the series converges on the interval $-1 < x \leqq 1$.

8.  arc tan $x$; powers of $x$.

$$f(x) \quad = \text{ arc tan } x \qquad\qquad\qquad\qquad f(0) \quad = 0$$

$$f'(x) \quad = \frac{1}{1 + x^2} = 1 - x^2 + x^4 - x^6 + \cdots \qquad f'(0) \quad = 1$$

$$f''(x) \quad = -2x + 4x^3 - 6x^5 + \cdots \qquad\qquad f''(0) \quad = 0$$
$$f'''(x) \quad = -2 + 12x^2 - 30x^4 + \cdots \qquad\qquad f'''(0) \quad = -2!$$
$$f^{iv}(x) \quad = 24x - 120x^3 + \cdots \qquad\qquad f^{iv}(0) \quad = 0$$
$$f^{v}(x) \quad = 24 - 360x^2 + \cdots \qquad\qquad f^{v}(0) \quad = 4!$$
$$f^{vi}(x) \quad = -720x + \cdots \qquad\qquad f^{vi}(0) \quad = 0$$
$$f^{vii}(x) \quad = -720 + \cdots \qquad\qquad f^{vii}(0) \quad = -6!$$

$$\text{arc tan } x \quad = \quad x - \frac{2!}{3!}x^3 + \frac{4!}{5!}x^5 - \frac{6!}{7!}x^7 + \cdots$$

$$= \quad x - \frac{x^3}{3} + \frac{x^5}{5} - \frac{x^7}{7} + \cdots + (-1)^{n-1}\frac{x^{2n-1}}{2n-1} + \cdots$$

From Problem 7, Chapter 51, the interval of convergence is $-1 \leqq x \leqq 1$.

9.  $e^{x/2}$; powers of $x - 2$.

$$f(x) \quad = e^{x/2} \qquad\qquad f(2) \quad = e$$
$$f'(x) \quad = \tfrac{1}{2}e^{x/2} \qquad\qquad f'(2) \quad = \tfrac{1}{2}e$$
$$f''(x) \quad = \tfrac{1}{4}e^{x/2} \qquad\qquad f''(2) \quad = \tfrac{1}{4}e$$

$$e^{x/2} \quad = \quad e\left\{ 1 + \frac{1}{2}(x - 2) + \frac{1}{4}\frac{(x-2)^2}{2!} + \cdots + \frac{1}{2^{n-1}} \cdot \frac{(x-2)^{n-1}}{(n-1)!} + \cdots \right\}$$

$$\lim_{n \to +\infty} \left| \frac{(x-2)^n}{2^n n!} \cdot \frac{2^{n-1}(n-1)!}{(x-2)^{n-1}} \right| \quad = \quad \frac{1}{2}|x - 2| \lim_{n \to +\infty} \frac{1}{n} \quad = \quad 0$$

The series converges for every value of $x$.

10.  $\ln x$; powers of $x - 2$.

$$f(x) \quad = \ln x \qquad\qquad f(2) \quad = \ln 2$$
$$f'(x) \quad = x^{-1} \qquad\qquad f'(2) \quad = \tfrac{1}{2}$$
$$f''(x) \quad = -x^{-2} \qquad\qquad f''(2) \quad = -\tfrac{1}{4}$$
$$f'''(x) \quad = 2x^{-3} \qquad\qquad f'''(2) \quad = \tfrac{1}{4}$$
$$f^{iv}(x) \quad = -6x^{-4} \qquad\qquad f^{iv}(2) \quad = -\tfrac{3}{8}$$

$$\cdots\cdots\cdots\cdots\qquad\qquad\cdots\cdots\cdots\cdots$$
$$\cdots\cdots\cdots\cdots\qquad\qquad\cdots\cdots\cdots\cdots$$

$$\ln x \quad = \quad \ln 2 + \frac{1}{2}(x - 2) - \frac{1}{4}\frac{(x-2)^2}{2!} + \frac{1}{4}\frac{(x-2)^3}{3!} - \frac{3}{8}\frac{(x-2)^4}{4!} + \cdots$$

$$= \quad \ln 2 + \frac{1}{2}(x - 2) - \frac{1}{8}(x - 2)^2 + \frac{1}{24}(x - 2)^3 - \frac{1}{64}(x - 2)^4 + \cdots$$

Since $\quad \lim_{n \to +\infty} \left| \frac{(x-2)^{n+1}}{2^{n+1}(n+1)} \cdot \frac{2^n n}{(x-2)^n} \right| = \frac{1}{2}|x - 2| \lim_{n \to +\infty} \frac{n}{n+1} = \frac{1}{2}|x - 2|$

the series converges for $|x - 2| < 2$ or $0 < x < 4$.

For $x = 0$, the series is $\ln 2 - $ (harmonic series) and diverges; for $x = 4$, the series is $\ln 2 + 1 - \frac{1}{2} + \frac{1}{3} - \frac{1}{4} + \cdots$ and converges. Thus the series converges on the interval $0 < x \leqq 4$.

11.  Obtain the Maclaurin's expansion for $\sqrt{1 + \sin x} = \sin \frac{1}{2}x + \cos \frac{1}{2}x$.

Replace $x$ by $\frac{1}{2}x$ in the expansion for $\sin x$ (Problem 6) to obtain

$$\sin \frac{1}{2}x \quad = \quad \frac{1}{2}x - \frac{x^3}{2^3 \cdot 3!} + \frac{x^5}{2^5 \cdot 5!} - \frac{x^7}{2^7 \cdot 7!} + \cdots$$

Differentiate this expansion to obtain

$$\cos\frac{1}{2}x \;=\; 2\left\{\frac{1}{2} - \frac{x^2}{2^3\cdot2!} + \frac{x^4}{2^5\cdot4!} - \frac{x^6}{2^7\cdot6!} + \cdots\right\}$$

$$=\; 1 - \frac{x^2}{2^2\cdot2!} + \frac{x^4}{2^4\cdot4!} - \frac{x^6}{2^6\cdot6!} + \cdots$$

Then

$$\sqrt{1+\sin x} \;=\; \sin\frac{1}{2}x + \cos\frac{1}{2}x \;=\; 1 + \frac{x}{2} - \frac{x^2}{2^2\cdot2!} - \frac{x^3}{2^3\cdot3!} + \frac{x^4}{2^4\cdot4!} + \frac{x^5}{2^5\cdot5!} - \cdots,$$

all values of $x$.

12. Obtain the Maclaurin's expansion for $e^{\cos x} = e \cdot e^{(\cos x - 1)}$.

Using $e^u = 1 + u + \dfrac{u^2}{2!} + \dfrac{u^3}{3!} + \cdots$ and $u = \cos x - 1 = -\dfrac{x^2}{2!} + \dfrac{x^4}{4!} - \dfrac{x^6}{6!} + \cdots$, we find

$$e^{\cos x} \;=\; e\left\{1 + \left(-\frac{x^2}{2!} + \frac{x^4}{4!} - \frac{x^6}{6!} + \cdots\right) + \frac{1}{2!}\left(\frac{x^4}{(2!)^2} - \frac{2x^6}{2!\,4!} + \cdots\right)\right.$$

$$\left. + \frac{1}{3!}\left(-\frac{x^6}{(2!)^3} + \cdots\right) + \cdots\right\}$$

$$=\; e\left\{1 - \frac{x^2}{2} + \frac{x^4}{6} - \frac{31}{720}x^6 + \cdots\right\}$$

13. Under the assumption that all necessary operations are valid, show (a) $e^{ix} = \cos x + i\sin x$, (b) $e^{-ix} = \cos x - i\sin x$, (c) $\sin x = (e^{ix} - e^{-ix})/2i$, (d) $\cos x = (e^{ix} + e^{-ix})/2$, where $i = \sqrt{-1}$.

$$e^z \;=\; 1 + z + \frac{z^2}{2!} + \frac{z^3}{3!} + \frac{z^4}{4!} + \frac{z^5}{5!} + \cdots$$

(a)  $$e^{ix} \;=\; 1 + (ix) + \frac{(ix)^2}{2!} + \frac{(ix)^3}{3!} + \frac{(ix)^4}{4!} + \frac{(ix)^5}{5!} + \cdots \;=\; 1 + ix - \frac{x^2}{2!} - i\frac{x^3}{3!} + \frac{x^4}{4!} + i\frac{x^5}{5!} + \cdots$$

$$=\; \left(1 - \frac{x^2}{2!} + \frac{x^4}{4!} - \cdots\right) + i\left(x - \frac{x^3}{3!} + \frac{x^5}{5!} - \cdots\right) \;=\; \cos x + i\sin x$$

(b)  $e^{-ix} = \cos(-x) + i\sin(-x) = \cos x - i\sin x$

(c)  $e^{ix} - e^{-ix} = 2i\sin x$; hence, $\sin x = (e^{ix} - e^{-ix})/2i$.

(d)  $e^{ix} + e^{-ix} = 2\cos x$; hence, $\cos x = (e^{ix} + e^{-ix})/2$.

# Supplementary Problems

14. Verify: (a) The series (i) and (ii) of Problem 2 converge for $|x| < 1$;  (b) (iii) converges for $-1 \leqq x < 1$; (c) (iv) converges for $-1 \leqq x \leqq 1$.

15. Verify: (a) The series obtained by adding (i) and (ii) of Problem 2 converges for $|x| < 1$;  (b) that obtained by adding (iii) and (iv) converges for $-1 \leqq x < 1$.

16. Find the power series $y = \Sigma c_n x^n$ satisfying the conditions  (i) $y = 2$ when $x = 0$,  (ii) $y' = 0$ when $x = 0$, and  (iii) $y'' - y = 0$.    Ans. $y = 2 + x^2 + \dfrac{x^4}{12} + \cdots + \dfrac{2x^{2n}}{(2n)!} + \cdots$.

17. Find the power series $y = \Sigma c_n x^n$ satisfying the conditions  (i) $y = 1$ when $x = 0$,  (ii) $y' = 1$ when $x = 0$, and  (iii) $y'' + y = 0$.    Ans. $y = 1 + x - \dfrac{x^2}{2!} - \dfrac{x^3}{3!} + \dfrac{x^4}{4!} + \dfrac{x^5}{5!} - \cdots$

**18.** Obtain the expansion in Maclaurin's series:

(a)  $\cos^2 x = 1 - \dfrac{2}{2!}x^2 + \dfrac{2^3}{4!}x^4 - \cdots + (-1)^n \dfrac{2^{2n-1}}{(2n)!}x^{2n} + \cdots,$  all values of $x$

(b)  $\sec x = 1 + \dfrac{1}{2}x^2 + \dfrac{5}{24}x^4 + \dfrac{61}{720}x^6 + \cdots,$  $-\pi/2 < x < \pi/2$

(c)  $\tan x = x + \dfrac{1}{3}x^3 + \dfrac{2}{15}x^5 + \dfrac{17}{315}x^7 + \cdots,$  $-\pi/2 < x < \pi/2$

(d)  $\arcsin x = x + \dfrac{1}{2}\dfrac{x^3}{3} + \dfrac{1 \cdot 3}{2 \cdot 4}\dfrac{x^5}{5} + \dfrac{1 \cdot 3 \cdot 5}{2 \cdot 4 \cdot 6}\dfrac{x^7}{7} + \cdots,$  $-1 < x < 1$

(e)  $\sin^2 x = \dfrac{2}{2!}x^2 - \dfrac{2^3}{4!}x^4 + \dfrac{2^5}{6!}x^6 - \cdots + (-1)^{n+1}\dfrac{2^{2n-1}}{(2n)!}x^{2n} + \cdots,$  all values of $x$

**19.** Obtain the expansion in Taylor's series:

(a)  $e^x = e^a\left[1 + (x-a) + \dfrac{(x-a)^2}{2!} + \dfrac{(x-a)^3}{3!} + \cdots + \dfrac{(x-a)^{n-1}}{(n-1)!} + \cdots\right],$  all values of $x$

(b)  $\sin x = \sin a + (x-a)\cos a - \dfrac{(x-a)^2}{2!}\sin a - \dfrac{(x-a)^3}{3!}\cos a + \cdots,$  all values of $x$

(c)  $\cos x = \dfrac{1}{\sqrt{2}}\left[1 - (x - \tfrac{1}{4}\pi) - \dfrac{(x - \tfrac{1}{4}\pi)^2}{2!} + \dfrac{(x - \tfrac{1}{4}\pi)^3}{3!} + \cdots\right],$  all values of $x$

**20.** Differentiate the expansion for $\sin x$ (Problem 6) to obtain the expansion for $\cos x$. Then identify the solution of Problem 17 as  $y = \sin x + \cos x$.

**21.** Replace $x$ by $\tfrac{1}{2}x$ in the expansion for $e^{-2x}$ (Problem 5) to obtain the expansion for $e^{-x}$. In this latter series replace $x$ by $-x$ to obtain the expansion for $e^x$; then identify the solution of Problem 16 as $y = e^x + e^{-x}$.

**22.** Obtain the Maclaurin's expansion  $\sin^2 x = (\sin x)^2 = x^2 - \dfrac{2x^4}{3!} + \dfrac{32x^6}{3!\,5!} - \dfrac{96x^8}{3!\,7!} + \cdots,$  all values of $x$.

**23.** Show that  $\displaystyle\int_0^x e^{-y^2}\,dy = x - \dfrac{x^3}{3 \cdot 1!} + \dfrac{x^5}{5 \cdot 2!} - \dfrac{x^7}{7 \cdot 3!} + \cdots,$  all values of $x$.

**24.** Obtain by division the series expansion of $\dfrac{1}{1+x^2}$; then obtain

$$\arctan x = \int_0^x \frac{dx}{1+x^2} = x - \frac{1}{3}x^3 + \frac{1}{5}x^5 - \frac{1}{7}x^7 + \cdots$$

and compare with Problem 8.

**25.** By the binomial theorem establish  $\dfrac{1}{\sqrt{1-x^2}} = 1 + \dfrac{1}{2}x^2 + \dfrac{1 \cdot 3}{2 \cdot 4}x^4 + \dfrac{1 \cdot 3 \cdot 5}{2 \cdot 4 \cdot 6}x^6 + \cdots;$  then obtain

$$\arcsin x = \int_0^x \frac{dx}{\sqrt{1-x^2}} = x + \frac{1 \cdot x^3}{2 \cdot 3} + \frac{1 \cdot 3 \cdot x^5}{2 \cdot 4 \cdot 5} + \frac{1 \cdot 3 \cdot 5 \cdot x^7}{2 \cdot 4 \cdot 6 \cdot 7} + \cdots$$

**26.** Multiply the respective series expansions to obtain:

(a)  $e^x \sin x = x + x^2 + \dfrac{x^3}{3} - \dfrac{x^5}{30} - \dfrac{x^6}{90} + \cdots$    (b)  $e^x \cos x = 1 + x - \dfrac{x^3}{3} - \dfrac{x^4}{6} - \dfrac{x^5}{30} + \cdots$

**27.** Write  $\sec x = \dfrac{1}{\cos x} = \dfrac{1}{1 - x^2/2! + x^4/4! - \cdots} = c_0 + c_1 x + c_2 x^2 + c_3 x^3 + \cdots.$  Clear of fractions in the last equality and equate coefficients of like powers of $x$ to obtain the expansion of $\sec x$.

# Chapter 53

## Maclaurin's and Taylor's Formulas with Remainders

**MACLAURIN'S FORMULA:** If $f(x)$ and its first $n$ derivatives are continuous on an interval containing $x = 0$, then there are numbers $x_0$ and $x_0^*$ between 0 and $x$ such that

$$f(x) = f(0) + \frac{f'(0)}{1!}x + \frac{f''(0)}{2!}x^2 + \cdots + \frac{f^{(n-1)}(0)}{(n-1)!}x^{n-1} + R_n(x)$$

where

$$R_n(x) = \frac{f^{(n)}(x_0)}{n!}x^n, \quad \text{(Lagrange Form)}$$

or

$$R_n(x) = \frac{f^{(n)}(x_0^*)}{(n-1)!}(x - x_0^*)^{n-1}x, \quad \text{(Cauchy Form)}$$

**TAYLOR'S FORMULA:** If $f(x)$ and its first $n$ derivatives are continuous on an interval containing $x = a$, then there are numbers $x_0$ and $x_0^*$ between $a$ and $x$ such that

$$f(x) = f(a) + \frac{f'(a)}{1!}(x - a) + \frac{f''(a)}{2!}(x - a)^2 + \cdots + \frac{f^{(n-1)}(a)}{(n-1)!}(x - a)^{n-1} + R_n(x)$$

where

$$R_n(x) = \frac{f^{(n)}(x_0)}{n!}(x - a)^n, \quad \text{(Lagrange Form)}$$

or

$$R_n(x) = \frac{f^{(n)}(x_0^*)}{(n-1)!}(x - x_0^*)^{n-1}(x - a), \quad \text{(Cauchy Form)}$$

Maclaurin's formula is a special case ($a = 0$) of Taylor's formula. Taylor's formula with the Lagrange form of the remainder is a simple variation of the Extended Law of the Mean (see Chapter 21). For the derivation of the formula with the Cauchy form of the remainder, see Problem 10.

The Maclaurin's and Taylor's series expansion of a function $f(x)$ obtained in Chapter 52 represent that function for those values, and only those values, of $x$ for which

$$\lim_{n \to +\infty} R_n(x) = 0$$

**SERIES FOR REFERENCE.** The following series, with the intervals on which they represent the function, are listed for reference.

$e^{ax} = 1 + ax + \frac{(ax)^2}{2!} + \frac{(ax)^3}{3!} + \cdots + \frac{(ax)^{n-1}}{(n-1)!} + \cdots$  All values of $x$.

$\sin ax = ax - \frac{(ax)^3}{3!} + \frac{(ax)^5}{5!} - \frac{(ax)^7}{7!} + \cdots + (-1)^{n-1}\frac{(ax)^{2n-1}}{(2n-1)!} + \cdots$  All values of $x$.

$\cos ax = 1 - \frac{(ax)^2}{2!} + \frac{(ax)^4}{4!} - \frac{(ax)^6}{6!} + \cdots + (-1)^{n-1}\frac{(ax)^{2n-2}}{(2n-2)!} + \cdots$  All values of $x$.

$\ln(a + x) = \ln a + \frac{x}{a} - \frac{x^2}{2a^2} + \frac{x^3}{3a^3} - \cdots + (-1)^{n-1}\frac{x^n}{na^n} + \cdots$  $-a < x \le a$.

$\arcsin x = x + \frac{1 \cdot x^3}{2 \cdot 3} + \frac{1 \cdot 3 \cdot x^5}{2 \cdot 4 \cdot 5} + \cdots + \frac{1 \cdot 3 \cdot 5 \ldots (2n-3)x^{2n-1}}{2 \cdot 4 \cdot 6 \ldots (2n-2)(2n-1)} + \cdots$  $-1 \le x \le 1$

$$\arctan x \;=\; x - \frac{x^3}{3} + \frac{x^5}{5} - \frac{x^7}{7} + \cdots + (-1)^{n-1}\frac{x^{2n-1}}{2n-1} + \cdots \qquad -1 \leqq x \leqq 1$$

$$\ln x \;=\; \ln a + \frac{1}{a}(x-a) - \frac{1}{2a^2}(x-a)^2 + \frac{1}{3a^3}(x-a)^3 - \cdots + \frac{(-1)^n}{(n-1)a^{n-1}}(x-a)^{n-1} + \cdots$$
$$0 < x \leqq 2a$$

$$e^x \;=\; e^a\left\{1 + (x-a) + \frac{(x-a)^2}{2!} + \frac{(x-a)^3}{3!} + \cdots + \frac{(x-a)^{n-1}}{(n-1)!} + \cdots\right\} \qquad \text{All values of } x.$$

$$\sin x \;=\; \sin a + (x-a)\cos a - \frac{(x-a)^2}{2!}\sin a - \frac{(x-a)^3}{3!}\cos a + \cdots \qquad \text{All values of } x.$$

$$\cos x \;=\; \cos a - (x-a)\sin a - \frac{(x-a)^2}{2!}\cos a + \frac{(x-a)^3}{3!}\sin a + \cdots \qquad \text{All values of } x.$$

# Solved Problems

1.  Find the interval for which $e^x$ may be represented by its Maclaurin's series.

    $f^{(n)}(x) = e^x$; the Lagrange form of the remainder is $|R_n(x)| = \left|\dfrac{x^n}{n!}f^{(n)}(x_0)\right| = \dfrac{|x^n|}{n!}e^{x_0}$, where $x_0$ is between 0 and $x$.

    The factor $\dfrac{x^n}{n!}$ is a general term of $e^x = 1 + x + \dfrac{x^2}{2!} + \dfrac{x^3}{3!} + \cdots$ which is known to converge for every value of $x$. Thus, $\lim\limits_{n\to+\infty}\dfrac{|x^n|}{n!} = 0$. The factor $e^{x_0}$ is finite for every value of $x$. Hence, $\lim\limits_{n\to+\infty}R_n(x) = 0$ (finite number) $= 0$ and the series represents $e^x$ for all values of $x$.

2.  Find the interval for which $\sin x$ may be represented by its Maclaurin's series.

    Apart from sign, $f^{(n)}(x) = \sin x$ or $\cos x$, and $|R_n(x)| = \dfrac{|x^n|}{n!}|\sin x_0|$ or $\dfrac{|x^n|}{n!}|\cos x_0|$, where $x_0$ is between 0 and $x$.

    As in Problem 1, $\dfrac{x^n}{n!}\to 0$ as $n\to+\infty$. Since $|\sin x_0|$ and $|\cos x_0|$ are never greater than 1, $\lim\limits_{n\to+\infty}R_n(x) = 0$ and the series represents $\sin x$ for all values of $x$.

3.  Find the interval for which $\cos x$ may be represented by its Taylor series in powers of $(x-a)$.

    Using the Lagrange form of the remainder, $|R_n(x)| = \dfrac{|(x-a)^n|}{n!}|\sin x_0|$ or $\dfrac{|(x-a)^n|}{n!}|\cos x_0|$, where $x_0$ is between $a$ and $x$.

    Since $\dfrac{|(x-a)^n|}{n!}\to 0$ as $n\to+\infty$, while $|\sin x_0|$ and $|\cos x_0|$ are never greater than 1, $\lim\limits_{n\to+\infty}R_n(x) = 0$ and the series represents $\cos x$ for all values of $x$.

4.  Find the interval for which $\ln(1+x)$ may be represented by its Maclaurin's series.

    Here $f^{(n)}(x) = (-1)^{n-1}\dfrac{(n-1)!}{(1+x)^n}$; then with $x_0$ and $x_0^*$ between 0 and $x$,

    (a) the Lagrange form of the remainder is
    $$R_n(x) \;=\; (-1)^{n-1}\frac{x^n}{n!}\cdot\frac{(n-1)!}{(1+x_0)^n} \;=\; \frac{(-1)^{n-1}}{n}\left(\frac{x}{1+x_0}\right)^n \qquad \text{and}$$

    (b) the Cauchy form of the remainder is
    $$R_n(x) \;=\; (-1)^{n-1}\frac{(x-x_0^*)^{n-1}}{(n-1)!}\cdot\frac{(n-1)!}{(1+x_0^*)^n}x \;=\; (-1)^{n-1}\frac{x(x-x_0^*)^{n-1}}{(1+x_0^*)^n}$$

    When $0 < x_0 < x \leqq 1$, $0 < x < 1 + x_0$ and $\dfrac{x}{1+x_0} < 1$; then using (a),

    $$|R_n(x)| \;=\; \frac{1}{n}\left(\frac{x}{1+x_0}\right)^n \;<\; \frac{1}{n} \qquad \text{and} \qquad \lim\limits_{n\to+\infty}R_n(x) = 0$$

    When $-1 < x < x_0^* < 0$, then $0 < 1 + x < 1 + x_0^*$ and $\dfrac{1}{1+x_0^*} < \dfrac{1}{1+x}$. Using (b),

$$|R_n(x)| = \frac{|x - x_0^*|^{n-1}}{(1 + x_0^*)^n}|x| = \left|\frac{x_0^* - x}{1 + x_0^*}\right|^{n-1} \cdot \frac{|x|}{1 + x_0^*} = \left(\frac{x_0^* + |x|}{1 + x_0^*}\right)^{n-1} \cdot \frac{|x|}{1 + x_0^*} < \left(\frac{x_0^* + |x|}{1 + x_0^*}\right)^{n-1} \cdot \frac{|x|}{1 + x}$$

Now since $1 > |x|$, $x_0^* < x_0^*|x|$, $x_0^* + |x| < |x| + x_0^*|x|$ and $\frac{x_0^* + |x|}{1 + x_0^*} < |x|$. Thus,

$$|R_n(x)| < \frac{|x|^n}{1 + x} \quad \text{and} \quad \lim_{n \to +\infty} R_n(x) = 0$$

Hence, $\ln(1 + x)$ is represented by its Maclaurin's series on the interval $-1 < x \leqq 1$.

5.  For the Maclaurin's series representing $e^x$, show that

$$|R_n(x)| < \frac{|x^n|}{n!} \text{ when } x < 0 \quad \text{and} \quad R_n(x) < \frac{x^n e^x}{n!} \text{ when } x > 0$$

From Problem 1, $R_n(x) = \frac{x^n}{n!}e^{x_0}$, where $x_0$ is between $0$ and $x$. When $x < 0$, $e^{x_0} < 1$; hence, $|R_n(x)| < \frac{|x^n|}{n!}$. When $x > 0$, $e^{x_0} < e^x$; hence, $R_n(x) < \frac{x^n e^x}{n!}$.

6.  For the Maclaurin's series representing $\ln(1 + x)$, show that

$$R_n(x) < \frac{x^n}{n} \text{ when } 0 < x \leqq 1 \quad \text{and} \quad |R_n(x)| < \frac{|x^n|}{n(1+x)^n} \text{ when } -1 < x < 0$$

From Problem 4(a), $|R_n(x)| = \frac{1}{n}\left|\frac{x}{1 + x_0}\right|^n$, where $x_0$ is between $0$ and $x$. When $0 < x_0 < x \leqq 1$, $\frac{1}{1 + x_0} < 1$; hence, $|R_n(x)| < \frac{x^n}{n}$. When $-1 < x < x_0 < 0$, $1 + x_0 > 1 + x$ and $\frac{1}{1 + x_0} < \frac{1}{1 + x}$; hence, $|R_n(x)| < \frac{|x^n|}{n(1 + x)^n}$.

# Supplementary Problems

7.  Find the interval for which $\cos x$ may be represented by its Maclaurin's series.
    *Ans.* All values of $x$.

8.  Find the intervals for which (a) $e^x$ and (b) $\sin x$ may be represented by their Taylor's series in powers of $(x - a)$.     *Ans.* All values of $x$.

9.  Show that $\ln x$ may be represented by its Taylor's series in powers of $(x - a)$ on the interval $0 < x \leqq 2a$.
    *Hint.* $|R_n(x)| = \left|\frac{(x - a)(x - x_0^*)^{n-1}}{(x_0^*)^n}\right|$.   For $0 < x < a$ and for $a < x \leqq 2a$, $\left|\frac{x - x_0^*}{x_0^*}\right| < 1$.

10. Let $T$ be defined by
    $$f(b) = f(a) + \frac{f'(a)}{1!}(b - a) + \frac{f''(a)}{2!}(b - a)^2 + \cdots + \frac{f^{(n-1)}(a)}{(n-1)!}(b - a)^{n-1} + T(b - a)$$
    and define
    $$F(x) = -f(b) + f(x) + \frac{f'(x)}{1!}(b - x) + \frac{f''(x)}{2!}(b - x)^2 + \cdots + \frac{f^{(n-1)}(x)}{(n-1)!}(b - x)^{n-1} + T(b - x)$$
    Carry through as in Problem 15, Chapter 21, and obtain Taylor's formula with the Cauchy form of the remainder.

11. (a)  In the Cauchy form of the remainder of Taylor's formula put $x_0^* = a + \theta(x - a)$, where $0 < \theta < 1$. Show that
    $$R_n(x) = \frac{f^{(n)}[a + \theta(x - a)]}{(n - 1)!}(1 - \theta)^{n-1}(x - a)^n$$

    (b)  Show that $R_n(x) = \frac{f^{(n)}(\theta x)}{(n - 1)!}(1 - \theta)^{n-1}x^n$  in Maclaurin's formula.

12. Show that $\frac{1}{1 - x}$ is represented by its Maclaurin's series on the interval $-1 \leqq x < 1$.
    *Hint.* From Prob. 11(b), $R_n(x) = \frac{n(1 - \theta)^{n-1}x^n}{(1 - \theta x)^{n+1}}$, $0 < \theta < 1$. For $|x| < 1$, $\frac{1 - \theta}{1 - \theta x} < 1$ and $1 - \theta x > 1 - |x|$.

13. (a)  Show that $xe^x = \sum_{i=1}^{+\infty}\frac{n}{n!}x^n$, for all values of $x$, and $\sum_{i=1}^{+\infty}\frac{n}{n!} = e$; also, $(x^2 + x)e^x = \sum_{i=1}^{+\infty}\frac{n^2}{n!}x^n$ and $\sum_{i=1}^{+\infty}\frac{n^2}{n!} = 2e$.   (b)  Obtain $\sum_{i=1}^{+\infty}\frac{n^3}{n!} = 5e$ and $\sum_{i=1}^{+\infty}\frac{n^4}{n!} = 15e$.

# Chapter 54

## Computations Using Power Series

**TABLES OF LOGARITHMS,** trigonometric functions, etc., are computed by means of power series. Other uses of series are suggested in the problems below.

At all times, it is necessary to have some estimate of how well the sum of the first $n$ terms of a series represents the corresponding function for a given value of the variable. For this purpose two theorems from preceding chapters are given:

1. If $f(x)$ is represented by an alternating series and if $x = \xi$ is on its interval of convergence, the error introduced by using the sum of the values of the first $n$ terms as an approximate value of $f(\xi)$ does not exceed the numerical value of the first term discarded.

2. If $f(x)$ is represented by its Taylor's series and if $x = \xi$ is on its interval of convergence, the error introduced by using the sum of the values of the first $n$ terms as an approximate value of $f(\xi)$ does not exceed $\dfrac{M}{n!}|x - a|^n$, where $M$ is equal to or greater than the maximum value of $|f^{(n)}(x)|$ on the interval $a$ to $\xi$.

For a Maclaurin's series, $a = 0$.

## Solved Problems

1. Find the value of $1/e$ correct to five decimal places.

$$e^{-x} = 1 - x + \frac{x^2}{2!} - \frac{x^3}{3!} + \cdots + (-1)^{n-1}\frac{x^{n-1}}{(n-1)!} + \cdots$$

$$e^{-1} = 1 - 1 + \frac{1}{2!} - \frac{1}{3!} + \frac{1}{4!} - \frac{1}{5!} + \cdots$$

$$= 1 - 1 + 0.500000 - 0.166667 + 0.041667 - 0.008333 + 0.001389$$
$$- 0.000198 + 0.000025 - 0.000003 + \cdots$$

$$= 0.36788$$

2. Find the value of $\sin 62°$ correct to five decimal places.

The Taylor's series in powers of $(x - a)$ is

$$\sin x = \sin a + (x - a)\cos a - \frac{(x-a)^2}{2!}\sin a - \frac{(x-a)^3}{3!}\cos a + \cdots$$

Take $a = 60°$ since it is near $62°$ and its trigonometric functions are known. Then

$$x - a = 62° - 60° = 2° = \pi/90 = 0.034907$$

and 
$$\sin 62° = \tfrac{1}{2}\sqrt{3} + \tfrac{1}{2}(0.034907) - \tfrac{1}{4}\sqrt{3}\,(0.034907)^2 - \tfrac{1}{12}(0.034907)^3 + \cdots$$
$$= 0.866025 + 0.017454 - 0.000528 - 0.000004 + \cdots = 0.88295$$

3. Find the value of $\ln 0.97$ correct to seven decimal places.

$$\ln(a - x) = \ln a - \frac{x}{a} - \frac{x^2}{2a^2} - \frac{x^3}{3a^3} - \cdots - \frac{x^n}{na^n} - \cdots$$

Take $a = 1$ and $x = 0.03$; then

$$\ln 0.97 = -0.03 - \tfrac{1}{2}(0.03)^2 - \tfrac{1}{3}(0.03)^3 - \tfrac{1}{4}(0.03)^4 - \tfrac{1}{5}(0.03)^5 - \cdots = -0.0304592$$

4.  How many terms in the expansion of $\ln(1+x)$ must be used to insure finding $\ln 1.02$ with an error not exceeding 0.00000005?

$$\ln 1.02 \;=\; 0.02 - \frac{(0.02)^2}{2} + \frac{(0.02)^3}{3} - \frac{(0.02)^4}{4} + \cdots$$

Since it is an alternating series, the error introduced by discarding all terms after the first $n$ is never greater than the numerical value of the first term discarded. The problem here is to find the first term whose numerical value is less than 0.00000005. This must be done by trial.

$$\frac{(0.02)^3}{3} = 0.0000027 \quad \text{and} \quad \frac{(0.02)^4}{4} = 0.00000004$$

Hence the desired accuracy is obtained when the first three terms are used.

5.  For what values of $x$ can $\sin x$ be replaced by $x$, if the allowable error is 0.0005?

$\sin x = x - x^3/3! + \cdots$ is an alternating series. The error in using only the first term is less than $|x^3|/3!$  Now $|x^3|/3! = 0.0005$ requires $|x^3| = 0.003$ or $|x| = 0.1442$; thus, $|x| < 8°15'$.

6.  How large may the angle be taken if the values of $\cos x$ are to be computed using three terms of the Taylor's series in powers of $(x - \pi/3)$ and the error must not exceed 0.00005?

Since $f'''(x) = \sin x$, $|R_3| = \dfrac{|\sin x_0|}{3!}|x - \pi/3|^3$, where $x_0$ is between $\pi/3$ and $x$.

Since $|\sin x_0| \le 1$, $|R_3| \le \frac{1}{6}|x - \pi/3|^3 = 0.00005$.

Then $|x - \pi/3| \le \sqrt[3]{0.0003} = 0.0669 = 3°50'$. Thus $x$ may have any value between $56°10'$ and $63°50'$.

7.  Approximate the amount by which an arc of a great circle on the earth 100 miles long will recede from its chord.

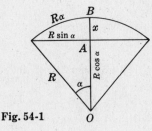

Let $x$ be the required amount. From Fig. 54-1, $x = OB - OA = R - R\cos\alpha$, where $R$ is the radius of the earth. Since angle $\alpha$ is small, $\cos\alpha = 1 - \frac{1}{2}x^2$, approximately, and

$$x = R\{1 - (1 - \tfrac{1}{2}\alpha^2)\} = \tfrac{1}{2}R\alpha^2 = (R\alpha)^2/2R = (50)^2/2R$$

Taking $R = 4000$ miles, $x = \frac{5}{16}$ mile.

**Fig. 54-1**

8.  Derive the approximation formula $\sin(\frac{1}{4}\pi + x) = \frac{1}{2}\sqrt{2}\,(1 + x)$ and use it to find $\sin 43°$.

Using the first two terms of the Taylor's expansion, we have

$$\sin(\tfrac{1}{4}\pi + x) = \sin\tfrac{1}{4}\pi + x\cos\tfrac{1}{4}\pi = \tfrac{1}{2}\sqrt{2} + \tfrac{1}{2}\sqrt{2}\,x = \tfrac{1}{2}\sqrt{2}\,(1 + x)$$

$$\sin 43° = \sin[\tfrac{1}{4}\pi + (-\pi/90)] = \tfrac{1}{2}\sqrt{2}\,(1 - 0.0349) = 0.6824$$

9.  Solve the equation $\cos x - 2x^2 = 0$.

Replace $\cos x$ by its first two terms $1 - \frac{1}{2}x^2$ of the Maclaurin's series.  Then

$$1 - \tfrac{1}{2}x^2 - 2x^2 = 0 \quad \text{or} \quad 2 - 5x^2 = 0$$

The roots are $\pm\sqrt{10}/5 = \pm 0.632$.  Newton's method gives the roots as $\pm 0.635$.

10.  Use power series expansions to evaluate $\lim\limits_{x\to 0} \dfrac{e^x - e^{-x}}{\sin x}$.

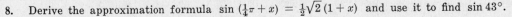

$$\lim_{x\to 0}\frac{e^x - e^{-x}}{\sin x} = \lim_{x\to 0}\frac{\left(1 + x + \dfrac{x^2}{2!} + \dfrac{x^3}{3!} + \cdots\right) - \left(1 - x + \dfrac{x^2}{2!} - \dfrac{x^3}{3!} + \cdots\right)}{x - \dfrac{x^3}{3!} + \dfrac{x^5}{5!} - \cdots}$$

$$= \lim_{x\to 0}\frac{2x + 2x^3/3! + \cdots}{x - x^3/3! + \cdots} = \lim_{x\to 0}\frac{2 + x^2/3 + \cdots}{1 - x^2/6 + \cdots} = 2$$

**11.** Expand $f(x) = x^4 - 11x^3 + 43x^2 - 60x + 14$ in powers of $(x-3)$ and find $\int_3^{3.2} f(x)\,dx$.

$f(3) = 5$, $f'(3) = 9$, $f''(3) = -4$, $f'''(3) = 6$, $f^{iv}(3) = 24$.   Hence,

$$f(x) = 5 + 9(x-3) - 2(x-3)^2 + (x-3)^3 + (x-3)^4$$

$$\int_3^{3.2} f(x)\,dx = 5x + \tfrac{9}{2}(x-3)^2 - \tfrac{2}{3}(x-3)^3 + \tfrac{1}{4}(x-3)^4 + \tfrac{1}{5}(x-3)^5 \Big|_3^{3.2} = 1.185$$

**12.** Evaluate $\int_0^1 \dfrac{\sin x}{x}\,dx$.

The difficulty here is that $\int \dfrac{\sin x}{x}\,dx$ cannot be expressed in elementary functions.  However,

$$\int_0^1 \frac{\sin x}{x}\,dx = \int_0^1 \frac{1}{x}\left(x - \frac{x^3}{3!} + \frac{x^5}{5!} - \frac{x^7}{7!} + \cdots\right)dx = \int_0^1\left(1 - \frac{x^2}{3!} + \frac{x^4}{5!} - \frac{x^6}{7!} + \cdots\right)dx$$

$$= x - \frac{x^3}{3\cdot 3!} + \frac{x^5}{5\cdot 5!} - \frac{x^7}{7\cdot 7!} + \cdots \Big|_0^1 = 0.946083$$

The error in using only four terms is $\leq \dfrac{1}{9\cdot 9!} = 0.0000003$.

# Supplementary Problems

**13.** Compute to four decimal places:
(a) $e^{-2} = 0.1353$, (b) $\sin 32° = 0.5299$, (c) $\cos 36° = 0.8090$, (d) $\tan 31° = 0.6009$.

**14.** For what range of $x$ can
(a) $e^x$ be replaced by $1 + x + \tfrac{1}{2}x^2$ if the allowable error is 0.0005?
(b) $\cos x$ be replaced by $1 - \tfrac{1}{2}x^2$ if the allowable error is 0.0005?
(c) $\sin x$ be replaced by $x - x^3/6 + x^5/120$ if the allowable error is 0.00005?
Ans. (a) $|x| < 0.1$, (b) $|x| < 18°57'$, (c) $|x| < 47°$

**15.** Use power series expansions to evaluate:
(a) $\lim\limits_{x\to 0} \dfrac{e - e^{\cos x}}{x^2} = \dfrac{1}{2}e$,  (b) $\lim\limits_{x\to 0} \dfrac{e^x - e^{\sin x}}{x^3} = \dfrac{1}{6}$,  (c) $\lim\limits_{x\to 0} \dfrac{\cosh x - \cos x}{\sinh x - \sin x} = \infty$.

**16.** Evaluate:
(a) $\int_0^{\pi/2} (1 - \tfrac{1}{2}\sin^2\phi)^{-1/2}\,d\phi = 1.854$,  (b) $\int_0^1 \cos\sqrt{x}\,dx = 0.76355$,  (c) $\int_0^{0.5} \dfrac{dx}{1+x^4} = 0.4940$.

**17.** Find the length of the curve $y = x^3/3$ from $x=0$ to $x=0.5$.     Ans. 0.5031

**18.** Find the area under the curve $y = \sin x^2$ from $x=0$ to $x=1$.     Ans. 0.3103

# Chapter 55

## Approximate Integration

**AN APPROXIMATE VALUE** of $\int_a^b f(x)\,dx$ may be obtained by means of certain formulas and by the use of mechanical integrators. Approximation procedures are necessary when ordinary integration is difficult, when the indefinite integral cannot be expressed in terms of elementary functions, or when the integrand $f(x)$ is defined by a table of values.

In Chapter 34, an approximation of $\int_a^b f(x)\,dx$ was obtained as the sum $S_n = \sum_{k=1}^{n} f(x_k)\,\Delta_k x$. In obtaining $S_n$ we interpreted the definite integral as an area, divided the area into $n$ strips, approximated the area of each strip by that of a rectangle, and summed the several approximations. The formulas developed below vary only as to the manner in approximating the areas of the strips.

**TRAPEZOIDAL RULE.** Let the area bounded above by the curve $y = f(x)$, below by the $x$-axis, and laterally by the ordinates $x = a$ and $x = b$, be divided into $n$ vertical strips each of width $h = (b-a)/n$, as in Fig. 55-1. Consider the $i$th strip bounded above by the arc $P_{i-1}P_i$ of $y = f(x)$. As an approximation of the area of this strip, we take

$$\tfrac{1}{2}h\{f[a + (i-1)h] + f(a + ih)\}$$

the area of the trapezoid obtained by replacing the arc $P_{i-1}P_i$ by the straight line segment $P_{i-1}P_i$. When each strip is so approximated, we have (where $\approx$ is to be read "is approximately"),

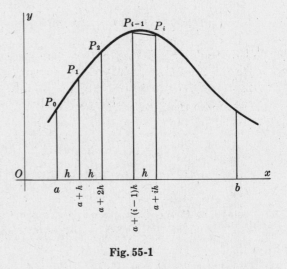

**Fig. 55-1**

$$\int_a^b f(x)\,dx \;\approx\; \frac{h}{2}\{f(a) + f(a+h)\} + \frac{h}{2}\{f(a+h) + f(a+2h)\}$$
$$+ \cdots + \frac{h}{2}\{f[a + (n-1)h] + f(b)\}$$

or
$$\int_a^b f(x)\,dx \;\approx\; \frac{h}{2}\{f(a) + 2f(a+h) + 2f(a+2h) \tag{1}$$
$$+ \cdots + 2f[a + (n-1)h] + f(b)\}$$

**PRISMOIDAL FORMULA.** Let the area defined by $\int_a^b f(x)\,dx$ be separated into two vertical strips of width $h = \tfrac{1}{2}(b-a)$ and let the arc $P_0P_1P_2$ of $y = f(x)$ be replaced by the arc of the parabola $y = Ax^2 + Bx + C$ through the points $P_0, P_1, P_2$, as in Fig. 55-2 below. Then after making certain changes of notation in the result of Problem 1, we have

$$\int_a^b f(x)\,dx \;\approx\; \frac{h}{3}\left\{f(a) + 4f\left(\frac{a+b}{2}\right) + f(b)\right\} \tag{2}$$

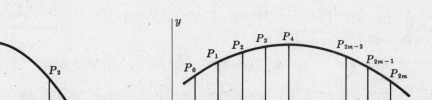

Fig. 55-2

Fig. 55-3

**SIMPSON'S RULE.**  Let the area under discussion be separated into $n = 2m$ strips each of width $h = (b - a)/n$, as in Fig. 55-3 above.  Using the prismoidal formula to approximate the area under each of the arcs $P_0P_1P_2$, $P_2P_3P_4$, ..., $P_{2m-2}P_{2m-1}P_{2m}$, we have

$$\int_a^b f(x)\,dx \;\approx\; \frac{h}{3}\,\{f(a) + 4f(a+h) + 2f(a+2h) + 4f(a+3h) + 2f(a+4h) \tag{3}$$

$$+ \cdots + 2f[a+(2m-2)h] + 4f[a+(2m-1)h] + f(b)\}$$

**POWER SERIES EXPANSION.**  The procedure here consists in replacing the integrand $f(x)$ by the first $n$ terms of its Maclaurin's or Taylor's series.  This method is available provided the integrand may be so expanded and the limits of integration fall within the interval of convergence of the series.  (See Chapter 54.)

# Solved Problems

1.  For the parabola $y = Ax^2 + Bx + C$, passing through the points
    $P_0(\xi, y_0)$, $P_1\!\left(\dfrac{\xi + \eta}{2},\, y_1\right)$, and $P_2(\eta, y_2)$, as shown in Fig. 55-4,

    show that $\displaystyle\int_\xi^\eta y\,dx = \frac{\eta - \xi}{6}(y_0 + 4y_1 + y_2)$.

    We have $\displaystyle\int_\xi^\eta y\,dx = \int_\xi^\eta (Ax^2 + Bx + C)\,dx$

    $$= \frac{\eta - \xi}{3}[A(\xi^2 + \xi\eta + \eta^2) + \tfrac{3}{2}B(\xi + \eta) + 3C].$$

    Now if $y = Ax^2 + Bx + C$ passes through the points $P_0, P_1, P_2$, then

    $$y_0 = A\xi^2 + B\xi + C$$
    $$y_1 = A\left(\frac{\xi + \eta}{2}\right)^2 + B\left(\frac{\xi + \eta}{2}\right) + C$$
    $$y_2 = A\eta^2 + B\eta + C$$

    and $$y_0 + 4y_1 + y_2 = 2[A(\xi^2 + \xi\eta + \eta^2) + \tfrac{3}{2}B(\xi + \eta) + 3C]$$

    Thus, $$\int_\xi^\eta y\,dx = \frac{\eta - \xi}{6}(y_0 + 4y_1 + y_2)$$

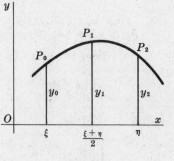

Fig. 55-4

2. Approximate $\displaystyle\int_0^{1/2} \frac{dx}{1+x^2}$ by each of the four methods and check by integration.

*Trapezoidal rule*, with $n = 5$.

Here, $h = \dfrac{\frac{1}{2}-0}{5} = 0.1$. Then $a = 0$, $a+h = 0.1$, $a+2h = 0.2$, $a+3h = 0.3$, $a+4h = 0.4$, $b = 0.5$.

$$\int_0^{1/2} \frac{dx}{1+x^2} \approx \frac{0.1}{2}[f(0) + 2f(0.1) + 2f(0.2) + 2f(0.3) + 2f(0.4) + f(0.5)]$$

$$\approx \frac{1}{20}\left(1 + \frac{2}{1.01} + \frac{2}{1.04} + \frac{2}{1.09} + \frac{2}{1.16} + \frac{1}{1.25}\right) = 0.4631$$

*Prismoidal formula.*

Here, $h = \dfrac{\frac{1}{2}-0}{2} = \dfrac{1}{4}$ and $f(a) = f(0) = 1$, $f\left(\dfrac{a+b}{2}\right) = f\left(\dfrac{1}{4}\right) = \dfrac{16}{17}$, $f(b) = f\left(\dfrac{1}{2}\right) = \dfrac{4}{5}$.

$$\int_0^{1/2} \frac{dx}{1+x^2} \approx \frac{1}{3}\cdot\frac{1}{4}\left(1 + \frac{64}{17} + \frac{4}{5}\right) = \frac{1}{12}(1 + 3.76471 + 0.8) = 0.4637$$

*Simpson's rule*, with $n = 4$.

Here, $h = \dfrac{\frac{1}{2}-0}{4} = \dfrac{1}{8}$. Then $a = 0$, $a+h = \frac{1}{8}$, $a+2h = \frac{1}{4}$, $a+3h = \frac{3}{8}$, $b = \frac{1}{2}$.

$$\int_0^{1/2} \frac{dx}{1+x} \approx \frac{1}{24}\left(1 + 4\frac{1}{1+(\frac{1}{8})^2} + 2\frac{1}{1+(\frac{1}{4})^2} + 4\frac{1}{1+(\frac{3}{8})^2} + \frac{1}{1+(\frac{1}{2})^2}\right)$$

$$\approx \frac{1}{24}\left(1 + \frac{256}{65} + \frac{32}{17} + \frac{256}{73} + \frac{4}{5}\right) = 0.4637$$

*Series expansion*, using 7 terms.

$$\int_0^{1/2}\frac{dx}{1+x^2} \approx \int_0^{1/2}(1 - x^2 + x^4 - x^6 + x^8 - x^{10} + x^{12})\,dx = \left[x - \frac{x^3}{3} + \frac{x^5}{5} - \frac{x^7}{7} + \frac{x^9}{9} - \frac{x^{11}}{11} + \frac{x^{13}}{13}\right]_0^{1/2}$$

$$\approx \frac{1}{2} - \frac{1}{3\cdot 2^3} + \frac{1}{5\cdot 2^5} - \frac{1}{7\cdot 2^7} + \frac{1}{9\cdot 2^9} - \frac{1}{11\cdot 2^{11}} + \frac{1}{13\cdot 2^{13}}$$

$$\approx 0.50000 - 0.04167 + 0.00625 - 0.00112 + 0.00022 - 0.00004 + 0.00001 = 0.4636$$

*Integration.*
$$\int_0^{1/2}\frac{dx}{1+x^2} = \arctan x\Big]_0^{1/2} = \arctan\frac{1}{2} = 0.4636$$

3. Find the area bounded by $y = e^{-x^2}$, the $x$-axis, and the lines $x = 0$ and $x = 1$ using (a) Simpson's rule with $n = 4$ and (b) series expansion.

(a) Here, $h = \frac{1}{4}$; $a = 0$, $a+h = \frac{1}{4}$, $a+2h = \frac{1}{2}$, $a+3h = \frac{3}{4}$, $b = 1$.

$$\int_0^1 e^{-x^2}\,dx \approx \frac{\frac{1}{4}}{3}(1 + 4e^{-1/16} + 2e^{-1/4} + 4e^{-9/16} + e^{-1})$$

$$\approx \frac{1}{12}\{1 + 4(0.9399) + 2(0.7788) + 4(0.5701) + 0.3679\} = 0.747 \text{ square units}$$

(b) $\displaystyle\int_0^1 e^{-x^2}\,dx \approx \int_0^1\left(1 - x^2 + \frac{x^4}{2!} - \frac{x^6}{3!} + \frac{x^8}{4!} - \frac{x^{10}}{5!} + \frac{x^{12}}{6!}\right)dx$

$$\approx \left[x - \frac{x^3}{3} + \frac{x^5}{5\cdot 2!} - \frac{x^7}{7\cdot 3!} + \frac{x^9}{9\cdot 4!} - \frac{x^{11}}{11\cdot 5!} + \frac{x^{13}}{13\cdot 6!}\right]_0^1$$

$$\approx 1 - \frac{1}{3} + \frac{1}{5\cdot 2!} - \frac{1}{7\cdot 3!} + \frac{1}{9\cdot 4!} - \frac{1}{11\cdot 5!} + \frac{1}{13\cdot 6!}$$

$$\approx 1 - 0.3333 + 0.1 - 0.0238 + 0.0046 - 0.0008 + 0.0001 = 0.747 \text{ square units}$$

4. A plot of land lies between a straight fence and a stream. At distances $x$ yd. from one end of the fence, the width of the plot ($y$ yd.) was measured as follows:

| $x$ | 0 | 20 | 40 | 60 | 80 | 100 | 120 |
|---|---|---|---|---|---|---|---|
| $y$ | 0 | 22 | 41 | 53 | 38 | 17 | 0 |

Use Simpson's rule to approximate the area of the plot.

Here, $h = 20$ and $\displaystyle\int_0^{120} f(x)\, dx \approx \frac{20}{3}(0 + 4 \cdot 22 + 2 \cdot 41 + 4 \cdot 53 + 2 \cdot 38 + 4 \cdot 17 + 0)$

$\approx 3507$ square yards.

5. A certain curve is given by the following pairs of rectangular coordinates:

| $x$ | 1 | 2 | 3 | 4 | 5 | 6 | 7 | 8 | 9 |
|---|---|---|---|---|---|---|---|---|---|
| $y$ | 0 | 0.6 | 0.9 | 1.2 | 1.4 | 1.5 | 1.7 | 1.8 | 2 |

(a) Approximate the area between the curve, the $x$-axis and the ordinates $x = 1$ and $x = 9$, using Simpson's rule.

(b) Approximate the volume generated by revolving the area in (a) about the $x$-axis, using Simpson's rule.

(a) Here, $h = 1$ and

$\displaystyle\int_1^9 y\, dx \approx \frac{1}{3}\{0 + 4(0.6) + 2(0.9) + 4(1.2) + 2(1.4) + 4(1.5) + 2(1.7) + 4(1.8) + 2\}$
$\approx 10.13$ square units

(b) $\displaystyle\pi\int_1^9 y^2\, dx \approx \frac{\pi}{3}\{0 + 4(0.6)^2 + 2(0.9)^2 + 4(1.2)^2 + 2(1.4)^2 + 4(1.5)^2 + 2(1.7)^2 + 4(1.8)^2 + 4\}$
$\approx 46.58$ cubic units

# Supplementary Problems

6. Derive Simpson's Rule.

7. Approximate $\displaystyle\int_2^6 \frac{dx}{x}$ using (a) the trapezoidal rule with $n = 4$, (b) the prismoidal formula, and (c) Simpson's rule with $n = 4$. Check by integration.  *Ans.* (a) 1.117, (b) 1.111, (c) 1.100; 1.099

8. Approximate $\displaystyle\int_1^5 \sqrt{35 + x}\, dx$ as in Problem 7.   *Ans.* (a) 24.654, (b) 24.655, (c) 24.655; 24.655

9. Approximate $\displaystyle\int_1^3 \ln x\, dx$ using (a) the trapezoidal rule with $n = 5$ and (b) Simpson's rule with $n = 8$. Check by integration.   *Ans.* (a) 1.2870, (b) 1.2958; 1.2958

10. Approximate $\displaystyle\int_0^1 \sqrt{1 + x^3}\, dx$ using (a) the trapezoidal rule with $n = 5$ and (b) Simpson's rule with $n = 4$.   *Ans.* (a) 1.115, (b) 1.111

11. Approximate $\displaystyle\int_0^\pi \frac{\sin x}{x}\, dx$ by Simpson's rule with $n = 6$.   *Ans.* 1.852

12. Use Simpson's rule to determine (a) the area under the curve and (b) the volume generated by revolving the area about the $x$-axis determined by the given data.

| $x$ | 1 | 2 | 3 | 4 | 5 |
|---|---|---|---|---|---|
| $y$ | 1.8 | 4.2 | 7.8 | 9.2 | 12.3 |

*Ans.* (a) 27.8, (b) $228.44\pi$

# Chapter 56

## Partial Derivatives

**FUNCTIONS OF SEVERAL VARIABLES.** If to each point $(x, y)$ of a part (region) of the $xy$-plane is assigned a real number $z$, then $z$ is said to be given as a function, $z = f(x, y)$, of the independent variables $x$ and $y$. The locus of all points $(x, y, z)$ satisfying $z = f(x, y)$ is a surface in ordinary space. In a similar manner functions $w = f(x, y, z, \ldots)$ of several independent variables may be defined but, at the moment, there is no geometric picture available.

There are a number of differences between the calculus of one and of two variables. Fortunately, the calculus of functions of three or more variables differs only slightly from that of functions of two variables. The study here will be limited largely to functions of two variables.

A function $f(x, y)$ is said to have a limit $A$ as $x \to x_0$ and $y \to y_0$ if for any $\epsilon > 0$, however small, there exists a $\delta > 0$ such that for all $(x, y)$ satisfying

**(i)** $$0 < \sqrt{(x - x_0)^2 + (y - y_0)^2} < \delta$$

then $|f(x, y) - A| < \epsilon$. Here (i) defines a deleted neighborhood of $(x_0, y_0)$, namely, all points except $(x_0, y_0)$ lying within a circle of radius $\delta$ and center $(x_0, y_0)$.

A function $f(x, y)$ is said to be continuous at $(x_0, y_0)$ provided $f(x_0, y_0)$ is defined and $\lim\limits_{\substack{x \to x_0 \\ y \to y_0}} f(x, y) = f(x_0, y_0)$.

See Problems 1-2.

**PARTIAL DERIVATIVES.** Let $z = f(x, y)$ be a function of the independent variables $x$ and $y$. Since $x$ and $y$ are independent, we may **(i)** allow $x$ to vary while $y$ is held fixed, **(ii)** allow $y$ to vary while $x$ is held fixed, **(iii)** permit $x$ and $y$ to vary simultaneously. In the first two cases, $z$ is in effect a function of a single variable and can be differentiated in accordance with the usual rules.

If $x$ varies while $y$ is held fixed, $z$ is a function of $x$ and its derivative with respect to $x$

$$f_x(x, y) = \frac{\partial z}{\partial x} = \lim_{\Delta x \to 0} \frac{f(x + \Delta x, y) - f(x, y)}{\Delta x}$$

is called *the (first) partial derivative of* $z = f(x, y)$ *with respect to* $x$.

If $y$ varies while $x$ is held fixed, $z$ is a function of $y$ and its derivative with respect to $y$

$$f_y(x, y) = \frac{\partial z}{\partial y} = \lim_{\Delta y \to 0} \frac{f(x, y + \Delta y) - f(x, y)}{\Delta y}$$

is called *the (first) partial derivative of* $z = f(x, y)$ *with respect to* $y$.

See Problems 3-8.

If $z$ is defined implicitly as a function of $x$ and $y$ by the relation $F(x, y, z) = 0$, the partial derivatives $\dfrac{\partial z}{\partial x}$ and $\dfrac{\partial z}{\partial y}$ may be found using the implicit differentiation rule of Chapter 6.

See Problems 9-12.

The partial derivatives defined above have simple geometric interpretations. Consider the surface $z = f(x, y)$ in Fig. 56-1. Let $APB$ and $CPD$ be sections of the surface by planes through $P$, parallel to $xOz$ and $yOz$ respectively. As $x$ varies while $y$ is held fixed, $P$ moves along the curve $APB$ and the value of $\dfrac{\partial z}{\partial x}$ at $P$ is the slope of the curve $APB$ at $P$.

Similarly, as $y$ varies while $x$ is held fixed, $P$ moves along the curve $CPD$ and the value of $\dfrac{\partial z}{\partial y}$ at $P$ is the slope of the curve $CPD$ at $P$.

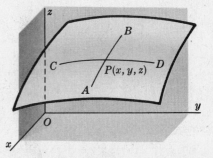

**Fig. 56-1**

See Problem 13.

**PARTIAL DERIVATIVES OF HIGHER ORDERS.** The partial derivative $\dfrac{\partial z}{\partial x}$ of $z = f(x, y)$ may in turn be differentiated partially with respect to $x$ and to $y$, yielding the second partial derivatives. $\dfrac{\partial^2 z}{\partial x^2} = f_{xx}(x, y) = \dfrac{\partial}{\partial x}\left(\dfrac{\partial z}{\partial x}\right)$ and $\dfrac{\partial^2 z}{\partial y\,\partial x} = f_{yx}(x, y) = \dfrac{\partial}{\partial y}\left(\dfrac{\partial z}{\partial x}\right)$.

Similarly, from $\dfrac{\partial z}{\partial y}$ may be obtained $\dfrac{\partial^2 z}{\partial x\,\partial y} = f_{xy}(x, y) = \dfrac{\partial}{\partial x}\left(\dfrac{\partial z}{\partial y}\right)$ and $\dfrac{\partial^2 z}{\partial y^2} = f_{yy}(x, y) = \dfrac{\partial}{\partial y}\left(\dfrac{\partial z}{\partial y}\right)$.

If $z = f(x, y)$ and its partial derivatives are continuous, the order of differentiation is immaterial, that is, $\dfrac{\partial^2 z}{\partial x\,\partial y} = \dfrac{\partial^2 z}{\partial y\,\partial x}$.

See Problems 14-15.

# Solved Problems

1.  Investigate $z = x^2 + y^2$ for continuity.

 For any set of finite values $(x, y) = (a, b)$, $z = a^2 + b^2$.
 As $x \to a$ and $y \to b$, $x^2 + y^2 \to a^2 + b^2$.
 Hence, the function is continuous everywhere.

2.  The following functions are continuous everywhere except at the origin $(0, 0)$ where they are not defined. Can they be made continuous there?

 (a) $z = \dfrac{\sin(x+y)}{x+y}$.

 Let $(x, y) \to (0, 0)$ over the line $y = mx$; then $z = \dfrac{\sin(x+y)}{x+y} = \dfrac{\sin(1+m)x}{(1+m)x} \to 1$. The function may be made continuous everywhere by redefining it as: $z = \dfrac{\sin(x+y)}{x+y}$, $(x, y) \neq (0, 0)$; $z = 1$, $(x, y) = (0, 0)$.

 (b) $z = \dfrac{xy}{x^2 + y^2}$.

 Let $(x, y) \to (0, 0)$ over the line $y = mx$; the limiting value of $z = \dfrac{xy}{x^2 + y^2} = \dfrac{m}{1 + m^2}$ depends on the particular line chosen. Thus, the function cannot be made continuous at $(0, 0)$.

In Problems 3-7, find the first partial derivatives.

3.  $z = 2x^2 - 3xy + 4y^2$.

 Treating $y$ as a constant and differentiating with respect to $x$, $\dfrac{\partial z}{\partial x} = 4x - 3y$.

 Treating $x$ as a constant and differentiating with respect to $y$, $\dfrac{\partial z}{\partial y} = -3x + 8y$.

4.   $z = \dfrac{x^2}{y} + \dfrac{y^2}{x}$.

Treating $y$ as a constant and differentiating with respect to $x$,   $\dfrac{\partial z}{\partial x} = \dfrac{2x}{y} - \dfrac{y^2}{x^2}$.

Treating $x$ as a constant and differentiating with respect to $y$,   $\dfrac{\partial z}{\partial y} = -\dfrac{x^2}{y^2} + \dfrac{2y}{x}$.

5.   $z = \sin(2x + 3y)$.         $\dfrac{\partial z}{\partial x} = 2\cos(2x + 3y)$,    $\dfrac{\partial z}{\partial y} = 3\cos(2x + 3y)$

6.   $z = \arctan x^2 y + \arctan xy^2$.     $\dfrac{\partial z}{\partial x} = \dfrac{2xy}{1 + x^4 y^2} + \dfrac{y^2}{1 + x^2 y^4}$,    $\dfrac{\partial z}{\partial y} = \dfrac{x^2}{1 + x^4 y^2} + \dfrac{2xy}{1 + x^2 y^4}$

7.   $z = e^{x^2 + xy}$.      $\dfrac{\partial z}{\partial x} = e^{x^2 + xy}(2x + y) = z(2x + y)$,    $\dfrac{\partial z}{\partial y} = e^{x^2 + xy}(x) = xz$

8.   The area of a triangle is given by $K = \frac{1}{2}ab\sin C$.   If $a = 20$, $b = 30$, and $C = 30°$, find:

    (a) The rate of change of $K$ with respect to $a$, when $b$ and $C$ are constant.

    (b) The rate of change of $K$ with respect to $C$, when $a$ and $b$ are constant.

    (c) The rate of change of $b$ with respect to $a$, when $K$ and $C$ are constant.

    (a)   $\dfrac{\partial K}{\partial a} = \frac{1}{2}b\sin C = \frac{1}{2}(30)(\sin 30°) = \dfrac{15}{2}$

    (b)   $\dfrac{\partial K}{\partial C} = \frac{1}{2}ab\cos C = \frac{1}{2}(20)(30)(\cos 30°) = 150\sqrt{3}$

    (c)   $b = \dfrac{2K}{a\sin C}$;    $\dfrac{\partial b}{\partial a} = -\dfrac{2K}{a^2\sin C} = -\dfrac{2(\frac{1}{2}ab\sin C)}{a^2\sin C} = -\dfrac{b}{a} = -\dfrac{3}{2}$

In Problems 9-11, find the first partial derivatives of $z$ with respect to the independent variables $x$ and $y$.

9.   $x^2 + y^2 + z^2 = 25$.

    *Solution 1.*   Solve for $z$ to obtain $z = \pm\sqrt{25 - x^2 - y^2}$.   Then

$$\frac{\partial z}{\partial x} = \frac{-x}{\pm\sqrt{25 - x^2 - y^2}} = -\frac{x}{z} \quad \text{and} \quad \frac{\partial z}{\partial y} = \frac{-y}{\pm\sqrt{25 - x^2 - y^2}} = -\frac{y}{z}$$

    *Solution 2.*   Differentiate implicitly with respect to $x$, treating $y$ as a constant.

$$2x + 2z\frac{\partial z}{\partial x} = 0 \quad \text{and} \quad \frac{\partial z}{\partial x} = -\frac{x}{z}$$

    Differentiate implicitly with respect to $y$, treating $x$ as a constant.

$$2y + 2z\frac{\partial z}{\partial y} = 0 \quad \text{and} \quad \frac{\partial z}{\partial y} = -\frac{y}{z}$$

10.   $x^2(2y + 3z) + y^2(3x - 4z) + z^2(x - 2y) = xyz$.

    The procedure of Solution 1, Problem 9, would be inconvenient here.

    Differentiating implicitly with respect to $x$,

$$2x(2y + 3z) + 3x^2\frac{\partial z}{\partial x} + 3y^2 - 4y^2\frac{\partial z}{\partial x} + 2z(x - 2y)\frac{\partial z}{\partial x} + z^2 = yz + xy\frac{\partial z}{\partial x}$$

and               $\dfrac{\partial z}{\partial x} = -\dfrac{4xy + 6xz + 3y^2 + z^2 - yz}{3x^2 - 4y^2 + 2xz - 4yz - xy}$

    Differentiating implicitly with respect to $y$,

$$2x^2 + 3x^2\frac{\partial z}{\partial y} + 2y(3x - 4z) - 4y^2\frac{\partial z}{\partial y} + 2z(x - 2y)\frac{\partial z}{\partial y} - 2z^2 = xz + xy\frac{\partial z}{\partial y}$$

and               $\dfrac{\partial z}{\partial y} = -\dfrac{2x^2 + 6xy - 8yz - 2z^2 - xz}{3x^2 - 4y^2 + 2xz - 4yz - xy}$

11. $xy + yz + zx = 1.$

     Differentiating with respect to $x$,   $y + y\dfrac{\partial z}{\partial x} + x\dfrac{\partial z}{\partial x} + z = 0$  and  $\dfrac{\partial z}{\partial x} = -\dfrac{y+z}{x+y}.$

     Differentiating with respect to $y$,   $x + y\dfrac{\partial z}{\partial y} + z + x\dfrac{\partial z}{\partial y} = 0$  and  $\dfrac{\partial z}{\partial y} = -\dfrac{x+z}{x+y}.$

12. Considering $x$ and $y$ as independent variables, find $\dfrac{\partial r}{\partial x}, \dfrac{\partial r}{\partial y}, \dfrac{\partial \theta}{\partial x}, \dfrac{\partial \theta}{\partial y}$ when $x = e^{2r}\cos\theta$, $y = e^{3r}\sin\theta$.

     Differentiate the relations partially with respect to $x$:

$$1 = 2e^{2r}\cos\theta\,\frac{\partial r}{\partial x} - e^{2r}\sin\theta\,\frac{\partial \theta}{\partial x} \quad\text{and}\quad 0 = 3e^{3r}\sin\theta\,\frac{\partial r}{\partial x} + e^{3r}\cos\theta\,\frac{\partial \theta}{\partial x}$$

     Solve simultaneously to obtain:   $\dfrac{\partial r}{\partial x} = \dfrac{\cos\theta}{e^{2r}(2 + \sin^2\theta)}$  and  $\dfrac{\partial \theta}{\partial x} = -\dfrac{3\sin\theta}{e^{2r}(2 + \sin^2\theta)}.$

     Differentiate the relations partially with respect to $y$:

$$0 = 2e^{2r}\cos\theta\,\frac{\partial r}{\partial y} - e^{2r}\sin\theta\,\frac{\partial \theta}{\partial y} \quad\text{and}\quad 1 = 3e^{3r}\sin\theta\,\frac{\partial r}{\partial y} + e^{3r}\cos\theta\,\frac{\partial \theta}{\partial y}$$

     Solve simultaneously to obtain:   $\dfrac{\partial r}{\partial y} = \dfrac{\sin\theta}{e^{3r}(2 + \sin^2\theta)}$  and  $\dfrac{\partial \theta}{\partial y} = \dfrac{2\cos\theta}{e^{3r}(2 + \sin^2\theta)}.$

13. Find the slopes of the curves cut from the surface $z = 3x^2 + 4y^2 - 6$ by planes through the point $(1, 1, 1)$ and parallel to the coordinate planes $xOz$ and $yOz$.

     The plane $x = 1$, parallel to the plane $yOz$, intersects the surface in the curve $z = 4y^2 - 3$, $x = 1$. Then $\partial z/\partial y = 8y = 8\cdot 1 = 8$ is the required slope.

     The plane $y = 1$, parallel to the plane $xOz$, intersects the surface in the curve $z = 3x^2 - 2$, $y = 1$. Then $\partial z/\partial x = 6x = 6$ is the required slope.

In Problems 14-15, find all second partial derivatives of $z$.

14. $z = x^2 + 3xy + y^2.$      $\dfrac{\partial z}{\partial x} = 2x + 3y,$     $\dfrac{\partial^2 z}{\partial x^2} = \dfrac{\partial}{\partial x}\left(\dfrac{\partial z}{\partial x}\right) = 2,$     $\dfrac{\partial^2 z}{\partial y\,\partial x} = \dfrac{\partial}{\partial y}\left(\dfrac{\partial z}{\partial x}\right) = 3$

                                     $\dfrac{\partial z}{\partial y} = 3x + 2y,$     $\dfrac{\partial^2 z}{\partial x\,\partial y} = \dfrac{\partial}{\partial x}\left(\dfrac{\partial z}{\partial y}\right) = 3,$     $\dfrac{\partial^2 z}{\partial y^2} = \dfrac{\partial}{\partial y}\left(\dfrac{\partial z}{\partial y}\right) = 2$

15. $z = x\cos y - y\cos x.$

     $\dfrac{\partial z}{\partial x} = \cos y + y\sin x,$      $\dfrac{\partial z}{\partial y} = -x\sin y - \cos x,$      $\dfrac{\partial^2 z}{\partial x^2} = \dfrac{\partial}{\partial x}\left(\dfrac{\partial z}{\partial x}\right) = y\cos x$

     $\dfrac{\partial^2 z}{\partial y\,\partial x} = \dfrac{\partial}{\partial y}\left(\dfrac{\partial z}{\partial x}\right) = -\sin y + \sin x = \dfrac{\partial^2 z}{\partial x\,\partial y},$      $\dfrac{\partial^2 z}{\partial y^2} = \dfrac{\partial}{\partial y}\left(\dfrac{\partial z}{\partial y}\right) = -x\cos y$

# Supplementary Problems

16. Investigate each of the following to determine whether or not it can be made continuous at $(0, 0)$.

    (a) $\dfrac{y^2}{x^2 + y^2}$,  (b) $\dfrac{x - y}{x + y}$,  (c) $\dfrac{x^3 + y^3}{x^2 + y^2}$,  (d) $\dfrac{x + y}{x^2 + y^2}$        *Ans.* (a) No, (b) No, (c) Yes, (d) No

17. For each of the following functions, find $\dfrac{\partial z}{\partial x}$ and $\dfrac{\partial z}{\partial y}$.

(a) $z = x^2 + 3xy + y^2$          Ans. $\dfrac{\partial z}{\partial x} = 2x + 3y$, $\dfrac{\partial z}{\partial y} = 3x + 2y$

(b) $z = \dfrac{x}{y^2} - \dfrac{y}{x^2}$          Ans. $\dfrac{\partial z}{\partial x} = \dfrac{1}{y^2} + \dfrac{2y}{x^3}$, $\dfrac{\partial z}{\partial y} = -\dfrac{2x}{y^3} - \dfrac{1}{x^2}$

(c) $z = \sin 3x \cos 4y$          Ans. $\dfrac{\partial z}{\partial x} = 3 \cos 3x \cos 4y$, $\dfrac{\partial z}{\partial y} = -4 \sin 3x \sin 4y$

(d) $z = \arctan \dfrac{y}{x}$          Ans. $\dfrac{\partial z}{\partial x} = \dfrac{-y}{x^2 + y^2}$, $\dfrac{\partial z}{\partial y} = \dfrac{x}{x^2 + y^2}$

(e) $x^2 - 4y^2 + 9z^2 = 36$          Ans. $\dfrac{\partial z}{\partial x} = -\dfrac{x}{9z}$, $\dfrac{\partial z}{\partial y} = \dfrac{4y}{9z}$

(f) $z^3 - 3x^2y + 6xyz = 0$          Ans. $\dfrac{\partial z}{\partial x} = \dfrac{2y(x - z)}{z^2 + 2xy}$, $\dfrac{\partial z}{\partial y} = \dfrac{x(x - 2z)}{z^2 + 2xy}$

(g) $yz + xz + xy = 0$          Ans. $\dfrac{\partial z}{\partial x} = -\dfrac{y + z}{x + y}$, $\dfrac{\partial z}{\partial y} = -\dfrac{x + z}{x + y}$

18. (a) If $z = \sqrt{x^2 + y^2}$, show that $x\dfrac{\partial z}{\partial x} + y\dfrac{\partial z}{\partial y} = z$.

(b) If $z = \ln \sqrt{x^2 + y^2}$, show that $x\dfrac{\partial z}{\partial x} + y\dfrac{\partial z}{\partial y} = 1$.

(c) If $z = e^{x/y} \sin\dfrac{x}{y} + e^{y/x} \cos\dfrac{y}{x}$, show that $x\dfrac{\partial z}{\partial x} + y\dfrac{\partial z}{\partial y} = 0$.

(d) If $z = (ax + by)^2 + e^{ax+by} + \sin(ax + by)$, show that $b\dfrac{\partial z}{\partial x} = a\dfrac{\partial z}{\partial y}$.

19. Find the equation of the tangent line

(a) to the parabola $z = 2x^2 - 3y^2$, $y = 1$ at the point $(-2, 1, 5)$.      Ans. $8x + z + 11 = 0$, $y = 1$
(b) to the parabola $z = 2x^2 - 3y^2$, $x = -2$ at the point $(-2, 1, 5)$.      Ans. $6y + z - 11 = 0$, $x = -2$
(c) to the hyperbola $z = 2x^2 - 3y^2$, $z = 5$ at the point $(-2, 1, 5)$.      Ans. $4x + 3y + 5 = 0$, $z = 5$

Show that these three lines lie in the plane $8x + 6y + z + 5 = 0$.

20. For each of the following functions, find $\dfrac{\partial^2 z}{\partial x^2}$, $\dfrac{\partial^2 z}{\partial x\, \partial y}$, $\dfrac{\partial^2 z}{\partial y\, \partial x}$, $\dfrac{\partial^2 z}{\partial y^2}$.

(a) $z = 2x^2 - 5xy + y^2$          Ans. $\dfrac{\partial^2 z}{\partial x^2} = 4$, $\dfrac{\partial^2 z}{\partial x\, \partial y} = \dfrac{\partial^2 z}{\partial y\, \partial x} = -5$, $\dfrac{\partial^2 z}{\partial y^2} = 2$

(b) $z = \dfrac{x}{y^2} - \dfrac{y}{x^2}$          Ans. $\dfrac{\partial^2 z}{\partial x^2} = -\dfrac{6y}{x^4}$, $\dfrac{\partial^2 z}{\partial x\, \partial y} = \dfrac{\partial^2 z}{\partial y\, \partial x} = 2\left(\dfrac{1}{x^3} - \dfrac{1}{y^3}\right)$, $\dfrac{\partial^2 z}{\partial y^2} = \dfrac{6x}{y^4}$

(c) $z = \sin 3x \cos 4y$          Ans. $\dfrac{\partial^2 z}{\partial x^2} = -9z$, $\dfrac{\partial^2 z}{\partial x\, \partial y} = \dfrac{\partial^2 z}{\partial y\, \partial x} = -12 \cos 3x \sin 4y$, $\dfrac{\partial^2 z}{\partial y^2} = -16z$

(d) $z = \arctan \dfrac{y}{x}$          Ans. $\dfrac{\partial^2 z}{\partial x^2} = -\dfrac{\partial^2 z}{\partial y^2} = \dfrac{2xy}{(x^2 + y^2)^2}$, $\dfrac{\partial^2 z}{\partial x\, \partial y} = \dfrac{\partial^2 z}{\partial y\, \partial x} = \dfrac{y^2 - x^2}{(x^2 + y^2)^2}$

21. (a) If $z = \dfrac{xy}{x - y}$, show that $x^2\dfrac{\partial^2 z}{\partial x^2} + 2xy\dfrac{\partial^2 z}{\partial x\, \partial y} + y^2\dfrac{\partial^2 z}{\partial y^2} = 0$.

(b) If $z = e^{\alpha x} \cos \beta y$ and $\beta = \pm\alpha$, show that $\dfrac{\partial^2 z}{\partial x^2} + \dfrac{\partial^2 z}{\partial y^2} = 0$.

(c) If $z = e^{-t}(\sin x + \cos y)$, show that $\dfrac{\partial^2 z}{\partial x^2} + \dfrac{\partial^2 z}{\partial y^2} = \dfrac{\partial z}{\partial t}$.

(d) If $z = \sin ax \sin by \sin kt\sqrt{a^2 + b^2}$, show that $\dfrac{\partial^2 z}{\partial t^2} = k^2\left\{\dfrac{\partial^2 z}{\partial x^2} + \dfrac{\partial^2 z}{\partial y^2}\right\}$.

22. For the gas formula $\left(p + \dfrac{a}{v^2}\right)(v - b) = ct$, where $a, b$, and $c$ are constants, show that

$$\dfrac{\partial p}{\partial v} = \dfrac{2a(v - b) - (p + a/v^2)v^3}{v^3(v - b)}, \quad \dfrac{\partial v}{\partial t} = \dfrac{cv^3}{(p + a/v^2)v^3 - 2a(v - b)}, \quad \dfrac{\partial t}{\partial p} = \dfrac{v - b}{c}, \quad \left(\dfrac{\partial p}{\partial v}\right)\left(\dfrac{\partial v}{\partial t}\right)\left(\dfrac{\partial t}{\partial p}\right) = -1$$

# Chapter 57

## Total Differentials and Total Derivatives

**TOTAL DIFFERENTIALS.** The differentials $dx$ and $dy$ for the function $y = f(x)$ of a single independent variable $x$ were defined in Chapter 23 as

$$dx = \Delta x, \qquad dy = f'(x)\, dx = \frac{dy}{dx}\, dx$$

Consider the function $z = f(x, y)$ of the two independent variables $x$ and $y$, and define $dx = \Delta x$ and $dy = \Delta y$. When $x$ varies while $y$ is held fixed, $z$ is a function of $x$ only and *the partial differential of $z$ with respect to $x$ is defined as* $d_x z = f_x(x, y)\, dx = \frac{\partial z}{\partial x}\, dx$. Similarly, *the partial differential of $z$ with respect to $y$ is defined by* $d_y z = f_y(x, y)\, dy = \frac{\partial z}{\partial y}\, dy$. The *total differential* $dz$ is defined as the sum of the partial differentials, e.g.,

$$dz = \frac{\partial z}{\partial x}\, dx + \frac{\partial z}{\partial y}\, dy \tag{1}$$

For a function $w = F(x, y, z, \dots t)$, the total differential $dw$ is defined as

$$dw = \frac{\partial w}{\partial x}\, dx + \frac{\partial w}{\partial y}\, dy + \frac{\partial w}{\partial z}\, dz + \cdots + \frac{\partial w}{\partial t}\, dt \tag{1'}$$

See Problems 1-2.

As in the case of a function of a single variable, the total differential of a function of several variables gives a good approximation of the total increment of the function when the increments of the several independent variables are small.

**Example:**

When $z = xy$, $dz = \frac{\partial z}{\partial x}\, dx + \frac{\partial z}{\partial y}\, dy = y\, dx + x\, dy$; and when $x$ and $y$ are given increments $\Delta x = dx$ and $\Delta y = dy$, the increment $\Delta z$ taken on by $z$ is

$$\begin{aligned}
\Delta z &= (x + \Delta x)(y + \Delta y) - xy \\
&= x\, \Delta y + y\, \Delta x + \Delta x\, \Delta y \\
&= x\, dy + y\, dx + dx\, dy
\end{aligned}$$

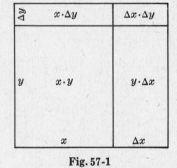

A geometric interpretation is given in Fig. 57-1. It will be seen that $dz$ and $\Delta z$ differ by the rectangle of area $\Delta x\, \Delta y = dx\, dy$.

See Problems 3-9.

**Fig. 57-1**

**THE CHAIN RULE FOR FUNCTIONS·OF FUNCTIONS.** If $z = f(x, y)$ is a continuous function of the variables $x, y$ with continuous partial derivatives $\partial z/\partial x$ and $\partial z/\partial y$, and if $x$ and $y$ are differentiable functions $x = g(t)$, $y = h(t)$ of a variable $t$, then $z$ is a function of $t$ and $dz/dt$, called the *total derivative* of $z$ with respect to $t$, is given by

$$\frac{dz}{dt} = \frac{\partial z}{\partial x}\frac{dx}{dt} + \frac{\partial z}{\partial y}\frac{dy}{dt} \tag{2}$$

Similarly, if $w = f(x, y, z, \ldots)$ is a continuous function of the variables $x, y, z, \ldots$, with continuous partial derivatives, and if $x, y, z, \ldots$ are differentiable functions of a variable $t$, the total derivative of $w$ with respect to $t$ is given by

$$\frac{dw}{dt} = \frac{\partial w}{\partial x}\frac{dx}{dt} + \frac{\partial w}{\partial y}\frac{dy}{dt} + \frac{\partial w}{\partial z}\frac{dz}{dt} + \cdots \tag{2'}$$

See Problems 10-16.

If $z = f(x, y)$ is a continuous function of the variables $x$ and $y$ with continuous partial derivatives $\partial z/\partial x$ and $\partial z/\partial y$, and if $x$ and $y$ are continuous functions $x = g(r, s)$, $y = h(r, s)$ of the independent variables $r$ and $s$, then $z$ is a function of $r$ and $s$ with

$$\frac{\partial z}{\partial r} = \frac{\partial z}{\partial x}\frac{\partial x}{\partial r} + \frac{\partial z}{\partial y}\frac{\partial y}{\partial r} \quad \text{and} \quad \frac{\partial z}{\partial s} = \frac{\partial z}{\partial x}\frac{\partial x}{\partial s} + \frac{\partial z}{\partial y}\frac{\partial y}{\partial s} \tag{3}$$

Similarly, if $w = f(x, y, z, \ldots)$ is a continuous function of the $n$ variables $x, y, z, \ldots$ with continuous partial derivatives $\partial w/\partial x, \partial w/\partial y, \partial w/\partial z, \ldots$, and if $x, y, z, \ldots$ are continuous functions of the $m$ independent variables $r, s, t, \ldots$, then

$$\frac{\partial w}{\partial r} = \frac{\partial w}{\partial x}\frac{\partial x}{\partial r} + \frac{\partial w}{\partial y}\frac{\partial y}{\partial r} + \frac{\partial w}{\partial z}\frac{\partial z}{\partial r} + \cdots$$

$$\frac{\partial w}{\partial s} = \frac{\partial w}{\partial x}\frac{\partial x}{\partial s} + \frac{\partial w}{\partial y}\frac{\partial y}{\partial s} + \frac{\partial w}{\partial z}\frac{\partial z}{\partial s} + \cdots \quad \text{etc.} \tag{3'}$$

See Problems 17-19.

## Solved Problems

In Problems 1-2, find the total differential.

1.  $z = x^3y + x^2y^2 + xy^3$.

$$\frac{\partial z}{\partial x} = 3x^2y + 2xy^2 + y^3, \qquad \frac{\partial z}{\partial y} = x^3 + 2x^2y + 3xy^2$$

Then $\qquad dz = \dfrac{\partial z}{\partial x}dx + \dfrac{\partial z}{\partial y}dy = (3x^2y + 2xy^2 + y^3)\,dx + (x^3 + 2x^2y + 3xy^2)\,dy$

2.  $z = x \sin y - y \sin x$.

$$\frac{\partial z}{\partial x} = \sin y - y \cos x, \qquad \frac{\partial z}{\partial y} = x \cos y - \sin x$$

Then $\qquad dz = \dfrac{\partial z}{\partial x}dx + \dfrac{\partial z}{\partial y}dy = (\sin y - y \cos x)\,dx + (x \cos y - \sin x)\,dy$

3.  Compare $dz$ and $\Delta z$, given $z = x^2 + 2xy - 3y^2$.

$$\frac{\partial z}{\partial x} = 2x + 2y, \qquad \frac{\partial z}{\partial y} = 2x - 6y, \qquad dz = 2(x+y)\,dx + 2(x-3y)\,dy$$

$$\Delta z = [(x+dx)^2 + 2(x+dx)(y+dy) - 3(y+dy)^2] - (x^2 + 2xy - 3y^2)$$
$$= 2(x+y)\,dx + 2(x-3y)\,dy + (dx)^2 + 2\,dx\,dy - 3(dy)^2$$

Thus $dz$ and $\Delta z$ differ by $(dx)^2 + 2\,dx\,dy - 3(dy)^2$.

4.  Approximate the area of a rectangle of dimensions 35.02 by 24.97 units.

For dimensions $x$ by $y$, the area is $A = xy$ so that $dA = \dfrac{\partial A}{\partial x}dx + \dfrac{\partial A}{\partial y}dy = y\,dx + x\,dy$. With $x = 35, dx = 0.02, y = 25, dy = -0.03$ we have $A = 35 \cdot 25 = 875$ and $dA = 25(.02) + 35(-.03) = -0.55$. The area is approximately $A + dA = 874.45$ square units.

5. Approximate the change in the hypotenuse of a right triangle of legs 6 and 8 inches, when the shorter leg is lengthened by $\frac{1}{4}$ in. and the longer leg is shortened by $\frac{1}{8}$ in.

Let $x, y, z$ be the shorter leg, the longer leg, and the hypotenuse of the triangle. Then

$$z = \sqrt{x^2 + y^2}, \quad \frac{\partial z}{\partial x} = \frac{x}{\sqrt{x^2 + y^2}}, \quad \frac{\partial z}{\partial y} = \frac{y}{\sqrt{x^2 + y^2}}, \quad dz = \frac{\partial z}{\partial x} dx + \frac{\partial z}{\partial y} dy = \frac{x\,dx + y\,dy}{\sqrt{x^2 + y^2}}$$

When $x = 6$, $y = 8$, $dx = 1/4$, and $dy = -1/8$, then $dz = \dfrac{6(1/4) + 8(-1/8)}{\sqrt{6^2 + 8^2}} = 1/20$ inch. Thus the hypotenuse is lengthened by approximately 1/20 inch.

6. The power consumed in an electrical resistor is given by $P = E^2/R$ watts. If $E = 200$ volts and $R = 8$ ohms, by how much does the power change if $E$ is decreased by 5 volts and $R$ is decreased by 0.2 ohm?

$$\frac{\partial P}{\partial E} = \frac{2E}{R}, \quad \frac{\partial P}{\partial R} = -\frac{E^2}{R^2}, \quad dP = \frac{2E}{R} dE - \frac{E^2}{R^2} dR$$

When $E = 200$, $R = 8$, $dE = -5$, and $dR = -0.2$, then

$$dP = \frac{2 \cdot 200}{8}(-5) - \left(\frac{200}{8}\right)^2 (-0.2) = -250 + 125 = -125 \text{ watts}$$

The power is reduced by approximately 125 watts.

7. In measuring a rectangular block of wood, the dimensions were found to be 10, 12, and 20 in. with a possible error of .05 in. in each of the measurements. Find approximately the greatest error in the surface area of the block and the percentage of error in the area caused by the errors in the individual measurements.

The surface area is $S = 2(xy + yz + zx)$; then

$$dS = \frac{\partial S}{\partial x} dx + \frac{\partial S}{\partial y} dy + \frac{\partial S}{\partial z} dz = 2(y + z)\,dx + 2(x + z)\,dy + 2(y + x)\,dz$$

The greatest error in $S$ will occur when the errors in the lengths are of the same sign, say positive. Then

$$dS = 2(12 + 20)(.05) + 2(10 + 20)(.05) + 2(12 + 10)(.05) = 8.4 \text{ in}^2$$

The percentage of error is (error/area)(100) = (8.4/1120)(100) = 0.75%.

8. In using the formula $R = E/C$, find the maximum error and the percentage of error if $C = 20$ with a possible error of 0.1 and $E = 120$ with a possible error of 0.05.

$$dR = \frac{\partial R}{\partial E} dE + \frac{\partial R}{\partial C} dC = \frac{1}{C} dE - \frac{E}{C^2} dC$$

The maximum error will occur when $dE = 0.05$ and $dC = -0.1$; then

$$dR = \frac{0.05}{20} - \frac{120}{400}(-0.1) = 0.0325 \text{ is the approximate maximum error}$$

The percentage of error is $\dfrac{dR}{R}(100) = \dfrac{0.0325}{8}(100) = 0.40625 = 0.41\%$.

9. Two sides of a triangle were measured as 150 and 200 feet and the included angle as 60°. If the possible errors are 0.2 foot in measuring the sides and 1° in the angle, what is the greatest possible error in the computed area?

$$A = \tfrac{1}{2}xy \sin\theta, \quad \partial A/\partial x = \tfrac{1}{2}y \sin\theta, \quad \partial A/\partial y = \tfrac{1}{2}x \sin\theta, \quad \partial A/\partial\theta = \tfrac{1}{2}xy \cos\theta$$

and

$$dA = \tfrac{1}{2}y \sin\theta\,dx + \tfrac{1}{2}x \sin\theta\,dy + \tfrac{1}{2}xy \cos\theta\,d\theta$$

When $x = 150$, $y = 200$, $\theta = 60°$, $dx = 0.2$, $dy = 0.2$, and $d\theta = 1° = \pi/180$, then

$$dA = \tfrac{1}{2}(200)(\sin 60°)(0.2) + \tfrac{1}{2}(150)(\sin 60°)(0.2) + \tfrac{1}{2}(150)(200)(\cos 60°)(\pi/180) = 161.21 \text{ ft}^2$$

10. Find $dz/dt$, given: $z = x^2 + 3xy + 5y^2$; $x = \sin t$, $y = \cos t$.

$$\frac{\partial z}{\partial x} = 2x + 3y, \quad \frac{\partial z}{\partial y} = 3x + 10y, \quad \frac{dx}{dt} = \cos t, \quad \frac{dy}{dt} = -\sin t$$

Then

$$\frac{dz}{dt} = \frac{\partial z}{\partial x}\frac{dx}{dt} + \frac{\partial z}{\partial y}\frac{dy}{dt} = (2x + 3y)\cos t - (3x + 10y)\sin t$$

**11.** Find $dz/dt$, given: $z = \ln(x^2 + y^2)$; $x = e^{-t}$, $y = e^t$.

$$\frac{\partial z}{\partial x} = \frac{2x}{x^2 + y^2}, \qquad \frac{\partial z}{\partial y} = \frac{2y}{x^2 + y^2}, \qquad \frac{dx}{dt} = -e^{-t}, \qquad \frac{dy}{dt} = e^t$$

Then $\qquad \dfrac{dz}{dt} = \dfrac{\partial z}{\partial x}\dfrac{dx}{dt} + \dfrac{\partial z}{\partial y}\dfrac{dy}{dt} = \dfrac{2x}{x^2 + y^2}(-e^{-t}) + \dfrac{2y}{x^2 + y^2}(e^t) = 2\dfrac{ye^t - xe^{-t}}{x^2 + y^2}$

**12.** Let $z = f(x, y)$ be a continuous function of $x$ and $y$ with continuous partial derivatives $\partial z/\partial x$ and $\partial z/\partial y$, and let $y$ be a differentiable function of $x$. Then $z$ is a differentiable function of $x$ and by $(2)$,

$$\frac{dz}{dx} = \frac{\partial f}{\partial x}\cdot\frac{dx}{dx} + \frac{\partial f}{\partial y}\cdot\frac{dy}{dx} = \frac{\partial f}{\partial x} + \frac{\partial f}{\partial y}\cdot\frac{dy}{dx}$$

The shift in notation from $z$ to $f$ is made here in order to avoid possible confusion arising in the use of $dz/dx$ and $\partial z/\partial x$ in the same expression.

**13.** Find $dz/dx$, given: $z = f(x, y) = x^2 + 2xy + 4y^2$, $y = e^{ax}$.

$$\frac{dz}{dx} = \frac{\partial f}{\partial x} + \frac{\partial f}{\partial y}\cdot\frac{dy}{dx} = (2x + 2y) + (2x + 8y)ae^{ax} = 2(x + y) + 2a(x + 4y)e^{ax}$$

**14.** Find (a) $dz/dx$ and (b) $dz/dy$, given: $z = f(x, y) = xy^2 + x^2y$, $y = \ln x$.

(a) Here $x$ is the independent variable.

$$\frac{dz}{dx} = \frac{\partial f}{\partial x} + \frac{\partial f}{\partial y}\cdot\frac{dy}{dx} = (y^2 + 2xy) + (2xy + x^2)\left(\frac{1}{x}\right) = y^2 + 2xy + 2y + x$$

(b) Here $y$ is the independent variable.

$$\frac{dz}{dy} = \frac{\partial f}{\partial x}\cdot\frac{dx}{dy} + \frac{\partial f}{\partial y} = (y^2 + 2xy)x + (2xy + x^2) = xy^2 + 2x^2y + 2xy + x^2$$

**15.** The altitude of a right circular cone is 15 in. and is increasing at 0.2 in/min. The radius of the base is 10 in. and is decreasing at 0.3 in/min. How fast is the volume changing?

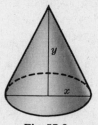

Let $x =$ radius and $y =$ altitude of cone. From $V = \frac{1}{3}\pi x^2 y$, considering $x$ and $y$ as functions of time $t$,

$$\frac{dV}{dt} = \frac{\partial V}{\partial x}\frac{dx}{dt} + \frac{\partial V}{\partial y}\frac{dy}{dt} = \frac{1}{3}\pi\left(2xy\frac{dx}{dt} + x^2\frac{dy}{dt}\right)$$

$$= \tfrac{1}{3}\pi[2\cdot 10\cdot 15\cdot(-0.3) + 10^2\cdot(0.2)] = -70\pi/3 \text{ in}^3/\text{min}$$

**Fig. 57-2**

**16.** A point $P$ is moving along the curve of intersection of the paraboloid $\dfrac{x^2}{16} - \dfrac{y^2}{9} = z$ and the cylinder $x^2 + y^2 = 5$, with $x$, $y$, and $z$ expressed in inches. If $x$ is increasing at 0.2 in/min, how fast is $z$ changing when $x = 2$?

From $z = \dfrac{x^2}{16} - \dfrac{y^2}{9}$, $\quad \dfrac{dz}{dt} = \dfrac{\partial z}{\partial x}\dfrac{dx}{dt} + \dfrac{\partial z}{\partial y}\dfrac{dy}{dt} = \dfrac{x}{8}\dfrac{dx}{dt} - \dfrac{2y}{9}\dfrac{dy}{dt}$.

Since $x^2 + y^2 = 5$, $y = \pm 1$ when $x = 2$; also, $x\dfrac{dx}{dt} + y\dfrac{dy}{dt} = 0$.

When $y = 1$, $\dfrac{dy}{dt} = -\dfrac{x}{y}\dfrac{dx}{dt} = -\dfrac{2}{1}(0.2) = -0.4$ and $\dfrac{dz}{dt} = \dfrac{2}{8}(0.2) - \dfrac{2}{9}(-0.4) = \dfrac{5}{36}$ in/min.

When $y = -1$, $\dfrac{dy}{dt} = -\dfrac{x}{y}\dfrac{dx}{dt} = 0.4$ and $\dfrac{dz}{dt} = \dfrac{2}{8}(0.2) - \dfrac{2}{9}(-1)(0.4) = \dfrac{5}{36}$ in/min.

**17.** Find $\partial z/\partial r$ and $\partial z/\partial s$, given: $z = x^2 + xy + y^2$; $x = 2r + s$, $y = r - 2s$.

$$\frac{\partial z}{\partial x} = 2x + y, \qquad \frac{\partial z}{\partial y} = x + 2y, \qquad \frac{\partial x}{\partial r} = 2, \qquad \frac{\partial x}{\partial s} = 1, \qquad \frac{\partial y}{\partial r} = 1, \qquad \frac{\partial y}{\partial s} = -2$$

Then $\qquad \dfrac{\partial z}{\partial r} = \dfrac{\partial z}{\partial x}\dfrac{\partial x}{\partial r} + \dfrac{\partial z}{\partial y}\dfrac{\partial y}{\partial r} = (2x + y)(2) + (x + 2y)(1) = 5x + 4y$

and $\qquad \dfrac{\partial z}{\partial s} = \dfrac{\partial z}{\partial x}\dfrac{\partial x}{\partial s} + \dfrac{\partial z}{\partial y}\dfrac{\partial y}{\partial s} = (2x + y)(1) + (x + 2y)(-2) = -3y$

18. Find $\dfrac{\partial u}{\partial \rho}, \dfrac{\partial u}{\partial \beta}, \dfrac{\partial u}{\partial \theta}$, given: $u = x^2 + 2y^2 + 2z^2$, $x = \rho \sin \beta \cos \theta$, $y = \rho \sin \beta \sin \theta$, $z = \rho \cos \beta$.

$$\frac{\partial u}{\partial \rho} = \frac{\partial u}{\partial x}\frac{\partial x}{\partial \rho} + \frac{\partial u}{\partial y}\frac{\partial y}{\partial \rho} + \frac{\partial u}{\partial z}\frac{\partial z}{\partial \rho} = 2x \sin \beta \cos \theta + 4y \sin \beta \sin \theta + 4z \cos \beta$$

$$\frac{\partial u}{\partial \beta} = \frac{\partial u}{\partial x}\frac{\partial x}{\partial \beta} + \frac{\partial u}{\partial y}\frac{\partial y}{\partial \beta} + \frac{\partial u}{\partial z}\frac{\partial z}{\partial \beta} = 2x \rho \cos \beta \cos \theta + 4y \rho \cos \beta \sin \theta - 4z \rho \sin \beta$$

$$\frac{\partial u}{\partial \theta} = \frac{\partial u}{\partial x}\frac{\partial x}{\partial \theta} + \frac{\partial u}{\partial y}\frac{\partial y}{\partial \theta} + \frac{\partial u}{\partial z}\frac{\partial z}{\partial \theta} = -2x \rho \sin \beta \sin \theta + 4y \rho \sin \beta \cos \theta$$

19. Find $du/dx$, given: $u = f(x, y, z) = xy + yz + zx$, $y = 1/x$, $z = x^2$.

Using $(3')$,

$$\frac{du}{dx} = \frac{\partial f}{\partial x} + \frac{\partial f}{\partial y}\cdot\frac{dy}{dx} + \frac{\partial f}{\partial z}\cdot\frac{dz}{dx} = (y + z) + (x + z)\left(-\frac{1}{x^2}\right) + (y + x)2x = y + z + 2x(x + y) - \frac{x + z}{x^2}$$

20. If $z = f(x, y)$ is a continuous function of $x$ and $y$ possessing continuous first partial derivatives $\partial z/\partial x$ and $\partial z/\partial y$, derive the basic formula

(A) $$\Delta z = \frac{\partial z}{\partial x}\Delta x + \frac{\partial z}{\partial y}\Delta y + \epsilon_1 \Delta x + \epsilon_2 \Delta y$$

where $\epsilon_1$ and $\epsilon_2 \to 0$ as $\Delta x$ and $\Delta y \to 0$.

When $x$ and $y$ are given increments $\Delta x$ and $\Delta y$ respectively, the increment given thereby to $z$ is

(i) $$\begin{aligned} \Delta z &= f(x + \Delta x, y + \Delta y) - f(x, y) \\ &= [f(x + \Delta x, y + \Delta y) - f(x, y + \Delta y)] + [f(x, y + \Delta y) - f(x, y)] \end{aligned}$$

In the first bracket expression, only $x$ changes; in the second, only $y$ changes. Thus, the Law of the Mean [(V) of Chapter 21] may be applied to each:

(ii) $$f(x + \Delta x, y + \Delta y) - f(x, y + \Delta y) = \Delta x \cdot f_x(x + \theta_1 \Delta x, y + \Delta y)$$

(iii) $$f(x, y + \Delta y) - f(x, y) = \Delta y \cdot f_y(x, y + \theta_2 \Delta y)$$

where $0 < \theta_1 < 1$ and $0 < \theta_2 < 1$. Note that here the derivatives involved are partial derivatives.

Since $\partial z/\partial x = f_x(x, y)$ and $\partial z/\partial y = f_y(x, y)$ are, by hypothesis, continuous functions of $x$ and $y$,

$$\lim_{\substack{\Delta x \to 0 \\ \Delta y \to 0}} f_x(x + \theta_1 \Delta x, y + \Delta y) = f_x(x, y) \quad \text{and} \quad \lim_{\substack{\Delta x \to 0 \\ \Delta y \to 0}} f_y(x, y + \theta_2 \Delta y) = f_y(x, y)$$

Then $$f_x(x + \theta_1 \Delta x, y + \Delta y) = f_x(x, y) + \epsilon_1, \quad f_y(x, y + \theta_2 \Delta y) = f_y(x, y) + \epsilon_2$$

where $\epsilon_1 \to 0$ and $\epsilon_2 \to 0$ as $\Delta x$ and $\Delta y \to 0$.

After making these replacements in (ii) and (iii) and then substituting in (i), we have, as required,

$$\begin{aligned} \Delta z &= \{f_x(x, y) + \epsilon_1\}\,\Delta x + \{f_y(x, y) + \epsilon_2\}\,\Delta y \\ &= f_x(x, y)\,\Delta x + f_y(x, y)\,\Delta y + \epsilon_1 \Delta x + \epsilon_2 \Delta y \end{aligned}$$

Note that the total derivative $dz$ is a fairly good approximation of the total increment $\Delta z$ when $|\Delta x|$ and $|\Delta y|$ are small.

# Supplementary Problems

**21.** Find the total differential, given:

(a) $z = x^3y + 2xy^3$     *Ans.* $dz = (3x^2 + 2y^2)y\,dx + (x^2 + 6y^2)x\,dy$

(b) $\theta = \arctan y/x$     *Ans.* $d\theta = \dfrac{x\,dy - y\,dx}{x^2 + y^2}$

(c) $z = e^{x^2 - y^2}$     *Ans.* $dz = 2z(x\,dx - y\,dy)$

(d) $z = x(x^2 + y^2)^{-1/2}$     *Ans.* $dz = \dfrac{y(y\,dx - x\,dy)}{(x^2 + y^2)^{3/2}}$

**22.** The fundamental frequency of vibration of a string or wire of circular section under tension $T$ is $n = \dfrac{1}{2rl}\sqrt{\dfrac{T}{\pi d}}$, where $l$ is the length, $r$ the radius, and $d$ the density of the string. Find  (a) the approximate effect of changing $l$ by a small amount $dl$,  (b) the effect of changing $T$ by a small amount $dT$, and  (c) the effect of changing $l$ and $T$ simultaneously.

*Ans.*  (a) $-\dfrac{n}{l}\,dl$,  (b) $\dfrac{n}{2T}\,dT$,  (c) $n\left(-\dfrac{dl}{l} + \dfrac{dT}{2T}\right)$

**23.** Use differentials to compute:

(a) the volume of a box of square base of side 8.005 and height 9.996 in.     *Ans.* 640.544 in³

(b) the diagonal of a rectangular box of dimensions 3.03 by 5.98 by 6.01 ft.     *Ans.* 9.003 ft

**24.** Approximate the maximum possible error and the percentage of error when $z$ is computed by the given formula:

(a) $z = \pi r^2 h$;  $r = 5 \pm 0.05$,  $h = 12 \pm 0.1$.     *Ans.* $8.5\pi$; 2.8%

(b) $1/z = 1/f + 1/g$;  $f = 4 \pm 0.01$,  $g = 8 \pm 0.02$.     *Ans.* 0.0067; 0.25%

(c) $z = y/x$;  $x = 1.8 \pm 0.1$,  $y = 2.4 \pm 0.1$.     *Ans.* 0.13; 10%

**25.** Find the approximate maximum percentage of error in

(a) $\omega = \sqrt[3]{g/b}$ if there is a possible 1% error in measuring $g$ and a possible $\frac{1}{2}$% error in measuring $b$.

*Hint.*  $\ln \omega = \frac{1}{3}(\ln g - \ln b)$;  $\dfrac{\partial\omega}{\omega} = \frac{1}{3}\left(\dfrac{dg}{g} - \dfrac{db}{b}\right)$,  $\left|\dfrac{dg}{g}\right| = 0.01$,  $\left|\dfrac{db}{b}\right| = 0.005$.     *Ans.* 0.005

(b) $g = 2s/t^2$ if there is a possible 1% error in measuring $s$ and $\frac{1}{4}$% error in measuring $t$.
*Ans.* 0.015

**26.** Find $du/dt$, given:

(a) $u = x^2y^3$, $x = 2t^3$, $y = 3t^2$.     *Ans.* $6xy^2\,t(2yt + 3x)$

(b) $u = x\cos y + y\sin x$, $x = \sin 2t$, $y = \cos 2t$.
*Ans.* $2(\cos y + y\cos x)\cos 2t - 2(-x\sin y + \sin x)\sin 2t$

(c) $u = xy + yz + zx$, $x = e^t$, $y = e^{-t}$, $z = e^t + e^{-t}$.     *Ans.* $(x + 2y + z)e^t - (2x + y + z)e^{-t}$

**27.** At a certain instant the radius of a right circular cylinder is 6 in. and is increasing at the rate 0.2 in/sec, while the altitude is 8 in. and is decreasing at the rate 0.4 in/sec. Find the time rate of change  (a) of the volume and  (b) of the surface at that instant.
*Ans.*  (a) $4.8\pi$ in³/sec,  (b) $3.2\pi$ in²/sec

**28.** A particle moves in a plane so that at any time $t$ its abscissa and ordinate are given by $x = 2 + 3t$, $y = t^2 + 4$ with $x$ and $y$ in feet and $t$ in minutes. How is the distance of the particle from the origin changing when $t = 1$?     *Ans.* $5/\sqrt{2}$ ft/min

29. A point is moving along the curve of intersection of $x^2 + 3xy + 3y^2 = z^2$ and the plane $x - 2y + 4 = 0$. When $x = 2$ and is increasing 3 units per sec, find (a) how $y$ is changing, (b) how $z$ is changing, (c) the speed of the point.

Ans. (a) inc. 3/2 units/sec, (b) inc. 75/14 units/sec at $(2, 3, 7)$ and dec. 75/14 units/sec at $(2, 3, -7)$, (c) 6.3 units/sec

30. Find $\partial z/\partial s$ and $\partial z/\partial t$, given:

(a) $z = x^2 - 2y^2$, $x = 3s + 2t$, $y = 3s - 2t$.        Ans. $6(x - 2y)$, $4(x + 2y)$

(b) $z = x^2 + 3xy + y^2$, $x = \sin s + \cos t$, $y = \sin s - \cos t$.        Ans. $5(x + y) \cos s$, $(x - y) \sin t$

(c) $z = x^2 + 2y^2$, $x = e^s - e^t$, $y = e^s + e^t$.        Ans. $2(x + 2y)e^s$, $2(2y - x)e^t$

(d) $z = \sin(4x + 5y)$, $x = s + t$, $y = s - t$.        Ans. $9 \cos(4x + 5y)$, $-\cos(4x + 5y)$

(e) $z = e^{xy}$, $x = s^2 + 2st$, $y = 2st + t^2$.        Ans. $2e^{xy}\{tx + (s + t)y\}$, $2e^{xy}\{(s + t)x + sy\}$

31. (a) If $u = f(x, y)$ and $x = r \cos \theta$, $y = r \sin \theta$, show that

$$\left(\frac{\partial u}{\partial x}\right)^2 + \left(\frac{\partial u}{\partial y}\right)^2 = \left(\frac{\partial u}{\partial r}\right)^2 + \frac{1}{r^2}\left(\frac{\partial u}{\partial \theta}\right)^2$$

(b) If $u = f(x, y)$ and $x = r \cosh s$, $y = r \sinh s$, show that

$$\left(\frac{\partial u}{\partial x}\right)^2 - \left(\frac{\partial u}{\partial y}\right)^2 = \left(\frac{\partial u}{\partial r}\right)^2 - \frac{1}{s^2}\left(\frac{\partial u}{\partial s}\right)^2$$

32. (a) If $z = f(x + \alpha y) + g(x - \alpha y)$, show that $\dfrac{\partial^2 z}{\partial x^2} = \dfrac{1}{\alpha^2}\dfrac{\partial^2 z}{\partial y^2}$.

Hint. Write $z = f(u) + g(v)$, $u = x + \alpha y$, $v = x - \alpha y$.

(b) If $z = x^n f\left(\dfrac{y}{x}\right)$, show that $x\dfrac{\partial z}{\partial x} + y\dfrac{\partial z}{\partial y} = nz$.

(c) If $z = f(x, y)$, $x = g(t)$, $y = h(t)$, show that subject to continuity conditions (Page 259)

$$\frac{d^2 z}{dt^2} = f_{xx}(g')^2 + 2f_{xy}g'h' + f_{yy}(h')^2 + f_x g'' + f_y h''$$

(d) If $z = f(x, y)$, $x = g(r, s)$, $y = h(r, s)$, show that subject to continuity conditions

$$\frac{\partial^2 z}{\partial r^2} = f_{xx}(g_r)^2 + 2f_{xy}g_r h_r + f_{yy}(h_r)^2 + f_x g_{rr} + f_y h_{rr}$$

$$\frac{\partial^2 z}{\partial r\, \partial s} = f_{xx}g_r g_s + f_{xy}(g_r h_s + g_s h_r) + f_{yy}h_r h_s + f_x g_{rs} + f_y h_{rs}$$

$$\frac{\partial^2 z}{\partial s^2} = f_{xx}(g_s)^2 + 2f_{xy}g_s h_s + f_{yy}(h_s)^2 + f_x g_{ss} + f_y h_{ss}$$

33. A function $f(x, y)$ is called homogeneous of order $n$ if $f(tx, ty) = t^n f(x, y)$. [For example, $f(x, y) = x^2 + 2xy + 3y^2$ is homogeneous of order 2; $f(x, y) = x \sin y/x + y \cos y/x$ is homogeneous of order 1.] Differentiate $f(tx, ty) = t^n f(x, y)$ with respect to $t$ and replace $t$ by 1 to show that $xf_x + yf_y = nf$. Verify this formula using the functions of the two examples. See also Problem 32(b).

34. If $z = \phi(u, v)$ where $u = f(x, y)$ and $v = g(x, y)$, and if $\dfrac{\partial u}{\partial x} = \dfrac{\partial v}{\partial y}$ and $\dfrac{\partial u}{\partial y} = -\dfrac{\partial v}{\partial x}$, show that

(a) $\dfrac{\partial^2 u}{\partial x^2} + \dfrac{\partial^2 u}{\partial y^2} = \dfrac{\partial^2 v}{\partial x^2} + \dfrac{\partial^2 v}{\partial y^2} = 0$

(b) $\dfrac{\partial^2 \phi}{\partial x^2} + \dfrac{\partial^2 \phi}{\partial y^2} = \left\{\left(\dfrac{\partial u}{\partial x}\right)^2 + \left(\dfrac{\partial v}{\partial x}\right)^2\right\}\left(\dfrac{\partial^2 \phi}{\partial u^2} + \dfrac{\partial^2 \phi}{\partial v^2}\right)$

35. Use (A) of Problem 20 to derive the chain rules (2) and (3). Hint. For (2), divide by $\Delta t$.

# Chapter 58

## Implicit Functions

**THE DIFFERENTIATION** of a function of one variable, defined implicitly by a relation $f(x, y) = 0$, was treated intuitively in Chapter 6. For this case, we state without proof:

**I.** If $f(x, y)$ is continuous in a region including a point $(x_0, y_0)$ for which $f(x_0, y_0) = 0$, if $\partial f/\partial x$ and $\partial f/\partial y$ are continuous throughout the region, and if $\partial f/\partial y \neq 0$ at $(x_0, y_0)$, then there is a neighborhood of $(x_0, y_0)$ in which $f(x, y) = 0$ can be solved for $y$ as a continuous differentiable function of $x$: $y = \phi(x)$ with $y_0 = \phi(x_0)$ and $\dfrac{dy}{dx} = -\dfrac{\partial f/\partial x}{\partial f/\partial y}$.

See Problems 1-3.

In extending this, we state

**II.** If $F(x, y, z)$ is continuous in a region including a point $(x_0, y_0, z_0)$ for which $F(x_0, y_0, z_0) = 0$, if $\dfrac{\partial F}{\partial x}, \dfrac{\partial F}{\partial y}, \dfrac{\partial F}{\partial z}$ are continuous throughout the region, and if $\partial F/\partial z \neq 0$ at $(x_0, y_0, z_0)$, then there is a neighborhood of $(x_0, y_0, z_0)$ in which $F(x, y, z) = 0$ can be solved for $z$ as a continuous differentiable function of $x$ and $y$: $z = \phi(x, y)$ with $z_0 = \phi(x_0, y_0)$ and $\dfrac{\partial z}{\partial x} = -\dfrac{\partial F/\partial x}{\partial F/\partial z}, \dfrac{\partial z}{\partial y} = -\dfrac{\partial F/\partial y}{\partial F/\partial z}$.

See Problems 4-5.

**III.** If $f(x, y, u, v)$ and $g(x, y, u, v)$ are continuous in a region including the point $(x_0, y_0, u_0, v_0)$ for which $f(x_0, y_0, u_0, v_0) = 0$ and $g(x_0, y_0, u_0, v_0) = 0$, if the several first partial derivatives of $f$ and of $g$ are continuous throughout the region, and if at $(x_0, y_0, u_0, v_0)$ the determinant $J\left(\dfrac{f, g}{u, v}\right) = \begin{vmatrix} \partial f/\partial u & \partial f/\partial v \\ \partial g/\partial u & \partial g/\partial v \end{vmatrix} \neq 0$, then there is a neighborhood of $(x_0, y_0, u_0, v_0)$ in which $f(x, y, u, v) = 0$ and $g(x, y, u, v) = 0$ can be solved simultaneously for $u$ and $v$ as continuous differentiable functions of $x$ and $y$: $u = \phi(x, y)$, $v = \psi(x, y)$. If at $(x_0, y_0, u_0, v_0)$ the determinant $J\left(\dfrac{f, g}{x, y}\right) \neq 0$, then there is a neighborhood of $(x_0, y_0, u_0, v_0)$ in which $f(x, y, u, v) = 0$ and $g(x, y, u, v) = 0$ can be solved for $x$ and $y$ as continuous differentiable functions of $u$ and $v$: $x = h(u, v)$, $y = k(u, v)$.

See Problems 6-7.

## Solved Problems

1. Use theorem I to show that $x^2 + y^2 - 13 = 0$ defines $y$ as a continuous differentiable function of $x$ in any neighborhood of the point $(2, 3)$ which does not include a point of the $x$-axis. Find the derivative at the point.

Set $f(x, y) = x^2 + y^2 - 13$. Then $f(2, 3) = 0$ while in any neighborhood of $(2, 3)$ described above in which the function is defined, its partial derivatives $\partial f/\partial x = 2x$ and $\partial f/\partial y = 2y$ are continuous, and $\partial f/\partial y \neq 0$. Then

$$\frac{\partial f}{\partial x} + \frac{\partial f}{\partial y} \cdot \frac{dy}{dx} = 0 \quad \text{and} \quad \frac{dy}{dx} = -\frac{\partial f/\partial x}{\partial f/\partial y} = -\frac{x}{y} = -\frac{2}{3} \quad \text{at } (2, 3)$$

2.  Find $\dfrac{dy}{dx}$, given  $f(x, y) = y^3 + xy - 12 = 0.$    $\dfrac{\partial f}{\partial x} = y,\ \dfrac{\partial f}{\partial y} = 3y^2 + x,$ and $\dfrac{dy}{dx} = -\dfrac{\partial f/\partial x}{\partial f/\partial y} = -\dfrac{y}{3y^2 + x}.$

3.  Find $dy/dx$, given  $e^x \sin y + e^y \sin x = 1.$

Put  $f(x, y) = e^x \sin y + e^y \sin x - 1.$  Then  $\dfrac{dy}{dx} = -\dfrac{\partial f/\partial x}{\partial f/\partial y} = -\dfrac{e^x \sin y + e^y \cos x}{e^x \cos y + e^y \sin x}.$

4.  Find $\partial z/\partial x$ and $\partial z/\partial y$, given  $F(x, y, z) = x^2 + 3xy - 2y^2 + 3xz + z^2 = 0.$

Treating $z$ as a function of $x$ and $y$ defined by the relation and differentiating partially with respect to $x$ and again with respect to $y$, we have

(i)  $$\frac{\partial F}{\partial x} + \frac{\partial F}{\partial z} \cdot \frac{\partial z}{\partial x} = (2x + 3y + 3z) + (3x + 2z)\frac{\partial z}{\partial x} = 0 \qquad \text{and}$$

(ii)  $$\frac{\partial F}{\partial y} + \frac{\partial F}{\partial z} \cdot \frac{\partial z}{\partial y} = (3x - 4y) + (3x + 2z)\frac{\partial z}{\partial y} = 0$$

From (i), $\dfrac{\partial z}{\partial x} = -\dfrac{\partial F/\partial x}{\partial F/\partial z} = -\dfrac{2x + 3y + 3z}{3x + 2z}$;  from (ii), $\dfrac{\partial z}{\partial y} = -\dfrac{\partial F/\partial y}{\partial F/\partial z} = -\dfrac{3x - 4y}{3x + 2z}.$

5.  Find $\partial z/\partial x$ and $\partial z/\partial y$, given  $\sin xy + \sin yz + \sin zx = 1.$

Set  $F(x, y, z) = \sin xy + \sin yz + \sin zx - 1$;  then

$$\frac{\partial F}{\partial x} = y \cos xy + z \cos zx, \quad \frac{\partial F}{\partial y} = x \cos xy + z \cos yz, \quad \frac{\partial F}{\partial z} = y \cos yz + x \cos zx$$

and

$$\frac{\partial z}{\partial x} = -\frac{\partial F/\partial x}{\partial F/\partial z} = -\frac{y \cos xy + z \cos zx}{y \cos yz + x \cos zx}, \quad \frac{\partial z}{\partial y} = -\frac{\partial F/\partial y}{\partial F/\partial z} = -\frac{x \cos xy + z \cos yz}{y \cos yz + x \cos zx}$$

6.  If $u$ and $v$ are defined as functions of $x$ and $y$ by the equations

$$f(x, y, u, v) = x + y^2 + 2uv = 0, \quad g(x, y, u, v) = x^2 - xy + y^2 + u^2 + v^2 = 0,$$

find  (i) $\partial u/\partial x,\ \partial v/\partial x$  and  (ii) $\partial u/\partial y,\ \partial v/\partial y.$

(i)  Differentiating $f$ and $g$ partially with respect to $x$, we have

$$1 + 2v\frac{\partial u}{\partial x} + 2u\frac{\partial v}{\partial x} = 0 \qquad \text{and} \qquad 2x - y + 2u\frac{\partial u}{\partial x} + 2v\frac{\partial v}{\partial x} = 0$$

Solving these relations simultaneously for $\partial u/\partial x$ and $\partial v/\partial x$, we find

$$\frac{\partial u}{\partial x} = \frac{v + u(y - 2x)}{2(u^2 - v^2)} \qquad \text{and} \qquad \frac{\partial v}{\partial x} = \frac{v(2x - y) - u}{2(u^2 - v^2)}$$

(ii)  Differentiating $f$ and $g$ partially with respect to $y$, we have

$$2y + 2v\frac{\partial u}{\partial y} + 2u\frac{\partial v}{\partial y} = 0 \qquad \text{and} \qquad -x + 2y + 2u\frac{\partial u}{\partial y} + 2v\frac{\partial v}{\partial y} = 0$$

Then  $\dfrac{\partial u}{\partial y} = \dfrac{u(x - 2y) + 2vy}{2(u^2 - v^2)}$  and  $\dfrac{\partial v}{\partial y} = \dfrac{v(2y - x) - 2uy}{2(u^2 - v^2)}$

7.  Given  $u^2 - v^2 + 2x + 3y = 0$ and $uv + x - y = 0$, find  (a) $\dfrac{\partial u}{\partial x}, \dfrac{\partial v}{\partial x}, \dfrac{\partial u}{\partial y}, \dfrac{\partial v}{\partial y}$, and  (b) $\dfrac{\partial x}{\partial u}, \dfrac{\partial y}{\partial u}, \dfrac{\partial x}{\partial v}, \dfrac{\partial y}{\partial v}.$

(a)  Here $x$ and $y$ are to be considered as independent variables.

Differentiate the given equations partially with respect to $x$:

$$2u\frac{\partial u}{\partial x} - 2v\frac{\partial v}{\partial x} + 2 = 0 \qquad \text{and} \qquad v\frac{\partial u}{\partial x} + u\frac{\partial v}{\partial x} + 1 = 0$$

Solve these relations simultaneously to obtain  $\dfrac{\partial u}{\partial x} = -\dfrac{u + v}{u^2 + v^2}$ and $\dfrac{\partial v}{\partial x} = \dfrac{v - u}{u^2 + v^2}.$

Differentiate the given equations partially with respect to $y$:

$$2u\frac{\partial u}{\partial y} - 2v\frac{\partial v}{\partial y} + 3 = 0 \qquad \text{and} \qquad v\frac{\partial u}{\partial y} + u\frac{\partial v}{\partial y} - 1 = 0$$

Solve simultaneously to obtain  $\dfrac{\partial u}{\partial y} = \dfrac{2v - 3u}{2(u^2 + v^2)}$ and $\dfrac{\partial v}{\partial y} = \dfrac{2u + 3v}{2(u^2 + v^2)}.$

(b) Here $u$ and $v$ are to be considered as independent variables.

Differentiate the given equations partially with respect to $u$:

$$2u + 2\frac{\partial x}{\partial u} + 3\frac{\partial y}{\partial u} = 0 \quad \text{and} \quad v + \frac{\partial x}{\partial u} - \frac{\partial y}{\partial u} = 0. \quad \text{Then} \quad \frac{\partial x}{\partial u} = -\frac{2u + 3v}{5} \quad \text{and} \quad \frac{\partial y}{\partial u} = \frac{2(v - u)}{5}.$$

Differentiate the given equations partially with respect to $v$:

$$-2v + 2\frac{\partial x}{\partial v} + 3\frac{\partial y}{\partial v} = 0 \quad \text{and} \quad u + \frac{\partial x}{\partial v} - \frac{\partial y}{\partial v} = 0. \quad \text{Then} \quad \frac{\partial x}{\partial v} = \frac{2v - 3u}{5} \quad \text{and} \quad \frac{\partial y}{\partial v} = \frac{2(u + v)}{5}.$$

# Supplementary Problems

8. Find $dy/dx$, given:

(a) $x^3 - x^2 y + xy^2 - y^3 = 1$      (b) $xy - e^x \sin y = 0$      (c) $\ln(x^2 + y^2) - \arctan y/x = 0$

Ans. (a) $\dfrac{3x^2 - 2xy + y^2}{x^2 - 2xy + 3y^2}$    (b) $\dfrac{e^x \sin y - y}{x - e^x \cos y}$    (c) $\dfrac{2x + y}{x - 2y}$

9. Find $\partial z/\partial x$ and $\partial z/\partial y$, given:

(a) $3x^2 + 4y^2 - 5z^2 = 60$                  Ans. $\partial z/\partial x = \dfrac{3x}{5z}$,   $\partial z/\partial y = \dfrac{4y}{5z}$

(b) $x^2 + y^2 + z^2 + 2xy + 4yz + 8zx = 20$      Ans. $\dfrac{\partial z}{\partial x} = -\dfrac{x + y + 4z}{4x + 2y + z}$,   $\dfrac{\partial z}{\partial y} = -\dfrac{x + y + 2z}{4x + 2y + z}$

(c) $x + 3y + 2z = \ln z$             Ans. $\dfrac{\partial z}{\partial x} = \dfrac{z}{1 - 2z}$,   $\dfrac{\partial z}{\partial y} = \dfrac{3z}{1 - 2z}$

(d) $z = e^x \cos(y + z)$            Ans. $\dfrac{\partial z}{\partial x} = \dfrac{z}{1 + e^x \sin(y + z)}$,   $\dfrac{\partial z}{\partial y} = \dfrac{-e^x \sin(y + z)}{1 + e^x \sin(y + z)}$

(e) $\sin(x + y) + \sin(y + z) + \sin(z + x) = 1$

Ans. $\dfrac{\partial z}{\partial x} = -\dfrac{\cos(x + y) + \cos(z + x)}{\cos(y + z) + \cos(z + x)}$,   $\dfrac{\partial z}{\partial y} = -\dfrac{\cos(x + y) + \cos(y + z)}{\cos(y + z) + \cos(z + x)}$

10. Find all the first and second partial derivatives of $z$, given: $x^2 + 2yz + 2zx = 1$.

Ans. $\dfrac{\partial z}{\partial x} = -\dfrac{x + z}{x + y}$,   $\dfrac{\partial z}{\partial y} = -\dfrac{z}{x + y}$,   $\dfrac{\partial^2 z}{\partial x^2} = \dfrac{x - y + 2z}{(x + y)^2}$,   $\dfrac{\partial^2 z}{\partial x \, \partial y} = \dfrac{x + 2z}{(x + y)^2}$,   $\dfrac{\partial^2 z}{\partial y^2} = \dfrac{2z}{(x + y)^2}$

11. If $F(x, y, z) = 0$ show that $\dfrac{\partial x}{\partial y} \cdot \dfrac{\partial y}{\partial z} \cdot \dfrac{\partial z}{\partial x} = -1$.

12. If $z = f(x, y)$ and $g(x, y) = 0$, show that $\dfrac{dz}{dx} = \dfrac{\dfrac{\partial f}{\partial x} \cdot \dfrac{\partial g}{\partial y} - \dfrac{\partial f}{\partial y} \cdot \dfrac{\partial g}{\partial x}}{\dfrac{\partial g}{\partial y}} = \dfrac{1}{\dfrac{\partial g}{\partial y}} J\left(\dfrac{f, g}{x, y}\right)$.

13. If $f(x, y) = 0$ and $g(z, x) = 0$, show that $\dfrac{\partial f}{\partial y} \cdot \dfrac{\partial g}{\partial x} \cdot \dfrac{\partial y}{\partial z} = \dfrac{\partial f}{\partial x} \cdot \dfrac{\partial g}{\partial z}$.

14. Find the first partial derivatives of $u$ and $v$ with respect to $x$ and $y$ and the first partial derivatives of $x$ and $y$ with respect to $u$ and $v$, given $2u - v + x^2 + xy = 0$, $u + 2v + xy - y^2 = 0$.

Ans. $\dfrac{\partial u}{\partial x} = -\dfrac{1}{5}(4x + 3y)$,   $\dfrac{\partial v}{\partial x} = \dfrac{1}{5}(2x - y)$,   $\dfrac{\partial u}{\partial y} = \dfrac{1}{5}(2y - 3x)$,   $\dfrac{\partial v}{\partial y} = \dfrac{4y - x}{5}$

$\dfrac{\partial x}{\partial u} = \dfrac{4y - x}{2(x^2 - 2xy - y^2)}$,   $\dfrac{\partial y}{\partial u} = \dfrac{y - 2x}{2(x^2 - 2xy - y^2)}$,   $\dfrac{\partial x}{\partial v} = \dfrac{3x - 2y}{2(x^2 - 2xy - y^2)}$,   $\dfrac{\partial y}{\partial v} = \dfrac{-4x - 3y}{2(x^2 - 2xy - y^2)}$

15. If $u = x + y + z$, $v = x^2 + y^2 + z^2$, $w = x^3 + y^3 + z^3$, show that

$$\dfrac{\partial x}{\partial u} = \dfrac{yz}{(x - y)(x - z)}, \quad \dfrac{\partial y}{\partial v} = \dfrac{x + z}{2(x - y)(y - z)}, \quad \dfrac{\partial z}{\partial w} = \dfrac{1}{3(x - z)(y - z)}$$

# Chapter 59

## Space Curves and Surfaces

**TANGENT LINE AND NORMAL PLANE TO A SPACE CURVE.** A space curve may be defined parametrically by the equations

$$x = f(t), \ y = g(t), \ z = h(t) \qquad (1)$$

At the point $P_0(x_0, y_0, z_0)$ of the curve (determined by $t = t_0$), the equations of the *tangent line* are

$$\frac{x - x_0}{\dfrac{dx}{dt}} = \frac{y - y_0}{\dfrac{dy}{dt}} = \frac{z - z_0}{\dfrac{dz}{dt}} \qquad (2)$$

and the equation of the *normal plane* (the plane through $P_0$ perpendicular to the tangent line there) is

$$\frac{dx}{dt}(x - x_0) + \frac{dy}{dt}(y - y_0) + \frac{dz}{dt}(z - z_0) = 0 \qquad (3)$$

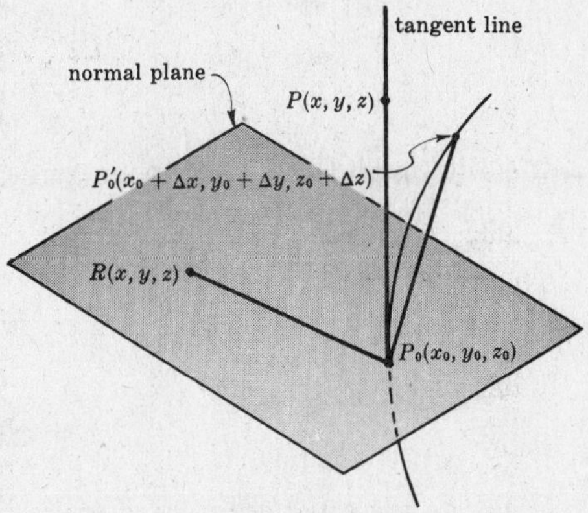

In both (2) and (3) it is understood that the derivatives have been evaluated at the point $P_0$.

See Problems 1-2.

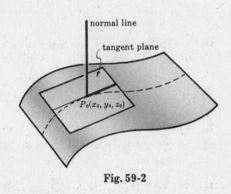

**Fig. 59-1**

**TANGENT PLANE AND NORMAL LINE TO A SURFACE.** The equation of the *tangent plane* to the surface $F(x, y, z) = 0$ at one of its points $P_0(x_0, y_0, z_0)$ is

$$\frac{\partial F}{\partial x}(x - x_0) + \frac{\partial F}{\partial y}(y - y_0) + \frac{\partial F}{\partial z}(z - z_0) = 0 \qquad (4)$$

and the equations of the *normal line* are

$$\frac{x - x_0}{\dfrac{\partial F}{\partial x}} = \frac{y - y_0}{\dfrac{\partial F}{\partial y}} = \frac{z - z_0}{\dfrac{\partial F}{\partial z}} \qquad (5)$$

with the understanding that the partial derivatives have been evaluated at the point $P_0$. Refer to Fig. 59-2.

See Problems 3-9.

**Fig. 59-2**

**A SPACE CURVE** may also be defined by the pair of equations

$$F(x, y, z) = 0, \ G(x, y, z) = 0 \qquad (6)$$

At the point $P_0(x_0, y_0, z_0)$ of the curve, the equations of the tangent line are

273

$$\frac{x - x_0}{\begin{vmatrix} \dfrac{\partial F}{\partial y} & \dfrac{\partial F}{\partial z} \\[2mm] \dfrac{\partial G}{\partial y} & \dfrac{\partial G}{\partial z} \end{vmatrix}} = \frac{y - y_0}{\begin{vmatrix} \dfrac{\partial F}{\partial z} & \dfrac{\partial F}{\partial x} \\[2mm] \dfrac{\partial G}{\partial z} & \dfrac{\partial G}{\partial x} \end{vmatrix}} = \frac{z - z_0}{\begin{vmatrix} \dfrac{\partial F}{\partial x} & \dfrac{\partial F}{\partial y} \\[2mm] \dfrac{\partial G}{\partial x} & \dfrac{\partial G}{\partial y} \end{vmatrix}} \tag{7}$$

and the equation of the normal plane is

$$\begin{vmatrix} \dfrac{\partial F}{\partial y} & \dfrac{\partial F}{\partial z} \\[2mm] \dfrac{\partial G}{\partial y} & \dfrac{\partial G}{\partial z} \end{vmatrix}(x - x_0) + \begin{vmatrix} \dfrac{\partial F}{\partial z} & \dfrac{\partial F}{\partial x} \\[2mm] \dfrac{\partial G}{\partial z} & \dfrac{\partial G}{\partial x} \end{vmatrix}(y - y_0) + \begin{vmatrix} \dfrac{\partial F}{\partial x} & \dfrac{\partial F}{\partial y} \\[2mm] \dfrac{\partial G}{\partial x} & \dfrac{\partial G}{\partial y} \end{vmatrix}(z - z_0) = 0 \tag{8}$$

In (7) and (8) it is to be understood that all partial derivatives have been evaluated at the point $P_0$.

See Problems 10-11.

# Solved Problems

1. Derive the equations (2) and (3) of the tangent line and normal plane to the space curve: $x = f(t)$, $y = g(t)$, $z = h(t)$ at the point $P_0(x_0, y_0, z_0)$ determined by the value $t = t_0$. Refer to Fig. 59-1.

   Let $P_0'(x_0 + \Delta x, y_0 + \Delta y, z_0 + \Delta z)$, determined by $t = t_0 + \Delta t$, be another point on the curve. As $P_0' \to P_0$ along the curve, the chord $P_0 P_0'$ approaches the tangent line to the curve at $P_0$ as limiting position.

   A simple set of direction numbers for the chord $P_0 P_0'$ is $[\Delta x, \Delta y, \Delta z]$ but we shall use $\left[\dfrac{\Delta x}{\Delta t}, \dfrac{\Delta y}{\Delta t}, \dfrac{\Delta z}{\Delta t}\right]$. Then as $P_0' \to P_0$, $\Delta t \to 0$ and $\left[\dfrac{\Delta x}{\Delta t}, \dfrac{\Delta y}{\Delta t}, \dfrac{\Delta z}{\Delta t}\right] \to \left[\dfrac{dx}{dt}, \dfrac{dy}{dt}, \dfrac{dz}{dt}\right]$, a set of direction numbers of the tangent line at $P_0$. Now if $P(x, y, z)$ is an arbitrary point on this tangent line, then $[x - x_0, y - y_0, z - z_0]$ is a set of direction numbers of $P_0 P$. Thus, since the sets of direction numbers are proportional, the equations of the tangent line at $P_0$ are

$$\frac{x - x_0}{dx/dt} = \frac{y - y_0}{dy/dt} = \frac{z - z_0}{dz/dt}$$

   If $R(x, y, z)$ is an arbitrary point in the normal plane at $P_0$ then, since $P_0 R$ and $P_0 P$ are perpendicular, the equation of the normal plane at $P_0$ is

$$(x - x_0)\frac{dx}{dt} + (y - y_0)\frac{dy}{dt} + (z - z_0)\frac{dz}{dt} = 0$$

2. Find the equations of the tangent line and normal plane to:

   (a) the curve $x = t$, $y = t^2$, $z = t^3$ at the point $t = 1$,

   (b) the curve $x = t - 2$, $y = 3t^2 + 1$, $z = 2t^3$ at the point where it pierces the $yz$-plane.

   (a) At the point $t = 1$ or $(1, 1, 1)$, $dx/dt = 1$, $dy/dt = 2t = 2$, and $dz/dt = 3t^2 = 3$. Using (2), the equations of the tangent line are $\dfrac{x - 1}{1} = \dfrac{y - 1}{2} = \dfrac{z - 1}{3}$ and, using (3), the equation of the normal plane is $(x - 1) + 2(y - 1) + 3(z - 1) = x + 2y + 3z - 6 = 0$.

   (b) The given curve pierces the $yz$-plane in the point where $x = t - 2 = 0$, that is, in the point $t = 2$ or $(0, 13, 16)$. At this point, $dx/dt = 1$, $dy/dt = 6t = 12$, and $dz/dt = 6t^2 = 24$. The equations of the tangent line are $\dfrac{x}{1} = \dfrac{y - 13}{12} = \dfrac{z - 16}{24}$ and the equation of the normal plane is $x + 12(y - 13) + 24(z - 16) = x + 12y + 24z - 540 = 0$.

3.  Derive the equations (4) and (5) of the tangent plane and normal line to the surface $F(x, y, z) = 0$ at the point $P_0(x_0, y_0, z_0)$.  Refer to Fig. 59-2.

Let $x = f(t)$, $y = g(t)$, $z = h(t)$ be the parametric equations of any curve on the surface $F(x, y, z) = 0$ and passing through the point $P_0$. Then at $P_0$,

$$\frac{\partial F}{\partial x} \cdot \frac{dx}{dt} + \frac{\partial F}{\partial y} \cdot \frac{dy}{dt} + \frac{\partial F}{\partial z} \cdot \frac{dz}{dt} = 0$$

with the understanding that all derivatives have been evaluated at $P_0$.

This relation expresses the fact that the line through $P_0$ with direction numbers (i) $\left[\dfrac{dx}{dt}, \dfrac{dy}{dt}, \dfrac{dz}{dt}\right]$ is perpendicular to the line through $P_0$ having direction numbers (ii) $\left[\dfrac{\partial F}{\partial x}, \dfrac{\partial F}{\partial y}, \dfrac{\partial F}{\partial z}\right]$. The set (i) belongs to the tangent to the curve which lies in the tangent plane of the surface. The set (ii) defines the normal line to the surface at $P_0$. The equations of this normal are

$$\frac{x - x_0}{\partial F/\partial x} = \frac{y - y_0}{\partial F/\partial y} = \frac{z - z_0}{\partial F/\partial z}$$

and the equation of the tangent plane at $P_0$ is

$$\frac{\partial F}{\partial x}(x - x_0) + \frac{\partial F}{\partial y}(y - y_0) + \frac{\partial F}{\partial z}(z - z_0) = 0$$

In Problems 4-5, find the equations of the tangent plane and normal line to the given surface at the given point.

4.  $z = 3x^2 + 2y^2 - 11$;  (2, 1, 3).

Put $F(x, y, z) = 3x^2 + 2y^2 - z - 11 = 0$. At $(2, 1, 3)$: $\dfrac{\partial F}{\partial x} = 6x = 12$, $\dfrac{\partial F}{\partial y} = 4y = 4$, and $\dfrac{\partial F}{\partial z} = -1$.

The equation of the tangent plane is $12(x - 2) + 4(y - 1) - (z - 3) = 0$ or $12x + 4y - z = 25$.

The equations of the normal line are $\dfrac{x - 2}{12} = \dfrac{y - 1}{4} = \dfrac{z - 3}{-1}$.

5.  $F(x, y, z) = x^2 + 3y^2 - 4z^2 + 3xy - 10yz + 4x - 5z - 22 = 0$;  (1, -2, 1).

At $(1, -2, 1)$: $\dfrac{\partial F}{\partial x} = 2x + 3y + 4 = 0$, $\dfrac{\partial F}{\partial y} = 6y + 3x - 10z = -19$, $\dfrac{\partial F}{\partial z} = -8z - 10y - 5 = 7$.

The equation of the tangent plane is $0(x - 1) - 19(y + 2) + 7(z - 1) = 0$ or $19y - 7z + 45 = 0$.

The equations of the normal line are $\dfrac{x - 1}{0} = \dfrac{y + 2}{-19} = \dfrac{z - 1}{7}$ or $x = 1$, $7y + 19z - 5 = 0$.

6.  Show that the equation of the tangent plane to the surface $\dfrac{x^2}{a^2} - \dfrac{y^2}{b^2} - \dfrac{z^2}{c^2} = 1$ at the point $P_0(x_0, y_0, z_0)$ is $\dfrac{xx_0}{a^2} - \dfrac{yy_0}{b^2} - \dfrac{zz_0}{c^2} = 1$.

At $P_0$: $\dfrac{\partial F}{\partial x} = \dfrac{2x_0}{a^2}$, $\dfrac{\partial F}{\partial y} = -\dfrac{2y_0}{b^2}$, $\dfrac{\partial F}{\partial z} = -\dfrac{2z_0}{c^2}$.

The equation of the tangent plane is $\dfrac{2x_0}{a^2}(x - x_0) - \dfrac{2y_0}{b^2}(y - y_0) - \dfrac{2z_0}{c^2}(z - z_0) = 0$.

This becomes $\dfrac{xx_0}{a^2} - \dfrac{yy_0}{b^2} - \dfrac{zz_0}{c^2} = \dfrac{x_0^2}{a^2} - \dfrac{y_0^2}{b^2} - \dfrac{z_0^2}{c^2} = 1$, since $P_0$ is on the surface.

7.  Show that the surfaces

$$F(x, y, z) = x^2 + 4y^2 - 4z^2 - 4 = 0 \quad \text{and} \quad G(x, y, z) = x^2 + y^2 + z^2 - 6x - 6y + 2z + 10 = 0$$

are tangent at point $(2, 1, 1)$.

It is to be shown that the two surfaces have the same tangent plane at the given point.

At $(2, 1, 1)$: $\dfrac{\partial F}{\partial x} = 2x = 4$, $\dfrac{\partial F}{\partial y} = 8y = 8$, $\dfrac{\partial F}{\partial z} = -8z = -8$ and

$\dfrac{\partial G}{\partial x} = 2x - 6 = -2$, $\dfrac{\partial G}{\partial y} = 2y - 6 = -4$, $\dfrac{\partial G}{\partial z} = 2z + 2 = 4$.

Since the sets of direction numbers $[4, 8, -8]$ and $[-2, -4, 4]$ of the normal lines of the two surfaces are proportional, the surfaces have the common tangent plane

$$1(x - 2) + 2(y - 1) - 2(z - 1) = 0 \quad \text{or} \quad x + 2y - 2z = 2$$

8. Show that the surfaces $F(x, y, z) = xy + yz - 4zx = 0$ and $G(x, y, z) = 3z^2 - 5x + y = 0$ intersect at right angles at the point $(1, 2, 1)$.

It is to be shown that the tangent planes to the surfaces at the point are perpendicular or, what is the same, that the normal lines at the point are perpendicular.

At $(1, 2, 1)$: $\dfrac{\partial F}{\partial x} = y - 4z = -2$, $\dfrac{\partial F}{\partial y} = x + z = 2$, and $\dfrac{\partial F}{\partial z} = y - 4x = -2$. A set of direction numbers for the normal line to $F(x, y, z) = 0$ is $[l_1, m_1, n_1] = [1, -1, 1]$.

At $(1, 2, 1)$: $\dfrac{\partial G}{\partial x} = -5$, $\dfrac{\partial G}{\partial y} = 1$, and $\dfrac{\partial G}{\partial z} = 6z = 6$. A set of direction numbers for the normal line to $G(x, y, z) = 0$ is $[l_2, m_2, n_2] = [-5, 1, 6]$.

Since $l_1 l_2 + m_1 m_2 + n_1 n_2 = 1(-5) + (-1)1 + 1(6) = 0$, these directions are perpendicular.

9. Show that the surfaces $F(x, y, z) = 3x^2 + 4y^2 + 8z^2 - 36 = 0$ and $G(x, y, z) = x^2 + 2y^2 - 4z^2 - 6 = 0$ intersect at right angles.

At any point $P_0(x_0, y_0, z_0)$ on the two surfaces:

$\dfrac{\partial F}{\partial x} = 6x_0$, $\dfrac{\partial F}{\partial y} = 8y_0$, $\dfrac{\partial F}{\partial z} = 16z_0$, and $[3x_0, 4y_0, 8z_0]$ is a set of direction numbers for the normal to the surface at $P_0$. Similarly, $[x_0, 2y_0, -4z_0]$ is a set of direction numbers for the normal line to $G(x, y, z) = 0$ at $P_0$.

Since $3x_0(x_0) + 4y_0(2y_0) + 8z_0(-4z_0) = 3x_0^2 + 8y_0^2 - 32z_0^2$
$$= 6(x_0^2 + 2y_0^2 - 4z_0^2) - (3x_0^2 + 4y_0^2 + 8z_0^2) = 6(6) - 36 = 0,$$

these directions are perpendicular.

10. Derive the equations (7) and (8) for the tangent line and normal plane to the space curve $C$: $F(x, y, z) = 0$, $G(x, y, z) = 0$ at one of its points $P_0(x_0, y_0, z_0)$.

At $P_0$ the directions $\left[\dfrac{\partial F}{\partial x}, \dfrac{\partial F}{\partial y}, \dfrac{\partial F}{\partial z}\right]$ and $\left[\dfrac{\partial G}{\partial x}, \dfrac{\partial G}{\partial y}, \dfrac{\partial G}{\partial z}\right]$ are normal respectively to the tangent planes of the surfaces $F(x, y, z) = 0$ and $G(x, y, z) = 0$. Now the direction

$$\left[\begin{vmatrix} \partial F/\partial y & \partial F/\partial z \\ \partial G/\partial y & \partial G/\partial z \end{vmatrix}, \begin{vmatrix} \partial F/\partial z & \partial F/\partial x \\ \partial G/\partial z & \partial G/\partial x \end{vmatrix}, \begin{vmatrix} \partial F/\partial x & \partial F/\partial y \\ \partial G/\partial x & \partial G/\partial y \end{vmatrix}\right]$$

being perpendicular to each of these directions is that of the tangent line to $C$ at $P_0$. Hence, the equations of the tangent line are

$$\frac{x - x_0}{\begin{vmatrix} \partial F/\partial y & \partial F/\partial z \\ \partial G/\partial y & \partial G/\partial z \end{vmatrix}} = \frac{y - y_0}{\begin{vmatrix} \partial F/\partial z & \partial F/\partial x \\ \partial G/\partial z & \partial G/\partial x \end{vmatrix}} = \frac{z - z_0}{\begin{vmatrix} \partial F/\partial x & \partial F/\partial y \\ \partial G/\partial x & \partial G/\partial y \end{vmatrix}}$$

and the equation of the normal plane is

$$\begin{vmatrix} \partial F/\partial y & \partial F/\partial z \\ \partial G/\partial y & \partial G/\partial z \end{vmatrix}(x - x_0) + \begin{vmatrix} \partial F/\partial z & \partial F/\partial x \\ \partial G/\partial z & \partial G/\partial x \end{vmatrix}(y - y_0) + \begin{vmatrix} \partial F/\partial x & \partial F/\partial y \\ \partial G/\partial x & \partial G/\partial y \end{vmatrix}(z - z_0) = 0$$

11. Find the equations of the tangent line and the normal plane to the curve $x^2 + y^2 + z^2 = 14$, $x + y + z = 6$ at the point $(1, 2, 3)$.

Set $F(x, y, z) = x^2 + y^2 + z^2 - 14 = 0$ and $G(x, y, z) = x + y + z - 6 = 0$. At $(1, 2, 3)$:

$$\begin{vmatrix} \partial F/\partial y & \partial F/\partial z \\ \partial G/\partial y & \partial G/\partial z \end{vmatrix} = \begin{vmatrix} 2y & 2z \\ 1 & 1 \end{vmatrix} = \begin{vmatrix} 4 & 6 \\ 1 & 1 \end{vmatrix} = -2,$$

$$\begin{vmatrix} \partial F/\partial z & \partial F/\partial x \\ \partial G/\partial z & \partial G/\partial x \end{vmatrix} = \begin{vmatrix} 6 & 2 \\ 1 & 1 \end{vmatrix} = 4, \qquad \begin{vmatrix} \partial F/\partial x & \partial F/\partial y \\ \partial G/\partial x & \partial G/\partial y \end{vmatrix} = \begin{vmatrix} 2 & 4 \\ 1 & 1 \end{vmatrix} = -2$$

With $[1, -2, 1]$ as a set of direction numbers of the tangent, its equations are $\dfrac{x - 1}{1} = \dfrac{y - 2}{-2} = \dfrac{z - 3}{1}$ and the equation of the normal plane is $(x - 1) - 2(y - 2) + (z - 3) = x - 2y + z = 0$.

# Supplementary Problems

12. Find the equations of the tangent line and the normal plane to each of the given curves at the given point:

(a) $x = 2t,\ y = t^2,\ z = t^3;\ t = 1$     Ans. $\dfrac{x-2}{2} = \dfrac{y-1}{2} = \dfrac{z-1}{3};\ \ 2x + 2y + 3z - 9 = 0$

(b) $x = te^t,\ y = e^t,\ z = t;\ t = 0$    Ans. $\dfrac{x}{1} = \dfrac{y-1}{1} = \dfrac{z}{1};\ \ x + y + z - 1 = 0$

(c) $x = t\cos t,\ y = t\sin t,\ z = t;\ t = 0$   Ans. $x = z,\ y = 0;\ x + z = 0$

13. Show that the curves (i) $x = 2 - t,\ y = -1/t,\ z = 2t^2$ and (ii) $x = 1 + \theta,\ y = \sin\theta - 1,\ z = 2\cos\theta$ intersect at right angles at $P(1, -1, 2)$. Obtain the equations of the tangent line and normal plane of each curve at $P$.

Ans. (i) $\dfrac{x-1}{-1} = \dfrac{y+1}{1} = \dfrac{z-2}{4};\ \ x - y - 4z + 6 = 0$  (ii) $x - y = 2,\ z = 2;\ \ x + y = 0$

14. Show that the tangents to the helix $x = a\cos t,\ y = a\sin t,\ z = bt$ meet the $xy$-plane at the same angle.

15. Show that the length of the curve (1) from the point $t = t_0$ to the point $t = t_1$ is given by

$$\int_{t_0}^{t_1} \sqrt{\left(\frac{dx}{dt}\right)^2 + \left(\frac{dy}{dt}\right)^2 + \left(\frac{dz}{dt}\right)^2}\ dt$$

Find the length of the helix of Problem 14 from $t = 0$ to $t = t_1$.   Ans. $\sqrt{a^2 + b^2}\ t_1$

16. Find the equations of the tangent line and the normal plane to each of the given curves at the given point:

(a) $x^2 + 2y^2 + 2z^2 = 5,\ 3x - 2y - z = 0;\ (1, 1, 1)$

(b) $9x^2 + 4y^2 - 36z = 0,\ 3x + y + z - z^2 - 1 = 0;\ (2, -3, 2)$    (c) $4z^2 = xy,\ x^2 + y^2 = 8z;\ (2, 2, 1)$

Ans. (a) $\dfrac{x-1}{2} = \dfrac{y-1}{7} = \dfrac{z-1}{-8};\ \ 2x + 7y - 8z - 1 = 0$

  (b) $\dfrac{x-2}{1} = \dfrac{z-2}{1},\ y + 3 = 0;\ \ x + z - 4 = 0$  (c) $\dfrac{x-2}{1} = \dfrac{y-2}{-1},\ z - 1 = 0;\ \ x - y = 0$

17. Find the equations of the tangent plane and normal line to the given surface at the given point:

(a) $x^2 + y^2 + z^2 = 14;\ (1, -2, 3)$    Ans. $x - 2y + 3z = 14;\ \dfrac{x-1}{1} = \dfrac{y+2}{-2} = \dfrac{z-3}{3}$

(b) $x^2 + y^2 + z^2 = r^2;\ (x_1, y_1, z_1)$    Ans. $x_1 x + y_1 y + z_1 z = r^2;\ \dfrac{x-x_1}{x_1} = \dfrac{y-y_1}{y_1} = \dfrac{z-z_1}{z_1}$

(c) $x^2 + 2z^2 = 3y^2;\ (2, -2, -2)$    Ans. $x + 3y - 2z = 0;\ \dfrac{x-2}{1} = \dfrac{y+2}{3} = \dfrac{z+2}{-2}$

(d) $2x^2 + 2xy + y^2 + z + 1 = 0;\ (1, -2, -3)$   Ans. $z - 2y = 1;\ x - 1 = 0,\ \dfrac{y+2}{2} = \dfrac{z+3}{-1}$

(e) $z = xy;\ (3, -4, -12)$    Ans. $4x - 3y + z = 12;\ \dfrac{x-3}{4} = \dfrac{y+4}{-3} = \dfrac{z+12}{1}$

18. (a) Show that the sum of the intercepts of the tangent plane of the surface $x^{1/2} + y^{1/2} + z^{1/2} = a^{1/2}$ at any of its points is $a$.

(b) Show that the square root of the sum of the squares of the intercepts of the tangent plane to the surface $x^{2/3} + y^{2/3} + z^{2/3} = a^{2/3}$ at any of its points is $a$.

19. Show that the pairs of surfaces are tangent at the given point:

(a) $x^2 + y^2 + z^2 = 18,\ xy = 9;\ (3, 3, 0)$

(b) $x^2 + y^2 + z^2 - 8x - 8y - 6z + 24 = 0,\ x^2 + 3y^2 + 2z^2 = 9;\ (2, 1, 1)$

20. Show that the pairs of surfaces are mutually perpendicular at the given point:

(a) $x^2 + 2y^2 - 4z^2 = 8,\ 4x^2 - y^2 + 2z^2 = 14;\ (2, 2, 1)$

(b) $x^2 + y^2 + z^2 = 50,\ x^2 + y^2 - 10z + 25 = 0;\ (3, 4, 5)$

21. Show that each of the surfaces (i) $14x^2 + 11y^2 + 8z^2 = 66$, (ii) $3z^2 - 5x + y = 0$, (iii) $xy + yz - 4zx = 0$ is perpendicular to the other two at the point $(1, 2, 1)$.

# Chapter 60

## Directional Derivatives
## Maximum and Minimum Values

**DIRECTIONAL DERIVATIVES.** Through $P(x, y, z)$, any point on the surface $z = f(x, y)$, pass planes parallel to the coordinate planes $xOz$ and $yOz$ cutting the surface in the arcs $PR$ and $PS$ and the plane $xOy$ in the lines $P'M$ and $P'N$, as shown in Fig. 60-1.

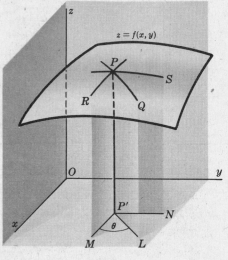

**Fig. 60-1**

The partial derivatives $\partial z/\partial x$ and $\partial z/\partial y$ evaluated at $P$ or at $P'(x, y)$ give respectively the rates of change of $z = P'P$ when $y$ and when $x$ is held fixed, that is, the rates of change of $z$ in directions parallel to the $x$- and $y$-axes or the slopes of the curves $PR$ and $PS$ at $P$.

Consider next a plane through $P$ perpendicular to the plane $xOy$ and making an angle $\theta$ with the $x$-axis. Let it cut the surface in the curve $PQ$ and the $xOy$ plane in the line $P'L$. The *directional derivative* of $f(x, y)$ at $P$ (or at $P'$) in the direction $\theta$ is given by

$$\frac{dz}{ds} = \frac{\partial z}{\partial x} \cos \theta + \frac{\partial z}{\partial y} \sin \theta \qquad (1)$$

The directional derivative gives the rate of change of $z = P'P$ in the direction of $P'L$ or the slope of the curve $PQ$ at $P$.

The directional derivative at a point $P$ is a function of $\theta$. If there is a direction for which the directional derivative at $P$ has a relative maximum value, this value is called the *gradient* of $f(x, y)$ at $P$. The gradient is the slope of the steepest tangent line which can be drawn to the surface at $P$.

See Problems 1-8.

For the function $w = F(x, y, z)$, the directional derivative at $P(x, y, z)$ in the direction $(\alpha, \beta, \gamma)$ is given by

$$\frac{dF}{ds} = \frac{\partial F}{\partial x} \cos \alpha + \frac{\partial F}{\partial y} \cos \beta + \frac{\partial F}{\partial z} \cos \gamma \qquad (2)$$

See Problem 9.

**RELATIVE MAXIMUM AND MINIMUM VALUES.** Suppose that $z = f(x, y)$ has a relative maximum (or minimum) value at $P_0(x_0, y_0, z_0)$. Any plane through $P_0$ perpendicular to the plane $xOy$ will cut the surface in a curve having a relative maximum (or minimum) point at $P_0$, that is, the directional derivative $\dfrac{\partial f}{\partial x} \cos \theta + \dfrac{\partial f}{\partial y} \sin \theta$ of $z = f(x, y)$ must equal 0 at $P_0$, for any whatever value of $\theta$. Thus, at $P_0$, $\partial f/\partial x = 0$ and $\partial f/\partial y = 0$.

278

The points, if any, at which $z = f(x, y)$ has a relative maximum (or minimum) value are among the point $(x_0, y_0)$ for which $\partial f/\partial x = 0$ and $\partial f/\partial y = 0$ simultaneously. To separate the cases, we quote without proof:

Let $z = f(x, y)$ have first and second partial derivatives in a certain region including the point $(x_0, y_0, z_0)$ at which $\partial f/\partial x = 0$ and $\partial f/\partial y = 0$. Then, if $\Delta = \left(\dfrac{\partial^2 f}{\partial x\, \partial y}\right)^2 - \left(\dfrac{\partial^2 f}{\partial x^2}\right)\left(\dfrac{\partial^2 f}{\partial y^2}\right) < 0$ at $P_0$, $z = f(x, y)$ has

a relative minimum at $P_0$ if $\dfrac{\partial^2 f}{\partial x^2} + \dfrac{\partial^2 f}{\partial y^2} > 0$

and

a relative maximum at $P_0$ if $\dfrac{\partial^2 f}{\partial x^2} + \dfrac{\partial^2 f}{\partial y^2} < 0$

If $\Delta > 0$, $P_0$ yields neither a maximum nor a minimum value; if $\Delta = 0$, the nature of the critical point $P_0$ is undetermined.

See Problems 10-15.

# Solved Problems

1.  In Fig. 60-1 let $P''(x + \Delta x, y + \Delta y)$ be a second point on $P'L$ and denote by $\Delta s$ the distance $P'P''$. Assuming that $z = f(x, y)$ possesses continuous first partial derivatives, we have by Problem 20, Chapter 57,

$$\Delta z = \frac{\partial z}{\partial x} \Delta x + \frac{\partial z}{\partial y} \Delta y + \epsilon_1 \Delta x + \epsilon_2 \Delta y$$

where $\epsilon_1$ and $\epsilon_2 \to 0$ as $\Delta x$ and $\Delta y \to 0$. The average rate of change of $z$ between the points $P'$ and $P''$ is

$$\frac{\Delta z}{\Delta s} = \frac{\partial z}{\partial x} \cdot \frac{\Delta x}{\Delta s} + \frac{\partial z}{\partial y} \cdot \frac{\Delta y}{\Delta s} + \epsilon_1 \cdot \frac{\Delta x}{\Delta s} + \epsilon_2 \cdot \frac{\Delta y}{\Delta s}$$

$$= \frac{\partial z}{\partial x} \cos \theta + \frac{\partial z}{\partial y} \sin \theta + \epsilon_1 \cos \theta + \epsilon_2 \sin \theta$$

where $\theta$ is the angle which the line $P'P''$ makes with the $x$-axis. Now let $P'' \to P'$ along $P'L$; the instantaneous rate of change of $z$ or the directional derivative at $P'$ is

$$\frac{dz}{ds} = \frac{\partial z}{\partial x} \cos \theta + \frac{\partial z}{\partial y} \sin \theta$$

2.  Find the directional derivative of $z = x^2 - 6y^2$ at point $P'(7, 2)$ in the direction (a) $\theta = 45°$, (b) $\theta = 135°$.

The directional derivative at any point $P'(x, y)$ in the direction $\theta$ is

$$\frac{dz}{ds} = \frac{\partial z}{\partial x} \cos \theta + \frac{\partial z}{\partial y} \sin \theta = 2x \cos \theta - 12y \sin \theta$$

(a)  At $P'(7, 2)$ in the direction $\theta = 45°$:    $dz/ds = 2 \cdot 7(\tfrac{1}{2}\sqrt{2}) - 12 \cdot 2(\tfrac{1}{2}\sqrt{2}) = -5\sqrt{2}$.

(b)  At $P'(7, 2)$ in the direction $\theta = 135°$:    $dz/ds = 2 \cdot 7(-\tfrac{1}{2}\sqrt{2}) - 12 \cdot 2(\tfrac{1}{2}\sqrt{2}) = -19\sqrt{2}$.

3.  Find the directional derivative of $z = ye^x$ at point $P'(0, 3)$ in the direction (a) $\theta = 30°$, (b) $\theta = 120°$.

$$\frac{dz}{ds} = ye^x \cos \theta + e^x \sin \theta$$

(a)  At point $(0, 3)$ in the direction $\theta = 30°$:    $dz/ds = 3 \cdot 1(\tfrac{1}{2}\sqrt{3}) + \tfrac{1}{2} = \tfrac{1}{2}(3\sqrt{3} + 1)$.

(b)  At point $(0, 3)$ in the direction $\theta = 120°$:    $dz/ds = 3 \cdot 1(-\tfrac{1}{2}) + \tfrac{1}{2}\sqrt{3} = \tfrac{1}{2}(-3 + \sqrt{3})$.

4.  The temperature $T$ of a heated circular plate at any of its points $(x, y)$ is given by $T = \dfrac{64}{x^2 + y^2 + 2}$, the origin being at the center of the plate. At the point $(1, 2)$ find the rate of change of $T$ in the direction $\theta = \pi/3$.

$$\frac{dT}{ds} = -\frac{64(2x)}{(x^2 + y^2 + 2)^2} \cos \theta - \frac{64(2y)}{(x^2 + y^2 + 2)^2} \sin \theta$$

At point $(1, 2)$ in the direction $\theta = \pi/3$:    $\dfrac{dT}{ds} = -\dfrac{128}{49}\left(\dfrac{1}{2}\right) - \dfrac{256}{49}\left(\dfrac{\sqrt{3}}{2}\right) = -\dfrac{64}{49}(1 + 2\sqrt{3})$.

5.   The electrical potential $V$ at any point $(x, y)$ is given by $V = \ln \sqrt{x^2 + y^2}$. Find the rate of change of $V$ at the point $(3, 4)$ in the direction toward the point $(2, 6)$.

$$\frac{dV}{ds} = \frac{x}{x^2 + y^2} \cos \theta + \frac{y}{x^2 + y^2} \sin \theta$$

Since $\theta$ is a second quadrant angle and $\tan \theta = \dfrac{6-4}{2-3} = -2$, $\cos \theta = -\dfrac{1}{\sqrt{5}}$ and $\sin \theta = \dfrac{2}{\sqrt{5}}$.

Hence, at $(3, 4)$ in the indicated direction, $\dfrac{dV}{ds} = \dfrac{3}{25}\left(-\dfrac{1}{\sqrt{5}}\right) + \dfrac{4}{25}\left(\dfrac{2}{\sqrt{5}}\right) = \dfrac{\sqrt{5}}{25}$.

6.   Find the gradient for the surface and point of Problem 2.

At point $(7, 2)$ in the direction $\theta$, $dz/ds = 14 \cos \theta - 24 \sin \theta$.

To find the value of $\theta$ for which $\dfrac{dz}{ds}$ is a maximum, set $\dfrac{d}{d\theta}\left(\dfrac{dz}{ds}\right) = -14 \sin \theta - 24 \cos \theta = 0$.

Then $\tan \theta = -24/14 = -12/7$ and $\theta$ is either a second or fourth quadrant angle. For the second quadrant angle, $\sin \theta = 12/\sqrt{193}$ and $\cos \theta = -7/\sqrt{193}$. For the fourth quadrant angle, $\sin \theta = -12/\sqrt{193}$ and $\cos \theta = 7/\sqrt{193}$.

Since $\dfrac{d^2}{d\theta^2}\left(\dfrac{dz}{ds}\right) = \dfrac{d}{d\theta}(-14 \sin \theta - 24 \cos \theta) = -14 \cos \theta + 24 \sin \theta$ is negative for the fourth quadrant angle, the gradient is $\dfrac{dz}{ds} = 14\left(\dfrac{7}{\sqrt{193}}\right) - 24\left(-\dfrac{12}{\sqrt{193}}\right) = 2\sqrt{193}$ and the direction is $\theta = 300°15'$.

7.   Find the gradient for the function and point of Problem 3.

At point $(0, 3)$ in the direction $\theta$, $dz/ds = 3 \cos \theta + \sin \theta$.

To find the value of $\theta$ for which $\dfrac{dz}{ds}$ is a maximum, set $\dfrac{d}{d\theta}\left(\dfrac{dz}{ds}\right) = -3 \sin \theta + \cos \theta = 0$.

Then $\tan \theta = 1/3$ and $\theta$ is either a first or third quadrant angle.

Since $\dfrac{d^2}{d\theta^2}\left(\dfrac{dz}{ds}\right) = \dfrac{d}{d\theta}(-3 \sin \theta + \cos \theta) = -3 \cos \theta - \sin \theta$ is negative for the first quadrant angle, the gradient is $\dfrac{dz}{ds} = 3\left(\dfrac{3}{\sqrt{10}}\right) + \dfrac{1}{\sqrt{10}} = \sqrt{10}$ and the direction is $\theta = 18°26'$.

8.   In Problem 5, show that $V$ changes most rapidly along the set of radial lines through the origin.

At any point $(x_1, y_1)$ in the direction $\theta$, $\dfrac{dV}{ds} = \dfrac{x_1}{x_1^2 + y_1^2} \cos \theta + \dfrac{y_1}{x_1^2 + y_1^2} \sin \theta$.

When $\dfrac{d}{d\theta}\left(\dfrac{dV}{ds}\right) = -\dfrac{x_1}{x_1^2 + y_1^2} \sin \theta + \dfrac{y_1}{x_1^2 + y_1^2} \cos \theta = 0$, $\tan \theta = \dfrac{y_1/(x_1^2 + y_1^2)}{x_1/(x_1^2 + y_1^2)} = \dfrac{y_1}{x_1}$.

Thus $\theta$ is the angle of inclination of the line joining the origin and point $(x_1, y_1)$.

9.   Find the directional derivative of $F(x, y, z) = xy + 2xz - y^2 + z^2$ at the point $(1, -2, 1)$ along the curve $x = t$, $y = t - 3$, $z = t^2$ in the direction of increasing $z$.

A set of direction numbers of the tangent to the curve at $(1, -2, 1)$ is $[1, 1, 2]$; the direction cosines are $[1/\sqrt{6}, 1/\sqrt{6}, 2/\sqrt{6}]$. The directional derivative is

$$\frac{\partial F}{\partial x} \cos \alpha + \frac{\partial F}{\partial y} \cos \beta + \frac{\partial F}{\partial z} \cos \gamma = 0 \cdot \frac{1}{\sqrt{6}} + 5 \cdot \frac{1}{\sqrt{6}} + 4 \cdot \frac{2}{\sqrt{6}} = \frac{13\sqrt{6}}{6}$$

10.  Examine $f(x, y) = x^2 + y^2 - 4x + 6y + 25$ for maximum and minimum values.

The conditions $\dfrac{\partial f}{\partial x} = 2x - 4 = 0$ and $\dfrac{\partial f}{\partial y} = 2y + 6 = 0$ are satisfied when $x = 2$, $y = -3$.

Since $f(x, y) = (x^2 - 4x + 4) + (y^2 + 6y + 9) + 25 - 4 - 9 = (x - 2)^2 + (y + 3)^2 + 12$, it is evident that $f(2, -3) = 12$ is a minimum value of the function.

Geometrically, $(2, -3, 12)$ is the minimum point of the surface $z = x^2 + y^2 - 4x + 6y + 25$.

**11.** Examine $f(x, y) = x^3 + y^3 + 3xy$ for maximum and minimum values.

The conditions $\dfrac{\partial f}{\partial x} = 3(x^2 + y) = 0$ and $\dfrac{\partial f}{\partial y} = 3(y^2 + x) = 0$ are satisfied when $x = 0$, $y = 0$ and when $x = -1$, $y = -1$.

At $(0,0)$: $\dfrac{\partial^2 f}{\partial x^2} = 6x = 0$, $\dfrac{\partial^2 f}{\partial x\, \partial y} = 3$, and $\dfrac{\partial^2 f}{\partial y^2} = 6y = 0$. Then $\left(\dfrac{\partial^2 f}{\partial x\, \partial y}\right)^2 - \dfrac{\partial^2 f}{\partial x^2} \cdot \dfrac{\partial^2 f}{\partial y^2} = 9 > 0$, and $(0, 0)$ yields neither a maximum nor minimum.

At $(-1, -1)$: $\dfrac{\partial^2 f}{\partial x^2} = -6$, $\dfrac{\partial^2 f}{\partial x\, \partial y} = 3$, $\dfrac{\partial^2 f}{\partial y^2} = -6$. Then $\left(\dfrac{\partial^2 f}{\partial x\, \partial y}\right)^2 - \dfrac{\partial^2 f}{\partial x^2} \cdot \dfrac{\partial^2 f}{\partial y^2} = -27 < 0$, and $\dfrac{\partial^2 f}{\partial x^2} + \dfrac{\partial^2 f}{\partial y^2} < 0$. Hence, $f(-1, -1) = 1$ is the maximum value of the function.

**12.** Divide 120 into 3 parts so that the sum of the products taken two at a time shall be a maximum.

Let $x$, $y$, and $120 - (x + y)$ be the three parts.

The function to be maximized is $S = xy + (x + y)(120 - x - y)$.

$\dfrac{\partial S}{\partial x} = y + (120 - x - y) - (x + y) = 120 - 2x - y$, $\quad$ $\dfrac{\partial S}{\partial y} = x + (120 - x - y) - (x + y) = 120 - x - 2y$.

When $\dfrac{\partial S}{\partial x} = \dfrac{\partial S}{\partial y} = 0$, we have $2x + y = 120$ and $x + 2y = 120$.

Then $x = 40$, $y = 40$, $120 - (x + y) = 40$ are the three parts, and $S = 3 \cdot 40^2 = 4800$.

For the division $1, 1, 118$, $S = 237$; clearly, $S = 4800$ is the maximum value.

**13.** Find the point in the plane $2x - y + 2z = 16$ nearest the origin.

Let $(x, y, z)$ be the required point; then the square of its distance from the origin is $D = x^2 + y^2 + z^2$. Since also $2x - y + 2z = 16$, $y = 2x + 2z - 16$ and $D = x^2 + (2x + 2z - 16)^2 + z^2$.

Then the conditions $\partial D/\partial x = 2x + 4(2x + 2z - 16) = 0$ and $\partial D/\partial z = 4(2x + 2z - 16) + 2z = 0$ are equivalent to $5x + 4z = 32$, $4x + 5z = 32$ and $x = z = 32/9$. Since it is known that a point for which $D$ is a minimum exists, $(32/9, -16/9, 32/9)$ is that point.

**14.** Show that a rectangular parallelepiped of maximum volume $V$ with constant surface area $S$ is a cube.

Let the dimensions be $x$, $y$, and $z$. Then $V = xyz$ and $S = 2(xy + yz + zx)$.

The second relation may be solved for $z$ and substituted in the first, thus expressing $V$ as a function of $x$ and $y$. We prefer to avoid this step by simply treating $z$ as a function of $x$ and $y$. Then

$$\frac{\partial V}{\partial x} = yz + xy\frac{\partial z}{\partial x}, \qquad \frac{\partial V}{\partial y} = xz + xy\frac{\partial z}{\partial y}, \qquad \frac{\partial S}{\partial x} = 0 = 2\left(y + z + x\frac{\partial z}{\partial x} + y\frac{\partial z}{\partial x}\right),$$

$$\frac{\partial S}{\partial y} = 0 = 2\left(x + z + x\frac{\partial z}{\partial y} + y\frac{\partial z}{\partial y}\right)$$

From the latter two of these, $\dfrac{\partial z}{\partial x} = -\dfrac{y + z}{x + y}$ and $\dfrac{\partial z}{\partial y} = -\dfrac{x + z}{x + y}$.

Then the conditions $\dfrac{\partial V}{\partial x} = yz - \dfrac{xy(y + z)}{x + y} = 0$ and $\dfrac{\partial V}{\partial y} = xz - \dfrac{xy(x + z)}{x + y} = 0$ reduce to $y^2(z - x) = 0$ and $x^2(z - y) = 0$. Thus $x = y = z$, as required.

**15.** Find the volume $V$ of the largest rectangular parallelepiped that can be inscribed in the ellipsoid $\dfrac{x^2}{a^2} + \dfrac{y^2}{b^2} + \dfrac{z^2}{c^2} = 1$.

Let $P(x, y, z)$ be the vertex in the first octant. Then $V = 8xyz$.

Consider $z$ defined as a function of the independent variables $x$ and $y$ by the equation of the ellipsoid. The necessary conditions for a maximum are

$$\frac{\partial V}{\partial x} = 8\left(yz + xy\frac{\partial z}{\partial x}\right) = 0 \qquad \text{and} \qquad \frac{\partial V}{\partial y} = 8\left(xz + xy\frac{\partial z}{\partial y}\right) = 0 \tag{1}$$

From the equation of the ellipsoid obtain $\dfrac{2x}{a^2} + \dfrac{2z}{c^2} \cdot \dfrac{\partial z}{\partial x} = 0$ and $\dfrac{2y}{b^2} + \dfrac{2z}{c^2} \cdot \dfrac{\partial z}{\partial y} = 0$.

Eliminate $\partial z/\partial x$ and $\partial z/\partial y$ between these relations and (1) to obtain

$$\frac{\partial V}{\partial x} = 8\left(yz - \frac{c^2 x^2 y}{a^2 z}\right) = 0 \qquad \text{and} \qquad \frac{\partial V}{\partial y} = 8\left(xz - \frac{c^2 x y^2}{b^2 z}\right) = 0$$

and finally

$$\frac{x^2}{a^2} = \frac{z^2}{c^2} = \frac{y^2}{b^2} \tag{2}$$

Combine (2) with the equation of the ellipsoid to get $x = a\sqrt{3}/3$, $y = b\sqrt{3}/3$, and $z = c\sqrt{3}/3$. Then $V = 8xyz = (8\sqrt{3}/9)abc$ cubic units.

# Supplementary Problems

16.  Find the directional derivative of the following functions at the given point in the indicated direction:
     (a) $z = x^2 + xy + y^2$, $(3, 1)$, $\theta = \pi/3$. (b) $z = x^3 + y^3 - 3xy$, $(2, 1)$, $\theta = \arctan 2/3$. (c) $z = y + x\cos xy$,
     $(0, 0)$, $\theta = \pi/3$. (d) $z = 2x^2 + 3xy - y^2$, $(1, -1)$, in the direction toward $(2, 1)$.
     Ans. (a) $\frac{1}{2}(7 + 5\sqrt{3})$    (b) $21\sqrt{13}/13$    (c) $\frac{1}{2}(1 + \sqrt{3})$    (d) $11\sqrt{5}/5$

17.  Find the gradient for each of the functions of Problem 16 at the given point.
     Ans. (a) $\sqrt{74}$, (b) $3\sqrt{10}$, (c) $\sqrt{2}$, (d) $\sqrt{26}$

18.  Show that the gradient of $V = \ln\sqrt{x^2 + y^2}$ of Problem 8 is constant along any circle $x^2 + y^2 = r^2$.

19.  On a hill represented by $z = 8 - 4x^2 - 2y^2$, find (a) the direction of the steepest grade at $(1, 1, 2)$ and (b) the direction of the contour line (direction for which $z = $ constant). Note that the directions are mutually perpendicular.    Ans. (a) $\arctan \frac{1}{2}$, 3rd quadrant; (b) $\arctan -2$

20.  Show that the sum of the squares of the directional derivatives of $z = f(x, y)$ at any of its points is constant for any two mutually perpendicular directions and is equal to the square of the gradient.

21.  Given $z = f(x, y)$ and $w = g(x, y)$ such that $\partial z/\partial x = \partial w/\partial y$ and $\partial z/\partial y = -\partial w/\partial x$. If $\theta_1$ and $\theta_2$ are two mutually perpendicular directions, show that at any point $P(x, y)$, $\partial z/\partial s_1 = \partial w/\partial s_2$ and $\partial z/\partial s_2 = -\partial w/\partial s_1$.

22.  Find the directional derivative of each given function at the given point in the indicated direction:
     (a) $xy^2 z$, $(2, 1, 3)$, $[1, -2, 2]$. (b) $x^2 + y^2 + z^2$, $(1, 1, 1)$, in the direction toward $(2, 3, 4)$. (c) $x^2 + y^2 - 2xz$,
     $(1, 3, 2)$, along $x^2 + y^2 - 2xz = 6$, $3x^2 - y^2 + 3z = 0$ in the direction of increasing $z$.
     Ans. (a) $-17/3$    (b) $6\sqrt{14}/7$    (c) $0$

23.  Examine each of the following functions for relative maximum and minimum values:

     (a) $z = 2x + 4y - x^2 - y^2 - 3$        Ans. Max. $= 2$ when $x = 1$, $y = 2$

     (b) $z = x^3 + y^3 - 3xy$                Ans. Min. $= -1$ when $x = 1$, $y = 1$

     (c) $z = x^2 + 2xy + 2y^2$               Ans. Min. $= 0$ when $x = 0$, $y = 0$

     (d) $z = (x - y)(1 - xy)$                Ans. Neither max. nor min.

     (e) $z = 2x^2 + y^2 + 6xy + 10x - 6y + 5$    Ans. Neither max. nor min.

     (f) $z = 3x - 3y - 2x^3 - xy^2 + 2x^2 y + y^3$    Ans. Min. $= -\sqrt{6}$ when $x = -\sqrt{6}/6$, $y = \sqrt{6}/3$;
                                                        max. $= \sqrt{6}$ when $x = \sqrt{6}/6$, $y = -\sqrt{6}/3$

     (g) $z = xy(2x + 4y + 1)$               Ans. Max. $= 1/216$ when $x = -1/6$, $y = -1/12$

24.  Find positive numbers $x, y, z$ such that:
     (a) $x + y + z = 18$ and $xyz$ is a maximum.     (c) $x + y + z = 20$ and $xyz^2$ is a maximum.
     (b) $xyz = 27$ and $x + y + z$ is a minimum.     (d) $x + y + z = 12$ and $xy^2 z^3$ is a maximum.
     Ans. (a) $x = y = z = 6$, (b) $x = y = z = 3$, (c) $x = y = 5$, $z = 10$, (d) $x = 2$, $y = 4$, $z = 6$

25.  Find the minimum value of the square of the distance from the origin to the plane $Ax + By + Cz + D = 0$.    Ans. $D^2/(A^2 + B^2 + C^2)$

26.  (a) The surface area of a rectangular box without top is to be 108 ft². Find the greatest possible volume. (b) The volume of a rectangular box without top is to be 500 ft³. Find the minimum surface area.    Ans. (a) 108 ft³, (b) 300 ft²

27.  Find the point on $z = xy - 1$ nearest the origin.    Ans. $(0, 0, -1)$

28.  Find the equation of the plane through $(1, 1, 2)$ which cuts off in the first octant the least volume.
     Ans. $2x + 2y + z = 6$

29.  Determine the values of $p$ and $q$ so that the sum $S$ of the squares of the vertical distances of the points $(0, 2)$, $(1, 3)$, and $(2, 5)$ from the line $y = px + q$ shall be a minimum.
     Hint. $S = (q - 2)^2 + (p + q - 3)^2 + (2p + q - 5)^2$.    Ans. $p = 3/2$, $q = 11/6$

# Chapter 61

## Space Vectors

**THE STUDY OF PLANE ANALYTIC GEOMETRY** using vector methods is complicated by the traditional dominance of the subject by the concept of the slope of a line. On the contrary, the study of Solid Analytic Geometry is greatly facilitated by the use of vectors.

Three vectors $\mathbf{a}, \mathbf{b}, \mathbf{c}$, not in the same plane and no two parallel, issuing from a common point are said to form a *right-handed system* or *triad* if $\mathbf{c}$ has the direction a right-threaded screw moves when rotated through the smaller angle in the direction from $\mathbf{a}$ to $\mathbf{b}$, as in Fig. 61-1. It will be seen that then $\mathbf{b}$ has the direction of the screw when rotated in the direction from $\mathbf{c}$ to $\mathbf{a}$ and $\mathbf{a}$ has the direction of the screw when rotated from $\mathbf{b}$ to $\mathbf{c}$.

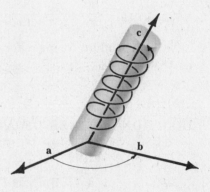

**Fig. 61-1**

We choose a right-handed rectangular coordinate system in space and let $\mathbf{i}, \mathbf{j}, \mathbf{k}$ be unit vectors along the positive $x$-, $y$-, $z$-axes respectively, as in Fig. 61-2. Then, paralleling Chapter 18, any free vector $\mathbf{a}$ may be written as

$$\mathbf{a} = a_1\mathbf{i} + a_2\mathbf{j} + a_3\mathbf{k}$$

while if $P(x, y, z)$ is a general point of space, the position vector $\mathbf{r}$ of $P$ is

$$\mathbf{r} = \mathbf{OP} = \mathbf{OB} + \mathbf{BP} = \mathbf{OA} + \mathbf{AB} + \mathbf{BP} \qquad (1)$$
$$= x\mathbf{i} + y\mathbf{j} + z\mathbf{k}$$

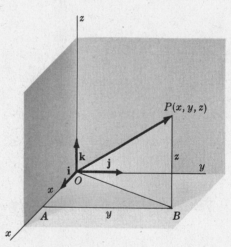

**Fig. 61-2**

Moreover, the algebra developed in Chapter 18 holds here with only such changes as the difference in dimension requires. For example, if $\mathbf{a} = a_1\mathbf{i} + a_2\mathbf{j} + a_3\mathbf{k}$ and $\mathbf{b} = b_1\mathbf{i} + b_2\mathbf{j} + b_3\mathbf{k}$, then

$k\mathbf{a} = ka_1\mathbf{i} + ka_2\mathbf{j} + ka_3\mathbf{k}$ for $k$, any scalar,

$\mathbf{a} = \mathbf{b}$ if and only if $a_1 = b_1$, $a_2 = b_2$, $a_3 = b_3$

$\mathbf{a} \pm \mathbf{b} = (a_1 \pm b_1)\mathbf{i} + (a_2 \pm b_2)\mathbf{j} + (a_3 \pm b_3)\mathbf{k}$

$\mathbf{a} \cdot \mathbf{b} = |\mathbf{a}|\,|\mathbf{b}|\cos\theta$, where $\theta$ is the smaller angle between $\mathbf{a}$ and $\mathbf{b}$

$\mathbf{i} \cdot \mathbf{i} = \mathbf{j} \cdot \mathbf{j} = \mathbf{k} \cdot \mathbf{k} = 1$; $\quad \mathbf{i} \cdot \mathbf{j} = \mathbf{j} \cdot \mathbf{k} = \mathbf{k} \cdot \mathbf{i} = 0$

$|\mathbf{a}| = \sqrt{\mathbf{a} \cdot \mathbf{a}} = \sqrt{a_1^2 + a_2^2 + a_3^2}$

$\mathbf{a} \cdot \mathbf{b} = 0$ if $\mathbf{a} = 0$, or $\mathbf{b} = 0$, or $\mathbf{a}$ and $\mathbf{b}$ are perpendicular.

From (1), we have

$$|\mathbf{r}| = \sqrt{\mathbf{r} \cdot \mathbf{r}} = \sqrt{x^2 + y^2 + z^2} \qquad (2a)$$

as the distance of the point $P(x, y, z)$ from the origin. Also, if $P_1(x_1, y_1, z_1)$ and $P_2(x_2, y_2, z_2)$ are any two points (see Fig. 61-3)

$$\mathbf{P_1P_2} = \mathbf{P_1B} + \mathbf{BP_2} = \mathbf{P_1A} + \mathbf{AB} + \mathbf{BP_2}$$
$$= (x_2 - x_1)\mathbf{i} + (y_2 - y_1)\mathbf{j} + (z_2 - z_1)\mathbf{k}$$

and

$$|\mathbf{P_1P_2}| = \sqrt{(x_2 - x_1)^2 + (y_2 - y_1)^2 + (z_2 - z_1)^2} \qquad (2b)$$

is the familiar formula for the distance between two points.

See Problems 1-3.

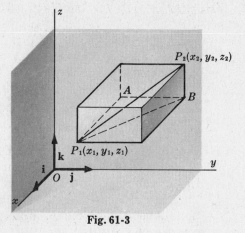

Fig. 61-3

**DIRECTION COSINES OF A VECTOR.** Let $\mathbf{a} = a_1\mathbf{i} + a_2\mathbf{j} + a_3\mathbf{k}$ make angles $\alpha, \beta, \gamma$ respectively with the positive $x$-, $y$-, $z$-axes, as in Fig. 61-4. From

$$\mathbf{i} \cdot \mathbf{a} = |\mathbf{i}| |\mathbf{a}| \cos \alpha = |\mathbf{a}| \cos \alpha$$
$$\mathbf{j} \cdot \mathbf{a} = |\mathbf{a}| \cos \beta, \qquad \mathbf{k} \cdot \mathbf{a} = |\mathbf{a}| \cos \gamma$$

we have

$$\cos \alpha = \frac{\mathbf{i} \cdot \mathbf{a}}{|\mathbf{a}|} = \frac{a_1}{|\mathbf{a}|}, \qquad \cos \beta = \frac{\mathbf{j} \cdot \mathbf{a}}{|\mathbf{a}|} = \frac{a_2}{|\mathbf{a}|}$$

$$\cos \gamma = \frac{\mathbf{k} \cdot \mathbf{a}}{|\mathbf{a}|} = \frac{a_3}{|\mathbf{a}|}$$

These are the *direction cosines* of $\mathbf{a}$. Since

$$\cos^2 \alpha + \cos^2 \beta + \cos^2 \gamma = \frac{a_1^2 + a_2^2 + a_3^2}{|\mathbf{a}|^2} = 1$$

the vector $\mathbf{u} = \mathbf{i} \cos \alpha + \mathbf{j} \cos \beta + \mathbf{k} \cos \gamma$ is a unit vector parallel to $\mathbf{a}$.

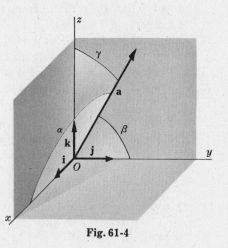

Fig. 61-4

**VECTOR PERPENDICULAR TO TWO VECTORS.** Let

$$\mathbf{a} = a_1\mathbf{i} + a_2\mathbf{j} + a_3\mathbf{k} \quad \text{and} \quad \mathbf{b} = b_1\mathbf{i} + b_2\mathbf{j} + b_3\mathbf{k}$$

be two non-parallel vectors taken with common initial point $P$. By an easy computation it can be shown that

$$\mathbf{c} = \begin{vmatrix} a_2 & a_3 \\ b_2 & b_3 \end{vmatrix} \mathbf{i} + \begin{vmatrix} a_3 & a_1 \\ b_3 & b_1 \end{vmatrix} \mathbf{j} + \begin{vmatrix} a_1 & a_2 \\ b_1 & b_2 \end{vmatrix} \mathbf{k} = \begin{vmatrix} \mathbf{i} & \mathbf{j} & \mathbf{k} \\ a_1 & a_2 & a_3 \\ b_1 & b_2 & b_3 \end{vmatrix} \qquad (3)$$

is perpendicular to (normal to) both $\mathbf{a}$ and $\mathbf{b}$ and, hence, to the plane of these vectors.

In Problems 5 and 6, we show

$$|\mathbf{c}| = |\mathbf{a}| |\mathbf{b}| \sin \theta$$
$$= \text{the area of the parallelogram having } \mathbf{a} \text{ and } \mathbf{b} \text{ as nonparallel sides.} \qquad (4)$$

If $\mathbf{a}$ and $\mathbf{b}$ are parallel, then $\mathbf{b} = k\mathbf{a}$ and (3) shows that $\mathbf{c} = 0$, that is, $\mathbf{c}$ is the zero vector. The zero vector, by definition, has magnitude 0 but no specified direction.

**VECTOR PRODUCT OF TWO VECTORS.**  Take

$$\mathbf{a} = a_1\mathbf{i} + a_2\mathbf{j} + a_3\mathbf{k} \quad \text{and} \quad \mathbf{b} = b_1\mathbf{i} + b_2\mathbf{j} + b_3\mathbf{k}$$

with initial point $P$ and denote by $\mathbf{n}$ the unit vector normal to the plane of $\mathbf{a}$ and $\mathbf{b}$, so directed that $\mathbf{a}, \mathbf{b}, \mathbf{n}$ in that order form a right-handed triad at $P$ as in Fig. 61-5. The *vector* or *cross product* of $\mathbf{a}$ and $\mathbf{b}$ is defined by

$$\mathbf{a} \times \mathbf{b} = |\mathbf{a}|\,|\mathbf{b}|\sin\theta\,\mathbf{n} \qquad (5)$$

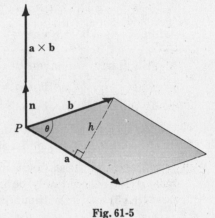

where $\theta$ is again the smaller angle between $\mathbf{a}$ and $\mathbf{b}$.  Thus, $\mathbf{a} \times \mathbf{b}$ is a vector perpendicular to both $\mathbf{a}$ and $\mathbf{b}$.

By Problem 6,

$$|\mathbf{a} \times \mathbf{b}| = |\mathbf{a}|\,|\mathbf{b}|\sin\theta$$

is the area of the parallelogram having $\mathbf{a}$ and $\mathbf{b}$ as nonparallel sides.

If $\mathbf{a}$ and $\mathbf{b}$ are parallel, $\theta = 0$ or $\pi$, and $\mathbf{a} \times \mathbf{b} = 0$.  Thus,

$$\mathbf{i} \times \mathbf{i} = \mathbf{j} \times \mathbf{j} = \mathbf{k} \times \mathbf{k} = 0 \qquad (6)$$

Fig. 61-5

If in (5) the order of $\mathbf{a}$ and $\mathbf{b}$ is reversed, $\mathbf{n}$ must be replaced by $-\mathbf{n}$; hence,

$$\mathbf{b} \times \mathbf{a} = -(\mathbf{a} \times \mathbf{b}) \qquad (7)$$

Since the coordinate axes were chosen as a right-handed system, it follows that

$$\begin{aligned}
\mathbf{i} \times \mathbf{j} &= \mathbf{k} & \mathbf{j} \times \mathbf{k} &= \mathbf{i} & \mathbf{k} \times \mathbf{i} &= \mathbf{j} \\
\mathbf{j} \times \mathbf{i} &= -\mathbf{k} & \mathbf{k} \times \mathbf{j} &= -\mathbf{i} & \mathbf{i} \times \mathbf{k} &= -\mathbf{j}
\end{aligned} \qquad (8)$$

In Problem 8, we prove for any vectors $\mathbf{a}, \mathbf{b}, \mathbf{c}$, the distributive law:

$$(\mathbf{a} + \mathbf{b}) \times \mathbf{c} = \mathbf{a} \times \mathbf{c} + \mathbf{b} \times \mathbf{c} \qquad (9)$$

Multiplying (9) by $-1$ and using (7), we have the companion distributive law:

$$\mathbf{c} \times (\mathbf{a} + \mathbf{b}) = \mathbf{c} \times \mathbf{a} + \mathbf{c} \times \mathbf{b} \qquad (9')$$

There follow

$$(\mathbf{a} + \mathbf{b}) \times (\mathbf{c} + \mathbf{d}) = \mathbf{a} \times \mathbf{c} + \mathbf{a} \times \mathbf{d} + \mathbf{b} \times \mathbf{c} + \mathbf{b} \times \mathbf{d} \qquad (10)$$

and

$$\mathbf{a} \times \mathbf{b} = \begin{vmatrix} \mathbf{i} & \mathbf{j} & \mathbf{k} \\ a_1 & a_2 & a_3 \\ b_1 & b_2 & b_3 \end{vmatrix} \qquad (11)$$

See Problems 9-10.

**TRIPLE SCALAR PRODUCT.**  In Fig. 61-6, let $\theta$ be the smaller angle between $\mathbf{b}$ and $\mathbf{c}$ and let $\phi$ be the smaller angle between $\mathbf{a}$ and $\mathbf{b} \times \mathbf{c}$.  Then the triple scalar product is by definition

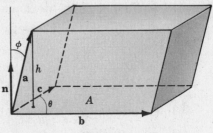

$$\begin{aligned}
\mathbf{a} \cdot (\mathbf{b} \times \mathbf{c}) &= \mathbf{a} \cdot |\mathbf{b}|\,|\mathbf{c}|\sin\theta\,\mathbf{n} = |\mathbf{a}|\,|\mathbf{b}|\,|\mathbf{c}|\sin\theta\cos\phi \\
&= (|\mathbf{a}|\cos\phi)(|\mathbf{b}|\,|\mathbf{c}|\sin\theta) = hA \\
&= \text{volume of parallelepiped}
\end{aligned}$$

It may be shown (see Problem 11) that

Fig. 61-6

$$\mathbf{a} \cdot (\mathbf{b} \times \mathbf{c}) = \begin{vmatrix} a_1 & a_2 & a_3 \\ b_1 & b_2 & b_3 \\ c_1 & c_2 & c_3 \end{vmatrix} = (\mathbf{a} \times \mathbf{b}) \cdot \mathbf{c} \tag{12}$$

Now $\qquad \mathbf{c} \cdot (\mathbf{a} \times \mathbf{b}) = \begin{vmatrix} c_1 & c_2 & c_3 \\ a_1 & a_2 & a_3 \\ b_1 & b_2 & b_3 \end{vmatrix} = \begin{vmatrix} a_1 & a_2 & a_3 \\ b_1 & b_2 & b_3 \\ c_1 & c_2 & c_3 \end{vmatrix} = \mathbf{a} \cdot (\mathbf{b} \times \mathbf{c})$

while $\qquad \mathbf{b} \cdot (\mathbf{a} \times \mathbf{c}) = \begin{vmatrix} b_1 & b_2 & b_3 \\ a_1 & a_2 & a_3 \\ c_1 & c_2 & c_3 \end{vmatrix} = -\begin{vmatrix} a_1 & a_2 & a_3 \\ b_1 & b_2 & b_3 \\ c_1 & c_2 & c_3 \end{vmatrix} = -\mathbf{a} \cdot (\mathbf{b} \times \mathbf{c})$

Similarly, we have

$$\mathbf{a} \cdot (\mathbf{b} \times \mathbf{c}) = \mathbf{c} \cdot (\mathbf{a} \times \mathbf{b}) = \mathbf{b} \cdot (\mathbf{c} \times \mathbf{a}) \tag{13}$$

and $\qquad \mathbf{a} \cdot (\mathbf{b} \times \mathbf{c}) = -\mathbf{b} \cdot (\mathbf{a} \times \mathbf{c}) = -\mathbf{c} \cdot (\mathbf{b} \times \mathbf{a}) = -\mathbf{a} \cdot (\mathbf{c} \times \mathbf{b}) \tag{14}$

From the definition of $\mathbf{a} \cdot (\mathbf{b} \times \mathbf{c})$ as a volume, it follows that if $\mathbf{a}, \mathbf{b}, \mathbf{c}$ are coplanar, then $\mathbf{a} \cdot (\mathbf{b} \times \mathbf{c}) \doteq 0$ and conversely.

The parentheses used in $\mathbf{a} \cdot (\mathbf{b} \times \mathbf{c})$ and $(\mathbf{a} \times \mathbf{b}) \cdot \mathbf{c}$ are not necessary. For example, $\mathbf{a} \cdot \mathbf{b} \times \mathbf{c}$ could only be interpreted as $\mathbf{a} \cdot (\mathbf{b} \times \mathbf{c})$ or $(\mathbf{a} \cdot \mathbf{b}) \times \mathbf{c}$. Now $\mathbf{a} \cdot \mathbf{b}$ is a scalar and $(\mathbf{a} \cdot \mathbf{b}) \times \mathbf{c}$ is without meaning.

See Problem 12.

**TRIPLE VECTOR PRODUCT.** In Problem 13, we show

$$\mathbf{a} \times (\mathbf{b} \times \mathbf{c}) = (\mathbf{a} \cdot \mathbf{c})\mathbf{b} - (\mathbf{a} \cdot \mathbf{b})\mathbf{c} \tag{15}$$

Similarly,

$$(\mathbf{a} \times \mathbf{b}) \times \mathbf{c} = (\mathbf{a} \cdot \mathbf{c})\mathbf{b} - (\mathbf{b} \cdot \mathbf{c})\mathbf{a} \tag{16}$$

Thus, except in the case $\mathbf{b}$ is perpendicular to both $\mathbf{a}$ and $\mathbf{c}$

$$\mathbf{a} \times (\mathbf{b} \times \mathbf{c}) \neq (\mathbf{a} \times \mathbf{b}) \times \mathbf{c}$$

and the use of parentheses is necessary.

**THE STRAIGHT LINE.** A line in space through a given point $P_0(x_0, y_0, z_0)$ will be defined as the locus of all points $P(x, y, z)$ such that $P_0P$ is parallel to a given direction $\mathbf{a} = a_1\mathbf{i} + a_2\mathbf{j} + a_3\mathbf{k}$. Let $\mathbf{r}_0$ and $\mathbf{r}$ be the position vectors of $P_0$ and $P$. Then

$$\mathbf{r} - \mathbf{r}_0 = k\mathbf{a} \qquad (k, \text{ a scalar variable}) \tag{17}$$

is the vector equation of the line $PP_0$. Refer to Fig. 61-7.

Writing (17) as

$$(x - x_0)\mathbf{i} + (y - y_0)\mathbf{j} + (z - z_0)\mathbf{k} = k(a_1\mathbf{i} + a_2\mathbf{j} + a_3\mathbf{k}),$$

separating components

$$x - x_0 = ka_1, \quad y - y_0 = ka_2, \quad z - z_0 = ka_3$$

and eliminating $k$, we have

$$\frac{x - x_0}{a_1} = \frac{y - y_0}{a_2} = \frac{z - z_0}{a_3} \tag{18}$$

as the equations in rectangular coordinates. Here, $[a_1, a_2, a_3]$ are a set of *direction numbers* for the line and $\left[\dfrac{a_1}{|\mathbf{a}|}, \dfrac{a_2}{|\mathbf{a}|}, \dfrac{a_3}{|\mathbf{a}|}\right]$ are a set of *direction cosines* of the line.

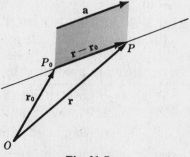

Fig. 61-7

If any one of the numbers $a_1, a_2, a_3$ is zero, the corresponding numerator in $(18)$ must be zero. For example, if $a_1 = 0$, $a_2a_3 \neq 0$, the equations of the line are

$$x - x_0 = 0, \qquad \frac{y - y_0}{a_2} = \frac{z - z_0}{a_3}$$

**THE PLANE.** A plane in space through the given point $P_0(x_0, y_0, z_0)$ will be defined as the locus of all lines through $P_0$ and perpendicular (normal) to a given line (direction) $\mathbf{a} = A\mathbf{i} + B\mathbf{j} + C\mathbf{k}$. Let $P(x, y, z)$ be any other point in the plane. Then $r - r_0 = \mathbf{P_0P}$ is perpendicular to $\mathbf{a}$, as in Fig. 61-8, and the required equation is

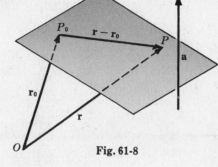

Fig. 61-8

$$(\mathbf{r} - \mathbf{r_0}) \cdot \mathbf{a} = 0 \qquad (19)$$

In rectangular coordinates, this becomes

$$\{(x - x_0)\mathbf{i} + (y - y_0)\mathbf{j} + (z - z_0)\mathbf{k}\} \cdot (A\mathbf{i} + B\mathbf{j} + C\mathbf{k}) = 0$$
or
$$A(x - x_0) + B(y - y_0) + C(z - z_0) = 0$$
or
$$Ax + By + Cz + D = 0 \qquad (20)$$

where $D = -(Ax_0 + By_0 + Cz_0)$.

Conversely, let $P_0(x_0, y_0, z_0)$ be a point on the surface

$$Ax + By + Cz + D = 0$$
Then
$$Ax_0 + By_0 + Cz_0 + D = 0$$
Subtracting,

$$A(x - x_0) + B(y - y_0) + C(z - z_0)$$
$$= (A\mathbf{i} + B\mathbf{j} + C\mathbf{k}) \cdot \{(x - x_0)\mathbf{i} + (y - y_0)\mathbf{j} + (z - z_0)\mathbf{k}\} = 0$$

and the constant vector $A\mathbf{i} + B\mathbf{j} + C\mathbf{k}$ is normal to the surface at each of its points. Thus, the surface is a plane.

# Solved Problems

1. Find the distance of the point $P_1(1, 2, 3)$ from $(a)$ the origin, $(b)$ the $x$-axis, $(c)$ the $z$-axis, $(d)$ the $xy$-plane, $(e)$ the point $P_2(3, -1, 5)$.

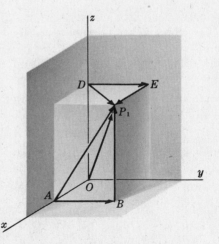

   $(a)$  $\mathbf{r} = \mathbf{OP_1} = \mathbf{i} + 2\mathbf{j} + 3\mathbf{k}$;  $|\mathbf{r}| = \sqrt{1^2 + 2^2 + 3^2} = \sqrt{14}$

   $(b)$  $\mathbf{AP_1} = \mathbf{AB} + \mathbf{BP_1} = 2\mathbf{j} + 3\mathbf{k}$;  $|\mathbf{AP_1}| = \sqrt{4 + 9} = \sqrt{13}$

   $(c)$  $\mathbf{DP_1} = \mathbf{DE} + \mathbf{EP_1} = 2\mathbf{j} + \mathbf{i}$;  $|\mathbf{DP_1}| = \sqrt{5}$

   $(d)$  $\mathbf{BP_1} = 3\mathbf{k}$;  $|\mathbf{BP_1}| = 3$

   $(e)$  $\mathbf{P_1P_2} = (3 - 1)\mathbf{i} + (-1 - 2)\mathbf{j} + (5 - 3)\mathbf{k} = 2\mathbf{i} - 3\mathbf{j} + 2\mathbf{k}$

   $\quad\ \ |\mathbf{P_1P_2}| = \sqrt{4 + 9 + 4} = \sqrt{17}$

2. Find the angle $\theta$ between the vectors joining $O$ to $P_1(1, 2, 3)$ and $P_2(2, -3, -1)$.

   $\mathbf{r_1} = \mathbf{OP_1} = \mathbf{i} + 2\mathbf{j} + 3\mathbf{k}$,   $\mathbf{r_2} = \mathbf{OP_2} = 2\mathbf{i} - 3\mathbf{j} - \mathbf{k}$

   $\cos \theta = \dfrac{\mathbf{r_1} \cdot \mathbf{r_2}}{|\mathbf{r_1}|\,|\mathbf{r_2}|} = \dfrac{1(2) + 2(-3) + 3(-1)}{\sqrt{14}\,\sqrt{14}} = -\dfrac{1}{2}$,   $\theta = 120°$

Fig. 61-9

3.  Find the angle $\alpha = \angle BAC$ of the triangle $ABC$ whose vertices are
    $A(1, 0, 1)$, $B(2, -1, 1)$, $C(-2, 1, 0)$.

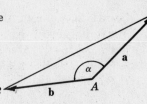

$$\mathbf{a} = \mathbf{AC} = -3\mathbf{i} + \mathbf{j} - \mathbf{k}, \quad \mathbf{b} = \mathbf{AB} = \mathbf{i} - \mathbf{j}$$

$$\cos \alpha = \frac{\mathbf{a} \cdot \mathbf{b}}{|\mathbf{a}|\,|\mathbf{b}|} = \frac{-3 - 1}{\sqrt{22}} = -0.85280, \quad \alpha = 148°31'$$

Fig. 61-10

4.  Find the direction cosines of $\mathbf{a} = 3\mathbf{i} + 12\mathbf{j} + 4\mathbf{k}$.

    The direction cosines are

$$\cos \alpha = \frac{\mathbf{i} \cdot \mathbf{a}}{|\mathbf{a}|} = \frac{3}{13}, \quad \cos \beta = \frac{\mathbf{j} \cdot \mathbf{a}}{|\mathbf{a}|} = \frac{12}{13}, \quad \cos \gamma = \frac{\mathbf{k} \cdot \mathbf{a}}{|\mathbf{a}|} = \frac{4}{13}$$

5.  If $\mathbf{a} = a_1\mathbf{i} + a_2\mathbf{j} + a_3\mathbf{k}$ and $\mathbf{b} = b_1\mathbf{i} + b_2\mathbf{j} + b_3\mathbf{k}$ are two vectors issuing from a point $P$ and if

$$\mathbf{c} = \begin{vmatrix} a_2 & a_3 \\ b_2 & b_3 \end{vmatrix} \mathbf{i} + \begin{vmatrix} a_1 & a_3 \\ b_1 & b_3 \end{vmatrix} \mathbf{j} + \begin{vmatrix} a_1 & a_2 \\ b_1 & b_2 \end{vmatrix} \mathbf{k}$$

show that $|\mathbf{c}| = |\mathbf{a}|\,|\mathbf{b}| \sin \theta$, where $\theta$ is the smaller angle between $\mathbf{a}$ and $\mathbf{b}$.

We have $\cos \theta = \dfrac{\mathbf{a} \cdot \mathbf{b}}{|\mathbf{a}|\,|\mathbf{b}|}$ and

$$\sin \theta = \sqrt{1 - \left\{ \frac{\mathbf{a} \cdot \mathbf{b}}{|\mathbf{a}|\,|\mathbf{b}|} \right\}^2} = \frac{\sqrt{(a_1^2 + a_2^2 + a_3^2)(b_1^2 + b_2^2 + b_3^2) - (a_1 b_1 + a_2 b_2 + a_3 b_3)^2}}{|\mathbf{a}|\,|\mathbf{b}|} = \frac{|\mathbf{c}|}{|\mathbf{a}|\,|\mathbf{b}|}$$

Hence, $|\mathbf{c}| = |\mathbf{a}|\,|\mathbf{b}| \sin \theta$ as required.

6.  Find the area of the parallelogram whose non-parallel sides are $\mathbf{a}$ and $\mathbf{b}$.

    From Fig. 61-11 below, $h = |\mathbf{b}| \sin \theta$ and the area is $h\,|\mathbf{a}| = |\mathbf{a}|\,|\mathbf{b}| \sin \theta$.

7.  Let $\mathbf{a}_1$ and $\mathbf{a}_2$ respectively be the components of $\mathbf{a}$ parallel and perpendicular to $\mathbf{b}$, as in Fig. 61-12
    below.  Show that
$$\mathbf{a}_2 \times \mathbf{b} = \mathbf{a} \times \mathbf{b} \quad \text{and} \quad \mathbf{a}_1 \times \mathbf{b} = 0$$

    If $\theta$ is the angle between $\mathbf{a}$ and $\mathbf{b}$, then $|\mathbf{a}_1| = |\mathbf{a}| \cos \theta$ and $|\mathbf{a}_2| = |\mathbf{a}| \sin \theta$. Since $\mathbf{a}, \mathbf{a}_2, \mathbf{b}$ are
    coplanar,
$$\mathbf{a}_2 \times \mathbf{b} = |\mathbf{a}_2|\,|\mathbf{b}| \sin \phi\, \mathbf{n} = |\mathbf{a}| \sin \theta\, |\mathbf{b}|\, \mathbf{n}$$
$$= |\mathbf{a}|\,|\mathbf{b}| \sin \theta\, \mathbf{n} = \mathbf{a} \times \mathbf{b}$$

    Since $\mathbf{a}_1$ and $\mathbf{b}$ are parallel, $\mathbf{a}_1 \times \mathbf{b} = 0$.

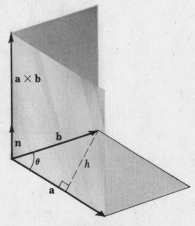

Fig. 61-11

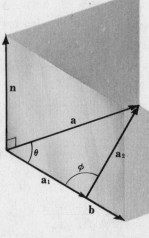

Fig. 61-12

8. Prove: $(\mathbf{a} + \mathbf{b}) \times \mathbf{c} = \mathbf{a} \times \mathbf{c} + \mathbf{b} \times \mathbf{c}$

In Fig. 61-13, the initial point $P$ of the vectors $\mathbf{a}, \mathbf{b}, \mathbf{c}$ is in the plane of the paper while their endpoints are above this plane. The vectors $\mathbf{a}_1$ and $\mathbf{b}_1$ are respectively the components perpendicular to $\mathbf{c}$ of $\mathbf{a}$ and $\mathbf{b}$. Then $\mathbf{a}_1, \mathbf{b}_1, \mathbf{a}_1 + \mathbf{b}_1, \mathbf{a}_1 \times \mathbf{c}, \mathbf{b}_1 \times \mathbf{c}$, and $(\mathbf{a}_1 + \mathbf{b}_1) \times \mathbf{c}$ all lie in the plane of the paper.

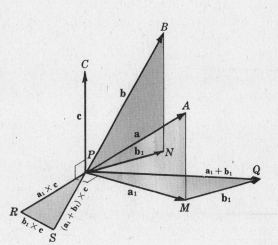

In the triangles $PRS$ and $PMQ$,

$$\frac{RS}{PR} = \frac{|\mathbf{b}_1 \times \mathbf{c}|}{|\mathbf{a}_1 \times \mathbf{c}|} = \frac{|\mathbf{b}_1|\,|\mathbf{c}|}{|\mathbf{a}_1|\,|\mathbf{c}|} = \frac{|\mathbf{b}_1|}{|\mathbf{a}_1|} = \frac{MQ}{PM};$$

thus, $PRS$ and $PMQ$ are similar. Now $PR$ is perpendicular to $PM$ and $RS$ is perpendicular to $MQ$; hence $PS$ is perpendicular to $PQ$ and $\mathbf{PS} = \mathbf{PQ} \times \mathbf{c}$. Then, since

$$\mathbf{PS} = \mathbf{PQ} \times \mathbf{c} = \mathbf{PR} + \mathbf{RS},$$

we have $(\mathbf{a}_1 + \mathbf{b}_1) \times \mathbf{c} = \mathbf{a}_1 \times \mathbf{c} + \mathbf{b}_1 \times \mathbf{c}$

By Problem 7, $\mathbf{a}_1$ and $\mathbf{b}_1$ may be replaced by $\mathbf{a}$ and $\mathbf{b}$ respectively to yield the required result

$$(\mathbf{a} + \mathbf{b}) \times \mathbf{c} = \mathbf{a} \times \mathbf{c} + \mathbf{b} \times \mathbf{c}$$

**Fig. 61-13**

9. When $\mathbf{a} = a_1\mathbf{i} + a_2\mathbf{j} + a_3\mathbf{k}$ and $\mathbf{b} = b_1\mathbf{i} + b_2\mathbf{j} + b_3\mathbf{k}$, show that

$$\mathbf{a} \times \mathbf{b} = \begin{vmatrix} \mathbf{i} & \mathbf{j} & \mathbf{k} \\ a_1 & a_2 & a_3 \\ b_1 & b_2 & b_3 \end{vmatrix}$$

We have by the distributive law

$$\begin{aligned}
\mathbf{a} \times \mathbf{b} &= (a_1\mathbf{i} + a_2\mathbf{j} + a_3\mathbf{k}) \times (b_1\mathbf{i} + b_2\mathbf{j} + b_3\mathbf{k}) \\
&= a_1\mathbf{i} \times (b_1\mathbf{i} + b_2\mathbf{j} + b_3\mathbf{k}) + a_2\mathbf{j} \times (b_1\mathbf{i} + b_2\mathbf{j} + b_3\mathbf{k}) + a_3\mathbf{k} \times (b_1\mathbf{i} + b_2\mathbf{j} + b_3\mathbf{k}) \\
&= (a_1 b_2 \mathbf{k} - a_1 b_3 \mathbf{j}) + (-a_2 b_1 \mathbf{k} + a_2 b_3 \mathbf{i}) + (a_3 b_1 \mathbf{j} - a_3 b_2 \mathbf{i}) \\
&= (a_2 b_3 - a_3 b_2)\mathbf{i} - (a_1 b_3 - a_3 b_1)\mathbf{j} + (a_1 b_2 - a_2 b_1)\mathbf{k} \\
&= \begin{vmatrix} \mathbf{i} & \mathbf{j} & \mathbf{k} \\ a_1 & a_2 & a_3 \\ b_1 & b_2 & b_3 \end{vmatrix}
\end{aligned}$$

10. Derive the law of sines of plane trigonometry.

Consider the triangle $ABC$ whose sides are $\mathbf{a}, \mathbf{b}, \mathbf{c}$ of magnitudes $a, b, c$ respectively and whose interior angles are $\alpha, \beta, \gamma$.

We have

$$\mathbf{a} + \mathbf{b} + \mathbf{c} = 0$$

Then     $\mathbf{a} \times (\mathbf{a} + \mathbf{b} + \mathbf{c}) = \mathbf{a} \times \mathbf{b} + \mathbf{a} \times \mathbf{c} = 0$     or     $\mathbf{a} \times \mathbf{b} = \mathbf{c} \times \mathbf{a}$

and      $\mathbf{b} \times (\mathbf{a} + \mathbf{b} + \mathbf{c}) = \mathbf{b} \times \mathbf{a} + \mathbf{b} \times \mathbf{c} = 0$     or     $\mathbf{b} \times \mathbf{c} = \mathbf{a} \times \mathbf{b}$

Thus                     $\mathbf{a} \times \mathbf{b} = \mathbf{b} \times \mathbf{c} = \mathbf{c} \times \mathbf{a}$

$$|\mathbf{a}|\,|\mathbf{b}| \sin \gamma = |\mathbf{b}|\,|\mathbf{c}| \sin \alpha = |\mathbf{c}|\,|\mathbf{a}| \sin \beta$$

$$ab \sin \gamma = bc \sin \alpha = ca \sin \beta$$

and

$$\frac{\sin \gamma}{c} = \frac{\sin \alpha}{a} = \frac{\sin \beta}{b}$$

**11.** If $\mathbf{a} = a_1\mathbf{i} + a_2\mathbf{j} + a_3\mathbf{k}$, $\mathbf{b} = b_1\mathbf{i} + b_2\mathbf{j} + b_3\mathbf{k}$, and $\mathbf{c} = c_1\mathbf{i} + c_2\mathbf{j} + c_3\mathbf{k}$, show that

$$\mathbf{a} \cdot (\mathbf{b} \times \mathbf{c}) = \begin{vmatrix} a_1 & a_2 & a_3 \\ b_1 & b_2 & b_3 \\ c_1 & c_2 & c_3 \end{vmatrix}$$

$$\mathbf{a} \cdot (\mathbf{b} \times \mathbf{c}) = (a_1\mathbf{i} + a_2\mathbf{j} + a_3\mathbf{k}) \cdot \begin{vmatrix} \mathbf{i} & \mathbf{j} & \mathbf{k} \\ b_1 & b_2 & b_3 \\ c_1 & c_2 & c_3 \end{vmatrix}$$

$$= (a_1\mathbf{i} + a_2\mathbf{j} + a_3\mathbf{k}) \cdot [(b_2c_3 - b_3c_2)\mathbf{i} + (b_3c_1 - b_1c_3)\mathbf{j} + (b_1c_2 - b_2c_1)\mathbf{k}]$$

$$= a_1(b_2c_3 - b_3c_2) + a_2(b_3c_1 - b_1c_3) + a_3(b_1c_2 - b_2c_1)$$

$$= \begin{vmatrix} a_1 & a_2 & a_3 \\ b_1 & b_2 & b_3 \\ c_1 & c_2 & c_3 \end{vmatrix}$$

**12.** Show that $\mathbf{a} \cdot (\mathbf{a} \times \mathbf{c}) = 0$.

By (12), $\mathbf{a} \cdot (\mathbf{a} \times \mathbf{c}) = (\mathbf{a} \times \mathbf{a}) \cdot \mathbf{c} = 0$.

**13.** If $\mathbf{a}, \mathbf{b}, \mathbf{c}$ are the vectors of Problem 11, show that $\mathbf{a} \times (\mathbf{b} \times \mathbf{c}) = (\mathbf{a} \cdot \mathbf{c})\mathbf{b} - (\mathbf{a} \cdot \mathbf{b})\mathbf{c}$.

$$\mathbf{a} \times (\mathbf{b} \times \mathbf{c}) = (a_1\mathbf{i} + a_2\mathbf{j} + a_3\mathbf{k}) \times \begin{vmatrix} \mathbf{i} & \mathbf{j} & \mathbf{k} \\ b_1 & b_2 & b_3 \\ c_1 & c_2 & c_3 \end{vmatrix}$$

$$= (a_1\mathbf{i} + a_2\mathbf{j} + a_3\mathbf{k}) \times [(b_2c_3 - b_3c_2)\mathbf{i} + (b_3c_1 - b_1c_3)\mathbf{j} + (b_1c_2 - b_2c_1)\mathbf{k}]$$

$$= \begin{vmatrix} \mathbf{i} & \mathbf{j} & \mathbf{k} \\ a_1 & a_2 & a_3 \\ b_2c_3 - b_3c_2 & b_3c_1 - b_1c_3 & b_1c_2 - b_2c_1 \end{vmatrix}$$

$$= \mathbf{i}(a_2b_1c_2 - a_2b_2c_1 - a_3b_3c_1 + a_3b_1c_3)$$
$$+ \mathbf{j}(a_3b_2c_3 - a_3b_3c_2 - a_1b_1c_2 + a_1b_2c_1)$$
$$+ \mathbf{k}(a_1b_3c_1 - a_1b_1c_3 - a_2b_2c_3 + a_2b_3c_2)$$

$$= \left\{ \begin{matrix} \mathbf{i}b_1(a_1c_1 + a_2c_2 + a_3c_3) \\ + \mathbf{j}b_2(a_1c_1 + a_2c_2 + a_3c_3) \\ + \mathbf{k}b_3(a_1c_1 + a_2c_2 + a_3c_3) \end{matrix} \right. - \left\{ \begin{matrix} \mathbf{i}c_1(a_1b_1 + a_2b_2 + a_3b_3) \\ + \mathbf{j}c_2(a_1b_1 + a_2b_2 + a_3b_3) \\ + \mathbf{k}c_3(a_1b_1 + a_2b_2 + a_3b_3) \end{matrix} \right.$$

$$= (b_1\mathbf{i} + b_2\mathbf{j} + b_3\mathbf{k})(\mathbf{a} \cdot \mathbf{c}) - (c_1\mathbf{i} + c_2\mathbf{j} + c_3\mathbf{k})(\mathbf{a} \cdot \mathbf{b})$$

$$= \mathbf{b}(\mathbf{a} \cdot \mathbf{c}) - \mathbf{c}(\mathbf{a} \cdot \mathbf{b}) = (\mathbf{a} \cdot \mathbf{c})\mathbf{b} - (\mathbf{a} \cdot \mathbf{b})\mathbf{c}$$

as required.

**14.** If $l_1$ and $l_2$ are two non-intersecting lines in space, the shortest distance $d$ between them is the distance from any point on $l_1$ to the plane through $l_2$ and parallel to $l_1$, that is, if $P_1$ is a point on $l_1$ and $P_2$ is a point on $l_2$ then, apart from sign, $d$ is the scalar projection of $\mathbf{P_1P_2}$ on a common perpendicular to $l_1$ and $l_2$.

Let $l_1$ pass through $P_1(x_1, y_1, z_1)$ in the direction $\mathbf{a} = a_1\mathbf{i} + a_2\mathbf{j} + a_3\mathbf{k}$, and $l_2$ pass through $P_2(x_2, y_2, z_2)$ in the direction $\mathbf{b} = b_1\mathbf{i} + b_2\mathbf{j} + b_3\mathbf{k}$.

Then $\mathbf{P_1P_2} = (x_2 - x_1)\mathbf{i} + (y_2 - y_1)\mathbf{j} + (z_2 - z_1)\mathbf{k}$ and the vector $\mathbf{a} \times \mathbf{b}$ is perpendicular to both $l_1$ and $l_2$. Thus,

$$d = \left| \frac{\mathbf{P_1P_2} \cdot (\mathbf{a} \times \mathbf{b})}{|\mathbf{a} \times \mathbf{b}|} \right| = \left| \frac{(\mathbf{r}_2 - \mathbf{r}_1) \cdot (\mathbf{a} \times \mathbf{b})}{|\mathbf{a} \times \mathbf{b}|} \right|$$

15. Write the equation of the line passing through $P_0(1, 2, 3)$ and parallel to $\mathbf{a} = 2\mathbf{i} - \mathbf{j} - 4\mathbf{k}$. Which of the points $A(3, 1, -1)$, $B(1/2, 9/4, 4)$, $C(2, 0, 1)$ are on this line?

From equation (17), the vector equation is

$$(x\mathbf{i} + y\mathbf{j} + z\mathbf{k}) - (\mathbf{i} + 2\mathbf{j} + 3\mathbf{k}) = k(2\mathbf{i} - \mathbf{j} - 4\mathbf{k})$$

or

(i) $$(x - 1)\mathbf{i} + (y - 2)\mathbf{j} + (z - 3)\mathbf{k} = k(2\mathbf{i} - \mathbf{j} - 4\mathbf{k})$$

and the rectangular equation is

(ii) $$\frac{x - 1}{2} = \frac{y - 2}{-1} = \frac{z - 3}{-4}$$

Using (ii), it is readily found that $A$ and $B$ are on the line while $C$ is not.

In the vector equation (i), a point $P(x, y, z)$ on the line is found by giving $k$ a value and comparing components. The point $A$ is on the line since

$$(3 - 1)\mathbf{i} + (1 - 2)\mathbf{j} + (-1 - 3)\mathbf{k} = k(2\mathbf{i} - \mathbf{j} - 4\mathbf{k})$$

when $k = 1$. Similarly $B$ is on the line since

$$-\tfrac{1}{2}\mathbf{i} + \tfrac{1}{4}\mathbf{j} + \mathbf{k} = k(2\mathbf{i} - \mathbf{j} - 4\mathbf{k})$$

when $k = -\tfrac{1}{4}$. The point $C$ is not on the line since

$$\mathbf{i} - 2\mathbf{j} - 2\mathbf{k} = k(2\mathbf{i} - \mathbf{j} - 4\mathbf{k})$$

for no value of $k$.

16. Write the equation of the plane

(a) passing through $P_0(1, 2, 3)$ and parallel to $3x - 2y + 4z - 5 = 0$.

(b) passing through $P_0(1, 2, 3)$ and $P_1(3, -2, 1)$, and perpendicular to the plane $3x - 2y + 4z - 5 = 0$.

(c) through $P_0(1, 2, 3)$, $P_1(3, -2, 1)$ and $P_2(5, 0, -4)$.

Let $P(x, y, z)$ be a general point in the required plane.

(a) Here $\mathbf{a} = 3\mathbf{i} - 2\mathbf{j} + 4\mathbf{k}$ is normal to the given plane and to the required plane. The vector equation of the latter is

$$(\mathbf{r} - \mathbf{r}_0) \cdot \mathbf{a} = 0$$

and the rectangular equation is

$$3(x - 1) - 2(y - 2) + 4(z - 3) = 0$$

or

$$3x - 2y + 4z - 11 = 0$$

(b) Here $\mathbf{r}_1 - \mathbf{r}_0 = 2\mathbf{i} - 4\mathbf{j} - 2\mathbf{k}$ and $\mathbf{a} = 3\mathbf{i} - 2\mathbf{j} + 4\mathbf{k}$ are parallel to the required plane; thus, $(\mathbf{r}_1 - \mathbf{r}_0) \times \mathbf{a}$ is normal to this plane. Its vector equation is

$$(\mathbf{r} - \mathbf{r}_0) \cdot [(\mathbf{r}_1 - \mathbf{r}_0) \times \mathbf{a}] = 0$$

The rectangular equation is

$$(\mathbf{r} - \mathbf{r}_0) \cdot \begin{vmatrix} \mathbf{i} & \mathbf{j} & \mathbf{k} \\ 2 & -4 & -2 \\ 3 & -2 & 4 \end{vmatrix} = [(x - 1)\mathbf{i} + (y - 2)\mathbf{j} + (z - 3)\mathbf{k}] \cdot [-20\mathbf{i} - 14\mathbf{j} + 8\mathbf{k}]$$

$$= -20(x - 1) - 14(y - 2) + 8(z - 3) = 0$$

or

$$20x + 14y - 8z - 24 = 0$$

(c) Here $\mathbf{r}_1 - \mathbf{r}_0 = 2\mathbf{i} - 4\mathbf{j} - 2\mathbf{k}$ and $\mathbf{r}_2 - \mathbf{r}_0 = 4\mathbf{i} - 2\mathbf{j} - 7\mathbf{k}$ are parallel to the required plane so that $(\mathbf{r}_1 - \mathbf{r}_0) \times (\mathbf{r}_2 - \mathbf{r}_0)$ is normal to it. The vector equation is

$$(\mathbf{r} - \mathbf{r}_0) \cdot [(\mathbf{r}_1 - \mathbf{r}_0) \times (\mathbf{r}_2 - \mathbf{r}_0)] = 0$$

and the rectangular equation is

$$(\mathbf{r} - \mathbf{r}_0) \cdot \begin{vmatrix} \mathbf{i} & \mathbf{j} & \mathbf{k} \\ 2 & -4 & -2 \\ 4 & -2 & -7 \end{vmatrix} = [(x - 1)\mathbf{i} + (y - 2)\mathbf{j} + (z - 3)\mathbf{k}] \cdot [24\mathbf{i} + 6\mathbf{j} + 12\mathbf{k}]$$

$$= 24(x - 1) + 6(y - 2) + 12(z - 3) = 0$$

or

$$4x + y + 2z - 12 = 0$$

17. Find the shortest distance $d$ between the point $P_0(1, 2, 3)$ and the plane $\text{II: } 3x - 2y + 5z - 10 = 0$.

A normal to the plane is $\mathbf{a} = 3\mathbf{i} - 2\mathbf{j} + 5\mathbf{k}$. Take $P_1(2, 3, 2)$ any convenient point in $\text{II}$. Then, apart from sign, $d$ is the scalar projection of $\mathbf{P_0 P_1}$ on $\mathbf{a}$. Hence,

$$d = \left| \frac{(\mathbf{r}_1 - \mathbf{r}_0) \cdot \mathbf{a}}{|\mathbf{a}|} \right| = \left| \frac{(\mathbf{i} + \mathbf{j} - \mathbf{k}) \cdot (3\mathbf{i} - 2\mathbf{j} + 5\mathbf{k})}{\sqrt{38}} \right| = \frac{2}{19}\sqrt{38}$$

# Supplementary Problems

18. Find the length of each of the vectors

   (a) $\mathbf{a} = 2\mathbf{i} + 3\mathbf{j} + \mathbf{k}$.

   (b) $\mathbf{b} = 3\mathbf{i} - 5\mathbf{j} + 9\mathbf{k}$.

   (c) $\mathbf{c}$, joining $P_1(3, 4, 5)$ to $P_2(1, -2, 3)$.        Ans. (a) $\sqrt{14}$, (b) $\sqrt{115}$, (c) $2\sqrt{11}$

19. For the vectors of Problem 18

   (a) show that $\mathbf{a}$ and $\mathbf{b}$ are perpendicular.

   (b) find the smaller angle between $\mathbf{a}$ and $\mathbf{c}$; also $\mathbf{b}$ and $\mathbf{c}$.

   (c) find the angles which $\mathbf{b}$ makes with the coordinate axes.
   Ans. (b) $165°14'$, $85°10'$; (c) $73°45'$, $117°47'$, $32°56'$

20. Prove: $\mathbf{i} \cdot \mathbf{i} = \mathbf{j} \cdot \mathbf{j} = \mathbf{k} \cdot \mathbf{k} = 1$; $\mathbf{i} \cdot \mathbf{j} = \mathbf{j} \cdot \mathbf{k} = \mathbf{k} \cdot \mathbf{i} = 0$.

21. Write a unit vector in the direction of $\mathbf{a}$ and a unit vector in the direction of $\mathbf{b}$ of Problem 18.

22. Find the interior angles $\beta$ and $\gamma$ of the triangle of Problem 3.
   Ans. $\beta = 22°12'$, $\gamma = 9°16'$

23. For the unit cube in the adjoining figure, find

   (a) the angle between its diagonal and an edge.

   (b) the angle between its diagonal and a diagonal of a face.
   Ans. (a) $54°44'$, (b) $35°16'$

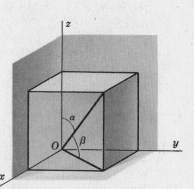

24. Show that the scalar projection of $\mathbf{b}$ onto $\mathbf{a}$ is given by $\dfrac{\mathbf{a} \cdot \mathbf{b}}{|\mathbf{a}|}$.

25. Show that the vector $\mathbf{c}$ of equation ($3$) is perpendicular to both $\mathbf{a}$ and $\mathbf{b}$.

**Fig. 61-14**

26. Given $\mathbf{a} = \mathbf{i} + \mathbf{j}$, $\mathbf{b} = \mathbf{i} - 2\mathbf{k}$, $\mathbf{c} = 2\mathbf{i} + 3\mathbf{j} + 4\mathbf{k}$,  find

   (a) $\mathbf{a} \times \mathbf{b} = -2\mathbf{i} + 2\mathbf{j} - \mathbf{k}$               (e) $\mathbf{a} \cdot (\mathbf{a} \times \mathbf{b}) = 0$

   (b) $\mathbf{b} \times \mathbf{c} = 6\mathbf{i} - 8\mathbf{j} + 3\mathbf{k}$               (f) $\mathbf{a} \cdot (\mathbf{b} \times \mathbf{c}) = -2$

   (c) $\mathbf{c} \times \mathbf{a} = -4\mathbf{i} + 4\mathbf{j} - \mathbf{k}$               (g) $\mathbf{a} \times (\mathbf{b} \times \mathbf{c}) = 3\mathbf{i} - 3\mathbf{j} - 14\mathbf{k}$

   (d) $(\mathbf{a} + \mathbf{b}) \times (\mathbf{a} - \mathbf{b}) = 4\mathbf{i} - 4\mathbf{j} + 2\mathbf{k}$               (h) $\mathbf{c} \times (\mathbf{a} \times \mathbf{b}) = -11\mathbf{i} - 6\mathbf{j} + 10\mathbf{k}$

27. Find the area of the triangle whose vertices are $A(1, 2, 3)$, $B(2, -1, 1)$, $C(-2, 1, -1)$.
   Hint. $|\mathbf{AB} \times \mathbf{AC}| = $ twice the area.        Ans. $5\sqrt{3}$

28. Find the volume of the parallelepiped whose edges are $OA, OB, OC$ when $A(1, 2, 3)$, $B(1, 1, 2)$, $C(2, 1, 1)$.
   Ans. $2$

29. If $\mathbf{u} = \mathbf{a} \times \mathbf{b}$, $\mathbf{v} = \mathbf{b} \times \mathbf{c}$, $\mathbf{w} = \mathbf{c} \times \mathbf{a}$, show that

    (a) $\mathbf{u} \cdot \mathbf{c} = \mathbf{v} \cdot \mathbf{a} = \mathbf{w} \cdot \mathbf{b}$

    (b) $\mathbf{a} \cdot \mathbf{u} = \mathbf{b} \cdot \mathbf{u} = 0$, $\mathbf{b} \cdot \mathbf{v} = \mathbf{c} \cdot \mathbf{v} = 0$, $\mathbf{c} \cdot \mathbf{w} = \mathbf{a} \cdot \mathbf{w} = 0$

    (c) $\mathbf{u} \cdot (\mathbf{v} \times \mathbf{w}) = \{\mathbf{a} \cdot (\mathbf{b} \times \mathbf{c})\}^2$

30. Show that $(\mathbf{a} + \mathbf{b}) \cdot \{(\mathbf{b} + \mathbf{c}) \times (\mathbf{c} + \mathbf{a})\} = 2\mathbf{a} \cdot (\mathbf{b} \times \mathbf{c})$.

31. Find the smaller angle of intersection of the planes $5x - 14y + 2z - 8 = 0$ and $10x - 11y + 2z + 15 = 0$.
    *Hint.* Find the angle between their normals.    *Ans.* $22°25'$

32. Write the vector equation of the line of intersection of the planes $x + y - z - 5 = 0$ and $4x - y - z + 2 = 0$.
    *Ans.* $(x-1)\mathbf{i} + (y-5)\mathbf{j} + (z-1)\mathbf{k} = k(-2\mathbf{i} - 3\mathbf{j} - 5\mathbf{k})$, where $P_0(1, 5, 1)$ is a point on the line.

33. Find the shortest distance between the line through $A(2, -1, -1)$ and $B(6, -8, 0)$ and the line through $C(2, 1, 2)$ and $D(0, 2, -1)$.    *Ans.* $\sqrt{6}/6$

34. Define a line through $P_0(x_0, y_0, z_0)$ as the locus of all points $P(x, y, z)$ such that $\mathbf{P_0P}$ and $\mathbf{OP_0}$ are perpendicular. Show that its vector equation is $(\mathbf{r} - \mathbf{r_0}) \cdot \mathbf{r_0} = 0$.

35. Find the equation of the line through $P_0(2, -3, 5)$ and

    (a) perpendicular to $7x - 4y + 2z - 8 = 0$.

    (b) parallel to the line $x - y + 2z + 4 = 0$, $2x + 3y + 6z - 12 = 0$.

    (c) through $P_1(3, 6, -2)$.

    *Ans.* (a) $\dfrac{x-2}{7} = \dfrac{y+3}{-4} = \dfrac{z-5}{2}$,  (b) $\dfrac{x-2}{12} = \dfrac{y+3}{2} = \dfrac{z-5}{-5}$,  (c) $\dfrac{x-2}{1} = \dfrac{y+3}{9} = \dfrac{z-5}{-7}$

36. Find the equation of the plane

    (a) through $P_0(1, 2, 3)$ and parallel to $\mathbf{a} = 2\mathbf{i} + \mathbf{j} - \mathbf{k}$ and $\mathbf{b} = 3\mathbf{i} + 6\mathbf{j} - 2\mathbf{k}$.

    (b) through $P_0(2, -3, 2)$ and the line $6x + 4y + 3z + 5 = 0$, $2x + y + z - 2 = 0$.

    (c) through $P_0(2, -1, -1)$ and $P_1(1, 2, 3)$ and perpendicular to $2x + 3y - 5z - 6 = 0$.
    *Ans.* (a) $4x + y + 9z - 33 = 0$,  (b) $16x + 7y + 8z - 27 = 0$,  (c) $9x - y + 3z - 16 = 0$

37. If $r_0 = \mathbf{i} + \mathbf{j} + \mathbf{k}$, $r_1 = 2\mathbf{i} + 3\mathbf{j} + 4\mathbf{k}$, and $r_2 = 3\mathbf{i} + 5\mathbf{j} + 7\mathbf{k}$ are three position vectors, show that $\mathbf{r_0} \times \mathbf{r_1} + \mathbf{r_1} \times \mathbf{r_2} + \mathbf{r_2} \times \mathbf{r_0} = 0$. What can be said of the terminal points of these vectors?
    *Ans.* Collinear

38. If $P_0, P_1, P_2$ are three non-collinear points and $\mathbf{r_0}, \mathbf{r_1}, \mathbf{r_2}$ are their position vectors, what is the position of $\mathbf{r_0} \times \mathbf{r_1} + \mathbf{r_1} \times \mathbf{r_2} + \mathbf{r_2} \times \mathbf{r_0}$ with respect to the plane $P_0 P_1 P_2$?    *Ans.* Normal

39. Prove:  (a) $\mathbf{a} \times (\mathbf{b} \times \mathbf{c}) + \mathbf{b} \times (\mathbf{c} \times \mathbf{a}) + \mathbf{c} \times (\mathbf{a} \times \mathbf{b}) = 0$

        (b) $(\mathbf{a} \times \mathbf{b}) \cdot (\mathbf{c} \times \mathbf{d}) = (\mathbf{a} \cdot \mathbf{c})(\mathbf{b} \cdot \mathbf{d}) - (\mathbf{a} \cdot \mathbf{d})(\mathbf{b} \cdot \mathbf{c})$.

40. Prove:  (a) The perpendiculars erected at the midpoints of the sides of a triangle meet in a point.

        (b) The perpendiculars dropped from the vertices to the opposite sides (produced if necessary) of a triangle meet in a point.

41. Let $A(1, 2, 3)$, $B(2, -1, 5)$ and $C(4, 1, 3)$ be three vertices of the parallelogram $ABCD$. Find: (a) the coordinates of $D$, (b) the area of $ABCD$, (c) the area of the orthogonal projection of $ABCD$ on each of the coordinate planes.    *Ans.* (a) $D(3, 4, 1)$, (b) $2\sqrt{26}$, (c) $8, 6, 2$

42. Prove that the area of a parallelogram in space is the square root of the sum of the squares of the areas of projections of the parallelogram on the coordinate planes.

# Chapter 62

## Vector Differentiation and Integration

**DIFFERENTIATION.** Let

$$\begin{aligned}
\mathbf{r} &= \mathbf{i}\,f_1(t) + \mathbf{j}\,f_2(t) + \mathbf{k}\,f_3(t) = \mathbf{i}\,f_1 + \mathbf{j}\,f_2 + \mathbf{k}\,f_3 \\
\mathbf{s} &= \mathbf{i}\,g_1(t) + \mathbf{j}\,g_2(t) + \mathbf{k}\,g_3(t) = \mathbf{i}\,g_1 + \mathbf{j}\,g_2 + \mathbf{k}\,g_3 \\
\mathbf{u} &= \mathbf{i}\,h_1(t) + \mathbf{j}\,h_2(t) + \mathbf{k}\,h_3(t) = \mathbf{i}\,h_1 + \mathbf{j}\,h_2 + \mathbf{k}\,h_3
\end{aligned}$$

be vectors whose components are functions of a single scalar variable $t$ having continuous first and second derivatives.

We can show, as in Chapter 18 for plane vectors, that

$$\frac{d}{dt}(\mathbf{r} \cdot \mathbf{s}) = \frac{d\mathbf{r}}{dt} \cdot s + \mathbf{r} \cdot \frac{d\mathbf{s}}{dt} \tag{1}$$

The reader familiar with the differentiation of determinants whose elements are functions of a single variable will find easily

$$\frac{d}{dt}(\mathbf{r} \times \mathbf{s}) = \frac{d}{dt}\begin{vmatrix} \mathbf{i} & \mathbf{j} & \mathbf{k} \\ f_1 & f_2 & f_3 \\ g_1 & g_2 & g_3 \end{vmatrix} = \begin{vmatrix} \mathbf{i} & \mathbf{j} & \mathbf{k} \\ f_1' & f_2' & f_3' \\ g_1 & g_2 & g_3 \end{vmatrix} + \begin{vmatrix} \mathbf{i} & \mathbf{j} & \mathbf{k} \\ f_1 & f_2 & f_3 \\ g_1' & g_2' & g_3' \end{vmatrix} \tag{2}$$

$$= \frac{d\mathbf{r}}{dt} \times \mathbf{s} + \mathbf{r} \times \frac{d\mathbf{s}}{dt}$$

and

$$\frac{d}{dt}\{\mathbf{r} \cdot (\mathbf{s} \times \mathbf{u})\} = \frac{d\mathbf{r}}{dt} \cdot (\mathbf{s} \times \mathbf{u}) + \mathbf{r} \cdot \left(\frac{d\mathbf{s}}{dt} \times \mathbf{u}\right) + \mathbf{r} \cdot \left(\mathbf{s} \times \frac{d\mathbf{u}}{dt}\right) \tag{3}$$

Otherwise, these formulas may be established by expanding the products before differentiating.

From (2) follows

$$\frac{d}{dt}\{\mathbf{r} \times (\mathbf{s} \times \mathbf{u})\} = \frac{d\mathbf{r}}{dt} \times (\mathbf{s} \times \mathbf{u}) + \mathbf{r} \times \frac{d}{dt}(\mathbf{s} \times \mathbf{u}) \tag{4}$$

$$= \frac{d\mathbf{r}}{dt} \times (\mathbf{s} \times \mathbf{u}) + \mathbf{r} \times \left(\frac{d\mathbf{s}}{dt} \times \mathbf{u}\right) + \mathbf{r} \times \left(\mathbf{s} \times \frac{d\mathbf{u}}{dt}\right)$$

**SPACE CURVES.** Consider the space curve

$$x = f(t), \quad y = g(t), \quad z = h(t) \tag{5}$$

where $f(t), g(t), h(t)$ have continuous first and second derivatives. Let the position vector of a general variable point $P(x, y, z)$ of the curve be given by

$$\mathbf{r} = x\mathbf{i} + y\mathbf{j} + z\mathbf{k}$$

As in Chapter 18, $\mathbf{t} = \dfrac{d\mathbf{r}}{ds}$ is the unit tangent vector to the curve. If $\mathbf{R}$ is the position vector of a point $(X, Y, Z)$ on the tangent line at $P$, the vector equation of this line is (see Chapter 61)

$$\mathbf{R} - \mathbf{r} = k\mathbf{t} \qquad (k, \text{ a scalar variable}) \tag{6}$$

and the rectangular equations are

$$\frac{X-x}{\dfrac{dx}{ds}} = \frac{Y-y}{\dfrac{dy}{ds}} = \frac{Z-z}{\dfrac{dz}{ds}}$$

where $\left[\dfrac{dx}{ds}, \dfrac{dy}{ds}, \dfrac{dz}{ds}\right]$ is a set of direction cosines of the line. In the corresponding equation, (2) of Chapter 59, a set of direction numbers $\left[\dfrac{dx}{dt}, \dfrac{dy}{dt}, \dfrac{dz}{dt}\right]$ was used.

The vector equation of the normal plane to the curve at $P$ is given by

$$(\mathbf{R} - \mathbf{r}) \cdot \mathbf{t} = 0 \tag{7}$$

where $\mathbf{R}$ is the position vector of a general point of the plane.

Again, as in Chapter 18, $\dfrac{d\mathbf{t}}{ds}$ is a vector perpendicular to $\mathbf{t}$. If $\mathbf{n}$ is a unit vector having the direction of $\dfrac{d\mathbf{t}}{ds}$, then

$$\frac{d\mathbf{t}}{ds} = |K|\,\mathbf{n}$$

where $|K|$ is the magnitude of the curvature at $P$. The unit vector

$$\mathbf{n} = \frac{1}{|K|}\frac{d\mathbf{t}}{ds} \tag{8}$$

is called the *principal normal* to the curve at $P$.

The unit vector $\mathbf{b}$ at $P$ defined by

$$\mathbf{b} = \mathbf{t} \times \mathbf{n} \tag{9}$$

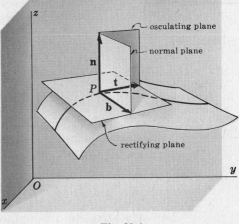

**Fig. 62-1**

is called the *binormal* at $P$. The three vectors $\mathbf{t}, \mathbf{n}, \mathbf{b}$ form at $P$ a right-handed triad of mutually orthogonal vectors which permit a *local* coordinate system in further investigation of a space curve near one of its points.          See Problems 1-2.

At a general point $P$ of a space curve, the vectors $\mathbf{t}, \mathbf{n}, \mathbf{b}$ determine three mutually perpendicular planes:

(i)   the *osculating plane*, containing $\mathbf{t}$ and $\mathbf{n}$, of equation

$$(\mathbf{R} - \mathbf{r}) \cdot \mathbf{b} = 0$$

(ii)  the *normal plane,* containing $\mathbf{n}$ and $\mathbf{b}$, of equation

$$(\mathbf{R} - \mathbf{r}) \cdot \mathbf{t} = 0$$

(iii) the *rectifying plane,* containing $\mathbf{t}$ and $\mathbf{b}$, of equation

$$(\mathbf{R} - \mathbf{r}) \cdot \mathbf{n} = 0$$

In each equation $\mathbf{R}$ is the position vector of a general point in the particular plane.

**SURFACES.** The oldest equation of a surface is $F(x, y, z) = 0$. (See Chapter 59.) A parametric representation results by writing $x, y, z$ as functions of two independent variables or parameters $u$ and $v$; for example,

$$x = f_1(u,v), \quad y = f_2(u,v), \quad z = f_3(u,v) \tag{10}$$

When $u$ is replaced by $u_0$, a constant, (10) becomes

$$x = f_1(u_0, v), \quad y = f_2(u_0, v), \quad z = f_3(u_0, v) \tag{11}$$

the equation of a space curve ($u$-curve) lying on the surface. Similarly, when $v$ is replaced by $v_0$, a constant, (10) becomes

$$x = f_1(u, v_0), \quad y = f_2(u, v_0), \quad z = f_3(u, v_0) \tag{12}$$

the equation of another space curve ($v$-curve) on the surface. The two curves intersect in a point of the surface obtained by setting $u = u_0$, $v = v_0$ simultaneously in (10).

The position vector of a general point $P$ on the surface is given by

$$\mathbf{r} = x\mathbf{i} + y\mathbf{j} + z\mathbf{k} = \mathbf{i}f_1(u, v) + \mathbf{j}f_2(u, v) + \mathbf{k}f_3(u, v) \tag{13}$$

Suppose (11) and (12) are the $u$- and $v$-curves through $P$. Then at $P$

$$\frac{\partial \mathbf{r}}{\partial v} = \mathbf{i}\frac{\partial}{\partial v} f_1(u_0, v) + \mathbf{j}\frac{\partial}{\partial v} f_2(u_0, v) + \mathbf{k}\frac{\partial}{\partial v} f_3(u_0, v)$$

is a vector tangent to the $u$-curve and

$$\frac{\partial \mathbf{r}}{\partial u} = \mathbf{i}\frac{\partial}{\partial u} f_1(u, v_0) + \mathbf{j}\frac{\partial}{\partial u} f_2(u, v_0) + \mathbf{k}\frac{\partial}{\partial u} f_3(u, v_0)$$

is a vector tangent to the $v$-curve there. The two tangents determine a plane which is the tangent plane to the surface at $P$. Clearly, a normal to this plane is given by

$$\frac{\partial \mathbf{r}}{\partial u} \times \frac{\partial \mathbf{r}}{\partial v} \tag{14}$$

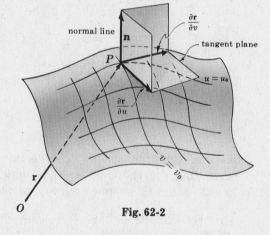

Fig. 62-2

The *unit normal* to the surface at $P$ is defined by

$$\mathbf{n} = \frac{\dfrac{\partial \mathbf{r}}{\partial u} \times \dfrac{\partial \mathbf{r}}{\partial v}}{\left| \dfrac{\partial \mathbf{r}}{\partial u} \times \dfrac{\partial \mathbf{r}}{\partial v} \right|} \tag{15}$$

If $\mathbf{R}$ is the position vector of a general point on the normal to the surface at $P$, its vector equation is

$$(\mathbf{R} - \mathbf{r}) = k\left( \frac{\partial \mathbf{r}}{\partial u} \times \frac{\partial \mathbf{r}}{\partial v} \right) \tag{16}$$

If $\mathbf{R}$ is the position vector of a general point on the tangent plane to the surface at $P$, its vector equation is given by

$$(\mathbf{R} - \mathbf{r}) \cdot \left( \frac{\partial \mathbf{r}}{\partial u} \times \frac{\partial \mathbf{r}}{\partial v} \right) = 0 \tag{17}$$

See Problem 3.

**THE OPERATOR** $\nabla$. In Chapter 60 the directional derivative of $z = f(x, y)$ at an arbitrary point $(x, y)$ and in a direction making an angle $\theta$ with the positive $x$-axis was given by

$$\frac{dz}{ds} = \frac{\partial f}{\partial x} \cos\theta + \frac{\partial f}{\partial y} \sin\theta$$

Let us write

$$\frac{\partial f}{\partial x}\cos\theta \,+\, \frac{\partial f}{\partial y}\sin\theta \;=\; \left(\mathbf{i}\frac{\partial f}{\partial x} + \mathbf{j}\frac{\partial f}{\partial y}\right)\cdot(\mathbf{i}\cos\theta + \mathbf{j}\sin\theta) \tag{18}$$

Now $\mathbf{a} = \mathbf{i}\cos\theta + \mathbf{j}\sin\theta$ is a unit vector whose direction makes the angle $\theta$ with the positive $x$-axis. The other factor, when written as

$$\left(\mathbf{i}\frac{\partial}{\partial x} + \mathbf{j}\frac{\partial}{\partial y}\right)f$$

suggest the definition of a vector differential operator $\nabla$ (del) defined by

$$\nabla \;=\; \mathbf{i}\frac{\partial}{\partial x} + \mathbf{j}\frac{\partial}{\partial y} \tag{19}$$

In vector analysis, $\nabla f = \mathbf{i}\dfrac{\partial f}{\partial x} + \mathbf{j}\dfrac{\partial f}{\partial y}$ is called the *gradient* of $f$ or *grad f*. From (18), we have that the component of $\nabla f$ in the direction of a *unit* vector $\mathbf{a}$ is the directional derivative of $f$ in the direction $\mathbf{a}$.

Let $\mathbf{r} = x\mathbf{i} + y\mathbf{j}$ be the position vector to $P(x, y)$. Since

$$\frac{df}{ds} \;=\; \frac{\partial f}{\partial x}\frac{dx}{ds} + \frac{\partial f}{\partial y}\frac{dy}{ds} \;=\; \left(\mathbf{i}\frac{\partial f}{\partial x} + \mathbf{j}\frac{\partial f}{\partial y}\right)\cdot\left(\mathbf{i}\frac{dx}{ds} + \mathbf{j}\frac{dy}{ds}\right)$$

$$\;=\; \nabla f \cdot \frac{d\mathbf{r}}{ds}$$

and

$$\left|\frac{df}{ds}\right| \;=\; |\nabla f|\cos\phi$$

where $\phi$ is the angle between the vectors $\nabla f$ and $\dfrac{d\mathbf{r}}{ds}$, we see that $\dfrac{df}{ds}$ is maximum when $\cos\phi = 1$, that is, when $\nabla f$ and $\dfrac{d\mathbf{r}}{ds}$ have the same direction. Thus, the maximum value of the directional derivative at $P$ is $|\nabla f|$ and its direction is that of $\nabla f$.

See Problem 4.

For $w = F(x, y, z)$, we define

$$\nabla F \;=\; \mathbf{i}\frac{\partial F}{\partial x} + \mathbf{j}\frac{\partial F}{\partial y} + \mathbf{k}\frac{\partial F}{\partial z}$$

and the directional derivative of $F(x, y, z)$ at an arbitrary point $P(x, y, z)$ in the direction $\mathbf{a} = a_1\mathbf{i} + a_2\mathbf{j} + a_2\mathbf{k}$ by

$$\frac{dF}{ds} \;=\; \nabla F \cdot \mathbf{a} \tag{20}$$

As in the case of functions of two variables, $|\nabla F|$ is the maximum value of the directional derivative of $F(x, y, z)$ at $P(x, y, z)$ and its direction is that of $\nabla F$.

See Problem 5.

Consider now the surface $F(x, y, z) = 0$. The equation of the tangent plane to the surface at one of its points $P_0(x_0, y_0, z_0)$ is given by

$$(x - x_0)\frac{\partial F}{\partial x} + (y - y_0)\frac{\partial F}{\partial y} + (z - z_0)\frac{\partial F}{\partial z}$$

$$\;=\; [(x - x_0)\mathbf{i} + (y - y_0)\mathbf{j} + (z - z_0)\mathbf{k}]\cdot\left[\mathbf{i}\frac{\partial F}{\partial x} + \mathbf{j}\frac{\partial F}{\partial y} + \mathbf{k}\frac{\partial F}{\partial z}\right] \;=\; 0 \tag{21}$$

with the understanding that the partial derivatives are evaluated at $P_0$. The first factor is an arbitrary vector through $P_0$ in the tangent plane; hence the second factor $\nabla F$, evaluated at $P_0$, is normal to the tangent plane, that is, is normal to the surface at $P_0$.

See Problems 6-7.

**DIVERGENCE AND CURL.** The *divergence* of a vector $\mathbf{F} = \mathbf{i} f_1(x, y, z) + \mathbf{j} f_2(x, y, z) + \mathbf{k} f_3(x, y, z)$, or *del dot* $\mathbf{F}$, is defined by

$$\operatorname{div} \mathbf{F} = \nabla \cdot \mathbf{F} = \frac{\partial}{\partial x} f_1 + \frac{\partial}{\partial y} f_2 + \frac{\partial}{\partial z} f_3 \tag{22}$$

The *curl* of the vector $\mathbf{F}$, or *del cross* $\mathbf{F}$, is defined by

$$\operatorname{curl} \mathbf{F} = \nabla \times \mathbf{F} = \begin{vmatrix} \mathbf{i} & \mathbf{j} & \mathbf{k} \\ \frac{\partial}{\partial x} & \frac{\partial}{\partial y} & \frac{\partial}{\partial z} \\ f_1 & f_2 & f_3 \end{vmatrix} \tag{23}$$

$$= \left( \frac{\partial}{\partial y} f_3 - \frac{\partial}{\partial z} f_2 \right) \mathbf{i} + \left( \frac{\partial}{\partial z} f_1 - \frac{\partial}{\partial x} f_3 \right) \mathbf{j} + \left( \frac{\partial}{\partial x} f_2 - \frac{\partial}{\partial y} f_1 \right) \mathbf{k}$$

See Problem 8.

**INTEGRATION.** The study of integration will be limited here to ordinary integration of vectors and to so-called line integrals. As an example of the former, let

$$\mathbf{F}(u) = \mathbf{i} \cos u + \mathbf{j} \sin u + au\mathbf{k}$$

be a vector depending upon the scalar variable $u$. Then

$$\mathbf{F}'(u) = -\mathbf{i} \sin u + \mathbf{j} \cos u + a\mathbf{k}$$

and

$$\int \mathbf{F}'(u)\, du = \int (-\mathbf{i} \sin u + \mathbf{j} \cos u + a\mathbf{k})\, du$$

$$= \mathbf{i} \int -\sin u\, du + \mathbf{j} \int \cos u\, du + \mathbf{k} \int a\, du$$

$$= \mathbf{i} \cos u + \mathbf{j} \sin u + au\mathbf{k} + \mathbf{c}$$

$$= \mathbf{F}(u) + \mathbf{c}$$

where $\mathbf{c}$ is an arbitrary constant vector independent of $u$. Moreover,

$$\int_{u=a}^{u=b} \mathbf{F}'(u)\, du = \Big[ \mathbf{F}(u) + \mathbf{c} \Big]_{u=a}^{u=b} = \mathbf{F}(b) - \mathbf{F}(a)$$

See Problems 9-10.

**LINE INTEGRALS.** Consider in space two points $P_0$ and $P_1$ joined by an arc $C$. The arc may be the segment of a straight line, a portion of a space curve $x = g_1(t)$, $y = g_2(t)$, $z = g_3(t)$ or may consist of several sub-arcs of curves. In any case, $C$ is assumed continuous at each of its points and does not intersect itself. Consider further a vector function

$$\mathbf{F} = \mathbf{F}(x, y, z) = \mathbf{i} f_1(x, y, z) + \mathbf{j} f_2(x, y, z) + \mathbf{k} f_3(x, y, z)$$

which at every point in a region about $C$ and, in particular, at every point of $C$ defines a vector of known magnitude and direction. Denote by

$$\mathbf{r} = x\mathbf{i} + y\mathbf{j} + z\mathbf{k} \tag{24}$$

the position vector of $P(x, y, z)$ on $C$. The integral

$$\int_C^{P_1}{}_{P_0} \left( \mathbf{F} \cdot \frac{d\mathbf{r}}{ds} \right) ds = \int_C^{P_1}{}_{P_0} \mathbf{F} \cdot d\mathbf{r} \tag{25}$$

is called a line integral, that is, an integral along a given path $C$.

As an example, let $\mathbf{F}$ denote a force. The work done by it in moving a particle over $d\mathbf{r}$ is given by (see Problem 9, Chapter 18)

$$|\mathbf{F}|\,|d\mathbf{r}|\cos\theta \;=\; \mathbf{F}\cdot d\mathbf{r}$$

and the work done in moving the particle from $P_0$ to $P_1$ along the arc $C$ is given by

$$\int_{C\,P_0}^{P_1} \mathbf{F}\cdot d\mathbf{r}$$

From (24),

$$d\mathbf{r} \;=\; \mathbf{i}\,dx \,+\, \mathbf{j}\,dy \,+\, \mathbf{k}\,dz$$

and (25) becomes

$$\int_{C\,P_0}^{P_1} \mathbf{F}\cdot d\mathbf{r} \;=\; \int_{C\,P_0}^{P_1} (f_1\,dx \,+\, f_2\,dy \,+\, f_3\,dz) \tag{26}$$

See Problem 11.

# Solved Problems

1. A particle moves along the curve $x = 4\cos t$, $y = 4\sin t$, $z = 6t$. Find the magnitude of the velocity and acceleration at time $t = 0$ and $t = \tfrac{1}{2}\pi$.

Let $P(x, y, z)$ be a point on the curve and

$$\mathbf{r} \;=\; x\mathbf{i} + y\mathbf{j} + z\mathbf{k} \;=\; 4\mathbf{i}\cos t + 4\mathbf{j}\sin t + 6\mathbf{k}\,t$$

be its position vector. Then

$$\mathbf{v} \;=\; \frac{d\mathbf{r}}{dt} \;=\; -4\mathbf{i}\sin t + 4\mathbf{j}\cos t + 6\mathbf{k}$$

$$\mathbf{a} \;=\; \frac{d^2\mathbf{r}}{dt^2} \;=\; -4\mathbf{i}\cos t - 4\mathbf{j}\sin t$$

At $t = 0$:   $\mathbf{v} = 4\mathbf{j} + 6\mathbf{k}$;   $|\mathbf{v}| = \sqrt{16+36} = 2\sqrt{13}$
$\mathbf{a} = -4\mathbf{i}$;   $|\mathbf{a}| = 4$

At $t = \tfrac{1}{2}\pi$:   $\mathbf{v} = -4\mathbf{i} + 6\mathbf{k}$;   $|\mathbf{v}| = \sqrt{16+36} = 2\sqrt{13}$
$\mathbf{a} = -4\mathbf{j}$;   $|\mathbf{a}| = 4$

2. At the point $(1, 1, 1)$ or $t = 1$ of the space curve $x = t$, $y = t^2$, $z = t^3$, find

(a) the equations of the tangent line and normal plane
(b) the unit tangent, principal normal and binormal
(c) the equations of the principal normal and binormal.

$$\mathbf{r} \;=\; t\mathbf{i} + t^2\mathbf{j} + t^3\mathbf{k}$$

$$\frac{d\mathbf{r}}{dt} \;=\; \mathbf{i} + 2t\mathbf{j} + 3t^2\mathbf{k}$$

$$\frac{ds}{dt} \;=\; \left|\frac{d\mathbf{r}}{dt}\right| \;=\; \sqrt{1 + 4t^2 + 9t^4}$$

$$\mathbf{t} \;=\; \frac{d\mathbf{r}}{ds} \;=\; \frac{d\mathbf{r}}{dt}\frac{dt}{ds} \;=\; \frac{\mathbf{i} + 2t\mathbf{j} + 3t^2\mathbf{k}}{\sqrt{1 + 4t^2 + 9t^4}}$$

At $t = 1$:   $\mathbf{r} = \mathbf{i} + \mathbf{j} + \mathbf{k}$ and $\mathbf{t} = \dfrac{1}{\sqrt{14}}(\mathbf{i} + 2\mathbf{j} + 3\mathbf{k})$.

(a) If $\mathbf{R}$ is the position vector of a general point $(X, Y, Z)$ on the tangent line, its vector equation is

$$\mathbf{R} - \mathbf{r} = k\mathbf{t}$$

or

$$(X-1)\mathbf{i} + (Y-1)\mathbf{j} + (Z-1)\mathbf{k} = \frac{k}{\sqrt{14}}(\mathbf{i} + 2\mathbf{j} + 3\mathbf{k})$$

and its rectangular equations are

$$\frac{X-1}{1} = \frac{Y-1}{2} = \frac{Z-1}{3}$$

If $\mathbf{R}$ is the position vector of a general point $(X, Y, Z)$ on the normal plane, its vector equation is

$$(\mathbf{R} - \mathbf{r}) \cdot \mathbf{t} = 0$$

or

$$[(X-1)\mathbf{i} + (Y-1)\mathbf{j} + (Z-1)\mathbf{k}] \cdot \frac{1}{\sqrt{14}}(\mathbf{i} + 2\mathbf{j} + 3\mathbf{k}) = 0$$

and its rectangular equation is

$$(X-1) + 2(Y-1) + 3(Z-1) = X + 2Y + 3Z - 6 = 0$$

[See Problem 2(a), Chapter 59.]

(b)

$$\frac{d\mathbf{t}}{ds} = \frac{d\mathbf{t}}{dt}\frac{dt}{ds} = \frac{(-4t - 18t^3)\mathbf{i} + (2 - 18t^4)\mathbf{j} + (6t + 12t^3)\mathbf{k}}{(1 + 4t^2 + 9t^4)^2}$$

At $t = 1$: $\dfrac{d\mathbf{t}}{ds} = \dfrac{-11\mathbf{i} - 8\mathbf{j} + 9\mathbf{k}}{98}$; $\left|\dfrac{d\mathbf{t}}{ds}\right| = \dfrac{1}{7}\sqrt{\dfrac{19}{14}} = |K|$. Then

$$\mathbf{n} = \frac{1}{|K|}\frac{d\mathbf{t}}{ds} = \frac{-11\mathbf{i} - 8\mathbf{j} + 9\mathbf{k}}{\sqrt{266}}$$

and

$$\mathbf{b} = \mathbf{t} \times \mathbf{n} = \frac{1}{\sqrt{14}\sqrt{266}}\begin{vmatrix} \mathbf{i} & \mathbf{j} & \mathbf{k} \\ 1 & 2 & 3 \\ -11 & -8 & 9 \end{vmatrix} = \frac{1}{\sqrt{19}}(3\mathbf{i} - 3\mathbf{j} + \mathbf{k})$$

(c) If $\mathbf{R}$ is the position vector of a general point $(X, Y, Z)$ on the principal normal, its vector equation is

$$\mathbf{R} - \mathbf{r} = k\mathbf{n}$$

or

$$(X-1)\mathbf{i} + (Y-1)\mathbf{j} + (Z-1)\mathbf{k} = k\frac{-11\mathbf{i} - 8\mathbf{j} + 9\mathbf{k}}{\sqrt{266}}$$

and the rectangular equations are

$$\frac{X-1}{-11} = \frac{Y-1}{-8} = \frac{Z-1}{9}$$

If $\mathbf{R}$ is the position vector of a general point $(X, Y, Z)$ on the binormal, its vector equation is

$$\mathbf{R} - \mathbf{r} = k \cdot \mathbf{b}$$

or

$$(X-1)\mathbf{i} + (Y-1)\mathbf{j} + (Z-1)\mathbf{k} = k\frac{3\mathbf{i} - 3\mathbf{j} + \mathbf{k}}{\sqrt{19}}$$

and the rectangular equations are

$$\frac{X-1}{3} = \frac{Y-1}{-3} = \frac{Z-1}{1}$$

3. Find the equation of the tangent plane and normal line to the surface $x = 2(u + v)$, $y = 3(u - v)$, $z = uv$ at the point $P(u = 2,\ v = 1)$.

$$\mathbf{r} = 2(u + v)\mathbf{i} + 3(u - v)\mathbf{j} + uv\mathbf{k}$$

$$\frac{\partial \mathbf{r}}{\partial u} = 2\mathbf{i} + 3\mathbf{j} + v\mathbf{k}, \qquad \frac{\partial \mathbf{r}}{\partial v} = 2\mathbf{i} - 3\mathbf{j} + u\mathbf{k}$$

At $P$: $\mathbf{r} = 6\mathbf{i} + 3\mathbf{j} + 2\mathbf{k}$, $\dfrac{\partial \mathbf{r}}{\partial u} = 2\mathbf{i} + 3\mathbf{j} + \mathbf{k}$, $\dfrac{\partial \mathbf{r}}{\partial v} = 2\mathbf{i} - 3\mathbf{j} + 2\mathbf{k}$ and

$$\frac{\partial \mathbf{r}}{\partial u} \times \frac{\partial \mathbf{r}}{\partial v} = 9\mathbf{i} - 2\mathbf{j} - 12\mathbf{k}$$

The vector and rectangular equations of the normal line are

$$\mathbf{R} - \mathbf{r} = k\frac{\partial \mathbf{r}}{\partial u} \times \frac{\partial \mathbf{r}}{\partial v}$$

or

$$(X-6)\mathbf{i} + (Y-3)\mathbf{j} + (Z-2)\mathbf{k} = k(9\mathbf{i} - 2\mathbf{j} - 12\mathbf{k})$$

and

$$\frac{X-6}{9} + \frac{Y-3}{-2} = \frac{Z-2}{-12}$$

The vector and rectangular equations of the tangent plane are

$$(\mathbf{R}-\mathbf{r}) \cdot \left(\frac{\partial \mathbf{r}}{\partial u} \times \frac{\partial \mathbf{r}}{\partial v}\right) = 0$$

or

$$[(X-6)\mathbf{i} + (Y-3)\mathbf{j} + (Z-2)\mathbf{k}] \cdot [9\mathbf{i} - 2\mathbf{j} - 12\mathbf{k}] = 0$$

and

$$9X - 2Y - 12Z - 24 = 0$$

4. (a) Find the directional derivative of $f(x,y) = x^2 - 6y^2$ at the point $(7,2)$ in the direction $\theta = \frac{1}{4}\pi$.
(b) Find the maximum value of the directional derivative at $(7,2)$.

(a)
$$\nabla f = \left(\mathbf{i}\frac{\partial}{\partial x} + \mathbf{j}\frac{\partial}{\partial y}\right)(x^2 - 6y^2) = \mathbf{i}\frac{\partial}{\partial x}(x^2 - 6y^2) + \mathbf{j}\frac{\partial}{\partial y}(x^2 - 6y^2)$$
$$= 2x\mathbf{i} - 12y\mathbf{j}$$

and
$$\mathbf{a} = \mathbf{i}\cos\theta + \mathbf{j}\sin\theta = \frac{1}{\sqrt{2}}\mathbf{i} + \frac{1}{\sqrt{2}}\mathbf{j}$$

At $(7,2)$: $\nabla f = 14\mathbf{i} - 24\mathbf{j}$ and

$$\nabla f \cdot \mathbf{a} = (14\mathbf{i} - 24\mathbf{j}) \cdot \left(\frac{1}{\sqrt{2}}\mathbf{i} + \frac{1}{\sqrt{2}}\mathbf{j}\right) = 7\sqrt{2} - 12\sqrt{2} = -5\sqrt{2}$$

is the directional derivative.

(b) At $(7,2)$, $\nabla f = 14\mathbf{i} - 24\mathbf{j}$ and $|\nabla f| = \sqrt{14^2 + 24^2} = 2\sqrt{193}$ is the maximum directional derivative. Since

$$\frac{\nabla f}{|\nabla f|} = \frac{7}{\sqrt{193}}\mathbf{i} - \frac{12}{\sqrt{193}}\mathbf{j} = \mathbf{i}\cos\theta + \mathbf{j}\sin\theta$$

the direction is $\theta = 300°15'$. (See Problems 2 and 6, Chapter 60.)

5. (a) Find the directional derivative of $F(x,y,z) = x^2 - 2y^2 + 4z^2$ at $P(1,1,-1)$ in the direction $\mathbf{a} = 2\mathbf{i} + \mathbf{j} - \mathbf{k}$.
(b) Find the maximum value of the directional derivative at $P$.

$$\nabla F = \left(\mathbf{i}\frac{\partial}{\partial x} + \mathbf{j}\frac{\partial}{\partial y} + \mathbf{k}\frac{\partial}{\partial z}\right)(x^2 - 2y^2 + 4z^2) = 2x\mathbf{i} - 4y\mathbf{j} + 8z\mathbf{k}$$

At $(1,1,-1)$, $\nabla F = 2\mathbf{i} - 4\mathbf{j} - 8\mathbf{k}$.

(a) $\nabla F \cdot \mathbf{a} = (2\mathbf{i} - 4\mathbf{j} - 8\mathbf{k}) \cdot (2\mathbf{i} + \mathbf{j} - \mathbf{k}) = 8$

(b) At $P$, $|\nabla F| = \sqrt{84} = 2\sqrt{21}$. The direction is $\mathbf{a} = 2\mathbf{i} - 4\mathbf{j} - 8\mathbf{k}$.

6. Given the surface $F(x,y,z) = x^3 + 3xyz + 2y^3 - z^3 - 5 = 0$ and one of its points $P_0(1,1,1)$, find
(a) a unit normal to the surface at $P_0$,
(b) the equations of the normal line at $P_0$,
(c) the equation of the tangent plane at $P_0$.

$$\nabla F = (3x^2 + 3yz)\mathbf{i} + (3xz + 6y^2)\mathbf{j} + (3xy - 3z^2)\mathbf{k}$$

At $P_0(1,1,1)$, $\nabla F = 6\mathbf{i} + 9\mathbf{j}$.

(a) $\dfrac{\nabla F}{|\nabla F|} = \dfrac{2}{\sqrt{13}}\mathbf{i} + \dfrac{3}{\sqrt{13}}\mathbf{j}$ is a unit normal at $P_0$; the other is $-\dfrac{2}{\sqrt{13}}\mathbf{i} - \dfrac{3}{\sqrt{13}}\mathbf{j}$.

(b) The equations of the normal line are $\dfrac{X-1}{2} = \dfrac{Y-1}{3}$, $Z = 1$.

(c) The equation of the tangent plane is $2(X-1) + 3(Y-1) = 2X + 3Y - 5 = 0$.

7. Find the angle of intersection of the surfaces

$$F_1 = x^2 + y^2 + z^2 - 9 = 0 \quad \text{and} \quad F_2 = x^2 + 2y^2 - z - 8 = 0$$

at the point $(2, 1, -2)$.

$$\nabla F_1 = \nabla(x^2 + y^2 + z^2 - 9) = 2x\mathbf{i} + 2y\mathbf{j} + 2z\mathbf{k}$$
$$\nabla F_2 = \nabla(x^2 + 2y^2 - z - 8) = 2x\mathbf{i} + 4y\mathbf{j} - \mathbf{k}$$

At $(2, 1, -2)$: $\nabla F_1 = 4\mathbf{i} + 2\mathbf{j} - 4\mathbf{k}$ and $\nabla F_2 = 4\mathbf{i} + 4\mathbf{j} - \mathbf{k}$.

Now $\nabla F_1 \cdot \nabla F_2 = |\nabla F_1|\,|\nabla F_2|\cos\theta$, where $\theta$ is the required angle. Thus,

$$(4\mathbf{i} + 2\mathbf{j} - 4\mathbf{k}) \cdot (4\mathbf{i} + 4\mathbf{j} - \mathbf{k}) = |4\mathbf{i} + 2\mathbf{j} - 4\mathbf{k}|\,|4\mathbf{i} + 4\mathbf{j} - \mathbf{k}|\cos\theta$$

from which $\cos\theta = \frac{14}{99}\sqrt{33} = 0.81236$, and $\theta = 35°40'$.

8. When $\mathbf{B} = xy^2\mathbf{i} + 2x^2yz\mathbf{j} - 3yz^2\mathbf{k}$, find (a) div $\mathbf{B}$, (b) curl $\mathbf{B}$.

(a)
$$\text{div } \mathbf{B} = \nabla \cdot \mathbf{B} = \left(\frac{\partial}{\partial x}\mathbf{i} + \frac{\partial}{\partial y}\mathbf{j} + \frac{\partial}{\partial z}\mathbf{k}\right) \cdot (xy^2\mathbf{i} + 2x^2yz\mathbf{j} - 3yz^2\mathbf{k})$$

$$= \frac{\partial}{\partial x}(xy^2) + \frac{\partial}{\partial y}(2x^2yz) + \frac{\partial}{\partial z}(-3yz^2)$$

$$= y^2 + 2x^2z - 6yz$$

(b)
$$\text{curl } \mathbf{B} = \nabla \times \mathbf{B} = \begin{vmatrix} \mathbf{i} & \mathbf{j} & \mathbf{k} \\ \dfrac{\partial}{\partial x} & \dfrac{\partial}{\partial y} & \dfrac{\partial}{\partial z} \\ xy^2 & 2x^2yz & -3yz^2 \end{vmatrix}$$

$$= \left[\frac{\partial}{\partial y}(-3yz^2) - \frac{\partial}{\partial z}(2x^2yz)\right]\mathbf{i} + \left[\frac{\partial}{\partial z}(xy^2) - \frac{\partial}{\partial x}(-3yz^2)\right]\mathbf{j} + \left[\frac{\partial}{\partial x}(2x^2yz) - \frac{\partial}{\partial y}(xy^2)\right]\mathbf{k}$$

$$= -(3z^2 + 2x^2y)\mathbf{i} + (4xyz - 2xy)\mathbf{k}$$

9. Given $\mathbf{F}(u) = u\mathbf{i} + (u^2 - 2u)\mathbf{j} + (3u^2 + u^3)\mathbf{k}$, find (a) $\int \mathbf{F}(u)\,du$ and (b) $\int_0^1 \mathbf{F}(u)\,du$.

(a)
$$\int \mathbf{F}(u)\,du = \int [u\mathbf{i} + (u^2 - 2u)\mathbf{j} + (3u^2 + u^3)\mathbf{k}]\,du$$

$$= \mathbf{i}\int u\,du + \mathbf{j}\int (u^2 - 2u)\,du + \mathbf{k}\int (3u^2 + u^3)\,du$$

$$= \frac{u^2}{2}\mathbf{i} + \left(\frac{u^3}{3} - u^2\right)\mathbf{j} + \left(u^3 + \frac{u^4}{4}\right)\mathbf{k} + \mathbf{c}$$

where $\mathbf{c} = c_1\mathbf{i} + c_2\mathbf{j} + c_3\mathbf{k}$ with $c_1, c_2, c_3$ arbitrary scalars.

(b)
$$\int_0^1 \mathbf{F}(u)\,du = \left[\frac{u^2}{2}\mathbf{i} + \left(\frac{u^3}{3} - u^2\right)\mathbf{j} + \left(u^3 + \frac{u^4}{4}\right)\mathbf{k}\right]_0^1 = \frac{1}{2}\mathbf{i} - \frac{2}{3}\mathbf{j} + \frac{5}{4}\mathbf{k}$$

10. The acceleration of a particle at any time $t \geq 0$ is given by $\mathbf{a} = \dfrac{d\mathbf{v}}{dt} = e^t\mathbf{i} + e^{2t}\mathbf{j} + \mathbf{k}$. If at $t = 0$, the displacement is $\mathbf{r} = 0$ and the velocity is $\mathbf{v} = \mathbf{i} + \mathbf{j}$, find $\mathbf{r}$ and $\mathbf{v}$ at any time $t$.

$$\mathbf{v} = \int \mathbf{a}\,dt = \mathbf{i}\int e^t\,dt + \mathbf{j}\int e^{2t}\,dt + \mathbf{k}\int dt$$

$$= e^t\mathbf{i} + \tfrac{1}{2}e^{2t}\mathbf{j} + t\mathbf{k} + \mathbf{c}_1$$

At $t = 0$: $\mathbf{v} = \mathbf{i} + \tfrac{1}{2}\mathbf{j} + \mathbf{c}_1 = \mathbf{i} + \mathbf{j}$ and $\mathbf{c}_1 = \tfrac{1}{2}\mathbf{j}$. Then

$$\mathbf{v} = e^t\mathbf{i} + \tfrac{1}{2}(e^{2t} + 1)\mathbf{j} + t\mathbf{k}$$

and
$$\mathbf{r} = \int \mathbf{v}\,dt = e^t\mathbf{i} + (\tfrac{1}{4}e^{2t} + \tfrac{1}{2}t)\mathbf{j} + \tfrac{1}{2}t^2\mathbf{k} + \mathbf{c}_2$$

At $t = 0$: $\mathbf{r} = \mathbf{i} + \tfrac{1}{4}\mathbf{j} + \mathbf{c}_2 = 0$ and $\mathbf{c}_2 = -\mathbf{i} - \tfrac{1}{4}\mathbf{j}$. Thus,

$$\mathbf{r} = (e^t - 1)\mathbf{i} + (\tfrac{1}{4}e^{2t} + \tfrac{1}{2}t - \tfrac{1}{4})\mathbf{j} + \tfrac{1}{2}t^2\mathbf{k}$$

11.  Find the work done by a force  $\mathbf{F} = (x + yz)\mathbf{i} + (y + xz)\mathbf{j} + (z + xy)\mathbf{k}$  in moving a particle from the origin to $C(1, 1, 1)$

(a)  along the straight line $OC$

(b)  along the curve  $x = t,\ y = t^2,\ z = t^3$

(c)  along the straight lines $O$ to $A(1, 0, 0)$, $A$ to $B(1, 1, 0)$, $B$ to $C$.

$$\mathbf{F} \cdot d\mathbf{r} = [(x + yz)\mathbf{i} + (y + xz)\mathbf{j} + (z + xy)\mathbf{k}] \cdot [\mathbf{i}\,dx + \mathbf{j}\,dy + \mathbf{k}\,dz]$$
$$= (x + yz)\,dx + (y + xz)\,dy + (z + xy)\,dz$$

(a)  For the line $OC$,  $x = y = z$ and $dx = dy = dz$.

The integral to be evaluated becomes

$$W = \int_{C\,(0,0,0)}^{(1,1,1)} \mathbf{F} \cdot d\mathbf{r} = 3 \int_0^1 (x + x^2)\,dx = \left[\left(\frac{3}{2}x^2 + x^3\right)\right]_0^1 = \frac{5}{2}$$

(b)  Along the curve:  $x = t,\ dx = dt;\ y = t^2,\ dy = 2t\,dt;\ z = t^3,\ dz = 3t^2\,dt$.  At $O$, $t = 0$; and at $C$, $t = 1$.

$$W = \int_0^1 (t + t^5)\,dt + (t^2 + t^4)2t\,dt + (t^3 + t^3)3t^2\,dt$$

$$= \int_0^1 (t + 2t^3 + 9t^5)\,dt = \left[\frac{1}{2}t^2 + \frac{1}{2}t^4 + \frac{3}{2}t^6\right]_0^1 = \frac{5}{2}$$

(c)  From $O$ to $A$:  $y = z = 0$, $dy = dz = 0$ and $x$ varies from 0 to 1.

From $A$ to $B$:  $x = 1$, $z = 0$, $dx = dz = 0$ and $y$ varies from 0 to 1.

From $B$ to $C$:  $x = y = 1$, $dx = dy = 0$ and $z$ varies from 0 to 1.

Now for the distance from $O$ to $A$,  $W_1 = \int_0^1 x\,dx = \frac{1}{2}$; for the distance from $A$ to $B$,

$W_2 = \int_0^1 y\,dy = \frac{1}{2}$; for the distance from $B$ to $C$,  $W_3 = \int_0^1 (z + 1)dz = \frac{3}{2}$. Thus, $W = W_1 + W_2 + W_3 = 5/2$.

In general, the value of a line integral depends upon the path of integration. Here is an example of one which does not, that is, one which is independent of the path. It can be shown that a line integral $\int_C (f_1\,dx + f_2\,dy + f_3\,dz)$ is independent of the path provided there exists a function $\phi(x, y, z)$ such that

$$d\phi = f_1\,dx + f_2\,dy + f_3\,dz$$

Note that the integrand of this problem is

$$(x + yz)\,dx + (y + xz)\,dy + (z + xy)\,dz = d\{\tfrac{1}{2}(x^2 + y^2 + z^2) + xyz\}$$

# Supplementary Problems

12.  Find $\dfrac{ds}{dt}$ and $\dfrac{d^2s}{dt^2}$, given:

(a)  $\mathbf{s} = (t + 1)\mathbf{i} + (t^2 + t + 1)\mathbf{j} + (t^3 + t^2 + t + 1)\mathbf{k}$

(b)  $\mathbf{s} = \mathbf{i}e^t \cos 2t + \mathbf{j}e^t \sin 2t + t^2\mathbf{k}$

Ans.   (a)  $\mathbf{i} + (2t + 1)\mathbf{j} + (3t^2 + 2t + 1)\mathbf{k};\ 2\mathbf{j} + (6t + 2)\mathbf{k}$

(b)  $e^t(\cos 2t - 2 \sin 2t)\mathbf{i} + e^t(\sin 2t + 2 \cos 2t)\mathbf{j} + 2t\mathbf{k}$
$e^t(-4 \sin 2t - 3 \cos 2t)\mathbf{i} + e^t(-3 \sin 2t + 4 \cos 2t)\mathbf{j} + 2\mathbf{k}$

13.  Given:  $\mathbf{a} = u\mathbf{i} + u^2\mathbf{j} + u^3\mathbf{k}$,  $\mathbf{b} = \mathbf{i} \cos u + \mathbf{j} \sin u$,  $\mathbf{c} = 3u^2\mathbf{i} - 4u\mathbf{k}$.   First compute  $\mathbf{a} \cdot \mathbf{b}$, $\mathbf{a} \times \mathbf{b}$, $\mathbf{a} \cdot (\mathbf{b} \times \mathbf{c})$, $\mathbf{a} \times (\mathbf{b} \times \mathbf{c})$ and find the derivative of each. Then find the derivatives using the formulas.

14. A particle moves along the curve $x = 3t^2$, $y = t^2 - 2t$, $z = t^3$, where $t$ is time. Find (a) the magnitudes of velocity and acceleration at time $t = 1$, (b) the components of velocity and acceleration at time $t = 1$ in the direction $\mathbf{a} = 4\mathbf{i} - 2\mathbf{j} + 4\mathbf{k}$.  *Ans.* (a) $|\mathbf{v}| = 3\sqrt{5}$, $|\mathbf{a}| = 2\sqrt{19}$; (b) 6, 22/3

15. Using vector methods find the equations of the tangent line and normal plane to the curves of Problem 11, Chapter 59.

16. Solve Problem 12, Chapter 59, using vector methods.

17. Show that the surfaces $x = u$, $y = 5u - 3v^2$, $z = v$ and $x = u$, $y = v$, $z = \dfrac{uv}{4u - v}$ are perpendicular at $P(1, 2, 1)$.

18. Using vector methods find the equations of the tangent plane and normal line to the surface.

  (a) $x = u$, $y = v$, $z = uv$ at the point $(u, v) = (3, -4)$.
  (b) $x = u$, $y = v$, $z = u^2 - v^2$ at the point $(u, v) = (2, 1)$.

  *Ans.* (a) $4X - 3Y + Z - 12 = 0$, $\dfrac{X - 3}{-4} = \dfrac{Y + 4}{3} = \dfrac{Z + 12}{-1}$

         (b) $4X - 2Y - Z - 3 = 0$, $\dfrac{X - 2}{-4} = \dfrac{Y - 1}{2} = \dfrac{Z - 3}{1}$

19. (a) Find the equations of the osculating and rectifying plane to the curve of Problem 2 at the given point.

  (b) Find the equations of the osculating, normal and rectifying plane to $x = 2t - t^2$, $y = t^2$, $z = 2t + t^2$ at $t = 1$.

  *Ans.* (a) $3X - 3Y + Z - 1 = 0$, $11X + 8Y - 9Z - 10 = 0$

         (b) $X + 2Y - Z = 0$, $Y + 2Z - 7 = 0$, $5X - 2Y + Z - 6 = 0$

20. Show that the equation of the osculating plane to a space curve at $P$ is also given by

$$(\mathbf{R} - \mathbf{r}) \cdot \left( \frac{d\mathbf{r}}{dt} \times \frac{d^2\mathbf{r}}{dt^2} \right) = 0$$

21. Solve Problems 16 and 17, Chapter 60, using vector methods.

22. Find $\displaystyle \int_a^b \mathbf{F}(u)\, du$,  given

  (a) $\mathbf{F}(u) = u^3\mathbf{i} + (3u^2 - 2u)\mathbf{j} + 3\mathbf{k}$;  $a = 0$, $b = 2$
  (b) $\mathbf{F}(u) = e^u\mathbf{i} + e^{-2u}\mathbf{j} + u\mathbf{k}$;  $a = 0$, $b = 1$
  *Ans.* (a) $4\mathbf{i} + 4\mathbf{j} + 6\mathbf{k}$,  (b) $(e - 1)\mathbf{i} + \frac{1}{2}(1 - e^{-2})\mathbf{j} + \frac{1}{2}\mathbf{k}$

23. The acceleration of a particle at any time $t$ is given by $\mathbf{a} = \dfrac{d\mathbf{v}}{dt} = (t + 1)\mathbf{i} + t^2\mathbf{j} + (t^2 - 2)\mathbf{k}$. If at $t = 0$, the displacement is $\mathbf{r} = 0$ and the velocity is $\mathbf{v} = \mathbf{i} - \mathbf{k}$, find $\mathbf{v}$ and $\mathbf{r}$ at any time $t$.

  *Ans.* $\mathbf{v} = (\frac{1}{2}t^2 + t + 1)\mathbf{i} + \frac{1}{3}t^3\mathbf{j} + (\frac{1}{3}t^3 - 2t - 1)\mathbf{k}$,  $\mathbf{r} = (\frac{1}{6}t^3 + \frac{1}{2}t^2 + t)\mathbf{i} + \frac{1}{12}t^4\mathbf{j} + (\frac{1}{12}t^4 - t^2 - t)\mathbf{k}$

24. In each of the following, find the work done by the given force $\mathbf{F}$ in moving a particle from $O(0, 0, 0)$ to $C(1, 1, 1)$ along (i) the straight line $x = y = z$, (ii) the curve $x = t$, $y = t^2$, $z = t^3$, (iii) the straight lines from $O$ to $A(1, 0, 0)$, $A$ to $B(1, 1, 0)$, $B$ to $C$.

  (a) $\mathbf{F} = x\mathbf{i} + 2y\mathbf{j} + 3z\mathbf{k}$
  (b) $\mathbf{F} = (y + z)\mathbf{i} + (x + z)\mathbf{j} + (x + y)\mathbf{k}$
  (c) $\mathbf{F} = (x + xyz)\mathbf{i} + (y + x^2z)\mathbf{j} + (z + x^2y)\mathbf{k}$
  *Ans.* (a) 3, (b) 3, (c) 9/4, 33/14, 5/2

25. If $\mathbf{r} = x\mathbf{i} + y\mathbf{j} + z\mathbf{k}$, show (a) div $\mathbf{r} = 3$, (b) curl $\mathbf{r} = 0$.

26. If $f = f(x, y, z)$ has partial derivatives of order at least two, show

  (a) $\nabla \times \nabla f = 0$,  (b) $\nabla \cdot (\nabla \times f) = 0$,  (c) $\nabla \cdot \nabla f = \left( \dfrac{\partial^2}{\partial x^2} + \dfrac{\partial^2}{\partial y^2} + \dfrac{\partial^2}{\partial z^2} \right) f$.

# Chapter 63

# Double and Iterated Integrals

**THE (SIMPLE) INTEGRAL** $\displaystyle\int_a^b f(x)\,dx$ of a function $y = f(x)$, continuous over the finite interval $a \le x \le b$ of the $x$-axis, was defined in Chapter 33. We recall

(a)  the interval $a \le x \le b$ was divided into $n$ subintervals $h_1, h_2, \ldots, h_n$ of respective lengths $\Delta_1 x, \Delta_2 x, \ldots, \Delta_n x$ with $\lambda_n$ the greatest of the $\Delta_k x$,

(b)  points $x_1$ in $h_1$, $x_2$ in $h_2$, $\ldots$, $x_n$ in $h_n$ were selected and the sum $\displaystyle\sum_{k=1}^{n} f(x_k)\,\Delta_k x$ formed,

(c)  the interval was further subdivided in such a manner that $\lambda_n \to 0$ as $n \to +\infty$,

(d)  $\displaystyle\int_a^b f(x)\,dx \;=\; \lim_{n \to +\infty} \sum_{k=1}^{n} f(x_k)\,\Delta_k x.$

**THE DOUBLE INTEGRAL.**  Consider a function $z = f(x, y)$ continuous over a finite region $R$ of the $xOy$-plane. Let this region be subdivided (see Fig. 63-1) into $n$ subregions $R_1, R_2, \ldots, R_n$ of respective areas $\Delta_1 A, \Delta_2 A, \ldots, \Delta_n A$. In each subregion $R_k$ select a point $P_k(x_k, y_k)$ and form the sum

$$\sum_{k=1}^{n} f(x_k, y_k)\,\Delta_k A \;=\; f(x_1, y_1)\,\Delta_1 A + f(x_2, y_2)\,\Delta_2 A + \cdots + f(x_n, y_n)\,\Delta_n A \qquad (1)$$

Now defining the diameter of a subregion to be the greatest distance between any two points within or on its boundary and denoting by $\lambda_n$ the maximum diameter of the subregions, suppose the number of subregions to be increased in such a manner that $\lambda_n \to 0$ as $n \to +\infty$. Then the *double integral* of the function $f(x, y)$ over the region $R$ is defined as

$$\iint\limits_{R} f(x, y)\,dA \;=\; \lim_{n \to +\infty} \sum_{k=1}^{n} f(x_k, y_k)\,\Delta_k A \qquad (2)$$

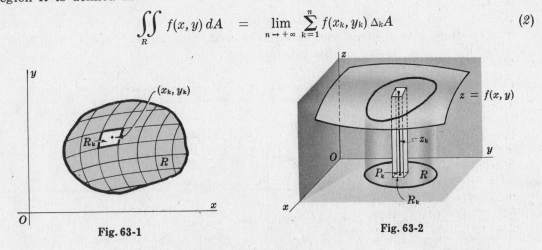

Fig. 63-1          Fig. 63-2

When $z = f(x, y)$ is non-negative over the region $R$, as in Fig. 63-2 above, the double integral (2) may be interpreted as a volume. Any term $f(x_k, y_k)\,\Delta_k A$ of (1) gives the volume of a vertical column whose parallel bases are of area $\Delta_k A$ and whose altitude is the distance $z_k$ measured along the vertical from the selected point $P_k$ to the surface $z = f(x, y)$. This, in turn, may be taken as an approximation of the volume of the vertical column whose lower base is the subregion $R_k$ and whose upper base is the projection of $R_k$ on the surface. Thus, $(1)$ is an approximation of the volume "under the surface", (that is, the volume with lower base in the $xOy$-plane and upper base in the surface generated by moving a line parallel to the $z$-axis along the boundary of $R$), and intuitively, at least, $(2)$ is the measure of this volume.

The evaluation of even the simplest double integral by direct summation is difficult and will not be attempted here.

**THE ITERATED INTEGRAL.** Consider a volume defined as above and assume that the boundary of $R$ is such that no line parallel to the $x$-axis or to the $y$-axis cuts it in more than two points. Draw in (see Fig. 63-3) the tangents $x = a$ and $x = b$ to the boundary with points of tangency $K$ and $L$, and the tangents $y = c$ and $y = d$ with points of tangency $M$ and $N$. Let the equation of the plane arc $LMK$ be $y = g_1(x)$ and that of the plane arc $LNK$ be $y = g_2(x)$.

Divide the interval $a \leqq x \leqq b$ into $m$ subintervals $h_1, h_2, \ldots, h_m$ of respective lengths $\Delta_1 x, \Delta_2 x, \ldots, \Delta_m x$ by the insertion of points $x = \xi_1, \ x = \xi_2, \ \ldots, \ x = \xi_{m-1}$ (as in Chapter 33) and the interval $c \leqq y \leqq d$ into $n$ subintervals $k_1, k_2, \ldots, k_n$ of respective lengths $\Delta_1 y, \Delta_2 y, \ldots, \Delta_n y$ by the insertion of points $y = \eta_1, \ y = \eta_2, \ \ldots, \ y = \eta_{n-1}$. Denote by $\lambda_m$ the greatest $\Delta_i x$ and by $\mu_n$ the greatest $\Delta_j y$. Draw in the parallel lines $x = \xi_1, \ x = \xi_2, \ \ldots, \ x = \xi_{m-1}$ and the parallel lines $y = \eta_1, \ y = \eta_2, \ \ldots, \ y = \eta_{n-1}$ thus dividing the region $R$ into a set of rectangles $R_{ij}$ of areas $\Delta_i x \cdot \Delta_j y$ and a set of non-rectangles which we shall ignore. On each subinterval $h_i$ select a point $x = x_i$ and on each subinterval $k_j$ select a point $y = y_j$, thereby determining in each subregion $R_{ij}$ a point $P_{ij}(x_i, y_j)$. With each subregion $R_{ij}$ associate by means of the equation of the surface a number $z_{ij} = f(x_i, y_j)$ and form the sum

$$\sum_{\substack{i = 1, 2, \ldots, m \\ j = 1, 2, \ldots, n}} f(x_i, y_j) \, \Delta_i x \cdot \Delta_j y \tag{3}$$

Now $(3)$ is merely a special case of $(1)$ so that if the number of rectangles is indefinitely increased in such a manner that both $\lambda_m$ and $\mu_n \to 0$, the limit of $(3)$ should be equal to the double integral $(2)$.

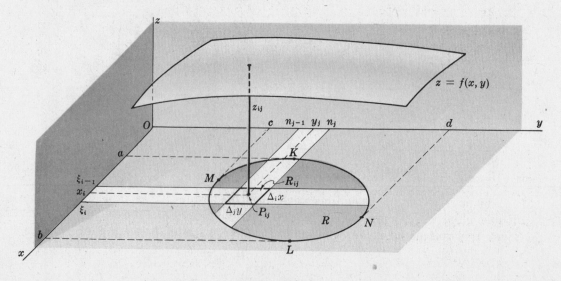

**Fig. 63-3**

In effecting this limit, let us first choose one of the subintervals, say $h_i$, and form the sum

$$\left\{ \sum_{j=1}^{n} f(x_i, y_j)\, \Delta_j y \right\} \Delta_i x, \qquad (i \text{ fixed})$$

of the contributions of all rectangles having $h_i$ as one dimension, that is, the contributions of all rectangles lying in the $i$th column. When $n \to +\infty$, $\mu_n \to 0$, and

$$\lim_{n \to +\infty} \left\{ \sum_{j=1}^{n} f(x_i, y_j)\, \Delta_j y \right\} \Delta_i x = \left\{ \int_{g_1(x_i)}^{g_2(x_i)} f(x_i, y)\, dy \right\} \Delta_i x$$
$$= \phi(x_i)\, \Delta_i x$$

Now summing over the $m$ columns and letting $m \to +\infty$, we have

$$\lim_{m \to +\infty} \sum_{i=1}^{m} \phi(x_i)\, \Delta_i x = \int_a^b \phi(x)\, dx \tag{4}$$
$$= \int_a^b \left[ \int_{g_1(x)}^{g_2(x)} f(x, y)\, dy \right] dx = \int_a^b \int_{g_1(x)}^{g_2(x)} f(x, y)\, dy\, dx$$

Although we shall not use the brackets hereafter, it must be clearly understood at all times that (4) calls for the evaluation of two simple definite integrals in a prescribed order: first, the integral of $f(x, y)$ with respect to $y$ (considering $x$ as a constant) from $y = g_1(x)$, the lower boundary of $R$, to $y = g_2(x)$, the upper boundary of $R$, and then the integral of this result with respect to $x$ from the abscissa $x = a$ of the left-most point of $R$ to the abscissa $x = b$ of the right-most point of $R$. The integral (4) is called an *iterated* or *repeated integral*.

It will be left as an exercise to sum first for the contributions of the rectangles lying in each row and then over all the rows to obtain the equivalent iterated integral

$$\int_c^d \int_{h_1(y)}^{h_2(y)} f(x, y)\, dx\, dy \tag{5}$$

where $x = h_1(y)$ and $x = h_2(y)$ are the equations of the plane arcs *MKN* and *MLN* respectively.

In Problem 1 it is shown by a different procedure that the iterated integral (4) measures the volume under discussion. For the evaluation of iterated integrals see Problems 2-6.

The principal difficulty in setting up the iterated integrals of the next several chapters will be that of inserting the limits of integration to cover the region $R$. The discussion here assumed the simplest of regions; more complex regions are considered in Problems 7-9.

# Solved Problems

1.  Let $z = f(x, y)$ be non-negative and continuous over the region $R$ of the plane $xOy$ whose boundary consists of the arcs of two curves $y = g_1(x)$ and $y = g_2(x)$ intersecting in the points $K$ and $L$, as in Fig. 63-4 below. Consider the volume $V$ under the surface.

    Let the section of this volume by a plane $x = x_i$, where $a < x_i < b$, meet the boundary of $R$ in the points $S[x_i, g_1(x_i)]$ and $T[x_i, g_2(x_i)]$ and the surface $z = f(x, y)$ in the arc $UV$ along which $z = f(x_i, y)$. The area of this section $STUV$ is given by

    $$A(x_i) = \int_{g_1(x_i)}^{g_2(x_i)} f(x_i, y)\, dy$$

Thus, the area of cross sections of the volume by planes parallel to the $yOz$ plane are known functions

$$A(x) = \int_{g_1(x)}^{g_2(x)} f(x, y)\, dy$$

of $x$, the distance of the sectioning plane from the origin. By Chapter 36, the required volume is given by

$$V = \int_a^b A(x)\, dx$$

$$= \int_a^b \left[ \int_{g_1(x)}^{g_2(x)} f(x, y)\, dy \right] dx$$

This is the iterated integral of equation (4).

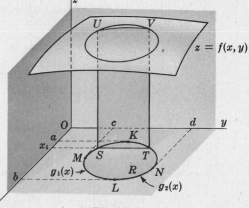

**Fig. 63-4**

2. $\displaystyle \int_0^1 \int_{x^2}^x dy\, dx = \int_0^1 \Big[ y \Big]_{x^2}^x dx = \int_0^1 (x - x^2)\, dx = \left[ \frac{x^2}{2} - \frac{x^3}{3} \right]_0^1 = \frac{1}{6}$

3. $\displaystyle \int_1^2 \int_y^{3y} (x + y)\, dx\, dy = \int_1^2 \left( \frac{1}{2} x^2 + xy \right)\Big]_y^{3y} dy = \int_1^2 6y^2\, dy = 2y^3 \Big]_1^2 = 14$

4. $\displaystyle \int_{-1}^2 \int_{2x^2-2}^{x^2+x} x\, dy\, dx = \int_{-1}^2 (xy)\Big]_{2x^2-2}^{x^2+x} dx = \int_{-1}^2 (x^3 + x^2 - 2x^3 + 2x)\, dx = \frac{9}{4}$

5. $\displaystyle \int_0^\pi \int_0^{\cos\theta} \rho \sin\theta\, d\rho\, d\theta = \int_0^\pi \left( \frac{1}{2}\rho^2 \sin\theta \right)\Big]_0^{\cos\theta} d\theta = \frac{1}{2} \int_0^\pi \cos^2\theta \sin\theta\, d\theta$

$$= \left( -\frac{1}{6}\cos^3\theta \right)\Big]_0^\pi = \frac{1}{3}$$

6. $\displaystyle \int_0^{\pi/2} \int_2^{4\cos\theta} \rho^3\, d\rho\, d\theta = \int_0^{\pi/2} \left( \frac{1}{4}\rho^4 \right)\Big]_2^{4\cos\theta} d\theta = \int_0^{\pi/2} (64\cos^4\theta - 4)\, d\theta$

$$= \left[ 64\left( \frac{3\theta}{8} + \frac{\sin 2\theta}{4} + \frac{\sin 4\theta}{32} \right) - 4\theta \right]_0^{\pi/2} = 10\pi$$

7. Evaluate $\displaystyle \iint_R dA$ where $R$ is the region in the first quadrant bounded by the semi-cubical parabola $y^2 = x^3$ and the line $y = x$.

The line and parabola intersect in the points $(0, 0)$ and $(1, 1)$ which establish the extreme values of $x$ and $y$ on the region $R$.

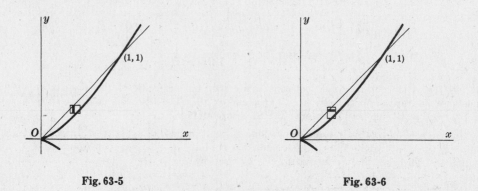

**Fig. 63-5**                    **Fig. 63-6**

*Solution 1.* See Fig. 63-5 above. Integrating first over a horizontal strip, that is, with respect to $x$ from $x = y$ (the line) to $x = y^{2/3}$ (the parabola) and then with respect to $y$ from $y = 0$ to $y = 1$,

$$\iint_R dA \;=\; \int_0^1 \int_y^{y^{2/3}} dx\,dy \;=\; \int_0^1 (y^{2/3} - y)\,dy \;=\; \left[\frac{3}{5}y^{5/3} - \frac{1}{2}y^2\right]_0^1 \;=\; \frac{1}{10}$$

*Solution 2.* See Fig. 63-6 above. Integrating first over a vertical strip, that is, with respect to $y$ from $y = x^{3/2}$ (the parabola) to $y = x$ (the line) and then with respect to $x$ from $x = 0$ to $x = 1$,

$$\iint_R dA \;=\; \int_0^1 \int_{x^{3/2}}^x dy\,dx \;=\; \int_0^1 (x - x^{3/2})\,dx \;=\; \left[\frac{1}{2}x^2 - \frac{2}{5}x^{5/2}\right]_0^1 \;=\; \frac{1}{10}$$

8. Evaluate $\displaystyle\iint_R dA$ where $R$ is the region between $y = 2x$ and $y = x^2$ lying to the left of $x = 1$.

Integrating first over the vertical strip (see Fig. 63-7), we have

$$\iint_R dA \;=\; \int_0^1 \int_{x^2}^{2x} dy\,dx \;=\; \int_0^1 (2x - x^2)\,dx \;=\; \frac{2}{3}$$

When horizontal strips are used (see Fig. 63-8), two iterated integrals are necessary. Let $R_1$ denote the part of $R$ lying below $AB$ and $R_2$ the part above $AB$. Then

$$\iint_R dA \;=\; \iint_{R_1} dA + \iint_{R_2} dA \;=\; \int_0^1 \int_{y/2}^{\sqrt{y}} dx\,dy + \int_1^2 \int_{y/2}^1 dx\,dy \;=\; \frac{5}{12} + \frac{1}{4} \;=\; \frac{2}{3}$$

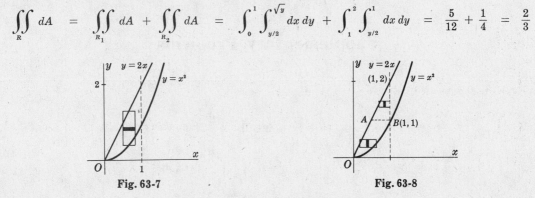

Fig. 63-7　　　　　　　　　　　　　Fig. 63-8

9. Evaluate $\displaystyle\iint_R x^2\,dA$ where $R$ is the region in the first quadrant bounded by the hyperbola $xy = 16$ and the lines $y = x$, $y = 0$, and $x = 8$. See Fig. 63-9 below.

It is evident from Fig. 63-9 that $R$ must be separated into two regions and an iterated integral evaluated for each. Let $R_1$ denote the part of $R$ lying above the line $y = 2$ and $R_2$ the part below that line. Then

$$\iint_R x^2\,dA \;=\; \iint_{R_1} x^2\,dA + \iint_{R_2} x^2\,dA \;=\; \int_2^4 \int_y^{16/y} x^2\,dx\,dy + \int_0^2 \int_y^8 x^2\,dx\,dy$$

$$=\; \frac{1}{3}\int_2^4 \left(\frac{16^3}{y^3} - y^3\right) dy + \frac{1}{3}\int_0^2 (8^3 - y^3)\,dy \;=\; 448$$

As an exercise, the reader will separate $R$ by the line $x = 4$ and obtain

$$\iint_R x^2\,dA \;=\; \int_0^4 \int_0^x x^2\,dy\,dx + \int_4^8 \int_0^{16/x} x^2\,dy\,dx$$

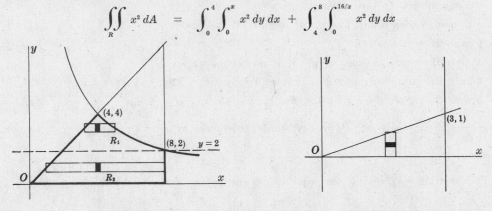

Fig. 63-9　　　　　　　　　　　　　Fig. 63-10

**10.** Evaluate $\displaystyle\int_0^1 \int_{3y}^3 e^{x^2}\, dx\, dy$ by first reversing the order of integration.

The given integral cannot be evaluated directly since $\displaystyle\int e^{x^2}\, dx$ is not an elementary function.

The region $R$ of integration is bounded by the lines $x = 3y$, $x = 3$, and $y = 0$. To reverse the order of integration, first integrate with respect to $y$ from $y = 0$ to $y = x/3$, and then with respect to $x$ from $x = 0$ to $x = 3$. Thus,

$$\int_0^1 \int_{3y}^3 e^{x^2}\, dx\, dy \;=\; \int_0^3 \int_0^{x/3} e^{x^2}\, dy\, dx$$

$$=\; \int_0^3 e^{x^2}\, y\,\Big]_0^{x/3}\, dx \;=\; \frac{1}{3}\int_0^3 e^{x^2}\, x\, dx \;=\; \frac{1}{6}\, e^{x^2}\,\Big]_0^3 \;=\; \frac{1}{6}(e^9 - 1)$$

# Supplementary Problems

**11.** Evaluate the iterated integrals:

(a) $\displaystyle\int_0^1 \int_1^2 dx\, dy \;=\; 1$

(b) $\displaystyle\int_1^2 \int_0^3 (x + y)\, dx\, dy \;=\; 9$

(c) $\displaystyle\int_2^4 \int_1^2 (x^2 + y^2)\, dy\, dx \;=\; \frac{70}{3}$

(d) $\displaystyle\int_0^1 \int_{x^2}^x xy^2\, dy\, dx \;=\; \frac{1}{40}$

(e) $\displaystyle\int_1^2 \int_0^{y^{3/2}} x/y^2\, dx\, dy \;=\; \frac{3}{4}$

(f) $\displaystyle\int_0^1 \int_x^{\sqrt{x}} (y + y^3)\, dy\, dx \;=\; \frac{7}{60}$

(g) $\displaystyle\int_0^1 \int_0^{x^2} xe^y\, dy\, dx \;=\; \frac{1}{2}e - 1$

(h) $\displaystyle\int_2^4 \int_y^{8-y} y\, dx\, dy \;=\; \frac{32}{3}$

(i) $\displaystyle\int_0^{\text{Arc}\tan 3/2} \int_0^{2\sec\theta} \rho\, d\rho\, d\theta \;=\; 3$

(j) $\displaystyle\int_0^{\pi/2} \int_0^2 \rho^2 \cos\theta\, d\rho\, d\theta \;=\; \frac{8}{3}$

(k) $\displaystyle\int_0^{\pi/4} \int_0^{\tan\theta\sec\theta} \rho^3 \cos^2\theta\, d\rho\, d\theta \;=\; \frac{1}{20}$

(l) $\displaystyle\int_0^{2\pi} \int_0^{1-\cos\theta} \rho^3 \cos^2\theta\, d\rho\, d\theta \;=\; \frac{49}{32}\pi$

**12.** Using an iterated integral, evaluate each of the following double integrals. When feasible, evaluate the iterated integral in both orders.

(a) $x$ over the region bounded by $y = x^2$ and $y \doteq x^3$.    *Ans.* 1/20

(b) $y$ over the region of (a).    *Ans.* 1/35

(c) $x^2$ over the region bounded by $y = x$, $y = 2x$, and $x = 2$.    *Ans.* 4

(d) $1$ over each first quadrant region bounded by $2y = x^2$, $y = 3x$, and $x + y = 4$.    *Ans.* 8/3; 46/3

(e) $y$ over the region above $y = 0$ bounded by $y^2 = 4x$ and $y^2 = 5 - x$.    *Ans.* 5

(f) $\displaystyle\frac{1}{\sqrt{2y - y^2}}$ over the region in the first quadrant bounded by $x^2 = 4 - 2y$.    *Ans.* 4

**13.** In Problems 11(a)-(h), reverse the order of integration and evaluate the resulting iterated integrals.

# Chapter 64

## Centroids and Moments of Inertia of Plane Areas

**PLANE AREA BY DOUBLE INTEGRATION.** If $f(x,y) = 1$, the double integral of Chapter 63 becomes $\iint\limits_{R} dA$. In cubic units, this measures the volume of a cylinder of unit height; in square units, it measures the area of the region $R$.

See Problems 1-2.

In polar coordinates, $A = \iint\limits_{R} dA = \int_{\alpha}^{\beta} \int_{\rho_1(\theta)}^{\rho_2(\theta)} \rho \, d\rho \, d\theta$, where $\theta = \alpha$, $\theta = \beta$, $\rho_1(\theta)$, and $\rho_2(\theta)$ are chosen to cover the region $R$.

See Problems 3-5.

**CENTROIDS.** The coordinates $(\bar{x}, \bar{y})$ of the centroid of a plane region $R$ of area $A = \iint\limits_{R} dA$ satisfy the relations

$$A \cdot \bar{x} = M_y \qquad \text{and} \qquad A \cdot \bar{y} = M_x$$

or

$$\bar{x} \cdot \iint\limits_{R} dA = \iint\limits_{R} x \, dA \qquad \text{and} \qquad \bar{y} \cdot \iint\limits_{R} dA = \iint\limits_{R} y \, dA$$

See Problems 6-9.

**THE MOMENT OF INERTIA** of a plane region $R$ with respect to the coordinate axes are given by

$$I_x = \iint\limits_{R} y^2 \, dA \qquad \text{and} \qquad I_y = \iint\limits_{R} x^2 \, dA$$

The polar moment of inertia (the moment of inertia with respect to a line through the origin and perpendicular to the plane of the area) of a plane region $R$ is given by

$$I_0 = I_x + I_y = \iint\limits_{R} (x^2 + y^2) \, dA$$

See Problems 10-12.

## Solved Problems

1.  Find the area bounded by the parabola $y = x^2$ and the line $y = 2x + 3$.

    Using vertical strips (see Fig. 64-1), we have

    $$A = \int_{-1}^{3} \int_{x^2}^{2x+3} dy \, dx$$

    $$= \int_{-1}^{3} (2x + 3 - x^2) \, dx$$

    $$= 32/3 \text{ sq. units}$$

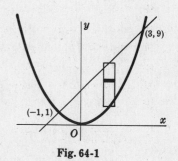

**Fig. 64-1**

2. Find the area bounded by the parabolas $y^2 = 4 - x$ and $y^2 = 4 - 4x$.

Using horizontal strips (see Fig. 64-2) and taking advantage of symmetry,

$$A = 2 \int_0^2 \int_{1-y^2/4}^{4-y^2} dx\, dy$$

$$= 2 \int_0^2 [(4 - y^2) - (1 - \tfrac{1}{4}y^2)]\, dy$$

$$= 6 \int_0^2 (1 - \tfrac{1}{4}y^2)\, dy$$

$$= 8 \text{ sq. units}$$

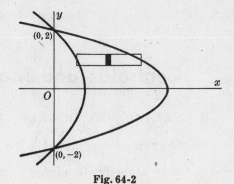

**Fig. 64-2**

3. Find the area outside the circle $\rho = 2$ and inside the cardioid $\rho = 2(1 + \cos\theta)$.

Due to symmetry, the required area is twice that swept over as $\theta$ varies from $\theta = 0$ to $\theta = \tfrac{1}{2}\pi$. Thus

$$A = 2 \int_0^{\pi/2} \int_2^{2(1+\cos\theta)} \rho\, d\rho\, d\theta = 2 \int_0^{\pi/2} \tfrac{1}{2}\rho^2 \Big]_2^{2(1+\cos\theta)} d\theta$$

$$= 4 \int_0^{\pi/2} (2\cos\theta + \cos^2\theta)\, d\theta$$

$$= 4 \left[ 2\sin\theta + \tfrac{1}{2}\theta + \tfrac{1}{4}\sin 2\theta \right]_0^{\pi/2} = (\pi + 8) \text{ sq. un.}$$

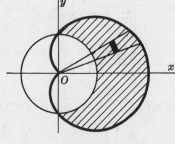

**Fig. 64-3**

4. Find the area inside the circle $\rho = 4\sin\theta$ and outside the lemniscate $\rho^2 = 8\cos 2\theta$.

The required area is twice that in the first quadrant bounded by the two curves and the line $\theta = \tfrac{1}{2}\pi$. Note that the arc $AO$ of the lemniscate is described as $\theta$ varies from $\theta = \pi/6$ to $\theta = \pi/4$, while the arc $AB$ of the circle is described as $\theta$ varies from $\theta = \pi/6$ to $\theta = \pi/2$. This area must then be considered as two regions, one below and one above the line $\theta = \pi/4$. Thus,

$$A = 2 \int_{\pi/6}^{\pi/4} \int_{2\sqrt{2\cos 2\theta}}^{4\sin\theta} \rho\, d\rho\, d\theta + 2 \int_{\pi/4}^{\pi/2} \int_0^{4\sin\theta} \rho\, d\rho\, d\theta$$

$$= \int_{\pi/6}^{\pi/4} (16\sin^2\theta - 8\cos 2\theta)\, d\theta + \int_{\pi/4}^{\pi/2} 16\sin^2\theta\, d\theta$$

$$= (\tfrac{8}{3}\pi + 4\sqrt{3} - 4) \text{ square units}$$

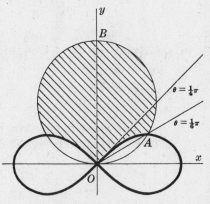

**Fig. 64-4**

5. Evaluate $N = \displaystyle\int_0^{+\infty} e^{-x^2}\, dx$.

Since $\displaystyle\int_0^{+\infty} e^{-x^2}\, dx = \int_0^{+\infty} e^{-y^2}\, dy$,

$$N^2 = \int_0^{+\infty} e^{-x^2}\, dx \cdot \int_0^{+\infty} e^{-y^2}\, dy$$

$$= \int_0^{+\infty} \int_0^{+\infty} e^{-(x^2+y^2)}\, dx\, dy = \iint_R e^{-(x^2+y^2)}\, dA$$

**Fig. 64-5**

Changing to polar coordinates $(x^2 + y^2 = \rho^2,\ dA = \rho\, d\rho\, d\theta)$,

$$N^2 = \int_0^{\pi/2} \int_0^{+\infty} e^{-\rho^2} \cdot \rho\, d\rho\, d\theta = \int_0^{\pi/2} \left\{ \lim_{a \to +\infty} (-\tfrac{1}{2}e^{-\rho^2}) \Big]_0^a \right\} d\theta = \frac{1}{2} \int_0^{\pi/2} d\theta = \frac{\pi}{4}$$

and $N = \sqrt{\pi}/2$.

6.  Find the centroid of the plane area bounded by the parabola $y = 6x - x^2$ and the line $y = x$.

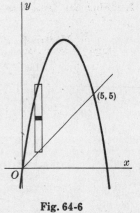

Fig. 64-6

$$A = \iint_R dA = \int_0^5 \int_x^{6x-x^2} dy\, dx = \int_0^5 (5x - x^2)\, dx = \frac{125}{6}$$

$$M_y = \iint_R x\, dA = \int_0^5 \int_x^{6x-x^2} x\, dy\, dx = \int_0^5 (5x^2 - x^3)\, dx = \frac{625}{12}$$

$$M_x = \iint_R y\, dA = \int_0^5 \int_x^{6x-x^2} y\, dy\, dx$$

$$= \frac{1}{2}\int_0^5 \{(6x - x^2)^2 - x^2\}\, dx = \frac{625}{6}$$

Hence, $\bar{x} = \dfrac{M_y}{A} = \dfrac{5}{2}$, $\bar{y} = \dfrac{M_x}{A} = 5$, and the coordinates of the centroid are $(\frac{5}{2}, 5)$.

7.  Find the centroid of the plane area bounded by the parabolas $y = 2x - x^2$ and $y = 3x^2 - 6x$.

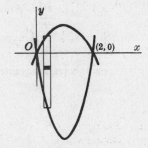

Fig. 64-7

$$A = \iint_R dA = \int_0^2 \int_{3x^2-6x}^{2x-x^2} dy\, dx = \int_0^2 (8x - 4x^2)\, dx = \frac{16}{3}$$

$$M_y = \iint_R x\, dA = \int_0^2 \int_{3x^2-6x}^{2x-x^2} x\, dy\, dx = \int_0^2 (8x^2 - 4x^3)\, dx = \frac{16}{3}$$

$$M_x = \iint_R y\, dA = \int_0^2 \int_{3x^2-6x}^{2x-x^2} y\, dy\, dx$$

$$= \frac{1}{2}\int_0^2 \{(2x - x^2)^2 - (3x^2 - 6x)^2\}\, dx = -\frac{64}{15}$$

Hence, $\bar{x} = \dfrac{M_y}{A} = 1$, $\bar{y} = \dfrac{M_x}{A} = -\dfrac{4}{5}$, and the centroid is $(1, -\frac{4}{5})$.

8.  Find the centroid of the plane area outside the circle $\rho = 1$ and inside the cardioid $\rho = 1 + \cos\theta$. See Fig. 64-8.

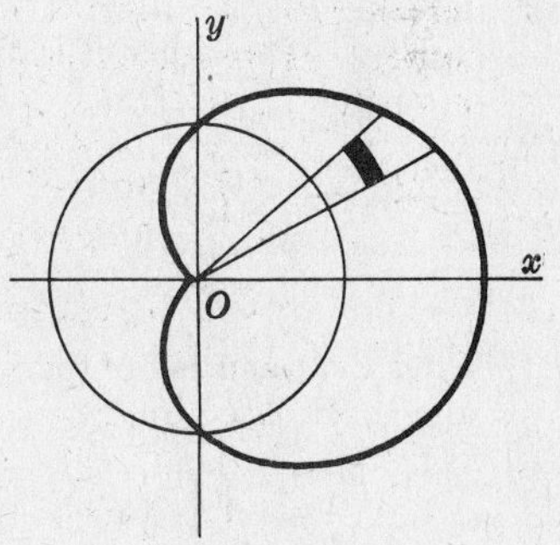

From Fig. 64-8 it is evident that $\bar{y} = 0$ and that $\bar{x}$ is the same whether computed for the given area or for the half lying above the polar axis.  For the latter area,

$$A = \iint_R dA = \int_0^{\pi/2} \int_1^{1+\cos\theta} \rho\, d\rho\, d\theta$$

$$= \frac{1}{2}\int_0^{\pi/2} \{(1 + \cos\theta)^2 - 1^2\}\, d\theta = \frac{\pi + 8}{8}$$

$$M_y = \iint_R x\, dA = \int_0^{\pi/2} \int_1^{1+\cos\theta} (\rho\cos\theta)\, \rho\, d\rho\, d\theta$$

Fig. 64-8

$$= \frac{1}{3}\int_0^{\pi/2} (3\cos^2\theta + 3\cos^3\theta + \cos^4\theta)\, d\theta$$

$$= \frac{1}{3}\left[\frac{3}{2}\theta + \frac{3}{4}\sin 2\theta + 3\sin\theta - \sin^3\theta + \frac{3}{8}\theta + \frac{1}{4}\sin 2\theta + \frac{1}{32}\sin 4\theta\right]_0^{\pi/2} = \frac{15\pi + 32}{48}$$

The coordinates of the centroid are $\left(\dfrac{15\pi + 32}{6(\pi + 8)}, 0\right)$.

9.  Find the centroid of the area inside $\rho = \sin\theta$ and outside $\rho = 1 - \cos\theta$.  See Fig. 64-9 below.

$$A = \iint_R dA = \int_0^{\pi/2} \int_{1-\cos\theta}^{\sin\theta} \rho\, d\rho\, d\theta = \frac{1}{2}\int_0^{\pi/2} (2\cos\theta - 1 - \cos 2\theta)\, d\theta = \frac{4 - \pi}{4}$$

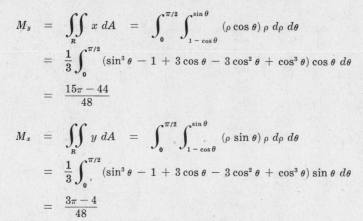

$$M_y = \iint_R x\,dA = \int_0^{\pi/2} \int_{1-\cos\theta}^{\sin\theta} (\rho\cos\theta)\,\rho\,d\rho\,d\theta$$

$$= \frac{1}{3}\int_0^{\pi/2} (\sin^3\theta - 1 + 3\cos\theta - 3\cos^2\theta + \cos^3\theta)\cos\theta\,d\theta$$

$$= \frac{15\pi - 44}{48}$$

$$M_x = \iint_R y\,dA = \int_0^{\pi/2} \int_{1-\cos\theta}^{\sin\theta} (\rho\sin\theta)\,\rho\,d\rho\,d\theta$$

$$= \frac{1}{3}\int_0^{\pi/2} (\sin^3\theta - 1 + 3\cos\theta - 3\cos^2\theta + \cos^3\theta)\sin\theta\,d\theta$$

$$= \frac{3\pi - 4}{48}$$

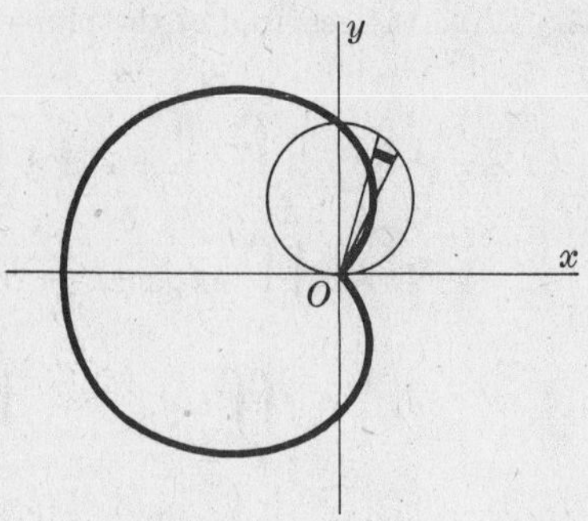

**Fig. 64-9**

The coordinates of the centroid are $\left(\dfrac{15\pi - 44}{12(4 - \pi)},\ \dfrac{3\pi - 4}{12(4 - \pi)}\right)$.

**10.** Find $I_x$, $I_y$ and $I_0$ for the area enclosed by the loop of $y^2 = x^2(2 - x)$.

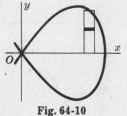

$$A = \iint_R dA = 2\int_0^2 \int_0^{x\sqrt{2-x}} dy\,dx = 2\int_0^2 x\sqrt{2-x}\,dx$$

$$= -4\int_{\sqrt2}^0 (2z^2 - z^4)\,dz = -4\left[\frac{2}{3}z^3 - \frac{1}{5}z^5\right]_{\sqrt2}^0 = \frac{32\sqrt2}{15}$$

using the transformation $2 - x = z^2$.

**Fig. 64-10**

$$I_x = \iint_R y^2\,dA = 2\int_0^2 \int_0^{x\sqrt{2-x}} y^2\,dy\,dx = \frac{2}{3}\int_0^2 x^3(2-x)^{3/2}\,dx$$

$$= -\frac{4}{3}\int_{\sqrt2}^0 (2-z^2)^3 z^4\,dz = -\frac{4}{3}\left[\frac{8}{5}z^5 - \frac{12}{7}z^7 + \frac{2}{3}z^9 - \frac{1}{11}z^{11}\right]_{\sqrt2}^0 = \frac{2048\sqrt2}{3465} = \frac{64}{231}A$$

$$I_y = \iint_R x^2\,dA = 2\int_0^2 \int_0^{x\sqrt{2-x}} x^2\,dy\,dx = 2\int_0^2 x^3\sqrt{2-x}\,dx$$

$$= -4\int_{\sqrt2}^0 (2-z^2)^3 z^2\,dz = -4\left[\frac{8}{3}z^3 - \frac{12}{5}z^5 + \frac{6}{7}z^7 - \frac{1}{9}z^9\right]_{\sqrt2}^0 = \frac{1024\sqrt2}{315} = \frac{32}{21}A$$

$$I_0 = I_x + I_y = \frac{13312\sqrt2}{3465} = \frac{416}{231}A.$$

**11.** Find $I_x$, $I_y$ and $I_0$ for the first quadrant area outside the circle $\rho = 2a$ and inside the circle $\rho = 4a\cos\theta$.

$$A = \iint_R dA = \int_0^{\pi/3} \int_{2a}^{4a\cos\theta} \rho\,d\rho\,d\theta$$

$$= \frac{1}{2}\int_0^{\pi/3} \{(4a\cos\theta)^2 - (2a)^2\}\,d\theta = \frac{2\pi + 3\sqrt3}{3}a^2$$

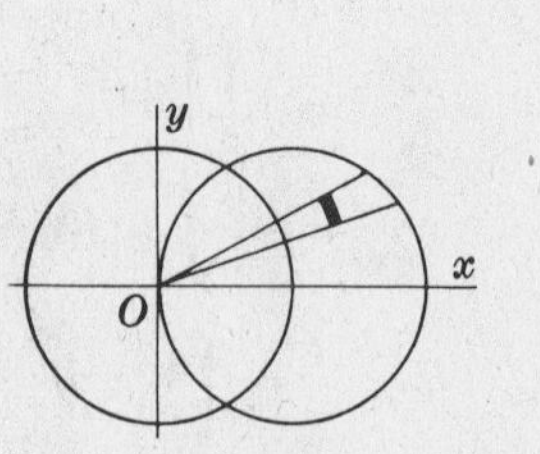

**Fig. 64-11**

$$I_x = \iint_R y^2\,dA = \int_0^{\pi/3} \int_{2a}^{4a\cos\theta} (\rho\sin\theta)^2 \rho\,d\rho\,d\theta = \frac{1}{4}\int_0^{\pi/3} \{(4a\cos\theta)^4 - (2a)^4\}\sin^2\theta\,d\theta$$

$$= 4a^4\int_0^{\pi/3} (16\cos^4\theta - 1)\sin^2\theta\,d\theta = \frac{4\pi + 9\sqrt3}{6}a^4 = \frac{4\pi + 9\sqrt3}{2(2\pi + 3\sqrt3)}a^2A$$

$$I_y = \iint_R x^2\,dA = \int_0^{\pi/3} \int_{2a}^{4a\cos\theta} (\rho\cos\theta)^2 \rho\,d\rho\,d\theta = \frac{12\pi + 11\sqrt3}{2}a^4 = \frac{3(12\pi + 11\sqrt3)}{2(2\pi + 3\sqrt3)}a^2A$$

$$I_0 = I_x + I_y = \frac{20\pi + 21\sqrt3}{3}a^4 = \frac{20\pi + 21\sqrt3}{2\pi + 3\sqrt3}a^2A$$

12. Find $I_x$, $I_y$ and $I_0$ of the area of the circle $\rho = 2(\sin \theta + \cos \theta)$.

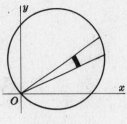

**Fig. 64-12**

Since $x^2 + y^2 = \rho^2$,

$$I_0 = \iint_R (x^2 + y^2)\, dA = \int_{-\frac{1}{4}\pi}^{\frac{3}{4}\pi} \int_0^{2(\sin \theta + \cos \theta)} \rho^2 \cdot \rho\, d\rho\, d\theta$$

$$= 4 \int_{-\frac{1}{4}\pi}^{\frac{3}{4}\pi} (\sin \theta + \cos \theta)^4\, d\theta$$

$$= 4 \left[ \frac{3}{2}\theta - \cos 2\theta - \frac{1}{8}\sin 4\theta \right]_{-1/4\,\pi}^{3/4\,\pi} = 6\pi = 3A$$

It is evident from Fig. 64-12 that $I_x = I_y$. Hence, $I_x = I_y = \frac{1}{2}I_0 = \frac{3}{2}A$.

# Supplementary Problems

13. Use double integration to find the following areas:

   (a) bounded by $3x + 4y = 24$, $x = 0$, $y = 0$.     *Ans.* 24 sq. un.

   (b) bounded by $x + y = 2$, $2y = x + 4$, $y = 0$.     *Ans.* 6 sq. un.

   (c) bounded by $x^2 = 4y$, $8y = x^2 + 16$.     *Ans.* 32/3 sq. un.

   (d) within $\rho = 2(1 - \cos \theta)$.     *Ans.* $6\pi$ sq. un.

   (e) bounded by $\rho = \tan \theta \sec \theta$ and $\theta = \pi/3$.     *Ans.* $\frac{1}{2}\sqrt{3}$ sq. un.

   (f) outside $\rho = 4$ and inside $\rho = 8 \cos \theta$.     *Ans.* $8(\frac{2}{3}\pi + \sqrt{3}\,)$ sq. un.

14. Locate the centroid of the following areas:

   (a) Problem 13(a).     *Ans.* (8/3, 2)

   (b) first quadrant area of Problem 13(c).     *Ans.* (3/2, 8/5)

   (c) in first quadrant bounded by $y^2 = 6x$, $y = 0$, $x = 6$.     *Ans.* (18/5, 9/4)

   (d) bounded by $y^2 = 4x$, $x^2 = 5 - 2y$, $x = 0$.     *Ans.* (13/40, 26/15)

   (e) in first quadrant bounded by $x^2 - 8y + 4 = 0$, $x^2 = 4y$, $x = 0$.     *Ans.* (3/4, 2/5)

   (f) Problem 13(e).     *Ans.* $(\frac{1}{2}\sqrt{3}, 6/5)$

   (g) first quadrant area of Problem 13(f).     *Ans.* $\left( \dfrac{16\pi + 6\sqrt{3}}{2\pi + 3\sqrt{3}}, \dfrac{22}{2\pi + 3\sqrt{3}} \right)$

15. Verify $\dfrac{1}{2} \int_\alpha^\beta [g_2^2(\theta) - g_1^2(\theta)]\, d\theta = \int_\alpha^\beta \int_{g_1(\theta)}^{g_2(\theta)} \rho\, d\rho\, d\theta = \iint_R dA$; then infer

$$\iint_R f(x, y)\, dA = \iint_R f(\rho \cos \theta, \rho \sin \theta)\, \rho\, d\rho\, d\theta$$

16. Find $I_x$ and $I_y$ for the following areas:

   (a) Problem 13(a).     *Ans.* $I_x = 6A$, $I_y = \frac{32}{3}A$

   (b) cut from $y^2 = 8x$ by it latus rectum.     *Ans.* $I_x = \frac{16}{5}A$, $I_y = \frac{12}{7}A$

   (c) bounded by $y = x^2$ and $y = x$.     *Ans.* $I_x = \frac{3}{14}A$, $I_y = \frac{3}{10}A$

   (d) bounded by $y = 4x - x^2$ and $y = x$.     *Ans.* $I_x = \frac{459}{70}A$, $I_y = \frac{27}{10}A$

17. Find $I_x$ and $I_y$ for one loop of $\rho^2 = \cos 2\theta$.     *Ans.* $I_x = \left( \dfrac{\pi}{16} - \dfrac{1}{6} \right) A$, $I_y = \left( \dfrac{\pi}{16} + \dfrac{1}{6} \right) A$

18. Find $I_0$ for the following areas:

   (a) loop of $\rho = \sin 2\theta$.     *Ans.* $\frac{3}{8}A$        (b) enclosed by $\rho = 1 + \cos \theta$.     *Ans.* $\frac{35}{24}A$

# Chapter 65

## Volume Under a Surface
### Double Integration

**THE VOLUME UNDER A SURFACE** $z = f(x, y)$ or $z = f(\rho, \theta)$, that is, the volume of a vertical column whose upper base is in the surface and whose lower base is in the $xOy$ plane is defined by the double integral $V = \iint\limits_{R} z\, dA$, the region $R$ being the lower base of the column.

## Solved Problems

1. Find the volume in the first octant between the planes $z = 0$ and $z = x + y + 2$, and inside the cylinder $x^2 + y^2 = 16$.

   From Fig. 65-1, below, it is evident that $z = x + y + 2$ is to be integrated over a quadrant of the circle $x^2 + y^2 = 16$ in the $xOy$ plane. Hence,

   $$V = \iint\limits_{R} z\, dA = \int_{0}^{4} \int_{0}^{\sqrt{16 - x^2}} (x + y + 2)\, dy\, dx = \int_{0}^{4} (x\sqrt{16 - x^2} + 8 - \tfrac{1}{2}x^2 + 2\sqrt{16 - x^2})\, dx$$

   $$= \left[ -\frac{1}{3}(16 - x^2)^{3/2} + 8x - \frac{x^3}{6} + x\sqrt{16 - x^2} + 16 \arcsin\frac{1}{4}x \right]_{0}^{4} = \left( \frac{128}{3} + 8\pi \right) \text{ cubic units}$$

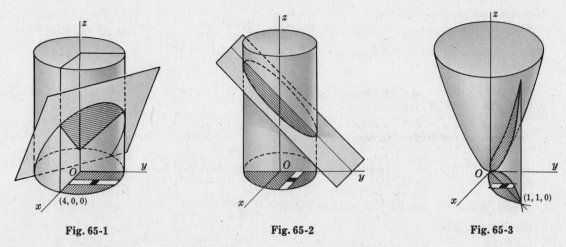

| Fig. 65-1 | Fig. 65-2 | Fig. 65-3 |

2. Find the volume bounded by the cylinder $x^2 + y^2 = 4$ and the planes $y + z = 4$ and $z = 0$.

   From Fig. 65-2, above, it is evident that $z = 4 - y$ is to be integrated over the circle $x^2 + y^2 = 4$ in the $xOy$ plane. Hence,

   $$V = \int_{-2}^{2} \int_{-\sqrt{4 - y^2}}^{\sqrt{4 - y^2}} (4 - y)\, dx\, dy = 2 \int_{-2}^{2} \int_{0}^{\sqrt{4 - y^2}} (4 - y)\, dx\, dy = 16\pi \text{ cubic units}$$

316

3. Find the volume bounded above by the paraboloid $x^2 + 4y^2 = z$, below by the plane $z = 0$, and laterally by the cylinders $y^2 = x$ and $x^2 = y$. See Fig. 65-3 above.

The required volume is obtained by integrating $z = x^2 + 4y^2$ over the region $R$ common to the parabolas $y^2 = x$ and $x^2 = y$ in the $xOy$ plane. Hence,

$$V = \int_0^1 \int_{x^2}^{\sqrt{x}} (x^2 + 4y^2)\, dy\, dx = \int_0^1 \left( x^2 y + \frac{4}{3} y^3 \right)\Big]_{x^2}^{\sqrt{x}} dx = \frac{3}{7} \text{ cubic units}$$

4. Find the volume of one of the wedges cut from the cylinder $4x^2 + y^2 = a^2$ by the planes $z = 0$ and $z = my$. See Fig. 65-4 below.

The volume is obtained by integrating $z = my$ over half the ellipse $4x^2 + y^2 = a^2$. Hence,

$$V = 2 \int_0^{a/2} \int_0^{\sqrt{a^2 - 4x^2}} my\, dy\, dx = m \int_0^{a/2} y^2 \Big]_0^{\sqrt{a^2 - 4x^2}} dx = \frac{ma^3}{3} \text{ cubic units}$$

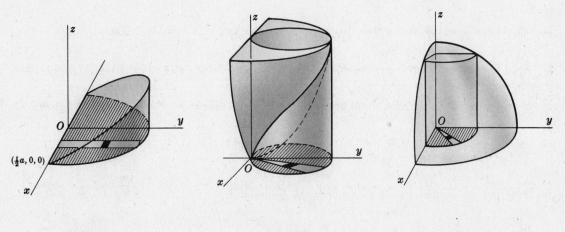

**Fig. 65-4**          **Fig. 65-5**          **Fig. 65-6**

5. Find the volume bounded by the paraboloid $x^2 + y^2 = 4z$, the cylinder $x^2 + y^2 = 8y$, and the plane $z = 0$. See Fig. 65-5 above.

The required volume is obtained by integrating $z = \frac{1}{4}(x^2 + y^2)$ over the circle $x^2 + y^2 = 8y$. Using cylindrical coordinates, the volume is obtained by integrating $z = \frac{1}{4}\rho^2$ over the circle $\rho = 8 \sin \theta$. Then,

$$V = \iint_R z\, dA = \int_0^\pi \int_0^{8 \sin \theta} z\rho\, d\rho\, d\theta = \frac{1}{4} \int_0^\pi \int_0^{8 \sin \theta} \rho^3\, d\rho\, d\theta$$

$$= \frac{1}{16} \int_0^\pi \rho^4 \Big]_0^{8 \sin \theta} d\theta = 256 \int_0^\pi \sin^4 \theta\, d\theta = 96\pi \text{ cubic units}$$

6. Find the volume removed when a hole of radius $a$ is bored through a sphere of radius $2a$, the axis of the hole being a diameter of the sphere. See Fig. 65-6 above.

From the figure, it is seen that the required volume is eight times the volume in the first octant bounded by the cylinder $\rho^2 = a^2$, the sphere $\rho^2 + z^2 = 4a^2$, and plane $z = 0$. The latter volume is obtained by integrating $z = \sqrt{4a^2 - \rho^2}$ over a quadrant of the circle $\rho = a$. Hence,

$$V = 8 \int_0^{\pi/2} \int_0^a \sqrt{4a^2 - \rho^2}\, \rho\, d\rho\, d\theta = \frac{8}{3} \int_0^{\pi/2} (8a^3 - 3\sqrt{3}\, a^3)\, d\theta = \frac{4}{3}(8 - 3\sqrt{3})a^3 \pi \text{ cubic units}$$

# Supplementary Problems

7.  Find the volume cut from $9x^2 + 4y^2 + 36z = 36$ by the plane $z = 0$.     *Ans.* $3\pi$ cubic units

8.  Find the volume under $z = 3x$ and above the first quadrant area bounded by $x = 0$, $y = 0$, $x = 4$, and $x^2 + y^2 = 25$.     *Ans.* 98 cubic units

9.  Find the volume in the first octant bounded by $x^2 + z = 9$, $3x + 4y = 24$, $x = 0$, $y = 0$, and $z = 0$.     *Ans.* 1485/16 cubic units

10. Find the volume in the first octant bounded by $xy = 4z$, $y = x$, and $x = 4$.     *Ans.* 8 cu. un.

11. Find the volume in the first octant bounded by $x^2 + y^2 = 25$ and $z = y$.     *Ans.* 125/3 cu. un.

12. Find the volume common to the cylinders $x^2 + y^2 = 16$ and $x^2 + z^2 = 16$.     *Ans.* 1024/3 cu. un.

13. Find the volume in the first octant inside $y^2 + z^2 = 9$ and outside $y^2 = 3x$.     *Ans.* $27\pi/16$ cu. un.

14. Find the volume in the first octant bounded by $x^2 + z^2 = 16$ and $x - y = 0$.     *Ans.* 64/3 cu. un.

15. Find the volume in front of $x = 0$ and common to $y^2 + z^2 = 4$ and $y^2 + z^2 + 2x = 16$.     *Ans.* $28\pi$ cu. un.

16. Find the volume inside $\rho = 2$ and outside the cone $z^2 = \rho^2$.     *Ans.* $32\pi/3$ cu. un.

17. Find the volume inside $y^2 + z^2 = 2$ and outside $x^2 - y^2 - z^2 = 2$.     *Ans.* $8\pi(4 - \sqrt{2})/3$ cu. un.

18. Find the volume common to $\rho^2 + z^2 = a^2$ and $\rho = a \sin \theta$.     *Ans.* $2(3\pi - 4)a^3/9$ cu. un.

19. Find the volume inside $x^2 + y^2 = 9$, bounded below by $x^2 + y^2 + 4z = 16$ and above by $z = 4$.     *Ans.* $81\pi/8$ cu. un.

20. Find the volume cut from the paraboloid $4x^2 + y^2 = 4z$ by the plane $z - y = 2$.     *Ans.* $9\pi$ cu. un.

21. Find the volume generated by revolving the cardioid $\rho = 2(1 - \cos \theta)$ about the polar axis.
    *Ans.* $V = 2\pi \iint y \rho \, d\rho \, d\theta = 64\pi/3$ cu. un.

22. Find the volume generated by revolving a petal of $\rho = \sin 2\theta$ about either axis.
    *Ans.* $32\pi/105$ cu. un.

23. A square hole 2 units on a side is cut symmetrically through a sphere of radius 2 units. Show that the volume removed is $\frac{4}{3}(2\sqrt{2} + 19\pi - 54 \text{ Arc tan } \sqrt{2})$ cubic units.

# Chapter 66

## Area of a Curved Surface
### Double Integration

**IN COMPUTING THE LENGTH OF AN ARC,** (1) the arc is projected on a convenient coordinate axis, thus establishing an interval on the axis, and (2) an integrand function, $\sqrt{1 + \left(\dfrac{dy}{dx}\right)^2}$ if the projection is on the $x$-axis or $\sqrt{1 + \left(\dfrac{dx}{dy}\right)^2}$ if the projection is on the $y$-axis, is integrated over the interval.

A similar procedure is used to compute the area $S$ of a portion $R'$ of a surface $z = f(x, y)$:

(1) $R'$ is projected on a convenient coordinate plane, thus establishing a region $R$ on the plane.

(2) An integrand function is integrated over $R$.

If $R'$ is projected on $xOy$, $\quad S = \displaystyle\iint\limits_R \sqrt{1 + \left(\frac{\partial z}{\partial x}\right)^2 + \left(\frac{\partial z}{\partial y}\right)^2}\, dA.$

If $R'$ is projected on $yOz$, $\quad S = \displaystyle\iint\limits_R \sqrt{1 + \left(\frac{\partial x}{\partial y}\right)^2 + \left(\frac{\partial x}{\partial z}\right)^2}\, dA.$

If $R'$ is projected on $zOx$, $\quad S = \displaystyle\iint\limits_R \sqrt{1 + \left(\frac{\partial y}{\partial x}\right)^2 + \left(\frac{\partial y}{\partial z}\right)^2}\, dA.$

## Solved Problems

1.  Consider a region $R'$ of area $S$ on the surface $z = f(x, y)$. Through the boundary of $R'$ pass a vertical cylinder (see Fig. 66-1) cutting the $xOy$ plane in the region $R$. Now divide $R$ into $n$ subregions $\Delta A_i$ (of areas $\Delta A_i$) and denote by $\Delta S_i$ the area of the projection of $\Delta A_i$ on $R'$. In each subregion $\Delta S_i$, choose a point $P_i$ and draw there the tangent plane to the surface. Let the area of the projection of $\Delta A_i$ on this tangent plane be denoted by $\Delta T_i$. We shall use $\Delta T_i$ as an approximation of the corresponding surface area $\Delta S_i$.

Now the angle between the $xOy$ plane and the tangent plane at $P_i$ is the angle $\gamma_i$ between the $z$-axis, $[0, 0, 1]$, and the normal, $\left[-\dfrac{\partial f}{\partial x}, -\dfrac{\partial f}{\partial y}, 1\right] = \left[-\dfrac{\partial z}{\partial x}, -\dfrac{\partial z}{\partial y}, 1\right]$, to the surface at $P_i$; thus

$$\cos \gamma_i = \frac{1}{\sqrt{\left(\dfrac{\partial z}{\partial x}\right)^2 + \left(\dfrac{\partial z}{\partial y}\right)^2 + 1}}$$

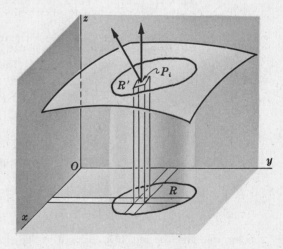

**Fig. 66-1**

319

Then (see Fig. 66-2),

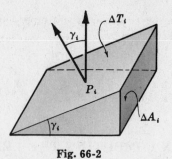

$$\Delta T_i \cdot \cos \gamma_i = \Delta A_i \qquad \text{and} \qquad \Delta T_i = \sec \gamma_i \cdot \Delta A_i$$

Hence, an approximation of $S$ is $\displaystyle\sum_{i=1}^{n} \Delta T_i = \sum_{i=1}^{n} \sec \gamma_i \cdot \Delta A_i$, and

$$S = \lim_{n \to +\infty} \sum_{i=1}^{n} \sec \gamma_i \cdot \Delta A_i = \iint_R \sec \gamma \cdot dA$$

$$= \iint_R \sqrt{\left(\frac{\partial z}{\partial x}\right)^2 + \left(\frac{\partial z}{\partial y}\right)^2 + 1}\, dA$$

**Fig. 66-2**

2. Find the area of the portion of the cone $x^2 + y^2 = 3z^2$ lying above the $xOy$ plane and inside the cylinder $x^2 + y^2 = 4y$.

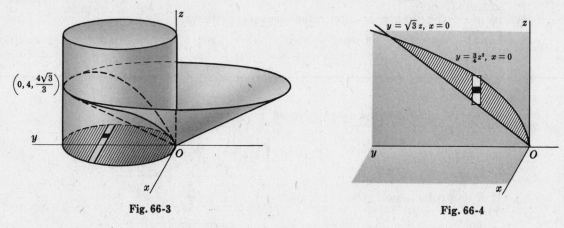

**Fig. 66-3**                                        **Fig. 66-4**

*Solution 1.* Refer to Fig. 66-3 above. The projection of the required area on the $xOy$ plane is the region $R$ enclosed by the circle $x^2 + y^2 = 4y$. For the cone,

$$\frac{\partial z}{\partial x} = \frac{1}{3} \cdot \frac{x}{z}, \quad \frac{\partial z}{\partial y} = \frac{1}{3} \cdot \frac{y}{z}, \quad \text{and} \quad 1 + \left(\frac{\partial z}{\partial x}\right)^2 + \left(\frac{\partial z}{\partial y}\right)^2 = \frac{9z^2 + x^2 + y^2}{9z^2} = \frac{12z^2}{9z^2} = \frac{4}{3}$$

$$S = \iint_R \sqrt{1 + \left(\frac{\partial z}{\partial x}\right)^2 + \left(\frac{\partial z}{\partial y}\right)^2}\, dA = \int_0^4 \int_{-\sqrt{4y-y^2}}^{\sqrt{4y-y^2}} \frac{2}{\sqrt{3}}\, dx\, dy = 2\left(\frac{2}{\sqrt{3}}\right) \int_0^4 \int_0^{\sqrt{4y-y^2}} dx\, dy$$

$$= \frac{4}{\sqrt{3}} \int_0^4 \sqrt{4y - y^2}\, dy = \frac{8\sqrt{3}}{3}\pi \text{ square units}$$

*Solution 2.* Refer to Fig. 66-4 above. The projection of one half the required area on the $yOz$ plane is the region $R$ bounded by the line $y = \sqrt{3}z$ and the parabola $y = \frac{3}{4}z^2$, the latter obtained by eliminating $x$ between the equations of the two surfaces. For the cone,

$$\frac{\partial x}{\partial y} = -\frac{y}{x}, \quad \frac{\partial x}{\partial z} = \frac{3z}{x}, \quad \text{and} \quad 1 + \left(\frac{\partial x}{\partial y}\right)^2 + \left(\frac{\partial x}{\partial z}\right)^2 = \frac{x^2 + y^2 + 9z^2}{x^2} = \frac{12z^2}{x^2} = \frac{12z^2}{3z^2 - y^2}$$

Hence,

$$S = 2\int_0^4 \int_{y/\sqrt{3}}^{2\sqrt{y}/\sqrt{3}} \frac{2\sqrt{3}\,z}{\sqrt{3z^2 - y^2}}\, dz\, dy = \frac{4\sqrt{3}}{3} \int_0^4 \sqrt{3z^2 - y^2}\Big]_{y/\sqrt{3}}^{2\sqrt{y}/\sqrt{3}}\, dy = \frac{4\sqrt{3}}{3} \int_0^4 \sqrt{4y - y^2}\, dy$$

*Solution 3.* Using cylindrical coordinates in *Solution 1*, $\sqrt{1 + \left(\frac{\partial z}{\partial x}\right)^2 + \left(\frac{\partial z}{\partial y}\right)^2} = \frac{2}{\sqrt{3}}$ is to be integrated over the region $R$ enclosed by the circle $\rho = 4\sin\theta$. Then,

$$S = \iint_R \frac{2}{\sqrt{3}}\, dA = \int_0^\pi \int_0^{4\sin\theta} \frac{2}{\sqrt{3}} \rho\, d\rho\, d\theta = \frac{1}{\sqrt{3}} \int_0^\pi \rho^2 \Big]_0^{4\sin\theta}\, d\theta$$

$$= \frac{16}{\sqrt{3}} \int_0^\pi \sin^2\theta\, d\theta = \frac{8\sqrt{3}}{3}\pi \text{ square units}$$

**3.** Find the area of the portion of the cylinder  $x^2 + z^2 = 16$  lying inside the cylinder  $x^2 + y^2 = 16$.

Fig. 66-5 shows one-eighth of the required area, its projection on the $xOy$ plane being a quadrant of the circle  $x^2 + y^2 = 16$.  For the cylinder  $x^2 + z^2 = 16$,

$$\frac{\partial z}{\partial x} = -\frac{x}{z}, \quad \frac{\partial z}{\partial y} = 0, \quad \text{and} \quad 1 + \left(\frac{\partial z}{\partial x}\right)^2 + \left(\frac{\partial z}{\partial y}\right)^2 = \frac{x^2 + z^2}{z^2} = \frac{16}{16 - x^2}$$

Then,

$$S = 8 \int_0^4 \int_0^{\sqrt{16 - x^2}} \frac{4}{\sqrt{16 - x^2}} \, dy \, dx = 32 \int_0^4 dx = 128 \text{ square units}$$

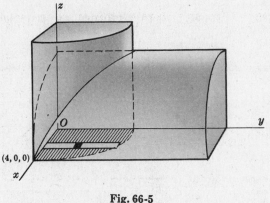

(4, 0, 0)

**Fig. 66-5**

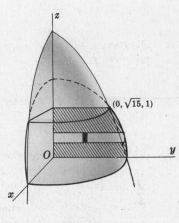

(0, √15, 1)

**Fig. 66-6**

**4.** Find the area of the portion of the sphere  $x^2 + y^2 + z^2 = 16$  outside the paraboloid  $x^2 + y^2 + z = 16$.

Fig. 66-6, above, shows one-fourth of the required area, its projection on the $yOz$ plane being the region $R$ bounded by the circle  $y^2 + z^2 = 16$,  the $y$ and $z$ axes, and the line $z = 1$.  For the sphere,

$$\frac{\partial x}{\partial y} = -\frac{y}{x}, \quad \frac{\partial x}{\partial z} = -\frac{z}{x}, \quad \text{and} \quad 1 + \left(\frac{\partial x}{\partial y}\right)^2 + \left(\frac{\partial x}{\partial z}\right)^2 = \frac{x^2 + y^2 + z^2}{x^2} = \frac{16}{16 - y^2 - z^2}$$

Then,

$$S = 4 \iint_R \sqrt{1 + \left(\frac{\partial x}{\partial y}\right)^2 + \left(\frac{\partial x}{\partial z}\right)^2} \, dA = 4 \int_0^1 \int_0^{\sqrt{16 - z^2}} \frac{4}{\sqrt{16 - y^2 - z^2}} \, dy \, dz$$

$$= 16 \int_0^1 \arcsin \frac{y}{\sqrt{16 - z^2}} \bigg]_0^{\sqrt{16 - z^2}} dz = 16 \int_0^1 \frac{1}{2}\pi \, dz = 8\pi \text{ square units}$$

**5.** Find the area of the portion of the cylinder  $x^2 + y^2 = 6y$  lying inside the sphere  $x^2 + y^2 + z^2 = 36$.

Fig. 66-7 shows one-fourth of the required area. Its projection on the $yOz$ plane is the region $R$ bounded by the $z$ and $y$ axes and the parabola $z^2 + 6y = 36$, the latter obtained by eliminating $x$ from the equations of the two surfaces.  For the cylinder,

$$\frac{\partial x}{\partial y} = \frac{3 - y}{x}, \quad \frac{\partial x}{\partial z} = 0$$

and

$$1 + \left(\frac{\partial x}{\partial y}\right)^2 + \left(\frac{\partial x}{\partial z}\right)^2 = \frac{x^2 + 9 - 6y + y^2}{x^2} = \frac{9}{6y - y^2}$$

Then,

$$S = 4 \int_0^6 \int_0^{\sqrt{36 - 6y}} \frac{3}{\sqrt{6y - y^2}} \, dz \, dy$$

$$= 12 \int_0^6 \frac{\sqrt{6}}{\sqrt{y}} \, dy = 144 \text{ square units}$$

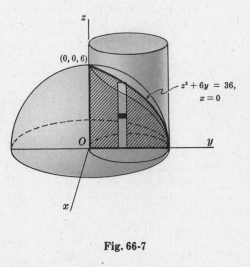

(0, 0, 6)

$z^2 + 6y = 36$, $x = 0$

**Fig. 66-7**

# Supplementary Problems

6. Find the area of the portion of the cone $x^2 + y^2 = z^2$ inside the vertical prism whose base is the triangle bounded by the lines $y = x$, $x = 0$, and $y = 1$ in the $xOy$ plane.      *Ans.* $\frac{1}{2}\sqrt{2}$ square units

7. Find the area of the portion of the plane $x + y + z = 6$ inside the cylinder $x^2 + y^2 = 4$.
   *Ans.* $4\sqrt{3}\,\pi$ sq. un.

8. Find the area of the portion of the sphere $x^2 + y^2 + z^2 = 36$ inside the cylinder $x^2 + y^2 = 6y$.
   *Ans.* $72(\pi - 2)$ sq. un.

9. Find the area of the portion of the sphere $x^2 + y^2 + z^2 = 4z$ inside the paraboloid $x^2 + y^2 = z$.
   *Ans.* $4\pi$ sq. un.

10. Find the area of the portion of the sphere $x^2 + y^2 + z^2 = 25$ between the planes $z = 2$ and $z = 4$.
    *Ans.* $20\pi$ sq. un.

11. Find the area of the portion of the surface $z = xy$ inside the cylinder $x^2 + y^2 = 1$.
    *Ans.* $2\pi(2\sqrt{2} - 1)/3$ sq. un.

12. Find the area of the surface of the cone $x^2 + y^2 - 9z^2 = 0$ above the plane $z = 0$ and inside the cylinder $x^2 + y^2 = 6y$.      *Ans.* $3\sqrt{10}\,\pi$ sq. un.

13. Find the area of that part of the sphere $x^2 + y^2 + z^2 = 25$ within the elliptic cylinder $2x^2 + y^2 = 25$.
    *Ans.* $50\pi$ sq. un.

14. Find the area of the surface of $x^2 + y^2 - az = 0$ which lies directly above the lemniscate
    $4\rho^2 = a^2 \cos 2\theta$.      *Ans.* $S = \dfrac{4}{a} \displaystyle\iint \sqrt{4\rho^2 + a^2}\,\rho\,d\rho\,d\theta = \dfrac{a^2}{3}\left\{\dfrac{5}{3} - \dfrac{\pi}{4}\right\}$ sq. un.

15. Find the area of the surface of $x^2 + y^2 + z^2 = 4$ which lies directly above the cardioid $\rho = 1 - \cos\theta$.
    *Ans.* $8[\pi - \sqrt{2} - \ln(\sqrt{2} + 1)]$ sq. un.

# Chapter 67

## Triple Integrals

**THE TRIPLE INTEGRAL** $\iiint\limits_{R} f(x,y,z)\, dV$ of a function of three independent variables over a closed region $R$ of points $(x,y,z)$, of volume $V$, on which the function is single-valued and continuous, is an extension of the notion of single and double integrals.

If $f(x,y,z) = 1$, $\iiint\limits_{R} f(x,y,z)\, dV$ may be interpreted as measuring the volume of the region $R$.

**EVALUATION OF THE TRIPLE INTEGRAL** $\iiint\limits_{R} f(x,y,z)\, dV$ in rectangular coordinates.

$$\iiint\limits_{R} f(x,y,z)\, dV \;=\; \int_{a}^{b}\int_{y_1(x)}^{y_2(x)}\int_{z_1(x,y)}^{z_2(x,y)} f(x,y,z)\, dz\, dy\, dx$$

$$=\; \int_{c}^{d}\int_{x_1(y)}^{x_2(y)}\int_{z_1(x,y)}^{z_2(x,y)} f(x,y,z)\, dz\, dx\, dy, \quad \text{etc.}$$

where the limits of integration are chosen to cover the region $R$.

**EVALUATION OF THE TRIPLE INTEGRAL** $\iiint\limits_{R} f(\rho,\theta,z)\, dV$ in cylindrical coordinates.

$$\iiint\limits_{R} f(\rho,\theta,z)\, dV \;=\; \int_{\alpha}^{\beta}\int_{\rho_1(\theta)}^{\rho_2(\theta)}\int_{z_1(\rho,\theta)}^{z_2(\rho,\theta)} f(\rho,\theta,z)\, \rho\, dz\, d\rho\, d\theta$$

where the limits of integration are chosen to cover the region $R$.

**EVALUATION OF THE TRIPLE INTEGRAL** $\iiint\limits_{R} f(\rho,\phi,\theta)\, dV$ in spherical coordinates.

$$\iiint\limits_{R} f(\rho,\phi,\theta)\, dV \;=\; \int_{\alpha}^{\beta}\int_{\phi_1(\theta)}^{\phi_2(\theta)}\int_{\rho_1(\phi,\theta)}^{\rho_2(\phi,\theta)} f(\rho,\phi,\theta)\, \rho^2 \sin\phi\, d\rho\, d\phi\, d\theta$$

where the limits of integration are chosen to cover the region $R$.

**CENTROIDS AND MOMENTS OF INERTIA.** The coordinates $(\bar{x},\bar{y},\bar{z})$ of the *centroid of a volume* satisfy the relations

$$\bar{x}\iiint\limits_{R} dV = \iiint\limits_{R} x\, dV, \quad \bar{y}\iiint\limits_{R} dV = \iiint\limits_{R} y\, dV,$$

$$\bar{z}\iiint\limits_{R} dV = \iiint\limits_{R} z\, dV$$

The *moments of inertia of a volume* with respect to the coordinate axes are given by

$$I_x = \iiint\limits_{R} (y^2 + z^2)\, dV, \quad I_y = \iiint\limits_{R} (z^2 + x^2)\, dV,$$

$$I_z = \iiint\limits_{R} (x^2 + y^2)\, dV$$

# Solved Problems

1. Consider the function $f(x, y, z)$, continuous over a region $R$ of ordinary space. After slicing $R$ by planes $x = \xi_i$ and $y = \eta_j$ as in Chapter 63, let these subregions be further sliced by planes $z = \zeta_k$. The region $R$ has now been separated into a number of rectangular parallelepipeds of volume $\Delta V_{ijk} = \Delta x_i \cdot \Delta y_j \cdot \Delta z_k$ and a number of partial parallelepipeds which we shall ignore. In each complete parallelepiped select a point $P_{ijk}(x_i, y_j, z_k)$; then compute $f(x_i, y_j, z_k)$ and form the sum

(i) $$\sum_{\substack{i=1,\ldots,m \\ j=1,\ldots,n \\ k=1,\ldots,p}} f(x_i, y_j, z_k) \cdot \Delta V_{ijk} \;=\; \sum_{\substack{i=1,\ldots,m \\ j=1,\ldots,n \\ k=1,\ldots,p}} f(x_i, y_j, z_k)\, \Delta x_i\, \Delta y_j\, \Delta z_k$$

The triple integral of $f(x, y, z)$ over the region $R$ is defined to be the limit of (i) as the number of parallelepipeds is indefinitely increased in such a manner that all dimensions of each $\to 0$.

In evaluating this limit, we may sum first each set of parallelepipeds having $\Delta_i x$ and $\Delta_j y$, for fixed $i$ and $j$, as two dimensions and consider the limit as each $\Delta_k z \to 0$. We have

$$\lim_{p \to +\infty} \sum_{k=1}^{p} f(x_i, y_j, z_k)\, \Delta_k z\, \Delta_i x\, \Delta_j y \;=\; \int_{z_1}^{z_2} f(x_i, y_j, z)\, dz\, \Delta_i x\, \Delta_j y$$

Now these are the columns, the basic sub-regions, of Chapter 63; hence,

$$\lim_{\substack{m \to +\infty \\ n \to +\infty \\ p \to +\infty}} \sum_{\substack{i=1,\ldots,m \\ j=1,\ldots,n \\ k=1,\ldots,p}} f(x_i, y_j, z_k) \cdot \Delta V_{ijk} \;=\; \iiint_R f(x, y, z)\, dz\, dx\, dy \;=\; \iiint_R f(x, y, z)\, dz\, dy\, dx$$

2. Evaluate:

(a) $$\int_0^1 \int_0^{1-x} \int_0^{2-x} xyz\, dz\, dy\, dx$$

$$= \int_0^1 \left[ \int_0^{1-x} \left\{ \int_0^{2-x} xyz\, dz \right\} dy \right] dx$$

$$= \int_0^1 \left[ \int_0^{1-x} \left\{ \frac{xyz^2}{2} \Big|_{z=0}^{z=2-x} \right\} dy \right] dx \;=\; \int_0^1 \left[ \int_0^{1-x} \frac{xy(2-x)^2}{2}\, dy \right] dx$$

$$= \int_0^1 \left[ \frac{xy^2(2-x)^2}{4} \Big|_{y=0}^{y=1-x} \right] dx \;=\; \frac{1}{4} \int_0^1 (4x - 12x^2 + 13x^3 - 6x^4 + x^5)\, dx \;=\; \frac{13}{240}$$

(b) $$\int_0^{\pi/2} \int_0^1 \int_0^2 z\rho^2 \sin\theta\, dz\, d\rho\, d\theta$$

$$= \int_0^{\pi/2} \int_0^1 \frac{z^2}{2} \Big|_0^2\, \rho^2 \sin\theta\, d\rho\, d\theta \;=\; 2 \int_0^{\pi/2} \int_0^1 \rho^2 \sin\theta\, d\rho\, d\theta$$

$$= \frac{2}{3} \int_0^{\pi/2} \rho^3 \Big|_0^1 \sin\theta\, d\theta \;=\; -\frac{2}{3} \cos\theta \Big|_0^{\pi/2} \;=\; 2/3$$

(c) $$\int_0^{\pi} \int_0^{\pi/4} \int_0^{\sec\phi} \sin 2\phi\, d\rho\, d\phi\, d\theta \;=\; 2 \int_0^{\pi} \int_0^{\pi/4} \sin\phi\, d\phi\, d\theta \;=\; 2 \int_0^{\pi} (1 - \tfrac{1}{2}\sqrt{2})\, d\theta \;=\; (2 - \sqrt{2})\pi$$

3. Compute the triple integral of $F(x, y, z) = z$ over the region $R$ in the first octant bounded by the planes $y = 0$, $z = 0$, $x + y = 2$, $2y + x = 6$, and the cylinder $y^2 + z^2 = 4$. See Fig. 67-1 below.

Integrate first with respect to $z$ from $z = 0$ (the $xOy$ plane) to $z = \sqrt{4 - y^2}$ (the cylinder), then with respect to $x$ from $x = 2 - y$ to $x = 6 - 2y$, and finally with respect to $y$ from $y = 0$ to $y = 2$. Then

$$\iiint_R z\, dV \;=\; \int_0^2 \int_{2-y}^{6-2y} \int_0^{\sqrt{4-y^2}} z\, dz\, dx\, dy \;=\; \int_0^2 \int_{2-y}^{6-2y} (\tfrac{1}{2}z^2) \Big]_0^{\sqrt{4-y^2}} dx\, dy$$

$$= \frac{1}{2} \int_0^2 \int_{2-y}^{6-2y} (4 - y^2)\, dx\, dy \;=\; \frac{1}{2} \int_0^2 (4 - y^2)x \Big]_{2-y}^{6-2y} dy \;=\; \frac{26}{3}$$

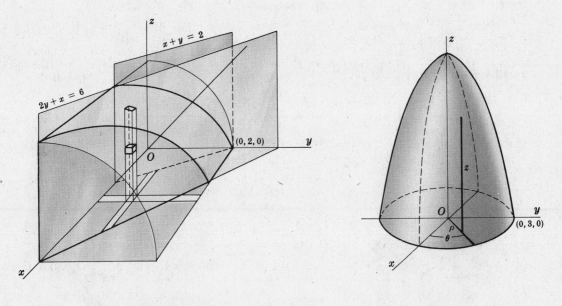

**Fig. 67-1**                                    **Fig. 67-2**

4.  Compute the triple integral of $f(\rho, \theta, z) = \rho^2$ over the region $R$ bounded by the paraboloid $\rho^2 = 9 - z$ and the plane $z = 0$. See Fig. 67-2 above.

   Integrate first with respect to $z$ from $z = 0$ to $z = 9 - \rho^2$, then with respect to $\rho$ from $\rho = 0$ to $\rho = 3$, and finally with respect to $\theta$ from $\theta = 0$ to $\theta = 2\pi$. Then

$$\iiint_R \rho^2 \, dV \;=\; \int_0^{2\pi} \int_0^3 \int_0^{9-\rho^2} \rho^2 (\rho \, dz \, d\rho \, d\theta) \;=\; \int_0^{2\pi} \int_0^3 \rho^3 (9 - \rho^2) \, d\rho \, d\theta$$

$$=\; \int_0^{2\pi} \left. (\tfrac{9}{4}\rho^4 - \tfrac{1}{6}\rho^6) \right]_0^3 \, d\theta \;=\; \int_0^{2\pi} \frac{243}{4} \, d\theta \;=\; \frac{243}{2}\pi$$

5.  Show that the integrals   $(a)$  $4 \displaystyle\int_0^4 \int_0^{\sqrt{16-x^2}} \int_{(x^2+y^2)/4}^4 dz \, dy \, dx$,  $(b)$  $4 \displaystyle\int_0^4 \int_0^{2\sqrt{z}} \int_0^{\sqrt{4z-x^2}} dy \, dx \, dz$,  and

   $(c)$  $4 \displaystyle\int_0^4 \int_{y^2/4}^4 \int_0^{\sqrt{4z-y^2}} dx \, dz \, dy$   give the same volume.

   $(a)$  Here $z$ ranges from $z = \frac{1}{4}(x^2 + y^2)$ to $z = 4$; that is, the volume is bounded below by the paraboloid $4z = x^2 + y^2$ and above by the plane $z = 4$. The ranges of $y$ and $x$ cover a quadrant of the circle $x^2 + y^2 = 16$, $z = 0$, the projection of the curve of intersection of the paraboloid and the plane $z = 4$, on the $xOy$ plane. Thus, the integral gives the volume cut from the paraboloid by the plane $z = 4$.

   $(b)$  Here $y$ ranges from $y = 0$ to $y = \sqrt{4z - x^2}$; that is, the volume is bounded on the left by the $zOx$ plane and on the right by the paraboloid $y^2 = 4z - x^2$. The ranges of $x$ and $z$ cover one-half the area cut from the parabola $x^2 = 4z$, $y = 0$, the curve of intersection of the paraboloid and the $zOx$ plane, by the plane $z = 4$. The region $R$ is that of $(a)$.

   $(c)$  Here the volume is bounded behind by the $yOz$ plane and in front by the paraboloid $4z = x^2 + y^2$. The ranges of $z$ and $y$ cover one-half the area cut from the parabola $y^2 = 4z$, $x = 0$, the curve of intersection of the paraboloid and the $yOz$ plane, by the plane $z = 4$. The region $R$ is that of $(a)$.

6.  Compute the triple integral of $F(\rho, \phi, \theta) = 1/\rho$ over the region $R$ in the first octant bounded by the cones $\phi = \frac{1}{4}\pi$ and $\phi = \arctan 2$, and the sphere $\rho = \sqrt{6}$. See Fig. 67-3 below.

Integrate first with respect to $\rho$ from $\rho = 0$ to $\rho = \sqrt{6}$, then with respect to $\phi$ from $\phi = \frac{1}{4}\pi$ to $\phi = \arctan 2$, and finally with respect to $\theta$ from $\theta = 0$ to $\theta = \frac{1}{2}\pi$. Then

$$\iiint_R \frac{1}{\rho}\, dV = \int_0^{\pi/2} \int_{\pi/4}^{\arctan 2} \int_0^{\sqrt{6}} \frac{1}{\rho} \rho^2 \sin\phi \, d\rho \, d\phi \, d\theta$$

$$= 3\int_0^{\pi/2} \int_{\pi/4}^{\arctan 2} \sin\phi \, d\phi \, d\theta = -3\int_0^{\pi/2} \left(\frac{1}{\sqrt{5}} - \frac{1}{\sqrt{2}}\right) d\theta = \frac{3\pi}{2}\left(\frac{1}{\sqrt{2}} - \frac{1}{\sqrt{5}}\right)$$

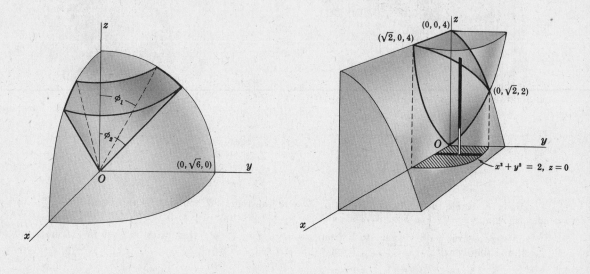

Fig. 67-3                                                      Fig. 67-4

7.  Find the volume bounded by the paraboloid $z = 2x^2 + y^2$ and the cylinder $z = 4 - y^2$. See Fig. 67-4.

Integrate first with respect to $z$ from $z = 2x^2 + y^2$ to $z = 4 - y^2$, then with respect to $y$ from $y = 0$ to $y = \sqrt{2 - x^2}$ (obtain $x^2 + y^2 = 2$ by eliminating $z$ between the equations of the two surfaces), and finally with respect to $x$ from $x = 0$ to $x = \sqrt{2}$ (obtained by setting $y = 0$ in $x^2 + y^2 = 2$) to obtain one-fourth of the required volume. Thus,

$$V = 4\int_0^{\sqrt{2}} \int_0^{\sqrt{2-x^2}} \int_{2x^2+y^2}^{4-y^2} dz \, dy \, dx = 4\int_0^{\sqrt{2}} \int_0^{\sqrt{2-x^2}} \{(4 - y^2) - (2x^2 + y^2)\}\, dy \, dx$$

$$= 4\int_0^{\sqrt{2}} \left(4y - 2x^2 y - \frac{2y^3}{3}\right)\Big]_0^{\sqrt{2-x^2}} dx = \frac{16}{3}\int_0^{\sqrt{2}} (2 - x^2)^{3/2}\, dx = 4\pi \text{ cubic units}$$

8.  Find the volume within the cylinder $\rho = 4\cos\theta$ bounded above by the sphere $\rho^2 + z^2 = 16$ and below by the plane $z = 0$. See Fig. 67-5 below.

Integrate first with respect to $z$ from $z = 0$ to $z = \sqrt{16 - \rho^2}$, then with respect to $\rho$ from $\rho = 0$ to $\rho = 4\cos\theta$, and finally with respect to $\theta$ from $\theta = 0$ to $\theta = \pi$ to obtain the required volume. Thus,

$$V = \int_0^{\pi} \int_0^{4\cos\theta} \int_0^{\sqrt{16-\rho^2}} \rho \, dz \, d\rho \, d\theta = \int_0^{\pi} \int_0^{4\cos\theta} \rho\sqrt{16 - \rho^2}\, d\rho \, d\theta$$

$$= -\frac{64}{3}\int_0^{\pi} (\sin^3\theta - 1)\, d\theta = \frac{64}{9}(3\pi - 4) \text{ cubic units}$$

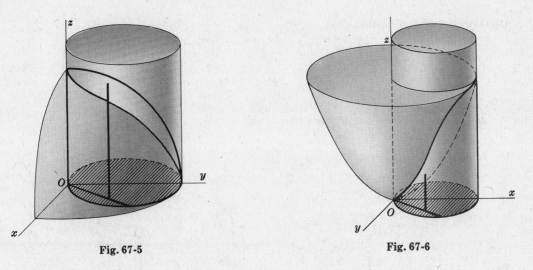

**Fig. 67-5**                                    **Fig. 67-6**

9.  Find the coordinates of the centroid of the volume within the cylinder $\rho = 2\cos\theta$, bounded above by the paraboloid $z = \rho^2$ and below by the plane $z = 0$.  See Fig. 67-6 above.

$$V = 2\int_0^{\pi/2}\int_0^{2\cos\theta}\int_0^{\rho^2}\rho\,dz\,d\rho\,d\theta = 2\int_0^{\pi/2}\int_0^{2\cos\theta}\rho^3\,d\rho\,d\theta$$

$$= \frac{1}{2}\int_0^{\pi/2}\rho^4\Big]_0^{2\cos\theta}\,d\theta = 8\int_0^{\pi/2}\cos^4\theta\,d\theta = \frac{3}{2}\pi$$

$$M_{yz} = \iiint_R x\,dV = 2\int_0^{\pi/2}\int_0^{2\cos\theta}\int_0^{\rho^2}\rho\cos\theta\cdot\rho\,dz\,d\rho\,d\theta$$

$$= 2\int_0^{\pi/2}\int_0^{2\cos\theta}\rho^4\cos\theta\,d\rho\,d\theta = \frac{64}{5}\int_0^{\pi/2}\cos^6\theta\,d\theta = 2\pi, \quad\text{and}\quad \bar{x} = \frac{M_{yz}}{V} = \frac{4}{3}$$

By symmetry, $\bar{y} = 0$.

$$M_{xy} = \iiint_R z\,dV = 2\int_0^{\pi/2}\int_0^{2\cos\theta}\int_0^{\rho^2}z\cdot\rho\,dz\,d\rho\,d\theta = \int_0^{\pi/2}\int_0^{2\cos\theta}\rho^5\,d\rho\,d\theta$$

$$= \frac{32}{3}\int_0^{\pi/2}\cos^6\theta\,d\theta = \frac{5}{3}\pi, \quad\text{and}\quad \bar{z} = \frac{M_{xy}}{V} = \frac{10}{9}$$

Thus, the centroid has coordinates (4/3, 0, 10/9).

10.  For the right circular cone of radius $r$ and height $h$, find (a) the centroid, (b) the moment of inertia with respect to its axis, (c) the moment of inertia with respect to any line through its vertex and perpendicular to its axis, (d) the moment of inertia with respect to any line through its centroid and perpendicular to its axis, (e) the moment of inertia with respect to any diameter of its base.

Take the cone as in Fig. 67-7, so that its equation is $\rho = \frac{r}{h}z$.  Then

$$V = 4\int_0^{\pi/2}\int_0^r\int_{\frac{h}{r}\rho}^h \rho\,dz\,d\rho\,d\theta$$

$$= 4\int_0^{\pi/2}\int_0^r\left(h\rho - \frac{h}{r}\rho^2\right)d\rho\,d\theta$$

$$= \frac{2}{3}hr^2\int_0^{\pi/2}d\theta = \frac{1}{3}\pi hr^2$$

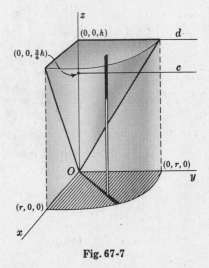

**Fig. 67-7**

(a) The centroid lies on the $z$-axis.

$$M_{xy} = \iiint_R z \, dV = 4 \int_0^{\pi/2} \int_0^r \int_{\frac{h}{r}\rho}^h z \, \rho \, dz \, d\rho \, d\theta$$

$$= 2 \int_0^{\pi/2} \int_0^r \left( h^2\rho - \frac{h^2}{r^2}\rho^3 \right) d\rho \, d\theta = \frac{1}{2} h^2 r^2 \int_0^{\pi/2} d\theta = \frac{1}{4}\pi h^2 r^2$$

and $\bar{z} = \dfrac{M_{xy}}{V} = \dfrac{3}{4} h$. Hence, the centroid has coordinates $(0, 0, \frac{3}{4}h)$.

(b) $I_z = \iiint_R (x^2 + y^2) \, dV = 4 \int_0^{\pi/2} \int_0^r \int_{\frac{h}{r}\rho}^h \rho^2 \cdot \rho \, dz \, d\rho \, d\theta = \frac{1}{10}\pi h r^4 = \frac{3}{10} r^2 V$

(c) Take the line as the $y$-axis.

$$I_y = \iiint_R (x^2 + z^2) \, dV = 4 \int_0^{\pi/2} \int_0^r \int_{\frac{h}{r}\rho}^h (\rho^2 \cos^2\theta + z^2) \rho \, dz \, d\rho \, d\theta$$

$$= 4 \int_0^{\pi/2} \int_0^r \left\{ \left( h\rho^3 - \frac{h}{r}\rho^4 \right) \cos^2\theta + \frac{1}{3}\left( h^3\rho - \frac{h^3}{r^3}\rho^4 \right) \right\} d\rho \, d\theta$$

$$= \frac{1}{5}\pi h r^2 \left( h^2 + \frac{1}{4}r^2 \right) = \frac{3}{5}\left( h^2 + \frac{1}{4}r^2 \right) V$$

(d) Let the line $c$ through the centroid be parallel to the $y$-axis. By the parallel axis theorem,

$$I_y = I_c + V(\tfrac{3}{4}h)^2 \quad\text{and}\quad I_c = \tfrac{3}{5}(h^2 + \tfrac{1}{4}r^2)V - \tfrac{9}{16}h^2 V = \tfrac{3}{80}(h^2 + 4r^2)V$$

(e) Let $d$ denote the diameter of the base of the cone parallel to the $y$-axis. Then

$$I_d = I_c + V(\tfrac{1}{4}h)^2 = \tfrac{3}{80}(h^2 + 4r^2)V + \tfrac{1}{16}h^2 V = \tfrac{1}{20}(2h^2 + 3r^2)V$$

11. Find the volume cut from the cone $\phi = \frac{1}{4}\pi$ by the sphere $\rho = 2a\cos\phi$. See Fig. 67-8 below.

$$V = 4 \iiint_R dV = 4 \int_0^{\pi/2} \int_0^{\pi/4} \int_0^{2a\cos\phi} \rho^2 \sin\phi \, d\rho \, d\phi \, d\theta$$

$$= \frac{32a^3}{3} \int_0^{\pi/2} \int_0^{\pi/4} \cos^3\phi \sin\phi \, d\phi \, d\theta = 2a^3 \int_0^{\pi/2} d\theta = \pi a^3 \text{ cubic units}$$

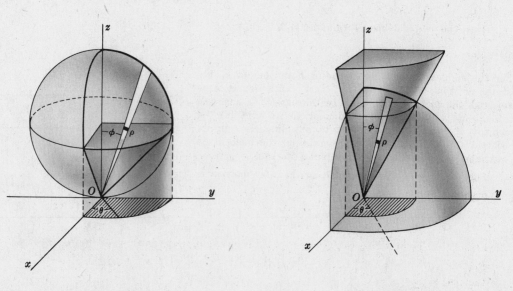

Fig. 67-8                                        Fig. 67-9

12. Locate the centroid of the volume cut from one nappe of a cone of vertex angle 60° by a sphere of radius 2 whose center is at the vertex of the cone. See Fig. 67-9 above.

Take the surfaces as in Fig. 67-9 above, so that $\bar{x} = \bar{y} = 0$. In spherical coordinates, the equation of the cone is $\phi = \pi/6$ and the equation of the sphere is $\rho = 2$.

$$V = \iiint_R dV = 4 \int_0^{\pi/2} \int_0^{\pi/6} \int_0^2 \rho^2 \sin\phi \, d\rho \, d\phi \, d\theta = \frac{32}{3} \int_0^{\pi/2} \int_0^{\pi/6} \sin\phi \, d\phi \, d\theta$$

$$= -\frac{32}{3}\left(\frac{\sqrt{3}}{2} - 1\right) \int_0^{\pi/2} d\theta = \frac{8\pi}{3}(2 - \sqrt{3})$$

$$M_{xy} = \iiint_R z \, dV = 4 \int_0^{\pi/2} \int_0^{\pi/6} \int_0^2 \rho \cos\phi \cdot \rho^2 \sin\phi \, d\rho \, d\phi \, d\theta$$

$$= 8 \int_0^{\pi/2} \int_0^{\pi/6} \sin 2\phi \, d\phi \, d\theta = \pi, \quad \text{and} \quad \bar{z} = \frac{M_{xy}}{V} = \frac{3(2 + \sqrt{3})}{8}.$$

13. Find the moment of inertia with respect to the $z$-axis of the volume of Problem 12.

$$I_z = \iiint_R (x^2 + y^2) \, dV = 4 \int_0^{\pi/2} \int_0^{\pi/6} \int_0^2 \rho^2 \sin^2\phi \cdot \rho^2 \sin\phi \, d\rho \, d\phi \, d\theta$$

$$= \frac{128}{5} \int_0^{\pi/2} \int_0^{\pi/6} \sin^3\phi \, d\phi \, d\theta = \frac{128}{5}\left(\frac{2}{3} - \frac{3}{8}\sqrt{3}\right) \int_0^{\pi/2} d\theta = \frac{8\pi}{15}(16 - 9\sqrt{3}) = \frac{5 - 2\sqrt{3}}{5} V$$

## Supplementary Problems

14. Evaluate the following triple integrals:

(a) $\displaystyle\int_0^1 \int_1^2 \int_2^3 dz \, dx \, dy = 1$

(b) $\displaystyle\int_0^1 \int_{x^2}^x \int_0^{xy} dz \, dy \, dx = 1/24$

(c) $\displaystyle\int_0^6 \int_0^{12-2y} \int_0^{4-2y/3-x/3} x \, dz \, dx \, dy = 144 = \int_0^{12} \int_0^{6-x/2} \int_0^{4-2y/3-x/3} x \, dz \, dy \, dx$

(d) $\displaystyle\int_0^{\pi/2} \int_0^4 \int_0^{\sqrt{16-z^2}} (16 - \rho^2)^{1/2} \rho \, z \, d\rho \, dz \, d\theta = \frac{256}{5}\pi$

(e) $\displaystyle\int_0^{2\pi} \int_0^\pi \int_0^5 \rho^4 \sin\phi \, d\rho \, d\phi \, d\theta = 2500\pi$

15. (a) Calculate the integral of Problem 14(b) after changing the order to $dz \, dx \, dy$.
    (b) Calculate the integral of Problem 14(c) changing the order to $dx \, dy \, dz$ and also to $dy \, dz \, dx$.

16. Find the following volumes, using triple integrals in rectangular coordinates:
    (a) inside $x^2 + y^2 = 9$, above $z = 0$, and below $x + z = 4$.     *Ans.* $36\pi$ cu. un.
    (b) bounded by the coordinate planes and $6x + 4y + 3z = 12$.     *Ans.* 4 cu. un.
    (c) inside $x^2 + y^2 = 4x$, above $z = 0$, and below $x^2 + y^2 = 4z$.     *Ans.* $6\pi$ cu. un.

17. Find the following volumes, using triple integrals in cylindrical coordinates:
    (a) Problem 5.
    (b) Problem 16(c).
    (c) inside $\rho^2 = 16$, above $z = 0$, and below $2z = y$.     *Ans.* 64/3 cu. un.

**18.** Find the centroid of each of the following volumes:

    (a) under $z^2 = xy$ and above the triangle $y = x$, $y = 0$, $x = 4$
       in the plane $z = 0$.                     *Ans.* (3, 9/5, 9/8)

    (b) Problem 16(b).                       *Ans.* (1/2, 3/4, 1)

    (c) first octant volume of Problem 16(a).    *Ans.* $\left( \dfrac{64 - 9\pi}{16(\pi - 1)}, \ \dfrac{23}{8(\pi - 1)}, \ \dfrac{73\pi - 128}{32(\pi - 1)} \right)$

    (d) Problem 16(c).                     *Ans.* (8/3, 0, 10/9)

    (e) Problem 17(c).                     *Ans.* $(0, 3\pi/4, 3\pi/16)$

**19.** Find the moments of inertia $I_x$, $I_y$, $I_z$ of the following volumes:

    (a) Problem 5.                      *Ans.* $I_x = I_y = \frac{32}{3}V$, $I_z = \frac{16}{3}V$

    (b) Problem 16(b).                  *Ans.* $I_x = \frac{5}{2}V$, $I_y = 2V$, $I_z = \frac{13}{10}V$

    (c) Problem 16(c).                  *Ans.* $I_x = \frac{55}{18}V$, $I_y = \frac{175}{18}V$, $I_z = \frac{80}{9}V$

    (d) cut from $z = \rho^2$ by the plane $z = 2$.   *Ans.* $I_x = I_y = \frac{7}{3}V$, $I_z = \frac{2}{3}V$

**20.** Show that, in cylindrical coordinates, the triple integral of a function $f(\rho, \theta, z)$ over a region $R$ may be represented by

$$\int_\alpha^\beta \int_{\rho_1(\theta)}^{\rho_2(\theta)} \int_{z_1(\rho, \theta)}^{z_2(\rho, \theta)} f(\rho, \theta, z) \, \rho \, dz \, d\rho \, d\theta$$

*Hint.* Consider in Fig. 67-10 below, a representative subregion of $R$ bounded by two cylinders having $Oz$ as axis and of radii $\rho$ and $\rho + \Delta\rho$ respectively, by two horizontal planes through $(0, 0, z)$ and $(0, 0, z + \Delta z)$ respectively, and by two vertical planes through $Oz$ making angles $\theta$ and $\theta + \Delta\theta$ respectively with the $xOz$ plane. Take $\Delta V = (\rho \, \Delta\theta) \, \Delta\rho \cdot \Delta z$ as an approximation of its volume.

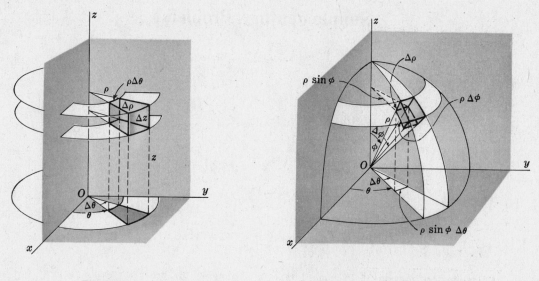

        **Fig. 67-10**                                     **Fig. 67-11**

**21.** Show that, in spherical coordinates, the triple integral of a function $f(\rho, \phi, \theta)$ over a region $R$ may be represented by

$$\int_\alpha^\beta \int_{\phi_1(\theta)}^{\phi_2(\theta)} \int_{\rho_1(\phi, \theta)}^{\rho_2(\phi, \theta)} f(\rho, \phi, \theta) \, \rho^2 \sin\phi \, d\rho \, d\phi \, d\theta$$

*Hint.* Consider in Fig. 67-11 above, a representative subregion of $R$ bounded by two spheres centered at $O$ of radii $\rho$ and $\rho + \Delta\rho$ respectively, by two cones having $O$ as vertex, $Oz$ as axis, and semi-vertical angles $\phi$ and $\phi + \Delta\phi$ respectively, and by two vertical planes through $Oz$ making angles $\theta$ and $\theta + \Delta\theta$ respectively with the $zOy$ plane. Take

$$\Delta V \ = \ (\rho \, \Delta\phi)(\rho \sin\phi \, \Delta\theta)(\Delta\rho) \ = \ \rho^2 \sin\phi \, \Delta\rho \, \Delta\phi \, \Delta\theta$$

as an approximation of its volume.

# Chapter 68

## Masses of Variable Density

**HOMOGENEOUS MASSES** have been treated in previous chapters as geometric figures by taking the density $\delta = 1$. The mass of a homogeneous body of volume $V$ and density $\delta$ is $m = \delta V$.

For a non-homogeneous mass whose density $\delta$ varies continuously from point to point, an element of mass $dm$ is given by:

$\delta(x, y)\, ds$     for a material curve (i.e., a piece of fine wire),

$\delta(x, y)\, dA$     for a material two-dimensional spread (i.e., a thin sheet of metal),

$\delta(x, y, z)\, dV$   for a material body.

## Solved Problems

1. Find the mass of a semicircular wire whose density varies as the distance from the diameter joining the ends. See Fig. 68-1 below.

   Take the wire as in Fig. 68-1 so that $\delta(x, y) = ky$. Then, from $x^2 + y^2 = r^2$,

   $$ds = \sqrt{1 + \left(\frac{dy}{dx}\right)^2}\, dx = \frac{r}{y}\, dx$$

   and
   $$m = \int \delta(x, y)\, ds = \int_{-r}^{r} ky \cdot \frac{r}{y}\, dx = kr \int_{-r}^{r} dx = 2kr^2 \text{ units}$$

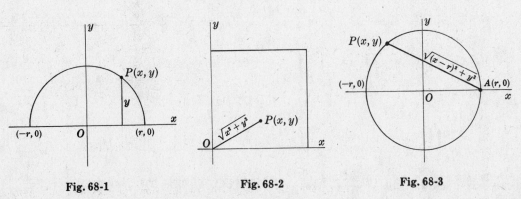

| Fig. 68-1 | Fig. 68-2 | Fig. 68-3 |

2. Find the mass of a square plate of side $a$ if the density varies as the square of the distance from a vertex. See Fig. 68-2 above.

   Take the square as in Fig. 68-2 and let the vertex from which distances are measured be at the origin. Then $\delta(x, y) = k(x^2 + y^2)$ and

   $$m = \iint_R \delta(x, y)\, dA = \int_0^a \int_0^a k(x^2 + y^2)\, dx\, dy = k \int_0^a (\tfrac{1}{3}a^3 + ay^2)\, dy = \tfrac{2}{3}ka^4 \text{ units}$$

331

**3.** Find the mass of a circular plate of radius $r$ if the density varies as the square of the distance from a point on the circumference. See Fig. 68-3 above.

Take the circle as in Fig. 68-3 and let $A(r, 0)$ be the fixed point on the circumference. Then $\delta(x, y) = k\{(x - r)^2 + y^2\}$ and

$$m = \iint_R \delta(x, y)\, dA = 2\int_{-r}^{r}\int_{0}^{\sqrt{r^2 - x^2}} k\{(x - r)^2 + y^2\}\, dy\, dx = \frac{3}{2} k\pi r^4 \text{ units}$$

**4.** Find the center of mass of a plate in the form of the segment cut from the parabola $y^2 = 8x$ by its latus rectum $x = 2$ if the density varies as the distance from the latus rectum. See Fig. 68-4 below.

Here, $\delta(x, y) = 2 - x$ and, by symmetry, $\bar{y} = 0$. For the upper half of the plate,

$$m = \iint_R \delta(x, y)\, dA = \int_0^4 \int_{y^2/8}^2 k(2 - x)\, dx\, dy = k\int_0^4 \left(2 - \frac{y^2}{4} + \frac{y^4}{128}\right) dy = \frac{64}{15}k,$$

$$M_y = \iint_R \delta(x, y)\, x\, dA = \int_0^4 \int_{y^2/8}^2 k(2 - x)\, x\, dx\, dy = k\int_0^4 \left(\frac{4}{3} - \frac{y^4}{64} + \frac{y^6}{24\cdot 64}\right) dy = \frac{128}{35}k$$

and $\bar{x} = \dfrac{M_y}{m} = \dfrac{6}{7}$. The center of mass has coordinates $(\frac{6}{7}, 0)$.

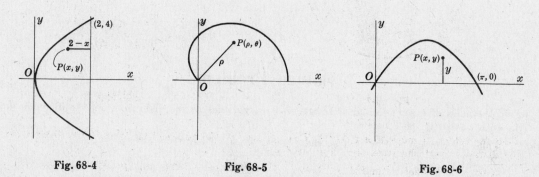

**Fig. 68-4**　　　　　　　　　　**Fig. 68-5**　　　　　　　　　　**Fig. 68-6**

**5.** Find the center of mass of a plate in the form of the upper half of the cardioid $\rho = 2(1 + \cos\theta)$ if the density varies as the distance from the pole. See Fig. 68-5 above.

$$m = \iint_R \delta(\rho, \theta)\, dA = \int_0^\pi \int_0^{2(1 + \cos\theta)} k\rho \cdot \rho\, d\rho\, d\theta = \frac{8}{3}k\int_0^\pi (1 + \cos\theta)^3\, d\theta = \frac{20}{3}k\pi,$$

$$M_x = \iint_R \delta(\rho, \theta)\, y\, dA = \int_0^\pi \int_0^{2(1 + \cos\theta)} k\rho \cdot \rho\sin\theta \cdot \rho\, d\rho\, d\theta$$

$$= 4k\int_0^\pi (1 + \cos\theta)^4 \sin\theta\, d\theta = \frac{128}{5}k,$$

$$M_y = \iint_R \delta(\rho, \theta)\, x\, dA = \int_0^\pi \int_0^{2(1 + \cos\theta)} k\rho \cdot \rho\cos\theta \cdot \rho\, d\rho\, d\theta = 14k\pi$$

Then, $\bar{x} = \dfrac{M_y}{m} = \dfrac{21}{10}$, $\bar{y} = \dfrac{M_x}{m} = \dfrac{96}{25\pi}$, and the center of mass has coordinates $\left(\dfrac{21}{10}, \dfrac{96}{25\pi}\right)$.

**6.** Find the moment of inertia with respect to the $x$-axis of a plate having for edges one arch of the curve $y = \sin x$ and the $x$-axis if its density varies as the distance from the $x$-axis. See Fig. 68-6 above.

$$m = \iint_R \delta(x, y)\, dA = \int_0^\pi \int_0^{\sin x} ky\, dy\, dx = \frac{1}{2}k\int_0^\pi \sin^2 x\, dx = \frac{1}{4}k\pi$$

and

$$I_x = \iint_R \delta(x, y)\, y^2\, dA = \int_0^\pi \int_0^{\sin x} ky \cdot y^2 \cdot dy\, dx = \frac{1}{4}k\int_0^\pi \sin^4 x\, dx = \frac{3}{32}k\pi = \frac{3}{8}m$$

7.   Find the mass of a sphere of radius $r$ if the density varies inversely as the square of the distance from the center. See Fig. 68-7 below.

Take the sphere as in Fig. 68-7. Then   $\delta(x, y, z) = \dfrac{k}{x^2 + y^2 + z^2} = \dfrac{k}{\rho^2}$  and

$$m = \iiint\limits_{R} \delta(x, y, z)\, dV = 8 \int_{0}^{\pi/2} \int_{0}^{\pi/2} \int_{0}^{r} \frac{k}{\rho^2} \cdot \rho^2 \sin\phi\, d\rho\, d\phi\, d\theta$$

$$= 8kr \int_{0}^{\pi/2} \int_{0}^{\pi/2} \sin\phi\, d\phi\, d\theta = 8kr \int_{0}^{\pi/2} d\theta = 4k\pi r \text{ units}$$

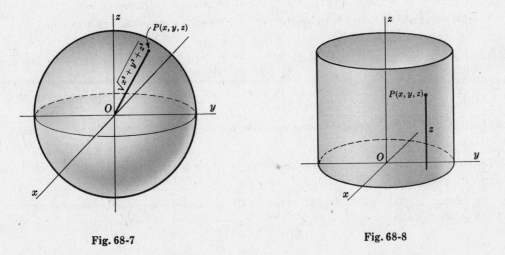

**Fig. 68-7**                                             **Fig. 68-8**

8.   Find the center of mass of a right circular cylinder of radius $r$ and height $h$ if the density varies as the distance from the base. See Fig. 68-8 above.

Take the cylinder as in Fig. 68-8, so that its equation is $\rho = r$ and the volume in question is that part of the cylinder between the planes $z = 0$ and $z = h$. Clearly, the center of mass lies on the $z$-axis.

$$m = \iiint\limits_{R} \delta(z, \rho, \theta)\, dV = 4 \int_{0}^{\pi/2} \int_{0}^{r} \int_{0}^{h} kz \cdot \rho\, dz\, d\rho\, d\theta = 2kh^2 \int_{0}^{\pi/2} \int_{0}^{r} \rho\, d\rho\, d\theta$$

$$= kh^2 r^2 \int_{0}^{\pi/2} d\theta = \frac{1}{2} k\pi h^2 r^2,$$

$$M_{xy} = \iiint\limits_{R} \delta(z, \rho, \theta)\, z\, dV = 4 \int_{0}^{\pi/2} \int_{0}^{r} \int_{0}^{h} kz^2 \cdot \rho\, dz\, d\rho\, d\theta = \frac{4}{3} kh^3 \int_{0}^{\pi/2} \int_{0}^{r} \rho\, d\rho\, d\theta$$

$$= \frac{2}{3} kh^3 r^2 \int_{0}^{\pi/2} d\theta = \frac{1}{3} k\pi h^3 r^2 \quad \text{and} \quad \bar{z} = \frac{M_{xy}}{m} = \frac{2}{3} h$$

Thus the center of mass has coordinates $(0, 0, \frac{2}{3} h)$.

# Supplementary Problems

9. Find the mass of:

(a) a straight rod of length $a$ whose density varies as the square of the distance from one end.
   Ans. $\frac{1}{3}ka^3$ units

(b) a plate in the form of a right triangle with legs $a$ and $b$ if the density varies as the sum of the distances from the legs.   Ans. $\frac{1}{6}kab(a+b)$ units

(c) a circular plate of radius $a$ whose density varies as the distance from the center.
   Ans. $\frac{2}{3}ka^3\pi$ units

(d) a plate in the form of an ellipse $b^2x^2 + a^2y^2 = a^2b^2$ if the density varies as the sum of the distances from its axes.   Ans. $\frac{4}{3}kab(a+b)$ units

(e) a circular cylinder of height $b$ and radius of base $a$ if the density varies as the square of the distance from its axis.   Ans. $\frac{1}{2}ka^4b\pi$ units

(f) a sphere of radius $a$ whose density varies as the distance from a fixed diametral plane.
   Ans. $\frac{1}{2}ka^4\pi$ units

(g) a circular cone of height $b$ and radius of base $a$ whose density varies as the distance from its axis.
   Ans. $\frac{1}{6}ka^3b\pi$ units

(h) a spherical surface whose density varies as the distance from a fixed diametral plane.
   Ans. $2ka^3\pi$ units

10. Find the center of mass of:

(a) one quadrant of 9(c).   Ans. $(3a/2\pi, 3a/2\pi)$

(b) one quadrant of a circular plane of radius $a$ if the density varies as the distance from a bounding radius ($x$-axis).   Ans. $(3a/8, 3a\pi/16)$

(c) a cube of edge $a$ if the density varies as the sum of the distances from three adjacent edges (coordinate axes).   Ans. $(5a/9, 5a/9, 5a/9)$

(d) an octant of a sphere of radius $a$ if the density varies as the distance from one of the plane faces.
   Ans. $(16a/15\pi, 16a/15\pi, 8a/15)$

(e) a right circular cone of height $b$ and radius of base $a$ if the density varies as the distance from its base.   Ans. $(0, 0, 2b/5)$

11. Find the moment of inertia of:

(a) a square plate of side $a$ with respect to a side if the density varies as the square of the distance from an extremity of that side.   Ans. $\frac{7}{15}a^2m$

(b) a plate in the form of a circle of radius $a$ with respect to its center if the density varies as the square of the distance from the center.   Ans. $\frac{2}{3}a^2m$

(c) a cube of edge $a$ with respect to an edge if the density varies as the square of the distance from one extremity of that edge.   Ans. $\frac{38}{45}a^2m$

(d) a right circular cone of height $b$ and radius of base $a$ with respect to its axis if the density varies as the distance from the axis.   Ans. $\frac{2}{5}a^2m$

(e) the cone of (d) if the density varies as the distance from the base.   Ans. $\frac{1}{5}a^2m$

# Chapter 69

## Differential Equations

**A DIFFERENTIAL EQUATION** is an equation which involves derivatives or differentials; for example, $\dfrac{d^2y}{dx^2} + 2\dfrac{dy}{dx} + 3y = 0$, $dy = (x+2y)\,dx$, etc.

The *order* of a differential equation is the order of the derivative of the highest order appearing in it. The first of the above equations is of order two and the second is of order one. Both are said to be of *degree* one.

A *solution* of a differential equation is any relation between the variables which is free of derivatives or differentials and which satisfies the equation identically. The *general solution* of a differential equation of order $n$ is that solution which contains the maximum number ($=n$) of essential arbitrary constants. See Problems 1-3.

**AN EQUATION OF THE FIRST ORDER AND DEGREE** has the form $M(x,y)\,dx + N(x,y)\,dy = 0$. If such an equation has the particular form $f_1(x) \cdot g_2(y)\,dx + f_2(x) \cdot g_1(y)\,dy = 0$, the variables are *separable* and the solution is obtained as

$$\int \frac{f_1(x)}{f_2(x)}\,dx + \int \frac{g_1(y)}{g_2(y)}\,dy = C$$

See Problems 4-6.

A function $f(x,y)$ is said to be *homogeneous of degree $n$* in the variables if $f(\lambda x, \lambda y) = \lambda^n f(x,y)$. The equation $M(x,y)\,dx + N(x,y)\,dy = 0$ is said to be *homogeneous* if $M(x,y)$ and $N(x,y)$ are homogeneous of the same degree. The substitution

$$y = vx, \qquad dy = v\,dx + x\,dv$$

will transform the homogeneous equation into one whose variables $x$ and $v$ are separable. See Problems 7-9.

**CERTAIN DIFFERENTIAL EQUATIONS** may be solved readily by taking advantage of the presence of integrable combinations of terms.

An equation, not immediately solvable by the above method, may be so solved after it is multiplied by a properly chosen function of $x$ and $y$. This multiplier is called an *integrating factor* of the equation. See Problems 10-14.

The so-called linear differential equation of the first order $\dfrac{dy}{dx} + Py = Q$, where $P$ and $Q$ are functions of $x$ alone, has $\xi(x) = e^{\int P\,dx}$ as integrating factor.

See Problems 15-17.

An equation of the form $\dfrac{dy}{dx} + Py = Qy^n$, where $n \neq 0, 1$, and $P$ and $Q$ are functions of $x$ alone, is reduced to the linear form by the substitution

$$y^{1-n} = z, \qquad y^{-n}\frac{dy}{dx} = \frac{1}{1-n}\frac{dz}{dx}$$

See Problems 18-19.

# Solved Problems

**1.** Show that (a) $y = 2e^x$, (b) $y = 3x$, and (c) $y = C_1 e^x + C_2 x$, where $C_1$ and $C_2$ are arbitrary constants, are solutions of the differential equation $y''(1 - x) + y'x - y = 0$.

(a) Differentiate $y = 2e^x$ twice to obtain $y' = 2e^x$ and $y'' = 2e^x$. Substitute in the differential equation to obtain the identity $2e^x(1 - x) + 2e^x x - 2e^x = 0$.

(b) Differentiate $y = 3x$ twice to obtain $y' = 3$ and $y'' = 0$. Substitute in the differential equation to obtain the identity $0(1 - x) + 3x - 3x = 0$.

(c) Differentiate $y = C_1 e^x + C_2 x$ twice to obtain $y' = C_1 e^x + C_2$ and $y'' = C_1 e^x$. Substitute in the differential equation to obtain the identity $C_1 e^x(1 - x) + (C_1 e^x + C_2)x - (C_1 e^x + C_2 x) = 0$.

Solution (c) is the *general solution* of the differential equation since it satisfies the equation and contains the proper number of essential arbitrary constants. Solutions (a) and (b) are called *particular solutions* since each may be obtained by assigning particular values to the arbitrary constants of the general solution.

**2.** Form the differential equation whose general solution is

$$(a) \quad y = Cx^2 - x \quad \text{and} \quad (b) \quad y = C_1 x^3 + C_2 x + C_3$$

(a) Differentiate $y = Cx^2 - x$ once to obtain $y' = 2Cx - 1$. Solve for $C = \dfrac{1}{2}\left(\dfrac{y' + 1}{x}\right)$ and substitute in the given relation (general solution) to obtain $y = \dfrac{1}{2}\left(\dfrac{y' + 1}{x}\right)x^2 - x$ or $y'x = 2y + x$.

(b) Differentiate $y = C_1 x^3 + C_2 x + C_3$ three times to obtain $y' = 3C_1 x^2 + C_2$, $y'' = 6C_1 x$, $y''' = 6C_1$. Then $y'' = xy'''$ is the required equation. Note that the given relation is a solution of the equation $y^{iv} = 0$ but is not the general solution since it contains only three arbitrary constants.

**3.** Form the differential equation of all parabolas with principal axis along the $x$-axis.

The system of parabolas has equation $y^2 = Ax + B$, where $A$ and $B$ are arbitrary constants.

Differentiate twice to obtain $2yy' = A$ and $2yy'' + 2y'^2 = 0$.

Then $2yy'' + 2y'^2 = 0$ is the required equation.

**4.** Solve $\dfrac{dy}{dx} + \dfrac{1 + y^3}{xy^2(1 + x^2)} = 0$.

Here $xy^2(1 + x^2)dy + (1 + y^3)dx = 0$ or $\dfrac{y^2}{1 + y^3}dy + \dfrac{1}{x(1 + x^2)}dx = 0$ and the variables are separated. Then

$$\frac{y^2\, dy}{1 + y^3} + \frac{dx}{x} - \frac{x\, dx}{1 + x^2} = 0,$$

$$\frac{1}{3}\ln|1 + y^3| + \ln|x| - \frac{1}{2}\ln(1 + x^2) = c,$$

$$2\ln|1 + y^3| + 6\ln|x| - 3\ln(1 + x^2) = 6c,$$

$$\ln\frac{x^6(1 + y^3)^2}{(1 + x^2)^3} = 6c \quad \text{and} \quad \frac{x^6(1 + y^3)^2}{(1 + x^2)^3} = e^{6c} = C.$$

**5.** Solve $\dfrac{dy}{dx} = \dfrac{1 + y^2}{1 + x^2}$.

Here $\dfrac{dy}{1 + y^2} = \dfrac{dx}{1 + x^2}$. Then $\arctan y = \arctan x + \arctan C$ and

$$y = \tan(\arctan x + \arctan C) = \frac{x + C}{1 - Cx}$$

**6.** Solve $\dfrac{dy}{dx} = \dfrac{\cos^2 y}{\sin^2 x}$.

$$\frac{dy}{\cos^2 y} = \frac{dx}{\sin^2 x}, \qquad \sec^2 y \, dy = \csc^2 x \, dx, \qquad \text{and} \qquad \tan y = -\cot x + C$$

**7.** Solve $2xy \, dy = (x^2 - y^2) \, dx$.

The equation is homogeneous of degree two. The transformation $y = vx$, $dy = v \, dx + x \, dv$ yields

$2x \cdot vx \, (v \, dx + x \, dv) = (x^2 - v^2 x^2) \, dx$ or $\dfrac{2v \, dv}{1 - 3v^2} = \dfrac{dx}{x}$.

Then $-\tfrac{1}{3} \ln |1 - 3v^2| = \ln |x| + \ln c$, $\quad \ln |1 - 3v^2| + 3 \ln |x| + \ln C' = 0$ or $\quad C' \, |x^3(1 - 3v^2)| = 1$.

Now $\pm C' x^3 (1 - 3v^2) = C x^3 (1 - 3v^2) = 1$ and, using $v = y/x$, $C(x^3 - 3xy^2) = 1$.

**8.** Solve $x \sin \dfrac{y}{x} \, (y \, dx + x \, dy) + y \cos \dfrac{y}{x} \, (x \, dy - y \, dx) = 0$.

The equation is homogeneous of degree two. The transformation $y = vx$, $dy = v \, dx + x \, dv$ yields

$$x \sin v \, (vx \, dx + x^2 \, dv + vx \, dx) + vx \cos v \, (x^2 \, dv + vx \, dx - vx \, dx) = 0$$

$$\sin v \, (2v \, dx + x \, dv) + xv \cos v \, dv = 0, \qquad \frac{\sin v + v \cos v}{v \sin v} \, dv + 2 \frac{dx}{x} = 0$$

Then $\ln |v \sin v| + 2 \ln |x| = \ln C'$, $\quad x^2 \cdot v \cdot \sin v = C$, and $\quad xy \sin \dfrac{y}{x} = C$.

**9.** Solve $(x^2 - 2y^2) \, dy + 2xy \, dx = 0$.

The equation is homogeneous of degree two and the standard transformation yields

$$(1 - 2v^2)(v \, dx + x \, dv) + 2v \, dx = 0, \qquad \frac{1 - 2v^2}{v(3 - 2v^2)} \, dv + \frac{dx}{x} = 0, \qquad \frac{dv}{3v} - \frac{4v \, dv}{3(3 - 2v^2)} + \frac{dx}{x} = 0$$

$$\tfrac{1}{3} \ln |v| + \tfrac{1}{3} \ln |3 - 2v^2| + \ln |x| = \ln c, \qquad \ln |v| + \ln |3 - 2v^2| + 3 \ln |x| = \ln C'$$

Then $vx^3(3 - 2v^2) = C$ and $y(3x^2 - 2y^2) = C$.

**10.** Solve $(x^2 + y) \, dx + (y^3 + x) \, dy = 0$.

Integrate $x^2 \, dx + (y \, dx + x \, dy) + y^3 \, dy = 0$, term by term, to obtain $\dfrac{x^3}{3} + xy + \dfrac{y^4}{4} = C$.

**11.** Solve $(x + e^{-x} \sin y) \, dx - (y + e^{-x} \cos y) \, dy = 0$.

Integrate $x \, dx - y \, dy - (e^{-x} \cos y \, dy - e^{-x} \sin y \, dx) = 0$, term by term, to obtain

$$\tfrac{1}{2} x^2 - \tfrac{1}{2} y^2 - e^{-x} \sin y = C$$

**12.** Solve $x \, dy - y \, dx = 2x^3 \, dx$.

The combination $x \, dy - y \, dx$ suggests $d \left( \dfrac{y}{x} \right) = \dfrac{x \, dy - y \, dx}{x^2}$. Hence, multiplying the given

equation by $\xi(x) = \dfrac{1}{x^2}$, $\quad \dfrac{x \, dy - y \, dx}{x^2} = 2x \, dx$ and $\dfrac{y}{x} = x^2 + C$ or $y = x^3 + Cx$.

**13.** Solve $x \, dy + y \, dx = 2x^2 y \, dx$.

The combination $x \, dy + y \, dx$ suggests $d(\ln xy) = \dfrac{x \, dy + y \, dx}{xy}$. Hence, multiplying the given

equation by $\xi(x, y) = \dfrac{1}{xy}$, $\quad \dfrac{x \, dy + y \, dx}{xy} = 2x \, dx$ and $\ln |xy| = x^2 + C$.

**14.** Solve $x \, dy + (3y - e^x) \, dx = 0$.

Multiply the equation by $\xi(x) = x^2$ to obtain $x^3 \, dy + 3x^2 y \, dx = x^2 e^x \, dx$.

Then $x^3 y = \displaystyle\int x^2 e^x \, dx = x^2 e^x - 2x e^x + 2 e^x + C$.

**15.** Solve $\dfrac{dy}{dx} + \dfrac{2}{x}y = 6x^3$.

Here $P(x) = \dfrac{2}{x}$, $\displaystyle\int P(x)\,dx = \ln x^2$, and $\xi(x) = e^{\ln x^2} = x^2$.

Multiply the given equation by $\xi(x) = x^2$ to obtain $x^2\,dy + 2xy\,dx = 6x^5\,dx$. Then $x^2 y = x^6 + C$.

*Note 1.* After multiplying by the integrating factor, the terms on the left side of the resulting equation are an *integrable combination.*

*Note 2.* The integrating factor of a given equation is not unique. In this problem $x^2$, $3x^2$, $\tfrac{1}{2}x^2$, etc., are all integrating factors. Hence, we write the simplest particular integral of $P(x)\,dx$ rather than the general integral, $\ln x^2 + \ln C = \ln Cx^2$.

**16.** Solve $\tan x\,\dfrac{dy}{dx} + y = \sec x$.

Since $\dfrac{dy}{dx} + y\cot x = \csc x$, $\displaystyle\int P(x)\,dx = \int \cot x\,dx = \ln|\sin x|$ and $\xi(x) = e^{\ln|\sin x|} = |\sin x|$.

Then $\sin x\left(\dfrac{dy}{dx} + y\cot x\right) = \sin x\csc x$, $\sin x\,dy + y\cos x\,dx = dx$, and $y\sin x = x + C$.

**17.** Solve $\dfrac{dy}{dx} - xy = x$.

Here $P(x) = -x$, $\displaystyle\int P(x)\,dx = -\tfrac{1}{2}x^2$, and $\xi(x) = e^{-\frac{1}{2}x^2}$.

Then $e^{-\frac{1}{2}x^2}\,dy - xye^{-\frac{1}{2}x^2}\,dx = xe^{-\frac{1}{2}x^2}\,dx$, $ye^{-\frac{1}{2}x^2} = -e^{-\frac{1}{2}x^2} + C$, and $y = Ce^{\frac{1}{2}x^2} - 1$.

**18.** Solve $\dfrac{dy}{dx} + y = xy^2$.

The equation is of the form $\dfrac{dy}{dx} + Py = Qy^n$ with $n = 2$.

Use the substitution $y^{1-n} = y^{-1} = z$, $y^{-2}\dfrac{dy}{dx} = -\dfrac{dz}{dx}$. (For convenience, write the equation in the form $y^{-2}\dfrac{dy}{dx} + y^{-1} = x$.) Then $-\dfrac{dz}{dx} + z = x$ or $\dfrac{dz}{dx} - z = -x$.

The integrating factor is $\xi(x) = e^{\int P\,dx} = e^{-\int dx} = e^{-x}$. Then $e^{-x}dz - ze^{-x}dx = -xe^{-x}dx$ and $ze^{-x} = xe^{-x} + e^{-x} + C$. Finally, since $z = y^{-1}$, $\dfrac{1}{y} = x + 1 + Ce^x$.

**19.** Solve $\dfrac{dy}{dx} + y\tan x = y^3\sec x$.

Write the equation in the form $y^{-3}\dfrac{dy}{dx} + y^{-2}\tan x = \sec x$.

Use the substitution $y^{-2} = z$, $y^{-3}\dfrac{dy}{dx} = -\dfrac{1}{2}\dfrac{dz}{dx}$ to obtain $\dfrac{dz}{dx} - 2z\tan x = -2\sec x$.

The integrating factor is $\xi(x) = e^{-2\int\tan x\,dx} = \cos^2 x$. Then $\cos^2 x\,dz - 2z\cos x\sin x\,dx = -2\cos x\,dx$, $z\cos^2 x = -2\sin x + C$, and $\dfrac{\cos^2 x}{y^2} = -2\sin x + C$.

**20.** When a bullet is fired into a sand bank, it will be assumed that its retardation is equal to the square root of its velocity on entering. For how long will it travel if its velocity on entering the bank is 144 ft/sec?

Let $v$ represent the velocity $t$ sec after striking the bank.

Then the retardation $= -\dfrac{dv}{dt} = \sqrt{v}$ or $\dfrac{dv}{\sqrt{v}} = -dt$ and $2\sqrt{v} = -t + C$.

When $t = 0$, $v = 144$ and $C = 2\sqrt{144} = 24$. Thus, $2\sqrt{v} = -t + 24$ is the law governing the motion of the bullet. When $v = 0$, $t = 24$; the bullet will travel 24 sec before coming to rest.

**21.** A tank contains 100 gallons of brine holding 200 pounds of salt in solution. Water containing 1 pound of salt per gallon flows into the tank at the rate of 3 gallons per minute and the mixture, kept uniform by stirring, flows out at the same rate. Find the amount of salt in the tank at the end of 90 minutes.

Let $q$ denote the number of pounds of salt in the tank at the end of $t$ minutes. Then $\dfrac{dq}{dt}$ is the rate of change of the amount of salt at time $t$.

Three pounds of salt enters the tank each minute and $.03q$ pounds is removed. Thus, $\dfrac{dq}{dt} = 3 - .03q$,

$\dfrac{dq}{3 - .03q} = dt$, and $\dfrac{\ln(.03q - 3)}{.03} = -t + C$.

When $t = 0$, $q = 200$ and $C = \dfrac{\ln 3}{.03}$ so that $\ln(.03q - 3) = -.03t + \ln 3$, $.01q - 1 = e^{-.03t}$, and $q = 100 + 100e^{-.03t}$. When $t = 90$, $q = 100 + 100e^{-2.7} = 106.72$ pounds.

22. Under certain conditions cane sugar in water is converted into dextrose at a rate which is proportional to the amount unconverted at any time. If, of 75 grams at time $t = 0$, 8 grams are converted during the first 30 minutes, find the amount converted in $1\frac{1}{2}$ hours.

Let $q$ denote the amount converted in $t$ minutes.

Then $\dfrac{dq}{dt} = k(75 - q)$, $\dfrac{dq}{75 - q} = k\,dt$, and $\ln(75 - q) = -kt + C$.

When $t = 0$, $q = 0$ and $C = \ln 75$ so that $\ln(75 - q) = -kt + \ln 75$.

When $t = 30$, $q = 8$, $30k = \ln 75 - \ln 67$, and $k = .0038$. Thus, $q = 75(1 - e^{-.0038t})$.

When $t = 90$, $q = 75(1 - e^{-.34}) = 21.6$ grams.

# Supplementary Problems

23. Form the differential equation whose general solution is:

(a) $y = Cx^2 + 1$     (c) $y = Cx^2 + C^2$     (e) $y = C_1 + C_2x + C_3x^2$     (g) $y = C_1 \sin x + C_2 \cos x$

(b) $y = C^2x + C$     (d) $xy = x^3 - C$     (f) $y = C_1e^x + C_2e^{2x}$     (h) $y = C_1e^x \cos(3x + C_2)$

Ans.    (a) $xy' = 2(y - 1)$    (c) $4x^2y = 2x^3y' + (y')^2$    (e) $y''' = 0$       (g) $y'' + y = 0$

         (b) $y' = (y - xy')^2$    (d) $xy' + y = 3x^2$         (f) $y'' - 3y' + 2y = 0$    (h) $y'' - 2y' + 10y = 0$

24. Solve:

(a) $y\,dy - 4x\,dx = 0$               Ans. $y^2 = 4x^2 + C$

(b) $y^2\,dy - 3x^5\,dx = 0$           Ans. $2y^3 = 3x^6 + C$

(c) $x^3y' = y^2(x - 4)$             Ans. $x^2 - xy + 2y = Cx^2y$

(d) $(x - 2y)\,dy + (y + 4x)\,dx = 0$    Ans. $xy - y^2 + 2x^2 = C$

(e) $(2y^2 + 1)y' = 3x^2y$          Ans. $y^2 + \ln|y| = x^3 + C$

(f) $xy'(2y - 1) = y(1 - x)$       Ans. $\ln|xy| = x + 2y + C$

(g) $(x^2 + y^2)\,dx = 2xy\,dy$        Ans. $x^2 - y^2 = Cx$

(h) $(x + y)\,dy = (x - y)\,dx$       Ans. $x^2 - 2xy - y^2 = C$

(i) $x(x + y)\,dy - y^2\,dx = 0$      Ans. $y = Ce^{-y/x}$

(j) $x\,dy - y\,dx + xe^{-y/x}\,dx = 0$    Ans. $e^{y/x} + \ln|Cx| = 0$

(k) $dy = (3y + e^{2x})\,dx$         Ans. $y = (Ce^x - 1)e^{2x}$

(l) $x^2y^2\,dy = (1 - xy^3)\,dx$      Ans. $2x^3y^3 = 3x^2 + C$

25. The tangent and normal to a curve at point $P(x, y)$ meet the $x$-axis in $T$ and $N$ respectively and the $y$-axis in $S$ and $M$ respectively. Determine the family of curves satisfying the condition:

(a) $TP = PS$    (b) $NM = MP$    (c) $TP = OP$    (d) $NP = OP$

Ans. (a) $xy = C$    (b) $2x^2 + y^2 = C$    (c) $xy = C$, $y = Cx$    (d) $x^2 \pm y^2 = C$

26. Solve Problem 21, assuming that pure water flows into the tank at the rate 3 gallons per minute and the mixture flows out at the same rate.     Ans. 13.44 lb

27. Solve Problem 26 assuming that the mixture flows out at the rate 4 gallons per minute.

Hint.   $dq = -\dfrac{4q}{100 - t}\,dt$       Ans. 0.02 lb

# Chapter 70

## Differential Equations of Order Two

**THE FOLLOWING TYPES** of second-order differential equations will be considered:

$(1)$   $\dfrac{d^2y}{dx^2} = f(x)$                                                  See Problem 1.

$(2)$.  $\dfrac{d^2y}{dx^2} = f\!\left(x, \dfrac{dy}{dx}\right)$                               See Problems 2-3.

$(3)$   $\dfrac{d^2y}{dx^2} = f(y)$                                                See Problems 4-5.

$(4)$   $\dfrac{d^2y}{dx^2} + P\dfrac{dy}{dx} + Qy = R,$    where $P$ and $Q$ are constants and $R$ is a constant or function of $x$ only.                   See Problems 6-11.

If the equation $m^2 + Pm + Q = 0$ has two *distinct* roots $m_1$ and $m_2$, $y = C_1 e^{m_1 x} + C_2 e^{m_2 x}$ is the general solution of the equation $\dfrac{d^2y}{dx^2} + P\dfrac{dy}{dx} + Qy = 0$. If the two roots are identical, $m_1 = m_2 = m$, $y = C_1 e^{mx} + C_2 x e^{mx}$ is the general solution.

The general solution of $\dfrac{d^2y}{dx^2} + P\dfrac{dy}{dx} + Qy = 0$ is called the *complementary function* of the equation $\dfrac{d^2y}{dx^2} + P\dfrac{dy}{dx} + Qy = R(x)$. If $y = f(x)$ satisfies the latter equation, then $y =$ complementary function $+ f(x)$ is its general solution.

## Solved Problems

**1.**   Solve   $\dfrac{d^2y}{dx^2} = xe^x + \cos x.$

    **Here**   $\dfrac{d}{dx}\!\left(\dfrac{dy}{dx}\right) = xe^x + \cos x,$   $\dfrac{dy}{dx} = \displaystyle\int (xe^x + \cos x)\,dx = xe^x - e^x + \sin x + C_1,$   and

$$y = xe^x - 2e^x - \cos x + C_1 x + C_2$$

**2.**   Solve   $x^2\dfrac{d^2y}{dx^2} + x\dfrac{dy}{dx} = a.$

    Let $p = \dfrac{dy}{dx}$; then $\dfrac{d^2y}{dx^2} = \dfrac{dp}{dx}$ and the given equation becomes $x^2\dfrac{dp}{dx} + xp = a$ or $x\,dp + p\,dx = \dfrac{a}{x}\,dx.$

Then $xp = a\ln|x| + C_1,$ $x\dfrac{dy}{dx} = a\ln|x| + C_1,$ $dy = a\ln|x|\dfrac{dx}{x} + C_1\dfrac{dx}{x},$ and $y = \tfrac{1}{2}a\ln^2|x| + C_1\ln|x| + C_2.$

**3.**   Solve $xy'' + y' + x = 0.$

    Let $p = \dfrac{dy}{dx}.$ Then $\dfrac{d^2y}{dx^2} = \dfrac{dp}{dx}$ and the given equation becomes

$$x\dfrac{dp}{dx} + p + x = 0 \quad \text{or} \quad x\,dp + p\,dx = -x\,dx$$

Then $xp = -\tfrac{1}{2}x^2 + C_1,$ $\dfrac{dy}{dx} = -\dfrac{1}{2}x + \dfrac{C_1}{x},$ and $y = -\tfrac{1}{4}x^2 + C_1\ln|x| + C_2.$

**4.** Solve $\dfrac{d^2y}{dx^2} - 2y = 0$.

Since $\dfrac{d}{dx}(y'^2) = 2y'y''$, multiply the given equation by $2y'$ to obtain

$$2y'y'' = 4yy', \qquad y'^2 = 4\int yy'\,dx = 4\int y\,dy = 2y^2 + C_1$$

Then $\qquad \dfrac{dy}{dx} = \sqrt{2y^2 + C_1}, \qquad \dfrac{dy}{\sqrt{2y^2 + C_1}} = dx, \qquad \ln|\sqrt{2}\,y + \sqrt{2y^2 + C_1}| = \sqrt{2}\,x + \ln C_2'$

and $\qquad\qquad\qquad\qquad \sqrt{2}\,y + \sqrt{2y^2 + C_1} = C_2 e^{\sqrt{2}\,x}$

**5.** Solve $y'' = -\dfrac{1}{y^3}$.

Multiply by $2y'$ to obtain $2y'y'' = -\dfrac{2y'}{y^3}$. Then

$$(y')^2 = \frac{1}{y^2} + C_1, \qquad \frac{dy}{dx} = \frac{\sqrt{1 + C_1 y^2}}{y}, \qquad \frac{y\,dy}{\sqrt{1 + C_1 y^2}} = dx, \qquad \sqrt{1 + C_1 y^2} = C_1 x + C_2$$

and $\qquad\qquad\qquad\qquad (C_1 x + C_2)^2 - C_1 y^2 = 1$

**6.** Solve $\dfrac{d^2y}{dx^2} + 3\dfrac{dy}{dx} - 4y = 0$.

Here $m^2 + 3m - 4 = 0$ and $m = 1, -4$. The general solution is $y = C_1 e^x + C_2 e^{-4x}$.

**7.** Solve $\dfrac{d^2y}{dx^2} + 3\dfrac{dy}{dx} = 0$.

Here $m^2 + 3m = 0$ and $m = 0, -3$. The general solution is $y = C_1 + C_2 e^{-3x}$.

**8.** Solve $\dfrac{d^2y}{dx^2} - 4\dfrac{dy}{dx} + 13y = 0$.

Here $m^2 - 4m + 13 = 0$ and the roots are $m_1 = 2 + 3i$ and $m_2 = 2 - 3i$. The general solution is

$$y = C_1 e^{(2+3i)x} + C_2 e^{(2-3i)x} = e^{2x}(C_1 e^{3ix} + C_2 e^{-3ix})$$

Since $e^{iax} = \cos ax + i \sin ax$, then $e^{3ix} = \cos 3x + i \sin 3x$, $e^{-3ix} = \cos 3x - i \sin 3x$, and the solution may be put in the form

$$\begin{aligned}
y &= e^{2x}\{C_1(\cos 3x + i \sin 3x) + C_2(\cos 3x - i \sin 3x)\} \\
&= e^{2x}\{(C_1 + C_2)\cos 3x + i(C_1 - C_2)\sin 3x\} \\
&= e^{2x}(A \cos 3x + B \sin 3x)
\end{aligned}$$

**9.** Solve $\dfrac{d^2y}{dx^2} - 4\dfrac{dy}{dx} + 4y = 0$.

Here $m^2 - 4m + 4 = 0$ and $m = 2, 2$. The general solution is $y = C_1 e^{2x} + C_2 x e^{2x}$.

**10.** Solve $\dfrac{d^2y}{dx^2} + 3\dfrac{dy}{dx} - 4y = x^2$.

From Problem 6, the complementary function is $y = C_1 e^x + C_2 e^{-4x}$.

To find a particular integral of the equation, note that the right hand member is $R(x) = x^2$. This suggests that the particular integral will contain a term in $x^2$ and perhaps other terms obtained by successive differentiation. We shall assume it to be of the form $y = Ax^2 + Bx + C$ where the constants $A, B, C$ are to be determined.

Substitute $y = Ax^2 + Bx + C$, $y' = 2Ax + B$, $y'' = 2A$, in the differential equation to obtain

$$2A + 3(2Ax + B) - 4(Ax^2 + Bx + C) = x^2, \quad -4Ax^2 + (6A - 4B)x + (2A + 3B - 4C) = x^2$$

Since this is an identity in $x$, $-4A = 1$, $6A - 4B = 0$, $2A + 3B - 4C = 0$.

Then $A = -\frac{1}{4}$, $B = -\frac{3}{8}$, $C = -\frac{13}{32}$, and $y = -\frac{1}{4}x^2 - \frac{3}{8}x - \frac{13}{32}$ is a particular integral.

Thus, the general solution is $y = C_1 e^x + C_2 e^{-4x} - \frac{1}{4}x^2 - \frac{3}{8}x - \frac{13}{32}$.

11. Solve $\dfrac{d^2y}{dx^2} - 2\dfrac{dy}{dx} - 3y = \cos x$.

Here $m^2 - 2m - 3 = 0$, $m = -1, 3$, and the complementary function is $y = C_1 e^{-x} + C_2 e^{3x}$. The right hand member of the differential equation suggests that a particular integral is of the form $A \cos x + B \sin x$.

Substitute $y = A \cos x + B \sin x$, $y' = B \cos x - A \sin x$, $y'' = -A \cos x - B \sin x$, in the differential equation to obtain

$$(-A \cos x - B \sin x) - 2(B \cos x - A \sin x) - 3(A \cos x + B \sin x) = \cos x$$
$$-2(2A + B) \cos x + 2(A - 2B) \sin x = \cos x$$

Then $-2(2A + B) = 1$, $A - 2B = 0$ and $A = -\frac{1}{5}$, $B = -\frac{1}{10}$.

The general solution is $C_1 e^{-x} + C_2 e^{3x} - \frac{1}{5} \cos x - \frac{1}{10} \sin x = y$.

12. A weight attached to a spring moves up and down, so that the equation of motion is $\dfrac{d^2s}{dt^2} + 16s = 0$, where $s$ is the stretch of the spring at time $t$. If $s = 2$ and $\dfrac{ds}{dt} = 1$ when $t = 0$, find $s$ in terms of $t$.

Here $m^2 + 16 = 0$, $m = \pm 4i$, and the general solution is $s = A \cos 4t + B \sin 4t$.

When $t = 0$, $s = 2 = A$, so that $s = 2 \cos 4t + B \sin 4t$.

When $t = 0$, $ds/dt = 1 = -8 \sin 4t + 4B \cos 4t = 4B$ and $B = \frac{1}{4}$.

Thus, the required equation is $s = 2 \cos 4t + \frac{1}{4} \sin 4t$.

13. The electric current in a certain circuit is given by $\dfrac{d^2I}{dt^2} + 4\dfrac{dI}{dt} + 2504I = 110$. If $I = 0$ and $\dfrac{dI}{dt} = 0$ when $t = 0$, find $I$ in terms of $t$.

$m^2 + 4m + 2504 = 0$, $m = -2 + 50i$, $-2 - 50i$; the complementary function is $e^{-2t}(A \cos 50t + B \sin 50t)$.

The particular integral is $I = 110/2504 = .044$.

Thus, the general solution is $I = e^{-2t}(A \cos 50t + B \sin 50t) + .044$.

When $t = 0$, $I = 0 = A + .044$; then $A = -.044$.

When $t = 0$, $\dfrac{dI}{dt} = 0 = e^{-2t}[(-2A + 50B) \cos 50t - (2B + 50A) \sin 50t] = -2A + 50B$.

Then $B = -.0018$ and the required relation is $I = -e^{-2t}(0.044 \cos 50t + 0.0018 \sin 50t) + 0.044$.

14. A chain 4 ft long starts to slide off a flat roof with 1 ft hanging over the edge. Discounting friction, find (a) the velocity with which it slides off and (b) the time required to slide off.

Let $s$ denote the length of the chain hanging over the edge of the roof at time $t$.

(a) The force $F$ causing the chain to slide off the roof is the weight of the part hanging over the edge.

Force $= $ mass $\times$ acceleration $= ms'' = \frac{1}{4}mgs$ or $s'' = \frac{1}{4}gs$.

$2s's'' = \frac{1}{2}gss'$ and $(s')^2 = \frac{1}{4}gs^2 + C_1$.

When $t = 0$, $s = 1$ and $s' = 0$. Then $C_1 = -\frac{1}{4}g$ and $s' = \frac{1}{2}\sqrt{g} \cdot \sqrt{s^2 - 1}$.

When $s = 4$, $s' = \frac{1}{2}\sqrt{15g}$ ft/sec.

(b) $\dfrac{ds}{\sqrt{s^2 - 1}} = \frac{1}{2}\sqrt{g}\, dt$ and $\ln|s + \sqrt{s^2 - 1}| = \frac{1}{2}\sqrt{g}\, t + C_2$.

When $t = 0$, $s = 1$. Then $C_2 = 0$ and $\ln(s + \sqrt{s^2 - 1}) = \frac{1}{2}\sqrt{g}\, t$.

When $s = 4$, $t = \dfrac{2}{\sqrt{g}} \ln(4 + \sqrt{15})$ sec.

15. A boat of mass 1600 lb has a speed of 20 ft/sec when the engine is cut off (at $t = 0$). The resistance of the water is proportional to its speed and is 200 lb when $t = 0$. How far will the boat have moved when its speed is reduced to 5 ft/sec?

Let $s$ denote the distance traveled by the boat $t$ sec after the engine is cut off.

$$ms'' = -Ks' \qquad \text{or} \qquad s'' = -ks'$$

To determine $k$:  At $t = 0$, $s' = 20$, $s'' = \dfrac{\text{force}}{\text{mass}} = -\dfrac{200g}{1600} = -4$  and  $k = \dfrac{1}{5}$.

$s'' = \dfrac{dv}{dt} = -\dfrac{v}{5}$,  $\ln v = -\dfrac{1}{5}t + C_1$,  and  $v = C_1 e^{-t/5}$.

When $t = 0$, $v = 20$.  Then  $C_1 = 20$,  $v = \dfrac{ds}{dt} = 20e^{-t/5}$  and  $s = -100e^{-t/5} + C_2$.

When $t = 0$, $s = 0$.  Then  $C_2 = 100$  and  $s = 100(1 - e^{-t/5})$.

When $v = 5 = 20e^{-t/5}$,  $s = 100(1 - \tfrac{1}{4}) = 75$ ft.

# Supplementary Problems

Solve:

16.  $\dfrac{d^2y}{dx^2} = 3x + 2$        *Ans.*  $y = \tfrac{1}{2}x^3 + x^2 + C_1 x + C_2$

17.  $e^{2x}\dfrac{d^2y}{dx^2} = 4(e^{4x} + 1)$     *Ans.*  $y = e^{2x} + e^{-2x} + C_1 x + C_2$

18.  $\dfrac{d^2y}{dx^2} = -9\sin 3x$      *Ans.*  $y = \sin 3x + C_1 x + C_2$

19.  $x\dfrac{d^2y}{dx^2} - 3\dfrac{dy}{dx} + 4x = 0$    *Ans.*  $y = x^2 + C_1 x^4 + C_2$

20.  $\dfrac{d^2y}{dx^2} - \dfrac{dy}{dx} = 2x - x^2$     *Ans.*  $y = x^3/3 + C_1 e^x + C_2$

21.  $x\dfrac{d^2y}{dx^2} - \dfrac{dy}{dx} = 8x^3$     *Ans.*  $y = x^4 + C_1 x^2 + C_2$

22.  $\dfrac{d^2y}{dx^2} - 3\dfrac{dy}{dx} + 2y = 0$     *Ans.*  $y = C_1 e^x + C_2 e^{2x}$

23.  $\dfrac{d^2y}{dx^2} + 5\dfrac{dy}{dx} + 6y = 0$     *Ans.*  $y = C_1 e^{-2x} + C_2 e^{-3x}$

24.  $\dfrac{d^2y}{dx^2} - \dfrac{dy}{dx} = 0$      *Ans.*  $y = C_1 + C_2 e^x$

25.  $\dfrac{d^2y}{dx^2} - 2\dfrac{dy}{dx} + y = 0$     *Ans.*  $y = C_1 x e^x + C_2 e^x$

26.  $\dfrac{d^2y}{dx^2} + 9y = 0$       *Ans.*  $y = C_1 \cos 3x + C_2 \sin 3x$

27.  $\dfrac{d^2y}{dx^2} - 2\dfrac{dy}{dx} + 5y = 0$     *Ans.*  $y = e^x(C_1 \cos 2x + C_2 \sin 2x)$

28.  $\dfrac{d^2y}{dx^2} - 4\dfrac{dy}{dx} + 5y = 0$     *Ans.*  $y = e^{2x}(C_1 \cos x + C_2 \sin x)$

29.  $\dfrac{d^2y}{dx^2} + 4\dfrac{dy}{dx} + 3y = 6x + 23$    *Ans.*  $y = C_1 e^{-x} + C_2 e^{-3x} + 2x + 5$

30.  $\dfrac{d^2y}{dx^2} + 4y = e^{3x}$      *Ans.*  $y = C_1 \sin 2x + C_2 \cos 2x + e^{3x}/13$

31.  $\dfrac{d^2y}{dx^2} - 6\dfrac{dy}{dx} + 9y = x + e^{2x}$    *Ans.*  $y = C_1 e^{3x} + C_2 x e^{3x} + e^{2x} + \dfrac{x}{9} + \dfrac{2}{27}$

32.  $\dfrac{d^2y}{dx^2} - y = \cos 2x - 2\sin 2x$    *Ans.*  $y = C_1 e^x + C_2 e^{-x} - \tfrac{1}{5}\cos 2x + \tfrac{2}{5}\sin 2x$

33.  A particle of mass $m$, moving in a medium which offers a resistance proportional to the velocity, is subject to an attracting force proportional to the displacement. Find the equation of motion if at time $t = 0$, $s = 0$ and $s' = v_0$.

  *Hint.*  Here  $m\dfrac{d^2s}{dt^2} = -k_1\dfrac{ds}{dt} - k_2 s$  or  $\dfrac{d^2s}{dt^2} + 2b\dfrac{ds}{dt} + c^2 s = 0$,  $b > 0$.

  *Ans.* If  $b^2 = c^2$, $s = v_0 t e^{-bt}$.

   If  $b^2 < c^2$,  $s = \dfrac{v_0}{\sqrt{c^2 - b^2}}e^{-bt}\sin\sqrt{c^2 - b^2}\,t$.

   If  $b^2 > c^2$,  $s = \dfrac{v_0}{2\sqrt{b^2 - c^2}}(e^{(-b+\sqrt{b^2-c^2})t} - e^{(-b-\sqrt{b^2-c^2})t})$

# INDEX

Catalog

If you are interested in a list of SCHAUM'S
OUTLINE SERIES send your name
and address, requesting your free catalog, to:

SCHAUM'S OUTLINE SERIES, Dept. C
McGRAW-HILL BOOK COMPANY
1221 Avenue of Americas
New York, N.Y. 10020